Humanoid Robots

Human-like Machines

Edited by
Matthias Hackel

Humanoid Robots - Human-like Machines
Edited by Matthias Hackel

This is a reprint of the book published in 2007 by I-Tech

Republished by InTech
Janeza Trdine 9, 51000 Rijeka, Croatia

Technical Editor Teodora Smiljanic

Cover Designer Dino Smrekar

Additional hard copies can be obtained from orders@intechopen.com

Humanoid Robots - Human-like Machines, Edited by Matthias Hackel
p. cm.
ISBN 978-3-902613-07-3

For EU product safety concerns:
IN TECH d.o.o., Prolaz Marije Krucifikse Kozulić 3, 51000 Rijeka, Croatia,
info@intechopen.com or visit our website at intechopen.com.

Preface

As the world at our time has to face developments like keen climate change and globalisation, one may ask about the significance of building human-like machines, which probably will never operate as effective as humans in their diversity of potentialities. The benefits from tumbling two-legged mechatronic creatures and multiprocessor-based question-and-answer games seem hard to discover for non-involved persons. In general the benefits from fundamental research are not evident – and humanoid robotics research means fundamental research in the field of robotics. It is an enormous challenge for all humanoid researchers that evolution has originated such effective sensors, controllers and actuators. Building humanoid robots involves the development of lightweight and energy-saving actuators, fast and intelligent sensors and exceptional complex control systems. By merging these technologies we investigate the cooperation of complex sensor-actor systems as well as and human-machine interaction. In analogy to space research humanoid robotics research, driven by the goal to copy and serve the pride of creation, will have a strong impact in daily live products.

In this book the variety of humanoid robotic research can be obtained. The first chapter deals with remarkable hardware developments, whereby complete humanoid robotic systems are as well described as partial solutions.
In the second chapter diverse results around the biped motion of humanoid robots are presented. The autonomous, efficient and adaptive two-legged walking is one of the main challenge in humanoid robotics. The two-legged walking will enable humanoid robots to enter our environment without rearrangement.
Developments in the field of visual sensors, data acquisition, processing and control are to be observed in third chapter. In the fourth chapter some "mind building" and communication technologies are presented.

Editor
Matthias Hackel

Contents

Mind Organisation: Learning and Interaction

1

Design of an Assistive Gait Device for Strength Endurance and Rehabilitation

K. H. Low[1], Xiaopeng Liu[1] and Haoyong Yu[2]
[1]School of MAE, Nanyang Technological University
[2]Dept. of ME, National University of Singapore
Singapore

1. Introduction

1.1 Background

Exoskeletons for human performance augmentation (EHPA) are controlled and wearable devices and machines that can increase the speed, strength, and endurance of the operator. EHPA is expected to do that by increasing the physical performance of the solider wearing it, including:

- Increased payload: more fire power, more supplies, and thicker and heavier armor increasing the soldier chance of surviving a direct hit or even an explosion.
- Increased speed and extended range: enhanced ground reconnaissance and battle space dominance.
- Increased strength: larger caliber weapons, obstacle clearance, repairing heavy ma chinery such as tack.

Besides these, "Imagine the psychological impact upon a foe when encountering squads of seemingly invincible warriors protected by armor and endowed with superhuman capabilities, such as the ability to leap over 20-foot walls, " said Ned Thomas, the director of Institute for Soldier Nanotechnologies (ISN) (Wakefield, 2002). Another use of exoskeletons is that they could help integrate women into combat situations. John Knowles, publisher of the defense industry newsletter, The Knowles Report, said that in terms of marksmanship and other combat skills, "Women have proven themselves very equal." The prevailing argument against women in combat is that most can't meet the job's physical requirements. Exoskeletons, could "radically equalize that," thus enabling a 130-pound woman to lift, carry and be as effective as a 180-pound man (Hembree, 2001). In non-military areas, one of the most possible applications of exoskeletons may be used to help aged or disabled people whose lower extremities have locomotor deficiencies due to various reasons: polio, paraparesis, paralysis, dystrophia, etc. They are unable to walk without assistance and may lose muscular strength in their legs and become bedridden. They can only move around by a wheelchair or by using a wheeled walker. Unfortunately, barriers such as bumps and steps restrict the area that these people have access to. It is hoped that the lower exoskeleton can enhance their muscular strength and enable them walk as normal people.

1.2 Objective and Scopes

In light of the recent surge of interest in exoskeletons, much research has been devoted to developing exoskeleton systems. However, most of these studies only focus on upper extremity exoskeletons, which are ground based. Lacking the locomotion capabilities for walking with the user, their application is inevitably limited. With fund from Ministry of Defense, Singapore, we are considering to develop a lower extremity exoskeleton (LEE). The LEE is aiming to help the user carry heavy loads by transferring the load weight to the ground (not to the wearer). It could provide soldiers the ability to carry heavy loads such as food, communications gear, and weaponry, without the strain typically associated with demanding labor. We also hope with improvements it might provide a versatile transport platform for mission-critical equipment. The scopes of the present work are:

- Study and examination of the key technologies required to successfully build an ex oskeleton
- Investigation of the control strategy and the exoskeleton to user interface
- Design and construction of an experimental prototype
- Implementation of the control algorithm on the constructed prototype
- Realization of normal walking

2. Literature Review

Exoskeleton has been an active field of research in recent years (Guizzo & Goldstein, 2005). This section briefly describe some exoskeleton systems and assistive devices developed around the world.

2.1 UC Berkeley's Exoskeletons

In 2000, the Defense Advanced Research Projects Agency (DARPA) launched a program over five years on developing EHAP to give infantry soldiers an extra edge *(Exoskeletons for Human Performance Augmentation Projects,* 2002). One of the facilities that received the research founds from DAPRA is located at the UC Berkeley.

In 2004, Berkeley Lower Extremity Exoskeleton (BLEEX) was first unveiled. The BLEEX is designed to have the same degrees of freedom similar to those of the pilot: three degrees at the ankle and the hip, and one degree at the knee. However, it is hydraulically actuated only at the hips, knees and ankles to allow flexion and extension of the hip joints and knee joints as well as dorsiflexion and plantarflexion of the ankle joints. The other non-actuated degrees of movements are then spring loaded to a default standing posture. The exoskeleton connects to the user at the foot by strapping onto the user's boots. A bendable sole allows for bending of the users toes; and ankle abduction and vertical rotation are allowed for better flexibility. A torso connects the user's back and hips to the exoskeleton legs. A full-body vest is incorporated onto the torso to avoid discomfort or abrasion to the user. The BLEEX employs a high-tech compact Hydraulic Power Unit (HPU). The stand-alone hybrid power source is able to deliver hydraulic power for actuation and electric power for the computers and sensors for long hours. The HPU is the synthesis of a gasoline engine, a three-phase brushless generator and a hydraulic gear pump. Exclusive designing by UC Berkeley enables the HPU to regulate the hydraulic pressure and engine speed via an engine throttle and a hydraulic valve. This unique employment of power supply enables the BLEEX to operate more efficiently and lightweight for a longer period of time. The control scheme needs no direct measurements from the pilot or the human-machine interface, instead, the controller estimates, based on measurements from the exoskeleton only. The basic principle for the control of BLEEX rests on the notion that the exoskeleton needs to shadow the wearers voluntary and involuntary movements quickly, and without delay (Kazerooni, Racine, Huang, & Steger, 2005). This requires a high level of sensitivity in response to all forces and torques on the exoskeleton. However, an exoskeleton with high sensitivity to external forces and torques would respond to other external forces not initiated by its pilot, the pilot would receive motion from the exoskeleton unexpectedly and would have to struggle with it to avoid unwanted movement. The other concern is that systems with high sensitivity to external forces and torques are not robust to variations and therefore the precision of the system performance will be proportional to the precision of the exoskeleton dynamic model. Hence the model accuracy is crucial to exoskeleton control. The dynamics of the exoskeleton should be understand quite well and the controller is heavily model based. In together, that maximizing system sensitivity leads to a loss of robustness in the system. However, inventive or state-of-the-art the BLEEX may sound, actual videos of tests done on the BLEEX shows certain drawbacks. First of all, the user seemed hampered and unnatural during walking. Also, the backpack looks bulky and makes the whole system rather unbalanced. The costs of building a BLEEX prototype were also not mentioned by the developers. One would suspect that such a complex system, especially with its high-tech HPU, would cost quite considerably.

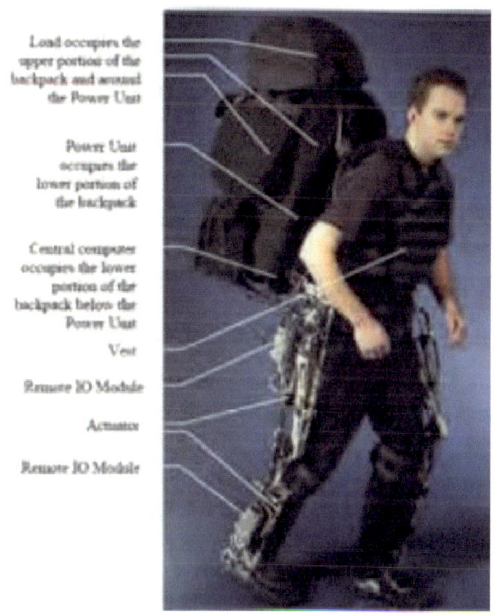

Figure 1. BLEEX prototype *(Berkeley Robotics Laboratory,* Dec. 2004)

2.2 Assistive Devices

Many institutions around the world have carried out research and development on exoskeletons and assistive devices to empower or aid the human limbs. Yobotics Incorporation's RoboKnee *(Robo Walker,* Dec. 2003), Japan's hybrid assistive limbs (HAL) *(Powered Suit HAL (Hybrid Assistive Leg),* Dec. 2003), Northeastern University's Active Knee Rehabilitation Device (AKROD) (Mavroidis, 2005) are some of the leading developments in the area of assistive devices to aid the human limb.

2.3 RoboWalker

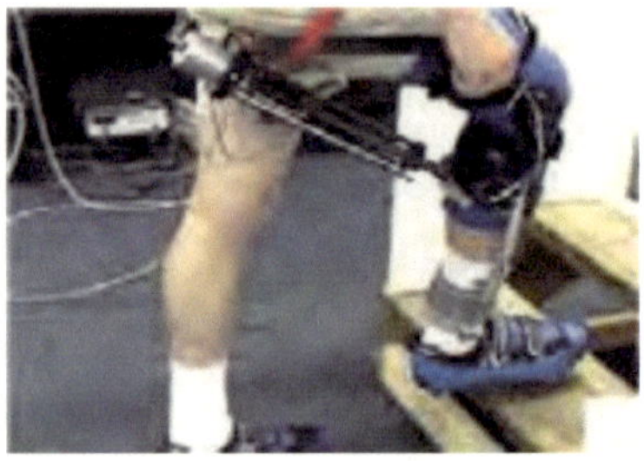

Figure 2. RoboKnee developed by Yobotics, Inc. *(RoboWalker,* Dec. 2003)

To help people who are suffering from weakness in their lower extremities, Yobotics, Inc., is developing a powered, wearable device called the RoboWalker. The RoboWalker augments or replaces muscular functions of the lower extremities. In 2001, they produced a prototype powered knee orthotic, called the RoboKnee, shown in Figure 2. With the computer, amplifiers, and batteries in a backpack, while not super impressive, the RoboKnee did provide for super-human capabilities by allowing the user to perform deep knee bends almost indefinitely. According to their homepage *(RoboWalker,* Dec. 2003), they are still looking for funds for the next stage of development.

2.4 Hybrid Assistive Leg

As shown in Figure 3, Hybrid Assistive Leg (HAL) (Kasaoka & Sankai, 2001; Kawamoto & Sankai, 2002; Tomohiro Hayashi & Sankai, 2005) is an exoskeleton type power assist system developed by Tsukuba University to realize the walking aid for aged people or gait-disorder persons.

2.5 AKROD

The AKROD-v2 (Mavroidis, 2005) developed by the Northeastern University consists of a robotic knee brace that is portable and programmable. As shown in Figure 4, the device contains electro-rheological fluid on one side of its structure such that it can turn solid in less than a millisecond with the application of a voltage. This would provide resistance to motion on a healing joint and it aims to help the muscle regain strength. The purpose of this device is to retrain the gait of a stroke patient.

Figure 3. HAL-3 developed by Japan's Tsukuba University *(Powered-suit gives aged a leg up,* Dec. 2003)

2.6 NTU Assistive Device

For the purpose of assisted walking, another prototype with the footpad design, as shown in Figure 5, has bee developed and tested for the walking and stair-climbing (Low & Yin, 2007).

Figure 4. Northeastern University's AKROD (*Mavroidis*, 2005)

Figure 5. NTU's Assistive Device (*Low & Yin*, 2007)

3. Gait Dynamics

Biped locomotion has been at the focus of researchers for decades. It is well known that biped gait can be divided into two phases: *single support phase* and *double support phase* (Whittle, 1991). In the single support phase, one leg is moving in the air, while the other leg is in contact with the ground. In the double support phase, both feet are with contact with the ground. The two support phases take place in sequence during walking. All of the biped mechanism joints are powered and directly controllable, except for the contact area between the foot and the ground. Foot behavior cannot be controlled directly, it is controlled indirectly by ensuring appropriate dynamics of the mechanism above the foot. To account for this, the concept of zero moment point (ZMP) (Vukobratović & Juricic, 1969), which is denned as the point on the ground at which the net moment of the inertial forces and the gravity forces has no component along the horizontal axes, has been used. The gait is balanced when and only when the ZMP trajectory remains within the support area. In this case, the system dynamics is perfectly balanced by the ground reaction force and overturning will not occur. In the single-support phase, the support polygon is identical to the foot surface. In the double support phase, however, the size of the support polygon is denned by the size of the foot surface and by the distance between them (the convex hulls of the two supporting feet).

The ZMP concept provides a useful dynamic criterion for the analysis and synthesis of biped locomotion. The ZMP ensures the gait balance during the entire gait cycle and provides a quantitative measure for the unbalanced moment about the support foot and for the robustness (balancing margin) of the dynamic gait equilibrium. Another term is center of pressure (CoP) (Vukobratović, Borovać, Šurdilović, & Stokić, 2001), which is commonly used in biped gait analysis based on force or pressure measurements. CoP represents the point on the support foot polygon at which the resultant of distributed foot ground reaction force acts. According to their definitions, it is obviously that in the considered single-support phase and for balanced dynamic gait equilibrium, the ZMP coincides with the CoP. However, in the dynamically unbalanced single-support situation that is characterized by a moment about CoP that could not be balanced by the sole reaction forces, the CoP and the ZMP do not coincide. The ZMP location outside the support area provides useful information for gait balancing (Low, Liu, Goh, & Yu, 2006). The fact that ZMP is instantaneously on the edge or outside of the support polygon indicates the occurrence of an unbalanced moment that cannot be compensated for by foot reaction forces. The distance of ZMP from the foot edge provides the measure for the unbalanced moment that tends to rotate the biped around the supporting foot and, possibly, to cause a fall. As depicted in Figure 6, the exoskeleton is composed of the trunk, the pelvis, two shanks, two thighs and two feet, will be considered. The trunk carries the payload, which can be seen as a part of the trunk. The vertical Z-axis and horizontal X-axis lie in the sagittal plane, as shown in Figure 6. By observing typical human joint trajectories, it is noted that the motion range is greater in sagittal plane than in other planes (Marchese, Muscato, & Virk, 2001) and most movements happen in the sagittal plane during walking. Hence, at the first stage, only the joints rotating around the Y-axis are actuated and movements in the sagittal plane are studied. The dynamical equation that describes the movement of a biped (or exoskeleton) has the following form:

$$M(\vec{q})\ddot{\vec{q}} = f(\vec{q}, \dot{\vec{q}}) + \vec{Q} \tag{1}$$

where $\vec{q} = (q_1, \dots, q_7)^T$ is the vector of generalized coordinates, which are the joint angles.

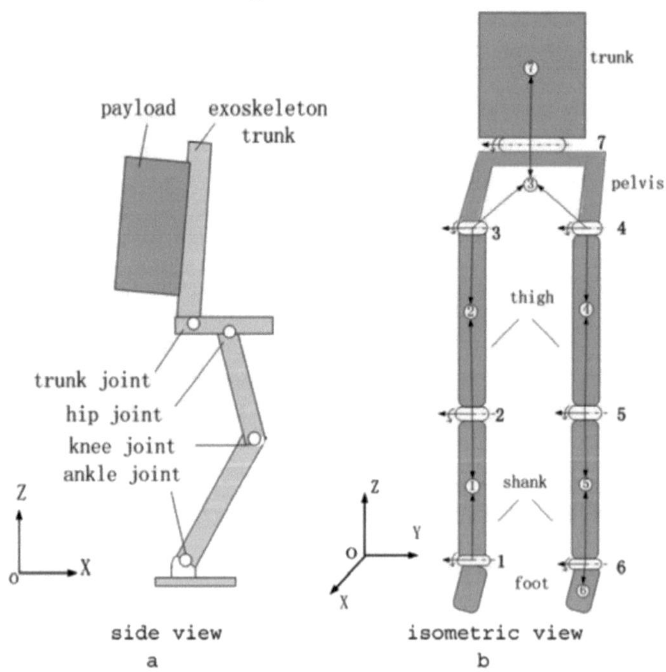

Figure 6. The model of the exoskeleton

The matrix function $M(\vec{q})$ takes into account the mass distribution, and the vector function $f(\vec{q}, \dot{\vec{q}})$ describes the effect of both inertial and gravity forces. The elements of the vector $\vec{Q}$ are generalized forces applied to the system, while the dots denote the time derivatives. Applying the basic theorems of rigid body kinematics, we obtain the following recursive (Low et al., 2006):

$$\begin{aligned} \vec{\omega}_i &= \vec{\omega}_{i-1} + \dot{q}_i \vec{e}_i \\ \vec{v}_i &= \vec{v}_{i-1} + \vec{\omega}_{i-1} \times \vec{r}_{i-1,i} + \vec{\omega}_i \times \vec{r}_{ii} \\ \vec{\alpha}_i &= \vec{\alpha}_{i-1} + \dot{q}_i \vec{\omega}_{i-1} \times \vec{e}_i + \ddot{q}_i \vec{e}_i \\ \vec{a}_i &= \vec{a}_{i-1} + \vec{\alpha}_{i-1} \times \vec{r}_{i-1,i} + \vec{\omega}_{i-1} \times (\vec{\omega}_{i-1} \times \vec{r}_{i-1,i}) \\ &\quad + \vec{\alpha}_i \times \vec{r}_{ii} + \vec{\omega}_i \times (\vec{\omega}_i \times \vec{r}_{ii}) \end{aligned} \tag{2}$$

where $\vec{\omega}_i$, $\vec{v}_i$, $\vec{\alpha}_i$ and $\vec{a}_i$ are the angular velocity, linear velocity of the center of mass, angular acceleration, and linear acceleration of the center of mass of the i-th link, respectively. The inertial force of the center of mass of the i-th link $\vec{F}_i$ and moment of the i-th link $\vec{M}_i$ can then be obtained by using Newton-Euler equations, respectively,

$$\vec{F}_i = m_i \vec{a}_i \tag{3}$$

$$\vec{M}_i = I_i \vec{\alpha}_i + \vec{\omega}_i \times I_i \vec{\omega}_i \tag{4}$$

4. Control Strategy of the Exoskeleton System

An important feature of the exoskeleton system, which is also the main difference between exoskeleton and biped robot, is the participation role of human in the process of control and decision-making. By introducing human as part of the control system, some intractable tasks for robots such as navigation, path planning, obstacle crossing and gait selection can be easily undertaken by the pilot instead of robot's complex artificial controller and vision system. However, two problems remain for the exoskeleton controller to solve: how to transfer the pilot's intention to the exoskeleton and how to keep the stability of the exoskeleton. Accordingly, the proposed control strategy can be divided into two parts, namely *Locomotion control* and *ZMP control.*

4.1 Locomotion Control

During the single support phase, the trajectory of the swinging foot determines the gait parameters such as step length, step height, etc. To ensure that the exoskeleton and the wearer can walk together, the trajectory of the exoskeleton's swing foot should trace that of the user in time. To do that, a mechanical linkage named *inner exoskeleton* is attached to the wearer operator (Low et al., 2006). Accordingly, the exoskeleton that is controllable and carrying payloads is named *outer exoskeleton.* The inner exoskeleton equipped with encoders is to capture the joint information of the pilot.

4.2 Control of the ZMP

If the ZMP of the exoskeleton is within the support area, it implies that the exoskeleton can keep the stability only by using the ground reaction force without adding any force to the user. In other words, the user will not feel any extra burden from the exoskeleton. Hence the purpose of the ZMP control is to make sure the ZMP remain within the support polygon. From the definition of the ZMP, we have

$$(\vec{M}_G + \vec{M}_F) \cdot \vec{e}_X = 0 \tag{5}$$

$$(\vec{M}_G + \vec{M}_F) \cdot \vec{e}_Y = 0 \tag{6}$$

where $\vec{M}_G$ is the total movement of gravity forces with respect to ZMP, $\vec{M}_F$ is the total moment of inertial forces of all the links with respect to ZMP, while $\vec{e}_X$ and $\vec{e}_Y$ denote unit vectors of the X and Y axes of the absolute coordinate frame. Equation (6) can be further replaced with

$$\sum_{i=1}^{7}[(\vec{p}_i - \vec{p}_z) \times (G_i + \vec{F}_i) + \vec{M}_i].\vec{e}_Y = 0 \tag{7}$$

where $\vec{p}_z$ is the ZMP coordinates in the global coordinate frame, $\vec{p}_i$ is the position vector of the center of mass of the i-th link,

$$\vec{p}_i = \vec{p}^* + \sum_{j=1}^{i-1}(\vec{r}_{jj} - \vec{r}_{j,j+1}) + \vec{r}_{ii} \tag{8}$$

where $G_i = m_i g$ is the gravity force of link i, $\vec{p}^*$ is the position vector of joint 1 with respect to the global coordinate system. Substitute Eqs. (2), (3), (4), (8) into Eq. (7), one can obtain

$$\sum_{i=1}^{7} a_i\ddot{q}_i + \sum_{i=1}^{7}\sum_{j=1}^{7} b_{ij}\dot{q}_i\dot{q}_j + \sum_{i=1}^{7} c_i G_i = 0 \tag{9}$$

where the coefficients a_i, b_{ij} and c_i are the functions of the generalized coordinates q_i. The trajectories of q_1 to q_6 are determined by the signals measured from the pilot's legs, as mentioned before, while q_7 is determined according to Eq. (9) to ensure the ZMP in the support polygon. Such a ZMP is the desired ZMP. However, the actual ZMP may be different from the desired ZMP due to all kinds of reasons such as disturbance from the environment or error of the actuators. A footpad that can online measure the actual ZMP is thus designed.

4.3 Measurement of ZMPs

In a stable gait, during the single support phase, the CoP of the supporting foot is also the ZMP of the whole exoskeleton; during the double support phase, the relationship between the ZMP and the CoP is described by

$$X_p = \frac{f_{LZ}X_L + f_{RZ}X_R}{f_{LZ} + f_{RZ}}, \quad Y_p = \frac{f_{LZ}Y_L + f_{RZ}Y_R}{f_{LZ} + f_{RZ}} \tag{10}$$

where

ZMP= (X_p, Y_p, Z_p): ZMP of the whole biped

CoP_L, = (X_L, Y_L, Z_L): CoP of the left foot

CoP_R = (X_R, Y_R, Z_R): CoP of the right foot

$f_L = (f_{LX}, f_{LY}, f_{LZ})$: ground reaction force at CoP_L

$f_R = (f_{RX}, f_{RY}, f_{RZ})$: ground reaction force at CoP_R

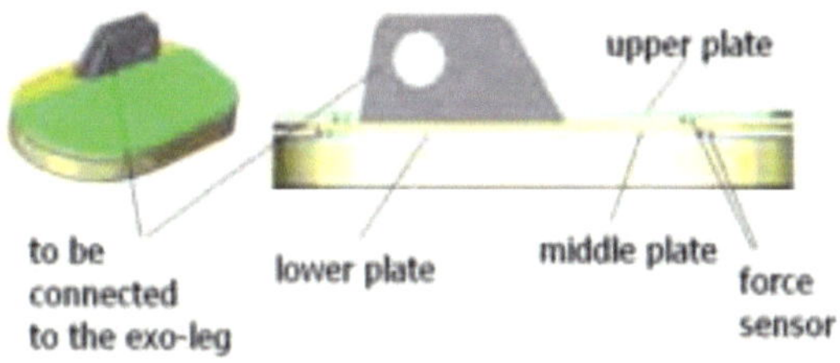

Figure 7. Design of the exoskeleton foot

To measure the ZMPs of the wearer and the exoskeleton, a footpad is designed as shown in Figure 7. The wearer's foot will be on the upper plate, and the exoskeleton leg will be connected to the middle plate. There are four force sensors between the upper plate and middle plate, the middle plate and lower plate, respectively. The sensors are distributed as shown in Figure 8. During the single support phase, Sensors 1-4 measure the ground reaction force under the human foot, and the ZMP coordinates of the human in the foot's local coordinate frame can be obtained according to

$$ZMP_h = \frac{\sum_{i=1}^{4} F_i r_i}{\sum_{i=1}^{4} F_i} \tag{11}$$

where F_i is the force measured by sensor i at the distance (r_i) from O, as denned in Figure 8. Sensors 5-8 measure the ground reaction force under the whole system (the human and the exoskeleton). Similarly, the ZMP of the whole system can be calculated by

$$ZMP_w = \frac{\sum_{i=5}^{8} F_i r_i}{\sum_{i=5}^{8} F_i} \tag{12}$$

The ZMP of the exoskeleton is on the radial distance from the human ZMP to the whole system's ZMP, and its position can be obtained by

$$ZMP_e = \frac{\sum_{i=1}^{4} F_i (r_w - r_h)}{\sum_{i=5}^{8} F_i - \sum_{i=1}^{4} F_i} + r_w \tag{13}$$

as we take the moment about the point of ZMPW,

$$\left(\sum_{i=5}^{8} F_i - \sum_{i=1}^{4} F_i\right)(ZMP_e - r_w) = \sum_{i=1}^{4} F_i (r_w - r_h) \tag{14}$$

in which r_h and r_w are the coordinates of the human's ZMP and the ZMP of the whole system (human plus exoskeleton), respectively, as shown in Figure 9. Note that the ZMP is expressed in terms of X, Y coordinates. During the double support phase, instead of the ZMPs, the CoPs of each foot are obtained from Eqs. (11) (13). By substituting those CoPs of the human and the exoskeleton into Eq. (10), respectively, ZMP of the human and that of the exoskeleton can be obtained accordingly.

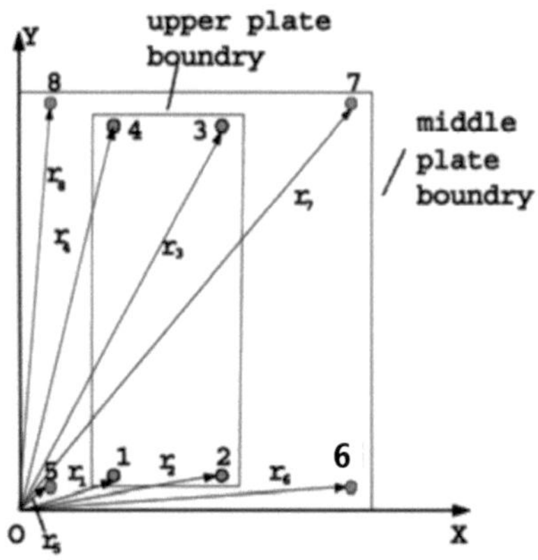

Figure 8. Location of the distributed sensors

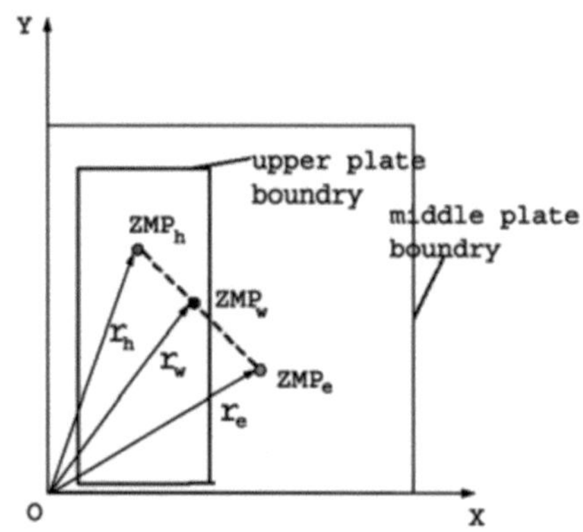

Figure 9. Relationship between the human ZMP and the exoskeleton's ZMP

4.4 Trunk Compensation

If the actual (currently measured) ZMP of the exoskeleton differs from the desired ZMP, trunk compensation will be applied to shift the actual ZMP to an appropriate position. Without losing generality, only motion in the sagittal plane during single support phase is discussed here. The trunk compensation in the frontal plane or during double support phase can be performed in the similar way. As shown in Figure 10, the actual ZMP differs from the desired ZMP in the direction of $\boldsymbol{X}$ axis by Δx. Note that the ground reaction force $\boldsymbol{F}_z$ acts on the exoskeleton can be derived from $F_z = \sum_{i=5}^{8} F_i - \sum_{i=1}^{4} F_i$. For simplicity, we assume that the action of the trunk joint k will not cause a change in the motion at any other joint. The system will then behave as if it was composed of two rigid links connected at trunk joint k, as depicted in Figure 10. The payload and the exoskeleton trunk as shown in the figure are considered as an upper part of total mass m and inertia moment $\boldsymbol{I}_k$ for the axis of joint k (as $I_k = I_c + ml_1^2$). Point c is the mass center of the upper part, and the distance from k to $\boldsymbol{C}$ is denoted by l_1. The lower part, representing the sum of all the links below the trunk joint k, including another leg that is not drawn in the figure, is also considered as a rigid body, which is standing on the ground surface and does not move. The distance from $\boldsymbol{O}$ to k is denoted by l_2. Note that ΔT_k stands for the correctional actuator torque, applied at joint k.

Assuming that the additional torque ΔT_k will cause change in acceleration of the upper part $\Delta\ddot{\beta}$, while velocities will not change due to the action of ΔT_k, $\Delta\dot{\beta} \approx 0$. Next the following equations are derived:

$$\Delta T_k = I_k \Delta\ddot{\beta} \tag{15}$$

$$F_z \Delta x = \Delta T_k + m l_1 l_2 (\cos\beta\cos\alpha + \sin\beta\sin\alpha)\Delta\ddot{\beta} \tag{16}$$

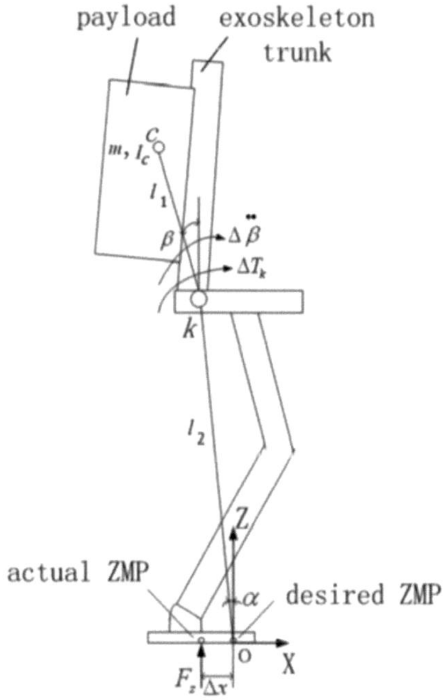

Fig. 10. Adjusting ZMP by trunk compensation

By virtue of Eq. (15), we have

$$\Delta\ddot{\beta} = \frac{\Delta T_k}{I_k} \tag{17}$$

Substituting Eq. (17) into Eq. (16), we obtain

$$\Delta T_k = \frac{F_z \Delta x}{1 + \frac{m l_1 l_2 (\cos\beta\cos\alpha + \sin\beta\sin\alpha)}{I_k}} \tag{18}$$

Taking into account that ΔT_k is derived by introducing certain simplifications, an additional feedback gain K_{zmp} is introduced into Eq. (18) as

$$\Delta T_k - K_{zmp} \frac{F_z \Delta x}{1 + \frac{m l_1 l_2 (\cos\beta\cos\alpha + \sin\beta\sin\alpha)}{I_k}} \tag{19}$$

where K_{zmp} can be determined by the feedback in the actual walking. Equation (19) shows how to drive the actual ZMP towards the desired ZMP by controlling the torque output of the trunk joint.

5. Simulation

It is necessary to verify a control algorithm using a software simulator before it is put into practice, because unsuccessful control method may cause great damages to the user or/and the exoskeleton. On the other hand, to develop a hardware exoskeleton with the desired physical capability, it is desired to clarify the required specifications of components in advance. An exoskeleton simulator, consisting of component parameters, is a useful tool to solve this problem. In this chapter, a human and an exoskeleton simulators are established using ADAMS (*Homepage of the Adams University*, Jan. 2005) and MATLAB (*MATLAB - the language of technical computing*, Jan. 2005) and walking simulations are performed to verify the control algorithms. Moreover, the moving range of each joint and the power needed by each actuator are analyzed.

5.1 Simulation Environment

To predict the physical biped (human/exoskeleton) motion, it is necessary to accurately formulate and solve the kinematic and dynamic equations of the mechanisms. However, the biped is a system with multiple bodies and many degrees of freedom. The analytical complexity of nonlinear algebraic equations of kinematics and nonlinear differential equations of dynamics makes it practically impossible to obtain closed-form solutions. On the other hand, there are some commercial dynamic software packages such as ADAMS, 3D Working Model, DADS, etc., whose usefulness has been illustrated in many areas. These software packages can automatically formulate the equations of kinematics and dynamics, solve the nonlinear equations, and provide computer graphics output of the simulation results (Haug, 1989). After compared with other commercially available simulation software, ADAMS is selected due to its high numerical accuracy. Another benefit of ADAMS is that it incorporates easily with control software packages such as MATLAB and MATLAB accepts user defined subfunctions. It is therefore possible to model detailed functions, define special constraints, and develop original control algorithms.

5.2 Exoskeleton Model

This section describes the kinematic structure of the human model. Figure 11 shows the human model established in ADAMS, and the human consists of legs, a waist, a trunk, and arms and a head. The CAD model adopts measured human data and parameters are set according to an adult with a 65 kg weight and 1.7 m height. The lengths of the leg's parts, which are crucial parameters of walking, are listed in Table 1. There are three degrees of freedom (DoF) in the hip joint, one in the ankle joint and one in the knee joint. There is also a DoF between the trunk and the waist. Also shown in Figure 11, an exoskeleton model is established, and a payload is carried by the exoskeleton. The links of the exoskeleton is parallel to those of the human, and the height of each joint (ankle, knee and hip) is equal to the corresponding one of the human. The exoskeleton model is set to have the same size and mass as those of the actual exoskeleton constructed. The mass of each part of the exoskeleton is listed in Table 2.

Height of Ankle	Length of Calf	Length of Thigh	Length of Foot	Width of Foot
80 mm	420 mm	440 mm	200 mm	100 mm

Table 1. Parameters of the human model's lower limbs

Foot and Ankle Joint	Calf	Thigh and Knee Joint	Waist Frame and Two Hip Joints	Back Frame and Linear Actuators
3 kg	1.68 kg	4.77 kg	9.54 kg	6.12 kg
Total	(3+1. 68+4.77)*2+9. 54+6.12= 34.56 kg			

Table 2. Mass of the exoskeleton

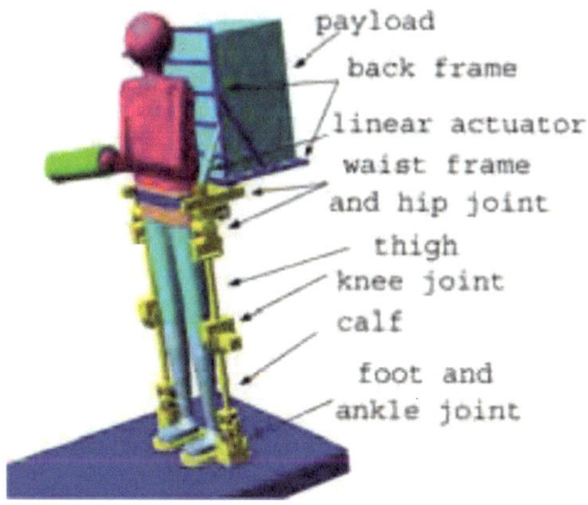

Figure 11. The NTU exoskeleton model

5.3 Walking Simulation

To run the walking simulation, there are three problems should be solved, namely contact model, locomotion control and trunk compensation control.

5.3.1 Contact Model

The reaction force between the feet and the ground strongly affects the behavior of the exoskeleton and human model. For example, if the impact force arising when the feet hit the ground is too large, the burden on joints of supporting the exoskeleton becomes large, and the exoskeleton will easily vibrate and become unstable. ADAMS adopts a spring-damper element to model the normal force between two contact objects. The IMPACT function activates when the distance between the first object and the second object falls below a nominal free length (x_1), that is, when two parts collide. As long as the distance between the first object and second object is greater than x_1, the force is zero. The force has two components, a spring or stiffness component and a damping or viscous component. The stiffness component is proportional to the stiffness k, and is a function of the penetration of the first object within the free length distance from the second object. The stiffness component opposes the penetration. The damping component of the force is a function of the speed of penetration. The damping opposes the direction of relative motion. To prevent a discontinuity in the damping force at contact, the damping coefficient is, by definition, a cubic step function of the penetration. Thus, at zero penetration, the damping coefficient is always zero. The damping coefficient achieves a maximum, c_{max}, at a user-defined penetration, d. The equation defining IMPACT is:

$$IMPACT = \begin{cases} \max(0, k(x_1 - x)^e - STEP(x, x_1 - d, c_{\max}, x_1, 0) * \dot{x}) & : x < x_1 \\ 0 & : x > x_1 \end{cases}$$

To absorb the shock when the exoskeleton's feet contact the ground, we propose to install elastomeric material on the feet of the exoskeleton. This situation can be simulated to investigate the effect of this measure on the ground reaction force. Figure 12 shows the two ground reaction force curves of the exoskeleton's right foot corresponding to the two walking with different footpad stiffness. It can be seen that with the stiffness decreasing, not only the impact force is smaller, the force curve is also more smooth. The simulation demonstrates that the elastomeric material can low-pass filter shock impulse and reduce the impact, therefore allowing the structure interacting with the environment at a pretty high speed.

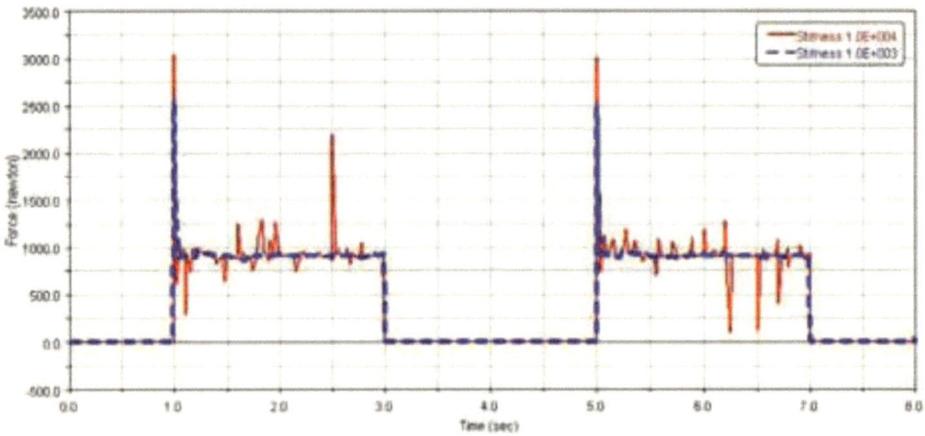

Fig. 12. Effect of foot's stiffness on the contact force

5.3.2 Locomotion Control

The human motion capture data (HMCD) is used to generate human-like walking motions of the human and exoskeleton models. As mentioned above, the human model's parameters are set according to an adult. To obtain the HMCD, encoders are attached to the joints of the wearer's lower limbs and the walking motion is thus captured. The human limbs are soft due to the muscles and tendons, while the model's limbs are firm. Besides, human has much more DoFs than the model. Hence the human motion data cannot be used directly to drive the human model. The recorded data is processed and edited based on the inverse kinematics of the human model and interpolated using Akima spline before being input into the ADAMS. Adams will calculate the human model's joints move according to those trajectories (Low et al., 2006).

5.3.3 Trunk Compensation

As mentioned earlier, the control of the exoskeleton is divided into leg trajectories control and ZMP control. In this simulation, the exoskeleton model's leg trajectories is controlled by human model's leg trajectories. To keep the exoskeleton walk stably, trunk should be controlled according to the ZMP of the exoskeleton. The trunk compensation function is implemented in MATLAB.

During the simulations using the MATLAB toolbox Simulink, at each time interval, the angular values, joint velocities and accelerations are sent to MATLAB, and those data is used by a trunk compensation module to calculate the torque needed at the trunk joint and such a torque will be applied in Adams so that the exoskeleton model can walk stably. Figure 13 shows the torque applied to the exoskeleton trunk joint for the changing masses of the load (5 kg and 20 kg).

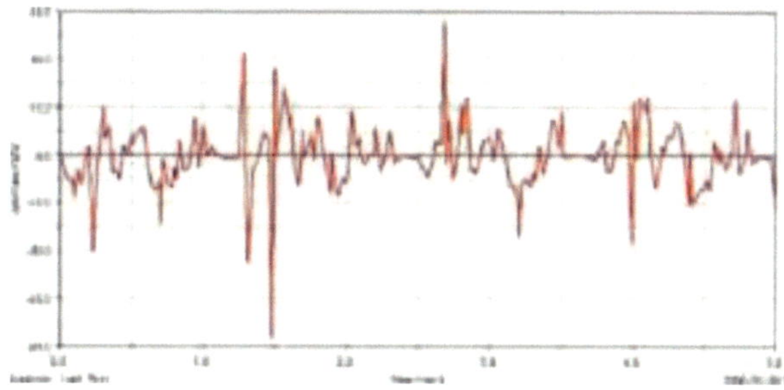

mass of load = 5kg

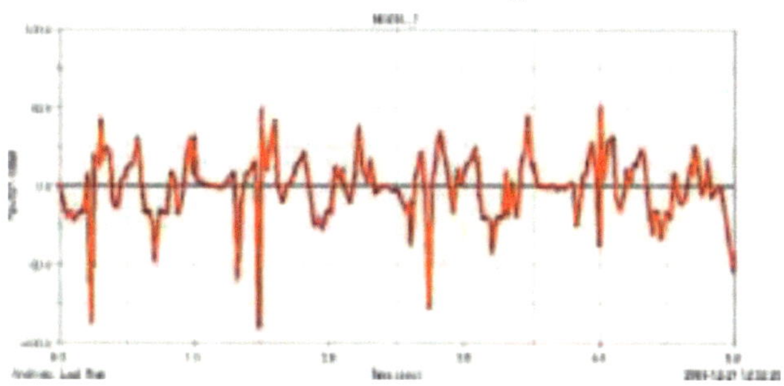

mass of load - 20kg

Figure 13. Torque at the trunk joint

5.4 Simulation Results

Without losing generality, we present only the results of the joints of the exoskeleton's right leg. The 30-kg payload is assumed and used in our simulation.

Ankle

Figure 14 shows the data of the ankle flexion/extension torque. The ankle torque is almost zero when the right leg is in swing phase. The torque is largest when the right foot is in its heel off phase. The instantaneous ankle mechanical power is calculated by multiplying the joint angular velocity and the instantaneous joint torque (Figure 14), as shown in Figure 15

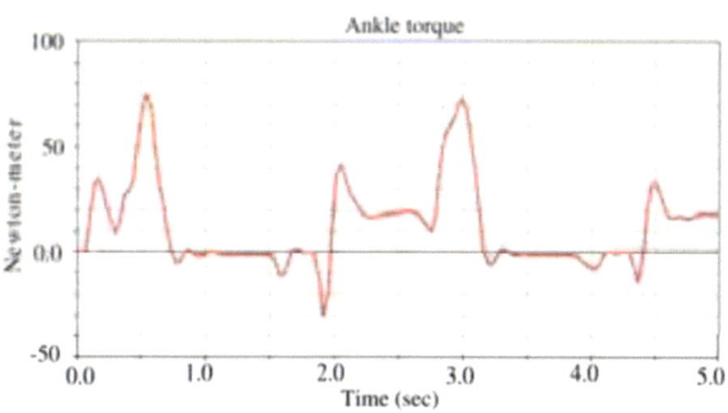

Figure 14. Ankle flexion/extension torque

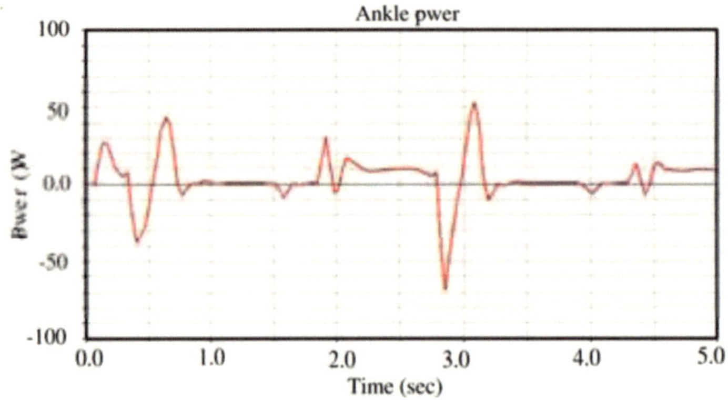

Figure 15. Ankle instantaneous mechanical power

Knee

The knee buckles momentarily in early stance to absorb the impact of heel strike then undergoes a large flexion during swing. This knee flexion decreases the effective leg length, allowing the foot to clear the ground when swinging forward. Although the walking knee flexion is up to approximately 0.6 rad or 34°, the human has significantly more flexibility-up to 160° when kneeling (Kapandji, 1987). The LEE knee motion range was set to from 0° to 110°. Figure 16 shows the data of the knee flexion/extension torque. The highest peak torque is in early stance (from heel contact to foot flat). Figure 17 shows the instantaneous mechanical power required by the knee joint.

Hip

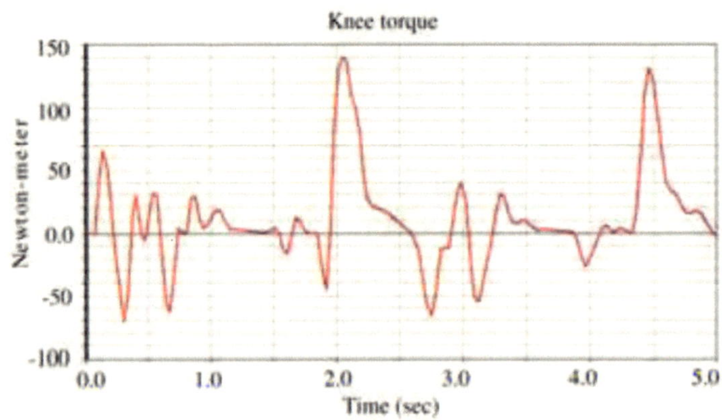

Figure 16. Knee flexion/extension torque

The thigh moves in a sinusoidal pattern with the thigh flexed upward at heel contact to allow foot contact the ground in front of the person. This is followed by an extension of the hip through most of stance phase and a flexion through swing. An average person can flexion reaches 90° with the knee extended. With the knee flexed, flexion can reach up to 120° or even beyond. When the knee is in extension, extension of the hip reaches 20° , when the knee is flexed, extension of the hip reaches 10°. The LEE hip angle is designed to 10° extension and 80° flexion. Figure 18 shows the data of the hip flexion/extension torque and Figure 19 shows the instantaneous mechanical power required by the hip joint.

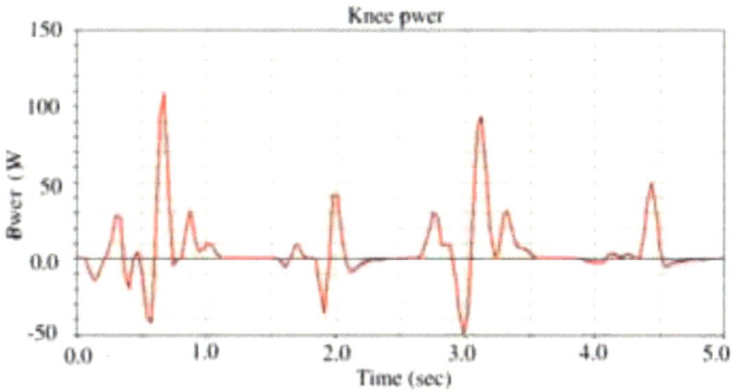

Figure 17. Knee instantaneous mechanical power

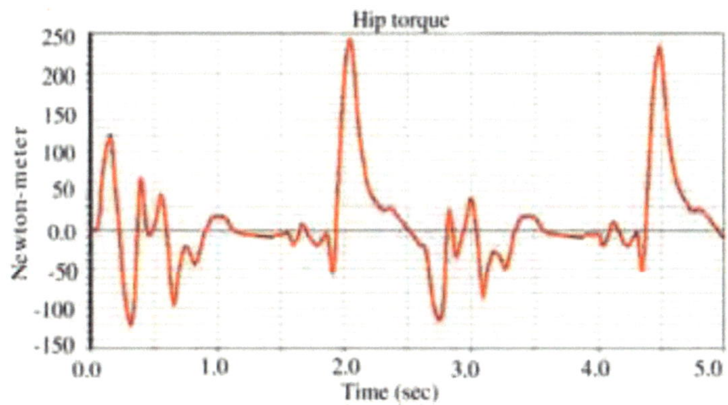

Figure 18. Hip flexion/extension torque

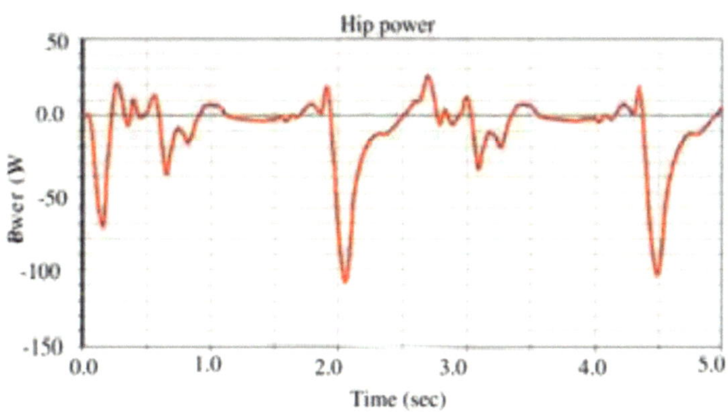

Figure 19. Hip instantaneous mechanical power

6. Design and Construction of Prototypes

This section introduces the design of the lower extremity exoskeleton system, including the inner lower exoskeleton and the outer lower exoskeleton.

6.1 Inner Lower Exoskeleton

The inner exoskeleton is used to attach the encoders to the human limbs. It exists between the human limbs and the outer lower exoskeleton and its weight is carried by the user. Hence, it must be as light and compact as possible. The conceptual design is shown in Figure 20. A minimum protrusion is achieved by designing the brackets to allow the tip of the encoder spindle to actually touch the surface of the user. This feat was achieved by the use of a larger but extra thin bearing which allows the rotary encoder to reside within the bearing itself. In this design, linkages are separated from the encoder housing. This modular concept of the housing allows the same housing unit to be repeated for the hip unit, the knee unit and the ankle unit. This modular design also cuts down on both design and manufacturing time. To position the rotary encoder brackets firmly on the respective bodily positions on the lower extremity of the human body, a need arise to design a harness system which is versatile, reliable, light and durable. Nylon being a material that economical, durable and easily worked with was selected as the choice material for the harness. For the harness used in the knee brackets, Velcro was chosen due to its ability to conform readily to the shape of the user's thigh. Also, with the Velcro, we can easily compensate for slight variations the variant dimensions of the various user thigh heights without having to adjust the buckle position.

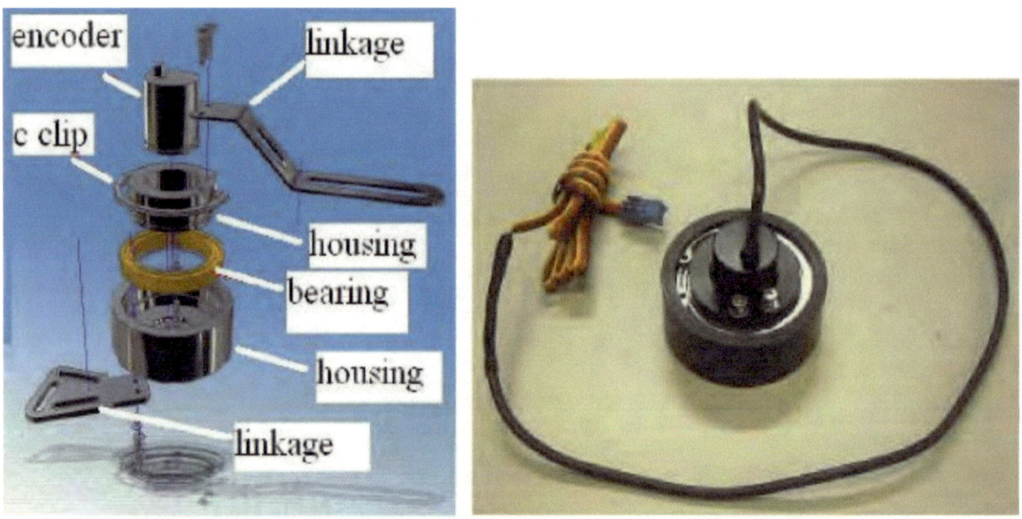

(a) (b)

Figure 20. The encoder bracket

6.2 Outer Lower Exoskeleton

The outer lower exoskeleton (OLE) provides the mechanical hardware means for the exoskeleton system. Therefore, it must be made of structures and mechanisms strong enough to carry a load and travel by itself. The OLE is a powered lower exoskeleton, which means that it is to be worn by the user from the user's waist down to the user's feet. The OLE has to be able to carry its own weight as well as a payload and should not hinder the walking movements of the user. The fully OLE CAD model is shown in Figure 21, while Figure 22 shows the overview of the whole OLE prototype.

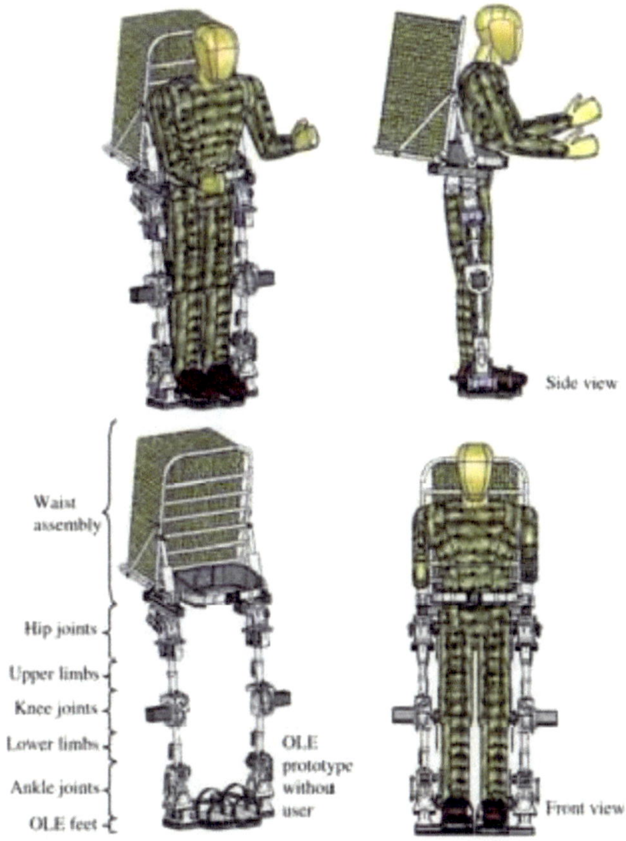

Figure 21. The overall design

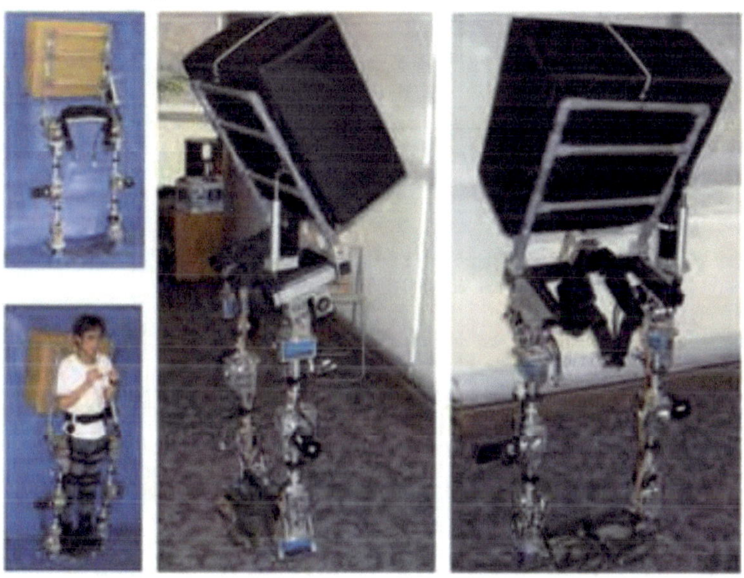

Figure 22. Photos of the OLE prototype

7. Walking Implementation of the Exoskeleton

The important function of the inner exoskeleton is to read necessary input data from the operator. These data will be analyzed, and transformed into corresponding commanded signals for the outer exoskeleton using certain mapping algorithms, and then transferred to the outer exoskeleton through a communication means. An operation of a dynamic system is called a real-time operation if the combined reaction- and operation-time of a task is shorter than the maximum delay that is allowed. When the human operator performs one action, the OLE should react correspondingly as fast and precise as possible, i.e., the OLE should be controlled real-time.

7.1. Real-Time Operating System Used in this Work

A Real Time Operating System or RTOS is an operating system that has been developed to serve for real-time applications. Some examples of RTOSes are QNX, RMX, RT-11, LynxOS, MicroC/OS-II, etc. In this work, xPC Target toolbox in MATLAB is employed (Mosterman et al., 2005). xPC Target provides a high-performance, host-target prototyping environment that enables developers to connect system simulation models (which can be created using Simulink and Stateflow toolbox from MATLAB) to physical systems and execute them in real time on PC-compatible hardware. xPC Target provides comprehensive software capabilities for rapid prototyping and hardware-in-the-loop simulation of control and signal processing systems. It can also enables users to add I/O interface blocks to their models (system models and controller designs, etc.), automatically generate code with R,eal-Time Workshop and Stateflow Coder, and download the code to a second PC running the xPC

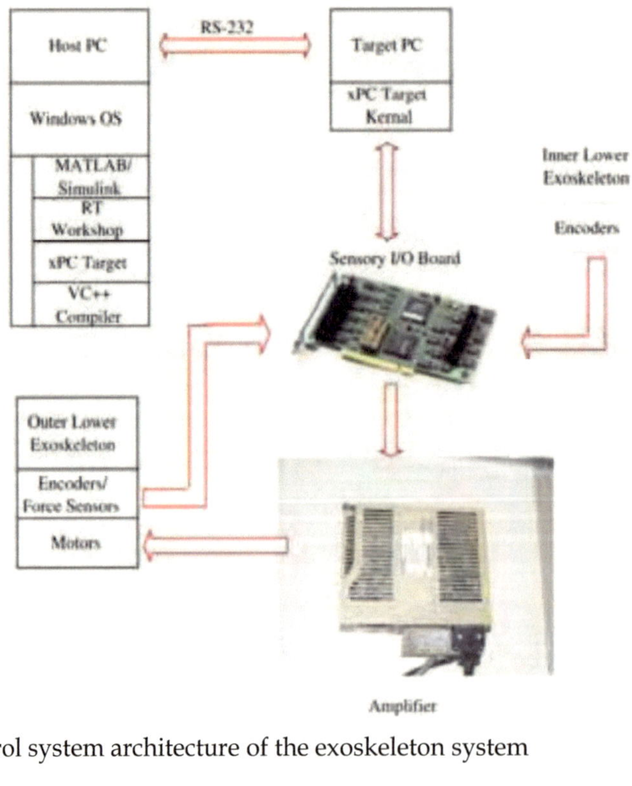

Figure 23. Control system architecture of the exoskeleton system

Target realtime kernel. This capability to rapidly produce code directly from verified system results allows many different techniques to be tried on hardware in a more convenient way compared to other software, operating systems. Developers, who use the other systems, most of the time, have to deal with low-level programming languages directly. These kinds of low-level programming languages are somewhat not very efficient in the sense that they take developers long time to deal with coding and debugging. The system includes softwares such as Matlab, xPC Target etc., which run in the PCs, and I/O boards that receive commands from xPC Target and control the motors via amplifies. The overall architecture of the host-target hardware-in-the-loop exoskeleton control system established in this work is shown in Figures 23 and 24.

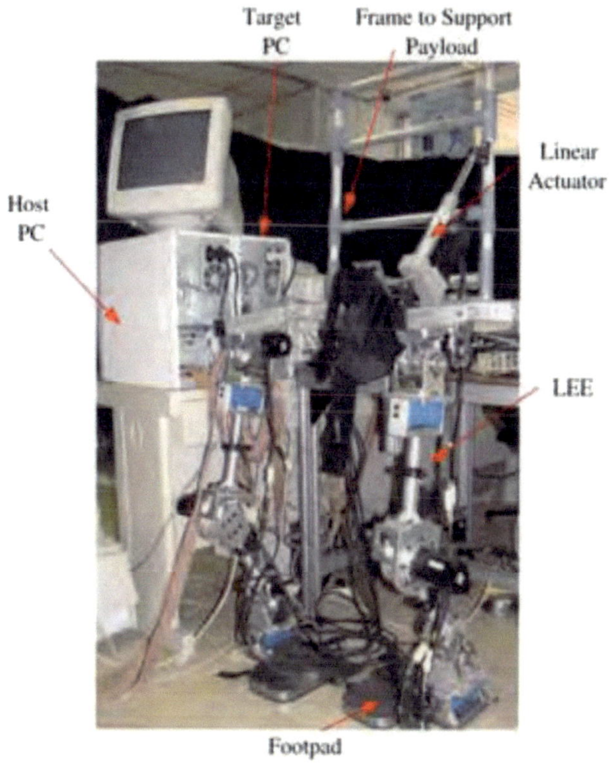

Figure 24. Set-up of developed control system

7.2 Control of the Developed Exoskeleton

Quite often, robots have the two lowest control levels. The tactical level generates the trajectories of each DoF, which perform the desired functional movement, while the executive level executes these trajectories by means of appropriate actuators, incorporated in each DoF. The two upper control levels are generally recognized as intelligence level. R.obots with the two upper levels such as ASIMO (*Asimo. the humanoid robot by Honda,* Dec. 2003; Hirose, Haikawa, Takenaka, & Hirai, 2001), QRIO (*QRIO. the humanoid entertainment robot by SONY,* Dec. 2003), HRP-2 (Kanehiro et al., 2003; Kaneko et al., 2004), etc. can walk on a terrain of unknown profile. While those robots walking in a known environment whose trajectories of each link can be pre-defined off-line do not need the two upper levels. In stead of artificial intelligence, the highest level of the exoskeleton's control structure is implemented by the human user. The user recognizes the obstacles using his/her natural sensors, decide where to go and how to go. The second level is to transfer the human intention into the exoskeleton's controller and the controller divides the imposed operation into elementary movements. The two lowest control levels are similar to those of the robots. In other words, the exoskeleton can be seen as a robot integrated with the human intelligence. Figure 25 shows the snapshots of a trial with a patient whose left leg is very weak after strokes. The patient confirmed that the exoskeleton's leg responds well and he could hardly feel any extra burden.

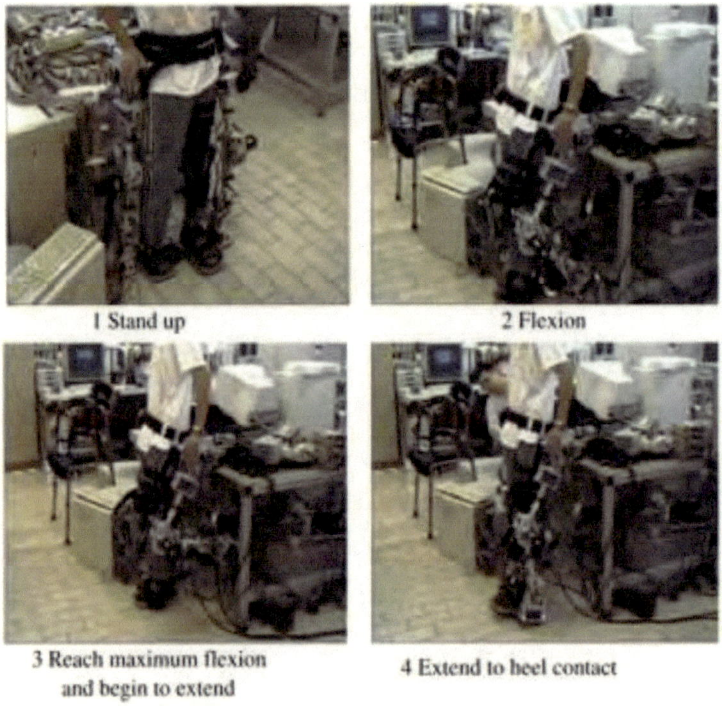

Figure 25. Exoskeleton trial with post-stroke patient

7.3 Measurement of the ZMP

As mentioned earlier, the ZMP has been introduced into the control strategy for stable exoskeleton walking. In this section, the experiment of measuring the human ZMP during walking is presented. In this experiment, a human walks as described in Figure 26. First, the human stands up, the width between his left foot and right foot is 300 mm. He then walks forward, and his right foot moves first, from position Rl to R2. Next, his left foot moves from position LI to position L2. Thirdly, his right foot moves from position R2 to position R3. At last, his left foot moves from position L2 to position L3 and he recovers stand posture. The step length is 300 mm. There are four force sensors placed below the human each foot. The reading in Figure 27 shows the timing of the ground reaction force below the two feet. At the first second, the human begins to walk. From the fifth second the human recovers stand posture. Using Eqs. (10) and (11), the ZMP of the human is calculated, as shown in Figure 28. Figure 27 and Figure 28 show that when the human begins to walk, the ground reaction force below his right foot decreases while the force below his left foot increases. The ZMP shifts from the middle of two feet to the left foot, i.e., the supporting foot area. During the walking, the ZMP shifts between the two feet. At last, when the human recovers stand posture, the ZMP stays between the two feet.

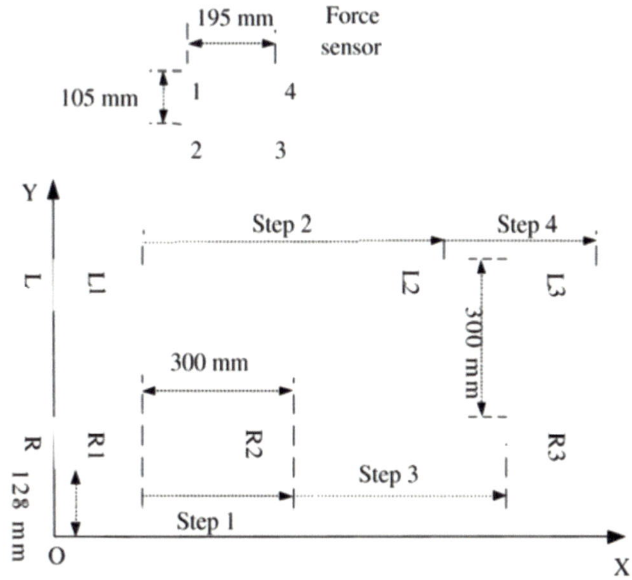

Figure 26. Steps of the walking experiment

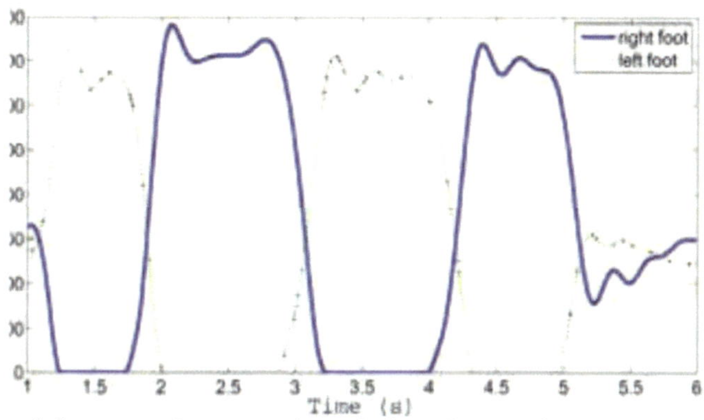

Figure 27. Timing of the ground reaction force below the two feet

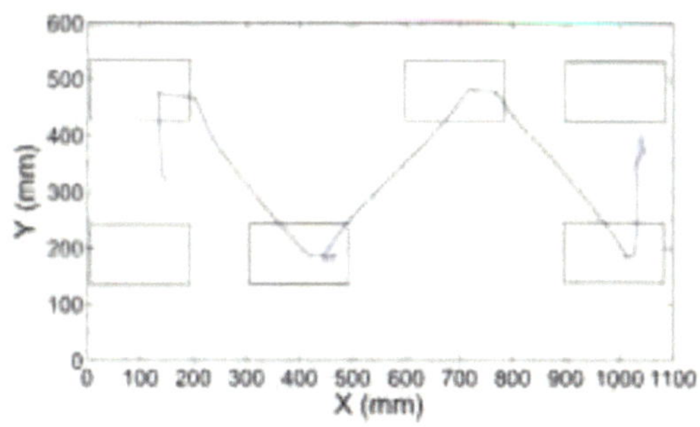

Figure 28. ZMP trajectory in the X-Y coordinate system

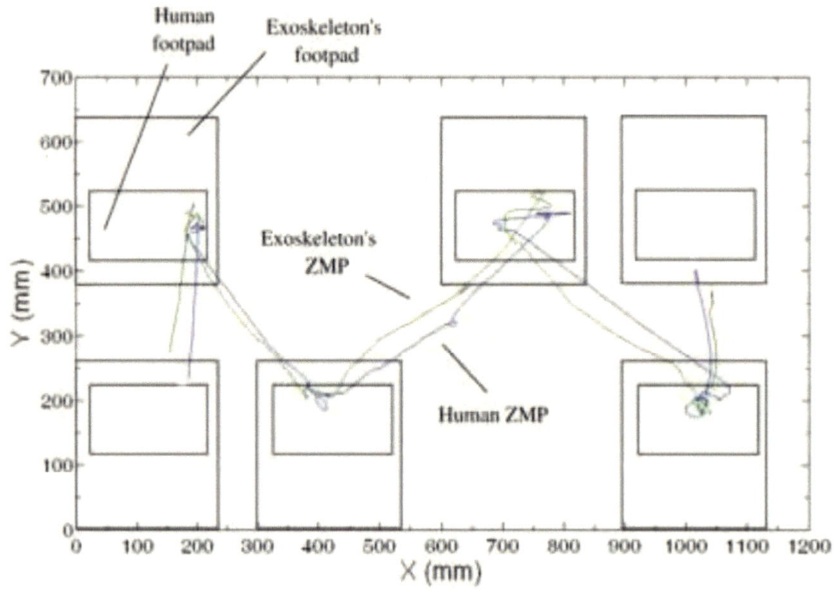

Figure 29. Human and exoskeleton's ZMP trajectories in the X-Y coordinate system

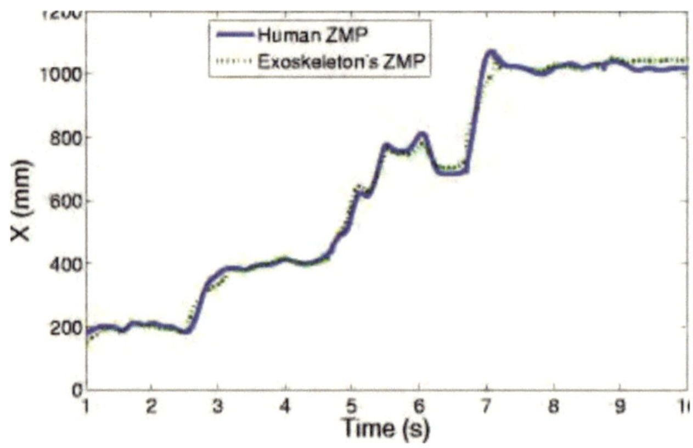

Figure 30. X coordinate of the human and exoskeleton ZMPs with respect to time

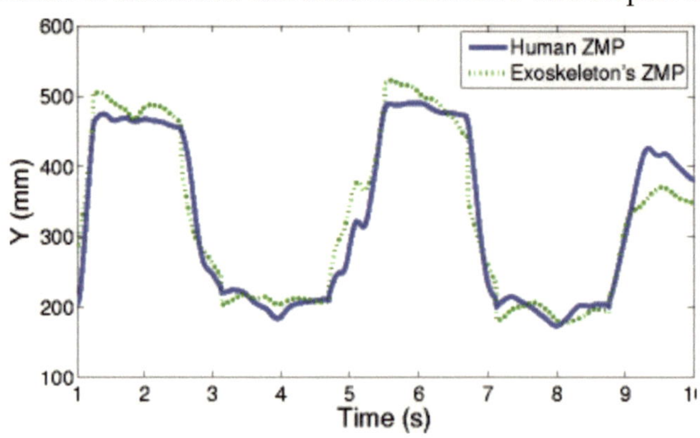

Figure 31. Y coordinate of the human and exoskeleton ZMPs with respect to time

7.4 Exoskeleton's Walking Experiments

An online balancer is added into the control scheme (Low et al., 2006). One more I/O board is employed, which reads the encoder of the trunk joint, signals from force sensors and control the actuator of the trunk joint. The balancer module reads the current joint angles and measured human ZMP then calculates the desired OLE's ZMP. If the measured OLE's ZMP differs from the desired one, suitable command signal will be applied to the actuator of the trunk joint and trunk compensation will shift the actual ZMP toward the desired position, as described previously. The ZMP will stay within the supporting area and the OLE will walk stably. A walking experiment is performed to check the balancer module. The walk procedure is similar to that in Figure 26. Figure 29 shows human and exoskeleton's ZMP trajectories in the *X-Y* coordinate system. This figure shows that the ZMP is adjusted during single support phase and the ZMPs are always kept in support area. Hence the exoskeleton can walk with the user stably. To see the relationship between the human ZMP and exoskeleton's ZMP more clearly, Figures 30 and 31 show the *X* and *Y* coordinates of the human and exoskeleton ZMPs with respect to time, respectively. These figures show that in *X* direction, the exoskeleton's ZMP is close to the human ZMP due to the trunk compensation, which ensures that the exoskeleton can walk stably following the human. In *Y* direction, the distance between exoskeleton's ZMP and the human ZMP is bigger. This is because the exoskeleton's footpad is much wider than human footpad (see Figure 29), therefore the allowable range of the exoskeleton's ZMP in *Y* direction is bigger. More experiments are performed including walking forward and backward with a payload of 20 kg. Figure 32 shows some snapshots of one of those experiments.

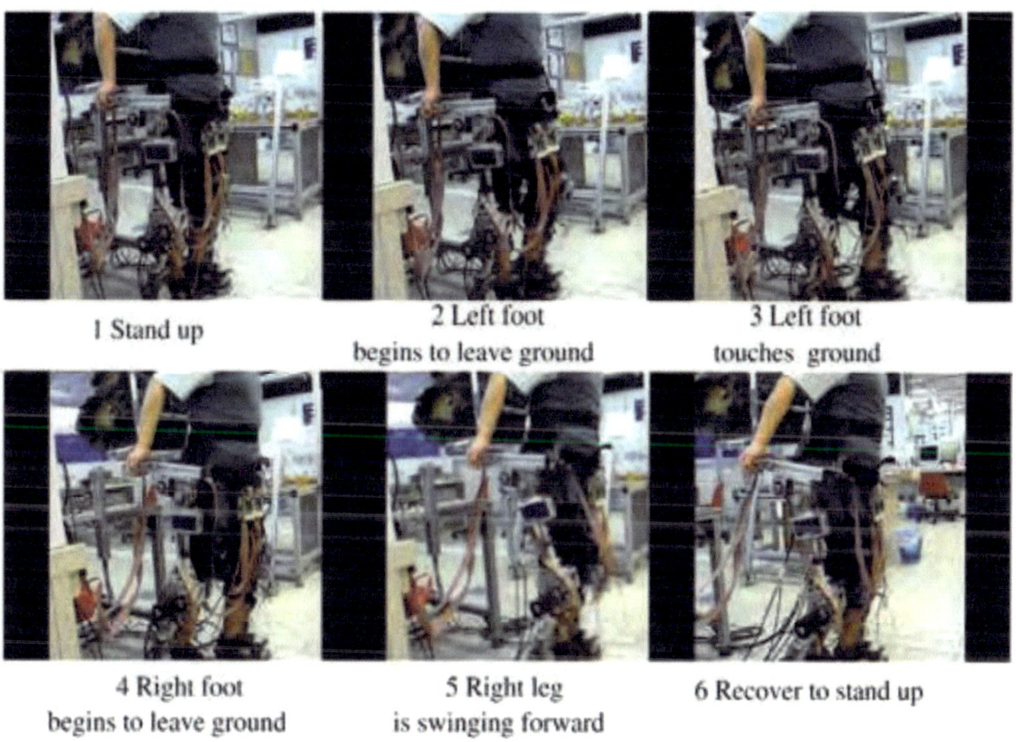

Figure 32. Snapshots of a walking test

8. Conclusion and Recommendations

8.1 Conclusion

This chapter has presented the development and control of a wearable lower exoskeleton system for augmentation of human walking ability, which incorporates human as the integral part of the control system and can relieve humans physical fatigue caused by excessive walking and heavy payloads. In this work, xPC Target, together with other toolboxes from MATLAB have been used so as to provide a real-time operating system and an integrated development environment for controlling the exoskeleton. Real-time control of the exoskeleton is implemented in this environment. At last, walking experiments are performed and demonstrated.

8.2 Future Work

The first prototype is only a test bed to verify the control algorithms. It looks bulky and rough. Though it has been verified that it can walk, there are some topics need to improve before one will see a reliable and useful exoskeleton that can effectively enhance humans strength and endurance. Another area of improvement may be the addition of more passive joints to the OLE prototype. These passive joints may be spring loaded such that it allows for passive movements without affecting the structural characteristics of the OLE. Examples of passive joints can be at positions like the feet where the passive joint can allow the bending of the OLE foot when walking. This will pertain to the users extension of the toes while walking. This addition of passive joints at the OLE feet will enable the OLE to be able to reciprocate the human movements better and at the same time, allow the user to have a more natural walking gait while wearing the OLE. Other areas where passive joints can be added are the hip joints and the ankle joints to allow them rotate. Now it is the inner exoskeleton who senses the user movements. However, the inner exoskeleton increases the distance between the user and the OLE. They can be replaced with goniometers *(BIOPAC systems. Inc.,* n.d.), which are more compact and light. Besides, human joints are not simple pin joints but complex joints with a changing instantaneous center of rotation. Compared to the hard linkages of the inner exoskeleton, goniometer has a telescopic end block that compensates for changes in distance between the two mounting points as the limb moves. The gauge mechanism allows for more accurate measurement of polycentric joints. To save energy and consequently reduce the size of the power source (e.g. batteries) carried by the OLE, ways to help decrease the torque required at the OLE joints could be developed. Such torque compensation mechanisms are not easy to construct, especially on the OLE. Issues like the amount of torque that will be reduced versus the weight increased by the addition of such a mechanism are widely debatable. Also, such a mechanism should not hinder the movements of the user or the OLE. With the addition of such torque compensation mechanisms, the controlling of the actuators will become more complicated. This is because of the additional parameters to controlling the torque output of the actuators. Nevertheless, a few torque compensation mechanisms can be a possible solution for the OLE. Firstly, in the development of the Saika-4 by Tohoku University (Shirata, Konno, & Uchiyama, 2004), a method of torque compensation was mentioned. Using a stainless-steel wire and a contrive spring layout, the mechanism is able to compensate without depending on joint angles. This is important as methods that depend on joint angles compensate with changing forces at every joint angle position. This increases the robustness of the control of the system. The mechanism in Saika-4 is reported to be able to reduce the torque requirements of the motors

at the robot joints because the robot legs always tend to return to the initial standing position by the spring force provided in the mechanism.
By virtue of the inner-outer exoskeleton systems, the proposed assistive gait device has been developed and tested initially for strength endurance and rehabilitation. The extension of the present study is currently applied to the rehabilitation of people with SCI (spinal cord injury) and post-stroke.

9. Acknowledgements

The authors would like to thank Miss Yunyun Huang, Mr. Tien-Boon Tan, Mr. Sui-Nan Lim, Mr. Tze-Kuang Ng, and Mr. Rene Wachter for their assistance in programming and experiments. Also, the help of Mr. Leong-Yeo Wai on the word processing is greatly appreciated. This research is supported by Research Grant MINDEF-NTU/02/01, a grant from the Ministry of Defence, Singapore. Robotics Research Center of the Nanyang Technological University provides technical helps, space and facilities.

10. References

Low K. H. and Yin Y., An integrated lower exoskeleton system towards design of a portable active orthotic device, *International Journal of Robotics and Automation*, 22 (1), pp. 32-42.

Berkeley robotics laboratory. (http://bleex.me.berkeley.edu/), Dec. 2004.

Exoskeletons for human performance augmentation projects. (http://www.darpa.gov/dso/thrust/md/Exoskeletons), 2002, Dec.

Guizzo, E., & Goldstein, H. The rise of the body bots. *IEEE Spectrum*, 42-48. 2005, Oct.

Haug, E. J. *Computer aided kinematics and dynamics of mechanical systems.* vol. 1: basic methods. Allyn & Bacon, Inc. 1989

Hembree, A. *Armor: the things they'll carry.* Wired News Online. (http://www.wired.eom/news/culture/0,1284,41216,00.html) 2001, Jan.

Hirose, M., Haikawa, Y., Takenaka, T., & Hirai, K. *Development of humanoid robot ASIMO.* 2001.

Homepage of the Adams University. (http://university.adams.com/tutorials.htm) Jan. 2005.

Kanehiro, F., Kaneko, K., Fujiwara, K., Harada, K., Kajita, S., Yokoi, K., et al. *The first humanoid robot that has the same size and that can lie down and get up.* 2003, Sep.

Kaneko, K., Kanehiro, F., KajitaK, S., Hirukawa, H., Kawasaki, T., Hirata, M., et al.. *Humanoid robot HRP-2.* 2004.

Kapandji, I. A. (1987). *The physiology of the joints* (5th ed., Vol. 2). Edinburgh London Melbourne and New York: Churchill Livingstone. (Translation of: Physiologic articulaire. v. 1. Upper limb - v. 2. Lower limb - v. 3. The trunk and the vertebral column (2nd ed., 1974))

Kasaoka, K., & Sankai, Y. *Predictive control estimating operator's intention for stepping-up motion by exoskeleton type power assist system HAL.* 2001, Oct..

Kawamoto, H., & Sankai, Y. *Comfortable power assist control method for walking aid by HAL-3.* 2002, Oct.)

Kazerooni, H., Racine, J.-L., Huang, L., & Steger, R. *On the control of the berkeley lower extremity exoskeleton* (BLEEX). 2005, Apr.

Low, K. H., Liu, X., Goh, C. H., & Yu, H. Locomotive control of a wearable lower exoskeleton for walking enhancement. *Journal of Vibration and Control*, 12(12), 1311-1336. 2006.

Low, K. H., & Yin, Y. An integrated lower exoskeleton system towards design of a portable active orthotic device. *International Journal of Robotics and Automation*, 22(1) in press. 2007.

Marchese, S., Muscato, G., & Virk, G. S. *Dynamically stable trajectory synthesis for a biped robto during the single-support phase*. Como, Italy. 2001, Jul.

Mavroidis, C. Smart portable rehabilitation devices. *Journal of NeuroEngineering and Rehabilitation*, 2(18), 8 pages. 2005.

Asimo, the humanoid robot by Honda. (http://www.honda.co.jp/ASIMO), Dec. 2003.

BIOPAC systems, Inc. (n.d.). (http://biopac.com/fr_prod.htm)

MATLAB - the language of technical computing. (http://www.mathworks.com/products/matlab/), Jan. 2005.

QRIO, the humanoid entertainment robot by SONY. (http:// www. sony.net/Sony Info/QRIO/story/index_nf. html), Dec. 2003.

RoboWalker. (http://yobotics.com/robowalker/robowalker.html), Dec. 2003.

Mosterman, P., Prabhu, S., Dowd, A., Glass, J., Erkkinen, T., Kluza, J., et al. *Handbook on networked and embedded systems.* In D. Hristu-Varsakelis & W. S. Levine (Eds.), (chap. Section 3.4: Embedded Real-Time Control via MATLAB, Simulink, and xPC Target). Birkhauser, Boston. 2005.

Powered-suit gives aged a leg up. (http://newpaper.asia!.com.sg/top/story/0,4136,31694,00.html), Dec. 2003

Powered suit HAL (Hybrid Assistive Leg). (http://sanlab.kz.tsukuba.ac.jp/HAL/indexE.html), Dec. 2003.

Shirata, S., Konno, A., & Uchiyama, M. Design and development of a light-weight biped humanoid robot Saika-4-, 2004, Sep.

Tomohiro Hayashi, H. K., & Sankai, Y. *Control method of robot suit HAL working as operator's muscle using biological and dynamical information.* 2005, Aug.

Vukobratovic, M., Borovac, B., Surdilovic, D., & Stokic, D. *The mechanicl systems design handbook*: modeling, measurement, and control. In (chap. Humanoid Robots). Osita D.I. Nwokah, Yildirim Hurmuzlu, Southern Methodist University Dallas, Texas. 2001.

Vukobratovic, M., & Juricic, D. *Contribution to the synthesis of biped gait.* 1969.

Wakefield, J. *US looks to create robo-soldier.* BBC News Online. (http://news.bbc.co.Uk/2/hi/science/nature/1908729.stm), 2002, Apr.

Whittle, M. W. *Gait analysis*: an introduction. Oxford; Boston: Butterworth-Heinemann. 1991.

2

A Novel Anthropomorphic Robot Hand and its Master Slave System

Tetsuya Mouri and Haruhisa Kawasaki
Gifu University
Japan

1. Introduction

Future humanoid robots will execute various complicated tasks based on communication with human users. These humanoid robots will be equipped with anthropomorphic robot hands much like the human hand. The robots will eventually supplant human labor in the execution of intricate and dangerous tasks in areas such as manufacturing, space exploration and the seabeds.

Many multi-fingered robot hands (Salisbury & Craig, 1982) (Jacobsen et al., 1984) (Jau, 1995) (Kyriakopoulos et al., 1997) have been developed. These robot hands are driven by actuators in a location remote from the robot hand frame and are connected by tendon cables. Elasticity in the tendon cable causes inaccurate joint angle control, and the long wiring of tendon cables may obstruct the motion of the robot when the hand is attached to the tip of a robot arm. To solve these problems, robot hands in which the actuators are built into the hand (Bekey et al., 1990) (Rosheim, 1994) (Lin & Huang, 1996) (Butterfass et al., 2001) (Namiki et al., 2003) (Yamano et al., 2003) have been developed. However, these hands have the problem that their movement differs from that of the human hand because both the number of fingers and the number of joints in the fingers are insufficient. Recently, many reports (Fearing, 1990) (Howe, 1994) (Shimojo et al., 1995) (Johnston et al., 1996) (Jockusch et al., 1997) have been presented on the use of tactile sensors that attempt to realize adequate object manipulation through contact with the fingers and palm. There has been only the slight development of a hand that combines a 6-axes force sensor attached to the fingertips with a distributed tactile sensor mounted on the hand surface.

To provide a standard robot hand used to study grasping and dexterous manipulation, our group has developed the Gifu Hand I (Kawasaki & Komatsu, 1998) (Kawasaki & Komatsu, 1999), the Gifu Hand II (Kawasaki et al., 1999), the Gifu Hand III (Mouri et al., 2002), and the kinetic humanoid hand (Kawasaki et al., 2004).

This paper presents a novel robot hand called the KH (Kinetic Humanoid) Hand type S for sign language, which requires a high degree of fingertip velocity. In addition, we construct a PC-based master slave system to demonstrate effectiveness in grasping and manipulating objects. An experiment involving grasping and manipulating objects by the master slave control is shown. Our results show that the KH Hand type S has a higher potential than previous robot hands in rendering a picturesque hand shape and performing dexterous object manipulations like the human hand.

2. An Anthropomorphic Robot Hand

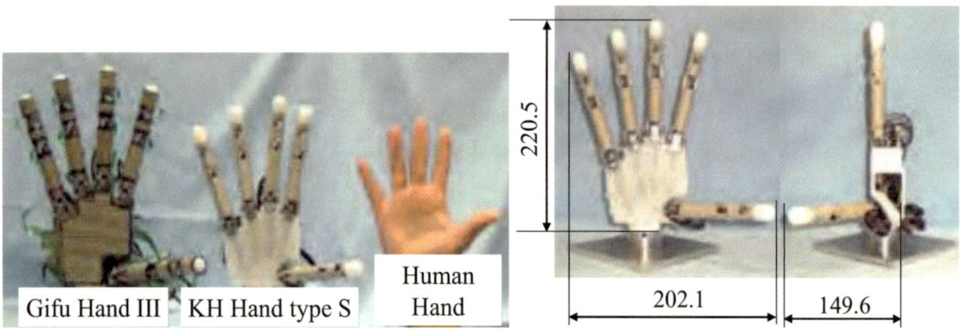

Figure 1. KH Hand type S

Recently, various robot hands with built-in actuators have been developed. However, these hands present the problem that their movement differs from that of the human hand because they have had an insufficient number of fingers and finger joints. To provide a standard robot hand to be used to study grasping and dexterous manipulation, our group developed the Gifu Hand I (Kawasaki & Komatsu, 1998) (Kawasaki & Komatsu, 1999), the Gifu Hand II (Kawasaki et al., 1999), and the Gifu Hand III (Mouri et al., 2002). The design concept for these robot hands was as follows.

1. **Size**: It is desirable that for skillful manipulation the robot hand resemble the human hand in size.
2. **Finger DOF**: The number of joints and the DOF motion in the robot hand are similar to those of the human hand.
3. **Opposability of the thumb**: The robot hand thumb is opposed to the four other fingers, enabling the hand to manipulate objects dexterously like the human hand.
4. **Force sensor**: The robot hand grasps and manipulates objects dexterously with the help of tactile sensors and force sensors in the fingers.
5. **Built-in servomotor method**: The motion of the robot arm is not disturbed by the robot hand, and the robot hand is easily attached to the robot arm.
6. **Unit design of the finger**: Each joint must be modular, and each finger must be a unit in order to realize easy maintenance and easy manufacture of the robot hand.

The Gifu Hand series are 5-finger hands driven by built-in servomotors that have 20 joints with 16 DOF. These hands use commercial motors. The length of the robot hand, in which the actuators are built, depends on the size of the motors. On this basis, we have developed and are presenting the smaller kinetic humanoid hand, which uses prototype brushless motors (Kawasaki et al., 2004). In the older robot hands, the fingertip velocity was slow because their motors had high reduction ratio gears. The shape and freedom of motion of our robot hands are almost equivalent to those of human hands. Therefore, we can use the robot hands not only for grasping and manipulating objects but also as communication tools for such as sign language. Because the drivers for the brushless motor are large and have much hardwiring, it has been difficult to miniaturize and make practicable the kinetic humanoid hand. Therefore, the new robot hand, which can be driven at same speed as a human hand, has been developed based on the use of a commercial DC motor with the kinetic humanoid hand.

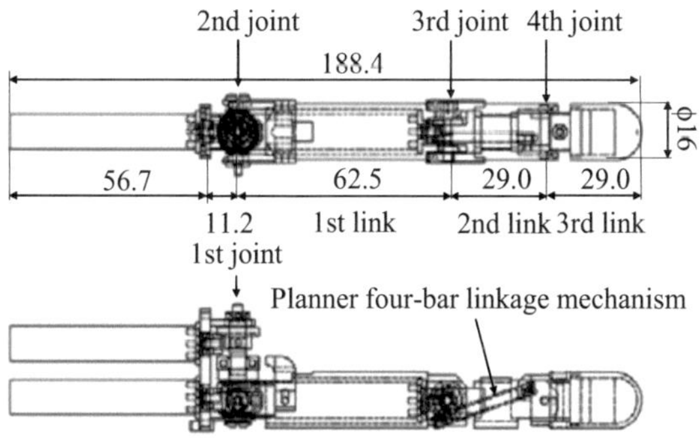

Figure 2. Design of fingers

Weight [kg]	Finger	0.097
	Total	0.619
Length [mm]	Finger	188.4
	Total	220.5
Operating angle of joints [deg]	1st	-20 ~ 20
	2nd	-10 ~ 90
	3rd	-10 ~ 90
	4th	-10 ~ 90
Fingertip force [N]		0.86
Gear ratio	1st	614.1 : 1
	2nd	307.2 : 1
	3rd	134.4 : 1

Table 1. Specifications

2.1 Characteristics

An overview of the developed KH Hand type S is shown in Figure 1. The hand has five fingers. The finger mechanism is shown in Figure 2. The servomotors and the joints are numbered from the palm to the fingertip. Each of the fingers has 4 joints, each with 3 DOF. The movement of the first finger joint allows adduction and abduction; the movement of the second to the fourth joints allows anteflexion and retroflexion. The third servomotor actuates the fourth joint of the finger through a planar four-bar linkage mechanism. The fourth joint of the robot finger can engage the third joint almost linearly in the manner of a human finger. All five fingers are used as common fingers because the hand is developed for the purpose of expressing sign language. Thus, the hand has 20 joints with 15 DOF. Table 1 summarizes the characteristics of KH Hand type S. The weight of the hand is 0.656 kg, and the bandwidth for the velocity control of the fingers is more than 15 Hz, which gives them a faster response than human fingers. The dexterity of the robot hand in manipulating an object is based on thumb opposability. The thumb opposability (Mouri et al., 2002) of the robot hand is 3.6 times better than that of the Gifu Hand III. To enable compliant pinching, we designed each finger to be equipped with a six-axes force sensor, a commercial item. Tactile sensors (distributed tactile sensors made by NITTA Corporation) are distributed on the surface of the fingers and palm. The hand is compact, lightweight, and anthropomorphic in terms of geometry and size so that it is able to grasp and manipulate like the human hand. The mechanism of KH Hand type S is improved over that of the kinetic humanoid hand, as described in the next section.

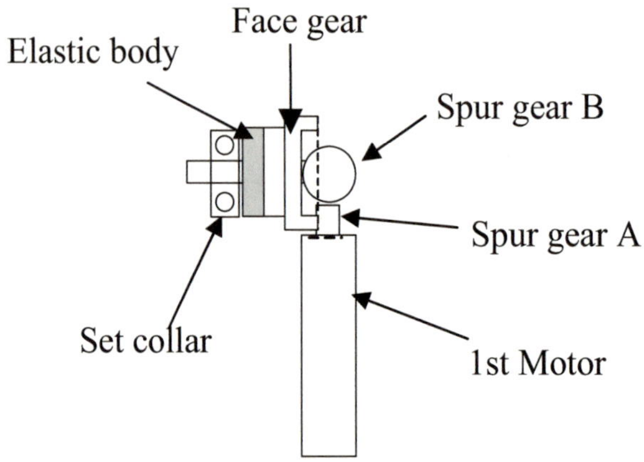

Figure 3. Reduction of backlash

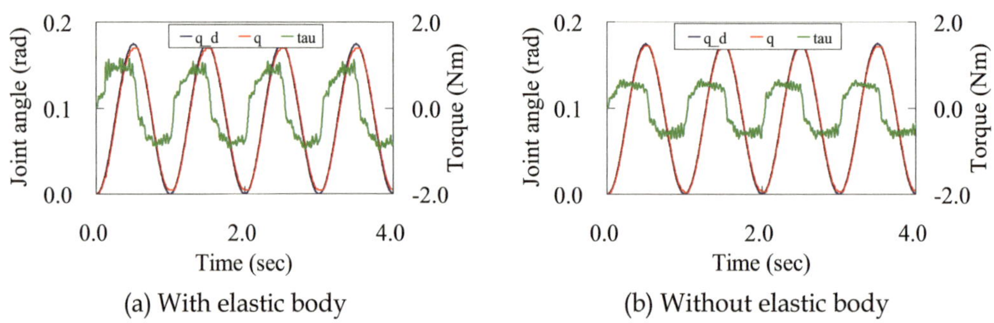

Figure 4. Effects of elastic body

2.2 Weight Saving

The weight of Gifu Hand III and the kinetic humanoid hand are 1.4 and 1.09 kg, respectively. The major part of the weight of Gifu Hand III is the titanium frame of the fingers. Therefore, the new KH Hand type S uses a plastic frame for the fingers and palm, and its weight is 0.61 times lighter than that of the older kinetic humanoid hand.

2.3 Motors

The Gifu Hand III has been developed with an emphasis on fingertip forces. High output motors have been used, with the hand's size being rather larger than that of the human hand. In order to miniaturize the robot hand, compact DC motors (the Maxson DC motor, by Interelectric AG), which have a magnetic encoder with 12 pulses per revolution, are used in the new robot hand. The diameter of servomotors was changed from 13 to 10 mm. The fingertip force of KH Hand type S is 0.48 times lower than that of the Gifu Hand III and has a value of 0.86 N. At the same time, its fingertip velocity is higher.

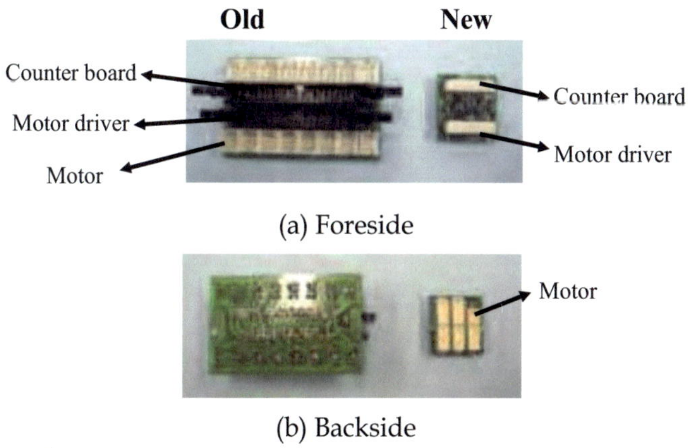

Figure 5. Transfer substrate

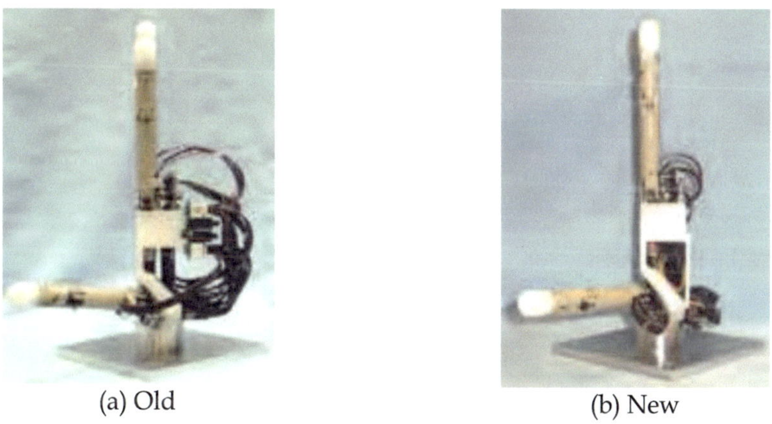

Figure 6. Over view with transfer substrate

2.4 Reduction of Backlash in the Transmission

The rotation of the first and second joints is controlled independently through an asymmetrical differential gear by the first and second servomotors. The backlash of the first and second joints depends on the adjustment of the gears shown in Figure 3. The lower the backlash we achieve, the higher becomes the friction of the gears transmission. An elastic body, which keeps a constant contact pressure, was introduced between the face gear and spur gears to guarantee a low friction. The effects of the elastic body were previously tested in Gifu Hand III, with the experimental results shown in Figure 4. Both the transmissions with and without the elastic body were accommodated at the same level. A desired trajectory is a sine wave, and for that the joint torque is measured. Figure 4 shows that the root mean joint torques without and with the elastic bodies were 0.72 and 0.49 Nm, respectively. Hence, the elastic body helps to reduce the friction between the gears.

2.5 Transfer Substrate

The robot hand has many cables, which are motors and encoders. The transfer substrate works the cables of counter boards and a power amp of the driving motors that are connected to the motors that are built in the fingers. Therefore, a new transfer substrate was developed for downsizing. Figure 5 shows the foreside and backside of the developed transfer substrate, which is a double-sided printed wiring board. The pitch of the connectors was changed from 2.5 to 1.0 mm. Compared with the previous transfer substrate, the weight is 0.117 times lighter and the occupied volume is 0.173 times smaller. Figure 6 shows an overview of a KH Hand type S equipped with each transfer substrate. As a result of the change, the backside of the robot hand became neat and clean, and the hand can now be used for the dexterous grasping and manipulation of objects, such as an insertion into a gap in objects.

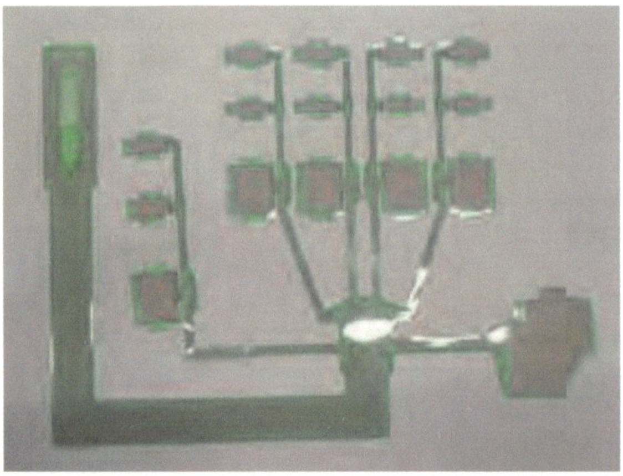

Figure 7. Distributed tactile sensor

Number of detecting points	
Total	895
Palm	321
Thumb	126
Finger	112
Maximum load [N/m^2]	2.2x10^5
Electrode column width [mm]	2.55
Electrode row width [mm]	3.35
Column pitch [mm]	3.40
Row pitch [mm]	4.20

Table 2. Characteristic of distributed tactile sensor

2.6 Distributed Tactile Sensor

Tactile sensors for the kinetic humanoid hand to detect contact positions and forces are mounted on the surfaces of the fingers and palm. The sensor is composed of grid-pattern electrodes and uses conductive ink in which the electric resistance changes in proportion to the pressure on the top and bottom of a thin film. A sensor developed in cooperation with the Nitta Corporation for the KH Hand is shown in Figure 7, and its characteristics are shown in Table 2. The numbers of sensing points on the palm, thumb, and fingers are 321,

126 and 112, respectively, with a total number of 895. Because the KH Hand has 36 tactile sensor points more than the Gifu Hand III, it can identify tactile information more accurately.

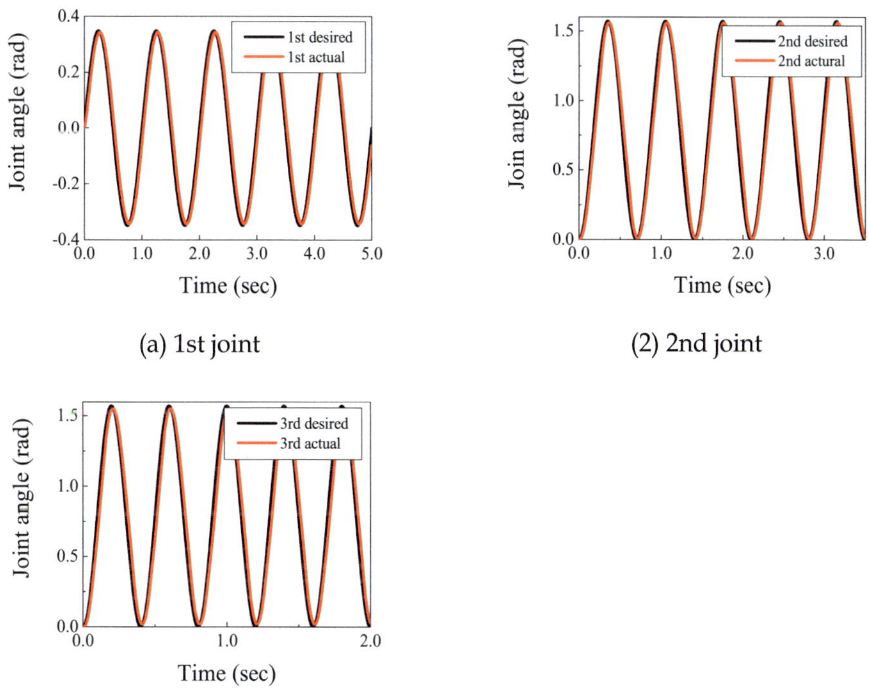

(a) 1st joint

(2) 2nd joint

(c) 3rd joint

Figure 8. Trajectory control

2.7 Sign Language

To evaluate the new robot hand, we examined control from branching to clenching. Figure 8 shows the experiment results. The result means that the angle velocity of the robot hand is sufficient for a sign language.

Sign language differs from country to country. Japanese vocals of the finger alphabet using the KH Hand type S are shown in Figure 9. The switching time from one finger alphabet sign to another one is less than 0.5 sec, a speed which indicates a high hand shape display performance for the robot hand.

3. Master Slave System

In order to demonstrate effectiveness in grasping and manipulating objects, we constructed a PC-based master slave system, shown in Figure 10. An operator and a robot are the master and slave, respectively. The operator controls the robot by using a finger joint angle, hand position and orientation. The fingertip force of the robot is returned to the operator, as shown in Figure 11. This is a traditional bilateral controller for teleoperations, but to the best of our knowledge no one has previously presented a bilateral controller applied to a five fingers anthropomorphic robot hand. In general, in a master slave system, a time delay in communications must be considered (Leung et al., 1995), but since our system is installed in a single room, this paper takes no account of the time delay.

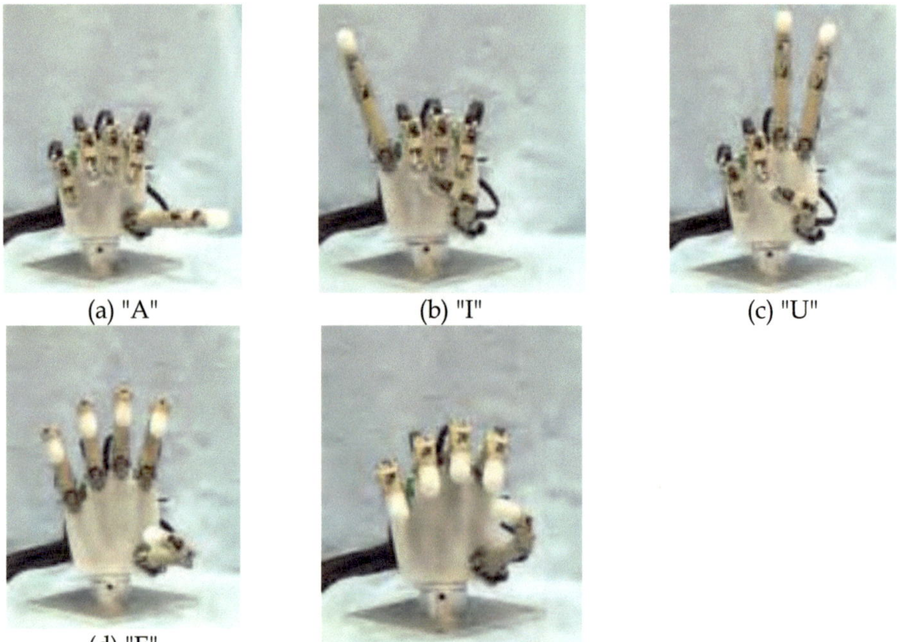

(a) "A" (b) "I" (c) "U" (d) "E" (e) "O"

Figure 9. Japanese finger alphabet

3.1 Master System

The master system to measure the movement of the operator and to display the force feeling is composed of four elements. The first element, a force feedback device called a FFG, displays the force feeling, as will be described in detail hereinafter. The second is a data glove (CyberGlove, Immersion Co.) for measuring the joint angle of the finger. The third is a 3-D position measurement device (OPTOTRAK, Northern Digital Inc.) for the hand position of operator and has a resolution of 0.1 mm and a maximum sampling frequency of 1500 Hz. The fourth element is an orientation tracking system (InertiaCube2, InterSense Inc.) for the operator's hand posture; the resolution of this device is 3 deg RMS, and its maximum sampling frequency is 180 Hz. The operating system of the PCs for the master system is Windows XP. The sampling cycle of the FFG controller is 1 ms. The measured data is transported through a shared memory (Memolink, Interface Co.). The hand position is measured by a PC with a 1 ms period. The sampling cycle of the hand orientation and the joint angle is 15 ms. The FFG is controlled by a PI force control. Since sampling cycles for each element are different, the measured data are run through a linear filter.

The developed robot hand differs geometrically and functionally from a human hand. A method of mapping from a human movement to the command of the robot is required, but our research considers that the operator manipulates the system in a visceral manner. The joint angle can be measured by the data glove, so that this system directly transmits the joint data and the hand position to the slave system, as we next describe.

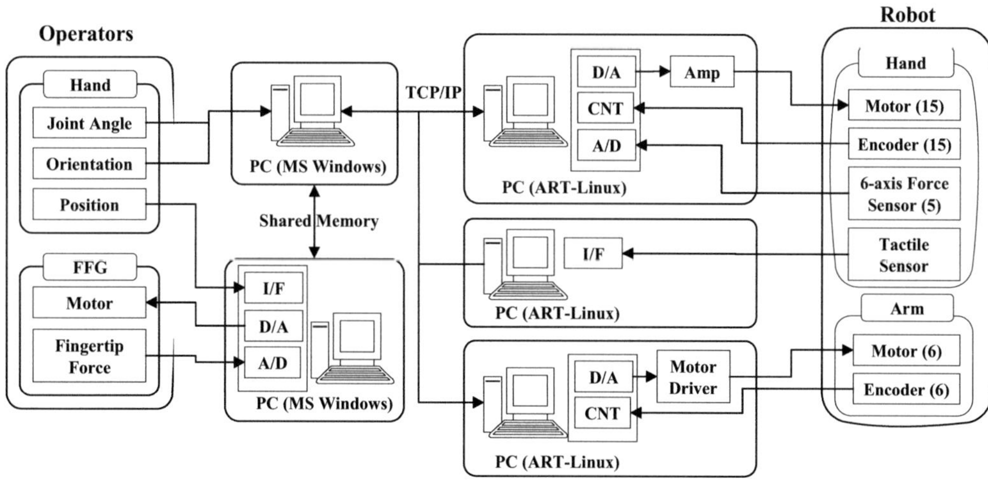

Figure 10. Control system

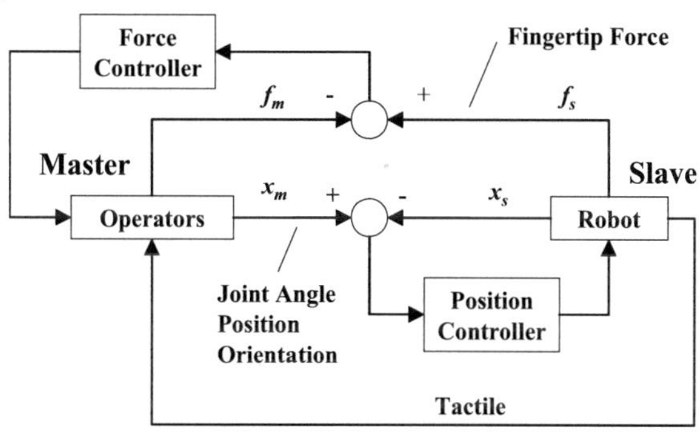

Figure 11. Master slave system

3.2 Slave System

The slave system consists of a hand and an arm. The robot hand is the developed KH Hand type S equipped with the 6-axes force sensor (NANO 5/4, BL. AUTOTEC Co.) at each fingertip and the developed tactile sensor. The robot arm is the 6-DOF robot arm (VS6354B, DENSO Co.). The operating system of the PCs for the slave system is ART-Linux, a real-time operating system (Movingeye, 2001). The tactile sensor output is processed by a PC with a 10 ms period. The measured tactile data is transported to a FFG control PC through TCP/IP. The sampling cycle of the hand and arm controller is 1 ms. Both the robot arm and hand are controlled by a PD position control.

3.3 Force Feed Back Glove

The forces generated from grasping an object are displayed to the human hand using the force feedback glove (FFG), as shown in Figure 12 (Kawasaki et al., 2003). The operator attaches the FFG on the backside of the hand, where a force feedback mechanism has 5 servomotors. Then the torque produced by the servomotor is transmitted to the human fingertips through a wire rope. The fingertip force is measured by a pressure sensitive conductive elastomer sensor (Inaba Co). A human can feel the forces at a single point on each finger, or on a total of 5 points on each hand. The resolution of the grasping force generated by the FFG is about 0.2 N. The force mechanism also has 11 vibrating motors located in finger surfaces and on the palm to present the feeling at the moment that objects are contacted. A person can feel the touch sense exactly at two points on each finger and at one point on the palm, or at a total of 11 points on each hand.

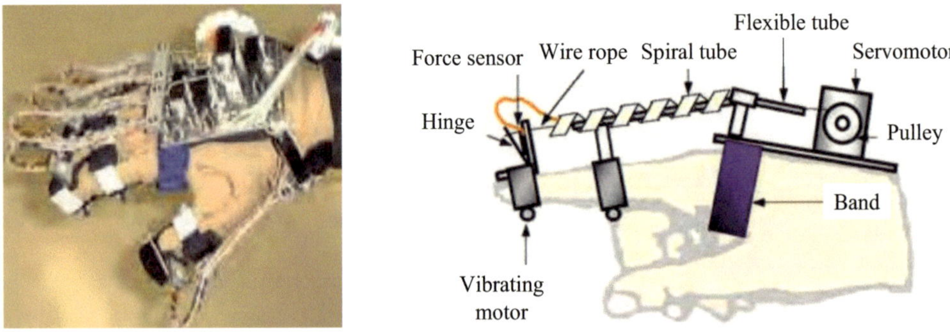

(a) Overview (b) Mechanism

Figure 12. Force feedback glove

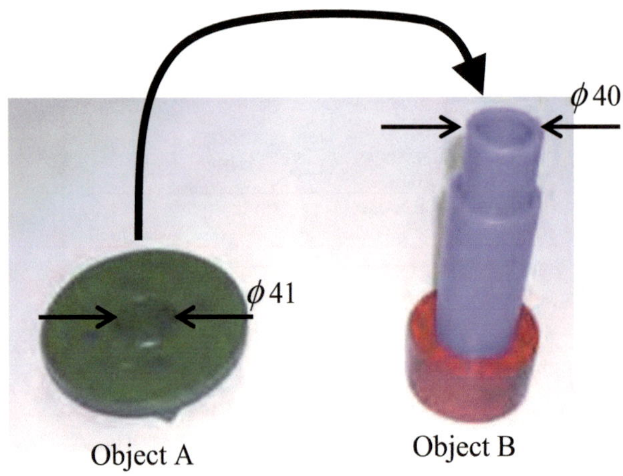

Figure 13. Peg-in-hole task

4. Experiment

4.1 Peg-in-hole task

As shown in Figure 13, a peg-in-hole task was conducted because it is the most fundamental assembly operation. We used two objects: a disk with a hole (A) and a cylinder (B). The weight of object A is 0.253 kg, the outer diameter is 0.13 m, and the hole's diameter is 0.041 m. The weight of object B is 0.198 kg, and the diameter is 0.040 m. The clearance between object A and B is 0.001 m.

The peg-in-hole task sequence is as follows. The robot (operator) approaches an object A, grasps the object, translates it closely to object B, and inserts it into object B.

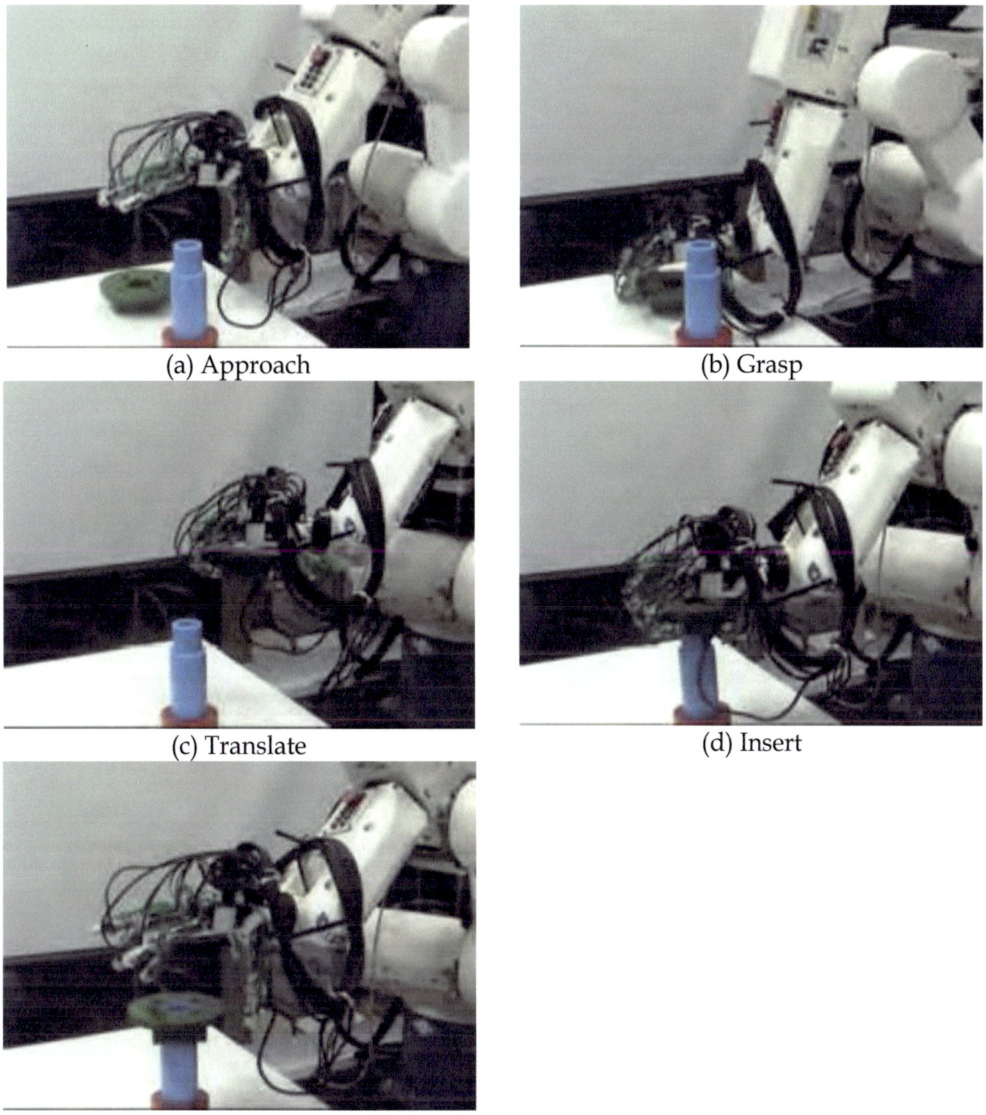

(a) Approach

(b) Grasp

(c) Translate

(d) Insert

(e) Completion

Figure 14. Sequence of peg-in-hole task

4.2 Experimental Result

Experimental results of the peg-in-hole task controlled by the master slave system are shown in Figure 14. Figures 15 and 16 show the joint angles and the position and orientation of the robot hand. We used the KH Hand types with the previous transfer substrate in this experiment. They indicate that the controlled variables are close to the desired ones. These results show that the KH Hand type S can perform dexterous object grasping and manipulation like the human hand.

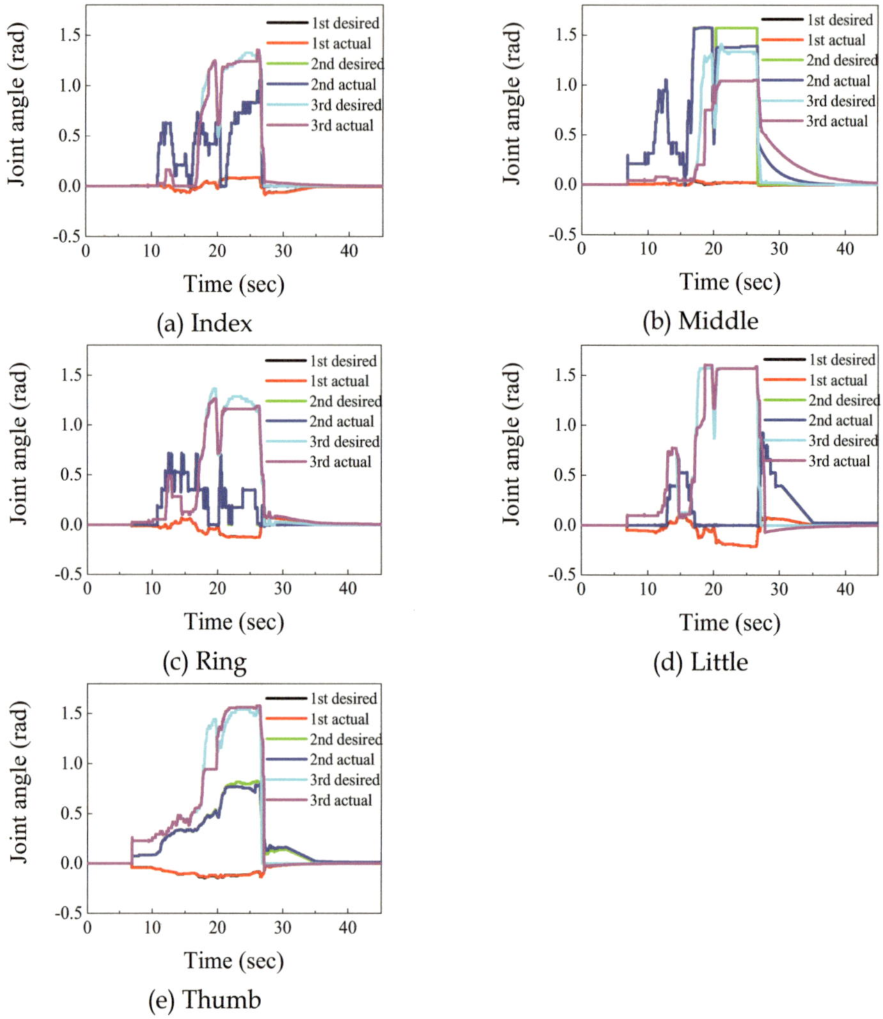

Figure 15. Joint angle of robot hand

5. Conclusion

We have presented the newly developed anthropomorphic robot hand named the KH Hand type S and its master slave system using the bilateral controller. The use of an elastic body has improved the robot hand in terms of weight, the backlash of the transmission, and friction between the gears. We have demonstrated the expression of the Japanese finger alphabet. We have also shown an experiment of a peg-in-hole task controlled by the bilateral controller. These results indicate that the KH Hand type S has a higher potential than previous robot hands in performing not only hand shape display tasks but also in grasping and manipulating objects in a manner like that of the human hand. In our future work, we are planning to study dexterous grasping and manipulation by the robot.

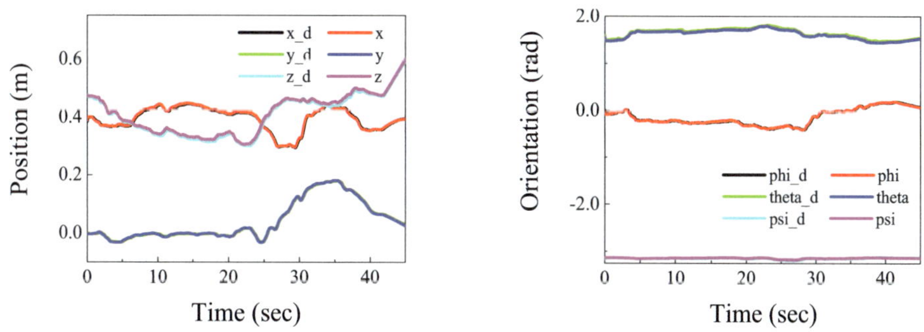

(a) Position (b) Orientation

Figure 16. Joint angle of robot arm

6. Acknowledgment

We would like to express our thanks to the Gifu Robot Hand Group for their support and offer special thanks to Mr. Umebayashi for his helpful comments.

7. References

Salisbury, J. K. & Craig, J. J. (1982). Articulated Hands: Force Control and Kinematic Issues, *International Journal Robotics Research*, Vol. 1, No. 1, pp. 4-17.

Jacobsen, S. C.; Wood, J. E.; Knutti, D. F. & Biggers, K. B. (1984). The Utah/MIT dexterous hand: Work in progress, *International Journal of Robotics Research*, Vol. 3, No. 4, pp. 21-50.

Jau, B. M. (1995). Dexterous Telemanipulation with Four Fingered Hand System, *Proceedings of IEEE Robotics and Automation*, pp. 338-343.

Kyriakopoulos, K. J.; Zink, A. & Stephanou, H. E. (1997). Kinematic Analysis and Position/Force Control of the Anthrobot Dextrous Hand, *Transaction on System, Man, and Cybernetics-Part B: cybernetics*, Vol. 27, No. 1, pp. 95-104.

Bekey, G. A.; Tomovic, R. & Zeljkovic, I. (1990). Control Archtecture for the Bergrade/USC hand, In S. T. Venkataraman and T. Iberall(Editors), *Dexterous Robot Hand*, Springer Verlay, pp.136-149.

Rosheim, M. (1994). *Robot Evolution*, John Wiley & Sons Inc., pp. 216-224.

Lin, L. R. & Huang, H. P. (1996). Integrating Fuzzy Control of the Dexterous National Taiwan University (NTU) Hand, *IEEE/ASME Transaction on Mechatronics*, Vol. 1, No. 3, pp. 216-229.

Butterfass, J.; Grebenstein, M.; Liu, H. & Hirzinger, G. (2001). DLR-Hand II: Next Generation of a Dextrous Robot Hand, *Proceedings of IEEE International Conference on Robotics and Automation*, pp. 109-114.

Namiki, A.; Imai, Y.; Ishikawa, M. & Kanneko, M. (2003). Development of a High-speed Multifingered Hand System and Its Application to Catching, *Proceedings of the 2003 IEEE/RSJ International Conference on Intelligent Robots and Systems*, pp. 2666-2671.

Yamano, I.; Takemura, K. & Maeno, T. (2003). Development of a Robot Finger for Five-fingered Hand using Ultrasonic Motors, *Proceedings of the 2003 IEEE/RSJ International Conference on Intelligent Robots and Systems*, pp. 2648-2653.

Fearing, R. S. (1990). Tactile Sensing Mechanisms, *International Journal of Robotics Research*, Vol. 9, No. 3, pp. 3-23.

Howe, R. D. (1994). Tactile Sensing and control of robotic manipulation, *Advanced Robotics*, Vol. 8, No. 3, pp. 245-261.

Shimojo, M.; Sato, S.; Seki, Y. & Takahashi, A. (1995). A System for Simulating Measuring Grasping Posture and Pressure Distribution, *Proceedings of IEEE International Conference on Robotics and Automation*, pp. 831-836.

Johnston, D.; Zhang, P.; Hollerbach, J. & Jacobsen, S. (1996). A Full Tactile Sensing Suite for Dextrous Robot Hands and Use In Contact Force Control, *Proceedings of IEEE International Conference on Robotics and Automation*, pp. 3222-3227.

Jockusch, J.; Walter, J. & Ritter, H. (1997). A Tactile Sensor System for a Three-Fingered Robot Manipulator, *Proceedings of IEEE International Conference on Robotics and Automation*, pp. 3080-3086.

Kawasaki, H. & Komatsu, T. (1998). Development of an Anthropomorphic Robot Hand Driven by Built-in Servo-motors, *Proceedings of the 3rd International Conference on ICAM*, Vol. 1, pp. 215-220.

Kawasaki, H. & Komatsu, T. (1999). Mechanism Design of Anthropomorphic Robot Hand: Gifu Hand I, *Journal of Robotics and Mechatronics*, Vol. 11, No.4, pp. 269-273.

Kawasaki, H.; Komatsu, T.; Uchiyama, K. & Kurimoto, T. (1999). Dexterous Anthropomorphic Robot Hand with Distributed tactile Sensor: Gifu Hand II, *Proceedings of 1999 IEEE ICSMC*, Vol. II, pp. 11782-11787.

Mouri, T.; Kawasaki, H.; Yoshikawa, K.; Takai, J. & Ito, S. (2002). Anthropomorphic Robot Hand: Gifu Hand III, *Proceedings of 2002 International Conference on Control, Automation and Systems*, pp. 1288-1293.

Kawasaki, H.; Mouri, T. & Ito, S. (2004). Toward Next Stage of Kinetic Humanoid Hand, *CD-ROM of World Automation Congress 10th International Symposium on Robotics with Applications*.

Leung, G. M. H.; Francis, B. A. & Apkarian, J. (1995). Bilateral Controller for Teleoperators with Time Delay via μ-Synthesis, *IEEE Transaction on Robotics and Automation*, Vol. 11, No. 1, pp. 105-116.

Movingeye Inc. (2001). http://www.movingeye.co.jp/mi6/artlinux_feature.html

Kawasaki, H.; Mouri, T.; Abe, T. & Ito, S. (2003). Virtual Teaching Based on Hand Manipulability for Multi-Fingered Robots, *Journal of the Robotics Society of Japan*, Vol. 21, No.2, pp. 194-200 (in Japanese).

3

Development of Biped Humanoid Robots at the Humanoid Robot Research Center, Korea Advanced Institute of Science and Technology (KAIST)

Ill-Woo Park, Jung-Yup Kim, Jungho Lee, Min-Su Kim, Baek-Kyu Cho and Jun-Ho Oh
Korea Advanced Institute of Science and Technology (KAIST)
Korea

1. Introduction

Recently, many studies have focused on the development of humanoid biped robot platforms. Some of the well-known humanoid robots are Honda's humanoid robots, the WABIAN series of robots from Waseda University, Partner, QRIO, H6 and H7, HRP and JOHNNIE. Given that humanoids are complex, expensive and unstable, designers face difficulties in constructing the mechanical body, integrating the hardware system, and realizing real-time motion and stability control based on human-like sensory feedback. Among the robots, HRP and ASIMO are the most well known humanoid robot platforms.

HRP-3P is a humanoid robot developed jointly by the National Institute of Advanced Industrial Science and Technology and Kawada Industries, Inc in Japan. It stands 1.6 m tall, weighs 65 kg, and has 36 degrees of freedom (DOF). Upgraded from HRP-2, the new platform is protected against dust and water. In addition, Honda has unveiled a new type of ASIMO, termed the ASIMO Type-R, which stands 1.3 m tall, weighs 54 kg, and has 34 DOF. With the i-WALK technology, this robot has an impressive walking feature: it can walk at 2.7 km/h, and run at 6 km/h.

HUBO is essentially an upgraded version of KHR-2. The objective of the development of HUBO was to develop a reliable and handsome humanoid platform that enables the implementation of various theories and algorithms such as dynamic walking, navigation, human interaction, and visual and image recognition. With the focus on developing a human-friendly robot that looks and moves like humans, one focus was on closely aligning the mechanical design with an artistic exterior design. This chapter also discusses the development of control hardware and the system integration of the HUBO platform. Numerous electrical components for controlling the robot have been developed and integrated into the robot. Servo controllers, sensors, and interface hardware in the robot have been explained. Electrical hardware, mechanical design, sensor technology and the walking algorithm are integrated in this robot for the realization of biped walking. This system integration technology is very important for the realization of this biped humanoid.

The technologies utilized in HUBO are the basis of the development of other HUBO series robot such as Albert HUBO and HUBO FX-1.

Albert HUBO is the only humanoid robot that has an android head and is able to walk with two legs. The face, which resembles Albert Einstein, can imitate human facial expressions such as surprise, disgust, laughter, anger, and sadness. The body, comprising the arms, hands, torso, and legs, is that of HUBO. The body of HUBO was modified to have the natural appearance despite the disproportionate sizes of the head and the body. It can be described as Albert Einstein in a space suit. The realization of a biped walking robot with an android head is a first-in-the-world achievement. The design and system integration between the head and the body are discussed. RC motors are used for the head mechanism, enabling facial expressions. The head and body are controlled by different controllers. The head controller generates facial motions and recognizes voices and images using a microphone and CCD cameras.

HUBO FX-1 is human-riding biped robot. There are a few research results on the subject of practical uses for human-like biped robots. HUBO FX-1 was developed for carrying humans or luggage. This is very useful in the construction or entertainment industries. As HUBO FX-1 uses two legs as transportation method, it offsets the limitations in the use of a wheel and caterpillar. The robot uses AC motors and harmonic drives for joints. As it should sustain heavy weight in the region of 100kg, it requires high power actuators and transmissible high-torque reduction gears.

2. HUBO

2.1 Overall Description

HUBO (Project name: KHR-3) is a biped walking humanoid robot developed by the Humanoid Robot Research Center at KAIST. It is 125cm tall and weights 55kg. The inside frame is composed of aluminum alloy and its exterior is composite plastic. A lithium-polymer battery located inside of HUBO allows the robot to be run for nearly 90 minutes without external power source. All electrical and mechanical parts are located in the body, and the operator can access HUBO using wireless communications. HUBO can walk forward, backward, sideways, and it can turn around. Its maximum walking speed is 1.25km/h and it can walk on even ground or on slightly slanted ground. HUBO has enough degrees of freedom (DOF) to imitate human motions. In particular, with five independently moving fingers, it can imitate difficult human motions such as sign language for deaf people. Additionally, with its many sensors HUBO can dance with humans. It has two CCD cameras in its head that approximate human eyes, giving it the ability to recognize human facial expressions and objects. It can also understand human conversation, allowing it to talk with humans.

HUBO is an upgraded version of KHR-2. The mechanical stiffness in the links was improved through modifications and the gear capacity of the joints was readjusted. The increased stiffness improves the stability of the robot by minimizing the uncertainty of the joint positions and the link vibration control. In the design stage, features of the exterior, such as the wiring path, the exterior case design and assembly, and the movable joint range were critically reconsidered, all of which are shown in Fig. 1. In particular, strong efforts were made to match the shape of the joints and links with the art design concept, and the joint controller, the motor drive, the battery, the sensors, and the main controller (PC) were designed in such a way that they could be installed in the robot itself. Table 1 lists the specifications of the robot. The following are the design concepts and their strategies in the design of the HUBO platform.

1. Low development cost
- Rather than using custom-made mechanical parts, commercially available components such as motors and harmonic gears were used in the joints.
2. Light weight and compact joints
- The power capacity of the motors and reduction gears enables short periods of overdrive due to the weight and size problem of the actuators.
3. Simple kinematics
- For kinematic simplicity, the joint axis was designed to coincide at one point or at one axis.
4. High rigidity
- To maintain rigidity, the cantilever-type joint design was avoided.
5. Slight uncertainty of the joints
- Harmonic drive reduction gears were used at the output side of the joints, as they do not have backlash.
6. Self-contained system
- All of the components, including the battery, controllers and sensors, are enclosed inside the robot. There are no external power cables or cables for operating the robot.

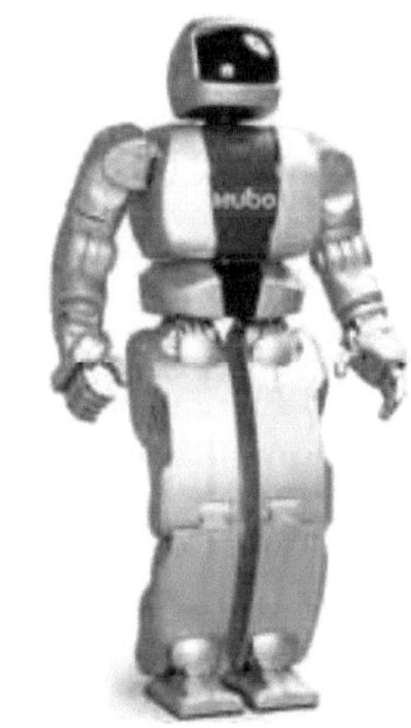

Figure 1. Humanoid Robot, HUBO

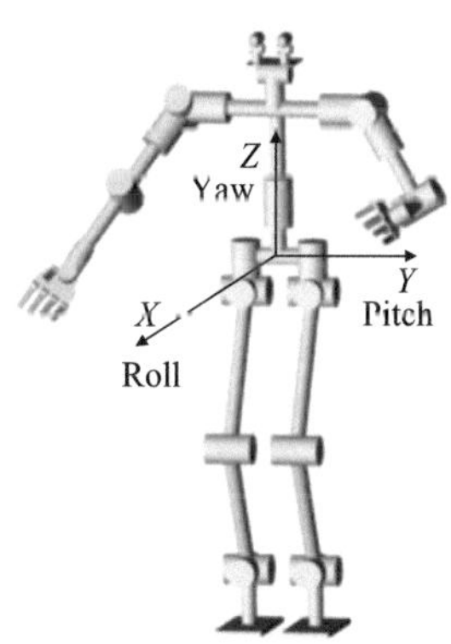

Figure 2. Schematic of the joints and links

Research period		January 2004 up to the present
Weight		55 kg
Height		1.25 m
Walking speed		0 ~ 1.25 km/h
Walking cycle, stride		0.7 ~ 0.95 s, 0 ~ 64 cm
Grasping force		0.5 kg/finger
Actuator		Servomotor + harmonic reduction gear
Control unit		Walking control unit, servo control unit, sensor unit, power unit, and etc.
Sensors	Foot	3-axis force torque sensor; accelerometer
	Torso	Inertial sensor system
Power section	Battery	24 V - 20 Ah (Lithium polymer)
	External power	24 V (battery and external power changeable)
Operation section		Laptop computer with wireless LAN
Operating system		Windows XP and RTX
Degree of Freedom		41 DOF

Table 1. Overall Specifications of HUBO

2.2 Mechanical Design

Degrees of Freedom and Movable Joint Angles

Table 2 shows the degrees of freedom of HUBO. Attempts were made to ensure that HUBO had enough degrees of freedom to imitate various forms of human motion, such as walking, hand shaking, and bowing. It has 12 DOF in the legs and 8 DOF in the arms. Furthermore, it can independently move its fingers and eyeballs as it has 2 DOF for each eye (for panning and tilting of the cameras), 1 DOF for the torso yaw, and 7 DOF for each hand (specifically, 2 DOF for the wrist and 1 DOF for each finger). As shown in Fig. 2, the joint axis of the shoulder (3 DOF/arm), hip (3 DOF/leg), wrist (2 DOF/wrist), neck (2 DOF) and ankle (2 DOF/ankle) cross each other for kinematic simplicity and for a dynamic equation of motion.

Head	Torso	Arm	Hand	Leg	Total
2 neck 2/eye (pan-tilt)	1/torso (yaw)	3/shoulder 1/elbow	5/hand 2/wrist	3/hip 1/knee 2/ankle	
6 DOF	1 DOF	8 DOF	14 DOF	12 DOF	41 DOF

Table 2. Degrees of Freedom of HUBO

Table 3 shows the movable angle range of the lower body joints. The ranges are from the kinematic analysis of the walking. The maximum and normal moving angle ranges of the joints are related to the exterior artistic design in Fig. 3. While determining the ranges, a compromise was reached in terms of the angle range and the appearance of the robot.

Joint		Angle range
Hip	Yaw	0 ~ +45°
	Roll	-31° ~ +28°
	Pitch	-90° ~ +90°
Knee	Pitch	-10° ~ +150°
Ankle	Pitch	-90° ~ +90°
	Roll	-23° ~ +23°

Table 3. Movable lower body joint angle ranges of HUBO

Figure 3. Artistic design of HUBO

Actuator (Reduction Gear and DC Motor)

Two types of reduction gears are used: a planetary gear and a harmonic gear. A planetary gear is used for joints such as finger joints, wrist-pan joints, neck-pan joints and eyeball joints, where small errors (such as backlash) are allowable. Errors in the finger and wrist-pan joints do not affect the stability of the entire body or the overall motion of the arms and legs. Harmonic gears are used for the leg and arm, as well as for neck tilt and wrist tilt joints. As a harmonic gear has little backlash on its output side and only a small amount of friction on its input side, it is particularly useful for leg joints, where errors can affect the stability of the entire system and the repeatability of the joint position. This harmonic type of reduction gear is connected to the motor in two ways: through a direct connection and through an indirect connection. The indirect connection requires various power transmission mechanisms (such as a pulley belt or a gear mechanism) between the reduction gear unit and the motor. HUBO has an indirect type of connection for the neck tilt, the shoulder pitch, the hip, the knee, and the ankle joints.

	Joint		**Reduction gear type**	**Input gear ratio**	**Motor power**
Hand	Finger		Planetary gear (256:1)	1.56:1 (pulley belt)	2.64 W
	Wrist	Pan	Planetary gear (104:1)	None	11 W
		Tilt	Harmonic drive (100:1)	2:1 (pulley belt)	
Head	Neck	Pan	Planetary gear (104:1)	None	
		Tilt	Harmonic drive (100:1)	2:1 (pulley belt)	
	Eye	Pan	Planetary gear (256:1)	None	2.64 W
		Tilt		1.56:1(pulley belt)	
Arm	Elbow	Pitch	Harmonic drive (100:1)	None	90 W
	Shoulder	Roll			
		Pitch		1:1	
		Yaw		None	
Trunk		Yaw			

Table 4. Upper body actuators of HUBO

Joint		Harmonic drive reduction ratio	Input gear ratio	Motor power
Hip	Roll	120:1	Gear (2.5:1)	150 W
	Pitch	160:1	Pulley belt (1.78:1)	150 W
	Yaw	120:1	Pulley belt (2:1)	90 W
Knee	Pitch	120:1	Pulley belt (1:1)	150 W*2
Ankle	Roll	100:1	Pulley belt (2:1)	90 W
	Pitch	100:1	Pulley belt (1.93:1)	

Table 5. Lower body actuators of HUBO

The choice of gear types and harmonic drive types was limited by specific design constraints (such as the space, shape, permissible power, and weight). With flexibility in designing the size, shape and wiring, it was easier to develop brushed DC motor drivers compared to other types of motors (such as brushless DC motors or AC motors). The brushed DC motors also have a suitable thermal property. When they are driven in harsh conditions, for example at a high speed and severe torque, the generated heat is less than if brushless DC motors were used. Hence, there is less of a chance that heat will be transferred from the motors to devices such as the sensors or the controller.

There are trade-offs in terms of the voltage of the motor. If the motor has a high voltage, it cannot drive a high current, and vice versa. The voltage of the motors is related to the size and weight of the battery. A high-voltage source requires more battery cells to be connected serially. The number of battery cells is directly related to the weight of the battery system and the weight distribution of the robot.

Weight Distribution

The main controller (PC), the battery, and the servo controller and drivers for the upper body are in the torso. The mass, except for the actuators, was concentrated in the torso due to the need to reduce the load of the actuators in frequently moving parts such as the arms and legs; in addition, it was desired that the torso have sufficiently large inertia for a small amplitude fluctuation. With this approach, the robot achieves low power consumption while swinging its arms and legs; moreover, the control input command ensured a zero moment point with a small positioning of the torso. When the inverted pendulum model is used for gait generation and control, making the legs lighter is important for the realization of biped walking because the model does not consider the weight and the moment of inertia of the lifting leg.

Mechanical Component of Force Torque Sensor (F/T Sensor)

Shaped like a Maltese cross, the F/T sensors can detect 1-force and 2-moment. As shown in Fig. 4, the sensors are attached the wrist (Φ50) and ankle (80 mm x 80 mm). To sense the magnitude of a beam deflection, strain gages are glued onto the points where the load causes the largest strain. These points were located at the ends of the beam but the gages were glued 5 mm apart to minimize the problems of stress concentration and physical space. The ankle sensor was designed for a maximum normal force (F_Z) of 100 kg and maximum moments (M_X, M_Y) of 50 Nm.

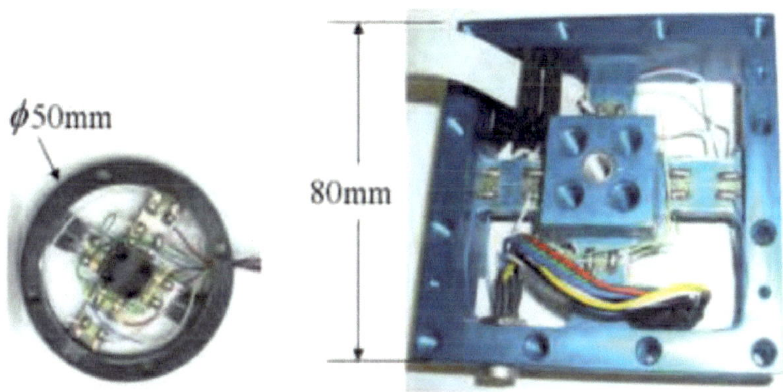

Figure 4. Three-axis F/T sensor

It can be physically assumed that the distance between the sole and the sensor is negligible and that the transversal forces in the x-y plane are small. From the principle of equivalent force-torque, the sensor-detected moment is then

$$M_{Sensor} = M_{ZMP} + r \times F_{ZMP} \tag{1}$$

where $F_{ZMP} = \begin{bmatrix} F_x \\ F_y \\ F_z \end{bmatrix}$, $M_{Sensor} = \begin{bmatrix} M_{s,x} \\ M_{s,y} \\ M_{s,z} \end{bmatrix}$, $r = \begin{bmatrix} r_x \\ r_y \\ r_z \end{bmatrix}$.

This 3-axis F/T sensor can only sense F_z, $M_{s,x}$, $M_{s,y}$. By the definition of ZMP, the moment at ZMP is $M_{ZMP} = 0$. It can be assumed that the F/T sensor is on the sole and that the transversal forces in the x-y plain are small. In this case, $r_z F_x$ and $r_z F_y$ are negligible. Through a simple calculation, the relationship between the ZMP and the detected force/moment are

$$r_x \approx -\frac{M_y}{F_z},\ r_y \approx \frac{M_x}{F_z} \tag{2}$$

2.3 Control Hardware System

The hardware architecture of the control system is shown in Fig. 5, and the location of the hardware components is displayed in Fig. 6. A Pentium III-933MHz embedded PC with the Windows XP operating system (OS) is used as the main computer. Other devices such as servo controllers (joint motor controller) and sensors are connected to the controller area network (CAN) communication lines to the main computer. The robot can be operated via a PC through a wireless LAN communications network. The main computer serves as the master controller. The master controller calculates the feedback control algorithm after receiving the sensor data, generates trajectories of the joints, and sends the control command of the robot to the servo controller of the joints via CAN communication.

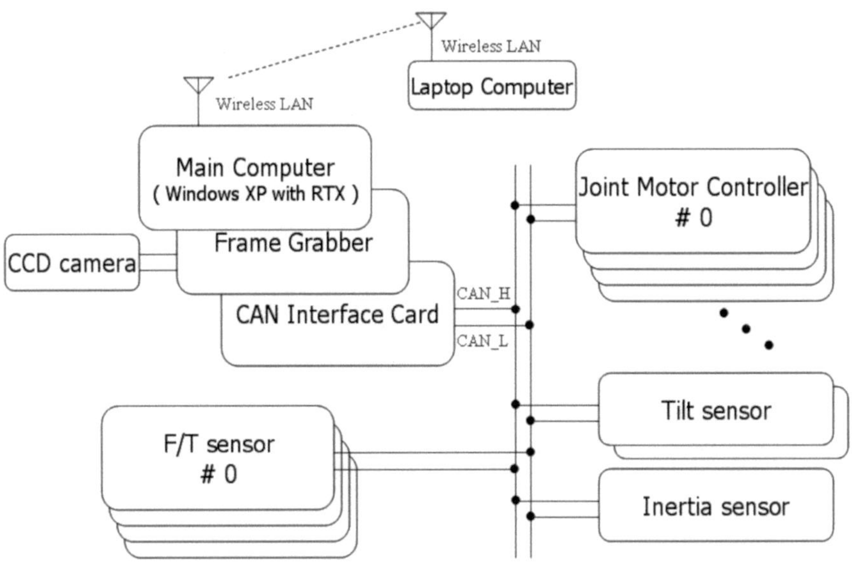

Figure 5. Control System Hardware of HUBO

The software architecture of the OS is shown in Fig. 7. Windows XP operates the main controller for the convenience of software development and for system management. Windows XP is a common OS, which is easy for the developer to access and handle. This widespread OS made it possible to develop the robot control algorithm more effectively, as it is easy to use with free or commercial software and with hardware and drivers. A graphical user interface (GUI) programming environment shortened and clarified the development time of the control software. However, the OS is not feasible for real-time control. Real-time extension (RTX) software is the solution for this situation. The operational environment and the GUI of the robot software were developed in the familiar Windows XP, and a real-time control algorithm including the CAN communications was programmed in RTX.

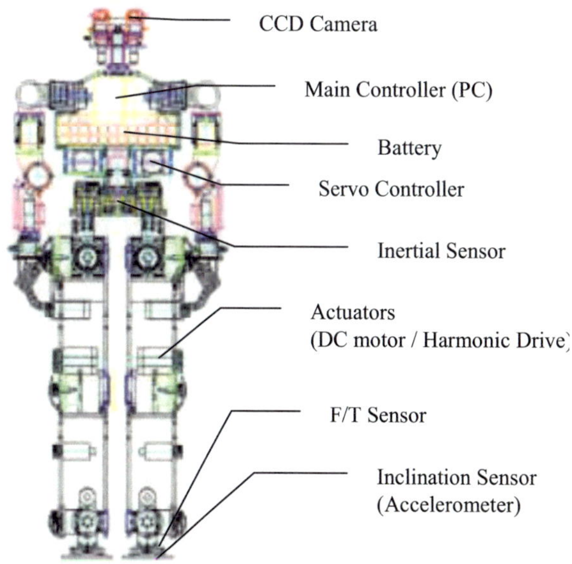

Figure 6. Hardware System Structure of HUBO

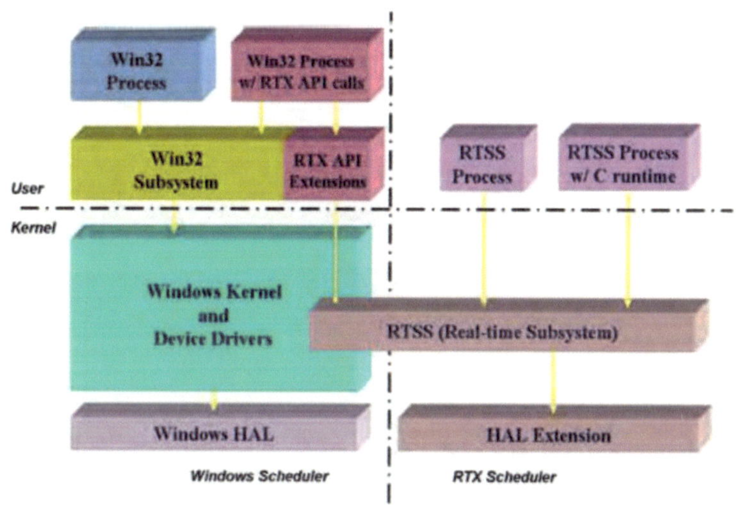

Figure 7. Software Architecture of the OS

Brushed DC motors were used for joint actuators. The motors, as used in this robot, are divided by their power capacity. High-power motors are used for joints such as the hip, knee, ankle, shoulder, elbow and torso. These joint actuators require high levels of torque, speed and reliability. For example, the motors used in the lower limb are directly related to the walking performance and stability of the robot. In addition, the arm joint motors also require high power as it was desired that the robot could imitate human motions such as bowing, sign language, and simple dancing. Low-power motors are used for joints such as the fingers, wrists, neck, and eyes. These motors have little connection to the overall walking stability of the robot; they were added for the decoration of motion. Two types of servo controllers were developed: a low-power controller and a high- power controller.

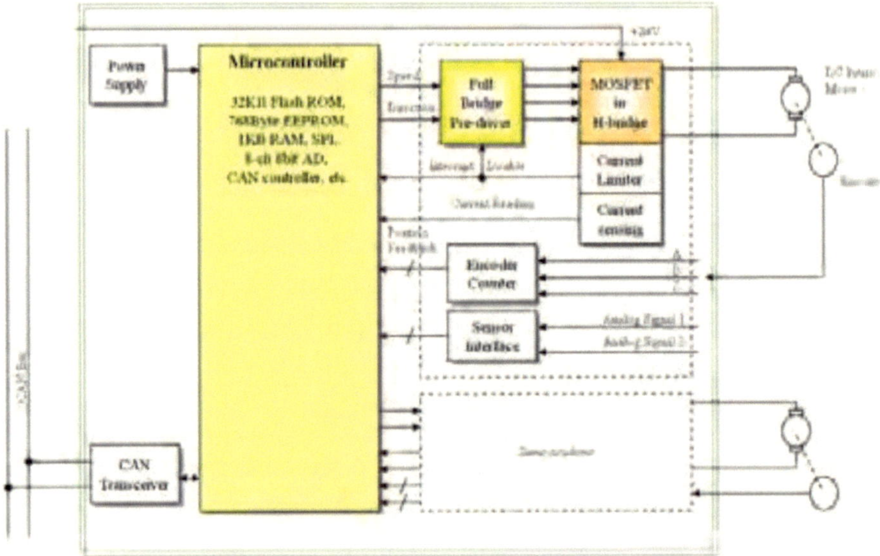

Figure 8. Hardware Configuration of the Servo controller

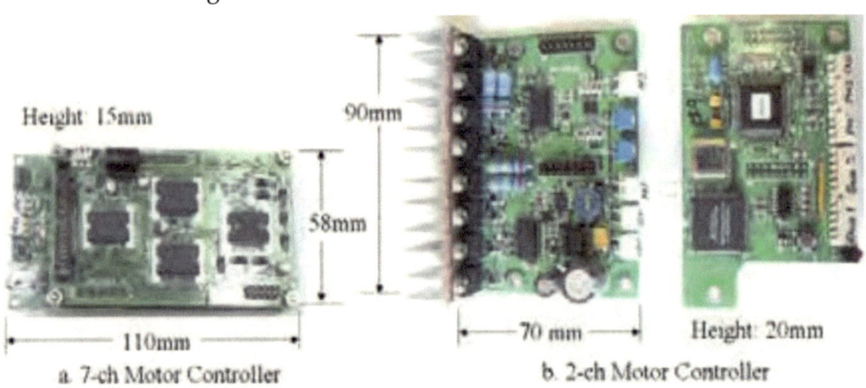

Figure 9. Servomotor Controllers

The controllers operate at 1000Hz, which interpolates linearly the position command issued by the main controller at a frequency of 100Hz. The detailed hardware configuration and the features of the controllers are shown in Figs. 8 and 9. As mentioned above, two types of servo controllers were used; these are shown in Fig. 9. Both are composed of a microcontroller module and power amplifier module. The microprocessor module receives the joint position commands, controls a real DC motor using given commands and encoder counting, and sends the actual position or current data of the motors. Fig. 9a shows low-power servo controllers that control the low-powered joints. These controllers can control 7-channel motors. There is also a 5-channel A/D port for additional sensors such as the pressure sensors in the fingertips. The power capacity is 40W/ch for the head and hand joints, which requires low power, a small space, and multiple motors. The other type of servomotor, as shown in Fig. 9, controls the high-power DC motors. It can handle 2-channel motors and a 2-channel A/D port for additional sensors such as accelerometers. It has a channel power capacity of 480W, allowing it to control the high-power motors.

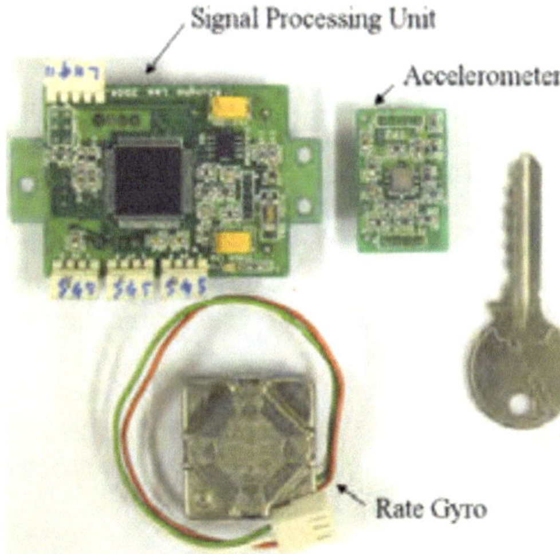

Figure 10. Inertia Sensor System

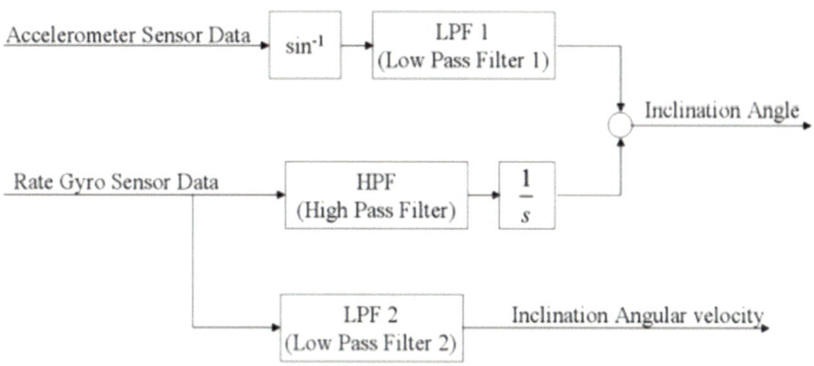

Figure 11. Signal Processing Block Diagram of the Inertia Sensor System

HUBO has an inertia sensor system enclosed in its chest. The walking control algorithm of the robot uses the attitude sensor actively. The inertia sensor system is composed of a 2-channel accelerometer, a 2-channel rate gyro and a signal-condition processor board, as shown in Fig. 10. In practice, the accelerometer can sense the robot's inclination using an arcsine function. However, it is very sensitive to unwanted acceleration resulting from a shock or a jerk. The rate gyro is good for sensing the angular velocity, but it drifts under a low frequency. Therefore, it is necessary to utilize signal-processing methods. As shown above in Fig. 11, the robot's attitude and its rate of change can be used instead. The sensor measures the inclination of the torso; the angle control is very important for the robot's stability and in terms of repeatability.

3. Albert HUBO

The design concept of the android-type humanoid robot Albert HUBO is described as follows:

1. A human-like head with a famous face
2. Can hear, see, speak, and express various facial expressions.

3. Can walk dynamically
4. Spacesuit-type exterior
5. Long battery life per single charge
6. Self-contained system
7. Two independent robotic systems of a head and a body

Albert HUBO is an android-type humanoid robot with a human face, as shown in Fig. 12. It has a height of 137 cm, a weight of 57 Kg, and 66 degrees of freedom. Essentially, its frame structures and systems are based on HUBO, which is a biped humanoid robot explained in a previous section. Based on HUBO, the control system architecture, battery capacity, and head system were upgraded. The head and body are independent robotic systems; accordingly, they have different roles. The head system manages intelligent human-robot interactions while the body system performs movements such as biped walking. Hence, the battery capacity was enlarged in order to power these two robotic systems sufficiently.

Figure 12. Albert HUBO

	Albert HUBO	HUBO
Height	137 cm	125 cm
Weight	57 kg	55 kg
DOF	66	41
Actuator	Brushed DC motor + RC servomotor	Brushed DC motor

Table 6. Mechanical Specifications of Albert HUBO and HUBO

Computer	Main computer 1	Pentium III 1.1GHz
	Main computer 2	Pentium III 933MHz
Operating System	General OS	Windows XP
	Real time OS	RTX
Communication	Internal	CAN
	External	IEEE 802.11g
Vision	Vision system using CCD Cameras	
Voice	Voice recognition and voice synthesis	

Table 7. System Specifications of Albert HUBO

Table 6 and 7 present the simple and overall specifications of Albert HUBO. The robot uses two PCs. The first of these is termed main computer 1, which mainly handles the role of head motion control. The controller generates head motions such as the facial expressions. It also processes vocal expression data and CCD camera image data from the microphone and CCD camera, respectively. The second PC is termed main computer 2, which mainly handles motions and the walking control of the entire robot system apart from the head. It controls walking and motions analogous to the main computer of HUBO.

Android Head Design

Historically, the entertainment industry has most aggressively explored realistic and nearly realistic robotic hardware for use in movies and theme parks. The field of such entertainment devices is known as "animatronics." These machines have taken a wide diversity of form, from the realistic "Abe" Lincoln of Disneyland, to the bizarre aliens of the "Men in Black" movies.

In the field of animatronics, problems of costliness, low expressivity, and power consumption all result from the physical dissimilarity of the simulated facial soft tissues of animatronics relative to real human tissues. Human facial soft tissues are mostly liquid, filling cellular membranes approximating billions of tiny water balloons. The liquid molecules will slide into any geometry that the membranes can tolerate. In this way, the human face is like a sealed wet sponge. Animatronic facial tissue on the other hand, is made of solid elastomers (e.g. rubber), which are typically composed of tangled, spring-like polymer molecules that unwind when elongated but are geometrically interlocked. Thus, these molecules are fundamentally restricted from reproducing the geometric plasticity of human facial tissues. In effect, the force required to move animatronics materials expressively are orders of magnitude above that required by facial soft tissues.

To resolve these issues, Hanson Robotics developed a series of methods for creating sponge-like elastomer materials that move more like facial soft-tissues. These materials, known collectively as "Frubber" (a contraction of "flesh" and "rubber"), wrinkle, crease, and amass much more like skin than do animatronics materials. They also consume very little power – less than 10W – while affecting a full range of facial expressions and speech-related mouth motions. In tests, the material requires less than 1/22nd the force and energy to move into facial expressions relative to animatronics materials.

The reduced energy consumption enables battery-powered biped walking. Being porous, Frubber also weighs much less, which is also a benefit for untethered walking robots. Such integration with a walking gesture-capable body is significant as it allows an exploration of the aesthetics of the entire, integrated humanoid figure as an autonomous social being.

As shown in Fig. 13, Frubber is used for Albert HUBO's facial skin. In addition, twenty-eight RC servomotors for the facial movements and three RC servomotors for the neck movements are used in order to generate a full range of facial actions and expressions, such as laughs, sadness, anger, or surprise. The servomotors are linked with the various points of Albert HUBO's face through strings. Therefore, the face motions are generated by drawing or releasing strings linked to various points on the face such as the eyebrows, eyes, jaws, lips, and other points.

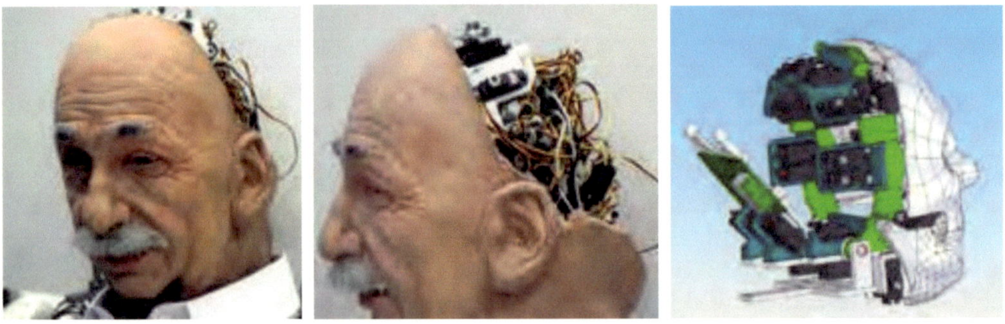

Figure 13. The head of Albert HUBO

To control the thirty-one RC servomotors, the mini SSC II (Scott Edwards Electronics) is used as the head motor controller. The main controller sends the position data via RS-232 signals to the head motor controller. The head motor controller then converts the RS-232 signals to PWM signals in order to drive the thirty-one RC servomotors. The head motor controller sends the PWM signals to the RC servomotors without feedback, as the RC servomotors incorporate feedback control circuits in their systems.

Body Design

The most important part of the design of Albert HUBO is the torso, as numerous important parts are densely crowded into this space. Thus, a high level of space efficiency is required. For example, the speaker and microphone systems, two main computers, a Li-Polymer battery, a Ni-MH battery, joint motor controllers, an inertial sensor, and a switch panel are located in the torso. To support this hardware, the torso is composed of a rectangular-shaped chest frame and two supporting columns between the chest and the pelvis. Fig. 14 shows a 2D drawing of Albert HUBO's body. By locating many parts into the torso, which does not have joints, energy consumption can be reduced during dynamic motions. The dimensions of robot body were determined in consideration of a human body. However, the distance between the right and left shoulders was designed to be greater than that of a human due to a spacesuit-like exterior.

The body frames and covers of HUBO were modified to connect them with an android head. The main design problem of the Albert HUBO project was the natural connection of the machine-shaped body and the realistic human-like head. More specifically, it was important to provide the best image to people without an unusual appearance. To match the robotic body with the human-like head, the authors conceived of Albert Einstein's face and coupled it with a spacesuit-type body exterior, as the spacesuit gives the robot a scientific and robotic appearance. This was felt to be well matched with the image of Albert Einstein in this regard.

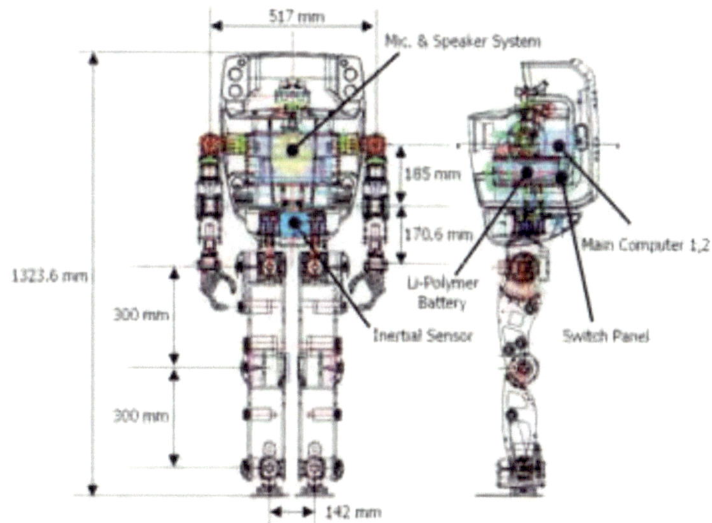

Figure 14. 2D drawing of Albert HUBO's body

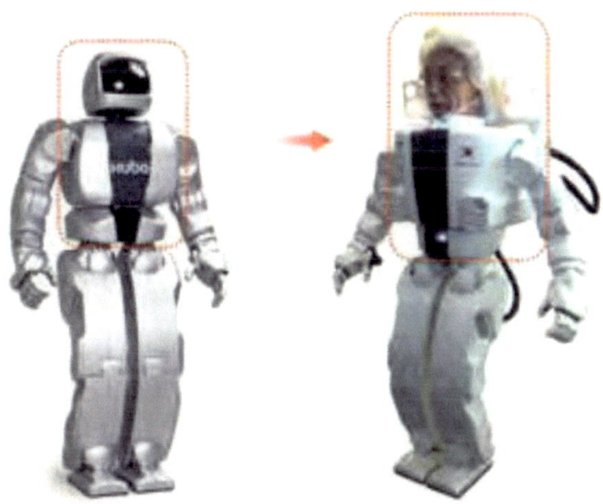

Figure 15. Modified parts of the HUBO Platform

During the body styling process, there was an unexpected problem: the head was out of proportion with the body. This problem was serious, as the robot may have appeared overly unusual. The first design alternative was an inverted triangle shape for the body, applying a perspective effect. As ordinary people understand the perspective view well, they do not recognize distortions regarding the depth of perspective until these distortions are large. As modern people are familiar with different camera views or screen angles used with movie cameras, they naturally recognize distortion from a lens as well. A normal adult looks down Albert HUBO from a high vantage point, because a person of normal height is taller than the robot. Thus, to someone looking at the robot, the upper body appears wider and the lower body narrower. This effect becomes greater as the distance between the viewer and the robot becomes closer and as the view angle becomes higher. If the perspective obtained from a high viewpoint is applied to the styling, an inverted-triangle robot shape results, as shown in Fig. 16.

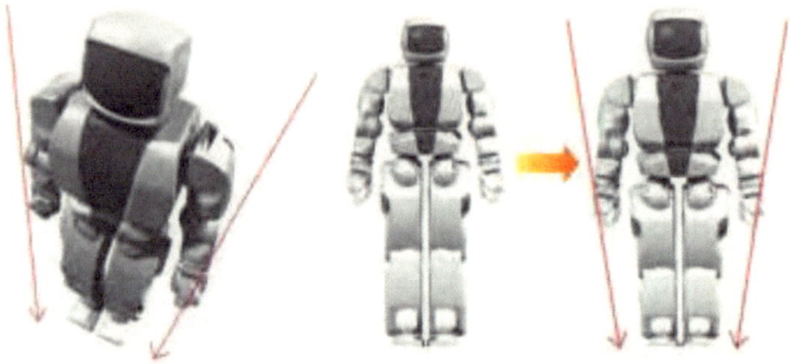

Figure 16. Distorted image with the perspective view and its application

In spite of various points of view, people are reminded of the shape from the viewpoint of a camera from a high viewing angle. This case represents the natural situation shown in Fig. 17.

Figure 17. Seeing the distorted robot, people imagine the image from the perspective of a camera

For the upper body cover design, an inverted conic shape was applied in order to prevent distortion from the side quarter view. Hence, the arm parts were moved approximately 45 mm more to the outside from the body center in order to expand the upper body. The pelvis cover was contracted compared to the original HUBO platform. To minimize the expansion of the upper body cover, it was made to be round so that it would not interfere with shoulder and neck, as shown in Fig. 18.

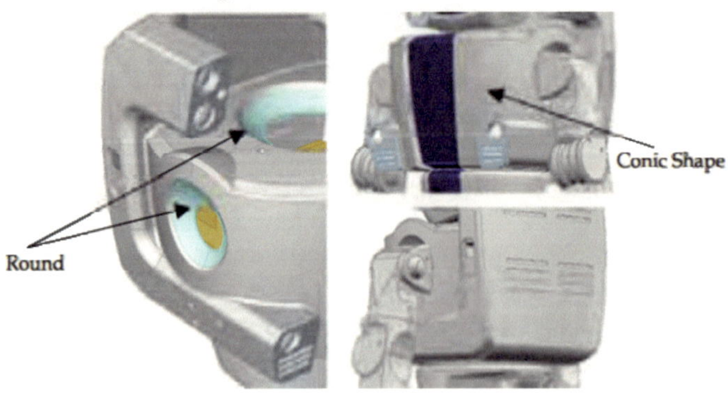

Figure 18. Design of the upper body cover

The second design alternative was a reduction of the exposure of the head (except for face, as it is the most important part in Albert HUBO body styling). To shade the back of the head, the backpack was expanded to the ears. The backpack protects the head from impacts, adds room for inner mechanic parts and controllers, and acts as a base for the attachment of various types of equipment as shown in fig. 19.

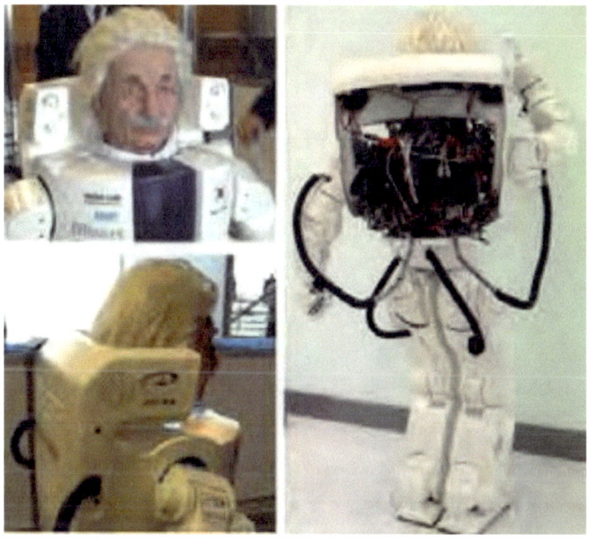

Figure 19. The backpack of Albert HUBO

The third design alternative was the application of effective lighting (shown in Fig. 20.). The sidelights on the head can disperse a spectator's sight when concentrated on the head. Furthermore, these powerful halogen lights make a dramatic effect when the robot presents on the stage in an artificial fog. Additional blue LED lights in the neck and the shoulder connection parts offer a feeling of mystique or a fantasy mood. In particular, the LED light in the neck shades the gap in the connection between the head and body and drives a spectator's sight to the face.

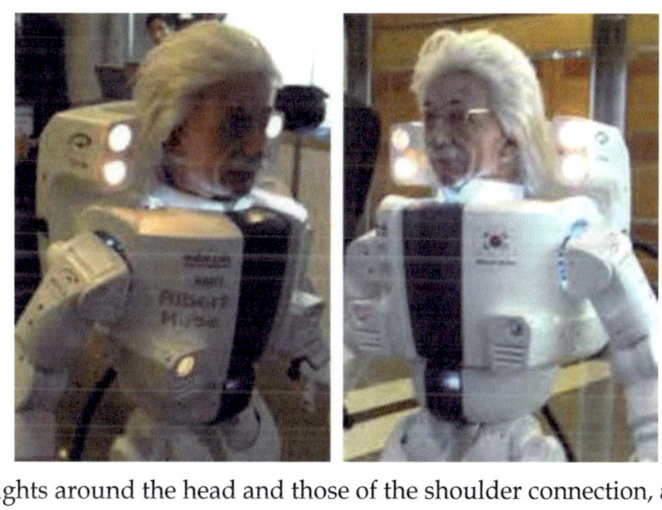

Figure 20. The lights around the head and those of the shoulder connection, and the front of the waist

System Integration

The motion control structure is shown in Fig. 21. Albert HUBO behaves with intelligent activities and body motions. The intelligent activities are the visual processing, voice recognition, and facial expressions of the robot. The body motions are upper body motions, gait pattern planning, and posture stabilization. The upper body motions are for interactions with humans or for the accomplishment of tasks. Gait pattern planning generates the biped walking motions, and posture stabilization based on sensory feedback is combined with other motions in order to maintain the balance of the robot continually. Both of the intelligent motions and body motions interact with the environment through the use of CCD cameras, a microphone, a speaker, force/torque sensors, an inertial sensor, and tilt sensors.

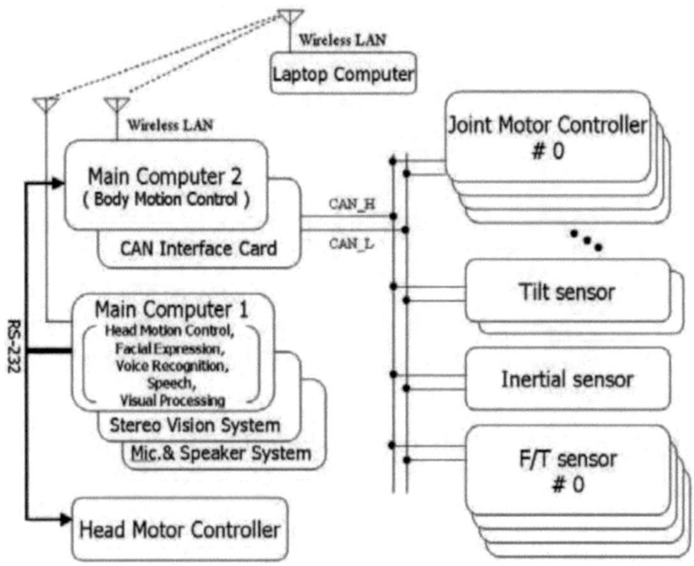

Figure 21. Schematic of Albert HUBO system

Using this motion control structure, the robot achieves human-robot interactive functions. For example, if a man commands Albert HUBO to bring a certain item, Albert HUBO initially judges whether the commander is his master by means of the voice and face. If the commander is his master, Albert HUBO replies with a positive answer and with a smile. Following this, Albert HUBO searches for the target item using the biped walking and vision systems. When the target item is found, Albert HUBO grasps it using an upper body motion and CCD cameras. Finally, Albert HUBO passes the target item to his commander. In this way, a simple intelligent job can be realized.

	Main Computer 1 (Head)	Main Computer 2 (Body: Arms and Legs)
Expansion	PC104+	PC104+ and PC104
Power Consumption	Typical 5V@ 2.3A, 12V@0.5mA Max 5V@ 3.68 A, 12V@0.5mA	Typical 5V@3.5A, 12V@0.02A Max 5V@3.99 A, 12V@0.08A
Size/Weight	108 mm x 115 mm, 0.279 Kg	96 mm x 115 mm, 0.2 Kg
Operating System	Windows XP	Windows XP + RTX

Table 8. Specifications of the main computers

4. HUBO FX-1

HUBO FX-1 was developed for research concerning applications of a biped walking robot. HUBO FX-1 is a scaled-up version of the original HUBO's lower body, and is useful in industrial fields. Heavy industrial companies need robots that can carry heavy materials, thus this type of robot can be seen as a milestone in industrial robot history. HUBO FX-1 will be used for a transportation system that can carry heavy materials or a human. HUBO FX-1 is a transportation system based on HUBO's lower body. It can transport luggage in an industrial field or a human by modification of the seat or upper body. HUBO FX-1 is scaled up to 1.5 times of the original HUBO, and is capable of generating larger power movements. It can carry luggage or a human weighing up to 100kg and can walk stably. To walk with 100kg, additional parts were added in an effort to increase its stiffness. An operator can control HUBO FX-1 via wireless communications or using a joystick located in its upper body (seat).

Overview of HUBO FX-1

HUBO FX-1 has 12 DOF, including 2 DOF in its ankle, 1 DOF in its knee and 3 DOF in the hip for each leg. Fig. 22 shows HUBO FX-1 with a seat for carrying a human. Except for its upper body, the height of HUBO FX-1 is 1.393m and its weight is about 130Kg. It can walk at 1.25Km/h. All of the main controllers and other devices are embedded in the robot. Therefore, operators can control HUBO FX-1 through a wireless network and a joystick. The former HUBO's walking algorithm was applied, and the algorithm was verified as very stable. Table 9 shows overall specifications of HUBO FX-1.

Figure 22. HUBO FX-1 (Human Riding Biped Walking Robot)

Research Term		2005.04~
Height and Weight		1.393m(1.988 with chair) and 130Kg(150Kg with chair)
Walking Speed		1.25Km/h
Actuator		AC Servomotor + Harmonic Reduction Gear + Drive Unit
Control Unit		Main controller, sub-controller and AC servo controller
Sensor	Foot	3-Axis Force-Torque Sensor and Inclinometer
	Torso	Rate-Gyro and Inclinometer
Power		External AC power(220V)
Operation		Windows XP and RTX with Wireless network and joystick
DOF		12 DOF

Table 9. Specifications of HUBO FX-1

Actuator and Reduction Gear

An AC servomotor is used for actuating HUBO FX-1 joints. The type of DC motor that is generally used for the other type of biped robots, such as HUBO, is not sufficient for the expected design parameters in this case. Each motor and reduction gear for the joints was selected using computer simulation to withstand its own weight and a 100Kg payload (maximum). In addition, to minimize backlash phenomenon on the output side of the joint reduction gear, a Harmonic Drive reduction gear was utilized in each joint. These are connected to the motor by pulley-belt system to adjust the reduction ratio easily. Table 10 outlines the AC motor and the Harmonic Drive reduction gears.

AC Servomotor	400Watt	Max. torque	1.27 Nm
AC Servomotor	400Watt	Inertia	$0.34\ gf \cdot cm \cdot s^2$
AC Servomotor	400Watt	RPM	5000 rpm
AC Servomotor	800Watt	Max. torque	2.39 Nm
AC Servomotor	800Watt	Inertia	$1.08\ gf \cdot cm \cdot s^2$
AC Servomotor	800Watt	RPM	5000 rpm
Harmonic Drive	CSF-25	Reduction ratio	100:1
Harmonic Drive	CSF-25	Max. torque	108 Nm
Harmonic Drive	CSF-32	Reduction ratio	100:1
Harmonic Drive	CSF-32	Max. torque	212 Nm

Table 10. Actuators and reduction gears used in the robot joints

Control Hardware

The electrical parts of HUBO FX-1 differ from the former HUBO series. As HUBO FX-1 uses AC servomotors and the former HUBO series use small DC motors, there are additional electrical devices. Fig. 23 shows overall structure of the system.

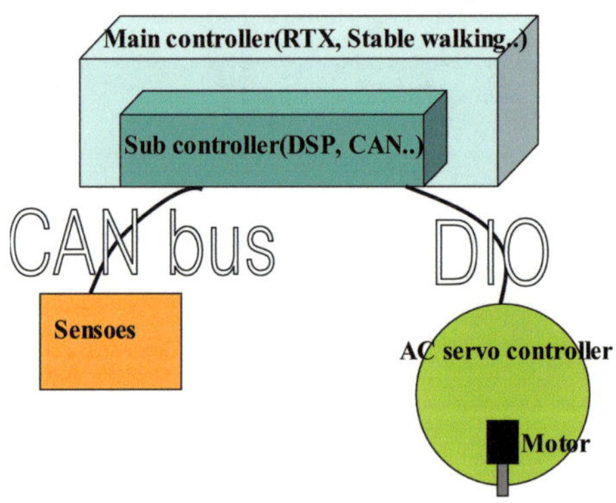

Figure 23. Overview of control system

The Windows XP operating system is utilized as the main operating system, while RTX is used for real time control. Controllers are made up of a main controller (PC), a sub-controller (controller board using DSP and PLX) and an AC servo controller (Servomotor controller of the AC motor), and sensors are made up of a 3-axis force-torque sensor, a rate gyro sensor and an inclinometer. Sensors communicate with the main controller using the CAN communication protocol.

The main controller has a PCI bus-type sub-controller, which the former HUBO series robots do not have. The main controller conducts the control algorithm using sensor signals and offers reference position information for each motor to the sub-controller during a 0.01s time period. The sub-controller offers reference position and direction signals for each motor at 50 KHz. These are created by a DDA (Digital Differential Analysis) algorithm and are sent to the AC servo controller based on this information (Fig. 24.).

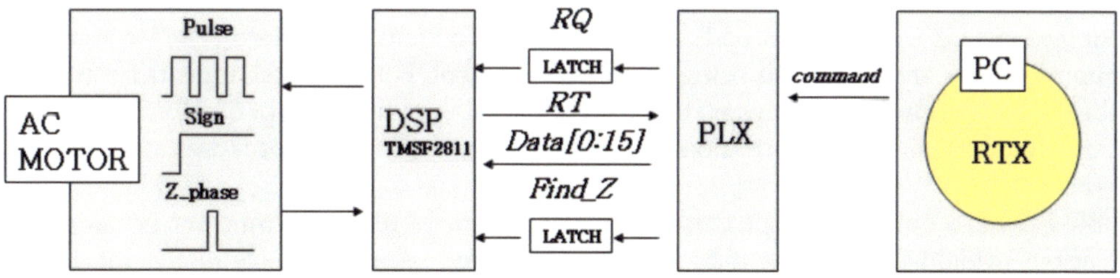

Figure 24. Control system diagram of the sub-controller

Using this procedure the AC servo controller controls each motor. Fig. 25 shows a timing diagram of the sub-controller. The PLX sestets the RQ signal to high in order to inform the DSP that the data is valid. The DSP then receives the data and returns an RT signal as an acknowledgement to the PLX. This procedure is conducted twelve times every 10msec, and all information necessary for the control of the motors are transmitted to DSP.
The sub-controller has a CAN communication module. It is used for communicating with the sensors. SJA1000 is a stand-alone CAN controller manufactured by Philips, which has inner Rx/Tx buffers, a bit stream processor and an acceptance filter. It supports the CAN 2.0B protocol and has a maximum 1Mbps communication speed. PCA82C250 as a CAN transceiver has a 1Mbps baud rate and a maximum 50nsec propagation delay.

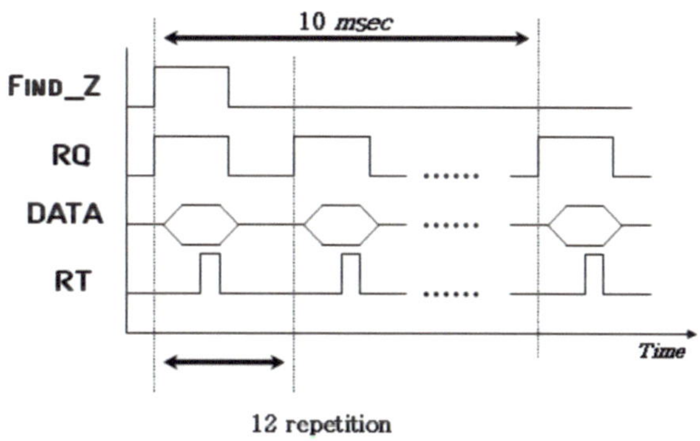

Figure 25 Timing diagram of the sub-controller

5. Conclusion and Future Work

In this chapter, The HUBO-series robot platform development is introduced. HUBO, Albert HUBO, and HUBO FX-1 are biped humanoids. Each robot has its own character. HUBO is a biped humanoid robot. The major function of this robot is to walk with its two legs and imitate the human's motions such as hand shaking, bowing, and communicating in sign language. The system components of the control hardware are connected using a CAN communication line. The control system of HUBO is based on the distributed control

architecture; all of the joints and sensors have their own microprocessors. The main controller sends a control command to the servo controller and receives environmental data from the sensor-signal condition board through CAN. This system architecture is similarly applied to the Albert HUBO and HUBO FX-1 robots.
Albert HUBO is a biped humanoid robot, which has a human-like (appearing as the famous physicist Albert Einstein) android head. It can be regarded as a robot with two independent robot systems i.e. a head and a body system. There is a main controller in each system, and communications are conducted using the RS232 protocol. For a natural appearance, in spite of the disproportion in size between the head and the body, the concept of Albert Einstein in a space suit was adopted. In addition, a number of artistic design approaches are applied to the robot.
HUBO FX-1 is a human-riding robot that can carry nearly 100kg. It can carry humans with of various weights. To enable it to walk with a heavy object, aspects of the mechanical design of the robot were thoroughly considered, including the stiffness of the robot frame structure, as well as the actuators and the reduction gears. The control system hardware differs from HUBO and Albert HUBO. It uses a PCI bus-based sub-controller containing a DSP processor, and CAN is used for sensor data communications.

6. References

K. Hirai, M. Hirose, Y. Haikawa, and T. Takenaka, The Development of Honda Humanoid Robot, in *Proc. IEEE Int. Conf. on Robotics and Automations*, pp.1321-1326, 1998.

Y. Sakagami, R. Watanabe, C. Aoyama, S. Matsunaga, N. Higaki, and K. Fujimura, The Intelligent ASIMO: System Overview and Integration, in *Proc. IEEE/RSJ Int. Conf. on Intelligent Robots and Systems*, pp. 2478-2483, 2002.

J. Yamaguchi, A. Takanishi, and I. Kato, Development of a Biped Walking Robot Compensating for Three-Axis Moment by Trunk Motion, in *Proc. IEEE/RSJ Int. Conf. on Intelligent Robots and Systems*, pp.561-566, 1993.

K. Nishiwaki, T. Sugihara, S. Kagami, F. Kanehiro, M. Inaba, and H. Inoue, Design and Development of Research Platform for Perception-Action Integration in Humanoid Robot: H6, in *Proc. IEEE/RJS Int. Conf. on Intelligent Robots and Systems*, pp.1559-1564, 2000.

K. Kaneko, F. Kanehiro, S. Kajita, K. Yokoyama, K. Akachi, T. Kawasaki, S. Ota, and T. Isozumi, Design of Prototype Humanoid Robotics Platform for HRP, in *Proc. IEEE Int. Conf. on Intelligent Robots and Systems*, pp.2431-2436, 1998.

M. Gienger, K. Loffler, and F. Pfeiffer, Towards the Design of Biped Jogging Robot, in *Proc. IEEE Int. Conf. on Robotics and Automation*, pp.4140-4145, 2001.

K. Akachi, K. Kaneko, N. Kanehira, S. Ota, G. Miyamori, M. Hirata, S. Kajita, and F. Kanehiro, Development of Humanoid Robot HRP-3P, in *Proc. of IEEE-RAS Int. Conf. on Humanoid Robots*, pp.50-55, 2005.

I.W. Park, J. Y. Kim, S.W. Park, and J.H. Oh, Development of Humanoid Robot Platform KHR-2 (KAIST Humanoid Robot - 2), *Int. Conf. on Humanoid* 2004.

Multipurpose Low-Cost Humanoid Platform and Modular Control Software Development

Filipe Silva and Vítor Santos
University of Aveiro
Portugal

1. Introduction

Humanoid robotics is becoming quite popular especially after Honda's robots in the 90s' and their evolution ever since (Hirai *et al.*, 1998; Sakami *et al.*, 2002). In the meantime, other research groups have developed their own humanoid platforms making them available commercially for both the edutainment or academic communities (Furuta *et al.*, 2001; Lohmeier et al., 2004; Kaneko et al., 2004; Kim *et al.*, 2004; Nagasaka *et al.*, 2004). Valuable and versatile platforms for research, however, still imply prohibitive costs for most small research groups wanting to perform activity on humanoid balance, walk and advanced perception, plus the difficulty associated to full autonomy, which is the ultimate requisite. Several initiatives to develop commonly accessible and open platforms for research and development have appeared, such as the well-known OpenPino project (Yamasaki *et al.*, 2000). Nonetheless, for several reasons, possibly technological issues, these open platforms did not surge as massively as would be expected. Another possible reason is that some academic groups prefer rather to develop a platform themselves since that creates opportunities to teach engineering for students. The combination of these reasons led the authors to gradually start a project, in early 2004, to develop a humanoid platform with off-the shelf low-cost components, but baring in mind the needs for versatility and full autonomy so as to comply with the RoboCup regulations and, eventually, enter the competition when possible. The topics covered in this chapter are largely inspired by the research work behind this project, namely the major design concerns and the development of the modular low-level control architecture. Special emphasis is given to the distributed local control to achieve proper adaptation to real-world uncertainties in dynamic environments. The text seeks a balance between the application of various technical solutions to a practical engineering system, and a mathematically rigorous theoretical development.

In what concerns the humanoid system, the actuators, sensors, and mechanical components were selected and configured to devise the platform main structure (the hardware includes 22 DOFs); that accounts for mechanical axles and links, servomotor typology, gear units, adequate Ion-Li battery packs, among others. For enhanced control power and versatility, the entire system is based on a distributed architecture supported on a Controller Area Network (CAN) arrangement of PIC microcontrollers with local decision capabilities, but able to comply with global directives issued by a central unit based on an embedded PC system. Still sticking to low cost, proper electronics and signal conditioning were also developed; that includes force sensors for the feet that were custom-developed and calibrated for the force-based control developed latter. RC servomotors are a low-cost full integrated solution, but they offer too simple control schemes allowing position control only; for them too, microsecond-tuned velocity and trajectory control algorithms were conceived by fine PWM dynamic generation. The adoption of an outer motion control loop, in cascade with the servo own controller, to overcome the performance degradation has been experimentally demonstrated. These techniques settled at the PIC software level also allow to measure average current consumption, by integrating instantaneous readings at 50 Hz duly synchronized with PWM pulses generation. From the above statements, it is clear that a door remains open to motor torque control, at least in static or pseudo-static approaches, but also in dynamic motion with additional inertial information from gyros and accelerometers.

Although being the first prototype still requiring some tuning and rectifications to reach a mature platform for research and development, interesting results on local control and pure force based balance of one leg have been already achieved. The system architecture allows for several paradigms of control, ranging from the fully centralized (inadequate from our point of view) to exclusively local to each robot joint (too complex and possibly unrealizable from our point of view) and passing by the local control capabilities modulated by global directives from the main control unit, which is the envisaged approach. Local control techniques for force, torque, position or velocity control may be based on simple traditional controllers (P, PI, PID) or fractional order controllers (Silva & Santos, 2005), fuzzy or neural nets.

From the control point of view, the difficulty of bipedal locomotion lies in the uncertainty of the environment and the limitations of the contact between the robot and the environment (Popovic *et al.*, 2005). Despite recent advances, accepted definitions of stability with application to the gait and posture of biped robots remain an unsolved problem and the subject of much work in robotics. The expertise of today, however, agrees on the importance of the ability to adapt efficiently to the environment as the key to enhance stability and reliability of humanoid walking. Although many papers have been published on the reflexive control in robotics, the issue of enhancing biped locomotion stability and robustness by exploiting reflexive actions has rarely been studied in the literature (Huang *et al.*, 2005). This work explores designing and implementing fundamental local actions to deal with such events as unexpected disturbances or unknown ground irregularities. More concretely, an approach referred as adaptive leg balancing is proposed to stabilise the motion of a single leg that stands on a platform with unknown slope changes, while the control system relies on the ground reaction forces as measured by the force sensors inserted in each foot corner. The algorithm is based on the computation of the COG Jacobian transpose that provides a computationally efficient posture control algorithm less sensitive to parameter uncertainties. In order to validate the theoretical findings, most of the control schemes presented throughout the chapter are tested in several numerical simulations and experiments for representative motion tasks.

2. The platform hardware

Before starting the actual platform construction, several issues were considered as initial paradigms to follow. These have been: *low-cost* off-the-shelf components for affordability and easier reproducibility, *full autonomy* to comply with some contests' rules and also to provide independence of testing environments, *upgradeability* to permit initial simple approaches but with time allow the re-usage of developed modules to increase complexity, *versatile control* possibilities to allow for conventional but also non-conventional control methods and algorithms, *distributed processing* for robustness to failures and to allow growing overall computational power and also, but not the least, *motion flexibility* for more locomotion possibilities and approaches. Naturally, *efficient power* usage was also a concern and adequate power autonomy had to be sought. The global expectation was that a platform for research on control and humanoid locomotion could be developed.

2.1 The structure design

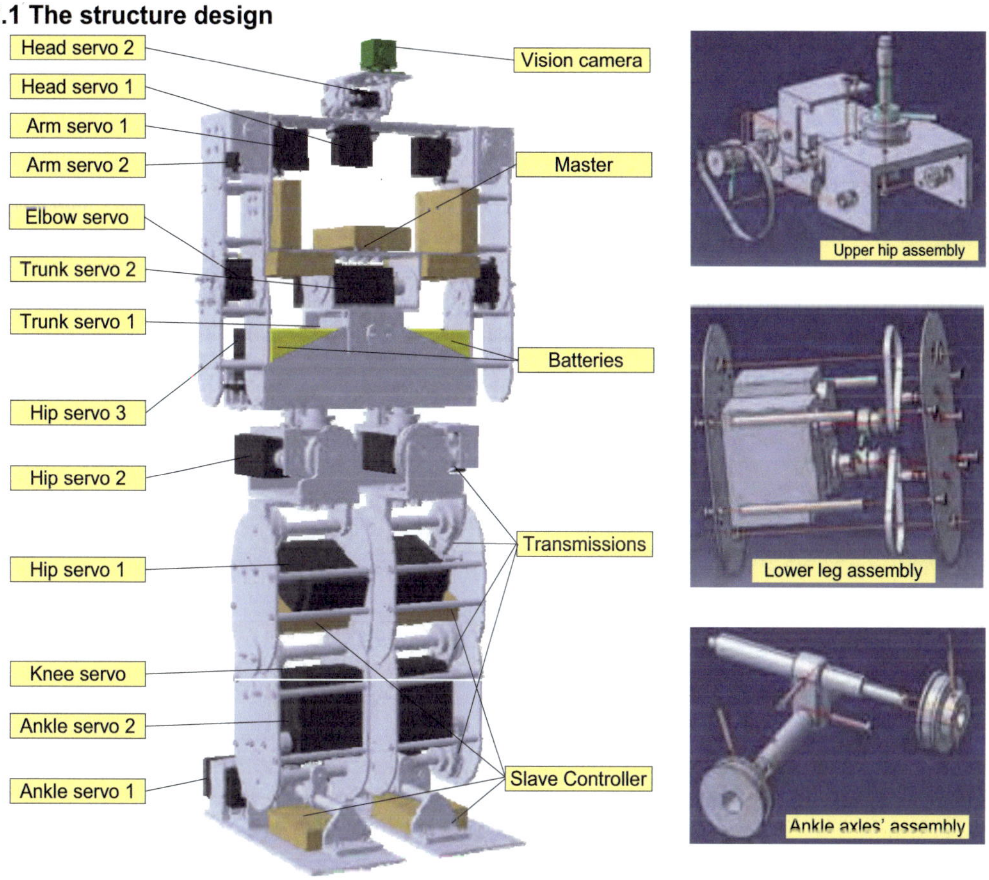

Figure 1. Three dimensional model for the humanoid structure and some assembly details

The very first concern on the platform structure was the number of degrees of freedom (DOFs). Legs were the central issue on those matters, and comparing activities and results from other research groups (mainly those present in RoboCup competitions since 2003), 6 DOFs for each leg was immediately decided. Attaching a trunk upon the hip was decided to be done with 2 DOFs. The main explanation for this decision was the expected enhanced flexibility for the moment when locomotion of the full body would arrive: having lateral and sagital compensation capabilities appeared better than only one of them. The arms were not

given as much attention as the rest since it was expected that later on much better supported decisions could be made based on the expected know-how to come. Two degrees of freedom for the shoulder plus one for the elbow were a sufficient start. Finally, two DOFs would allow independent pan and tilt of the head, that is, of the camera expected to place on it.
The structural links to build up the platform could be done hollow, exoskeleton-like, or based on a structural internal skeleton - animal bone endoskeleton-like. Both types of solutions have been explored by other authors, but the exoskeleton approach was decided taking into account the type of actuators by-then already decided: RC servomotors; due to their volumetric geometry, they would fit better "inside" hollow links rather than "around" bone-like structures. The robot dimensions were the next concern and here the expectation was "the smallest possible" to fit all the required DOFs and remainder components. The RoboCup rules established the 60 cm height as the boundary between the two categories defined at the time. So it was decided: the robot had to fit the 60 cm height, but keeping the right proportions to be declared a compliant humanoid by RoboCup regulations. A structure was conceived from scratch and initially it was only partial. Light materials such as nylons were first tried to build the first leg but it was soon realized that aluminium would be preferable for better linkage and fixation of components along it with greater resistance to mechanical stress and strain. After some months of layout studies, and simulations, a full 3D model of the platform was reached as shown in Figure 1.
The 3D model was developed in the CATIA software and the last version accounted for circa 600 parts covering everything from bolts to belts. The model allowed testing the kinematics properties and also component clearance during locomotion. For example, one relevant issue, among several others, was the horizontal separation of the legs which imposed a trade-off between manoeuvrability when rotating upon one leg at the hip level (for which large clearance will help) and maximum torque when swapping the stance and swing legs in locomotion (for which leg horizontal proximity is an advantage). Kinematics and geometry were the first challenges to be modelled and equated. After the decision of which actuators and associated gearing would be selected, construction issues when then to the shop-floor.

2.2 Actuators, gearing and power

Actuation for this kind of platform is quite demanding (high torques, local feedback control needed, etc.) and off-the-shelf components do not offer much more than the popular RC servomotors. The reasons to select this type of device are commonly invoked (Ruas *et al.*, 2006) but being sure that the servos comply with the project requisites may be difficult to predict due to their fully encapsulated behaviour and limited external controllability, since they possess a single digital control input for position setting, remaining their velocity totally beyond the control of the driving system. Hence, besides velocity control, the other concerning issue was whether their power and torque was enough to support the robot and provide locomotion. As initially only pseudo-static operation was expected, servomotors power was not a concern (velocity is small) and the attention focused on torque. Several static simulations were carried out based on the 3D model and using the real weights of motors and batteries. To compute the static torques on motors a basic formula was used (1):

$$\tau_k = \left\| \sum_{i=k}^{N} m_i \mathbf{r}_{ki} \times \mathbf{g} \right\| \tag{1}$$

where τ_k is the torque exerted by actuator k, m_i is the mass of link i and r_{ki} is the vector with the relative coordinates of i's centre of mass (CoM) relative to joint k reference frame. Link numbering starts on the foot. Figure 2 illustrates the procedure with a partial example.

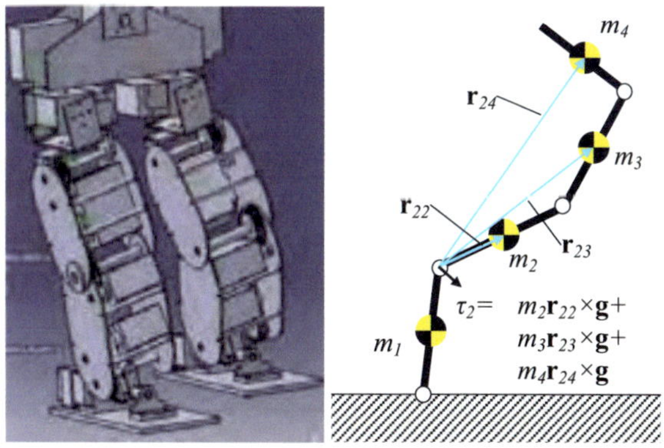

Figure 2. Calculating static torques on the Humanoid and generic example for joint 2

Table 1 shows he maximal torques predicted for each joint across the entire range of angles. The greatest torques occur on one of the foot joints and on one of the hip's joints.

	Motor/ Joint	θmin [º]	θmax [º]	Tmax [N.m]
Stance leg	Foot 1 roll	-35	+35	2.37
	Foot 1 tilt	-30	+60	0.30
	Knee 1	-45	+55	1.17
	Hip 1 tilt	-60	+60	0.35
	Hip 1 roll	-70	+21	2.57
Swing leg	Foot 2 roll	-35	+35	0.00
	Foot 2 tilt	-30	+60	0.12
	Knee 2	-45	+55	0.30
	Hip 2 tilt	-60	+60	0.14
	Hip 2 roll	-70	+21	0.30

Table 1. Maximal static torques on joints for the expected range of angular displacement

Off-the-shelf RC servomotors include several well-known brands, but it was found that the HITEC manufacturer implies normally lower costs. Even the strongest servomotors models from HITEC do not yield torques as large as 2.57 N.m, even when overpowered at 6.0 Volts. This implies that not all joints will be able to be direct driven by the servos. Mechanically this is a drawback because it requires gearing to increase torque (and lower the velocity). The solution adopted was to use toothed belts to connect gears with a reduction ratio of 2.5:1 (Figure 3). This increase in torque boosts the HITEC HS805BB announced 2.42 N.m to more that 5 N.m, which satisfies all static torque requirements. Moreover, it should be reminded that the torque calculations did not take advantage of trunk mobility. The trunk (and arms) was modelled as a fixed mass; with the future ability to pitch and roll the trunk, and also changing arms geometry, the CoM can be dynamically replaced and relief part of the torque applied upon some of the actuators.

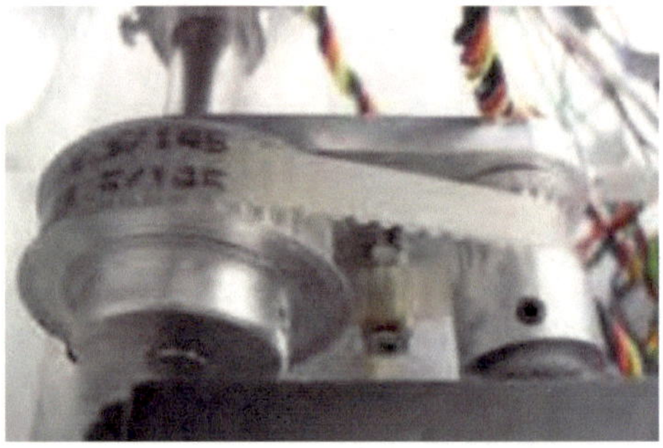

Figure 3. Transmission with a 2.5:1 gear ratio; an adjustable tensile roller holds the belt tight

As mention earlier, power to drive the motors is a most relevant issue. Servos absorb high currents, especially at start-up. Two ion-lithium batteries were installed and the system counts with a 7.2 V/9600 mAh pack, with maximal sustained current specified by the vendor at more than 19A. Each one of the two batteries weights circa 176 g. Proper fusing and charge monitoring were also implemented. Figure 4 shows the selected batteries and their location on the robot.

Figure 4. Batteries and their location on the robot

2.3 Sensors and perception

Autonomy demands perception in large extent, both of proprioceptive and exteroceptive types. For self-control and responsiveness to external perturbations due to contact, actuators and links must be monitored continuously; for awareness of the surroundings, intended for action planning or deliberate action upon the environment, some kind of remote or proximity detection is required.

The first category includes the static and dynamic status of joints which cover for angular position, but also velocity and, hopefully, torque. Moreover, measuring body interaction (force) against environment contacts can give relevant information to modify or adjust joint action to minimize some cost function during locomotion. A most meaningful example is the reaction of the floor on the feet. Indeed, if the appropriate monitoring of reaction forces on the foot sole is done, then balancing control against irregular floor may be an easier task. For this purpose, 4 force sensors were custom designed and installed on each foot. After calibration, the unbalanced force distribution among the 4 sensors on each foot may be used as input to joint actuation and permit some local control laws to rule robot behaviour, as described further. Another important issue would be to monitor the actuator torque; since that is not directly measurable (no appropriate sensors) a last attempt is to measure servos' current consumption; on servo motors that is not a simple task and further on some insight is given about the process.

Still on the proprioceptive sensors, inertial perception is believed to be an advantage on yet-to-come developments such as dynamic locomotion patterns. Integrated Micro-Electro-Mechanical (MEMS) accelerometers (ADXL202E, Analog Device) and a gyroscope (ENJ03JA, Murata) were selected and even installed, although they are not yet used to influence the system control. Finally, for the extraceptive perception, a vision system using a camera is to be mounted, although it is not yet used for decisions.

2.3.1 Measuring joint position

Servos have internal position feedback which is physically accessible by reading the internal potentiometer used by the device controller. However a connection must be wired to the output of the encapsulating box. Besides the tree default wires for power and input PWM pulse, a fourth wire is added to provide a voltage level between the power ground which is related to the potentiometer current position. The procedure, applicable in both the Futaba or HITEC brands (and possibly others), is sketched in Figure 5.

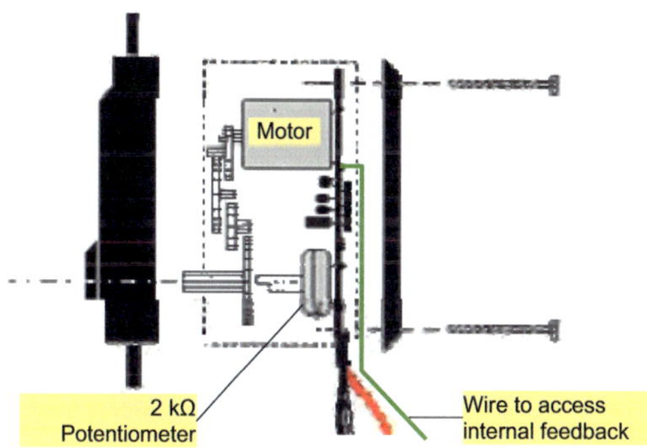

Figure 5. Wiring the servo to fetch the internal position feedback

Conditions seem now to exist to obtain joint angular position, except for the fact that voltage at the potentiometer output is not stable! Indeed, it depends on which part of the response to the PWM pulse the servo is at each moment. It was verified that the potentiometer output is only reliable during the duration of the pulse itself; once the pulse finishes, the servo internal controller enters the period of power application to the motor windings and interference will occurs as illustrated in Figure 6. A fine tuned software program had to be developed to perform potentiometer reading duly synchronized with PWM generation to ensure correct angular position evaluation.

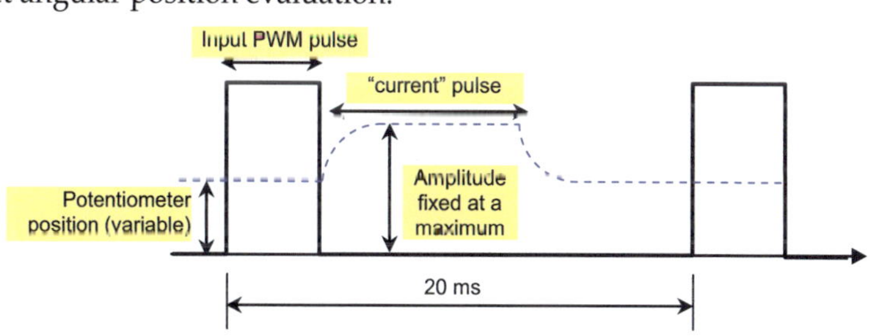

Figure 6. Value of the servomotor potentiometer along the PWM pulse sequence

2.3.2 Measuring current

The procedure described in the preceding section yielded another interesting outcome which was the possibility to assess the current consumption by the servo. The "current pulse" depicted in Figure 6 appears as a "pulse" occurring immediately after PWM falling edge and its width has been related to the power applied to the motor windings; the larger the pulse the more power is applied, or, as applied voltage is constant, the more instantaneous current is being absorbed by the motor, or finally, more torque is yielded by the motor. Measuring the width of the "current pulse" was done also in synchronization with PWM pulse generation by a sampling process at a much higher frequency than the PWM pulse itself which is 50 Hz.

2.3.3 Measuring foot reaction forces

To pursue a versatile platform that is expected to interact with the environment and be reactive to it, it is a must to measure contact reaction forces with the floor. This is needed to comply with floor irregularities and sloped paths, but ultimately it will provide direct feedback for balance, also in standard floor conditions. The idea is then to include the reaction forces in the control loop. Since miniature good quality load cells present prohibitive costs (hundreds of dollars), the team decided to develop low-cost strain gauge-based force sensors (Figure 7).

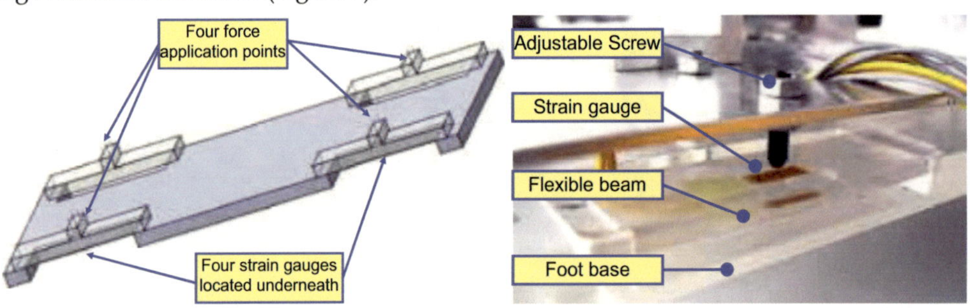

Figure 7. Model of sensitive foot and detail view of a force sensor on the foot

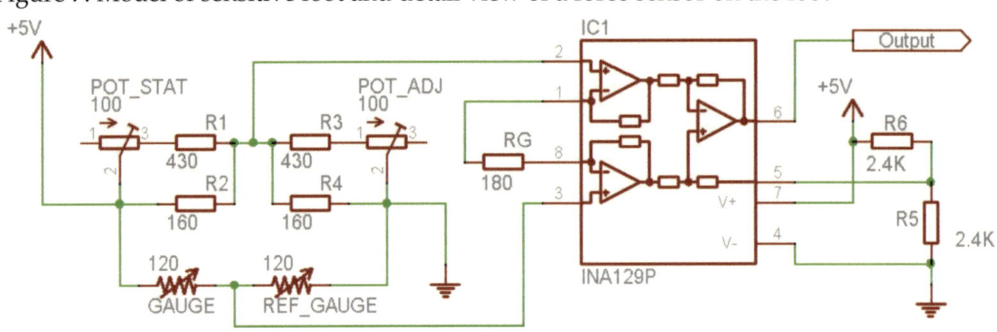

Figure 8. Electrical conditioning and amplification for the force sensor

Each foot possesses four of these sensors which allow for locomotion manoeuvres and control, either to keep the platform "upright" or even to perform dynamic balance when moving. The supporting material was made entirely of Plexiglas for greater flexibility and easier manufacture. A flexible beam of thinner Plexiglas holds the gauge which is connected to a Wheatstone bridge and an instrumentation amplifier with electrical conditioning and fine tuning components as shown in Figure 8. To compensate for asymmetric variations (temperature, shot noise, etc.) measuring bridge presents several points of symmetry, including a static strain gauge just for electric balancing purposes. Higher resistances than shown can be later used to low power consumption by the bridge.

3. Distributed control architecture

As mentioned before, the intended platform should have distributed computational capabilities either for modular and updatable software or for decentralized control capabilities in order to be more robust. Distributed architectures are not new but some systems (namely some smaller robots from those appearing in Robocup competitions) still attach to one central and global controller. The proposed approach wants to be scalable, and for so many degrees of freedom, one central controller simply is not practical.
The solution has then been to conceive a three level hierarchy of controllers: the lowest level, named Slave Units (SU), is responsible for actuator direct control and sensor reading and immediate processing; SUs may be in number of several. The second level comprises the so-called Master Unit (MU) whose role is to gather and maintain the status of the system as well as establish the communication protocols between levels. Finally, the Main Control Computer (MCC) will communicate with the MU and external interfaces, and will be responsible for the high level directives to be issued to the platform as a whole.
The Main Control, implemented as an embedded PC, will have computational power enough to acquire images and process vision. It will run a high level operating system and interfaces the MU by RS232 serial line. It issues orders of any level but does not ensure any type of real time control loop. Orders are dispatched to the Master Unit as well as querying status and other system variables, however not for real time control since there isn't even enough channel bandwidth for it. A general lay-out of the architecture appears in Figure 9.
The Slave Units are indeed in charge of system motion or static activity. Each slave unit is capable of controlling three servomotors as well as acquiring sensorial data form up to 16 sensors. All slave units are connected by a CAN bus which also includes the MU.

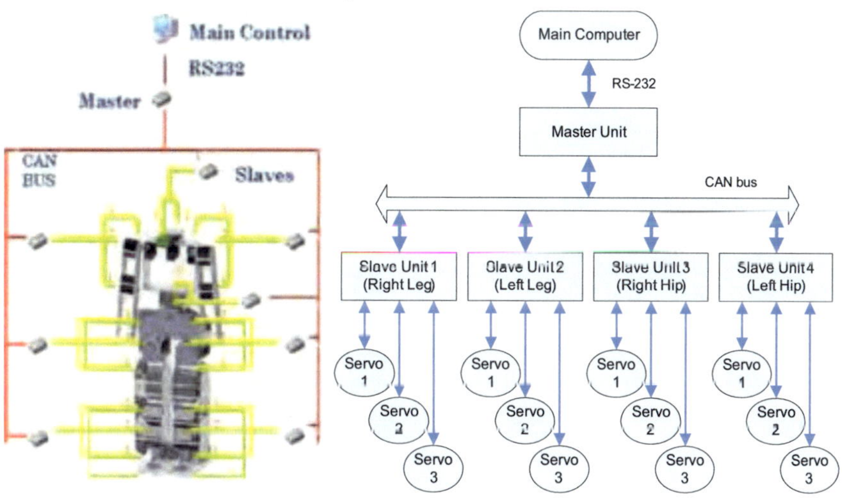

Figure 9. Concept of the distributed control architecture and one partial lay-out

The Slave Units only send data to the bus when asked to. Their main task is to maintain some local control law for each of the 3 servos, and possibly with variable feedback depending on local sensors and/or other directives that might have reached the unit. A PIC microcontroller is the centre of the processing unit and its main components are depicted in Figure 10. All SU have conceptually the same program with variations that can dynamically be imposed depending on the SU address which is hard-coded by on-board dip-switches.

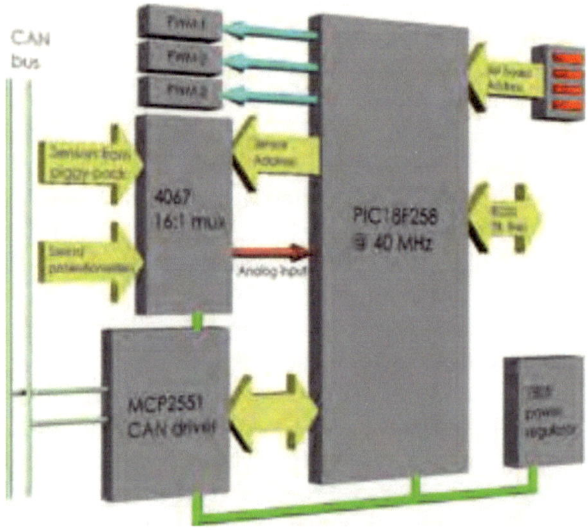

Figure 10. Functional layout of a slave controller

The Master Unit has a similar layout as slave units, but it holds a completely different program. The MU has normally no need to read any kind of sensors but it can do it if necessary. From the point of view of the hardware implementation, the basic board is similar, but a piggy-back board may be added where special sensors or other functions may be attached to a particular board (Figure 11).

Figure 11. Complete Slave Unit (left); base board and two cases of piggy-back boards (right)

The current stage of development of the humanoid robot designed and built in the scope of this project is shown in Figure 12. All the proposed ideas and particular control algorithms have been tested and verified on this real robot to form a critical hypothesis-and-test loop.

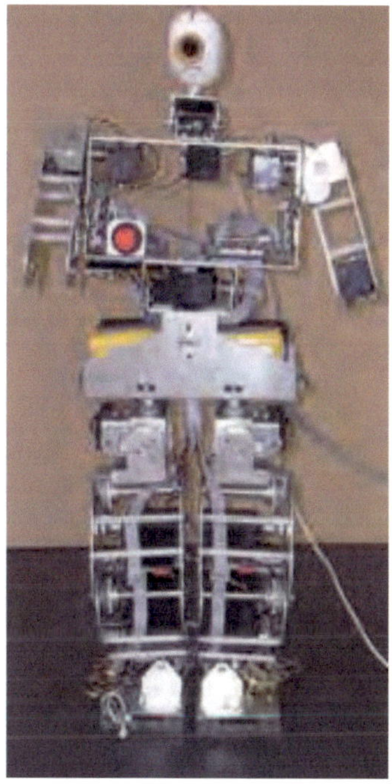

Figure 12. Biped humanoid robot with 22 DOFs

4. Low level joint control

Besides the computational resources, a major concern in building low-cost humanoid platforms is the implementation of the low level controllers, together with the constraints on the actuator systems. The success relies on the development of joint control algorithms and sensing devices to achieve proper performance when tracking a commanded trajectory. In this section, we will concentrate on the design and implementation of the local joint controllers. First, we review the limitations of RC servomotors with pulse-width control and how this affects our decisions. Then, we describe the implementation of an external position control loop closed around each slave unit. Adopting an outer loop, we establish a new control structure that introduces suitable compensation actions, significantly improving the system's performance and responsiveness.

4.1 Advantages and limitations of RC servomotors

The selected servomotors have themselves a built-in motor, gearbox, position feedback and controlling electronics, making them practical and robust devices. The control input is based on a digital signal whose pulse width indicates the desired position to be reached by the motor shaft. The internal position controller decodes this input pulse and tries to drive the motor up to the reference target based on the actual position determined by the potentiometer attached to each motor. However, the controller is not aware of the motor load and its velocity may vary rapidly and substantially. By design, servos drive to their commanded position fairly rapidly depending on the load, usually faster (slower) if the difference in position is larger (smaller). As the control task becomes more demanding, involving time-varying desired position (i.e., tracking control), the performance of the internal controller begins to deteriorate.

In order to validate the practical findings and gain insight into these problems, an entire system was set up intended to evaluate the actuator's performance. The experimental arrangement comprises several calibrated loads that will be applied to the servo shaft through a linkage 10 cm long (Figure 13). The servo is fixed in a mechanical lathe such that its zero position corresponds to the perpendicular between the link and the gravity vector.

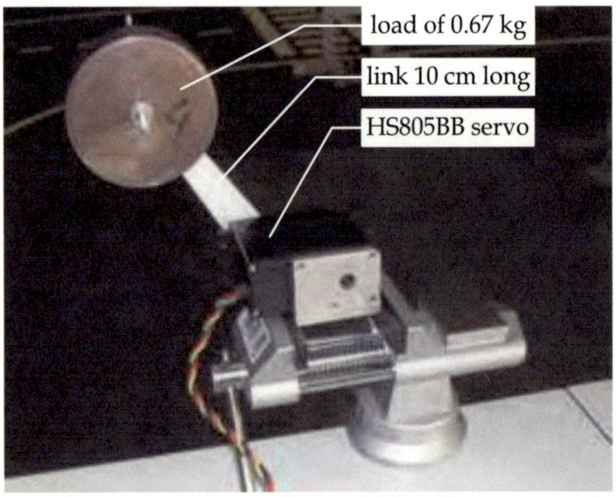

Figure 13. Experimental setup to assess servomotor response to variable loads

The setup used for experimental testing includes a master and a slave unit controlling a servomotor properly fixed and loaded. On the one side, the master unit is connected to a computer through a RS-232 link, using MatLab software as the user's interface. On the other side, the slave unit is connected to the servo mechanism in two ways: (*i*) by sending the desired servo position command as a pulse train with a given width; and (*ii*) by reading the potentiometer feedback signal (the only feedback available). In the experiments conducted below, the servo's internal controller is the only responsible for the resulting performance.

In the following, results of two experiments are described: the first experiment is performed with "large" steps (equivalent to 90º) for several loads and, then, a second experiment is carried out with smaller steps (few degrees each) in order to simulate some kind of ramp input and launching the basis for velocity control.

The results of applying a step input from -45º to +45º are presented in Figure 14 in terms of the desired and the actual response for two loads (258g and 1129g). The first notorious observation is the unstable dynamic behaviour on position reading, which shows at the beginning a sudden jump to a position below -45º and some oscillations during the path up to the final set point. Instead, the motor shaft presented a continuous and fast motion to the final position without speed inversions or any kind of oscillations. This seems to indicate that this process also requires care since the internal controller may interfere with the voltage drop on the potentiometer that can affect external readings of the shaft position. Another problem arising from the servo response, which may be critical as the load increases, is the considerable steady-state errors. Notice the presence of an appreciable value of steady-sate error for the larger load (about 8° error remains after the transient phase).

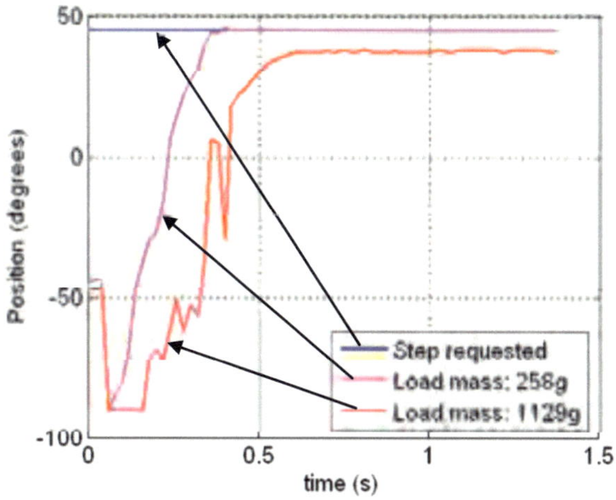

Figure 14. Response to a step input (– 45° to +45°) in the reference

In order to carry out a fair comparison with the previous case the joint has been placed in the same initial position (-45°) and should move to the same final position (+45°). However, to implement some sort of velocity control, the experiment was carried out in a manner that small position steps are successively requested to the servo controller. Their magnitude and rate will dictate some sort of desired "average velocity". This approach will generate an approximately linear increase for the position, which is to say, some constant velocity.
The results are presented in Figure 15 in terms of the desired and the actual response to a slope input. As above, and although the transient response has a very improved behaviour, the steady state error still exists. An experiment was carried out to stress this effect: the servo is requested to successively move a given load to some positions; for each position, after motion completion, the potentiometer is sampled to obtain the real position that the servo achieved. Relating the positional error with the static torque exerted in the joint, a direct conclusion can be drawn: the higher the torque, the higher is the steady state error.

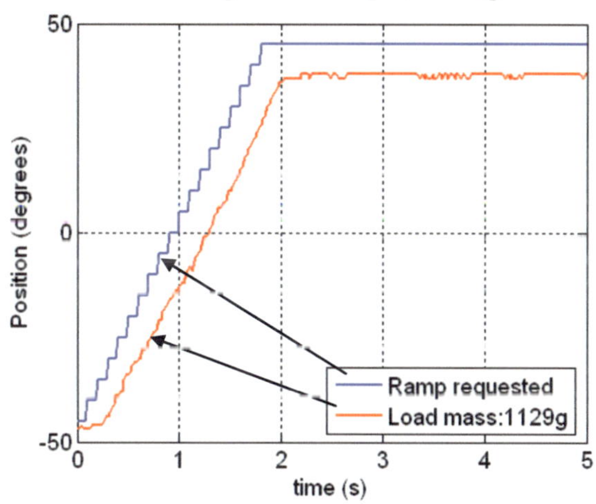

Figure 15. Response to a slope input in the reference

In conclusion, dynamic effects and improper servo's control turns the device into a highly non-linear actuator with limited performance, which restricts the scope of their application. Two common approaches can be devised to achieve higher performance: hardware modification or software compensation. The price to pay following the first direction is, often, the replacement of the electronics unit of the motor package by dedicated control boards. On the other hand, it is expected that enhanced performance can also be achieved by software compensation, provided that position and/or torque measurements are available.

4.2 Outer feedback control loop

The servo circuit has a narrow input control range and it is difficult to control accurately, though it has adequate speed and torque characteristics. In most practical situations, an effective strategy to improve the servo's operation is using an external controller where an outer position control loop is closed around the inner loop available in the servomotor. Figure 16 illustrates the block diagram of the servo controller proposed to achieve enhanced performance in terms of steady-state behaviour and trajectory tracking capabilities. The algorithm is based on dynamic PWM tracking using the servo own potentiometer for position feedback. For that purpose, the slave units have to track the motor positions (up to 3 motors) with time and adjust the PWM in order to accelerate or decelerate the joint motions. Practical issues like computation time or lack of speed measurements are challenged by devising the distributed architecture approach.

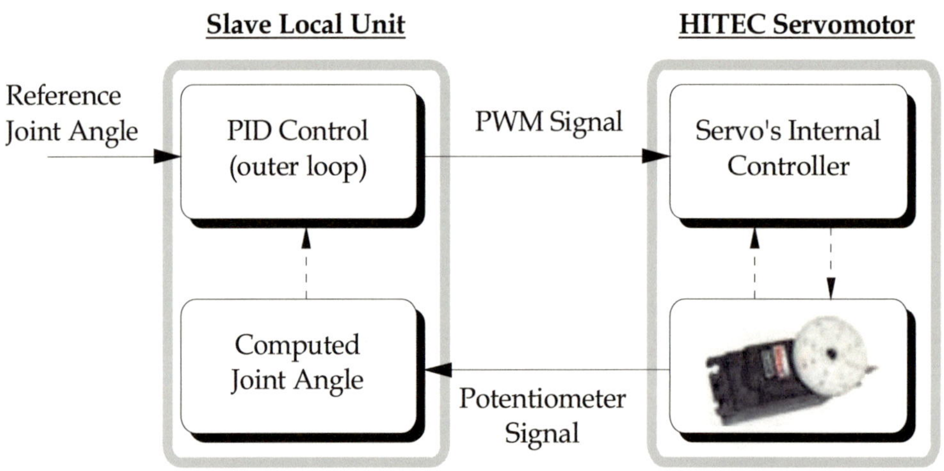

Figure 16. Block diagram of the joint control: the inner loop consists of the servo's own controller; the outer control loop generates a dynamic PWM using feedback from the servo's potentiometer

The potential offered by the external control strategy to ensure an improved behaviour is now investigated experimentally. For that purpose, several control schemes could be implemented in the PIC microcontroller. The focus of the present study is on digital PID-controller or any of its particular cases. The proposed control schemes are implemented in discrete time at 20 ms sampling interval and, then, tested in a number of experiments using the same setup as described before.

Two main considerations were made to guide the selection of the control structure. First, the system to control is formed by a single joint axis driven by an actuator with pulse-width control. Second, it is worth noting that an effective rejection of the steady-state errors is ensured by the presence of an integral action so as to cancel the effect of the gravitational component on the output. These facts suggest that the control problem can be solved by an incremental algorithm in which the output of the controller represents the increments of the control signal. In this line of thought, it is irrelevant the main drawback with this algorithm that cannot be used directly for a controller without integral action (P or PD). One advantage with the incremental algorithm is that most of the computation is done using increments only and short word-length calculations can often be used.

The first experiment is aimed at verify the effectiveness of the integral plus proportional actions. In this case, it is chosen a demanding specification for the desired slope: each new step position is update at the maximum rate of 50 Hz (corresponds to the PWM period) with amplitude of 5 degrees. Let the desired initial and final angular positions of the joint to be -90 and 50 degrees, respectively, with time duration of 1.12 seconds. The results are presented in Figure 17 in terms of the time history of the desired and actual angular positions, together with the trajectory errors for the full motion. It demonstrates the effect of increasing K_I for a fixed proportional term (K_P = 0.04): it reduces the lag time improving tracking accuracy, but at the expense of overshoot. Changing K_P to a higher value (K_P = 0.30) minimises the overshoot, maintaining the lag time as for K_I = 0.10. From these observations, the role of each component can be deduced: (*i*) integral action reduces time lag at the expense of an increased overshoot; and (*ii*) proportional action reduces overshoot, deteriorating the establishment time for very high gains.

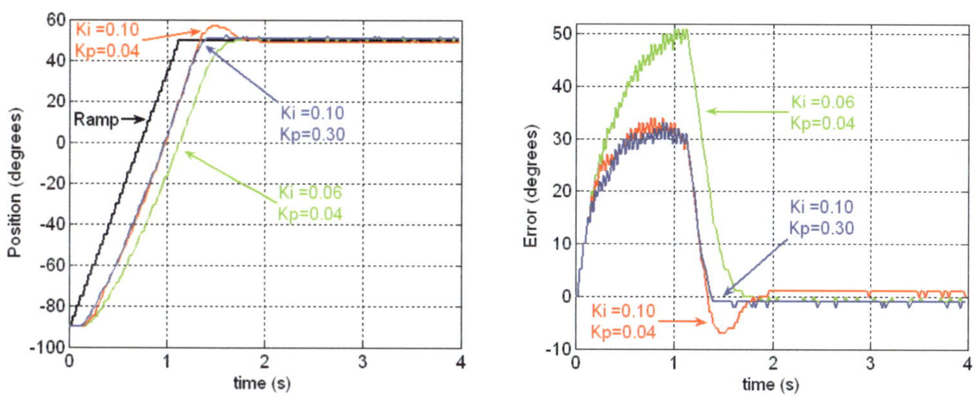

Figure 17. Behaviour of closed loop system with PI controller: the left graph shows the response to a slope input in the reference with different values of the control parameters; the right graph shows the trajectory errors for the full motion

Improvement of the position tracking accuracy might be achieved by increasing the position gain constant K_I, while controlling the overshoot effects by adjusting K_P. However, for high demands in terms of lag time, compensation tuning becomes very hard due to the presence of unstable oscillations during transient response. A solution to this drawback can be devised by rewrite the control algorithm aimed to include the proportional, integral and derivative terms. At the same time, the second experiment includes a planning algorithm used to generate smooth trajectories that not violate the saturation limits and do not excite resonant modes of the system. In general, it is required that the time sequence of joint variables satisfy some constraints, such as continuity of joint positions and velocities. A common method is to generate a time sequence of values attained by a polynomial function interpolating the desired trajectory. A third-order polynomial function in joint space was used to generate the reference trajectories. As result, the velocity has a parabolic profile and the acceleration has a linear profile with initial and final discontinuities. Figure 18 illustrates the time evolution obtained with the following initial and final conditions: q_i = -45°, q_f = 45°, t_f = 1.12 s. The gains of the various control actions have been optimized by trial and error in such a way to limit tracking errors. As observed, significant improvements are achieved in the servo's response: zero steady-state error with no overshoot and limited tracking errors.

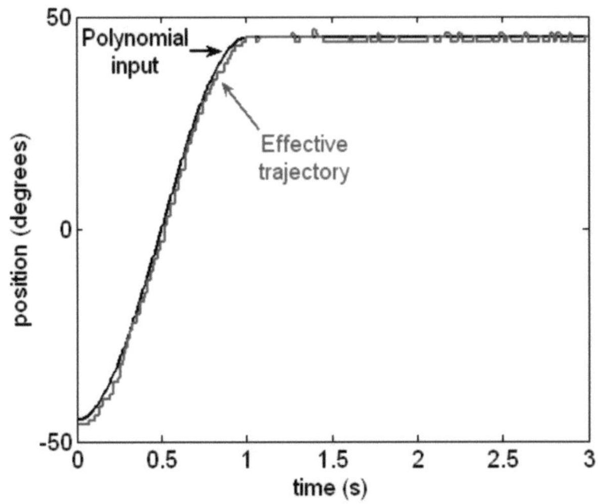

Figure 18. Behaviour of closed loop system with PID controller: the graph shows the response to a third-order polynomial joint trajectory in the reference

4.3 Dual leg behaviour

In this subsection, the previous control approach applied to the single-axis system is extended for the other robot's joints. Although this development phase may be facilitated by the reduction of performance demands and smaller joint excursions, the interpretation of the last results deserves attention given the influence of the driving system. The humanoid system is actuated by servomotors with reduction gears of low ratios for typically reduced joint velocities. The price to pay is the occurrence of joint friction, elasticity and backlash that contribute to the divergence between the commanded and the actual joint's position. At the lower level in the control system hierarchy lay the local controllers connected by a CAN bus to a master controller. These slave control units generate PWM waves to control three motors grouped by vicinity criteria (entire foot up to knee and hip joints) and monitor the joint angular positions by reading the servo own potentiometer. There are two servo loops for each joint control: the inner loop consists of the servo's internal controller as sold by the vendor; and the outer loop which provides position error information and is updated by the microprocessor every 20 ms.

We now compare the robotic system's behaviour when only the inner loop is present (hereinafter "open-loop control") and when the extra feedback loop is added (hereinafter "closed-loop control"). In the later case, the outer servo loop gains are constant and tuned to perform a well-damped behaviour at a predefined velocity. Once again, the joint trajectories along the path are generated according to a third-order interpolating polynomial with null initial and final velocities. The next trial demonstrates the behaviour of the legs in the double-support phase, while performing basic desired movements. More concretely, the desired movements to be performed consist of: (*i*) a vertical motion from an upright posture; and (*ii*) a lateral motion in which the leg leans sideways (±27 degrees). In both cases, an additional load of 2.1 kg is attached to the upper part of the leg to emulate the mass of other segments (Figure 19).

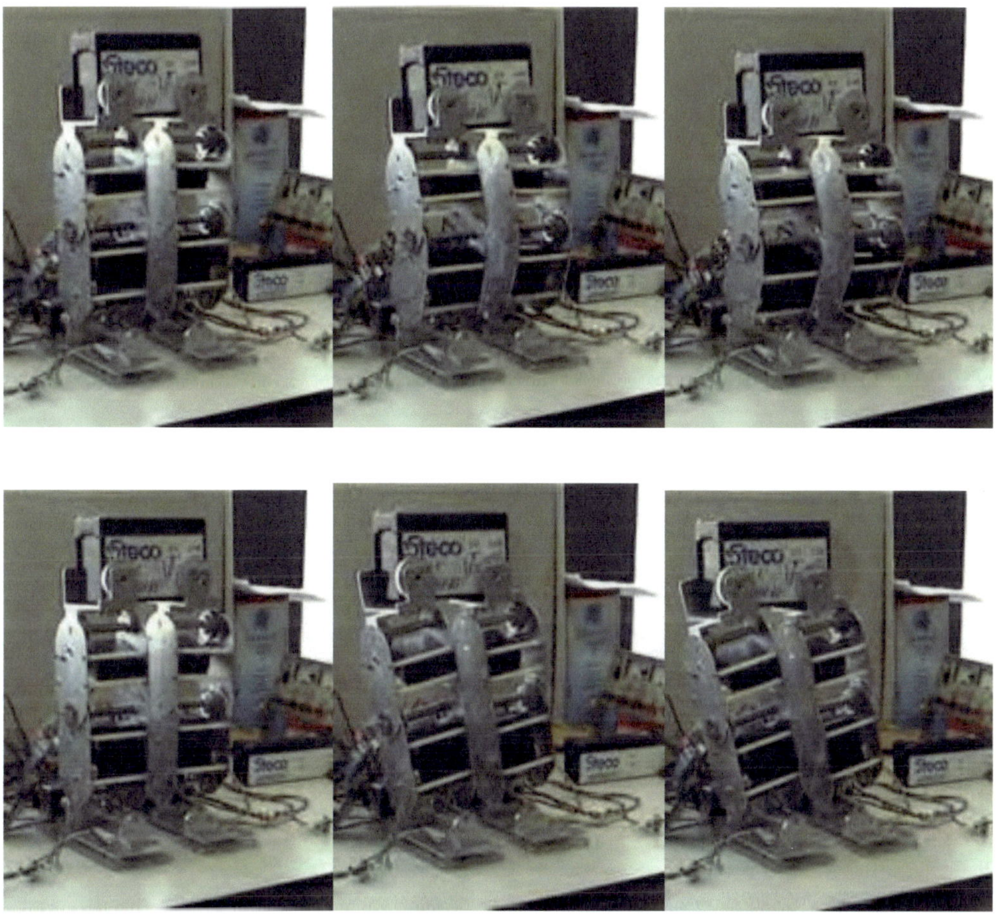

Figure 19. Snapshots of some stages in a motion sequence using two-legs and a load of 2.1 kg attached to the hip section: the top sequence shows the vertical motion; the bottom sequence shows the lateral motion

The experimental results in Figure 20 show the significant differences occurring in performance of the two control schemes (open-loop and the cascading close-loop controller). As expected, the open-loop control exhibits a poor performance, particularly for steady-state conditions. Due to the imposed vertical motion, the limitations of the open-loop scheme are more evident when observing the temporal evolution of the ankle (foot) joint. On the other hand, an improved performance is successfully achieved with the proposed outer control loop, both in terms of steady-state behaviour and enhanced trajectory tracking. Although further improvements could be possible by optimizing control gains, these results are adequate in demonstrating the efficacy of the external loop compensation approach. Finally, the performance of the servomotors is in accordance with theoretical considerations on the choice of a motor-gear combination.

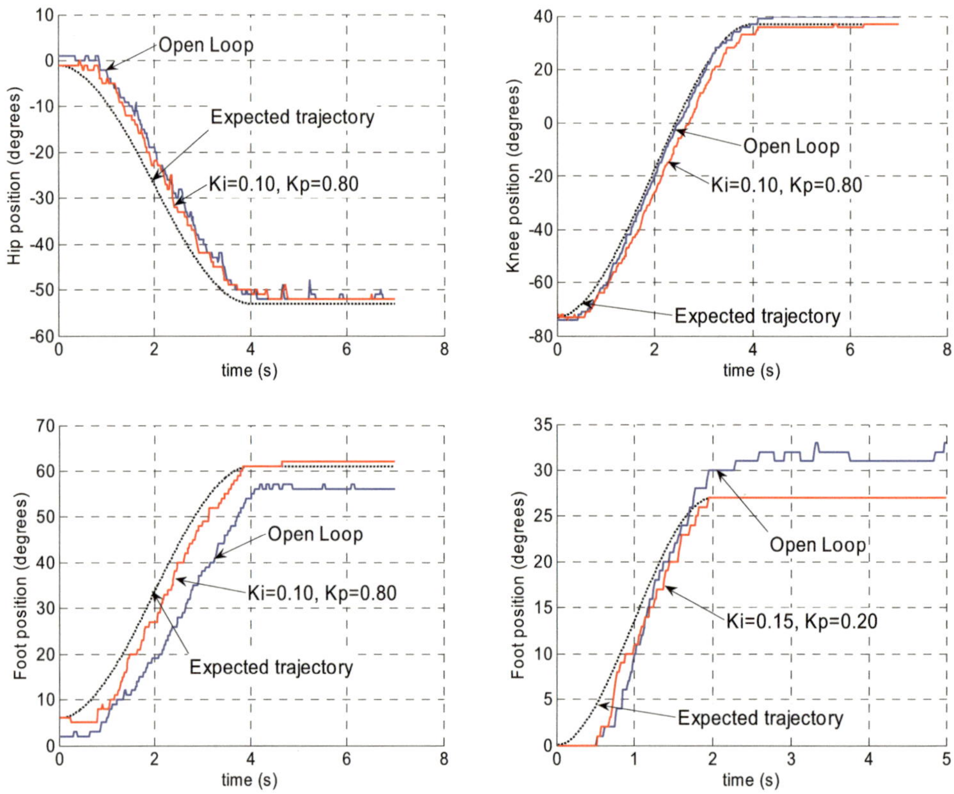

Figure 20. Comparison of performance between open and closed-loop control schemes: the top and left-bottom charts show the behaviour of the three joints during the vertical motion; the bottom-right chart shows the behaviour of the foot joint during the lateral motion

5. Force-driven local control

Balance maintenance is a core task for walking robots in order to engage useful tasks, ranging from standing upright posture to motion goals. The difficulty lies in the uncertainty of the environment and the limitations of the contact between the robot and the environment. Over the last years it becomes evident the dichotomy in the fundamental approaches of motion planning and control. On the one side, trajectory replaying approaches rely on accurate models of the walker being characterised by pre-planned trajectories that are played back during walking and, often, modified online through feedback (Sugihara et al., 2002; Yamasaki et al., 2002; Kajita et al., 2003). On the other side, realtime generation approaches ensure that planning and control are executed in a unified way. Gait trajectories are computed online feeding back the actual state of the system in accordance with the specified goal of the motion (Hirai et al., 1998; Denk & Schmidt, 2001;

Bourgeot et al., 2002). The combination of both approaches can be useful when adapting to deterministic but *a priori* unknown ground surfaces.
This section shows an example that is being developed to demonstrate the possibility of achieving proper humanoid leg balancing using a local control approach. To this purpose, it is considered feedback control from several sensors, including angular position in each joint and four force sensors inserted into the foot corners. The sensors provide information about the ground reaction forces and the location of the centre of pressure (COP). This opens up new avenues and possibilities for distributed architectures where centralised and local control co-exist and concur to provide robust full monitoring and efficient operation.

5.1 Adaptive leg balancing

The ability to balance in single support, while standing on one leg, is an important requirement for walking and other locomotion tasks. In the previous section, the approach to balance control assumed the presence of explicitly specified joint reference trajectories and calculations based on static configurations to derive the necessary PWM input signal. The goal of this section is to present the developed control algorithm that provides enhanced robustness in the control of balancing by accounting for the ground reaction forces. Thus, the system is able to stand on an uneven surface or one whose slope suddenly changes (Figure 21). In a similar way, the control system could sense that it has been pushed, using the force sensors in the soles of its foot, and acts to maintain the postural stability. The open challenge is to allow local controllers to perform control based on sensor feedback and possibly a general directive. Here, the global order is to keep balance in a desired COP location and, although all actuators can intervene, the ankle joints have the relevant role to keep an adequate force balance on each foot.

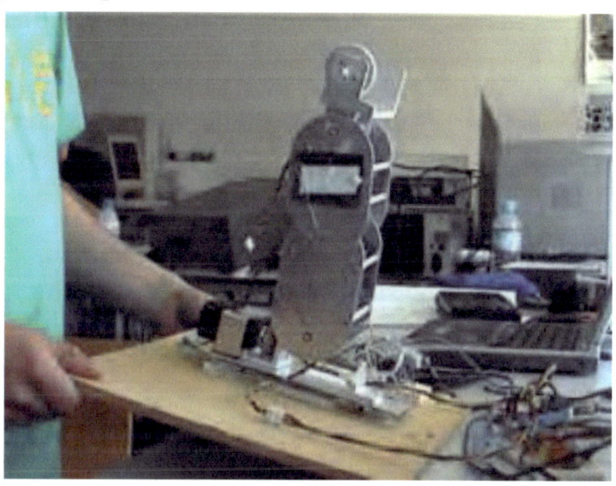

Figure 21. Single leg balancing on top of a surface with variable slope

The controller presents the following key features. First, the force sensors are used to measure the actual COP coordinates, instead of calculating other related variables, such as the centre of mass location. Second, the control system commands the joint actuators by relating the joint velocities ($\dot{q}$) to the error (e) between the desired and the current position of the COP. The choice of the relationship between $\dot{q}$ and e allows finding algorithms with different performances. The simplest method is the straightforward application of a proportional law, so that:

$$\dot{q} = Ke \tag{2}$$

The controller is independent of the robot's model or any nominal joint trajectory. This approach has the main advantage of its simplicity: each component of the error vector relates directly and in an independent way to the ankle joints (pitch and roll joints), due to their orthogonal relations. Alternatively, by interpreting a small displacement in the joint vector as a torque and the error vector as a force suggests the following update law:

$$\dot{q} = J^T Ke \tag{3}$$

Here, J^T is the transpose of the COG Jacobian matrix which transforms the differential variation in the joint space into the differential variation of the COG's position and K is a diagonal matrix properly chosen to ensure convergence. Another requirement is now imposed in order to stabilize the hip height: the error vector accounts for the operational space error between the desired and the actual end-effector position. Then, the Jacobian translates desired Cartesian motions of selected parts of the leg into corresponding joint space motions.

5.2 Experimental results

The following analysis illustrates the emergence of an appropriate behaviour when the system stands on a moving platform. The desired goal is to stand in an initial posture, while the control system relies on the reaction force data to estimate slope changes in the support surface. As stated before, the emphasis in this work is on procedures that allow the robot to calibrate itself with minimal human involvement. Thus, after an initial procedure in which the humanoid leg is displaced to the desired posture, the control system generates online the necessary joint adjustments in accordance with the pre-provided goal. The joint velocity values are computed in real time to modify dynamically the corresponding PWM signal. A joint velocity saturation function is used to avoid abrupt motions, while satisfying dynamic balance constraints.

The experimental results highlight the time evolution of the COP and the resulting ankle joint angles accordingly to the control laws presented above, while the humanoid leg adapts to unpredictable slope changes. Figure 22 and Figure 23 show the achieved behaviour for a predominant leg's motion in the sagittal plane, using both the proportional and the Jacobian-based control laws. Figure 24 and Figure 25 report the leg's behaviour for a predominant motion in the lateral plane. In both cases, the use of the proposed control algorithm gives rise to a tracking error which is bounded and tends to zero at steady state. This indicates that the posture was adjusted and the differences on the ground reaction forces become small. The algorithm based on the COG Jacobian provides a computationally efficient solution for simple models. For a practical humanoid, the Jacobian may be a complex non-linear matrix requiring fast and accurate calculations using a numerical approach. Ongoing work is exploiting the case when the reference COP is a time-varying function.

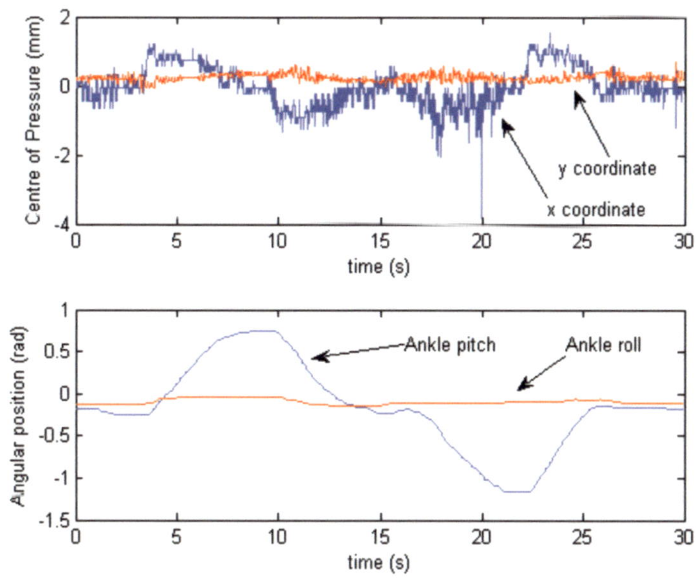

Figure 22. Leg's behaviour with predominant motion in the sagittal plane using the proportional law: temporal evolution of the centre of pressure (up) and joint angular positions (down)

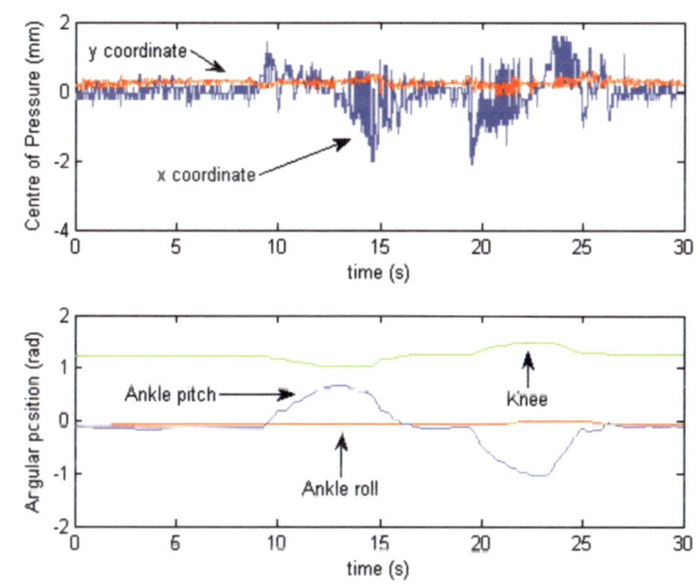

Figure 23. Leg's behaviour with predominant motion in the sagittal plane using the Jacobian-based method: temporal evolution of the centre of pressure (up) and joint angular positions (down)

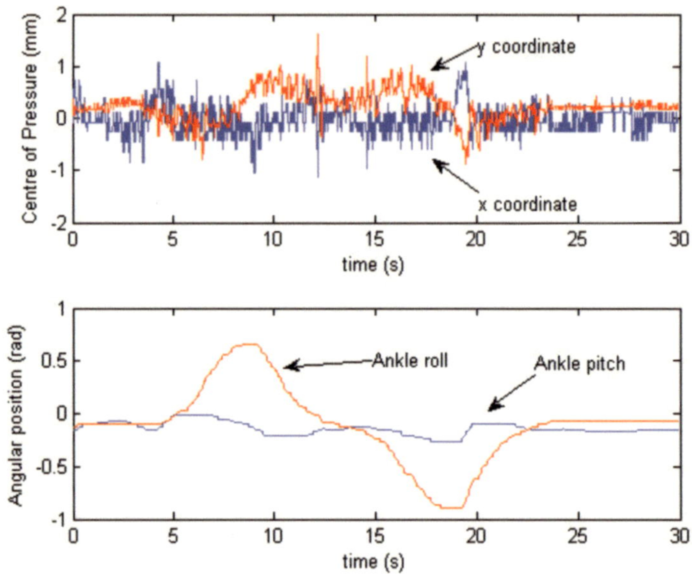

Figure 24. Leg's behaviour with predominant motion in the lateral plane using the proportional law: temporal evolution of the centre of pressure (up) and joint angular positions (down)

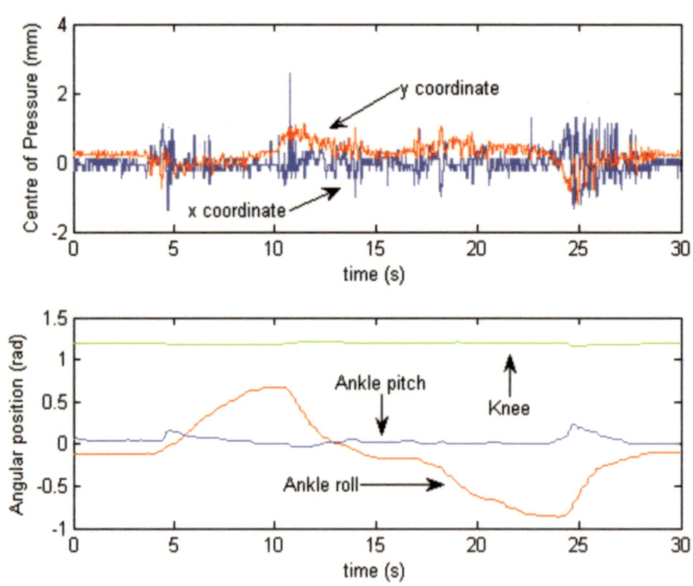

Figure 25. Leg's behaviour with predominant motion in the lateral plane using the Jacobian-based method: temporal evolution of the centre of pressure (up) and joint angular positions (down)

6. Conclusion

In this chapter we have described the development and integration of hardware and software components to build a small-size humanoid robot based on off-the-shelf technologies. A modular design is conceived to ensure easy maintenance and faster reproducibility. The most relevant feature of this implementation includes the distributed architecture in which independent and self-contained control units may allow either a cooperative or a standalone operation. The integration in these simpler control units of sensing, processing and acting capabilities play a key role towards localised control based on feedback from several sensors.

The adoption of an outer motion control loop to provide accurate trajectory tracking was presented and has been experimentally demonstrated. The strength of this approach lies in its performance, generality and overall simplicity. The humanoid platform reached a point where intermediate and high level control can now flourish. An example has been given for a kind of intermediate level control implemented as a local controller. From there, a force-driven actuation was successfully applied to demonstrate the possibility of keeping the humanoid robot in upright balance position using the ground reaction forces.

Ongoing developments on the humanoid platform cover the remainder hardware components, namely the inclusion of vision and its processing, possibly with a system based on PC104 or similar. A full autonomous humanoid robot for research is being developed that allows testing and evaluation of new ideas and concepts in both hardware and software modules. Future research, which has already started, will cover distributed control, alternative control laws and also deal with issues related to navigation of humanoids and, hopefully, cooperation. Force control techniques and more advanced algorithms such as adaptive and learning strategies will certainly be a key issue for the developments in periods to come in the near future.

7. Acknowledgments

The authors would like to thank the following students at the University of Aveiro for their support in the humanoid hardware and software development: David Gameiro, Filipe Carvalho, Luis Rego, Renato Barbosa, Mauro Silva, Luis Gomes, Ângelo Cardoso, Nuno Pereira and Milton Ruas.

8. References

Bourgeot, J.-M., Cislo, N. & Espiau, B. (2002). Path-planning and Tracking in a 3D Complex Environment for an Anthropomorphic Biped Robot, *Proceedings of the IEEE International Conference on Intelligent Robots and Systems*, pp. 2509-2514, October 2002, Lausanne, Switzerland.

Denk, J. & Schmidt, G. (2001). Synthesis of a Walking Primitive Database for a Humanoid Robot Using Optimal Control Techniques, *Proceedings of the IEEE International Conference on Humanoid Robots*, pp. 319-326, November, 2001, Waseda, Japan.

Furuta, T. et al. (2001). Design and Construction of a Series of Compact Humanoid Robots and Development of Biped Walk Control Strategies, *Robotics and Automation Systems*, Vol. 37, pp. 81-100.

Hirai, K. et al. (1998). The Development of Honda Humanoid Robot, *Proceedings of the IEEE International Conference on Robotics and Automation*, pp. 1321-1326, May 1998, Leuven, Belgium.

Huang, Q., Nakamura, Y. (2005). Sensory Reflex Control for Humanoid Walking, *IEEE Transactions on Robotics*, Vol. 21, n°. 5, pp. 977-984.

Lohmeier, S. et al. (2004). Computer System and Control of Biped "Johnnie", *Proceedings of the IEEE International Conference on Robotics and Automation*, pp. 4222-4227, April–May 2004, New Orleans, USA.

Kajita, S. et al. (2003). Resolved Momentum Control: Humanoid Motion Planning Based on the Linear Angular Momentum, *Proceedings of the IEEE International Conference on Intelligent Robots and Systems*, pp. 1644-1650, October 2003, Las Vegas, USA.

Kaneko, K et al. (2004). Humanoid Robot HRP-2, *Proceedings of the IEEE International Conference on Robotics and Automation*, pp. 1083-1090, April–May 2004, New Orleans, USA.

Kim, J.-H. et al. (2004). Humanoid Robot HanSaRam: Recent Progress and Developments, *Journal of Comp. Intelligence*, Vol 8, n°. 1, pp. 45-55.

Nagasaka, K. et al. (2004). Integrated Motion Control for Walking, Jumping and Running on a Small Bipedal Entertainment Robot, *Proceedings of the IEEE International Conference on Robotics and Automation*, pp. 3189-3194, April–May 2004, New Orleans, USA.

Popovic, M., Goswami, A. & Herr, H. (2005). Ground Reference Points in Legged Locomotion: Definitions, Biological Trajectories and Control Implications, *The International Journal of Robotics Research*, Vol. 24, n°. 12, pp. 1013-1032.

Ruas, M., Silva, F. & Santos, V. (2006). Techniques for Velocity and Torque Control of RC Servomotors for a Humanoid Robot, *Proceedings of the 9th International on Climbing and Walking Robots*, pp. 636-642, September 2006, Brussels, Belgium.

Sakagami, Y. et al. (2002). The Intelligent ASIMO: System Overview and Integration, *Proceedings of the IEEE International Conference on Intelligent Robots and Systems*, pp. 2478-2483, October 2002, Lausanne, Switzerland.

Sugihara, T., Nakamura, Y. & Inoue, H. (2002). Realtime Humanoid Motion Generation Through ZMP Manipulation Based on Inverted Pendulum Control, *Proceedings of the IEEE International Conference on Robotics and Automation*, pp. 1404-1409, May 2002, Washington, USA.

Silva, F. & Santos, V. (2005). Towards an Autonomous Small-Size Humanoid Robot: Design Issues and Control Strategies, *Proceedings of the IEEE International Symposium on Computational Intelligence in Robotics and Automation*, June 2005, Espoo, Finland.

Yamasaki, F. et al. (2000). PINO the Humanoid: A Basic Architecture, *Proceedings of the International Workshop on RoboCup*, August–September 2000, Melbourne, Australia.

Yamasaki, F. et al. (2002). A Control Method for Humanoid Biped Walking with Limited Torque, In: *RoboCup 2001*, A. Birk, S. Coradeshi & S. Tadokoro, (Ed.), pp. 60-70, Springer Verlag, Berlin Heidelberg.

5

Artificial Muscles for Humanoid Robots

Bertrand Tondu
LESIA, Institut National de Sciences Appliquées de Toulouse
Campus Universitaire de Rangueil, 31077 Toulouse
France

1. The question of actuator choice for humanoid robots

It is important to recall that humanoid robot technology derives from the technology of industrial robots. It is obvious that the developments of bipedal robots such as the integration of robot-upper limbs to complex anthropomorphic structures have benefited from progress in mechanical structures, sensors and actuators used in industrial robot-arms. A direct link is sometimes made between the technology of redundant robot-arms and humanoid robots as underlined in some technical documents of the Japanese AIST where it clearly appears that the HRP2 humanoid robot upper limb is directly derived from the Mitsubshi PA10 7R industrial robot-arm.
Due to its high number of degrees of freedom in comparison to industrial robots, a humanoid robot requires great compactness of all actuator and sensor components. This is why we believe that the harmonic drive technology associated with direct current electric motor technology has played a non-negligible part in humanoid robot development. The DC actuator offers the great advantage of being a straightforward technology, associated with simple and well-known physical models, its integration into mobile robots benefits from new developments in embedded batteries. However, its low maximum-torque-on-mass and maximum-torque-on- volume ratios are a serious drawback for its use in direct drive apparatuses. On the other hand, the ability of electric motors to generate very high velocities in comparison with moderate jointed velocities needed by industrial robot-arms and more by jointed anthropomorphic limbs, gives the possibility of using high ratio speed reducers to amplify motor-torque. Moreover, the choice of a high ratio speed reducer offers the advantage of masking inertial perturbations such as external torque perturbations. The technical achievement of such ratios induces specific mechanical difficulties due to the bulkiness of successive gears; harmonic drive technology - represented for example by Harmonic Drive AG - resolves this problem in a very elegant manner: the harmonic drive and the actuator fit together without excessive increase in mass and volume in comparison with the actuator alone. It can be considered that most of today's humanoid robots are actuated by DC motors with harmonic drives (this actuation mode is mentioned, for example, by Honda from its first paper about the P2 robot onwards (Hirai *et al.*, 1998) and then in the official ASIMO web site, as well as in papers concerning other Japanese and European humanoid robots). But if this technology simplifies actuator mechanical integration and leads to the use of simple joint linear control, despite the highly non-linear character of robot dynamics, it is well-known that the use of a speed reducer multiplies the

joint stiffness by the its ratio squared. A high joint stiffness contributes to joint accuracy and repeatability, but also induces a high danger level for users, which can be acceptable in the case of industrial robot-arms separated from factory staff by special safety devices, but becomes very problematical in the case of humanoid robots intended for working in public environments. The need to find an actuation mode which associates power, accuracy and a 'softness' adapted to human presence, that is the question of actuator choice in humanoid robotics. To address this problem we will first try to define the notion of artificial muscle in paragraph 2, then deal with the question of artificial muscle actuators for humanoid robots in paragraph 3, before analysing their integration within anthropomorphic limbs (paragraph 4) to finish with their control (paragraph 5).

2. Notion of artificial muscle

2.1 Performance criteria in the research of new actuators

A general theory of actuators does not exist; each actuator is defined according to the physical theory on which its legitimacy is founded. A comparison of actuators can as a consequence be delicate. This is why actuator designers have introduced a certain number of performance criteria aimed at making such comparisons easier. In general, actuation can be defined as a process of converting energy to mechanical forms, and an actuator as a device that accomplishes this conversion. **Power output per actuator mass**, and **per volume**, as actuator **efficiency** - defined as 'the ratio of mechanical work output to energy input during a complete cycle in cyclic operation ' (Huber *et al.*, 1997) - are three fundamental properties for characterizing actuators. However, artificial muscle technology considers more specific performance criteria so as to accurately specify new actuator technology in comparison with 'natural muscular motor' properties. The following terminology, justified by the linear actuator character of the artificial muscle, generally completes the power criteria - the definitions given in inverted commas are from (Madden *et al.*, 2004) :

- **Stress** : 'typical force per cross-sectional area under which the actuator materials are tested'; **maximum stress** corresponds to the maximum stress that can be generated in specified functioning conditions; as will be seen later, it is important for a given technology to specify the 'actuator materials' relating to stress: typical stresses of strips or fibres of a given technology is not obligatorily similar to that of the artificial muscle composed of a set of these strips or fibres;
- **Strain** : 'displacement normalized by the original material length in the direction of actuation'; **maximum strain** and maximum stress are according to Huber & others 'basic characteristics of an actuator [since] for a given size of actuator they limit the force and displacement' (Huber *et al.*, 1997, p. 2186); the terms **contraction ratio** and **maximum contraction ratio** will also be used;
- **Strain rate** : 'average change in strain per unit time during an actuator stroke'; the term **'response time'** - in the sense given by control theory, will also be used to characterize the speed of artificial muscle dynamic contraction;
- **Cycle life** : 'number of useful strokes that the material is known to be able to undergo'; this notion specifies the heavy-duty character of artificial muscle in given working conditions; in practice this is an important notion since artificial muscles are essentially made of 'soft' materials which can be weakened by shape changes imposed by the actuation mode;

- **Elastic modulus**: 'material stiffness multiplied by sample length and divided by cross-sectional area'; this is a typical material science notion; when the artificial muscle is considered as an actuator to be controlled, **stiffness** - and its inverse - **compliance** notions can appear more appropriate.

It is important to note that this criteria list is not exhaustive; depending on author, other performance criteria can be found: as an example, Huber *et al* consider two criteria not listed above: **actuator density** (the ratio of mass to initial volume of an actuator) and **strain resolution** (the smallest step increment of strain) (Huber *et al.*, 1997). The first directly concerns humanoid robotics: like skeletal muscles, artificial muscles are assumed to be located in moving robot links; the second criterion can be interesting in a control perspective to specify artificial muscle sensitivity.

Furthermore, we believe that the theoretical framework proposed by Hannaford and Winters (Hannaford & Winters, 1990) for the analysis of actuator properties based on Paynter's terminology of generalized dynamic systems (Paynter, 1961) can be particularly useful in our study: these authors propose characterizing any actuator by its two effort-flow and effort-displacement characteristics, where 'effort' represents the output force or torque, and 'flow' the linear velocity or angular velocity. For example, the DC motor is ideally characterized by a linear effort-flow curve and a constant effort-displacement curve, as illustrated in Figures 1.a and 1.b. Figures 1.c and 1.d give the typical characteristics of the skeletal muscle by the association of 'effort-flow' characteristics corresponding to the so-called tension-velocity curve with the 'effort-displacement' characteristics corresponding to the so-called tension-length curve. We will return to this in the modelling of skeletal muscle, but it is important to note the fundamental originality of skeletal muscle characteristics: while most of the actuators have constant or pseudo-constant 'effort-displacement' characteristics, this is not so for skeletal muscle. As a dynamic system in Paynter's sense, the 'effort-displacement' relationship defines passive element C (for compliance or capacitance). Classical actuators generally have infinite compliance; a dependence of force/torque on position can even appear as a drawback: it is up to the actuator control and not the actuator itself to impose this dependence. Conversely, living actuators aimed at a 'relationship life' have to combine the generation of the power necessary for body mobility with a non-infinite compliance making for easy contact with the environment - we will use later the term 'natural' compliance to characterize this compliance peculiar to the skeletal actuator. Research on artificial muscles can be understood as a new attempt to mimic the living so as to integrate it into a machine - the humanoid robot - an original power-softness combination, yet glaringly absent in machine technology.

2.2 The historical Kühn and Katchalsky notion of artificial muscle as a gel swelling and deswelling under the effect of a chemical agent

The origin of the artificial muscle notion must be found in the first works of chemists on certain materials whose swelling can be controlled in a reversible manner. At the end of the 1940s, Kühn & Katchalsky did indeed prove that an aqueous polymeric gel essentially composed of polyacrylic acid '… is found to swell enormously on addition of alkali and to contract rapidly when acid is added to the surrounding solution. Linear dilatations and contractions of the order of 300 per cent were observed. This cycle of swelling and de-swelling is reversible and can be repeated at will ' (Kühn *et al.*, 1950, p.515). At that time the authors had designed a device transforming chemical energy into mechanical working in the form of a 0.1 mm thick filament able to lift up a 360 mg load in some minutes when swollen. Thus the original Kühn & Katchalsky experiment historically created the present artificial muscle notion as a reversible contractile device. Katchalsky emphasized the natural tendency of artificial muscles to generate an 'equilibrium swelling' brought according to him about two opposing tendencies : first, '...the solution tendency of the polymeric molecules and the osmotic pressure of the cations of the alkali bound by the gel', secondly '...the caoutchouc-type contraction tendency of the stretched polymer molecules' (Katchalsky, 1949, p.V/8). More generally, we will go further and apply this natural tendency to an equilibrium state of any artificial muscle type as an open-loop stability in position. It is important to note, however, that this ability to pass from an equilibrium state to another state would be nothing if it were not associated to a reversibility whose life cycle is finally its measure. Note also that the life cycle of natural muscle is greater than 10^9 (Hunter & Lafontaine, 1992); no present-day artificial muscle is able to approach this value, which is linked to the ability of living tissues to self-repair. Kühn & Katchalsky's historical studies were reconsidered in the 1980s within the framework of a renewed interest for artificial muscles due to technological developments in robotics, and a demand for implantable artificial biological organs. However, from a practical point of view, the Kühn & Katchalsky actuator displays a major disadvantage: its excessively slow response time (in the order of minutes). More generally, it will be seen throughout this chapter that the major difficulty in the design of artificial muscles for robots consists of obtaining both quick response time and high-power-stress to mass-and-power to volume, adapted to the integration of artificial muscles to human-dimensions and mass anthropomorphic limbs. There now follows a brief overview of present-day artificial muscle technologies.

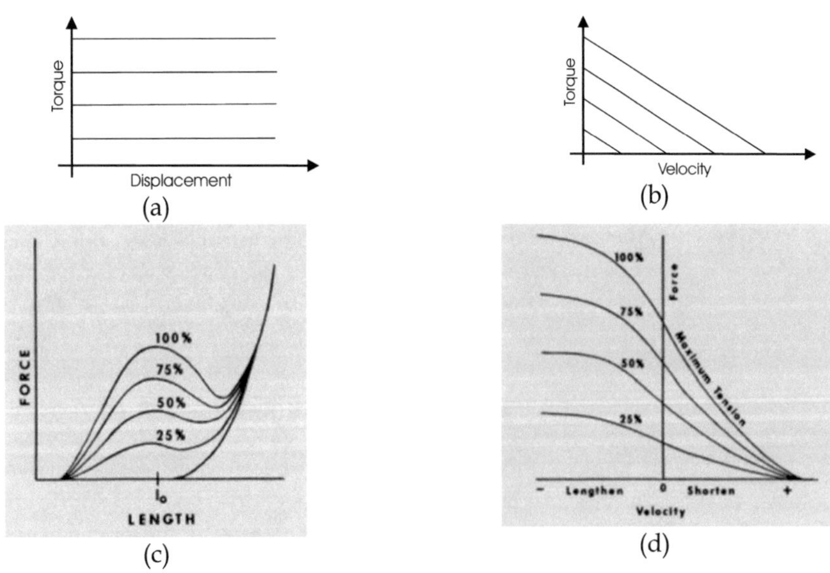

Figure 1. Characterization of artificial and natural actuators based on 'effort-flow' and on 'effort-displacement' relations (inspired from (Hannaford & Winters, 1990), (a) and (b) Case of the DC motor, (c) and (d) Case of the skeletal muscle (from (Winter, 1969))

2.3 The artificial muscle as any material structure exhibiting a reversible shape change by a chemical or a physical agent

a. Contractile polymer gels

Aqueous solutions of polymeric acids considered by Kühn & Katchalsky are a particular case of solvent-swollen crosslinked polymer gels that respond to changes in temperature, pH, solvent, illumination, or other physical or chemical stimuli. The choice of pH or a solvent as a stimulus applied to typically polyacrylamide (PAM), polyvinyl alcohol-polyacrylic acid (PVA-PAA) or polyacrylonitrile (PAN) fibers or strips (all commercially available) is particularly interesting due to the facility of producing the pH variation by adding of acid or alkali solutions (typically through the control of the hydrogen ion by means of chemical reactions with HCl and NaOH) or the facility of using cheap and readily available solvent like acetone. Parallel to a better understanding of gels physics it has been possible to develop micro sized gel fibers contracting in seconds and even tenths of seconds and to increase both the force produced by gel fibers as resistance fibers (some of them can support loads of about 100 N/cm^2 approximatively equal to that of human muscle. M. Suzuki, for example, has developed in the nineties a 12 cm length artificial muscle model by PVA-PAA-PA1Am gel film of 50μm thickness able to raise a ball of 2 g from a lower equilibrium position to an upper equilibrium position within 10 s (Suzuki, 1991). Although in these conditions, the maximum power density obtained with gel fibres can indeed be estimated as being close to that of skeletal muscle, it still appears difficult to apply this technology to the actuation of human-sized robot limbs. However, later we will analyse the interesting attempt made at the MIT in the 1990s to promote 'pH muscle'
The relative slowness of the complete contraction-elongation cycle of ion sensitive gel fibre is mainly due to the relative slowness of the ion sensitive mechanism itself. For this reason, other stimulus modes were considered, in particular heat transfer since heat diffusion is better than ion diffusion. Thermo-responsive polymer hydrogels are already used in drug release, and the study of tensile properties of thermo-responsive gels have been clearly established (Kim *et al.*, 2005). In the field of artificial muscles, this stimulus mode seems, however, to have been more studied in the case of fairly recent polymer types - in particular liquid crystal polymers - assumed to respond quicker than gel fibres. In any case, for gels as for other materials, thermal response always appears quicker in the heating than in the cooling phase (we will not deal in this chapter with the promising technology of liquid crystal polymers able to respond to a luminous flash in less than a second, but as noted by De Gennes (De Gennes *et al.*, 1997) which predicted their contractile possibilities and talked of them as 'semi-quick artificial muscles', the return to low temperatures is slow).

b. Shape memory alloys

Another form of reversible thermal response can be generated by means of thermally activated shape memory alloys based on the 'shape memory effect' discovered at the beginning of the 1950s. Among available materials nickel-titanium alloys (NiTi or nitinol) are particularly interesting for actuation use because of their low cost and electrical resistivity which heats the material by passing an electrical current. Due to this Joule heating, NiTi contraction and expansion cycles can occur based on transformations from martensitic to austenitic phases. Typically a shape is 'memorized' at a high temperature (600°) placing the nitinol in austenite phase. On decreasing the temperature, the material reverts to the martensite phase. When an external stress is applied, and as the material is reheated at a lower temperature the material, because of the instability of the martensite phase at these temperatures, returns to its well-defined high-temperature shape. In the shape memory effect, the material exhibits residual strains that can be used to generate a linear displacement. The possibility of developing NiTi fibres exhibiting both very high strain rates (300%/s) and very high stress (200 MPa) (Hunter & Lafontaine, 1992) naturally interested the designers of new robot actuators. Safak and Adam (Safak & Adams, 2002), for example, developed a lobster robot actuated by antagonistic nitinol artificial muscle pairs. Figure 2 shows an interesting application of nitinol to miniature humanoid robotics: the five fingers of the miniature robot hand can react at 0.2 s constant times to grasp light objects (Maeno & Hino, 2006). Besides the current need for small sizes to obtain quick responses, the main drawback of nitinol-type artificial muscles is its limited maximum strain: about 5%, which is very far from the natural skeletal muscle's 40 %. Moreover, the life cycle becomes more limited as the strain increases.

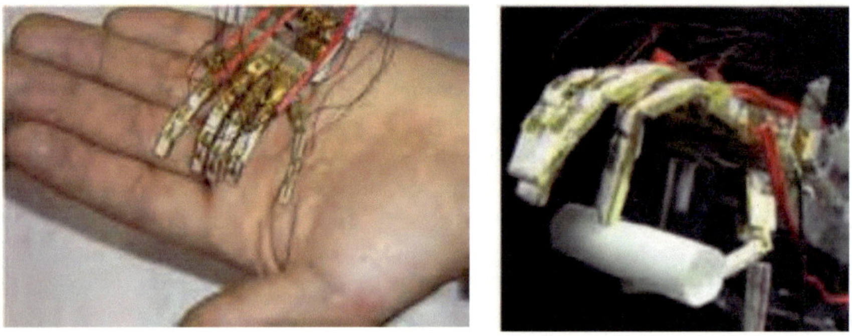

Figure 2. Miniature robot-hand actuated by Nitinol shape memory alloy artificial muscles controlled by heat transfer (from (Maeno & Hino, 2006))

The shape memory effect can be controlled not only by heat transfer, but also by means of an electric field which offers the advantage of being an easier control parameter. Ferromagnetic shape memory actuators have been largely studied and commercial versions exist able to develop maximum strains of 10%, typical strain rates of 10 000%/s and typical stresses of 1 Mpa to 9 MPa (Madden *et al.*, 2004). However, the requirement of high intensity magnets is a major drawback for human-sized robot-limbs. Finally, as for thermally activated shape memory alloys, the difficulty of accurately controlling muscle contraction is not a good point for applications to robotics. The same difficulty occurs with recent conducting shape memory composites.

c. Electroactive polymers : From ionic polymer gels to ionic polymer-metal composites

It was demonstrated in the 1960s that the swelling/shrinking of ionic polymeric gels generated by pH variation can also be obtained electrically. When an electric field is applied to a strip of PAM, PAMPS or PAA-PVA, for example suspended in a surfactant solution, the gel shows significant and quick bending towards the anode. According to Segalman and others (Segalman *et al.*, 1992) this is caused by the migration of unbound counter ions in the gel, and the impingement of solvent ions at its surface which produce the strip bending. The reversible phenomenon has been applied to the design of chemomechanical mobile systems such as Shahinpoor's swimming robot (Shahinpoor, 1992) or to polymer gel fingers (Hirose *et al.*, 1992). However, the application to linear actuators appears disappointing, as recently reported by Choe and Kin who studied polyacrylonitrile linear actuators (Choe & Kim, 2006): the tested fibre is a 10 mm long 'single strand' consisting of multiple filaments (about 2000) whose diameter in the contracted state is about 0.5 mm, and 1.2 mm in the elongated state; experimentally the fibre can produce a 0.1 N maximum force in both pH activation and electric activation cases, but while this static tension is generated in fewer than 10 s with a 1M HCl acid solution, it takes approximately 10 min for the same result with a 5V electric field..

If the electrostatic approach of ionic polymer-gel based linear actuators is unsatisfactory, it can be asked if more interesting results could be obtained using conductive polymers. Efforts to develop conductive polymer artificial muscles can be viewed as the search for associating polymer mechanical properties with the ease of electric field control. In general, conductive polymers are electronically conducting organic materials featuring conjugated structures, able to undergo dimensional changes in response to changes in the oxidation state. Polyaniline, trans-polyacetylene and polypyrrole are three examples of such employed structures. When, for example, a conducting polypyrole film is electrochemically oxidized, positive charges are generated along the chains and hydrated counter ions from the solution are forced to penetrate into the polymer in order to compensate for these positive charges. This induces an opening of the polymer structure and an increase in its volume. A reduction in polymer induces the reverse effect: positive charges are eliminated due the injected electrons: the polymer recovers its neutral state and the film volume decreases. This swelling-shrinking property can be applied to the design of artificial muscles such as the bilayer structure, consisting of a conducting and flexible polypyrolle layer adhering to an elastic and non-conducting film, proposed by Otero and Sansinena (Otero & Sansinena, 1995), or the monolayer structure by assembling two different polypyrroles (Ochoteco *et al.*, 2006). The resulting actuator is a bending type reacting in an aqeous solution under low intensity current variation. According to Madden & others a maximum 12% strain and a maximum 12%/s strain rate can be expected (Madden *et al*, 2004), consequently quicker than electrically-activated ionic polymer gels; however, conductive polymers suffer from the same major drawback as ionic polymer actuators: they need to be contained in a solvant bath.

The class of Ionic Polymer-Metal Composites (IPMC) can thus appear as a successful attempt to maintain the ionic polymer contractile principle without the need for a solvant bath. An IPMC is an ionic polymer in a composite form with a conductive metallic medium such as platinium. The nafion developed by DuPont de Nemours & Co. is generally used as a cation exchange thin membrane with metal electrode plated on both faces. Recently, liquid nafion has been used to manufacture IPMC strips in various thicknesses (Kim & Shahinpoor, 2002). Although they operate best in a humid environment, they can be designed as self-contained encapsulated actuators to operate in various environments. As theorized by Shahinpoor (Shahinpoor, 2002) an electric field induces a change in ion concentration which attracts water molecules to one side of the polymer. Non-uniform distribution of water produces a swelling on one side of the actuator, while the other side contracts. The bending movement of the strips toward the anode is obtained at a low voltage (typically 2V), and increases for higher voltages (typically up to 10 V) with a reaction speed between μs and s. Because the IPMC does not produce linear actuation, except in the case of fish-type mobile robots, its application to robotics is limited to gripping mechanisms able to lift a few grams (Shahinpoor *et al.*, 1998). The low stress generated by IPMC - 10 to 30 MPa (Shahinpoor *et al.* 1998) - is another drawback of this type of artificial muscle.

d. Electroactive polymers : Dielectric elastomers

Dielectric elastomer actuator technology appeared in the middle of the 1990s. As will be analysed in paragraph 3, this technology is considered as one of the most promising for developing artificial muscles adapted to robot-limbs (Bar-Cohen, 2004). As in IPMC technology, polymer shape change can be induced in dry environments, but at the expense of higher stresses. A dielectric elastomer actuator can be considered as a compliant capacitor inducing a stress when the capacitor is charged. According to Maxwell's theory applied to a constant volume elastomer, stress p and strains S_X , S_Y , S_Z of the dielectric elastomer (assuming small, e.g. < 10%), respectively, in X, Y and Z directions, as illustrated in Figures 3.a and 3.b, can be written as follows (Pelrine *et al.*, 1998), (Pelrine *et al.*, 2002) :

$$\begin{cases} p = ee_0\,(V/t^2) \\ S_Z = -ee_0\,V^2/(Yt^2) \\ (1+S_X)(1+S_Y)(1+S_Z)=1 \end{cases} \qquad (1)$$

where e_0 is the permittivity of free space, e the polymer dielectric constant, V applied voltage, t polymer thickness and Y elasticity modulus of the polymer-electrode composite film. These three equations constitute a simplified model (limited by the assumption of small strains) but which highlights the main parameters participating in the dimensioning of the actuator: polymer dielectric constant e - e is equal to 1 for air – varies between 3 to 10 according to elastomer choice (Mathew *et al.*, 2006) and electric field $E = V/t$ whose value, depending as it does on the elastomer, is an important factor in increasing actuation stress by a factor of 10 to 100 compared to conventional electrostatic actuators. The simplicity and relative efficiency of this model contrasts with the complexity of Flory's model which includes polymer gels swelling, in association with the difficulty of parametrizing the conductive polymers. Performances of dielectric elastomers are particularly interesting in comparison with other artificial muscle technologies as well as with natural muscle (Pelrine *et al.*, 2000) because they can associate high maximum strain, high maximum stress and low response times: in the case of a silicone elastomer, a maximum strain of 32 % can be generated with a maximum stress of 1.36 Mpa, and a time response of some msec. Furthermore, their ability to be configured in many ways is particularly important for robotics. The next paragraph analyses how linear actuators can be derived from this technology. Two main disadvantages, however, have to be highlighted: firstly, dielectric elastomers operate at relatively high voltages (100-4000 V), and secondly, due to their capacity, electrical energy remains after actuation, which in practice requires energy recovery circuits.

e. Elastomer inflatable structures

Physical elastomer properties are also at the origin of another large class of artificial muscles that recent advances in chemo-mechanical actuators have pushed into the background. They are not mentioned in numerous synthesis articles. We propose to call them 'elastomer inflatable structures' because they are based on the unique property of elastomers to support very large strains (up to 600%) without damage. Whereas the dielectric elastomer operation principle is based on the generation of a compression stress, the operation principle of elastomer inflatable structures is based on tensile stress. Thanks to a specific guiding structure, which will be specified in paragraph 3, the stress generated inside the elastomer inflatable structure is transformed into a linear force able to contract the artificial muscle, as illustrated in Figures 3.c and 3.d.

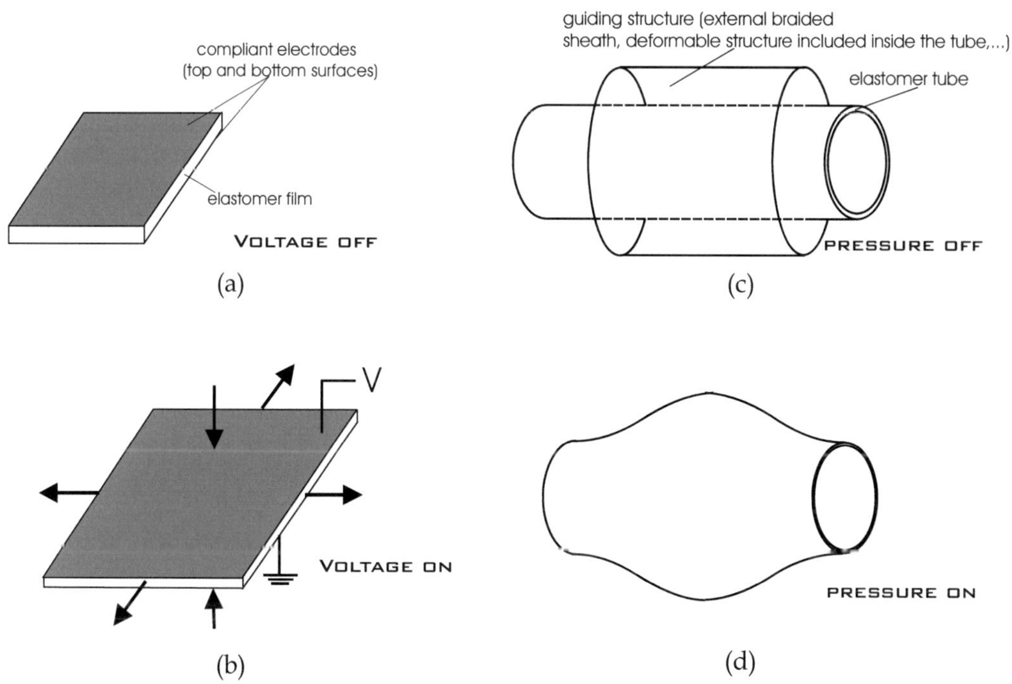

Figure 3. Use of elastomer elasticity in artificial muscle designing, (a) and (b) operation principles of dielectric elastomers (inspired from (Pelrine *et al.*, 2000)): the elastomer film placed between two compliant electrodes is compressed and expanded when an electric field is applied, (c) and (d) operation principles of contractile inflatable structures: a guiding structure is used to transform lateral stress generated by the pressure into a linear contraction stress

2.4 The artificial muscle as a linear actuator mimicking natural skeletal muscle behaviour: notion of skeletal artificial muscle

Until now the artificial muscle notion has been defined as a reversible contractile device, independently in some ways from any model of the skeletal muscle. In the framework of humanoid robots, based on a general anthropomorphism principle, the notion of artificial muscle can be considered from a more physiological point of view, as a linear actuator mimicking the natural skeletal muscle's static and dynamic behaviour. According to this approach, we suggest considering a fundamental abstract notion of artificial muscle inspired from the macroscopic modelling of skeletal muscle: the derived 'skeletal artificial muscle' notion will be used as a reference biomimetic model independent of the technology considered to achieve the abstract notion. According to the consideration developed by Hannaford & Winter on the actuator notion, referred to in paragraph 1, the skeletal artificial muscle notion can be specified by combining both tension-length and tension-velocity relationships, *i.e.* static and dynamic models of natural skeletal muscles. The resulting artificial muscle model will be, as a consequence, a phenomenological model which puts aside the microscopic phenomenon at the origin. It is well known that skeletal muscle contraction is induced both by the recruitment of motor units and their firing rate (Ghez, 1991). In a phenomelogical muscle model, these factors are set together in the form of a global variable which is traditionally called 'neural activation' and can be defined as a normalized variable u between 0 and 1. In isometric conditions, *i.e.* at a constant length, the typical skeletal muscle tension-length has already been illustrated in Figure 1.c. The static behaviour of a skeletal muscle of initial length l_0 is characterized both by an active tension corresponding to the contraction force produced when $l \leq l_0$, and by a passive tension due to the elasticity of muscle fibres beyond l_0,. as is also well known. If we focus on active tension, skeletal muscle behaves like a linear spring whose stiffness can be controlled by neural activation. The following linear model, illustrated in Figure 4.a, results from these considerations:

$$F_{stat} = uF_{max}(1 - \frac{\varepsilon}{\varepsilon_{max}}), \quad 0 \leq u \leq 1 \quad \text{and} \quad 0 \leq \varepsilon \leq \varepsilon_{max} \tag{2}$$

where F_{stat} corresponds to the static tension, ε to the muscle contraction ratio - defined as $\varepsilon = (l_0 - l)/l_0$ where l is the current length of the muscle - F_{max} the maximum tension and ε_{max} the maximum contraction ratio. This linear model captures the main static property of skeletal muscle : its 'natural compliance' C proportional to u which physically expresses the existence of a restoring force F_r which brings back the muscle to its equilibrium position when it is deviates of $\delta\varepsilon$; we get :

$$\begin{aligned} F_r &= -\frac{uF_{max}}{\varepsilon_{max}} \delta\varepsilon \\ \Rightarrow \quad C &= +\frac{\varepsilon_{max}}{uF_{max}} \end{aligned} \tag{3}$$

In our opinion the 'natural compliance' factor - or its inverse, the stiffness - has a greater importance than the 'young modulus' generally considered in artificial muscle technology, as defined in paragraph 1: young modulus reflects material characteristics, while compliance actuator characteristics.

Equation (2) is a purely static model; when applied to a dynamic contraction of the muscle, this muscle tension model results in purely oscillating behaviour. In order to take into account the natural damping of the natural muscle, it is consequently necessary to add a supplementary term to the static tension model, we note $F_{damp}(u, \varepsilon, \dot{\varepsilon})$ and which can depend on $\dot{\varepsilon}$ only if a pure viscous damping is assumed or on both u, ε and $\dot{\varepsilon}$ if the damping is associated to more complex friction phenomena :

$$F = uF_{max}(1 - \frac{\varepsilon}{\varepsilon_{max}}) - F_{damp}(u, \varepsilon, \dot{\varepsilon}), \quad 0 \leq u \leq 1 \quad \text{and} \quad 0 \leq \varepsilon \leq \varepsilon_{max} \tag{4}$$

How to define this damping function? No simple direct model can be derived from skeletal muscular physiology. However, Hill's model provides a particularly powerful and elegant indirect model . The Hill model is generally presented as the relationship between muscle shortening velocity V and its corresponding tension, denoted for reasons to be explained later, F_{Hill}, as follows :

$$(F_{Hill} + a)V = (F_0 - F_{Hill})b \tag{5}$$

where F_0 is the isometric contraction force at zero contraction ratio in given stimulation conditions – according to our static : $F_0 = uF_{max}$ – a is a constant having the dimensions of a force and b a constant having the dimensions of a velocity. Ratio (a/F_0) is typically between 0.2 and 0.3 which gives Hill's equation curve a hyperbola shape , as illustrated in Figure 4.b. Let us recall that the Hill equation was established in a specific isotonic quick-release test invented by Hill in his famous 1938 paper (Hill, 1938): during the almost constant speed phase of the quick-release in 'after-load' mode, F_{Hill} corresponds to the load driven by the muscle and captures the dynamic force produced by the muscle, including its damping component. As a consequence, the quick-release test applied to any artificial muscle allows an appreciation of its dynamic damping in comparison with that of skeletal muscle. Beyond the purely biomimetic argument, we have recently tried to show the interest for an artificial muscle to be in accordance with the Hill model in order to benefit from the natural load variation adaptivity which seems to be an integral part of this dynamic model (Tondu & Diaz, 2006).

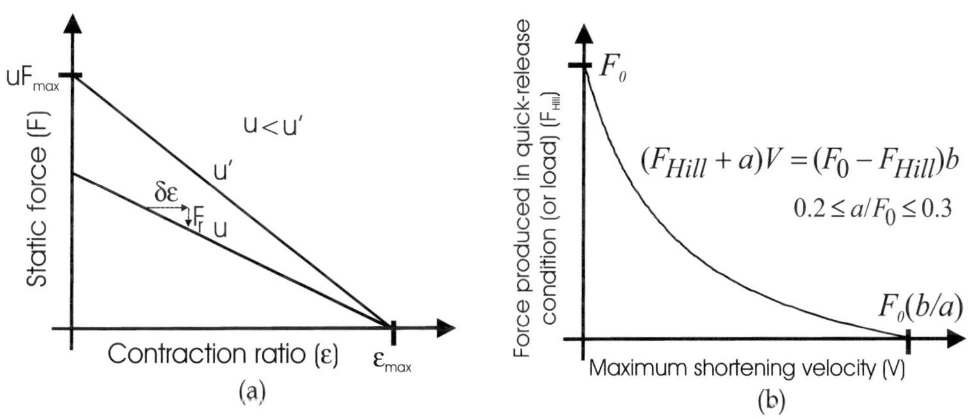

Figure 4. Physical model of skeletal artificial muscle based on both Force-Displacement and Force-Velocity curves of the natural muscle: a linear static force-displacement whose slope is proportional to the control variable is considered (a) with an additive damping term in such a way that the artificial muscle can be dynamically in accordance with (b) the fundamental Hill curve

Among all the technologies reported at the beginning of the paragraph, only the ones able to be adapted to this restrictive notion of skeletal artificial muscle would really be suitable for humanoid robotics. For example, the technology of certain conductive polymers essentially leads to bending artificial muscle difficulty applicable to linear actuators. Let us now analyse some attempts at artificial skeletal muscle.

3. Which artificial muscle technology for anthropomorphic limbs?

The three technologies that we present have all led to artificial muscles of skeletal muscle size; they illustrate the present diversity of the stimulus mode: chemical for pH muscle, electrical for roll actuator and fluidic for pneumatic muscles. It could be said that because nobody is currently able to mimic the complex sequence of biomechanical microscopic phenomena of natural skeletal muscle, the energy choice question of artificial muscle remains open.

3.1 pH-sensitive muscles and Solvent-sensitive muscles

The possibility of preparing relatively easily gel contractile fibres from commercially available polyacrylonitrile (PAN) or polyvinylalcohol (PVA) filaments led several researchers in the 1980s and 1990s to develop skeletal artificial muscles on roughly the same principle: a set of gel fibres is placed inside a rigid cell in which the circulation of two solutions alternatively generating the swelling and shrinking of the gel fibres is made easy. The resulting volume change can be used to pull and push a 'tendon'. Figure 5 illustrates an interesting prototype developed at MIT in the frame work of the *Artificial Muscle Project* by Brock and others (Brock *et al.*, 1994): acid and base are alternatively delivered to PAN gel fibre bundles through an array of perforated Teflon tubes

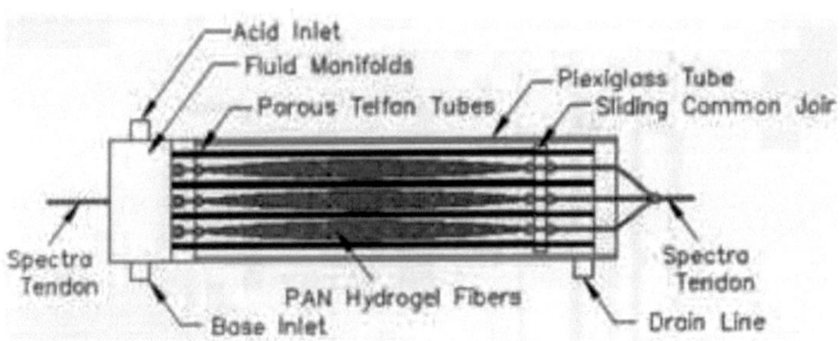

(a)

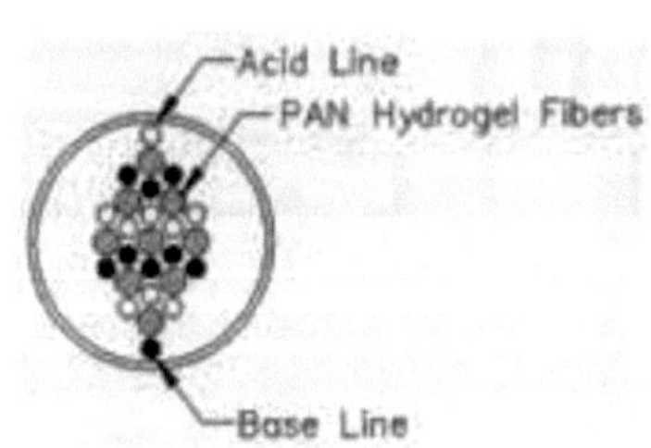

(b)

Figure 5. Cylindrical polymer gel 'pH muscle' using PAN gel fibers, (a) Gel bundles cell irrigated by perforated Teflon tubes connected to a common fluid manifold allowing alternatively acid (HCl) and base (NaOH) circulation, (b) Arrangement of gel fiber bundles, acid and base conduits in cross section (from (Brock et al., 1994))

The 'pH muscle' is used in antagonism with a spring and a modular pulley system: using a simple bang-bang control a rotation from 0 to 70° is generated in approximatively 45 seconds with a maximum tension of 0.35 N. The confrontation of this experimental device resulting from Caldwell's experiments leads to a better understanding of the difficulty of adapting this technology to powerful skeletal artificial muscles. The 'solvent-sensitive muscle' cell imagined by Caldwell (Caldwell & Taylor, 1990), (Caldwell *et al.*, 2000) is made of parallel PAA/PVA strips inside a container where an acetone and 4 molar NaCl solution can alternatively circulate. Originally (Caldwell & Taylor, 1990) 100 μm thick strips were used, but more recently it was shown (Caldwell *et al.*, 2000) that it was possible to optimize the thickness choice so as to highly increase contraction velocity without losing in stress: a 50 μm thickness gives a 43 %/s strain per s – instead of 11%/s for 100 μm strips and close to slow movements of the natural muscle – for a 30N/cm^2. With a 43% maximum contraction strain, the performances approach slow movements of the natural muscle. Ten strips are used in Caldwell's linear actuator, but the author highlights a major limit of this technology: the difficulty of use on a large scale. Because strain rate and stress evolve inversely with fibre thickness , it is necessary to constitute the artificial muscle of a great number of parallel fibres to generate useful forces for robot arms, thus inducing a discouraging complexity. To our knowledge, this interesting technology has not been considered for robot limb actuation. These attempts, however, highlight a central point for the development of artificial skeletal muscle: an mm^2 size scale technology is not necessarily efficient for a cm^2 size scale. The central notion of artificial muscle stress as defined in paragraph 2 can finally appear deceptive: in association with maximum strain and maximum strain rate it leads to making believe that a certain number of technologies are perfectly adapted to mimicking the human skeletal muscle, whereas their adaptation to a human size and weight robot has not been proved.

3.2 Electroelastomer rolls

SRI International (MenloPark, California) which claims the discovery of electroelastomers for new actuation has designed original linear actuators aimed at mimicking the shape and behaviour of natural skeletal muscles. These cylindrical actuators called *elastomer rolls* or simply *rolls* are composed of a central spring around which several layers of dielectric elastomers are rolled, as defined earlier. The actuator is closed by two caps, used as the electric poles between which the functioning high tension is placed. At rest, the central spring is maintained at compression by its surroundings and when tension is applied the compliant dielectric elastomers extend inducing actuator extension. Figure 11.a illustrates this actuator technology: the presented skeletal artificial muscle is 65 mm long with 45 mm active length, its highest strain is about 26% of active length for a 5 Hz response time, its diameter is 1.2 cm; composed of 20 layers of acrylic films generating a maximum force of 15 N. For this prototype there results a maximum stress of about 0.4 MPa significantly lower to the postulated value for dielectric elastomer artificial muscle technology (Pelrine *et al.*, 2000) – 1.36 Mpa as mentioned earlier - and also signficantly lower to that of natural skeletal muscle. Although the roll appears to be really more powerful than 'pH muscle' it suffers from the same drawback: the need to add layers – or fibres – in order to amplify its force, consequently bringing about an increase of its section like a greater design complexity : 35 layers are necessary, for example, to increase the muscle maximum force from 15 N to 21 N (Pei *et al.*, 2003) . For this reason, roll technology remains difficult to apply to a direct actuation of a robot-arm as the beautiful picture of the 'wrestling match between EAP actuated robotic arms and human' tends to prove (Figure 11.b) The roll actuators are too big to be placed on the skeleton arm and have been removed to an appendant box. Furthermore, unlike pH muscle (Caldwell *et al.*, 1990) it has not been proved that roll actuators accord with Hill's model : as mentioned by Pei, '...the stiffness of the roll is the sum of the central spring and the rolled polymer films' and it can be asked if the central restoring spring does not have a negative effect on the Hill-like damping behaviour which polymers seem to favour.

3.3 Pneumatic muscles

The roll actuator uses rubber stress as a force generator; in a complementary way, it can be said that rubber inflation-based actuators generate their contractile tension by means of the exceptional possibilities of rubber strain. McKibben artificial muscle is the most impressive example of this artificial muscle technological approach. After firstly analysing it , we will briefly discuss recent competitive approaches.

a. The McKibben artificial muscle

Without trying to accurately specify its origin which seems to lose itself in the golden age of rubber-industry derived technologies, McKibben muscle is a typical braided artificial muscle technology whose name comes from the American nuclear physicist Joseph L.McKibben who developed its prototype for bracing handicapped hands. As we have tried to justify (Tondu & Lopez, 2000), pneumatic McKibben artificial muscle is like a tyre whose braided sheath is free to open itself in order to allow the inflation of a pressurized inner tube. McKibben muscle is indeed composed of an inner tube surrounded by a double helix braid, characterized by its initial braid angle (α_0) as illustrated in Figure 6.a. If we assume that the inner tube integrally transmits its circumferential stress to the braided sheath, a simplified model of the resulting contractile linear tension can be deduced from the application of a virtual works theorem to an ideal cylinder the radius and length of which evolve according to the braid opening (Tondu & Lopez, 1995). Let us note l_0 as the cylinder initial length - *i.e.* initial active muscle length - and r_0 the cylinder initial radius - *i.e.* initial active muscle radius. The following static tension/strain model results :

$$F_{stat} = (\pi r_0^2) P[a(1-\varepsilon)^2 - b], \quad 0 \le \varepsilon \le \varepsilon_{\max} \quad \text{with} \quad a = 3/\tan^2\alpha_0 \text{ and } b = 1/\sin^2\alpha_0 \qquad (6)$$

where F_{stat} is the generated static tension for a control pressure P in function of the contraction ratio ε. This purely geometrical model does not take into account bound effects as force lose due to material effects whose major one is friction braid fibre on braid fibre - more accurate physical models can be found in (Chou & Hannaford, 1996), (Tondu & Lopez, 2000), (Colbrunn et al., 2001), (Davis et al., 2003)). It however captures three main characters which found the static analogy between McKibben muscle and natural skeletal muscle :

1. Static tension is proportional to P which can so play the part of the neural activation u (let us recall that static roll tension is proportional to the electric tension squared due to the equation (1) model);
2. Static tension is proportional to initial muscle section (πr_0^2) which is an important property for scaling robot-limb actuators as already discussed in the case of the two previous technologies;

3. Static tension continuously decreases when contraction ratio ε increases making the actuator a true contractile actuator as opposed to roll actuator.

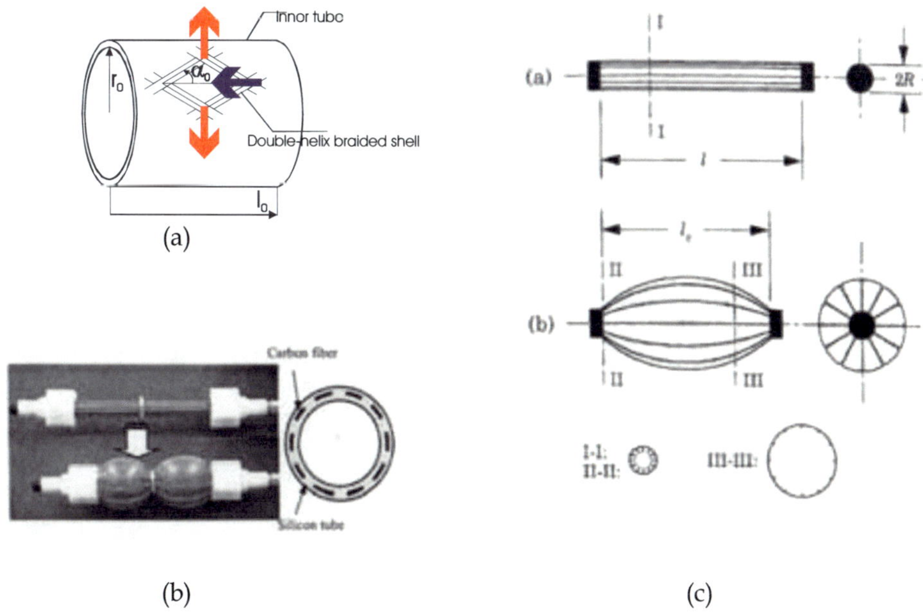

Figure 6. Pneumatic artificial muscle based on the use of a structure guiding the inflation of a elastomer chamber, (a) external structure case: operation principle of the McKibben muscle (see text), (b) internal structure case: longitudinal fibres included in a silicone tube (from (Saga *et al.*, 2005) , see also (Nakamura et al., 2003)) – this type of muscle is as 'old' as McKibben muscle (Matsushita, 1968) -(c) Typical membrane case : pleated muscle (from (Daerden & Lefeber, 2001)) (see text)

The static tension-contraction ratio curve parametrized in P offers a large linear range in a similar way as natural muscle (see Figures 1.c and 7.a). If the maximum contraction ratio, typically between 25% and 35% depending on the pressure, is relatively high in comparison with other artificial muscle technologies, it is still below the typical 40% maximum ratio of skeletal muscle (which even reaches 50% due to passive tension of skeletal muscle *in vivo*). Furthermore, if the fibre-on fibre-friction acting during muscle functioning necessarily induces a problematical dry and kinetic friction (which we discuss later) we also consider that kinetic friction caused by specific textile braided sheaths - like in rayon – can be responsible for a very good accordance of the dynamic McKibben muscle with Hill's model (Tondu & Diaz, 2006). Figure 7.b illustrates the tension-velocity curve obtained in quick-release conditions, at a constant 2 bar pressure, for a McKibben muscle characterized by initial length l_0=335 mm, initial radius r_0=10 mm, and of initial braid angle α_0=23°.
Finally, McKibben artificial muscle appears to have a close analogy with natural skeletal macroscopic static and dynamic behaviour. However, its purely mechanical principle implies an increase in artificial muscle volume while skeletal muscle works at constant volume: McKibben muscle volume typically doubles between its zero-contraction and full-contraction states. The maximum contraction ratio is less than the skeletal muscle 40% typical value, and which cannot be increased by its own passive tension, is a second

limitation for a perfect use of McKibben muscle in the actuation of anthropomorphic robot-limbs. Non-constant volume behaviour - which in fact is a characteristic of all fluidic muscles - induces another problem: fluid consumption of the artificial muscle, which implies a specific energy autonomy difficulty for its application to mobile humanoid robots. It is, however, possible in the case of an anthropomorphic musculature based on an antagonistic principle (see next paragraph) to imagine a recovering of consumed air from one elongated muscle to one contracted muscle but this is technologically delicate to put into work. This is the true paradox of McKibben artificial muscle which we will discuss at paragraph end: a unique phenomenological behaviour in analogy with skeletal muscle associated to a traditional and costly energy power source.

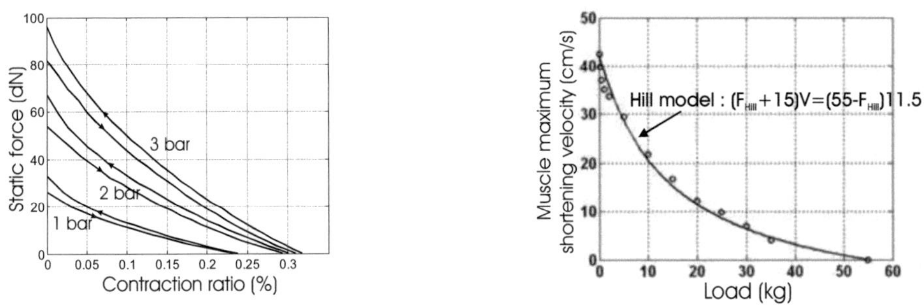

(a) (b)

Figure 7. Characteristics of a typical pneumatic McKibben muscle (of initial length l_0=335 mm, initial radius r_0=10 mm and initial braid angle α_0=23°), (a) Tension-Displacement curves at constant pressure 1, 2 and 3 bar, (b) Tension-Velocity curve at constant 2 bar pressure

b. Alternatives to McKibben muscle

McKibben muscle is one of the most known and used pneumatic artificial muscles in Robotics due to its ease of construction and its oblong shape which mimics the natural spindle shape of skeletal muscle, but a large number of other pneumatic muscles have also been developed. It is not easy to classify them; we propose to distinguish three main kinds of pneumatic artificial muscles according to the type of guiding structure of the elastomer inflatable chamber, as illustrated in Figure 6: firstly, the guiding structure is external to the inner tube, as in the case of McKibben muscle; secondly, the guiding structure is included inside the inner tube as considered when longitudinal fibres are embedded in the elastomer (Figure 6.b) and, thirdly, the elastomer chamber is specifically designed to inflate and contract (Figure 6.c). The first type is the easiest to elaborate while the third is the most sophisticated. 'Pleated muscles' developed at Brussels Vrije's University appear at the moment to be one of the most interesting examples of the third type applied to humanoid robotics. It was developed in opposition to McKibben muscle technology. If indeed we ignore the energy problem of pneumatic muscles to focus on the control problem it can be asked if the non-linear character of the actuator is not a too-difficult obstacle to surmount for stable and accurate control. In particular, in the case of an external guiding structure, a friction force generally appears inside the guiding 'mechanical structure' during muscle contraction as between the fibres of a McKibben braided sheath. Due to dry friction, a hysteresis phenomenon results, as illustrated in Figure 7.a for the muscle alone, and in Figure 15.b for the actuator made of two antagonistic muscles. Hysteresis is an undesirable non-linear effect; as will be analysed in the next paragraph; it is not, in our opinion, the major drawback to a robust control of anthropomorphic limbs actuated by artificial muscles. However, an artificial muscle without hysteresis is obviously a progress in the field; this is the case for the 'pleated muscle' which in the absence of any external braid rubbing against it, avoids the phenomenon. The 'pleated muscle' is a 'cylindrical membrane of high stiffness and high flexibility [folded] along its cylindrical axis like accordion bellows' (Daerden & Lefeber, 2001). When pressurized, inflation happens by unfolding and the artificial muscle shortens and bulges. As illustrated in Figure 12.a, static force characteristics - except for hysteresis - is globally similar to that of McKibben muscle, with a maximum contraction less dependent on pressure. The counterpart of braided sheath elimination is, however, a more complex geometrical shape of the artificial muscle (whereas the McKibben muscle globally keeps a cylindrical shape), pleated muscle offers a contracted complex 'flower'-like shape). What results, firstly, is an increase in the maximum muscle section which, at the same generated stress, can really be higher than that of McKibben muscle and which induces mechanical integration difficulties (see next paragraph). Secondly, muscle shape geometrical complexity implies a more complex mathematical model of muscle inflation. The three fundamental properties exhibited by McKibben muscle also appear present in pleated muscle technology, but in comparison to McKibben muscle no simplified corresponding model can be deduced. The following static model is indeed proposed (Daerden & Lefeber, 2001) :

$$F_{stat} = P l_0^2 f(\varepsilon, l_0/r_0, a) \tag{7}$$

where l_0 is the initial active length, ε the contraction ratio, r_0 the initial muscle radius, a is a parameter depending on the Young modulus of the membrane elastomer and f is a dimensionless force function for which no closed form seems easy to exhibit (Daerden & Lefeber, 2001). Furthermore, the concordance of the force-velocity curve with Hill's model has not yet been established. Lastly, it is important to note that 'pleated' artificial muscle can have a more limited lifespan than external guiding structure muscles due to the stress imposed to the membrane. As mentioned by Verrelst, the first generation pleated pneumatic artificial muscle has a limited lifespan due to the overlap used to make a cylindrical pleated membrane. The proposed second generation has a 400 000 cycle life but induces a complex mathematical model (Verrelst *et al.*, 2006).

In conclusion, the current field of available artificial muscles for humanoid robots is paradoxical: on the one hand, electroactive polymer-based roll actuators, although very promising, do not yet have sufficient power-to-volume to be embedded in robot-limbs and to actuate them; on the other hand, purely mechanical pneumatic artificial muscles appear to be at present the most biomimetic of muscles, although their energy mode is questionable. This paradoxical situation can, however, be settled in the following manner: even if given up in the future in favour of EAP-based muscles or other bio-inspired artificial muscles, fluidic artificial muscles, in our opinion, tackle the difficult question of mechanical integration and control of anthropomorphic limbs actuated by biomimetic artificial muscles, as considered in the two next paragraphs.

4. Specific design of anthropomorphic limbs actuated with skeletal artificial muscles

4.1 Antagonistic muscle revolute actuator

Skeletal artificial muscle applied to the actuation of anthropomorphic limbs is not aimed at acting alone. However, to mimic the complex organisation of human musculature, as explored by Washington's Biorobotics Laboratory in the upper limb case (Figure 8), appears at present to be an insurmountable task for the humanoid robot specialist. On the contrary, a basic muscular organization seems to be necessary for adapting linear skeletal artificial muscle to the revolute joint structure of anthropomorphic limbs: the revolute actuator made of two antagonistic artificial muscles illustrated in Figure 9. The two antagonistic muscles are assumed to be attached by means of a non-extensible pulley-chain system, where R denotes the pulley radius. We assume two identical artificial muscles of initial l_0 active length. In the general case, we will note, respectively, agonistic muscle control by u_1 and antagonistic muscle control by u_2, and the resulting actuator angle by θ.

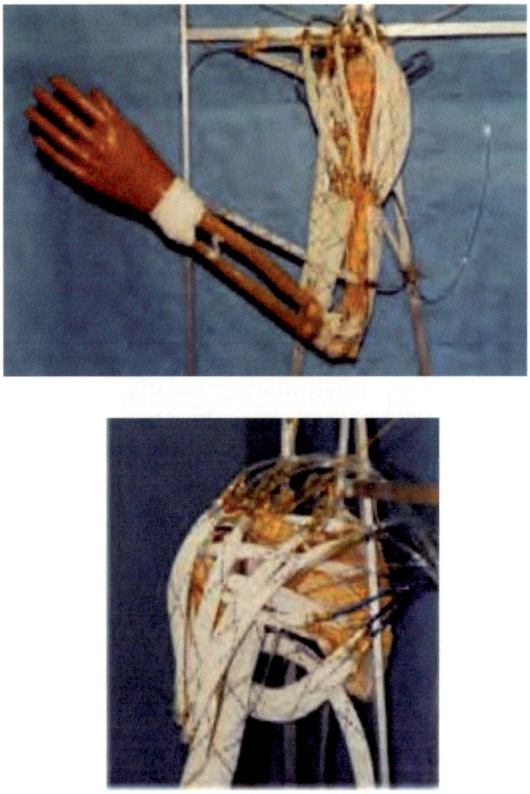

Figure 8. *Anthroform Biorobotic Arm* project of Washington's BioRobotics Laboratory (Hannaford *et al.*, 1995) composed of pneumatic McKibben artificial muscles mimicking the shoulder musculature attached to human bones (to our knowledge, no control of the arm has been developed)

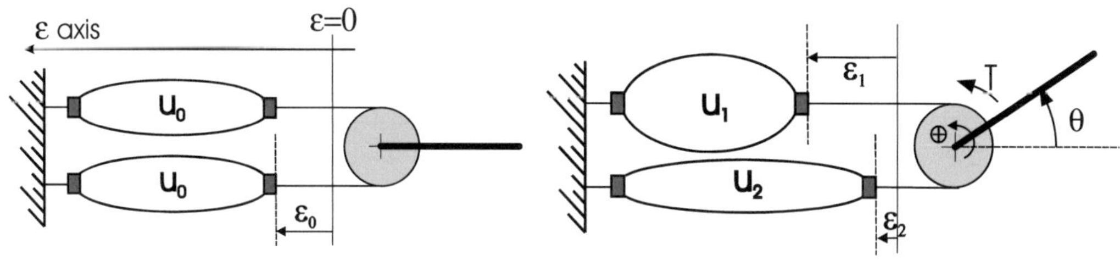

(a) (b)

Figure 9. Operation principle of the antagonistic muscle actuator, (a) Initial state, (b) Current state (see text)

If we limit our model to the use of the artificial muscles in purely active tension - i.e. without any form of passive tension - it is necessary to assume that in its rest angular position $\theta=0$, both muscles are equally contracted of ε_0 with a same control u_0 allowing the ε_0 initial contraction. In the current position θ the agonist and antagonist contraction ratios are respectively given by - let us number '1' the agonistic muscle and '2' the antagonistic muscle :

$$\varepsilon_1=\varepsilon_0+(R\theta/l_0) \text{ and } \varepsilon_2=\varepsilon_0-(R\theta/l_0) \quad (8)$$

The actuator torque T can be written as follows :

$$T=R(F_1-F_2) \quad (9)$$

where F_1 is the agonistic force and F_2 is the antagonistic force. The application of the fundamental model of equation (4) leads to the following expression :

$$T=K_1(u_1-u_2)-K_2(u_1+u_2)\theta-T_{damp}(u_1,u_2,\theta,\dot{\theta}) \quad (10)$$

with $K_1=RF_{\max}(1-\varepsilon_0/\varepsilon_{\max})$, $K_2=R^2F_{\max}/l_0\varepsilon_{\max}$ and $T_{damp}=R[F_{damp}(u_1,\varepsilon_1,\dot{\varepsilon}_1)-F_{damp}(u_2,\varepsilon_2,\dot{\varepsilon}_2)]$

Let us consider the static component of the actuation torque :

$$T_{stat}=K_1(u_1-u_2)-K_2(u_1+u_2)\theta \quad (11)$$

This general expression of the static torque - which is similar to the simplified Hogan's model of the biceps/triceps system (Hogan, 1984) - leads to highlight three fundamental properties of the artificial muscle antagonistic actuator in analogy with the fundamental biceps-triceps natural muscular system :

1.The actuator is stable in open-loop both in position and stiffness.

It is possible to define the following equilibrium position θ_{equ} corresponding to a null static torque :

$$\theta_{equ}=[K_1(u_1-u_2)]/[K_2(u_1+u_2)]. \quad (12)$$

In this equilibrium position, the stiffness actuator S defined as follows :

$$S=K_2(u_1+u_2) \tag{13}$$

can be changed by multiplying the u_1 and u_2 inputs by a same factor, although the conditions $u_1 \leq 1$ and $u_2 \leq 1$ are verified.

2. The actuator is both a torque and a stiffness generator.

The previous property is the consequence of the fact that the actuator can be considered as a double input-output system whose (u_1, u_2) are the inputs and (T_{stat}, S) are the outputs defined as follows :

$$\begin{bmatrix} T_{stat} \\ S \end{bmatrix} = \begin{bmatrix} K_1-K_2\theta & -K_1-K_2\theta \\ K_2 & K_2 \end{bmatrix} \begin{bmatrix} u_1 \\ u_2 \end{bmatrix} \Leftrightarrow \begin{bmatrix} u_1 \\ u_2 \end{bmatrix} = \frac{1}{2K_1K_2} \begin{bmatrix} +K_2 & K_1+K_2\theta \\ -K_2 & K_1-K_2\theta \end{bmatrix} \begin{bmatrix} T_{stat} \\ S \end{bmatrix} \tag{14}$$

We will use this double nature of the antagonistic artificial muscle actuator in next control paragraph.

3. The maximum actuator torque decreases continuously from an actuator angle limit to the other one.

Let us assume that each actuator muscle can fully contract of its ε_{max} maximum contraction ratio. If the two muscles are identical, the two actuator angle limits can be defined as follows:

$$[\theta_{\min}, \theta_{\max}] = [-(\varepsilon_{\max} - \varepsilon_0) l_0 / R, +(\varepsilon_{\max} - \varepsilon_0) l_0 / R] \tag{15}$$

For every actuator angle θ belonging to this range, we get the following expression of the maximum torque, corresponding to a maximum control of the agonistic muscle (u_1=1) and a null control of the antagonistic muscle (u_2=0) :

$$T_{\max}(\theta) = K_1 - K_2\theta \tag{16}$$

The maximum torque decreases in consequence from $\theta_{\min}$ to $\theta_{\max}$ with the $-K_2$ slope. In comparison with classical actuators, the presence of the restoring torque term ' $-K_2(u_1+u_2)\theta$ ' gives to the biomimetic actuator a natural compliance which can be actively controlled independently from the actuator position. However, a drawback results that is analysed in next paragraph: angular position actuator torque dependence induces a specific sensitivity of the actuator on gravity.

4.2 Gravity test with artificial muscle actuator

The first request for any actuation mode of a robot-limb is its ability to move each link all along the desired joint range. As for humans, the most important resistance of humanoid robot links is gravity. We consider that testing the ability of an artificial muscle actuator embedded on the kinematic chain to directly drive joints of a human size (in volume and weight) robot-limb could be a more rigorous test than the 'arm wrestling contest' organized by NASA (see the rules for the wrestling match between EAP actuated robotic arms and human on the website http://armwrestling.com/rulesandregulations.html) to test EAP-roll actuator performances, illustrated in Figure 11.b. In particular this 'gravity test' is made difficult by the specific nature of the artificial muscle actuator highlighted in the previous paragraph. Let us consider the case of a single actuator driving a link as illustrated in Figure 10.a: the two antagonistic muscles represent, for example, the biceps-triceps system actuating the arm placed in a vertical position, the considered joint the elbow flexion-extension and the link the set forearm + hand. Let us further consider a total joint range of 180° imposing the actuator zero-position when the moving link is at the horizontal position. If the positive direction corresponds to the flexion movement, gravity torque is always resistive and can be written as follows : $T_{gravity}=-mgl\cos\theta$, as illustrated in Figure 10.a. The maximum available actuator torque is given by equation (16) and subsequently the total torque can be deduced as simulated in Figure 10.b. It appears that in the [-90°,+90°] joint range, the total torque becomes minimum, and it is clear that if this minimum is negative, a special point results where the total torque becomes null and as a consequence, the actuator is no longer able to lift the link. In a more general manner, the gravity test appears as a quasi-static test which consists of checking that the total static joint torque keeps strictly positive on the considered actuator joint range. It is interesting to note that a load in hand can easily put the test in difficulty. The two shoulder movements in flexion-extension and abduction-adduction are particularly concerned by this test: the situation is analogous to the one illustrated in Figure 10 but now the link to lift represents the whole upper limb with a possible load in hand. Consequently, the actuation of a robot shoulder by artificial muscles can necessitate very strong muscles, notably if the flexion-extension and abduction-adduction movements are, respectively, performed by only one pair of antagonistic muscles. For example, the robot shoulder of our 7R anthropomorphic of the same size as a human upper limb - but of double weight – illustrated in Figure 13.b is actuated by McKibben pneumatic muscles able to develop more than 500 dN under a 5 bar pressure. Generally, this difficulty in actuating a human-like shoulder by artificial muscles reveals the need for future humanoid robots to mimic not only the antagonistic muscle principle, but also natural muscular system redundancy, as attempted in Figure 8.

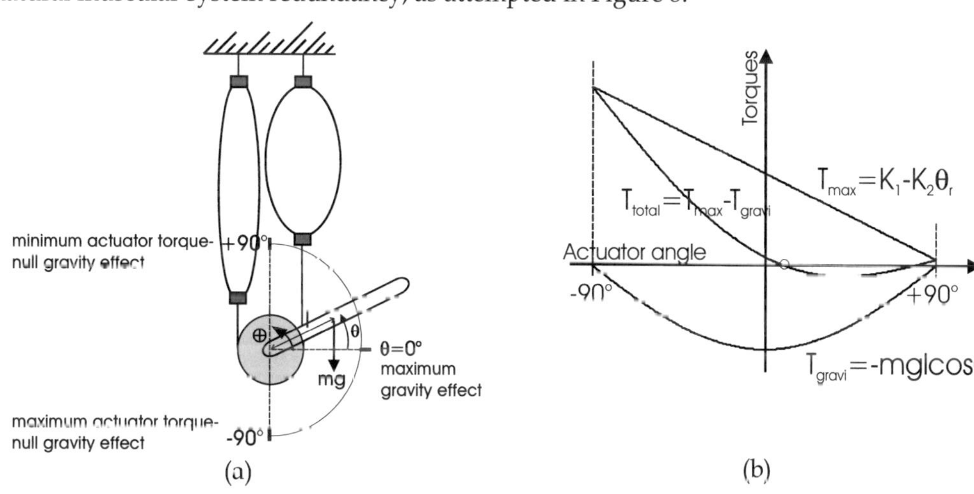

Figure 10. Effect of gravity on the antagonist artificial muscle actuator (a) Double joint angle dependence of the actuator when submitted to gravity, (b) Simulation of total torque highlighting a possible joint position beyond which the actuator is no longer able to lift the arm

4.3 Integrating artificial muscles to anthropomorphic kinematic structures

The integration of artificial muscle into a humanoid robot whose limbs, size and weight are those of a medium sized human being induces specific constraints in actuator joint range, actuator power-to-volume, actuator power-to-mass, and robot energy range. If we leave the latter constraint which depends on the development of embedded batteries, the first three are direct consequences of artificial muscle technology applied on a human-scale robot. Power-to-mass is generally not a difficulty due to the low mass density of polymer-type materials, but generating both actuator 'joint range' and 'power-to-volume' similar to skeletal muscle ones is a major challenge because of the difficulty in designing compact artificial muscle working at a constant volume as a skeletal muscle does. As already mentioned, EAP-roll actuators cannot yet fully satisfy these constraints. Pneumatic muscles are at present the most able ones within sight of these two criteria. This is the reason why all current anthropomorphic limbs actuated by artificial muscles use pneumatic ones, but it is important to emphasize that, even in the case of this technology choice, the integration of artificial muscle actuator to anthropomorphic robot-limbs is made difficult by their limited biomimetic behaviour :

- the global cylindrical shape of the McKibben pneumatic muscle helps its integration in spite of its volume increase, but its relatively limited maximum contraction is a real drawback for the actuation of large joint movements: in particular, flexion/extension and abduction/adduction human shoulder movements are particularly difficult to mimic with artificial muscles, without using excessively long artificial muscles; our 7R robot prototype, illustrated in Figure 13.b, has shoulder joint ranges equal to [0,+180°] in abduction-adduction and to [0,+100°] in flexion-extension thanks to transmission gears amplifying the actuator angular motion at a 1.5 ratio in order to limit shoulder width to about 35 cm (Tondu *et al.*, 2005). In the case of the Shadow Robot illustrated in Figure 13.a, it can be noted that the shoulder muscles are placed vertically in the robot waist, which is not adapted to humanoid architecture;
- the pleated muscles offer the advantage of high maximum contraction ratios weakly dependent on the control pressure, but its bulkiness when the muscle is contracted prevents the putting into parallel of two antagonistic muscles: as illustrated in Figure 12.b, the two pleated muscles have to be shifted in order to allow the antagonistic muscle actuator to get working which reduces the actuator range. Lucy robot (Verrelst *et al.* 2005), illustrated in Figure 12.c, is a successful application of pleated muscle actuation, but it can be asked if it is as well adapted to actuation by upper limbs necessitating larger joint ranges.

Analysis of these two points emphasizes the difficulty of actuating anthropomorphic robot-limbs by artificial muscles: the global 'optimal' character of the human muscular-skeletal system makes all attempts at mimicking the system difficult if one analogical term is missing.

(a)

(b)

Figure 11. Roll actuator composed of rolled dielectric elastomers (a) - from (Pei et al., 2003) - and application of this technology to the actuation of an anthropomorphic arm in the framework of the NASA's 'armwrestling contest' - from NASA web site (see the three big roll actuators inside the purple box in (b))

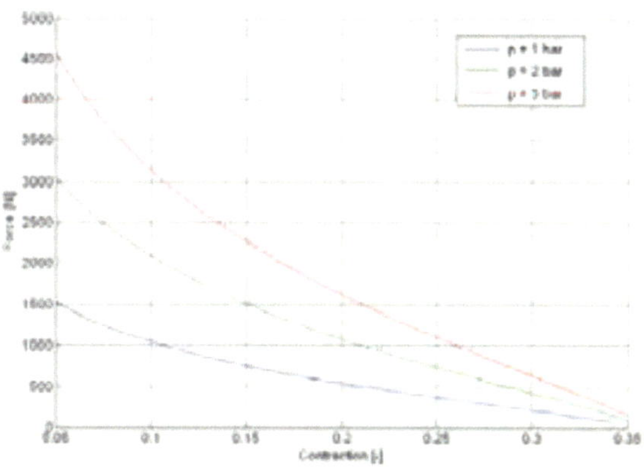

(a)

(b)

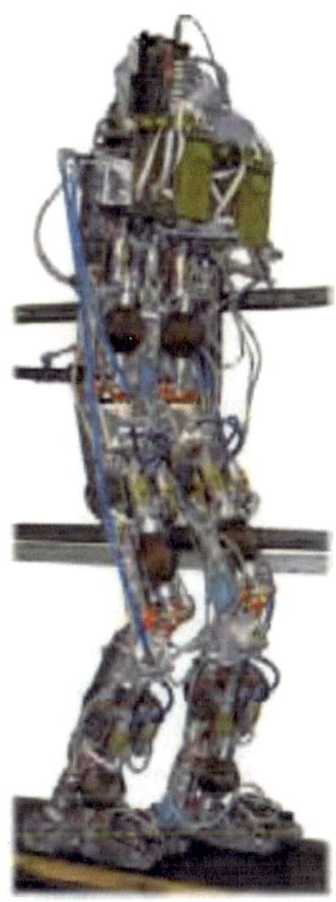

(c)

Figure 12. Pleated artificial muscle applied to biped robots, (a) Static characteristics of the pleated artificial muscle, (b) Corresponding antagonist actuator showing the difficulty to place simultaneously the two inflated muscles into antagonism, (c) Lucy biped robot moving in a vertical plane (from *Lucy* web site)

(a)

(b)

Figure 13. Two examples of anthropomorphic robot-arm design actuated by pneumatic McKibben muscles: (a) shadow robot-arm equipped with shadow hand showing artificial shoulder musculature placed in the robot's 'waist' (from *Shadow Robot Group* web site), (b) 7R anthropomorphic robot-arm prototype built in the laboratory with 30 cm horizontal shoulder muscles developing a maximum force exceeding 500 dN

5. Control of anthropomorphic limbs actuated by skeletal artificial muscles

5.1 Non-linearities of robot joints actuated by artificial muscle actuators

The use of flexible materials such as the recourse to original stimulus modes (pH, solvent, heat, etc.) are complexity factors of the physical models of any artificial muscle actuator. What results is a non-linear character generally more manifest than for other robotic actuators. In particular, it is well known that the more nonlinear the plant the more imprecise its physical or identified model on which its control can be based. Using Slotine & Li' terminology (Slotine & Li, 1991) the artificial muscle actuator is more concerned than others by 'structured (or parametric) uncertainties' as by 'unstructured uncertainties (or unmodelled dynamics)'. Furthermore, in the case of robot-limbs actuated by artificial muscles, the specific actuator non-linearities enter into combination with dynamic robot nonlinearities due to the direct drive character of robot joints. We emphasized in the previous paragraph the part played by gravitational forces but, as for any direct drive robot, jointed limbs actuated by artificial muscles have also to support dynamic perturbations due to inertial variations, or to velocity effects. Even if it is considered that a humanoid robot does not have to support the accelerations and velocities generated by the joints of high performance industrial robot-arms it is clear that the mimicking of certain sporting or daily-life gestures can induce torque perturbations due to the inertial, centrifugal or Coriolis terms of classical robot-limb dynamics. It seems important to us, however, to emphasize the following point: repeatability of the accuracy of the end-effector of a humanoid robot limb (hand, finger, foot, head, etc) can be defined in analogy with human gestures: they are, consequently, closer to the mm scale than to the 1/10 mm as required for a great number of tasks performed by industrial robot-arms. It can be roughly considered that an acceptable accuracy value for antagonistic artificial muscle actuators of a humanoid robot performing tasks at 'human accuracy' - the accuracy of drawing a line with a pencil - is about one or a bit less than one degree. From this point of view, the artificial muscle actuator can finally appear more adapted to humanoid robots mimicking human gestures than ultra-accurate, but non naturally compliant electric motors. This is true provided there is the possibility of being able to design a control mode of the artificial muscle as effective as the one resulting from the complex and badly known human movement learning principles. In the next paragraph we analyse some current or envisaged control modes of artificial muscle robot actuators: all results mentioned were obtained on pneumatic artificial muscles which as already emphasized, seem the only ones to have been actually tested on human-scale robot-limbs.

5.2. Single-variable position control

The antagonistic artificial muscle has been previously defined as a multiple input-multiple output (MIMO) system. Since the first target of the actuator control is a control position, it can be asked if it is possible to simplify the actuator functioning in order to consider it as a single input-single output (SISO) system whose output is reduced to the angular position. A simple way of doing this, initiated in our opinion by Bridgestone (Bridgestone, 1987), consists of a symmetrical control of agonist and antagonist muscles in the form of a 'Δu' input control added to the initial 'u_0' to control the agonist when antagonist control of 'Δu' decreases , as follows :

$$u_1 = u_0 + \Delta u \quad \text{and} \quad u_2 = u_0 - \Delta u \tag{17}$$

The new torque expression results :

$$T = 2K_1\Delta u - 2K_2 u_0 - T_{damp}(\Delta u, \theta, \dot{\theta}) \tag{18}$$

The relationship between input Δu and equilibrium position θ_{equ} is now $\theta_{equ}=(K_1/K_2u_0)\Delta u$ and actuator stiffness is now constant : $S=2K_2u_0$. The artificial muscle actuator can now be considered as a revolute actuator to which a linear or non-linear control approach can be applied. Furthermore, its open-loop position stability gives an original opportunity for facilitating joint control: it is indeed possible to identify the angular joint and to use the identification result as a feedforward term. We have demonstrated the advantage of this approach in controlling a 2R-SCARA-type robot actuated by four similar pneumatic McKibben muscles (Tondu & Lopez, 2000). In the framework of humanoid robotics, this kinematic architecture corresponds to a arm-forearm set performing horizontal shoulder and/or elbow flexion-extension movements - without the consequent gravity effect. In this case, a second-order linear model of the form :

$$\theta(p) = b/(p^2 + a_1 p + a_2)\Delta U(p) \tag{19}$$

appears to be very satisfactory to identify the step response. Physically, according to the torque model of equation (18), and assuming that the joint drives a constant inertia (forearm inertia or in the case of the shoulder joint, maximum forearm + arm inertia), the term 'a_2' can be interpreted as a specific actuator stiffness and 'a_1' as a linear viscous term approaching complex actuator damping. A typical linear controller illustrated in Figure 14.a results where the identified model is used as a feedforward term in association with a PID linear feedback, for example.

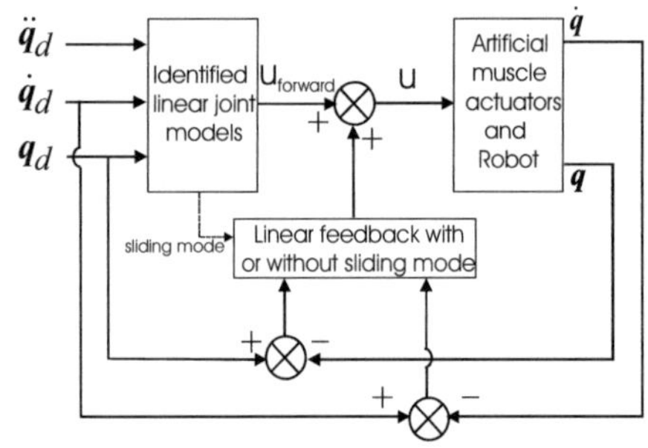

(a)

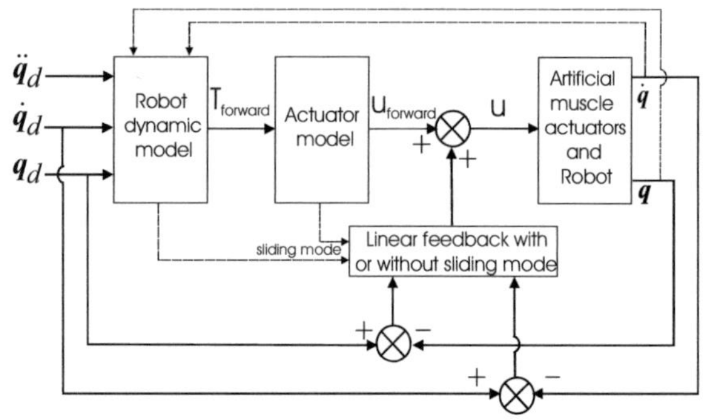

(b)

Figure 14. General scheme of position control of a robot-limb actuated by artificial muscle actuators, (a) control based on identified linear joint models, (b) control based on a robot dynamic model associated to a physical actuator model

However, as mentioned in paragraph 5.1, the artificial muscle actuator control has to face both actuator and robot non-linearities. A Simple linear control - even helped by the feedforward term - can appear not to be robust enough. Non-linear approaches are thus necessary to address the control problem of anthropomorphic limbs actuated by artificial muscles. Sliding mode control is one of these: it is particularly interesting because it integrates identified actuator models and/or robot dynamics. As emphasized by Slotine sliding control is one of the main approaches in robust control to deal with model uncertainties (Slotine & Li, 1991). Let us note $e=\theta_d-\theta$ and $\dot{e}=\dot{\theta}_d-\dot{\theta}$; if we limit our analysis to second order models, the sliding surface is the line defined by the equation :

$$S=\dot{e}+Ce \tag{20}$$

where C is the sliding line slope. Let us assume for example that the robot joint behaves like a second order linear model in the form of equation (19). The sliding condition $\dot{S}=0$ leads to the following expression of the perfect control $\hat{u}$:

$$\hat{u}=\frac{1}{b}[\ddot{\theta}_d+a_1\dot{\theta}_d+a_2\theta_d-a_2C+(C-a_1)\dot{e}] \tag{21}$$

Completed by a discontinuous component v chosen for example according to Harashima (Harashima et al., 1985) with α, β and γ parameters as :

$$v=[\alpha|e|+\beta|\dot{e}|+\gamma]\,\mathrm{sgn}(S) \tag{22}$$

the final actuator control is :

$$\Delta u=\hat{u}+v \tag{23}$$

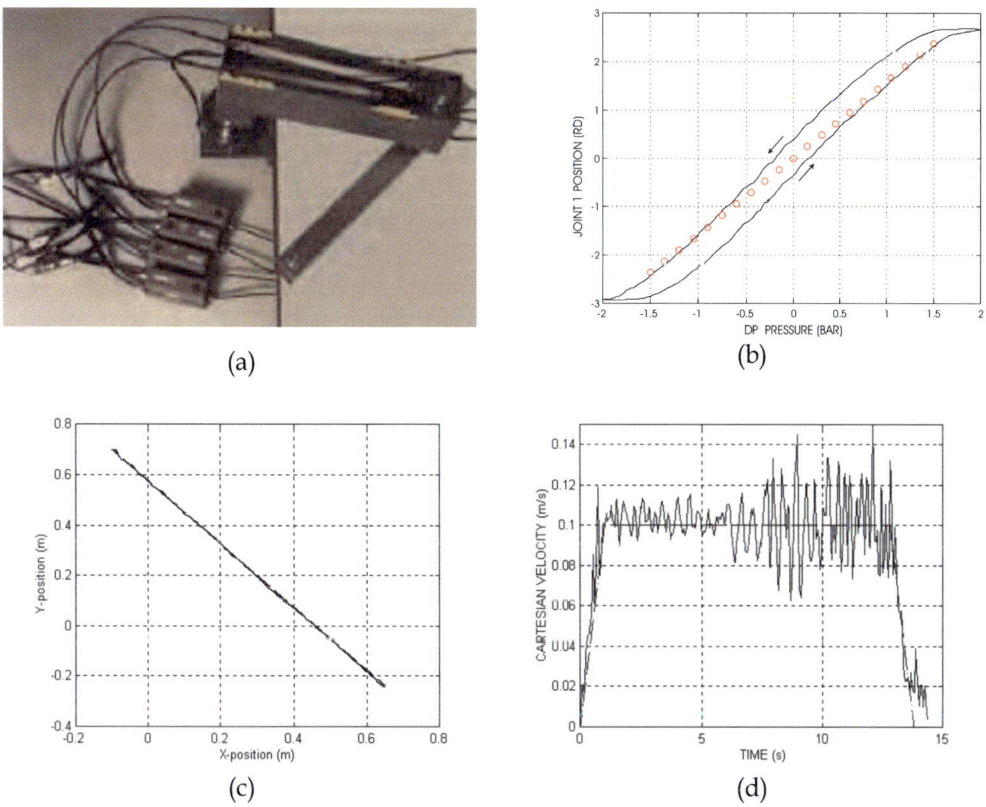

Figure 15. Experimental control of a 2R robot actuated by pneumatic McKibben muscles mimicking horizontal shoulder/elbow movements, (a) Photography of the robot, (b) Static characterics of the actuator, (c) and (d) Experimental tracking of a straight-line path according to a trapezoidal velocity profile with a sliding mode control (see text)

In comparison with the previous linear control, the feedforward model is now completed both by a continuous linear feedback component, and also by a discontinuous component aimed at giving robustness to the control, while sliding condition $S\dot{S} < 0$ is checked (see (Asada & Slotine, 1986) and (Slotine & Li, 1991) for theoretical analysis and application to robotics). Physically the controller's 'robustness' is preserved while the identified parameters of the model are kept to a limited range – typically 20%. This simple approach can be very adapted to robot limbs actuated by artificial muscles as emphasized in experiments performed in our laboratory on the arm-forearm prototype mimicking shoulder-elbow horizontal flexion-extension movements: the tracking of a straight-line at 0.1 m/s shows a mean path deviation of 2.5 mm with a maximum error of 8 mm (Figure 15 c) - occurring during acceleration/deceleration phases - and dynamic joint accuracy of +/- 0.5° for joint 2 and +/- 1.5° for joint 1 (see details in (Tondu & Lopez, 2000)). Note finally that sliding mode control has also to be applied to control 'pH muscle' (Brock. *et al.*, 1994) or shape memory alloy actuator (Grant & Hayward, 1997).

However, this approach has two limits: firstly, a moderate load at upper limb can induce large variations of inertial parameters; secondly, as previously emphasized, gravity has a large effect on the control: Figure 16 illustrates the identified linear model of the elbow joint of our 7R anthropomorphic arm moving on a vertical plane in response to pressure steps. A second order can be postulated as a first approximation, but a better result is obtained if this second order model is completed by a pure delay of 6 to 8 ms – thus leading to a third linear model approximation. Furthermore, the dynamic parameters now vary around their mean values of +/- 40% while their variation was limited to about +/- 15% in the case of non-gravity perturbed horizontal movements. Linear identified third order models have also be considered in the case of the antagonistic Rubbertuators - McKibben type muscles - actuated the Vanderbilt university's ISAC robot (Thongchai et al., 2001). These authors have proposed to use the joint identified model in the framework of a fuzzy controller using both linear quadratic regulator (LQR) and sliding mode techniques. Because a fuzzy controller has already appeared to us difficult to tune on the 2R robot of Figure 15.a (Tondu & Lopez, 2000), due to the actuator/robot system dynamics complexity, we are not sure that a fuzzy approach will be relevant to highly anthropomorphic robot limbs actuated with artificial muscles.

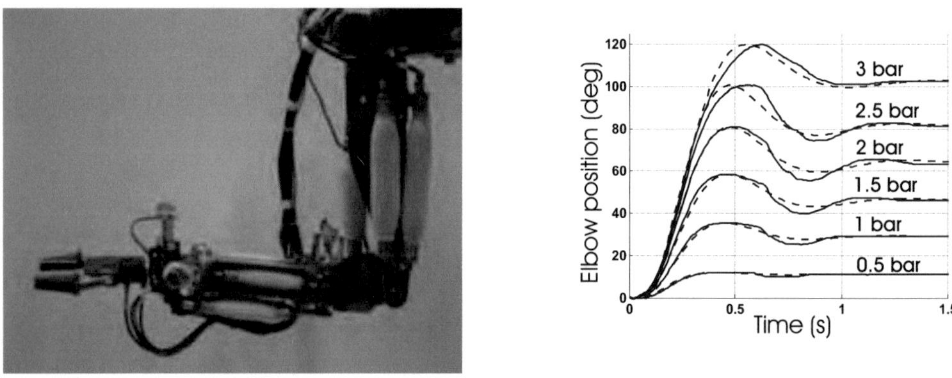

(a) (b)

Figure 16. Identification of the elbow joint of our 7R anthropomorphic robot-arm actuated by McKibben artificial muscle actuators, (a) Close-up on the elbow joint, (b) Open loop identification - model is in dotted line - in response to pressure differential steps

Consequently, it seems necessary so as to effectively control humanoid robots actuated by artificial muscles, to take into account both an actuator model and a robot dynamic model. In this framework, neural network control can constitute alternative bio-mimetic approaches (Hesselroth *et al.*, 1994), (Van der Smagt et al., 1996), (Tian *et al.*, 2004), (Thanh & Ahn, 2006) but their practical use in the programming of a large range of robot tasks is yet to be established. Adaptive methods can also be considered - see, for example, (Caldwell *et al.* 1995) - but to be effective they need a reference dynamic model and faced with the dynamic problem complexity, it seems necessary to add the knowledge of a complete dynamic robot model to the control scheme. Following the classic 'inverse torque method' a complete robot dynamic model is substituted to the linear identified joint model, but it is then necessary to associate it with a physical actuator model as emphasized in the control block scheme of Figure 14.b. This is a major drawback of the method when we are aware of the complexity of any artificial muscle physical model. Sliding mode control can always be applied to this dynamic model-based approach as developed by Cai and Dai (Cai & Dai, 2003) on the simulation of a vertical two-link manipulator using a 4 parameter McKibben muscle model (Cai & Yamaura, 1998). A dynamic limb model in association with an

antagonistic pleated muscle actuator model is also used to simulate the Lucy robot dynamic control (Verrelst *et al.*, 2005).
Control results based on this complete dynamic approach are still awaited to appreciate the possibilities of controlling humanoid robots actuated by artificial muscles. In this framework, it is clear that the simpler the actuator model, the easier is its control.

5.3. Multiple-variable position-compliance control

The linear or non-linear feedback component of the previous considered approaches introduces a 'servo-stiffness' which modifies the natural stiffness of the actuator. But if the feedback term stiffness is not too high - in particular by limiting proportional and integral components - the resulting global stiffness can be yet adapted to the achievement of tasks involving a contact of the robot with its environment as illustrated in Figure 17 : our 7R robot-arm prototype performs a straight-line against a solid wall fitted with a soft painting roller. A constant contact all along the trajectory is achieved by programming the end-effector tool slightly beyond the contact surface. This experiment proves that the SISO control of the artificial muscle actuator can also be adapted to contact.

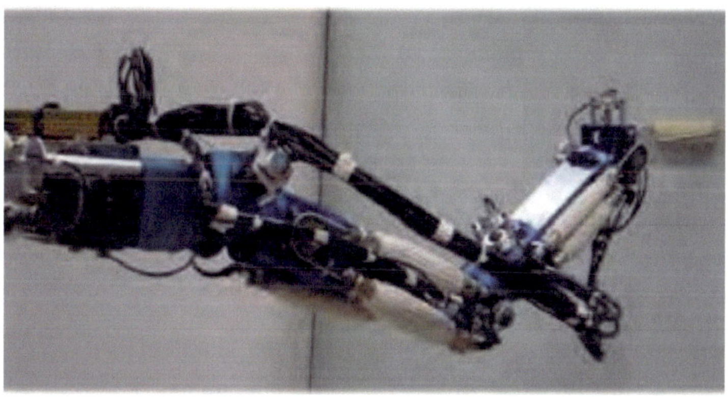

Figure17. Example of a task involving a permanent contact with the environment performed by our 7R anthropomorphic robot-arm actuated by pneumatic McKibben muscle actuators

However, the stiffness can be badly adapted to the task of producing, for example, Cartesian restoring force-torques varying in an inadequate manner with the imposed surface. By means of force-torque sensors the well-known hybrid position-force control approach can be naturally applied to robot-limbs actuated by artificial muscles. A more specific approach can, however, be highlighted: to use the MIMO nature of the artificial muscle actuator to control both position and stiffness in decoupling inputs 'u_1' and 'u_2'. The general MIMO scheme of Figure 18.a can be viewed as a generalization of Figure 14.b's SISO scheme, in which a actuator model in the form of the equation (14) model is introduced. The desired stiffness can now be imposed in accordance with Cartesian task requirements. Interesting preliminary results have been reported by Tonietti and Bicchi (Tonietti & Bicchi, 2002) based on a 2 d.o.f. robot-arm actuated by pneumatic McKibben muscles - Chou & Hannaford' McKibben muscle model was used - in which a time-varying stiffness was programmed. It is also possible to control the stiffness by estimating the real one assumed to be proportional to the sum of '$u_1 + u_2$' by means of muscle activation sensors -pressure sensors, for example, in the use of pneumatic muscles as made in the ambitious German Bielefeld University anthropomorphic grasping robot, without actually having resort to actuator and robot models. The association of this last basic scheme with the Figure 18.a scheme leads to a general approach of controlling both position and compliance in taking into account both robot dynamic model and actuator model for a global controller robustness.

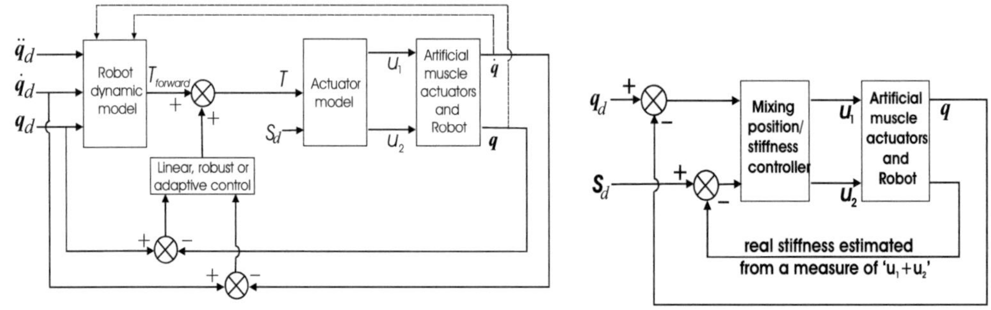

(a) (b)

Fig. 18. General schemes of position-compliance control of a robot-limb actuated by artificial muscle actuators : actuator compliance is imposed in the form of a desired stiffness (a) or controlled from a real stiffness estimation based on a measure of the actuator inputs sum (b)

As in the case of single position control approach, further experiments in hybrid position/stiffness control applied to anthropomorphic robot-limbs are necessary to prove the feasibility of this compliance-based approach. This is in opposition to the current and weakly biomimetic approach of controlling humanoid robot-limbs by means of wrist 6 axis force-torque sensors.

6. Conclusion

The functioning of natural skeletal muscle is based on microscopic phenomena that no technology is at present able to reproduce. The notion of artificial muscle is as a consequence mainly founded on a macroscopic model of the skeletal muscle. The mimicking of both tension-length and tension-velocity characteristics is aimed at giving future humanoid robots touch ability which is so fundamental in the 'relational life' of human beings. No definitive technology has as yet emerged in the design of artificial muscle. It is, however, interesting to note that the most promising ones are based on the use of polymers whose physical properties (responses to chemical or physical agents, elasticity, etc.) mimic some dynamic properties of animal tissues. In particular pH, temperature or electric field are now currently used to produce and control the shape changes of polymer fibres or polymer-based composite materials. These results are generally obtained on a small scale - typically a mm^2-section scale - and the application of these technologies to macroscopic skeletal muscle scales - typically a cm^2-section scale - generally induces a performance loose in power-to-volume and power-to-mass. Today the integration of artificial muscles to anthropomorphic limbs on a human-scale in volume and mass, necessitates power-to-mass and power-to-volume very close to human skeletal muscle.. Pneumatic artificial muscles, in the form of McKibben artificial muscles or alternative types such as pleated artificial muscles, are at present able to mimic these natural muscle dynamic properties. As a consequence, we consider that their use is interesting to test control approaches aimed at giving artificial muscle actuators speed, accuracy, robustness and compliance similar to human limb movements, while awaiting a more biomimetic technology able to supersede the dangerous DC motor/harmonic drive combination as a typical humanoid robot actuation mode.

7. References

Asada, H. & Slotine J.-J.E. (1986). *Robot Analysis and Control*, John Wiley & Sons, New-York.

Bar-Cohen, Y. Editor (2004). Electroactive Polymer (EAP) - Actuators as Artificial Muscles - Reality, Potential, and Challenges, SPIE society, Second edition, Bellingham, Washington, USA.

Bridgestone Corporation. (1987). Tokyo, Japan, *Soft Arm ACFAS Robot System*.

Brock, D.; Lee, W.; Segalman, D. & Witkowski, W. (1994). A Dynamic Model of a Linear Actuator Based on Polymer Hydrogel, *Journal of Intelligent Materials and Structures*, Vol. 5, N° 6, 764-771.

Cai, D. & Yamaura H. (1997). A VSS Control Method for a Manipulator Driven by an Artificial Muscle Actuator. *Electronics and Communications in Japan*, Vol. 80(3), 55-63.

Cai, D. & Dai, Y. (2003). A Sliding Mode Controller for Manipulator Driven by Artificial Muscle Actuator. *Electronics and Communications in Japan*, Vol. 86(11), 290-297.

Caldwell, D.G. & Taylor, P.M. (1990). Chemically Stimulated Pseudo-Muscular Actuation, *Int. J. Engng Sci* , Vol. 28, N°8, 797-808.

Caldwell, D.G.; Medrano-Cerda, G.A.; & Goodwin, M. (1995). Control of Pneumatic Muscle Actuators. *IEEE Control Systems Magazine* , Vol. 15, N° 1, 40-48.

Caldwell, D.G.; Tsagarakis, N.; & Medrano-Cerda, G.A. (2000). Bio-Mimetic Actuators: polymeric Pseudo Muscular Actuators and pneumatic Muscle Actuators for biological emulation, *Mechatronics* , Vol. 10, 499-530.

Choe, K. & Kim, K. (2006). Polyacrylonitrile Linear Actuators : Chemomechanical and Electro-Chemomechanical Properties, *Sensors and Actuators A*, Vol. 126, 165-172.

Chou, C.-P. & Hannaford, B. (1996). Measurement and Modeling of McKibben Pneumatic Artificial Muscles. *IEEE Trans. on Robotics and Automation* , Vol. 12(1), 90-102.

Colbrunn, R.W.; Nelson, G.M. & Quinn, R.D. (2001). Modeling of Braided Pneumatic Actuators for Robotic Control, *Proc. of the 2001 IEEE/RSJ Conference on Int. Robots and Systems*, Maui, Hawaii, USA, 1964-1970.

Daerden, F. & Lefeber, D. (2001). The Concept and Design of Pleated Pneumatic Artificial Muscles, *International Journal of Fluid Power 2*, N°3, 41-50.

Davis, S.; Tsagarakis, N.; Canderle, J. & Caldwell, D.C. (2003). Enhanced Modelling and Performance in Braided Pneumatic Muscle Actuators. *The Int. Journal of Robotics Research* Vol. 22(3-4), 213-227.

De Gennes, P.-G. (1997). Un muscle Artificiel semi-rapide, *C.R. Acad., Sci. Paris*, Vol. 324, Serie II, 343-348.

Ghez, C. (1991). Muscles : Effectors of the Motor System. In *Principles of Neural Science*, E.R. Kandel, J.H. Schwartz, T.M. Jessekk, Eds. Englewood Cliffs, 3rd edition,New-York, Prentice-Hall, 548-563.

Grant, D. & Hayward, V. (1997). Variable Structure Control of Shape Memory Alloy Actuators, *IEEE Control Systems Magazine*, Vol. 17(3), 80-88.

Hannaford, B. & Winters, J. (1990). Actuator Properties and Movement Control : Biological and Technological Models. In *Multiple Muscle Systems : Biomechanics and Movement Organization*, J.M.Winters and S.L.-Y.Woo (eds.), Springer Verlag, New-York.

Hannaford, B.; Winters, J.M; Chou, C.P. & Marbot, P. (1995). The Anthroform Biorobotic Arm: A System for the Study of Spinal Circuits *Ann. Biomed. Eng.* , Vol. 23, 399-408.

Harashima, F.; Ueshiba, T. & Hascimoto, H. (1985). Sliding Mode Control for Robotic Manipulator, Proc. of the EPE Conference, Brussels, 251-256.

Hesselroth, T.; Sarkar, K.; Van der Smagt, P. & Schulten, K. (1994). Neural Network Control of a Pneumatic Robot Arm. *IEEE Trans. on Systems, Man & Cyb.*, Vol. 24(1), 28-37.

Hill, A.V. (1938). The Heat of Shortening and the Dynamic Constants of Muscle, *Proc. Roy. Soc.*, Part B, Vol. 126, 136-195.

Hirai, K.; Hirose, M.; Haikawa, Y. & Takenaka, T. (1998). The Development of Honda Humanoid Robot, *Proc. of the 1998 IEEE Int. Conf. on Robotics & Automation*, Leuwen, Belgium, 1321-1326.

Hirose, Y.; Shiga, T.; Okada, A. & Kurauchi, T. (1992). Gel Actuators Driven by Electric Field, *Proc. on the Int. Symp. on Theory and Mechanisms*, Nagoya, Japan, 293-297

Hogan, N. (1984). Adaptive Control of Mechanical Impedance by Coactivation of Antagonistic Muscles, IEEE Trans. Automat. Contr., Vol. AC-29, N°8, 681-690.

Huber, J.E.; Fleck, N.A. & Ashby, M.F. (1997). The selection of Mechanical Actuators Based on Performance Indices, *Proc. Roy. Soc. London A*, Vol. 453, pp. 2185-2205.

Hunter, I.W. & Lafontaine, S. (1992). A Comparison of Muscle with Artificial Actuators, *Proceedings of the IEEE Solid-State Sensor and Actuator Workshop*, Hilton Head, SC (USA), 178-185.

Katchalsky, A. (1949). Rapid Swelling and Deswelling of Reversible Gels of Polymeric Acids by Ionization, *Experienta*, Vol. V/8, 319-320.

Kim, J.; Kim, B.; Ryu, J.; Jeong, Y.; Park, J.; Kim, H.C. & Chun, K. (2005). Potential of Thermo-Sensitive Hydrogel as an Actuator, *Japanese Journal of Applies Physics*, Vol. 44, N° 7B, 5764-5768.

Kim K.J. & Shahinpoor. (2002). M. A Novel Method of Manufacturing three-dimensional Ionic Polymer-metal Composites (IPMCs) Biomimetic Sensors, Actuators and Artificial Muscles, *Polymer*, Vol. 43, 797-802.

Kuhn, W. & Hargitay, B.; Katchalsky, A. & Eisenberg, H. (1950). Reversible Dilatation and Contraction by Changing the State of Ionization of High-Polymer Acid Networks, *Nature*, Vol. 165, pp.514-516.

Madden, J.D.W.; Vandesteeg, N.A.; Anquetil, A.; Madden, P.G.A.; Takshi, A.; Pytel, R.Z.; Lafontaine, S.R.; Wieringa, P.A. & Hunter, I.W. (2004). Artificial Muscle Technology: Physical Principles and Naval Prospects, *IEEE Journal of Oceanic Engineering*, Vol. 29, N° 3, 706-728.

Maeno, T. & Hino, T. (2006) Miniature Five-Fingered Robot Hand Driven by Shape Memory Alloy Actuators, *Proc. of the 12th IASTED Int. Conf. on Robotics and Applications*, Honolulu, Hawaï, USA, pp. 174-179.

Mathew, G.; Rhee, J.M.; Nah, C. & Leo D.J. (2006). Effects of Silicone Rubber on Properties of Dielectric Acrylate Elastomer Actuator, *Polymer Engineering and Science*, 1455-1460.

Matsushita, M. (1968). Synthesis of Rubber Artificial Muscle. *Journal of the Society of Instrument and Control Engineers* 7(12), 110-116. (in Japanese)

Nakamura, T.; Saga, N. & Yaegashi. (2003). Development of a Pneumatic Artificial Muscle based on Biomechanical Characteristics, *Proc. of the IEEE-ICIT 2003 Conference*, Maribor, Slovenia, 729-734.

Ochoteco, E.; Pomposo, J.A.; Bengoechea, M.; Grande, H. & Rodriguez, J. (2006). Assembled Cation-Exchange/Anion-Exchange Polypyrolle Layers as New Simplified Artificial Muscles, *Polymers for Advanced Technologies*, in press.

Otero, T.F. & Sansinena J.M. (1995). Artificial Muscles Based on Conducting Polymers, *Bioelectrochemistry and Bioenergetics*, Vol. 38, 411-414.

Paynter, H.M. (1961). *Analysis and Design of Engineering Systems*, MIT Press, Cambridge.

Pei, Q.; Pelrine, R.; Stanford, S.; Kornbluh, R. & Rosenthal, M. (2003). Electroelastomer Rolls and their ApplicaHigh-Speed Electrically Actuated Elastomers with Strain Greater tion for Biomimetic Robots, *Synthetic Metals*, Vol. 135-136, 129-131.

Pelrine, R.; Kornbluh, R. & Joseph, J. (1998). Electrostriction of Polymer Dielectrics with Compliant Electrodes as a Means of Actuation, *Sensors & Actuators A*, Vol. 64, 77-85.

Pelrine, R.; Kornbluh, R.; Joseph, J; Heydt, R; Pei Q. & Chiba, S. (2000). High-field Deformation of Elastomeric Dielectrics for Actuators, *Materials Science and Engineering C*, Vol. 11, 89-100.

Pelrine, R.; Kornbluh, R.; Pei, Q.; Stanford, S.; Oh, S. & Eckerle, J. (2002). Dielectric Elastomer Artificial Muscle Actuators : Toward Biomimetic Motion, Smart Structures and Materials 2002 : Electroactive Polymer Actuators and Devices (EAPAD), *Proc. of SPIE*, Vol. 4695, 126-137.

Safak, K. & Adams, G. (2002). Modeling and Simulation of an Artificial Muscle and its Application to Biomimetic Robot Posture Control, *Robotics and Autonomous Systems*, Vol. 41, 225-243.

Saga, N.; Saikawa, T. & Okano, H.(2005). Flexor Mechanism of Robot Arm Using Pneumatic Muscle Actuators, *Proc. of the IEEE Int. Conf. on Mechatronics & Automation*, Niagara Falls, Canada, 1261-1266.

Segalman, D.J.; Witkowski, W.R.; Adolf, D.B. & Shahinpoor, M. (1992). Theory and Application of Electrically Controlled Polymeric Gels, *Smart Material Structure*, Vol. 1, 95-100.

Shahinpoor, M. (1992). Conceptual Design, Kinematics and Dynamics of Swimming Robotic Structures Using Polymeric Gel Muscles, *Smart Mater. Struct.*, Vol. 1, 91-94.

Shahinpoor, M. (2002). Ionic Polymer-conductor Composites as Biomimetic Sensors, Robotic Actuators and Artificial Muscles - A Review, *Electrochimica Acta*, Vol. 48, 2343-2353.

Shahinpoor, M.; Bar-Cohen; Y.; Simpson; J.O. & Smith; J. (1998). Ionic Polymer-Metal Composites (IPMCs) as Biomimetic Sensors, Actuators and Artificial Muscles - A Review, *Smart Mater.* , Vol. 7, R15-R30.

Slotine, J.-J. & Li, W. (1991). *Applied Nonlinear Control,* Prenctice-Hall International Editions, Englewood Cliffs, USA.

Suzuki, M. (1991). Polymer Gels as a New Driving SourceforRobotics, Micromachines and Biomedical Applications, *Int. Journal of Japan Soc. Prec. Eng.*, Vol. 25, N°3, 169-174.

Thanh, T.D.C. & Ahn, K.K.(2006). Nonlinear PID Control to Improve the Control Performance of 2 Axes Pneumatic Artificial Muscle Manipulator Using Neural Network, *Mechatronics*, Vol. 16, 577-587.

Tian, S.; Ding, G.; Yan, D.; Lin, L. & Shi, M.(2004). Nonlinear Controlling of Artificial Muscle System with Neural Networks, *Proc. of the 2004 IEEE Int. Conf. on Robotics and Biomimetics,* Shenyang, China, 56-59.

Tonietti, G. & Bicchi, A. (2002). Adaptive Simultaneous Position and Stiffness Control for a Soft Robot Arm, *Proc. of the 2002 IEEE/RSJ Int. Conf. on Intelligent Robots and Systems,* Lausanne, 1992-1997.

Tondu, B. & Lopez, P. (1995). Theory of an Artificial Pneumatic Muscle and Application to the Modelling of McKibben Artificial Muscle *C.R.A.S. French National Academy of Sciences*, Series IIb, 320, 105-114. (in French with an abridged English version)

Tondu, B. & Lopez, P. (2000). Modeling and Control of McKibben Artificial Muscle Robot Actuators, *IEEE Control Systems Magazine* , Vol. 20, N°2, 15-38.

Tondu, B.; Daidie, A.; Ippolito, S. & Guiochet, J. (2005). A Seven-degrees-of-freedom Robot-arm Driven by Pneumatic Artificial Muscles for Humanoid Robots, *The International Journal of Robotics Research,* Vol. 24, N°4, 2005, 257-274.

Tondu, B. & Diaz Zagal, S. (2006). McKibben Artificial Muscle can be adapted to be in Accordance with the Hill Skeletal Muscle Model, *Proc. of BioRob2006,* Pisa, Italy, paper 276.

Thongchai, S.; Goldfarb, N.; Sarkar, N. & Kawamura, K. (2001). A Frequency Modeling Method of Rubbertuators for Control Application in an IMA Framework, *Proc. of the American Control Conference,* Arlington, VA, USA, 1710-1714.

Van der Smagt, P.; Groen, F. & Schulten, K. (1996). Analysis and Control of a Rubbertuator Arm, *Biological Cybernetics* , Vol. 75, 433-440.

Verrelst, B.; Van Ham, R.; Vanderborght, B.; Lefeber, D.; Daerden, F. & Van Damme, M. (2006). Second Generation Pleated Pneumatic Artificial Muscle and its Robotic Applications, *Advanced Robotics,* Vol. 20, N°7, 783-805.

Verrelst, B.; Van Ham, R.; Vanderborght, B.; Daerden, F.; Lefeber, D. & Vermeulen, J. (2005). The Pneumatic Biped "Lucy" Actuated with Pleated Pneumatic Artificial Muscles, *Autonomous Robots,* Vol. 18, 201-213.

Verrelst, B.; Vanderborght, B.; Vermeulen, J.; Van Ham, R.; Naudet, J. & Lefeber, D. (2005). Control Architecture for the Pneumatically Actuated Dynamic Walking Biped "Lucy", *Mechatronics,* Vol. 15, 703-729.

Winter, D.A. (1969). Chap.5. Mechanical Work, Energy, and Power, In *Biomechanics of Human Movement,* John Wiley & Sons, New-York.

6

Development of a CORBA-based Humanoid Robot and its Applications

Yasuo Nasu[1], Genci Capi[2], Hanafiah Yussof[3]
Mitsuhiro Yamano[1] and Masahiro Ohka[3]
[1]Faculty of Engineering, Yamagata University, [2] Faculty of Engineering, Toyama University, [3]Graduate School of Information Science, Nagoya University
Japan

1. Introduction

Recently, the research on humanoid robots has attracted many researchers. The research spans from stability and optimal control, gait generation, human-robot and robot-robot communication. In addition, humanoid robots have been also used to understand better human motion. Among humanoid robot prototypes the most well known is Honda humanoid robot (Hirai et. al., 1998). This robot has the ability to move forward and backward, sideways to the right or the left, as well as diagonally. In addition, the robot can turn in any direction, walk up and down stairs continuously. Other example is the 35 dof (degrees of freedom) Saika humanoid robot (Inaba et al. 1998). This robot can perform a reach-and-grasp motion through coordinating legs and arms. The key idea of the system architecture of this robot is a remote brained approach. In addition, the Waseda humanoid robot group has also developed an anthropomorphic dynamic biped walking robot adapting to the humans' living floor (Takanishi et. al., 1990). Fujitsu also has developed a commercial 25 dof miniature humanoid robot, named HOAP-1, for research purposes. Weighing 6 kg and standing 0.48 m tall, the light and compact HOAP-1 and accompanying simulation software can be used for developing motion control algorithms in such areas as two-legged walking, as well as in research on human-to-robot communication interfaces.

In our robotics laboratory at Yamagata University, we initialized the humanoid robot project. The goal of this project is to contribute to the research on humanoid robots. For this reason, we developed an anthropomorphic biped humanoid robot called Bonten-Maru. During the humanoid robot design, we tried to mimic as much as possible the human characteristics, from the viewpoints of links dimensions, body structure, as well as the number and configuration of the degrees of freedom. The high number of dofs helps the humanoid robot to realize complex motions in even and uneven terrains, like walking, going up and down stairs, crawling, etc. In this chapter, we present the development of Common Object Request Broker Architecture (CORBA) based humanoid robot control systems. Consequently, this chapter explains the application of real time generation of humanoid robot optimal gait by using soft computing techniques, and also teleoperation systems and its applications. Simulation and experimental results of the proposed system in

each research theme utilizing the humanoid robot Bonten-Maru are presented which reveals good performance of the robot control system.
This chapter is organized in the following sequence: Section 2 explains the motivations of this research. In Section 3, we briefly explain the development of research prototype humanoid robot *Bonten-Maru I* and *Bonten-Maru II* in term of hardware configurations and control systems. Next, Section 4 explains the CORBA-based Humanoid Robot Control Architecture (HRCA). This section includes concept, algorithm and experimental results of the HRCA. Next in Section 5, we present an optimal locomotion strategy using Genetic Algorithm (GA) and Neural Networks (NN) for bipedal humanoid robot. In this section we presents a GA gait synthesis method for biped robots during walking based on Consumed Energy (CE) and Torque Change (TC), and also application of Radial Basis Function Neural Networks (RBFNN). In Section 6, we explain development of teleoperation systems via internet and user-friendly Virtual Reality (VR) user interface. Consequently, experimental results are presented in each section. Finally, summary and conclusions are presented in Section 7, followed by future work in Section 8.

2. Motivations

At present, many robots are developed for particular industrial, entertainment and service applications. However, these robots cannot be applied for other application. Especially when different programming languages are used for other applications, the cooperation between different systems is needed and the programs must be converted to be the same. Furthermore, the application of tele-operated system in robotics field is highly demanded like nowadays telecommunication technology. Therefore, in order to build open robot control platforms in humanoid robot control system, CORBA has been proposed.
The used of CORBA for the humanoid robots has open new dimension in robotic research, for example in teleoperation operations via internet. The system can be apply not only for the humanoid robots systems, but also for other fields like tele-medical, industrial robots, service and security, and also in aerospace project. However, the accuracy issue and time delay problem are the main factor to be consider in order to make the project successful in common architecture applications. Therefore, we considered programming language built in network program like Java and Perl in the robot control programming which commonly used C or C++. The management in cooperation between network programs and robot control programs are expected to reduce the time delay and increase the accuracy of certain motion in the robot task. In addition, the design of robot hardware and control systems is also considered to obtain reliable and accurate motions in real time applications.
Eventually, we were one of the first humanoid robot groups, that proposed a humanoid robot control platform based on CORBA (Takeda et al., 2001). One of our main objectives is to make the control platform open for other researchers to test their results and also to conduct collaborative research. By using a CORBA based control platform, it is easy to add modules developed in different programming languages. In addition, the control of the humanoid robot is made in a distributed environment. Therefore, various humanoid robots in the world can share their own modules with each other via the internet.
Another direction of our research is the real time generation of humanoid robot optimal gait by using soft computing techniques (Capi et al. 2003). Genetic Algorithms (GA) was employed to minimize the energy for humanoid robot gait. For a real time gait generation, we used Radial Basis Function Neural Networks (RBFNN), which are trained based on

Genetic Algorithm (GA) data. Until now the walking and going up-stairs modules are created. The main goal is to create an autonomous humanoid robot that can operate in different environments. Based on the information received by the eye system, the appropriate module will be simulated to generate the optimal motion.
Control of humanoid robot in a long distance was also another research issue. In our research, the main objective is to create various sophisticated motions and new application fields for humanoid robots. For this reason, we considered accident site operations which are often unknown environments. In our work, we used a teleoperation system to control the humanoid robot via the internet. We carried out long distance experiments on the teleoperation system of the humanoid robot between Deakin University (Australia) and Yamagata University (Japan) (Nasu et al., 2003). Additionally, we have developed a user-friendly Virtual Reality (VR) user interface that is composed of ultrasonic 3D mouse system and a Head Mounted Display (HMD) (Kaneko et al., 2003). The real time experiments were conducted using the Bonten-Maru humanoid robot.

3. Development of Research Prototype Humanoid Robot

We have developed a research prototype humanoid robot system known as "Bonten-Maru". The earliest model was the 23 dof Bonten-Maru I. Next, we developed an advanced version called Bonten-Maru II which consists of 21 dof. The Bonten-Maru humanoid robot series are one of few research humanoid robots in the world that can be utilized in various aspects of studies.

3.1 Humanoid Robot Bonten-Maru I

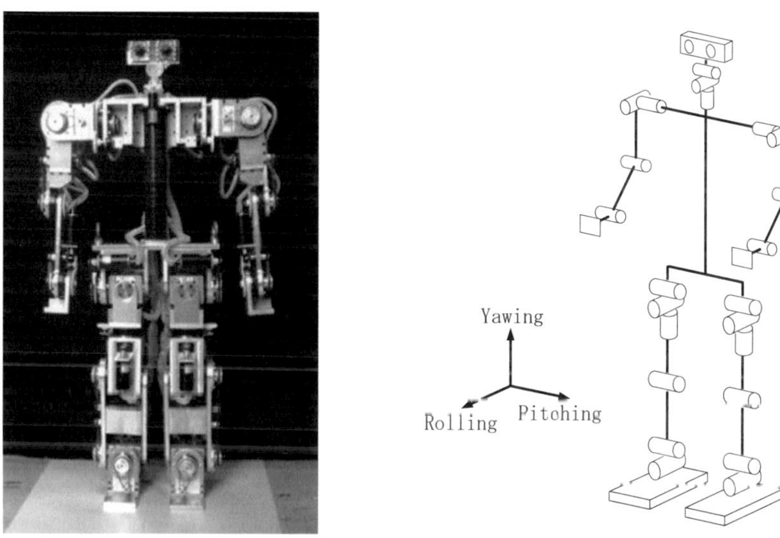

Figure 1. The Bonten-Maru I humanoid robot and its distribution of dofs

The Bonten-Maru I humanoid robot is consists of 23 dof. The appearance and distribution of dofs are shown in Fig. 1. Bonten-Maru I is 1.2 m high, weight about 35 kg. The properties of Bonten-Maru I is shown in Table 1, meanwhile Table 2 shows the configurations of dofs. The robot's each leg has six degree of freedom and is composed by three segments: upper leg,

lower leg and the foot. The hip is a ball-joint, permitting three dof; the knee joint one dof; the ankle is a double-axis design, permitting two. The shoulder has two dof, the elbow and wrist one dof. The distribution of dof is similar with the dof in human limbs. A DC servometer actuates each joint. The DC servomotors act across the three joints of the head, where is mounted the eye system, enabling a total of three dof.
Potentiometers are used to obtain the position and velocity of every link, interfaced to the computer via RIF-01 (ADC). The power is supplied to each joint b timing belt and harmonic drive reduction system. The gyro unit is connected to the PC by RS-232C interface. The head unit has 2 CCD cameras (542 x 492 pixels, Monochrome). The CCD camera units are connected to PC by video capture board. The control platform is based on CORBA. This allows an easy updating and addition of new modules. A Caleron based microcomputer (PC/AT compatible) is used to control the system. The OS are Linux 2.0.36 and FreeBSD2.2.8R.

	Body	***Lower leg***	***Upper leg***	***Lower leg + foot***
Moment of inertia [kg m²]	0.19	0.014	0.002	0.017
CoM from lower joint [m]	0.3	0.09	0.1	0.136
CoM from upper joint [m]	0.0	0.11	0.104	0.136
Length [m]	0.3	0.2	0.204	0.284
Mass of the link [kg]	12	2.93	3.89	4.09

CoM: Center of Mass

Table 1. Properties of Bonten-Maru I

JOINT		DIRECTION	DOF
Leg	Hip	Rolling, Pitching	2
	Thigh	Yawing	1
	Knee	Pitching	1
	Ankle	Rolling, Pitching ,	2
Arm	Shoulder	Rolling, Pitching	2
	Elbow	Yawing	1
	Wrist	Pitching	1
Head	Neck	Rolling, Pitching, Yawing	3

Table 2. Configurations of dofs in Bonten-Maru I

3.2 Humanoid Robot Bonten-Maru II

Bonten-Maru II is an anthropomorphic prototype humanoid robot. This robot is 1.25 m tall, weighting about 32.5 kg. Fig. 2 shows the appearance of Bonten-Maru II and distribution of dofs. The robot has a total of 21 dofs: six for each leg, three for each arm, one for the waist, and two for the head. Configurations of dofs at each joint and joint rotation range are shown in Table 3. The link dimensions are determined such that to mimic as much as possible the humans. The high number of dofs and configuration of joints that closely resemble those of humans provide Bonten-Maru II with the possibility of realizing complex trajectories to attain human-like motion. Each joint is driven by a DC servomotor with a rotary encoder and a harmonic drive-reduction system, and is controlled by a PC with the Linux OS. Under each foot are four force sensors, and the arms are equipped with six-axis force sensor. The head unit has two CCD cameras (542 x 492 pixels, Monochrome), which are connected to PC by video capture board. A Celeron based microcomputer (PC/AT compatible) is used to control the system.

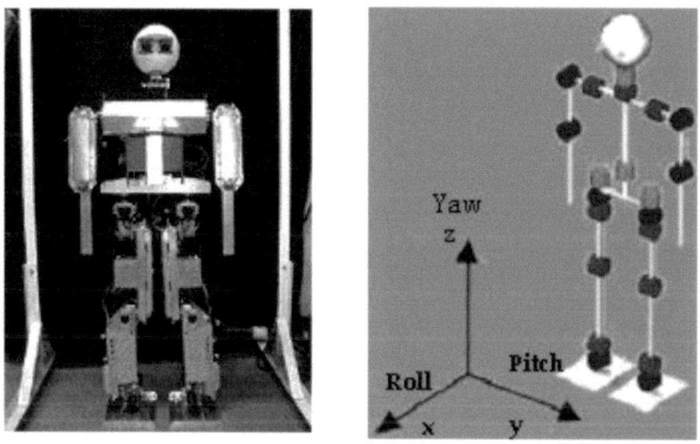

Figure 2. The Bonten-Maru II humanoid robot and its distribution of dofs

Axis	Quantity of dofs	Range of joint rotation angle (deg)
Neck (roll and pitch)	2	-90 ~ 90
Shoulder (pitch) right & left	2	-180 ~ 120
Shoulder (roll) right /left	2	-135 ~ 30/-30 ~ 135
Elbow (roll) right /left	2	0 ~ 90/0 ~ -90
Waist (yaw)	1	-45 ~ 45
Hip (yaw) right /left	2	-90 ~ 60/-60 ~ 90
Hip (roll) right/left	2	-90 ~ 25/-25 ~ 90
Hip (pitch) right & left	2	-130 ~ 45
Knee (pitch) right & left	2	-20 ~150
Ankle (pitch) right & left	2	-90 ~ 60
Ankle (roll) right/left	2	-20 ~ 90/-90 ~ 20

Table 3. Configurations of dofs and joint rotation range in Bonten-Maru II

4. CORBA-Based Humanoid Robot Control Architecture (HRCA)

The development of an efficient humanoid robot control system required the control modules to be developed independently and able to integrate easily to the system. Commonly, the control modules developed by many researchers are apart from OS and programming languages must be connected to the internet directly for the common use in the worldwide. For this reason, CORBA (Moubray et al., 1997) is a good platform for humanoid control system architecture. CORBA is a specification of message exchange among objects, which is specified by Open Management Group (OMG). CORBA has attracted many researchers. Vinoski (Vinoski, 1997) suggested the effectiveness of CORBA for heterogeneous systems. Pancerella (Pancerella et al., 1996) and Whiteside (Whiteside et al., 1997) have implemented a CORBA based distributed object software system for the Sandia Agile Manufacturing Testbed. Harrison (Harrison et al., 1997) has developed a real-time CORBA Event Service as a part of the TAO project at Washington University. CORBA is a useful distributed application platform, which can make the cooperation among distributed applications very easily. Also, CORBA is independent on a specific programming language or OS. Thus, it is able to communicate among objects developed in different programming languages and OS.

4.1 HRCA Concept

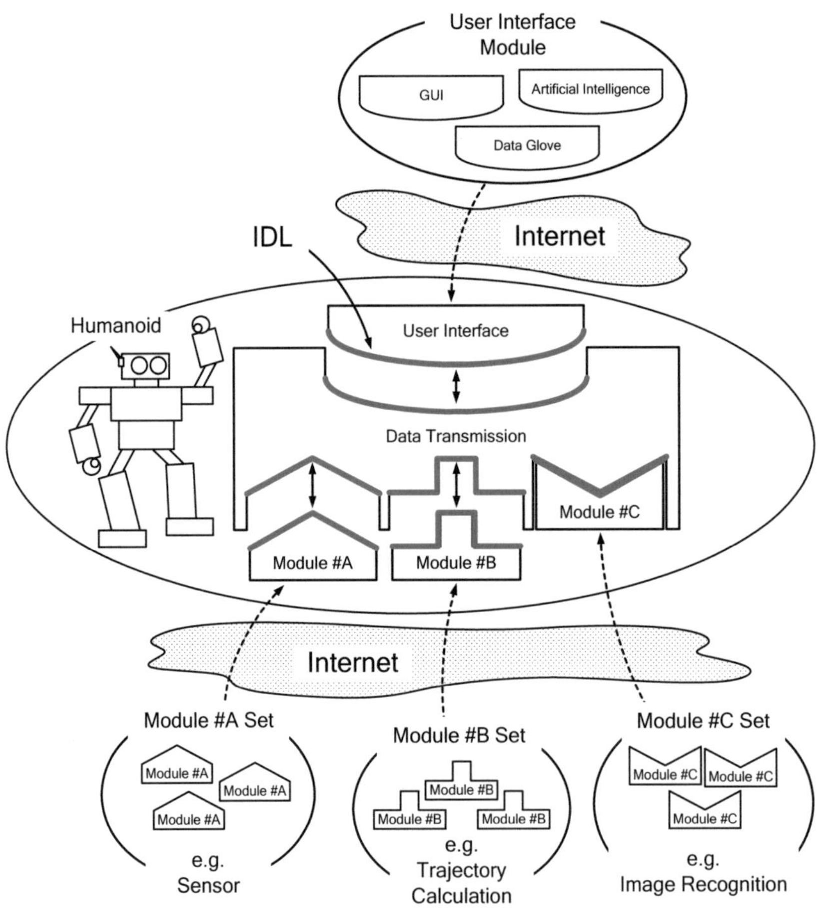

Figure 3. Concept of the proposed HRCA

We have proposed the HRCA to integrate the desired modules which developed by many researchers individually to control the motion of humanoid robot via the internet. The HRCA can share many modules among many users or researchers from remote distance through any computer by the internet communication. Figure 3 shows a basic concept of the proposed HRCA. The HRCA design is based on the Unified Modeling Language (UML), which includes the Use Case Diagram (UCD) and Class Diagram (CD).

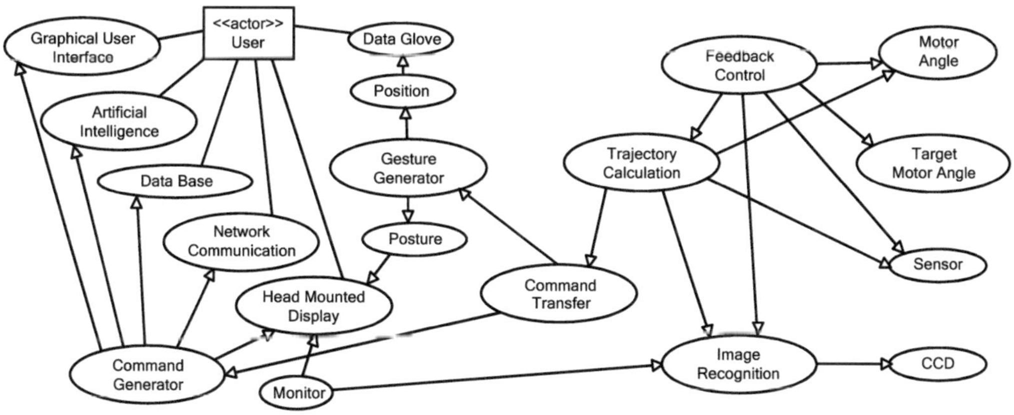

Figure 4. Relationships among the HRCA modules

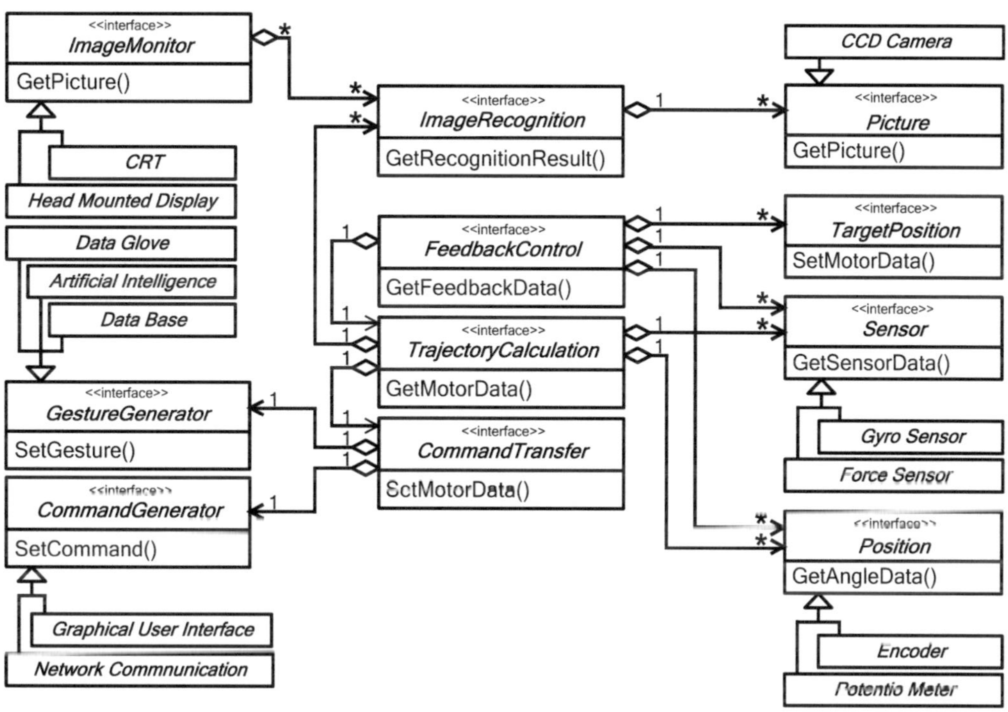

Figure 5. Class diagram of HRCA

The UML is used to define CORBA servers, clients, and it's IDL. Booch (Booch et al., 1999), Fowler (Fowler et al., 1997), and Open Manegement Group (OMG) proposed the UML for the design of the object-oriented programming. The HRCA modules are designed by using the UCD. The relationships among the HRCA modules are shown in Fig. 4. The HRCA is very complex, but in this figure we only show the highest level of the system. Each circle presents a Use Case (UC) and each arrow shows relationships among them. Eventually, when we design the CD, each UC is defined as a class.

The CD is shown in Fig. 5. There are many classes in each square, but in this study, we use only the interface class, because the IDL defines only the object's interface. In the CD, each square presents a class icon. Inside the square the stereotype, name, and method of the class are written. The symbol "◇——>" presents an association among classes.

The number written on the both ends of the symbols show how many classes are used. The symbol "*" shows that the class number is not limited. Finally, each class in Fig. 5 is implemented as HRCA modules, which correspond to CORBA servers and client. The IDL of each HRCA modules are obtained from CD, and convert to a programming language source code by IDL compiler.

4.2 Proposed HRCA Modules

The HRCA model is shown in Fig. 6. This figure presents some algorithms and devices, which can be implemented as HRCA modules. The HRCA is able to use these algorithms and devices by selecting the appropriate CORBA servers. Until now, we have implemented the following modules: DTCM, MCM, JTM, GSM, JAM, FCM, and UIM, which are shown in Fig. 7. Each module corresponds to "Data Transmission", "Target Position", "Angle Trajectory Calculation", "Sensor", "Position", "Feedback Control", "Command Generator", respectively, which are shown in Fig. 6. To implement CORBA servers and client, the Inter-Language Unification (ILU) is used, which has proposed by XEROX PARC. ILU supports many programming languages, such as C++, ANSI C, Python, Java, Common Lisp, Modula-3, Guile Scheme, and Perl 5. In our research, we used only C language for implement HRCA. But in the future, we would like to implement some HRCA modules using other languages.

In our HRCA, the DTCM controls the data flow of the modules. The DTCM communicates with MCM, JTM, GSM, JAM, and FCM by using their functions. But DTCM communicates with UIM by its own function. The data flow model is also shown in Fig. 7. Until now, the UIM is very simple, which is able to command "WALK", "OBJECT_OVERCOMING", and "GYRO_TEST" only.

The MCM controls the joint motors of the humanoid robot. The model of MCM and the IDL between MCM and DTCM, and are shown in Fig. 8 and Fig. 9, respectively. In Fig. 9, the MCM provides two functions, "SetMotorData()" and "SetMotorFeedbackData()". "SetMotorData()" is a function for the data input of the joint trajectories. "ROBOT_DEGREE_DATA" data type includes the time unit data and the joint trajectory data, which are named as "time_unit" and "degree_data", respectively. "SetMotorFeedbackData()" is a function for the feedback data input from FCM. "MOTOR_FEEDBACK_DEGREE_DATA" data type includes the joint feedback data, which is named as "feedback_degree_data". Using these arguments of the "SetMotorData()" and "SetMotorFeedbackData()", the MCM controls each joint motor. In addition, in this study, we used multi-threaded implementation for motor control routine because of the time delay, which is caused by controlling each joint motor, sequentially. Using multi-threaded implementation, the motors are controlled in parallel and the time delay is reduced.

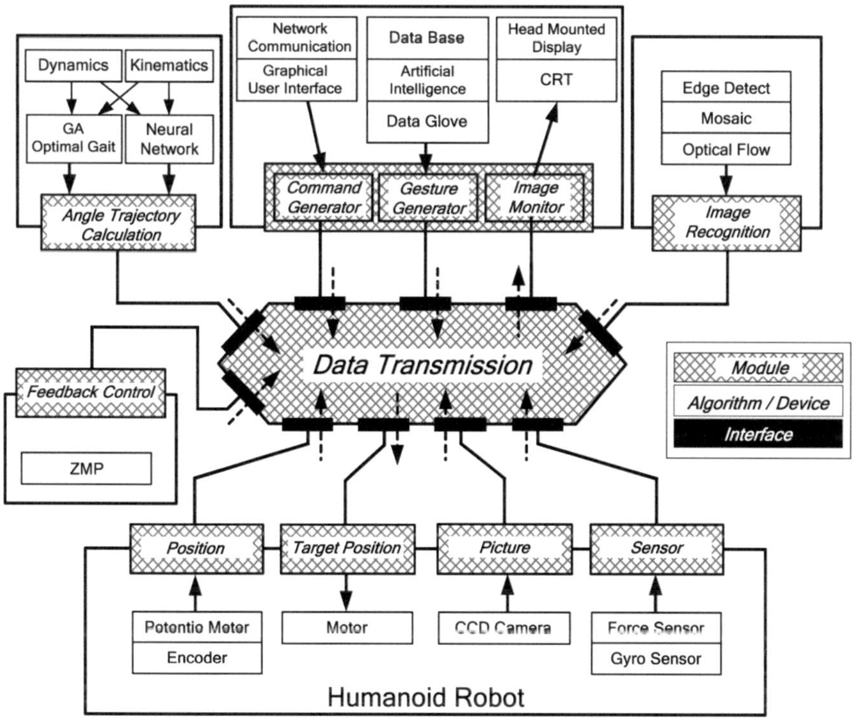

Figure 6. HRCA model

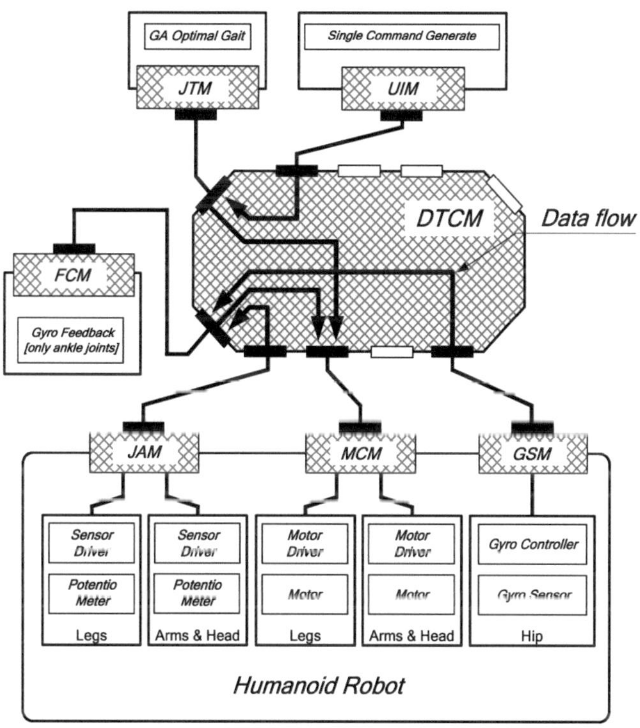

Figure 7. Implemented HRCA modules

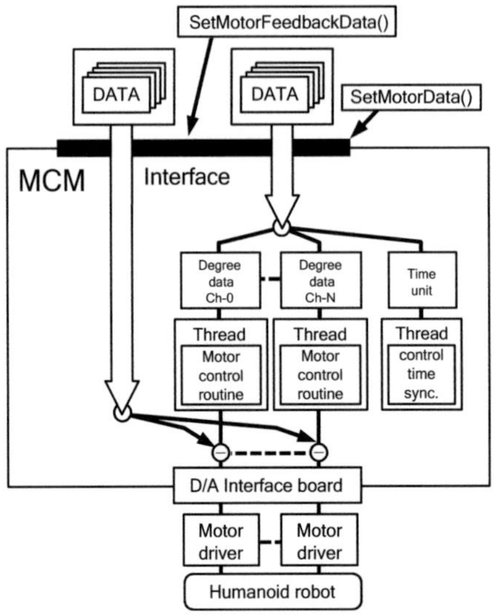

Figure 8. The model of Motor Control Module (MCM)

```
module MCM{

   // Define Array
   const long MOTOR = 25;              //max 25 motors
   const long TIMES = 100;             //max 100 set data

   typedef double            DEGREE[TIMES];

   typedef struct MotorDegreeData{
      DEGREE degrees;
   }MOTOR_DEGREE_DATA[MOTOR];

   typedef struct RobotDegreeData{
      long time_unit;
      MOTOR_DEGREE_DATA degree_data;
   }ROBOT_DEGREE_DATA;

   // For Robot Feedback
   typedef double    FEEDBACK_DEGREE[MOTOR];

   typedef struct MotorFeedbackDegreeData{
      FEEDBACK_DEGREE feedback_degree_data;
   }MOTOR_FEEDBACK_DEGREE_DATA;

   //Motor Interface
   interface MotorInterface{

      // Motor Angle Values Setting
      short
      SetMotorData(
           in ROBOT_DEGREE_DATA degree_order,
           in long data_size);

      short
      SetMotorFeedbackData(
           in MOTOR_FEEDBACK_DEGREE_DATA
                degree_feedback_order );

   };
};
```

Figure 9. MCM IDL and MCM module

The IDL of other modules are developed and implemented same as MCM. The JTM provides the joint trajectories data to DTCM. The joint trajectories data is defined same as the input data type of MCM. The joint trajectory data are calculated by a genetic algorithm program and are used in a data base. These data are provided from JTM to DTCM. The GSM provides the angle, angular velocity, and angular acceleration data of gyro sensor to DTCM. The JAM provides the joint angle data of humanoid robot to DTCM. The JAM is used for reading and recording the actual joint trajectory data of the humanoid robot by using multi-threaded implementation. The FCM provides the feedback joint angle data to MCM via DTCM. The FCM obtains the joint angle data from JAM and GSM via DTCM, which balance the humanoid robot. We use only gyro sensor data for ankle joints control, but in the future, we would like to add another sensor data for precise feedback control.

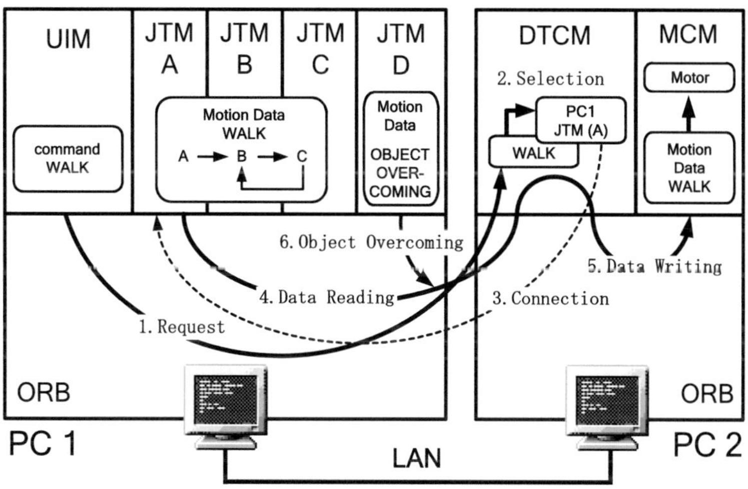

Figure 10. Data transfer flow in experimental system

4.3 Experiment and Result

Using the humanoid robot, we have carried out an experiment to show the effectiveness of the proposed HRCA. In this test, we are concentrated in the development of the humanoid robot control architecture, not in the control scheme itself or the robot response.

In order to measure the utilization of the proposed HRCA, the motion mode change tests are conducted on the Bonten-Maru I humanoid robot. The HRCA is applied to change the motion related to JTM as shown in Table 4. The static walking motion is divided into 3 parts in order to reuse the motion modules efficiently. The JTM (A, B, C, and D) and the UIM are running in PC1, the MCM and DTCM are running in PC2. The PC1 and PC2 are connected via LAN. The data transfer flow is shown in Fig. 10. The module changing procedure to control the motion of humanoid robot is explained as follows:

1. Request: The UIM sends an order sequence to DTCM (in this experiment it sends the "WALK" request);
2. JTM Selection: After receiving the "WALK" request from the UIM, the DTCM selects a JTM;
3. Connection: The DTCM is connected with JTM;
4. Data Reading: The DTCM reads the "WALK" data from JTM(A);

5. Data Writing: The DTCM transfers the data of JTM(A) to MCM and the MCM executes the data. When the humanoid robot is walking, the walking movement starts by JTM(A) and the normal walking is carried out by repeating in the round robin order JTM(B) and JTM(C);
6. Object Overcoming: The DTCM changes the JTM from "WALK" to "OBJECT_OVERCOMING", connects to JTM(D), and reads the "OBJECT_OVERCOMING" data from JTM(D). Then, the data is transferred to MCM, which moves the motor.

Figure 11 shows the joint angle trajectories, for right joint angles $(\theta_{R2}, \theta_{R3}, \text{and}\, \theta_{R6})$ and for left joint angles $(\theta_{L2}, \theta_{L3}, \text{and}\, \theta_{L6})$, during 4 steps JTM(A, B, C, and D). Figure 14 shows the continuous walking motion of humanoid robot by different modules of our proposed HRCA. The humanoid robot walks in smooth and static manner. Ideally, the time lag should be as short as possible at every step change. However, during the experiment, we measured that the time lag at every step change is about 200 milliseconds. But this time lag did not influence on the walking motion of the humanoid robot during every step because the humanoid robot walks in static condition. This experimental result shows that the proposed HRCA is able to control the static motion of humanoid robot accurately by anging the response of JTM. Figure 12 shows the continuous walking motion of humanoid robot by different modules of the proposed HRCA. The humanoid robot walks in smooth and static manner. This experiment result shows that the proposed HRCA is able to control the static motion of humanoid robot accurately by changing the respond of JTM.

1	JTM(A)	From stand position to right leg swing
2	JTM(B)	From right leg swing to left leg swing
3	JTM(C)	From left leg swing to right leg swing
4	JTM(D)	Object overcoming by using left leg

Table 4. Joint Trajectory Module (JTM) motions

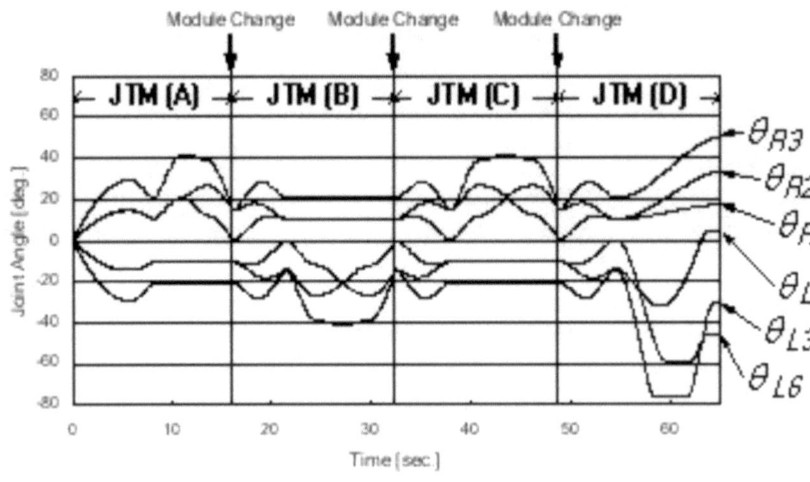

Figure 11. Joint angle trajectories

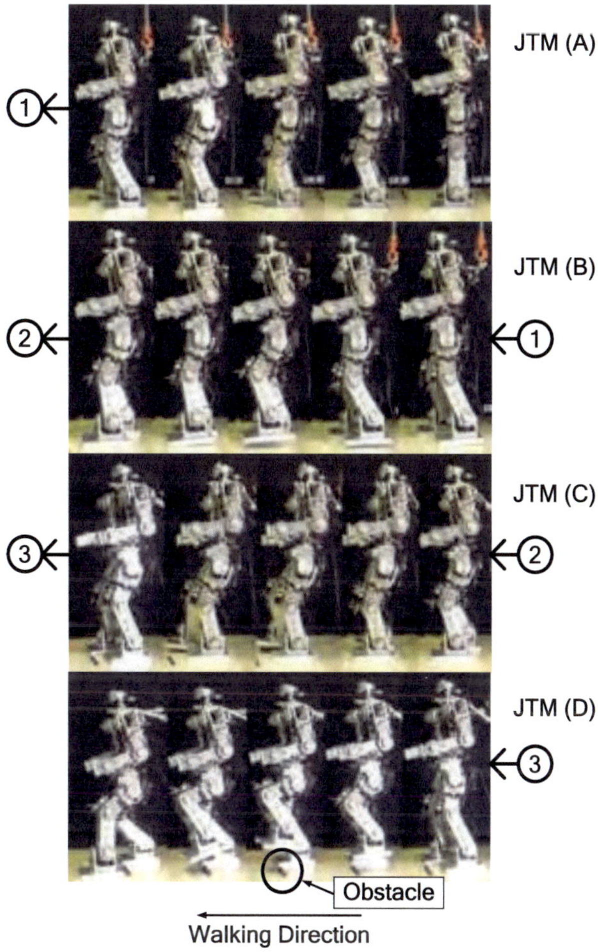

Figure 12. Continuous walking motion

5. Optimal gait strategy

We considered minimum Consumed Energy (CE) as criterion for humanoid robot gait generation, because autonomous humanoid robots make difficult to use external power supply. In order to have a long operation time when a battery actuates the motors, the energy must be minimized. From the viewpoint of energy consumption, one factor that has a great influence is the gait synthesis. In most of the previous papers related to biped robots (Vukobratovic et al., 1990) (Takanishi et al. 1990), the angle trajectories of the leg part are prescribed based on data taken from humans. The motion of upper body is calculated in order to have the ZMP inside the sole region. Some effort has been placed to analyze the effect of gait synthesis on consumed energy. In (Roussel et al., 1998) and (Silva et al., 1999) the minimum consumed energy gait synthesis during walking is treated. The body mass is considered concentrated on the hip of the biped robot (Roussel et al., 1998). In (Silva et al., 1999), the body link is restricted to the vertical position, the body forward velocity is considered to be constant and the tip motion of the swing leg is constrained to follow

sinusoidal functions. The effect of the walking velocity and step length on the consumed energy is treated in (Channon et al., 1996). A variation technique is used to minimize the cost function. However, in all these approaches related with optimal gait of biped robots, the stability and the real time implementation are not considered.
In this section, we describe a Genetic Algorithm (GA) gait synthesis method for biped robots during walking based on Consumed Energy (CE) and Torque Change (TC). The proposed method can easily be applied for other tasks like overcoming obstacles, going down-stairs, etc. The final aim of this research is to create modules for many tasks considered to be performed by Bonten-Maru I humanoid robot. Based on the information received by eye system, the appropriate module will be simulated to generate the optimal motion.
When solving for optimal gaits, some constrains must be considered. In our work, the most important constraint is the stability, which is verified through the ZMP concept. GA makes easy handling the constraints by using the penalty function vector, which transforms a constrained problem to an unconstrained one. GA has also been known to be robust for search and optimization problems (Channon et al., 1996), It has been used to solve difficult problems with objective functions that do not possess properties such as continuity, differentiability, etc. For a real time implementation, in (Roussel et al., 1998), the authors suggest to create in the future a database of pre-computed optimal gaits. This can generate the angle trajectories only for the step lengths and step times, which are included in the database. In order to cover all the interval of the pre-computed optimal gaits, we consider teaching a Radial Basis Function Neural Network (RBFNN) based on the GA results. In this section we present the results where input variables are the step length and step time. Simulations show good results generated by RBFNN in a very short time.

5.1 Genetic Algorithm (GA)

GA is a search algorithm based on the mechanics of natural selection and population genetics. The search mechanism is based on the interaction between individuals and the natural environment. GA comprises a set of individuals (the population) and a set of biologically inspired operators (the genetic operators). The individuals have genes, which are the potential solutions for the problem. The genetic operators are crossover and mutation. GA generates a sequence of populations by using genetic operators among individuals. Only the most suited individuals in a population can survive and generate offspring, thus transmitting their biological heredity to the new generation. The main steps are shown below:

1. Supply a population P_0 of N individuals and respective function values;
2. $i \leftarrow 1$;
3. $P_i' \leftarrow$ selection_function $(P_i - 1)$;
4. $P_i \leftarrow$ reproduction_function (P_i');
5. Evaluate (P_i);
6. $i \leftarrow i+1$;
7. Repeat step 3 until termination.

5.2 Biped Model

The arms of the humanoid robot will be fixed on the chest during performing motion. Therefore, it can be considered as a five-link biped robot in the saggital plane, as shown in Fig. 13. The motion of the biped robot is considered to be composed from a single support phase and an instantaneous double support phase. The friction force between the robot's feet and the ground is considered to be great enough to prevent sliding. During the single support phase, the ZMP must be within the sole length, so the contact between the foot and the ground will remain. To have a stable periodic motion, when the swing foot touches the ground, the ZMP must jump in its sole. This is realized by accelerating the body link. To have an easier relative motion of the body, the coordinate system from the ankle joint of the supporting leg is moved transitionally to the waist of the robot ($O_1X_1Z_1$). Referring to the new coordinate system, the ZMP position is written as follows:

$$\overline{X}_{ZMP} = \frac{\sum_{i=1}^{5} m_i(\ddot{\overline{z}}_i + \ddot{z}_w + g_z)\overline{x}_i - \sum_{i=1}^{5} m_i(\ddot{\overline{x}}_i + \ddot{x}_w)(\overline{z}_i + z_w)}{\sum_{i=1}^{5} m_i(\ddot{\overline{z}}_i + \ddot{z}_w + g_z)}, \tag{1}$$

where m_i is mass of the particle "*i*", x_w and z_w are the coordinates of the waist with respect to the coordinate system at the ankle joint of supporting leg, $\overline{x}_i$ and $\overline{z}_i$ are the coordinates of the mass particle "*i*" with respect to the $O_1X_1Z_1$, $\ddot{\overline{x}}_i$ and $\ddot{\overline{z}}_i$ are the acceleration of the mass particle "*i*" with respect to the $O_1X_1Z_1$ coordinate system. Based on the formula (1), if the position, $\overline{x}_i, \overline{z}_i$, and acceleration, $\ddot{\overline{x}}_i, \ddot{\overline{z}}_i$, of the leg part (i=1,2,4,5), the body angle, θ_3, and body angular velocity, $\dot{\theta}_3$, are known, then because $\ddot{\overline{x}}_3, \ddot{\overline{z}}_3$ are functions of $l_3, \theta_3, \dot{\theta}_3, \ddot{\theta}_3$, it is easy to calculate the body angular acceleration based on the ZMP position. Let $_{(0)}$ and $_{(f)}$ be the indexes at the beginning and at the end of the step, respectively. At the beginning of the step, $\ddot{\theta}_{30}$, causes the ZMP to be in the position ZMP_{jump}. At the end of the step, the angular acceleration, $\ddot{\theta}_{3f}$, is calculated in order to have the ZMP at the position ZMP_f, so that the difference between $\ddot{\theta}_{3f}$ and $\ddot{\theta}_{30}$ is minimal. Therefore, the torque necessary to change the acceleration of the body link will also be minimal.

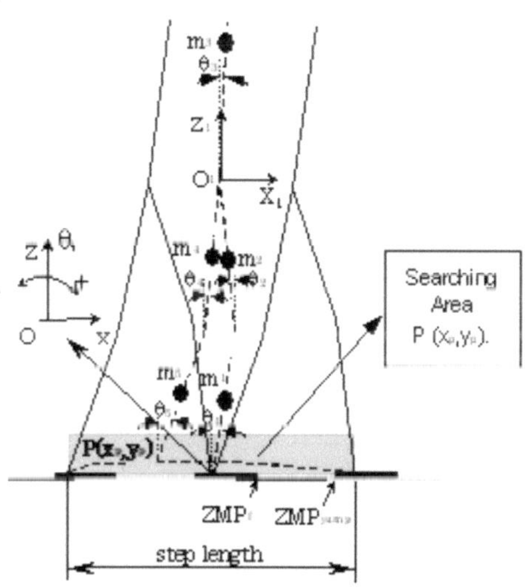

Figure 13. Five link biped robot

5.3 Problem Formulation

The problem consists of finding the joint angle trajectories, to connect the first and last posture of the biped robot for which the CE or TC is minimal. It can be assumed that the energy to control the position of the robot is proportional to the integration of the square of the torque with respect to the time. Because the joints of the manipulator are driven by torque, then the unit of torque, Nm, is equal to the unit of energy, joule. So, the cost function, J, can be defined as the following expression:

$$J = \frac{1}{2}\left(\int_0^{t_f} \tau^T \tau dt + \Delta\tau_{jump}^2 \Delta t + \int_0^{t_f} C dt\right) \tag{2}$$

where: t_f is the step time, τ is th e torque vector, $\Delta\tau_{jump}$ and Δt are the addition torque applied to the body link to cause the ZMP to jump and its duration time, and C is the constraint function, given as follows:

$$C = \begin{cases} 0 \text{ - if the constraints are satisfied,} \\ c_i \text{ - if the constraints are not satisfied,} \end{cases}$$

c denotes the penalty function vector. We consider the following constraints for our system.

- The walking to be stable or the ZMP to be within the sole length.
- The distance between the hip and ankle joint of the swing leg must not be longer then the length of the extended leg.
- The swing foot must not touch the ground prematurely.

The results generated for minimum CE cost function are compared with the angle trajectories that minimize the rate of change of the torque (Uno et al., 1989).
The cost function is as follows:

$$J_{\text{torque change}} = \frac{1}{2}\left(\int_0^{t_f} \left(\frac{d\tau}{dt}\right)^T \left(\frac{d\tau}{dt}\right) dt + \left(\frac{\Delta\tau}{\Delta t}\right)^2 + \int_0^{t_f} C dt\right). \tag{3}$$

5.4 Proposed Method

The block diagram of the proposed method is presented in Fig. 14. Based on the initial conditions and the range of searching variables, an initial population is generated. Every angle trajectory is presented as a polynomial of time. Its range is determined based on the number of angle trajectory constraints and the coefficients are calculated to satisfy these constraints. The torque vector is calculated from the inverse dynamics of the five-link biped robot (Mita et al., 1984) as follows:

$$J(\theta)\ddot{\theta} + X(\theta)\dot{\theta}^2 + Y\dot{\theta} + Z(\theta) = \tau \tag{4}$$

According to the formula (2) and (3), the cost function is calculated for minimum CE and minimum TC, respectively. The value of the cost function is attached to every individual in the population. The GA moves from generation to generation, selecting parents and producing offspring until the termination criterion (maximum number of generations

GN_{max}) is met. Based on the GA results, the gait synthesis is generated for minimum CE and minimum TC, respectively.

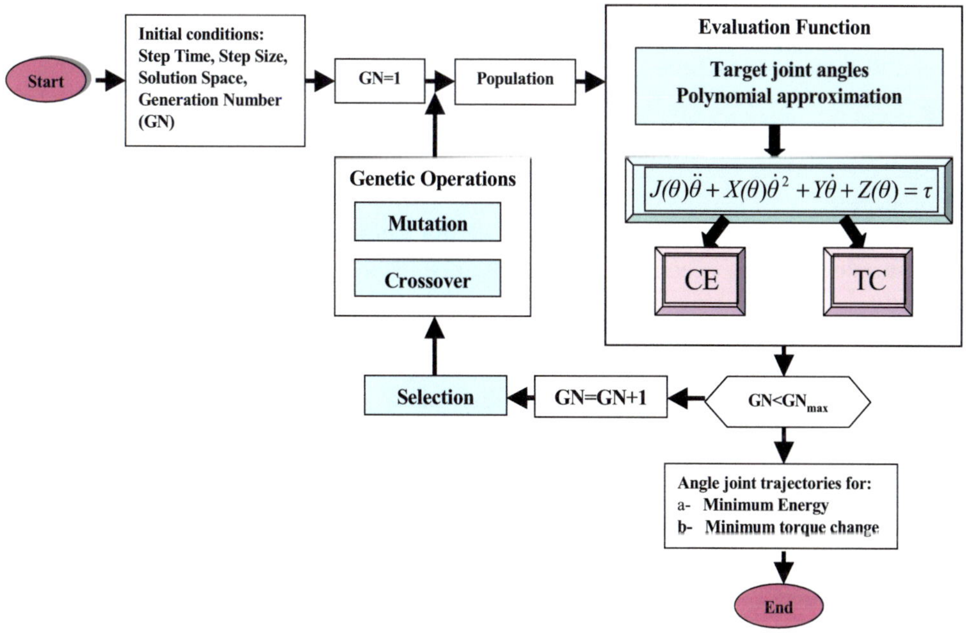

Figure 14. Proposed method

5.5 Boundary Conditions and CA Variables

To have a continuous periodic motion, the posture of the biped robot is considered to be the same at the beginning and at the end of the step. It must satisfy the following:

$$\theta_{10}=\theta_{5f},\ \theta_{20}=\theta_{4f},\ \theta_{1f}=\theta_{50},\ \theta_{2f}=\theta_{40},\ \theta_{30}=\theta_{3f}. \tag{5}$$

In order to find the best posture for walking, the optimum value of θ_{10}, θ_{20} and θ_{30} must be determined by GA. For a given step length during walking it is easy to calculate θ_{40} and θ_{50}. When referring to Fig. 13, it is clear that links 1, 2, 4 at the beginning of the step and links 2, 4, 5 at the end of the step, change the direction of rotation. Therefore, we can write:

$$\dot{\theta}_{10}=\dot{\theta}_{20}=\dot{\theta}_{40}=\dot{\theta}_{2f}=\dot{\theta}_{4f}=\dot{\theta}_{5f}=0. \tag{6}$$

The angular velocity of link 1 at the end of the step and link 5 at the beginning of the step is considered to be the same. This can be written in the form $\dot{\theta}_{1f}=\dot{\theta}_{50}$. In order to find the best value of angular velocity, we consider it as one variable of GA, because the rotation direction of these links does not change. GA will determine the optimal value of the angular velocity of the body link, which is considered to be the same at the beginning and at the end of the step. The following relations are considered for the angular acceleration:

$$\ddot{\theta}_{10}=\ddot{\theta}_{5f},\ddot{\theta}_{20}=\ddot{\theta}_{4f},\ddot{\theta}_{1f}=\ddot{\theta}_{50},\ddot{\theta}_{2f}=\ddot{\theta}_{40}. \tag{7}$$

In this way, during the instantaneous double support phase, we don't need to apply an extra torque to change the angular acceleration of the links. To find the upper body angle trajectory, an intermediate angle θ_{3p} and its passing time t_3 are considered as GA variables.

To determine the angle trajectories of the swing leg, the coordinates of an intermediate point $P(x_p,z_p)$ and their passing time t_p, are also considered as GA variables. The searching area for this point is shown in Fig. 13. Based on the number of constraints, the range of the time polynomial for $\theta_1, \theta_2, \theta_3, \theta_4$ and θ_5 are 3, 3, 7,6 and 6, respectively.

5.6 Simulation

In the simulations, we use the parameters of the Bonten-Maru I humanoid robot. The parameter values are presented in Table 1. We must produce many locomotion modules for humanoid robot, i.e. walking, going up-stairs, going down-stairs, obstacle overcoming, etc. In this chapter, we show the simulation results for a normal gait generation.
For the optimization of the cost function, a real-value GA was employed in conjunction with the selection, mutation and crossover operators. Many experiments comparing real value and binary GA has proven that the real value GA generates better results in terms of the solution quality and CPU time (Michalewich, 1994). To ensure a good result of the optimization problem, the best GA parameters are determined by extensive simulation that we have performed, as shown in Table 5. The maximum number of generations is used as the termination function. The GA converges within 40 generations (see Fig. 5). The average of the cost function En against the number of generations is shown in Fig. 6. The 33-th generation has the lowest value, which agrees with Fig. 5 results.

Function Name	Parameters
Arithmetic Crossover	2
Heuristic Crossover	[2 3]
Simple Crossover	2
Uniform Mutation	4
Non-Uniform Mutation	[4 GNmax 3]
Multi-Non-Uniform Mutation	[6 GNmax 3]
Boundary Mutation	4
Normalized Geometric Selection	0.08

Table 5. GA functions and parameters

Based on the Bonten-Maru I parameters, the step length can vary up to 0.5m. If the step length is smaller then 0.36 m, the ZMP can smoothly pass from one foot to the other during the instantaneous double support phase. The problem becomes more complex when the step length is larger then 0.36 m because the ZMP must jump to the new supporting foot. In the following, the optimal motion for the step length 0.42 m and step time 1.2 s is analyzed. The GA results are shown in Table 6. The joint angle trajectories (θ_i), the torque vector (τ_i) and the optimal motion are shown in Fig. 17 and Fig. 18, for minimum CE and minimum TC, respectively. Based on the simulation results, we see that the biped robot posture is straighter, like the human walking, when the minimum CE is used as cost function. Comparing Figs 17(b) and 18(b), the torques change more smoothly when minimum TC is used as a cost function. The swing foot does not touch the ground prematurely, and the

ZMP is always inside the sole length, as presented in Fig. 19. At the end of the step, the ZMP is at the position ZMP_f, as shown in Fig. 13. At the beginning of the step, the ZMP is not exactly at the position ZMP_{jump} because the foot's mass is not neglected. It should be noted that the mass of the lower leg is different when it is in supporting leg or swing leg. The values of CE, calculated by the equation (2) for minimum CE and minimum TC gait synthesis, are presented in Fig. 20. The minimum CE gait synthesis reduces the energy by about 30% compared with minimum TC.

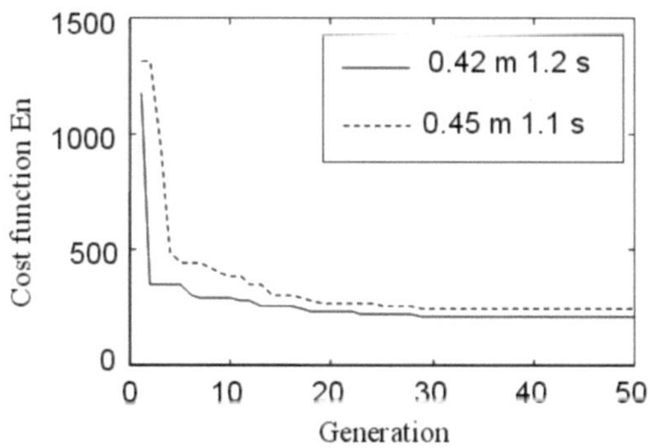

Figure 15. Cost function En vs generations

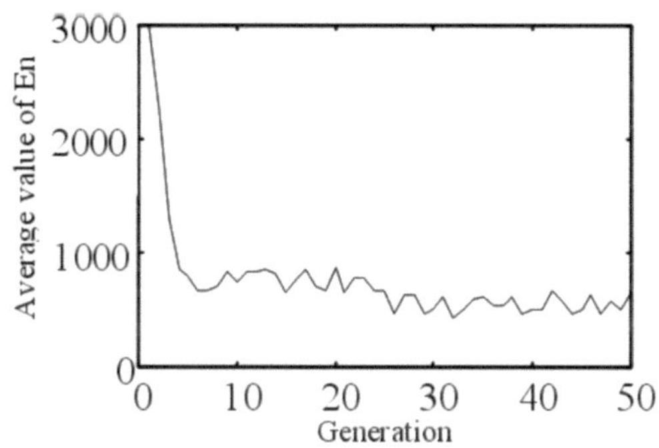

Figure 16. Average of the cost function En vs. generations

GA variables	**1.1.1. Limits**	**1.1.2. GA Results**	
		1.1.3. CE	TC
θ_{10}	-0.3~0	-0.1370	-0.0006
θ_{20}	-0.7~-0.3	-0.4374	-0.5199
θ_{30}	0 ~ 0.3	0.1309	0.1100
θ_{3p}	-0.1~0.2	0.0997	0.1001
t_3	0.3~0.7	0.5670	0.5900
x_{p1}	-0.2~0.2	-0.0861	-0.0585
y_{p1}	0.01~0.04	0.0142	0.0118
t_{p1}	0.1~0.9	0.4516	0.4471
$\dot{\theta}_{1f}$	0 ~ 2	0.6613	0.3935
$\dot{\theta}_{30}$	-1 ~ 1	0.0198	0.2514

Table 6. Variable space and GA results

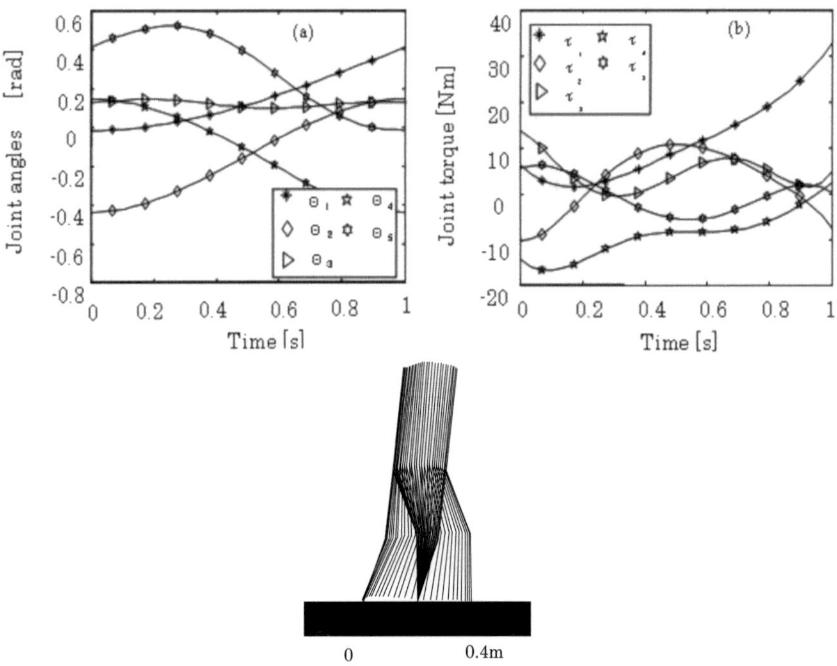

Figure 17. GA results and walking pattern of Joint angle (a) and Joint torque (b) for minimum CE cost function

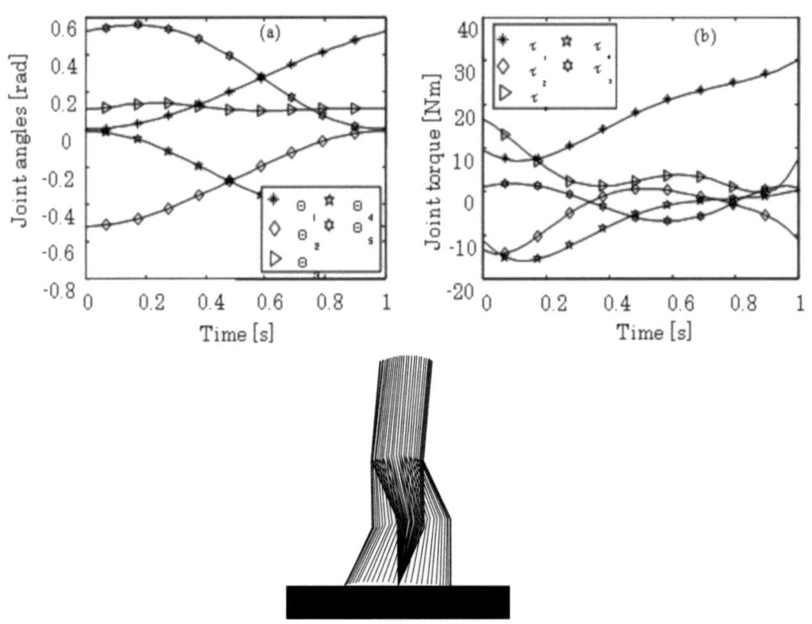

Figure 18. GA results and walking pattern of Joint angle (a) and Joint torque (b) for minimum TC cost function

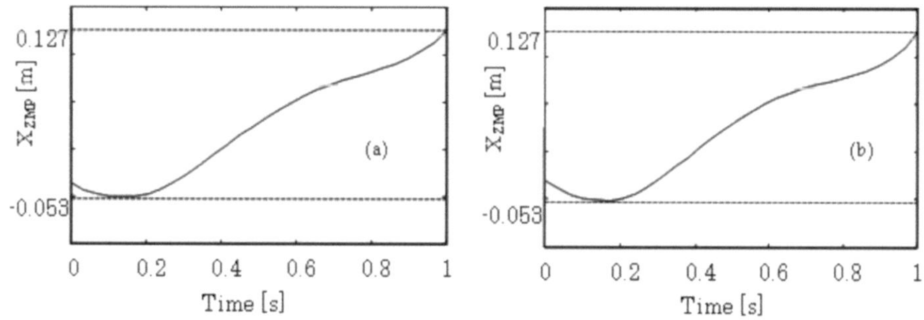

Figure 19. ZMP position for CE (a) and TC (b)

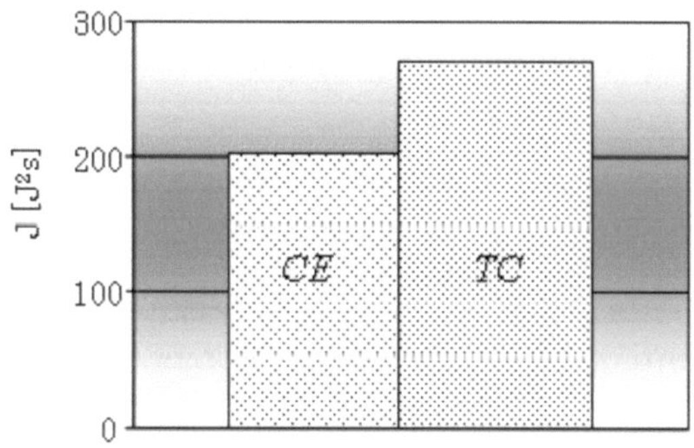

Figure 20. Energy comparison for normal step

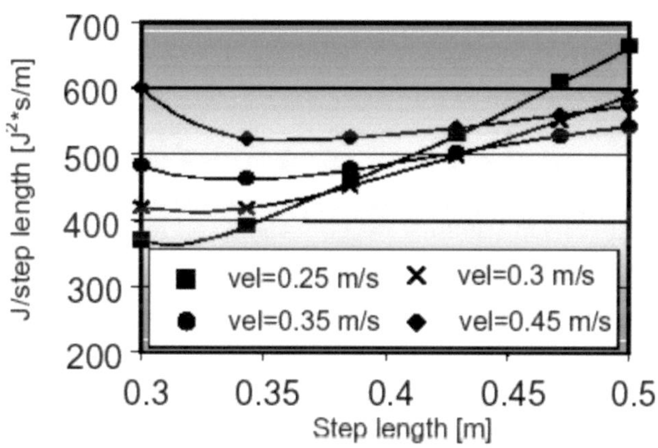

Figure 21. CE vs. the walking velocity

The energy required for 1 meter walking against the step length is presented in Fig. 21 for several walking velocities. In this case, the cost function is divided by step length. One result, which comes out from this figure, is that as the walking velocity gets higher, the optimal step length gets larger. The curves become more tended and don't intersect each other.

The energy required when the biped is moving slowly with a large step length is high. This makes that the curves of slow velocities to intersect the others. This suggests that low walking velocity doesn't mean low CE. In addition of walking velocity the step length must be also considered.

5.7 NN Implementation

In contrast to other optimization methods, GA needs more time to get the optimal solution. In our simulations, it needs about 10 minutes. However, in real time situations, based on the step length and step time, the angle trajectories must be generated in a very short time. In order to apply our method in real time, we considered teaching a NN based on the data generated by GA.

Our method employs the approximation abilities of a Radial Basis Function Neural Network (RBFNN) (Haykin, 1999). When the biped robot has to walk with a determined velocity and step length, the NN input variables would be the step length and step time. The output variables of the NN are the same as the variables generated by GA. From previous section, for a given step length it is a step time for which the CE is minimal. For this reason, when the walking velocity is not important the NN output will be the GA variables and the best step time. The NN input will be only the step length. In this chapter, we only show the NN simulation results, where the step length and step time are adapted as input variables.

RBFNN

The RBFNN involves three layers with entirely different roles, as shown in Fig. 22. The input layer connects the network to the environment. The second layer (hidden layer) applies a nonlinear transformation from the input space to the hidden space, which is of high dimensionality. We use as nonlinear transfer function the Gaussian function, because this is the most widely used function. The Gaussian function is expressed as follows:

$$h_i(x) = \exp\left(-\frac{\| x_i - c_i \|}{\sigma_i}\right), \tag{8}$$

where: h_i is the i-th output of the neuron, x_i is the input vector, c_i and σ_i are the center and the width of the i-th RBF neuron.

The width of Gaussian function is a positive constant that represents the standard deviation of the function. The output layer is linear, supplying the response of the network to the activation pattern (the signal applied to the input layer). Based on the number of nodes in the hidden layer the RBFNN are divided in generalized and regularization RBF networks. In our simulations we use a regularization RBF network.

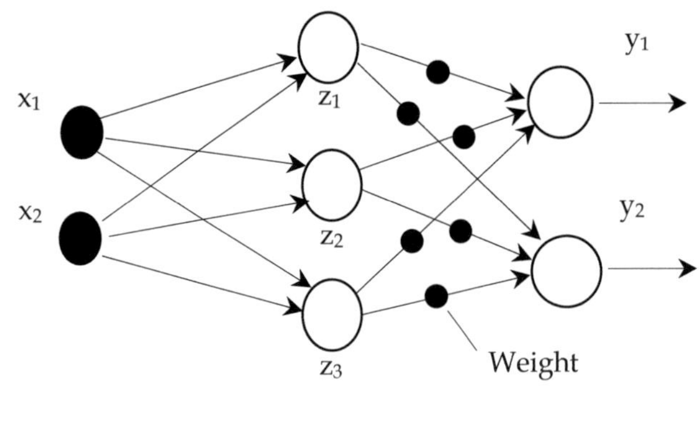

Figure 22. The structure of RBFNN

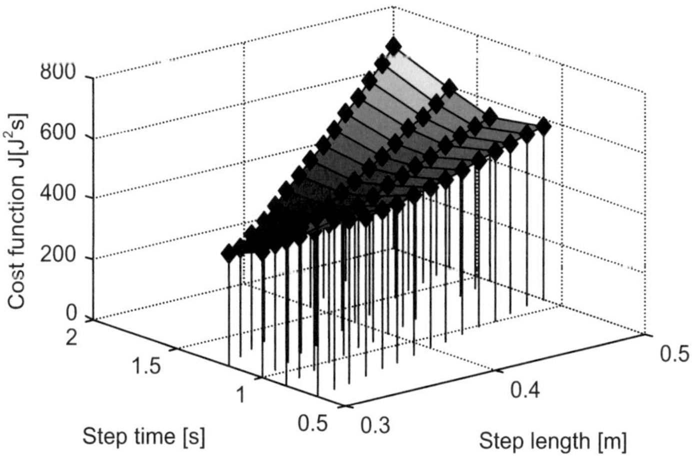

Figure 23. Cost function J vs step time and step length

5.8 RBFNN Results

The step length used to teach the NN varies from 0.3 to 0.5 and step time from 0.7s to 2s. Relation between cost Function J against the step length and step time is presented in Fig. 23. Because in our RBFNN the centers are the same with training data, determining the best value of the width σ is important in order to minimize the overall training error. The goal is to select the width in such way to minimize the overlapping of nearest neighbors, as well as maximize the generalization ability. The width selection depends on the distance between the two neighbor vectors. In our case the width σ is the same for all neurons, because the distances between the centers are the same. In Fig. 24 is shown the mean square error (mse) versus the width σ. The minimal value of mse is for $\sigma = 0.73$. We present the results generated by the NN for a set of step length and step time. It differs from training examples, for which the RBFNN output are the same with GA outputs. The input data of the NN have been [0.45 m 1.2 s]. The results generated by GA and NN are presented in Table 7. Based on these results, the angle trajectories are shown in Fig. 25. The difference between the NN and GA angle trajectories is very small. The time to generate the solution by the NN is 100ms, which is good for the real time implementation. The value of CE for NN gait is only 3.2 % more compared with GA one, as shown in Fig. 26.

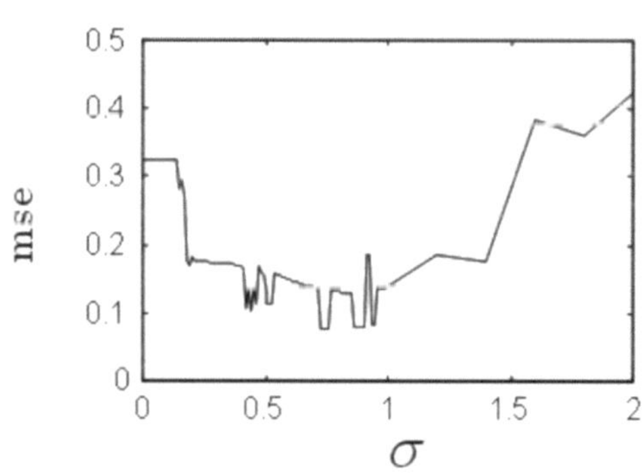

Figure 24. Mse vs the width σ

GA Var	Step length 0.42m, Step time 1.2s			St. length 0.45m, Step time 1.2s	
	Limits	CE	TC	GA	NN
θ_{10}	-.3~.0	-0.122	-0.0004	-0.124	-0.086
θ_{20}	-.7~-.3	-0.455	-0.57	-0.423	-0.447
θ_{30}	0~.3	0.107	0.370	0.258	0.264
$\dot{\theta}_{1f}$	0~2	0.523	0.399	0.395	0.259
$\dot{\theta}_{30}$	-1~1	-0.031	-0.11	-0.191	-0.167
θ_{3p}	-.1~.2	0.084	0.370	0.099	0.094
t_3	.2~.8	0.518	0.761	0.567	0.591
x_p	-.2~.2	-0.135	-0.132	-0.023	-0.006
y_p	.1~.4	0.016	0.017	0.018	0.014
t_p	0~1	0.441	0.432	0.613	0.632

Table 7. GA and NN results

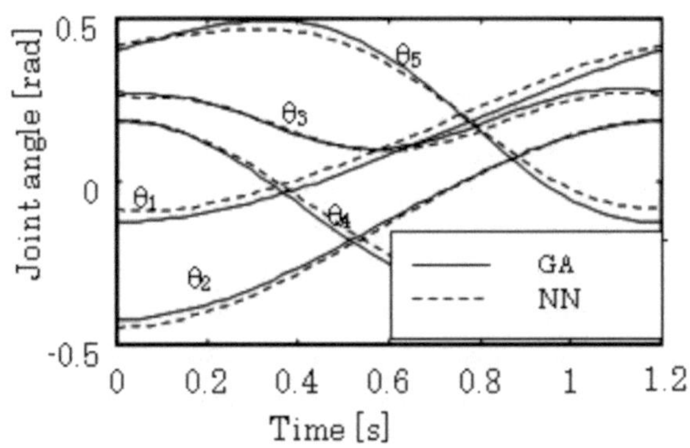

Figure 25. Comparison of GA and NN angle trajectories (step length 0.45m and step time 1.2s)

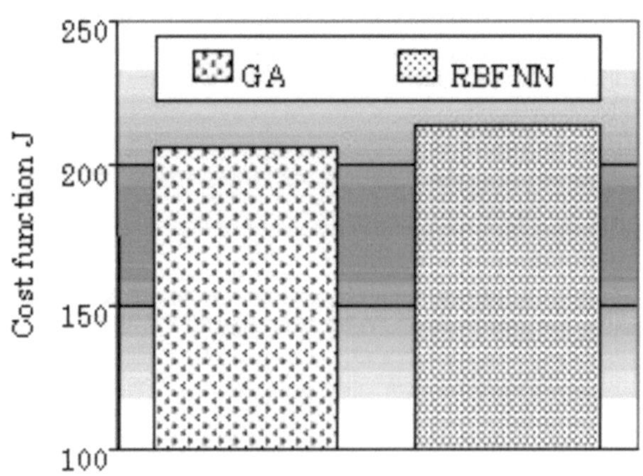

Figure 26. Comparison of values of J cost functions by GA and NN

6. Teleoperation Systems and its Applications

It is so effective to replace human being with humanoid robot for operation in disaster site and/or extreme environment (ex. Atomic power plants). These environments might change at every moment, thus, an operator must remotely control the robot. In order to remotely control the humanoid robot, several user interfaces have been developed. For example, remote control cockpit system is presented in (Hasunuma et al., 2002). In (Yokoi et al., 2003) a portable remote control device system with force feedback master arms was introduced. By using it, the operator can give his/her desired arm motion to the robot viscerally. But such device was very complex and expensive.
On the other hand, simple master device system (Neo et al., 2002) has two joysticks. Although the cost is reduced, because of a small number of degrees of freedom the system can not realize complex motions. In addition, it is hard to deal with environmental variations like sudden accidents. In order to overcome the shortcomings of previous systems, our objectives have been to: (1) develop a humanoid robot teleoperation system with a simple user interface; (2) able to reflect the operator's order; and (3) the cost for development and running to be low. Therefore, we first developed an on-line remote control of humanoid robot's arms. To carry out an easy operation, we developed an ultrasonic 3D mouse system as a master device for teleoperation system. In addition, we built a simple VR interface and a HMD equipped with a gyro sensor. In this chapter, we show the details of our teleoperation system and its user interface.
In this section, we present a new humanoid robot teleoperation system using the Internet/LAN and an easy user interface and a long distance teleoperation experiments.

6.1 Online Remote Control of Humanoid Robot using a Teleoperation System and User Interface

The schema of our teleoperation system is shown in Fig. 27. Our teleoperation system is a server-client system through the internet/LAN based on CORBA. There are two server PCs one to communicate and control the robot motion and the other for live streaming of CCD camera image. In addition, there are two client PCs for user interfaces, and for receiving live streaming vision images. While the operator sends the server his/her command including task commands and/or planned motion based on robot vision, the robot implements the order and returns results to the client, that is, current robot status (standing or crouching and so on) and robot vision. The communication between the operator and the humanoid robot is realized through TCP/IP with CORBA for motion operation and UDP for live streaming.

6.1.1 Operation Assist User Interface

Application of a joystick for giving commands to the robot based on images collected by robot vision system is a difficult task because of some troubles to manipulate the input device and robot camera at once. In addition, a joystick is not always suitable for quick 3D motion operation and to manipulate the input device and camera, separately. In order to overcome these difficulties, we decided to design the user interface as follow; 1) receive the operator's hand tip trajectory as order motion by a master device, 2) compose a VR interface with a HMD equipped with a gyro sensor to share the robot vision. The former needs to determine the space coordinates of operator's hand tip. Considering the environment, the operator's working area, the precision of measurement, and the cost of the system, we developed an ultrasonic 3D mouse system applying ultrasonic positioning system (Yu et al., 2001). The ultrasonic positioning system is applied to reduce the workload of manipulating the robot camera. Based on the visual information, the operator can synchronize the robot head motion.

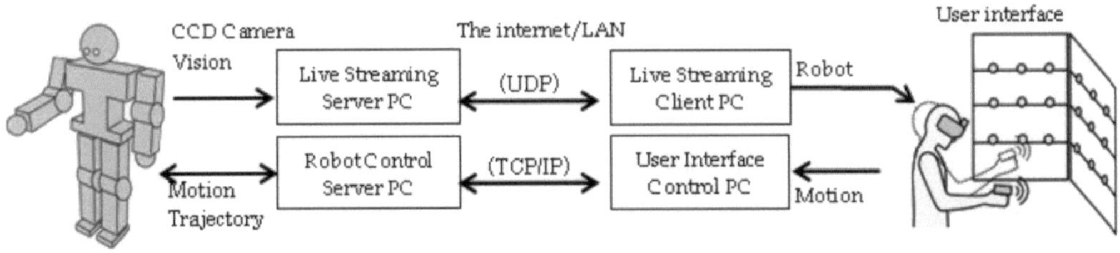

Figure 27. Schema of humanoid robot teleoperation system

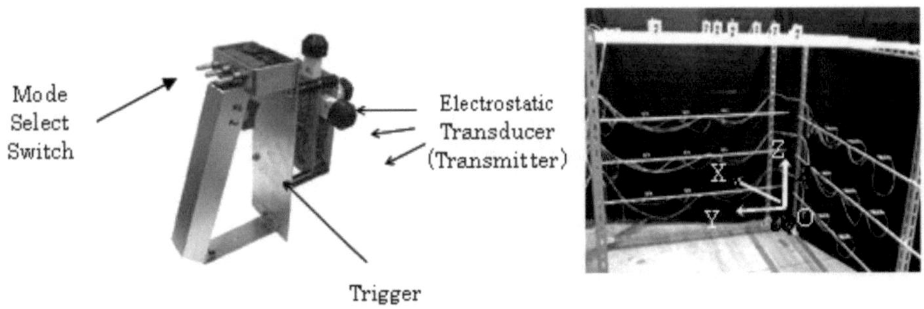

Figure 28. Ultrasonic 3D mouse and ultrasonic receiver net

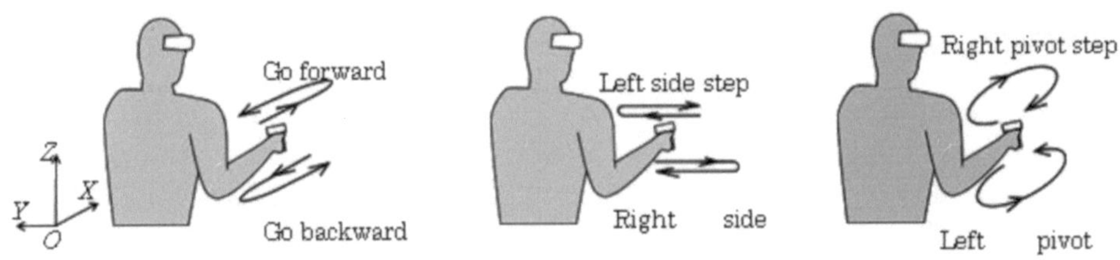

Figure 29. Gesture patterns corresponding to locomotion commands

6.1.2 Ultrasonic 3D Mouse System

This is a system to extract the operator's hand tip trajectory. The configuration is as follow; an ultrasonic 3D mouse; an ultrasonic receiver net cage and the system control PC (see Fig. 28). The 3D mouse has three electrostatic transducers (transmitter) and one trigger switch and three mode select switches. The receiver net cage has three planes that ultrasonic receivers are allocated by 300×300 mm regular interval on the frame of each plane. The origin of coordinate system is also shown in Fig. 28. The electrostatic transducers used in this study are MA40E7S/R made by Murata Manufacturing Co. Ltd, Japan. The specifications are shown in Table 8 . This system has two operating modes for manipulation of robot arms: the direct mode, which control the arm motion in real time, and the command mode to operate the locomotion by preset commands. The select switches on the 3D mouse are used to select the desired operating mode. Direct mode is used to operate one arm (right / left mode), or both arms (symmetrical / synchronized mode).

While the operator pulls the trigger, the system estimates 3D mouse position and extract the displacement vector of 3D mouse at every sampling. The vector is given to the robot as a reference motion data (reference motion vector). By using this system, the operator can generate in real time the robot's hand tip trajectory viscerally by dragging and dropping an icon on GUI desktop. In our system there is no need to consider the initial positioning between the 3D mouse and the robot hand tip at the start of operation, making easier to operate. On the other hand, the command mode is used to realize pre-designed motions like gesture input mode for walking motion. Here, gesture means an identification of the 3D mouse trajectory pattern. Preset commands for locomotion correspond with gesture patterns as shown in Fig. 29.

	Transmitter	Receiver
	MA40E7S	MA40E7R
Nominal Frequency	40 [kHz]	
Minimum Receiving Sensitivity	-74 [dB]	--
Minimum Transmitting Sensitivity	--	106 [dB]
Beam Angle	100 [deg]	
Capacitance	2200±20% [pF]	
Maximum Voltage	85 [V_{P-P}]	
Operating Conditions Temperature	-30 ~ 85 [°C]	
Measuring Range	0.2 ~ 3 [m]	
Resolving Power	9 [mm]	
Cover Size	18(D) x 12(H) [mm]	
Weight	4.5 [g]	
Character	Waterproof	

Table 8. Technical specifications for MA40E7R/S electrostatic transducer

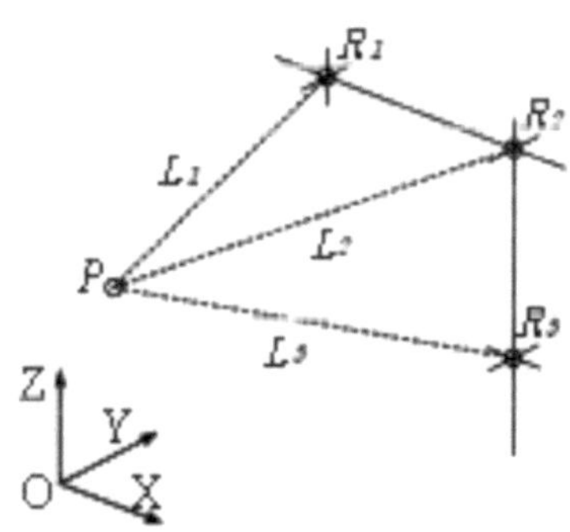

Figure 30. Position estimation

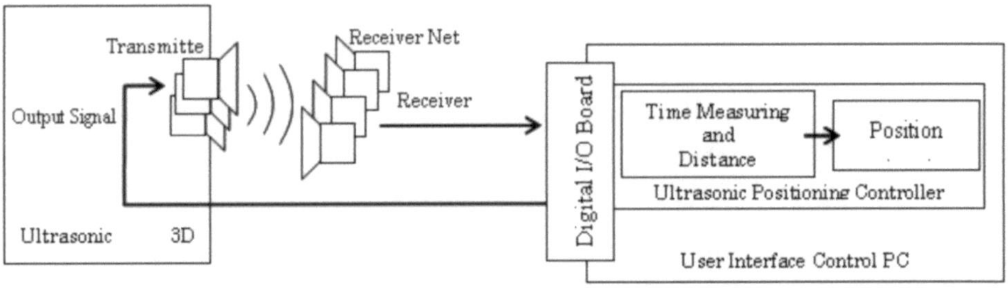

Figure 31. Diagram for position estimation

6.1.3 Ultrasonic Positioning Estimation

In our system, we know the speed of sonic wave at the air temperature and the propagation distance by measuring the wave propagation time. At least by knowing three distances between the 3D mouse and receivers, we can estimate the position of the 3D mouse in the sensor net by principle of triangulation. When the wave propagation time T_i [s] (i =1,2,3) is measured, the propagation distance L_i [m](i =1,2,3) is estimated as follow:

$$L_i = (331.5+0.6t)\, T_i \tag{9}$$

Here, t is room temperature [°C]. Assuming that receiver positions R_i (x_i, y_i, z_i), (i =1,2,3) are known and exist on same plane (not on a straight line) in an arbitrary Cartesian coordinate as shown in Fig. 30, the position of the 3D mouse P (x, y, z) is estimated by the following formulations:

$$\left.\begin{array}{l}(x_1 - x)^2 + (y_1 - y)^2 + (z_1 - z)^2 = L_1^2 \\ (x_3 - x)^2 + (y_3 - y)^2 + (z_3 - z)^2 = L_3^2 \\ (x_2 - x)^2 + (y_2 - y)^2 + (z_2 - z)^2 = L_2^2\end{array}\right\} \tag{10}$$

Figure 31 shows the diagram for the position estimation. When the ultrasonic positioning controller sends output signal to transmitter on the 3D mouse, it begins measuring the wave propagation time. In addition, a receiver detects ultrasonic waves and returns a receiver signal to the controller making possible to determine the time between the 3D mouse and the receiver. After sampling for 4 ms, the controller will calculate the distance between the 3D mouse and receivers, and estimate the position of 3D mouse.

The control PC used in this study is a DOS/V compatible PC with a Pentium III CPU (733MHz) and Linux OS (Red Hat 9). The time measuring process is executed by using CPU internal clock counter on the control PC. The precision for time measurement (the CPU frequency) depends on its operative conditions (power supply voltage, internal temperature and so on). But the sampling performance is about 250 kHz on the average, and the theoretical resolution for distance measurement is about 1.3mm at room temperature 20 °C. The total processing time is set 10 ms.

6.1.4 Live Streaming System

A live streaming system is applied to transmit robot camera vision to the operator. The robot camera vision is captured and it is encoded in real time to mpeg4 format data (QVGA (320x240 pixels)) on the live streaming server PC. Then it is transmitted to the client PC by UDP (Fig. 32). For the server and client application, we applied multicast application

"FocusShare", which is distributed at OpenNIME web site. The server PC used in this system is the DOS/V compatible PC with a Pentium IV CPU (2.53GHz) and Windows OS (Windows XP SP2). The live streaming data is decoded on the client PC (Notebook PC with Pentium M (900MHz) and Windows 2000 SP4), and projected on HMD. HMD used is i-Visor DH-4400VP made by Personal Display Systems, Inc., USA, and it has two 0.49inch, 1.44 million pixels LCD, and supports SVGA graphic mode. The gyro sensor used is InterTrax[2] is made by InterSense Inc. of USA, which can track roll, pitch, yaw direction angles (except for angular speed and acceleration), and its minimum resolution is 0.02deg.

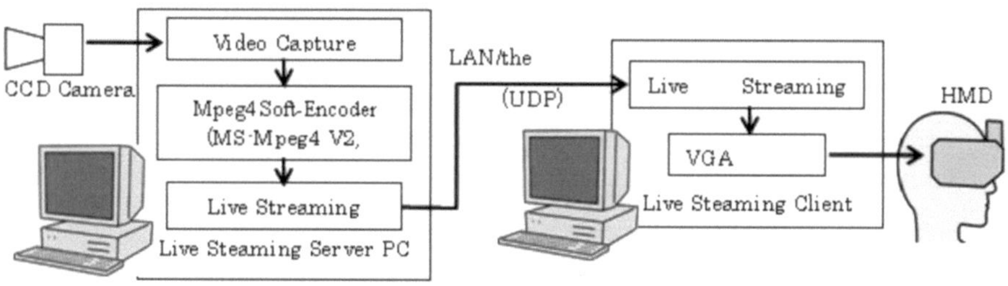

Figure 32. Live streaming system

6.1.5 Motion Trajectory Generation

For the motion trajectory generation we first added a reference motion vector given by the 3D mouse to current robot hand tip position. Therefore, the reference robot hand tip position is set. By linear interpolating the position and current robot hand, the reference hand tip trajectory is pre-released based on a given reference motion time (here, 10ms). At this moment, the trajectory is checked about collision and workspace of hand tip.
If there is any error, a new reference hand tip position will be set again, and a new reference hand tip trajectory will be released. Finally, it will be converted to reference arm joint angle trajectory by inverse kinematics. In Direct Mode, the reference motion vector is essentially handled as data for the right arm. Both reference hand tip positions are determined by adding same reference motion vector to each current robot hand. But in symmetrical mode, left reference hand tip position is determined by adding a reference motion vector that its Y direction element is reversed.

6.1.6 Experiments and Results

In order to evaluate the performance of the developed system, we completed experiments with Bonten-Maru II humanoid robot. In the following, we give the results of these experiments. First we discuss the results of right arm motion using the teleoperation system in a LAN environment. In this experiment the operator drew a simple quadrilateral hand tip trajectory on Y-Z plane in the ultrasonic receiver net with the 3D mouse. Fig. 33 (a) and (b) show an order trajectory given by 3D mouse and a motion trajectory of right robot hand tip. Note that in this experiment, the room temperature was 24 °C, and Fig. 33 (b) is viewed from the origin of right arm coordinate system located in the right shoulder. Although there is a difference in scaling that it is caused by feedback errors, each motion pattern matches well. And also, in Fig. 34 is shown the operation time in every communication. The horizontal axis is the number of communication times. There are some data spreads due to network traffics, but the operator could carry out the experiment in real time without serious time delay error.

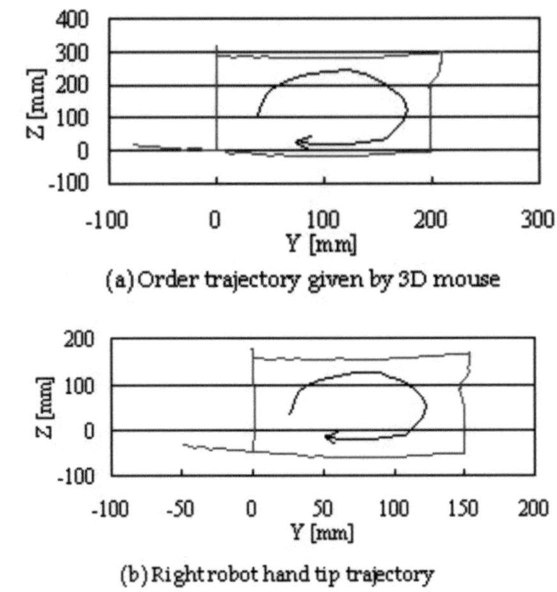

Figure 33. Results of teleoperation experiment

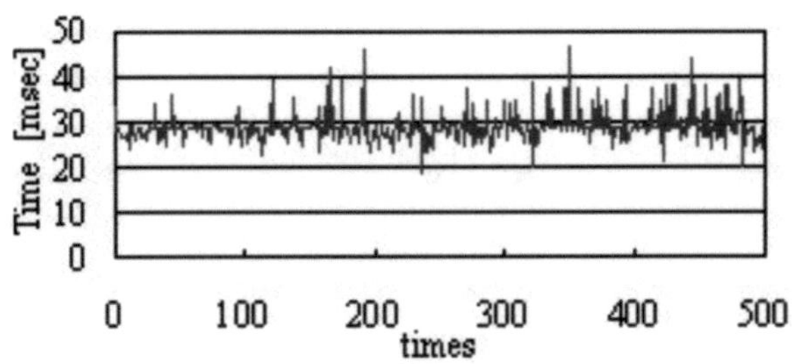

Figure 34. Operation time

In order to further verify the system performance, we performed an experiment to evaluate the ability to replicate the hand tip motion generated by the operator in *Y-Z* plane. In this experiment, the operator draws a quadrilateral hand tip trajectory on *Y-Z* plane. The operator cannot look his/her own hand because of the HMD. A stroboscopic photograph of the robot motion during the experiment is shown in Fig. 35. Fig. 36 (a) and (b) show an experimental measured operator's hand tip trajectory in the coordinate of receiver net and the right robot hand tip position viewed from the origin of right arm coordinates. Also in the Fig.11, the direction indicated by arrow shows the direction of motion. Each dot indicates the measured positions during the operation. The interval of each dot means one-operation cycle, which is about 1.5sec, including the sensing time in the receiver net, the robot motion time and the time-delay by the network traffics. The difference between Fig. 36 (a) and (b) originates in the decreasing reference data scale to 70%. In addition, this difference is exist because the robot hand tip trajectory is sometimes restricted due to the limitation of the workspace, the range of joint angles and change in trajectory to avoid the collision with the body. But both trajectory patterns are similar.

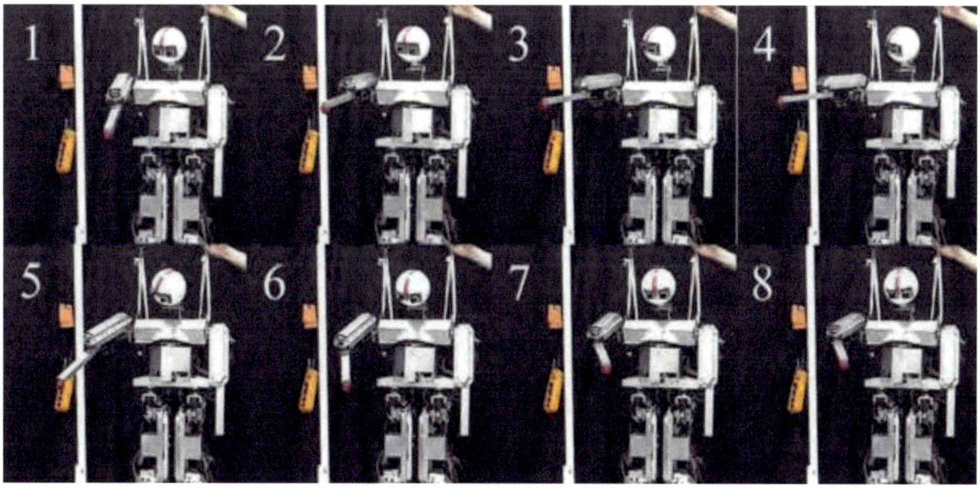

Figure 35. The robot motion during the experiment

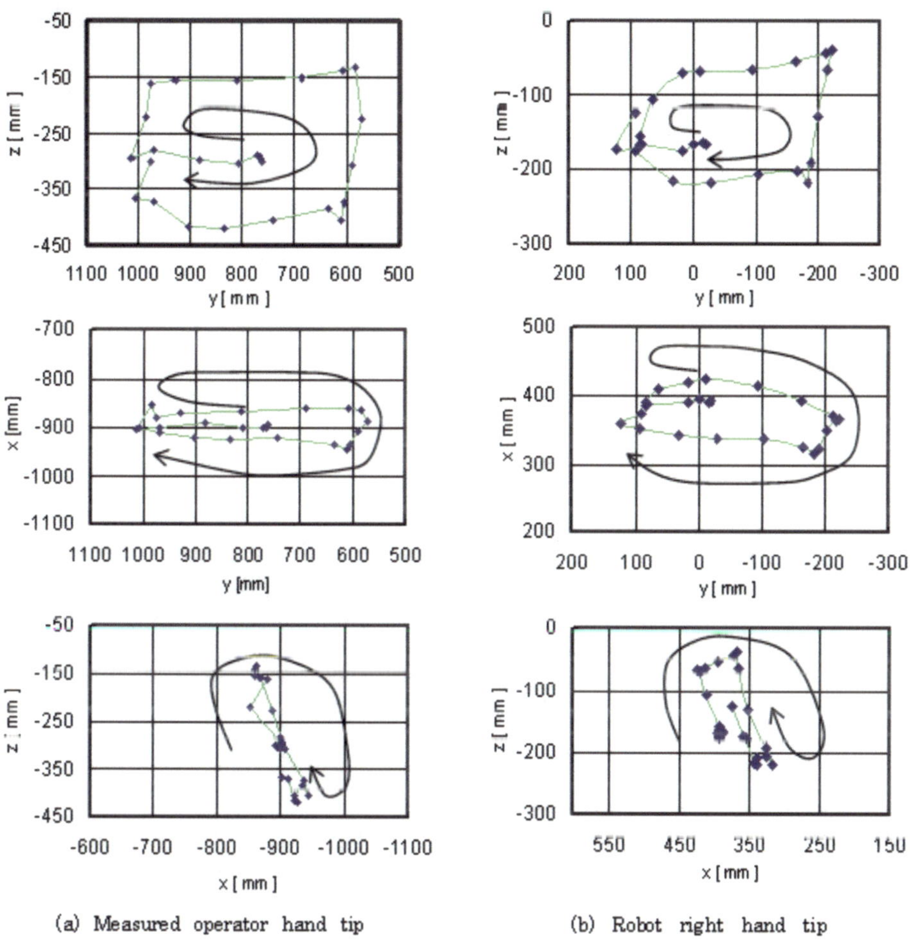

Figure 36. Results of the experiment

As previously mentioned, the operator cannot check on his/her own hand tip position. These mean that, the operator could correct his/her own hand tip position using the HMD vision and generate his/her planned motion. In other words, our user interface can function as a VR interface to share data with the robot. As the matter of fact, the communicating interval between the CORBA client and the CORBA server must be considered in order to minimize as much as possible.

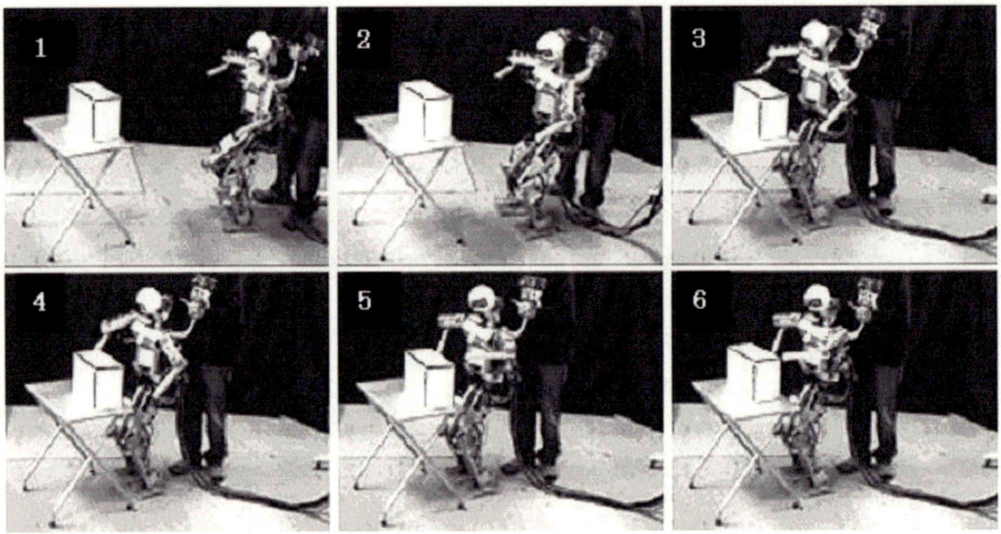

Figure 37. Video capture of teleoperation experiment

Next, we performed experiments using all the system. In this experiment, the operator gives locomotion commands by gesture input, in order to move the robot to a target box. Then the robot receives the command to touch the box. In Fig. 37 is shown a video capture of the robot. This experiment indicates that by using the developed teleoperation system we are able to communicate with the humanoid robot and realize complex motions. Fig. 38 shows a teleoperation demonstration to draw simple characters using the 3D mouse. The operator could draw simple characters easily.

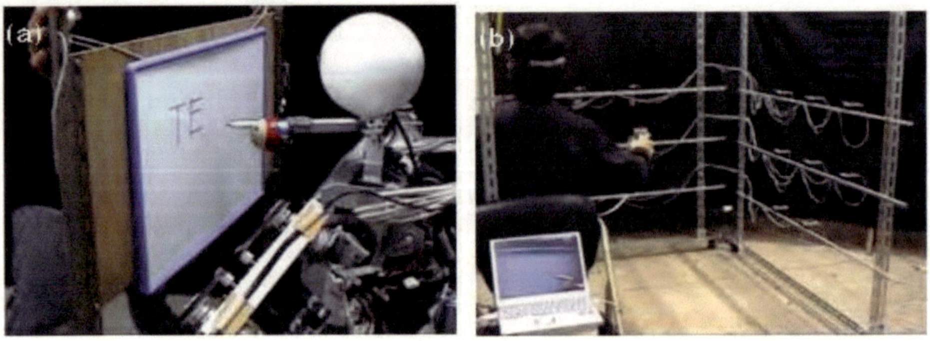

(a) Drawing simple characters (b) Operator with the 3D mouse

Figure 38. Demonstration test of the 3D mouse

6.2 Long Distance Teleoperation via the Internet

In this section, we explain a teleoperation system to control the humanoid robot through the internet. We carried out experiments on the teleoperation of the humanoid robot between Deakin University (Australia) and Yamagata University (Japan) (Nasu et al., 2003). The experimental results verified the good performance of the proposed system and control.

6.2.1 Teloperation system

Figure 39 shows the teleoperation schematic diagram. The operator uses this system as a CORBA client and commands several kinds of motions, i.e. walking, crouching, crawling, standing up, etc. Figure 40 shows the HRCA for Bonten-Maru II humanoid robot. We have implemented the following main modules: DTCM, MCM, JTM, GSM, JAM, FCM, CCM VCM and UIM in this figure. Each module corresponds to "Data Transmission", "Target Position", "Angle Trajectory Calculation", "Sensor", "Position", "Feedback Control", "CCD Camera", "Video Capture Control" and "Command Generator", respectively. Up to now, the operator can command the number of steps and humanoid robot walking direction.

The operator receives the camera image mounted in humanoid robot's head and based on the data displayed in PC1, measures the distance between the robot and objects. PC2 is used to read and manipulate the sensor data and send output commands to the actuators. PC3 is used to capture the CCD camera image. A notebook type computer with a Pentium III, 700 MHz processor running Red Hat Cygwin on the Windows XP was used as the client computer (PC1). Two different type computers were used as server computers: PC2 (Celeron, 433MHz), PC3 (Pentium II, 200 MHz) running Red Hat Linux 7.3.

6.2.2 Data Stream

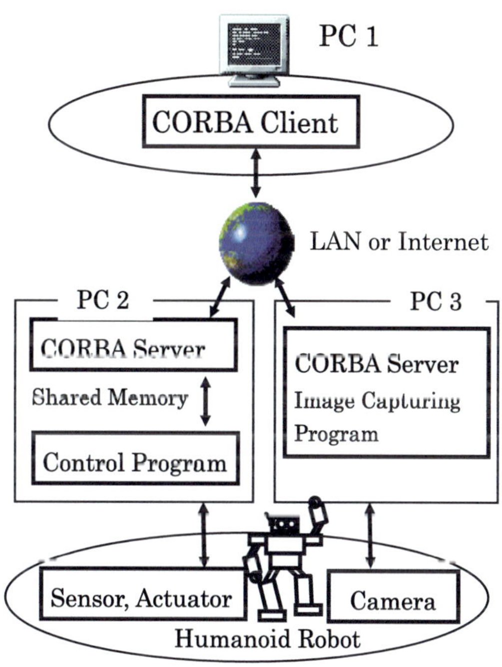

Figure 39. Teleoperation concept

CORBA server program receives a motion command from CORBA client and writes it on the shared memory of PC2. Sending and receiving the data between CORBA server program and control program are executed by using shared memory feature of UNIX OS. Among all programs on the LINUX, the control program OS implemented in accordance to highest-priority due to keep the control execution period. CORBA server program is implemented at default value. When the operator watches the camera image, PC1 and PC2 are used. When the operator executes CORBA client program of PC1, the image data, which is captured in PC3, is imported to PC1. The operator can use it to measure the object distance, to recognize the environment condition and make decision of the optimal motion.

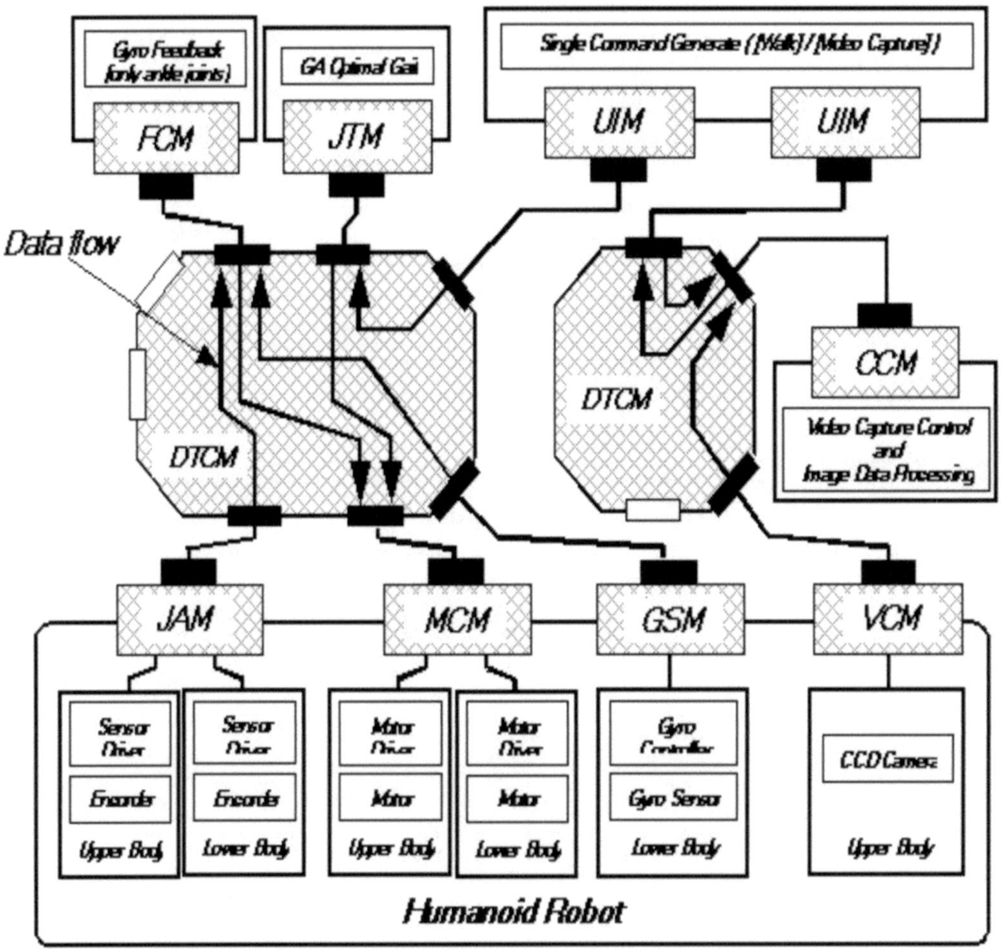

Figure 40. The HRCA for Bonten-Maru II humanoid robot

6.2.3 Experiments and Results

First, we measured the image capturing job time through the internet. The typical job time averaged about 13 second to a few minutes, because there are many communication traffic loads in the both universities LANs.

Second, using the humanoid robot, we have carried out two types of teleoperation obstacle avoidance experiments between Australia and Japan. The operator executed teleoperation program from Deakin University (Australia) through the internet.

Experiment 1: Obstacle avoidance by walk

At first, we set a box on the floor in front of humanoid robot. The operator recognized it in the image data from the humanoid robot. Fig. 41 shows a series of the obstacle avoidance walking motions and image data of the humanoid robot eyes. The humanoid robot received the following motion commands:

- Walk front (or back)
- Side step to left (or right)
- Spin left (or right)

The operator measures the distance between the robot and the obstacle, and plans a walk trajectory to avoid the obstacle. Because the measured obstacle data is not precious, the motion command is not always the best. But the operator can correct the walking trajectory by using the image information easily.

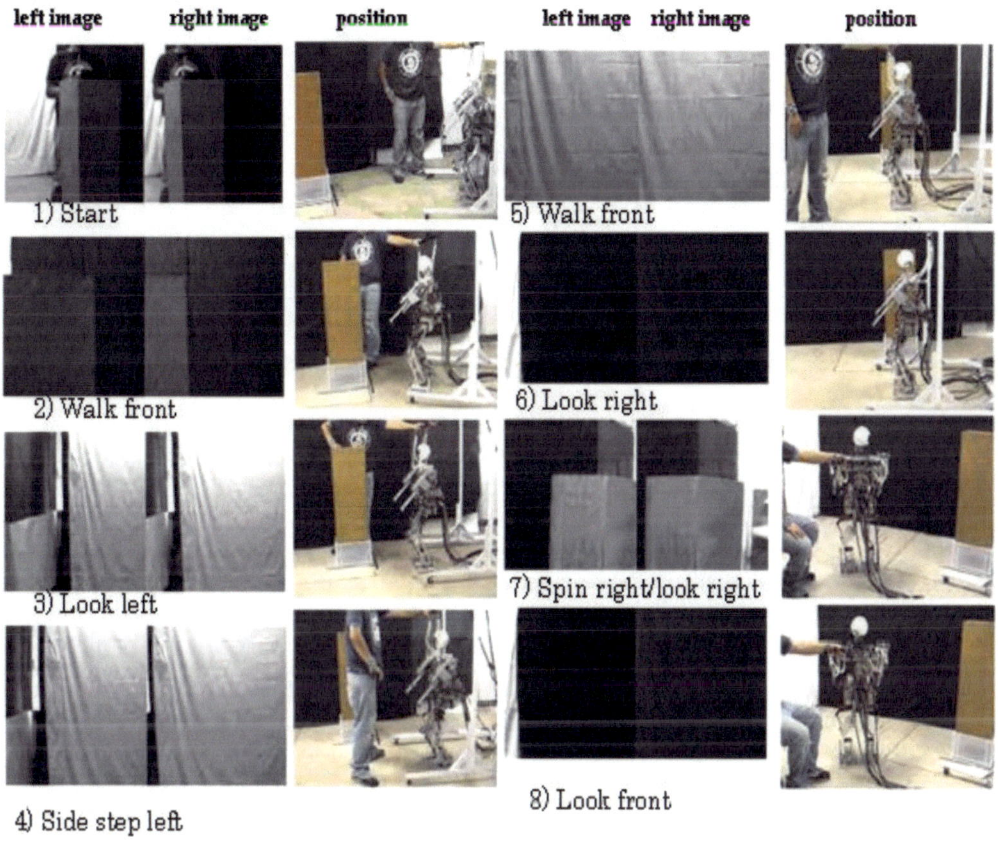

Figure 41. Walking and obstacle avoidance by teleoperation through the internet

Experiment 2: Sneaking under a low ceiling gate

At second, we set a low ceiling gate in front of the humanoid robot. The operator recognized it in the captured images data from the humanoid robot and judged that humanoid robot could not go through the gate having the body in upright position. Fig. 42 shows a series of the sneaking under a low ceiling gate (obstacle). The client commanded the following motion; 1) look front, 2) squat, 3) crawl start, 4)-8) crawl, 9) stand up, and 10) look front. The humanoid robot could go through the gate successfully.

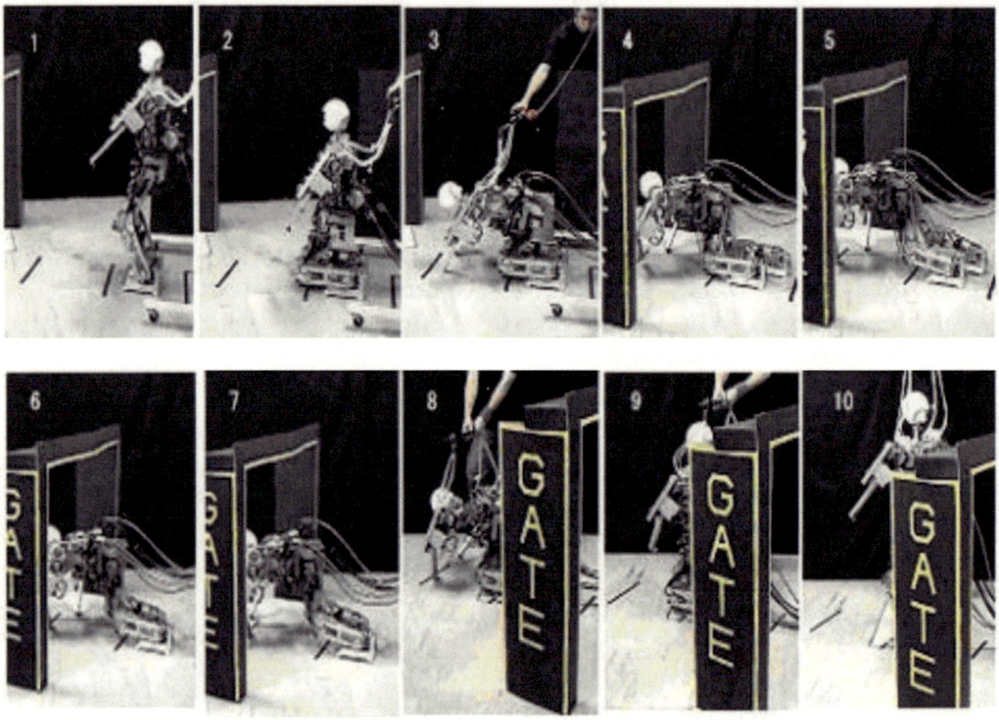

Figure 42. Sneaking and crawling under a low ceiling gate to avoid obstacle

7. Summary and Conclusions

We have developed anthropomorphic prototype humanoid robot; Bonten-Maru I and Bonten-Maru II. The Bonten-Maru humanoid robot series are one of few research prototype humanoid robots in the world which can be utilized in various aspects of studies. In this research, we utilized the Bonten-Maru in development of the CORBA-based humanoid robot control architecture, the optimal gait strategy and the teleoperation via internet.

7.1 CORBA-Based Humanoid Robot Control Architecture (HRCA)

In this section, we proposed a new robot control architecture called HRCA. The HRCA is developed as a CORBA client/server system and is implemented on the Bonten-Maru I humanoid robot. The HRCA allows easy addition, deletion, and upgrading of new modules. We have carried out simulations and experiments to evaluate the performance of the proposed HRCA. The experimental result shows that the proposed HRCA is able to control the static motion of humanoid robot accurately. By using the proposed HRCA various humanoid robots in the world can share their own modules each other via Internet.

7.2 Optimal Gait Strategy

This section presents the real time generation of humanoid robot optimal gait by using soft computing techniques. GA was employed to minimize the energy for humanoid robot gait. For a real time gait generation, we used the RBFNN, which are trained based on GA data. The performance evaluation is carried out by simulation, using the parameters of Bonten-Maru I humanoid robot. Based on the simulation results, we conclude:

- Each step length is optimal at a particular velocity;
- The stability is important to be considered when generating the optimal gait;
- The biped robot posture is straighter when minimum CE is used as the cost function, which is similar to the humans;
- The energy for CE is reduced 30% compared with TC cost function.

7.3 Teleoperation System and its Application

In this section, we described humanoid robot control architecture HRCA for teleoperation. The HRCA is developed as a CORBA client/server system and implemented on the new humanoid robot, which was designed to mimic as much as possible the human motion. Therefore, the humanoid robot can get several configurations, because each joint has a wide range rotation angle. A long distance teleoperation experiments between Japan and Australia were carried out through the internet. By using the image data from the humanoid robot, the operator judged and planned a series of necessary motion trajectories for obstacle avoidance.

This section also presented the teleoperation system for a humanoid robot and the operation assistance user interface. We developed an ultrasonic 3D mouse system for the user interface. In order to evaluate the system performance, we performed some teleoperation experiments the Bonten-Maru II humanoid robot. The results show that our system gives good results for control of humanoid robot in real time. However, there are still some problems which need to be considered in the future such as:

- The communication of live streaming system beyond network rooters.
- 3D mouse operation of robot hand postures.

Up to now we have applied the developed teleoperation system and the user interface on humanoid robot motion generation in simple environments. However, in complex environments the humanoid robot must generate skillful motions in a short time based on the visual information and operator's desired motion

The experimental results conducted with Bonten-Maru humanoid robot show a good performance of the system, whereby the humanoid robot replicates in real time the operators desired arm motion with high accuracy. The experimental results also verified the good performance of the proposed system and control.

8. Future Works

Recently, we focus in the development of contact interaction-based humanoid robot navigation (Hanafiah et al., 2006). Eventually, it is inevitable that the application of humanoid robots in the same workspace as humans will result in direct physical-contact interaction. We have proposed intelligent algorithm called groping locomotion (Hanafiah et al., 2005) to navigate humanoid robot locomotion by grasping using its arm and also avoiding obstacle. This method is useful during operation in dark area and also hazardous site. In addition, for the humanoid robot to work along human effectively, especially for object handling tasks, the robot will require additional sensory abilities. Besides sensor systems that help the robot to structure their environment, like cameras, radar sensors, etc., a system on the robot's surface is needed that enables to detect physical contact with its environment. A tactile sensor system is essential as a sensory device to support the robot control system. This tactile sensor is capable of sensing normal force, shearing force, and slippage, thus offering exciting possibilities for application in the field of robotics for determining object shape, texture, hardness, etc. In current research, we are developing tactile sensor that capable to define normal and shearing force, with the aim to install it on the humanoid robot arm (Ohka et al., 2006). This sensor is based on the principle of an optical waveguide-type tactile sensor. The tactile sensor system is combined with 3-DOF robot finger system where the tactile sensor in mounted on the fingertip. We believe that the demand for tactile sensing devices will grow in parallel with rapid progress in robotics research and development.

9. Acknowledgement

A part of this research was supported by fiscal 2006 grants from the Japan Ministry of Education, Culture, Sports, Science and Technology (Grant-in-Aid for Scientific Research in Exploratory Research, No. 18656079). The authors would like to thank all Nasu Lab. members, Ohka Lab. members and all individual involved in this research for their contribution, work and effort towards successful of this project.

10. References

Booch, G.; Rumbaugh, J. & Jacobson, I. (1999). *The Unified Modeling Language User Guide*, Addison-Wesley

Capi, G.; Nasu, Y.; Mitobe, K. & Barolli, L. (2003). Real time gait generation for autonomous humanoid robots: A case study for walking, *Journal Robotics and Autonomous Systems*, Vol. 42, No.2, (2003), pp. 169-178

Channon, P.H.; Pham, D.T. & Hopkins, S.H. (1996). A variational approach to the optimization of gait for a bipedal robot, *Journal of Mechanical Engineering Science*, Vol. 210, (1996), pp. 177-186

Fowler, M. & Scott, K. (1997). *UML Distilled: Applying the Standard Object Modeling Language*, Addison-Wesley

Hanafiah, Y.; Yamano, M.; Nasu, Y. & Ohka, M. (2005). Obstacle avoidance in groping locomotion of a humanoid robot, *Journal of Advanced Robotic Systems*, Vol.2 No. 3, (September 2005) pp. 251-258, ISSN 1729-8806

Hanafiah, Y.; Ohka, M.; Kobayashi, H.; Takata, J.; Yamano, M. & Nasu, Y. (2006). Contribution to the development of contact interaction-based humanoid robot navigation system: Application of an optical three-axis tactile sensor, *Proceeding of 3rd International Conference on Autonomous Robots and Agents (ICARA2006)*, pp. 63-68, ISBN-10: 0-473-11566-2, ISBN-13: 978-0-473-11566-1, Palmerston North, Dec. 2006, Massey Univ. Palmerston North, New Zealand

Harrison, T. H.; Levine, D. L. & Schmidt, D. C. (1997). The design and performance of a real-time CORBA event service, *Proceeding of the OOPSLA'97 Conference*, 1997

Hasunuma, H. (2002). A tele-operated humanoid robot drives a lift truck, *Proceeding of 2002 IEEE Int. Conf. on Robotics and Automation*, pp. 2246–2252, 2002

Haykin, S. (1999). *Neural Networks a Comprehensive Foundation*, Toronto, Prentice Hall International

Hirai, K.; Hirose, M.; Haikawa, Y. & Takenaka, T. (1998). The development of Honda humanoid robot, *Proceeding of IEEE Int. Conf. on Robotics & Automation*, pp. 1321-1326, Leuven, Belgium, 1998

Inaba, M.; Igarashi, T.; Kagami, S. & Inoue, H. (1998). Design and implementation of a 35 d.o.f full-Body humanoid robot that can sit, stand up, and grasp an object, *Journal Advanced Robotics*, Vol. 12, No.1, pp. 1-14

Kaneko, S.; Nasu, Y.; Yamano, M.; Mitobe, K. & Capi, G. (2005). Online remote control of humanoid robot using a teleoperation system and user interface, *WSEAS Transaction on Systems*, Issue 5, Vol. 4, May 2005, pp.561-568, ISSN 1109-2777

Michalewich, Z. (1994). *Genetic Algorithms + Data Structures = Evaluation Programs*, Springer-Verlag

Mita, T.; Yamaguchi, T.; Kashiwase, T. & Kawase, T. (1984). Realization of high speed biped using modern control theory, *Int. Journal Control*, Vol. 40, (1984), pp. 107 119

Mowbray, T. J. & Ruh, W. A. (1997). *Inside CORBA: Distributed Object Standards and Applications*, Addison-Wesley, 1997

Nasu, Y.; Kaneko, S.; Yamano, M.; Capi, G. & Nahavandi, S. (2003). Application of a CORBA-based humanoid robot system for accident site inspection through the internet, Proceeding of *7th WSEAS International Conference on Systems*, CD-ROM Proceedings, 6 pages, Corfu Island, Greece, July 7-10, 2003, *Computational Methods in Circuits and Systems Applications*, WSEAS Press, pp.177-184

Neo, E. S.; Yokoi, K.; Kajita, S.; Kanehiro, F. & Tanie, K. (2002). Whole body teleoperation of a humanoid robot -Development of a simple master device using joysticks-, *Proceeding of Int. Conf. on Intelligent Robotics and Systems (IROS)*, 2002

Ohka, M.; Kobayashi, H and Mitsuya, Y. (2006). Sensing precision of an optical three-axis tactile sensor for a robotic finger", *Proceeding of 15th RO-MAN2006*, pp 220-225, ISBN 1-4244-0565-3, Hatfield, U.K, 2006

Open Management Group, "UML Resource Page", http://www.omg.org/uml/

Open Management Group, Welcome to the OMG's CORBA Website, http://www.corba.org/

Open Management Group, The Object Management Group, http://www.omg.org/

Pancerella, C. M. & Whiteside, R. A. (1996). Using CORBA to integrate manufacturing cells to a virtual enterprise, *Proceeding of Plag and Play Software for Agile Manufacturing*, November 1996

Roussel, L.; Canudas-de-Wit, C. & Goswami, A. (1998). Generation of energy optimal complete gait cycles for biped robots, *Proceeding of IEEE Int. Conf. on Robotics and Automation*, 1998, pp. 2036-2041

Silva, F. M. & Machado, J. A. T. (1999). Energy analysis during biped walking, *Proceeding of IEEE Int. Conf. On Robotics and Automation*, pp. 59-64, 1999

Takanishi, A.; Ishida, M.; Yamazaki, Y. & Kato, I. (1990). A control method for dynamic biped walking under unknown external force, *Proceeding of IEEE Int. Workshop on Intelligent Robots and Systems*, pp.795-801, 1990

Takeda, K.; Nasu, Y.; Capi, G.; Yamano, M.; Barolli, L. & Mitobe, K. (2001). A CORBA-based approach for humanoid robot control, *Industrial Robot: An International Journal*, Vol. 28, No. 3, pp. 242-250

Uno, Y.; Kawato, M. & Suzuki, R. (1989). Formulation and control of optimal trajectory in human multijoint arm movement, *Journal Biol. Cybernet*, Vol. 61, (1989), pp. 89-101

Vinoski, S. (1997). CORBA: Integrating diverse applications within distributed heterogeneous environments, *IEEE Communications Magazine*, Vol.14, No.2, pp. 1-12, February 1997

Vukobratovic, M.; Borovac, B.; Surla, D. & Stokic, D. (1990). *Biped Locomotion, Dynamics, Stability, Control and Application*. Berlin, Springer-Verlag

Whiteside, R. A.; Pancerella, C. M. & Klevgard, P. A. (1997). A CORBA-based manufacturing environment, *Proc. of the Hawaii International Conference on Systems Sciences*, Jan. 1997

XEROX PARC, http://www.parc.xerox.com/parc-go.html

XEROX PARC, "Inter-Language unification", ftp://ftp.parc.xerox.com/pub/ilu/

Yokoi, K.; Nakashima, K.; Kobayashi, M.; Mihune, H.; Hasunuma, H.; Yanagihara, Y.; Ueno, T.; Gokyuu, T. & Endou, K. (2003). A tele-operated humanoid robot drives a backhoe in the open air, *Proceedings of the 2003 IEEE/RSJ Intl. Conference on Intelligent Robots and Systems*, 2003

Yu, Z.; Nasu, Y.; Nakajima, S. & Mitobe, K. (2001). Development of position measurement system in wide-area using ultrasonic receiver Net, *Journal of Japanese Society of Precision Engineering*, vol.67, no.5, 2001,pp. 764-769, (in Japanese)

7

Stability Analysis of a Simple Active Biped Robot with a Torso on Level Ground Based on Passive Walking Mechanisms

Terumasa Narukawa, Masaki Takahashi and Kazuo Yoshida
Keio University
Japan

1. Introduction

This study focuses on the passive dynamic walking to enable a biped robot on level ground to walk efficiently with simple mechanisms. To build an efficient bipedal robot, utilizing the dynamical property of the robot system is a useful approach. McGeer studied passive-dynamic walking, and showed a biped robot without actuators and controllers can walk stably down a shallow slope in simulations and experiments (McGeer, 1990). The simplest passive walker, which has only two mass-less links with hip mass, still can walk (Garcia et al., 1998). Collins et al. built the three-dimensional passive-dynamic walker which has knees and arms (Collins et al., 2001).

Passive-dynamic walking is useful to study efficient level-ground walking robots, (e.g. Collins et al. 2005), but passive walking has some limitations. The walking motion of the passive walker depends on the slope angle. The walking speed decreases with the slope angle. On the other hand, increasing the slope angle brings about a period doubling bifurcation leading to chaotic gaits and there are only unstable gaits in high speed region (Garcia et al., 1998). Biped robots based on the passive walking mechanisms were proposed (e.g. Goswami et al., 1997; Asano at al., 2004; Asano at al., 2005; Spong & Bullo, 2005), but the robots are mainly controlled by ankle torque, which has drawback from the viewpoints of Zero Moment Point (ZMP) condition, discussed in (Asano et al., 2005). The limitations of the passive-dynamic walkers and the ankle-torque controlled walkers should be addressed.

We propose the level-ground walking by using a torso and swing-leg control. Although using a torso for energy supply replacing potential energy, used in the case of the passive-dynamic walking, was proposed by McGeer (McGeer, 1988), there are few studies to use a torso explicitly for energy supply. Wisse et al. showed that the swing-leg motion is important to avoid falling forward (Wisse et al. 2005). From this viewpoint, we introduce a swing-leg control depending on the stance-leg motion. To modify the pendulum motion of the swing-leg by using the swing-leg control, the impact condition between the swing-leg and the ground will be satisfied before falling down.

In this paper, we study a knee-less biped robot with a torso on level ground. This paper presents a stability analysis of the biped robot to demonstrate the effectiveness of the swing-leg control. We use a Poincaré map to analyze walking motions which is a common tool in the study of the passive walking (McGeer, 1990; Goswami et al., 1996; Garcia et al., 1998;

Garcia et al., 2000). Walking analysis is as follows. First, using Newton-Raphson method, we search a periodic gait. Even though we know the existence of the periodic gait, we should know whether it is stable or unstable. Then we numerically approximate the Jacobian matrix of the Poincaré map of the periodic gait. If the Jacobian has all of its eigenvalues inside the unit circle, the gait is stable. Furthermore we search a set of initial conditions leading to stable walking. The stability analysis shows that the swing-leg control enables the robot to walk stably over a wide range of speed.

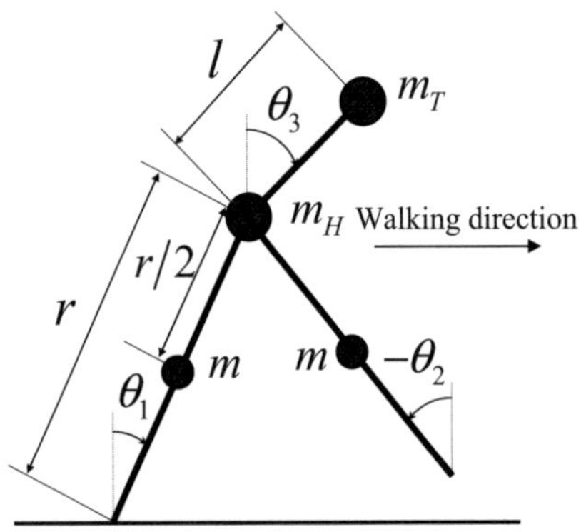

Figure 1. Biped knee-less walking robot

2. Biped walking model

2.1 Biped walking robot and model assumptions

The level-ground walking based on passive walk proposed in this paper needs a torso. In this paper, a simple biped robot with a torso shown in Fig. 1, is considered. This walking model is adding compass-like walking model (Goswami et al., 1996) to a torso, and has been studied in (Grizzle et al., 2001). The robot is composed of a torso, hips, and two legs. All masses are lumped. Dynamic variable values are measured from ground normal. Two torques u_1 and u_2, between the torso and the stance-leg, and between the torso and the swing-leg are applied, respectively. The motion of the robot is constrained to the sagittal plane. The scuffing problem of the swing-leg, which is inevitable in the case of a biped knee-less robot of which motion is constrained to the sagittal plane, is neglected during the swing phase, see in detail (McGeer, 1990; Grizzle et al., 2001).

2.2 Swing phase model

During the swing phase, the stance-leg acts as a pivot joint. By the method of Lagrange, the swing phase model is written as (Grizzle et al., 2001)

$$M(\boldsymbol{\theta})\ddot{\boldsymbol{\theta}} + C\left(\boldsymbol{\theta},\dot{\boldsymbol{\theta}}\right)\dot{\boldsymbol{\theta}} + G(\boldsymbol{\theta}) = B\mathbf{u} \tag{1}$$

where $\boldsymbol{\theta} = [\theta_1\ \theta_2\ \theta_3]^T$, $\mathbf{u} = [u_1\ u_2]^T$. The details of the matrices are

$$M(\boldsymbol{\theta}) = \begin{bmatrix} M_{11} & M_{12} & M_{13} \\ M_{12} & M_{22} & 0 \\ M_{13} & 0 & M_{33} \end{bmatrix}$$

$$M_{11} = \left(\tfrac{5}{4}m + m_H + m_T\right)r^2,\ M_{12} = -\tfrac{1}{2}mr^2\cos(\theta_1 - \theta_2)$$

$$M_{13} = m_T rl\cos(\theta_1 - \theta_3),\ M_{22} = \tfrac{1}{4}mr^2,\ M_{33} = m_T l^2.$$

$$C(\boldsymbol{\theta},\dot{\boldsymbol{\theta}}) = \begin{bmatrix} 0 & -\dot{\theta}_2 C_{12} & \dot{\theta}_3 C_{13} \\ \dot{\theta}_1 C_{12} & 0 & 0 \\ -\dot{\theta}_1 C_{13} & 0 & 0 \end{bmatrix}.$$

$$C_{12} = \tfrac{1}{2}mr^2\sin(\theta_1 - \theta_2),\ C_{13} = m_T rl\sin(\theta_1 - \theta_3).$$

$$G(\boldsymbol{\theta}) = \begin{bmatrix} -\tfrac{1}{2}g(3m + 2m_H + 2m_T)r\sin\theta_1 \\ \tfrac{1}{2}gmr\sin\theta_2 \\ -gm_T l\sin\theta_3 \end{bmatrix}.$$

$$B = \begin{bmatrix} -1 & 0 \\ 0 & -1 \\ 1 & 1 \end{bmatrix}.$$

2.3 Impact phase model

An impact occurs when the swing-leg touches the ground, which is called heel-strike. The condition of the impact, heel-strike, is given by

$$\theta_1 + \theta_2 = 0 \tag{2}$$

The impact is assumed to be inelastic and without slipping, and the stance-leg lifts from the ground without interaction (Hurmuzlu & Marghitu, 1994; Grizzle et al., 2001), and the actuators cannot generate impulses. Angular momentum is conserved at the impact for the whole robot about the new stance-leg contact point, for the torso about the hip, and for the new swing-leg about the hip. The conservation law of the angular momentum leads to the following compact equation between the pre- and post-impact angular velocities (Goswami et al. 1996) :

$$Q^+(\boldsymbol{\theta}^+)\dot{\boldsymbol{\theta}}^+ = Q^-(\boldsymbol{\theta}^-)\dot{\boldsymbol{\theta}}^- \tag{3}$$

The superscripts "-" and "+" respectively denote pre- and post-impact. During the impact phase, the configuration remains unchanged. The pre- and post-impact angles are identified with

$$\boldsymbol{\theta}^+ = J\boldsymbol{\theta} \tag{4}$$

where

$$J = \begin{bmatrix} 0 & 1 & 0 \\ 1 & 0 & 0 \\ 0 & 0 & 1 \end{bmatrix}$$

From (3) and (4) we have

$$\dot{\boldsymbol{\theta}}^+ = H(\boldsymbol{\theta}^-)\dot{\boldsymbol{\theta}}^- \tag{5}$$

The detail of the matrix is

$$H(\boldsymbol{\theta}^-) = \frac{1}{H_d}\begin{bmatrix} H_{11} & H_{12} & 0 \\ H_{21} & H_{22} & 0 \\ H_{31}/l & H_{32}/l & H_d \end{bmatrix}$$

$$H_d = -3m + 2m\cos(2\theta_1^- - 2\theta_2^-) - 4m_H - 2m_T + 2m_T\cos(2\theta_2^- - 2\theta_3^-)$$

$$H_{11} = (-2m - 4m_H - 2m_T)\cos(\theta_1^- - \theta_2^-) + 2m_T\cos(\theta_1^- + \theta_2^- - 2\theta_3^-),\ H_{12} = m$$

$$H_{21} = m - (4m + 4m_H + 2m_T)\cos(2\theta_1^- - 2\theta_2^-) + 2m_T\cos(2\theta_1^- - 2\theta_3^-),\ H_{22} = 2m\cos(\theta_1^- - \theta_2^-)$$

$$H_{31} = (-m - m_H - m_T)2r\cos(\theta_1^- - \theta_3^-) + (m + m_H + m_T)2r\cos(\theta_1^- - 2\theta_2^- + \theta_3^-) + mr\cos(3\theta_1^- - 2\theta_2^- - \theta_3^-)$$

$$H_{32} = -mr\cos(\theta_2^- - \theta_3^-)$$

Equation (5) can be also obtained by another method (Grizzle et al., 2001).

3. Simple control scheme

3.1 Torso and Swing-leg control

To hold the torso around a desired angle, the simple PD control scheme given by

$$T_T = -k_T^P(\theta_3 - \theta_3^d) - k_T^D\dot{\theta}_3 \tag{6}$$

is considered (McGeer, 1988). θ_3^d is the desired torso angle, k_T^P and k_T^D are control gain. k_T^P and k_T^D are determined as follows (McGeer, 1988). If the legs are firmly planted on the ground, the linearized equation of the torso motion about $\theta_3 = 0$ with the PD control becomes

$$m_T l^2\ddot{\theta}_3 + k_T^D\dot{\theta}_3 + (k_T^P - m_T gl)\theta_3 = 0 \tag{7}$$

The frequency of the torso is

$$\hat{\omega}_T = \sqrt{\frac{k_T^P - m_T gl}{m_T l^2}} \tag{8}$$

The damping ratio is

$$\hat{\zeta}_T = \frac{k_T^D}{2m_T l^2\hat{\omega}_T} \tag{9}$$

On the other hand, if the stance-leg is firmly planted on the ground, the linearized equation of the swing-leg motion about $\theta_2 = 0$ becomes

$$mr\ddot{\theta}_2 + 2mg\theta_2 = 0 \tag{10}$$

The natural frequency of the swing-leg is

$$\omega_S = \sqrt{\frac{2g}{r}} \tag{11}$$

In this paper, we determine the torso control parameters, k_T^P and k_T^D, to satisfy

$$\hat{\omega}_T = 3\omega_S \tag{12}$$

$$\hat{\zeta}_T = 0.7 \tag{13}$$

In order to satisfy the transition condition (Eq. (2)) before the robot falls down, we apply the simple control law given by

$$T_S = -k_S^P\left(\theta_2 - (-\theta_1)\right) \tag{14}$$

In the control law, the desired angle of the swing-leg depends on the stance-leg angle. $-\theta_1$ is the desired angle of the swing-leg which is opposed to the spring model between the legs (Kuo, 2002; Wisse et al., 2004). The swing-leg control will result in modifying the natural motion of the swing-leg. If the stance-leg angle is constant, the linearized equation of the swing-leg motion about $\theta_2 = 0$ with the swing-leg control becomes

$$mr^2\ddot{\theta}_2 + \left(2mgr + 4k_S^P\right)\theta_2 = 0 \tag{15}$$

The frequency is

$$\hat{\omega}_S = \sqrt{\frac{4k_S^P + 2mgr}{mr^2}} \tag{16}$$

k_S^P is determined by

$$\hat{\omega}_S = K\omega_S \tag{17}$$

K is a new swing-leg control parameter which shows the ratio between the frequencies of the swing-leg with the swing-leg control and without the swing-leg control. Then, we have

$$k_S^P = \tfrac{1}{4}mr^2K^2{\omega_S}^2 - \tfrac{1}{2}mgr \tag{18}$$

3.2 Control inputs for biped robot

From the torso control and the swing-leg control mentioned in the previous section, the control inputs are given by

$$u_1 = T_T + T_S \tag{19}$$

$$u_2 = -T_S \tag{20}$$

4. Stability analysis

4.1 Poincaré map

Poincaré map is commonly used to study the passive walking and quite useful to analysis biped locomotion. We follow the procedure to analysis the active biped robot on level ground.

The state just after the impact, heel-strike, is usually used as the Poincaré section. The Poincaré section removes one state. The Poincaré map is denoted as

$$P\left({}^{i}\mathbf{q}^{+}\right) = {}^{i+1}\mathbf{q}^{+} \tag{21}$$

where the superscript " i " denotes step number, and " $+$ " denotes post-impact between the swing-leg and the ground. Then ${}^{i}\mathbf{q}^{+}$ is the state just after the heel-strike of step i . A fixed point of the Poincaré map, $\mathbf{q}^{*}$, satisfies

$$P\left(\mathbf{q}^{*}\right) = \mathbf{q}^{*} \tag{22}$$

The fixed point represents a periodic (period-one) gait.

4.2 Periodic gaits

We can obtain a periodic gait to find the fixed point which is not only stable but also unstable, as follows (Garcia, 1999). Equation (22) corresponds to

$$\mathbf{g}\left({}^{i}\mathbf{q}\right) = 0 \tag{23}$$

where

$$\mathbf{g}\left({}^{i}\mathbf{q}\right) = P\left({}^{i}\mathbf{q}\right) - {}^{i}\mathbf{q} \tag{24}$$

To search for $\mathbf{q}^{*}$ such that $\mathbf{g}\left(\mathbf{q}^{*}\right) = 0$, Newton-Raphson method is used.

Given an initial guess at a fixed point, $\mathbf{q}_0$, the Jacobian of $\mathbf{g}$ is found numerically to perturb one state, i th element of $\mathbf{q}$ by ε and evaluate $\mathbf{g}_{\varepsilon}^{i}$. An estimate of the i th column of Jacobian is given by

$$\frac{\mathbf{g}_{\varepsilon}^{i} - \mathbf{g}(\mathbf{q}_0)}{\varepsilon} \tag{25}$$

Repeating this procedure find a numerical approximation to the Jacobian of $\mathbf{g}$.

Assuming that $\mathbf{g}\left(\mathbf{q}^{*}\right) = 0$, Newton-Raphson method provides the next approximation, $\mathbf{q}_1$, which is given by

$$\mathbf{q}_1 = \mathbf{q}_0 - \frac{\partial \mathbf{g}}{\partial \mathbf{q}}^{-1} \mathbf{g}(\mathbf{q}_0) \tag{26}$$

If a periodic gait exists and initial guess is sufficiently close, this search will converge to the fixed point $\mathbf{q}^*$.

4.3 Stability of the gait

By adding a small perturbation $\hat{\mathbf{q}}$ from the fixed point $\mathbf{q}^*$, Poincaré map P can be expressed as

$$P(\mathbf{q}^* + \hat{\mathbf{q}}) = P(\mathbf{q}^*) + J(\mathbf{q}^*)\hat{\mathbf{q}} \tag{27}$$

where J is the Jacobian of the Poincaré map.

$$J = \frac{\partial P(\mathbf{q})}{\partial \mathbf{q}} \tag{28}$$

J is determined approximately by performing the procedure described in Section 4.2. Note that instead of evaluating J, we can use the relationship Eq. (24). From Eq. (24), we obtain (Garcia et al., 1998)

$$\frac{\partial P(\mathbf{q})}{\partial \mathbf{q}} = \frac{\partial g(\mathbf{q})}{\partial \mathbf{q}} + \mathbf{I} \tag{29}$$

where $\mathbf{I}$ is the identity matrix.
Since $P(\mathbf{q}^*) = \mathbf{q}^*$, we can rewrite Eq. (27) as

$$P(\mathbf{q}^* + \hat{\mathbf{q}}) - \mathbf{q}^* = J(\mathbf{q}^*)\hat{\mathbf{q}} \tag{30}$$

Then we obtain

$${}^{i+1}\hat{\mathbf{q}} = J(\mathbf{q}^*)\, {}^{i}\hat{\mathbf{q}} \tag{31}$$

If all of its eigenvalues of the Jacobian are inside the unit circle, all sufficiently small perturbations $\hat{\mathbf{q}}$ will converge to 0, and the gait is asymptotically stable. If any eigenvalues of the Jacobian are outside the unite circle, the gait is unstable.

5. Simulation results

5.1 Simulation method

Values of the system parameters for the biped robot (Fig.1) are shown in Table 1.
To analysis the walking motion, we use numerical simulations. In swing phase, tha angular accelerations are solved as functions of the angles and the angular velocities to invert M in Eq. (1).

$$\ddot{\boldsymbol{\theta}} = M^{-1}(\boldsymbol{\theta})\left(-C(\boldsymbol{\theta},\dot{\boldsymbol{\theta}})\dot{\boldsymbol{\theta}} - G(\boldsymbol{\theta}) + B\mathbf{u}\right) \tag{32}$$

The simulations were run by using MATLAB®/SIMULINK®. We use ODE45 in MATLAB®/SIMULINK®, and specify a scalar relative error tolerance of 1e-8 and an absolute error tolerance of 1e-8. The heel strike of the biped robot was detected by zero-crossing detection in SIMULINK®. At the heel strike, the post-impact angular velocities and angles are calculated by Eq. (4) and (5).

Parameter	Unit	Value	Parameter	Unit	Value
m	kg	5	r	m	1
m_H	kg	10	l	m	0.5
m_T	kg	10	g	m/s^2	9.80665

Table 1. Values of the system parameters

In addition to the search of periodic gaits by using the Newton-Raphson method as mentioned in Section 4.2, by increasing the torso angle from 0.01rad in steps of 0.001rad, we find period-doubling bifurcations and chaotic gaits, which are demonstrated with the simplest model (Garcia et al., 1998), the compass-like model (Goswami et al., 1996) kneed models (Garcia et al., 2000), and level-ground walking (Howell & Baillieul, 1998)

5.2 Stability results

First, we search a stable walking while the swing-leg is left free, that is $K = 1.0$, where $\theta_3^d = 0.01[\text{rad}]$. In our search, a periodic gait could not be found. Then we introduce the swing-leg control.

Figure 2 shows the evolution of the walking speed as a function of the desired torso angle where the swing-leg control parameter $K = 1.4, 1.7\ 1.95$ and 2.0. Figure 2 demonstrates the walking speed increases with the desired torso angle and the maximum walking speed of the stable gait increases with the swing-leg control parameter. But when we increase further the swing-leg control parameter K, period-doubling bifurcations occur and we didn't find stable gaits. Figure 2 shows that the maximum walking speed of stable gaits doesn't necessarily increase with the swing-leg control parameter K. Figure 3 shows the evolution of the absolute eigengvalues of the Jacobian J as a function of the desired torso angle where $K = 1.95$ and 2.0.

5.3 Stable region

Even if a stable periodic walking is achieved, we want to know when the biped robot keeps walking, and when it falls down from a disturbance and incorrect launch. To answer it, yet partially, a set of initial conditions leading to stable walking is searched for. The more initial conditions result in the fixed point without falling down, the more stable the walking is. To find the initial conditions, we perturb one state of $\mathbf{q}^*$. For example, we perturb θ_1^+ from the fixed point and the other states $\left(\theta_3^+, \dot{\theta}_1^+, \dot{\theta}_2^+, \dot{\theta}_3^+\right)$ remain unchanged at the equilibrium position. Figure 4 shows the initial conditions of the perturbed state leading to continuous walking where $\theta_3^d = 0.2$.

From Fig. 4, increasing the swing-leg control parameter K results in the increase of the range of the initial conditions leading to stable walking. The simulation results show that the swing-leg control enlarges the stable region.

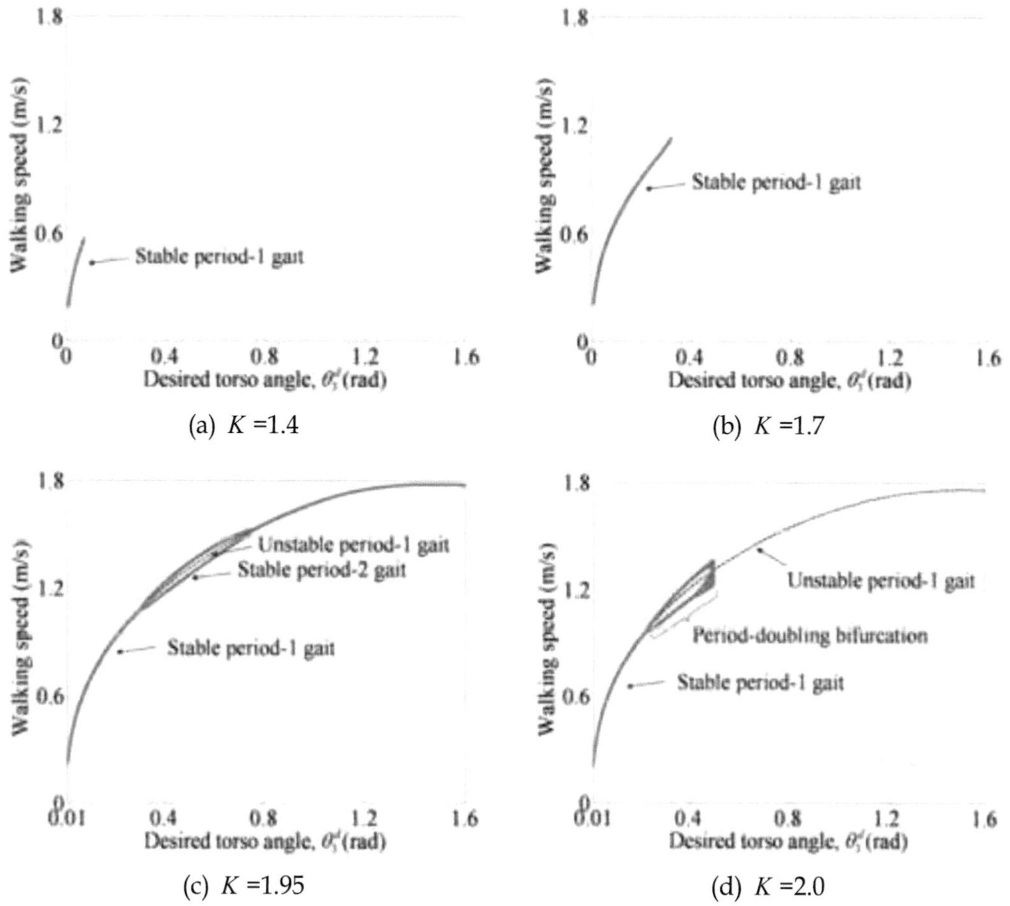

(a) K =1.4 (b) K =1.7

(c) K =1.95 (d) K =2.0

Figure 2. Walking speed versus desired torso angle

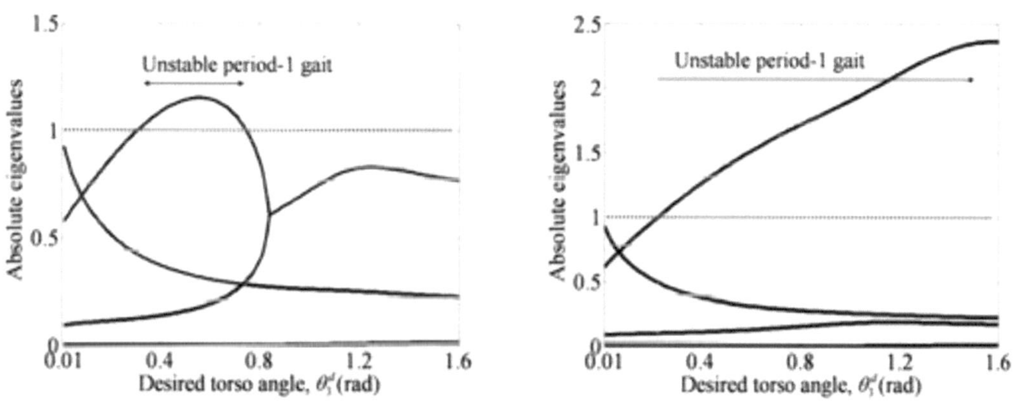

(a) K =1.95 (b) K =2.0

Figure 3. Absolute eigenvalues versus desired torso angle

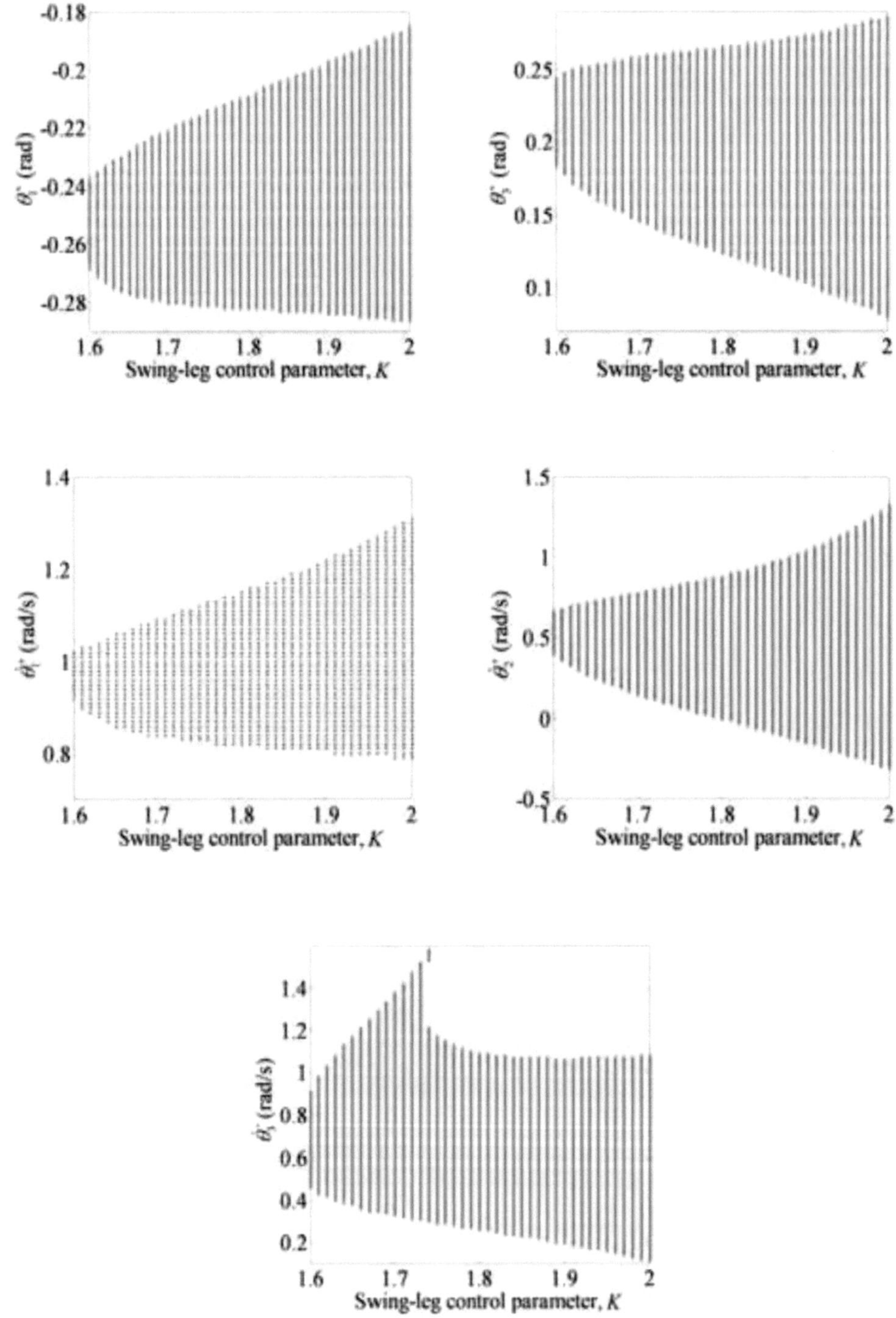

Figure 4. Stable range as a function of the swing-leg control parameter K where $\theta_3^d = 0.2$

6. Conclusion

We study a simple active biped robot on level ground to walk over a wide range of speed. The use of the torso for energy supply and the swing-leg control for stable waking are introduced. Numerical simulations show that the swing-leg control enables the biped robot on level ground to walk stably over a wide range of speed, and enlarges the stable region.

7. Acknowledgement

This work was supported in part by Grant-in-Aid for JSPS Fellows, 1853073.

8. References

Asano, F.; Yamakita, M., Kamamichi, N. & Luo, Z.-W. (2004). A novel gait generation for biped walking robots based on mechanical energy constraint, *IEEE Trans. Robotics and Automation*, Vol. 20, No. 3, pp. 565-573, 2004.

Asano, F.; Luo, Z.-W. & Yamakita, M. (2005). Biped gait generation and control based on a unified property of passive dynamic walking, *IEEE Trans. Robotics*, Vol. 21, No. 4, pp. 754-762, 2005.

Collins, S. H.; Ruina, A., Tedrake, R. & Wisse, M. (2005). Efficient bipedal robots based on passive-dynamic walkers, *Science*, 307, pp. 1082-1085, 2005.

Collins, S. H.; Wisse, M. & Ruina, A. (2001). A three-dimensional passive-dynamic walking robot with two legs and knees, *Int. J. Robotics Research*, Vol. 20, No. 7, pp. 607-615, 2001.

Garcia, M. (1999). Stability, scaling and chaos in passive-dynamic gait models, PhD thesis, Cornell University, 1999.

Garcia, M.; Chatterjee, A. & Ruina, A. (2000). Efficiency, speed, and scaling of two-dimensional passive-dynamic walking, *Dynamics and Stability of Systems*, Vol. 15, No. 2, pp. 75-99, 2000.

Garcia, M.; Chatterjee, A., Ruina, A. & Coleman, M. (1998). The simplest walking model: stability, complexity, and scaling, *J. Biomechanical Engineering*, Vol. 120, No. 2, pp. 281-288, 1998.

Goswami, A.; Espiau, B. & Keramane, A. (1997). Limit cycles in a passive compass gait biped and passivity-mimicking control laws, *Autonomous Robots*, Vol. 4, No. 3, pp. 273-286, 1997.

Goswami, A.; Thuilot, B. & Espiau, B. (1996). Compass-like biped robot-part I: stability and bifurcation of passive gaits, Technical Report 2996, INRIA, 1996.

Grizzle, J. W., Abba, G. & Plestan, F. (2001). Asymptotically stable walking for biped robots: analysis via systems with impulse effects, *IEEE Trans. Automatic Control*, Vol. 46, No. 1, pp. 51-64, 2001.

Howell, G. W. & Baillieul, J. (1998). Simple controllable walking mechanisms which exhibit bifurcations, *Proceedings of 37th IEEE Conf. Decision and Control*, Tampa, FL, Dec. pp. 3027-3032, 1998.

Hurmuzlu, Y. & Marghitu, D. B. (1994). Rigid body collisions of planar kinematic chains with multiple contact points, *Int. J. Robotics Research*, Vol. 13, No. 1, pp. 82-92, 1994.

Kuo, A. D. (2002). Energetics of actively powered locomotion using the simplest walking model, *J. Biomechanical Engineering*, Vol. 124, pp. 113-120, 2002.

McGeer, T. (1988). Stability and control of two-dimensional biped walking, Technical Report CSS-IS TR 88-01, Simon Fraser University, 1988.

McGeer, T. (1990). Passive dynamic walking, *Int. J. Robotics Research*, Vol. 9, No. 2, pp. 62-82, 1990.

McGeer, T. (1993). Dynamics and control of bipedal locomotion, *J. Theoretical Biology*, Vol. 163, No. 3, pp. 277-314, 1993.

Spong, M. W. & Bullo, F. (2005). Controlled symmetries and passive walking, *IEEE Trans. Automatic Control*, Vol. 50, No. 7, pp. 1025-1031, 2005.

Wisse, M.; Schwab, A. L. & van der Helm, F. C. T. (2004). Passive dynamic walking model with upper body, *Robotica*, Vol. 22, pp. 681-688, 2004.

Wisse, M.; Schwab, A. L., van der Linde, R. Q. & van der Helm, F. C. T. (2005). How to keep from falling forward: elementary swing leg action for passive dynamic walkers, *IEEE Trans. Robotics*, Vol. 21, No. 3, pp. 393-401, 2005.

8

Inertial Forces Posture Control for Humanoid Robots Locomotion

Victor Nunez[1], Nelly Nadjar-Gauthier[1], Kazuhito Yokoi[2], Pierre Blazevic[1] and Olivier Stasse[2]
[1]*Laboratoire d'Ingénierie des Systèmes de Versailles (LISV), Université de Versailles*
[2]*ISRI/AIST-STIC/CNRS Joint Japanese-French Robotics Laboratory (JRL)*
[1]*France,* [2]*Japan-France*

1. Introduction

In order to evolve in an environment designed for humans, a humanoid robot (or simply humanoid) is supposed capable of performing different motions depending on the given situation. With the walking paradigm in a mature stage, in recent years, the attention of many researchers has passed to more complicated locomotion modes. Some examples are: climbing stairs (Harada et al., 2004), falling down in a less damaging manner (Fujiwara et al., 2004), jumping (Nunez et al. 2005, Sakka et al. 2005) and running (Nagasaka et al., 2004; Kajita et al., 2005), crawling (Kanehiro et al., 2004), etc.
If different control strategies are used for each kind of locomotion, the autonomy and/or versatility of the humanoid can be affected by complicating the problem of implementing each control algorithm and switching between them.
To treat this issue it has been proposed (Tiong & Romyaldy, 2004) to differentiate two important parts on the humanoid locomotion system: the *motion generator* (MG) and the *posture controller* (PoCtr), see Fig. 1. The objective of the former is to define the desired locomotion using some specific parameters and the later will find the angles of all the actuated articulations of the robot such that this motion is achieved.
In this work, we will present a new posture controller which finds the angles of all the articulations of the humanoid ($\boldsymbol{\theta}$) which produces the desired generalized inertial forces ($\mathbf{Q}^{inertial,\mathbf{Ref}}$) and motion of the extremities of the humanoid ($\xi_i^{\mathbf{Ref}}$), see Fig 1.

1.1 Description of the approach

The motion generation section describes the input parameters for the PoCtr presented on section 3, and discusses the relationship between the inertial forces and the zmp and angular momentum will be presented.
The inertial force PoCtr is the main result presented in this work. It is based on the Lagrangian equations of motion which relates the accelerations of the actuated articulations with the generalized external forces (force and torque) acting at the support foot of the humanoid. The generalized inertial forces are the derivatives of the linear and angular momentum of the whole humanoid. By controlling these parameters it is possible to consider the *zero moment point* (zmp) stability of the robot.

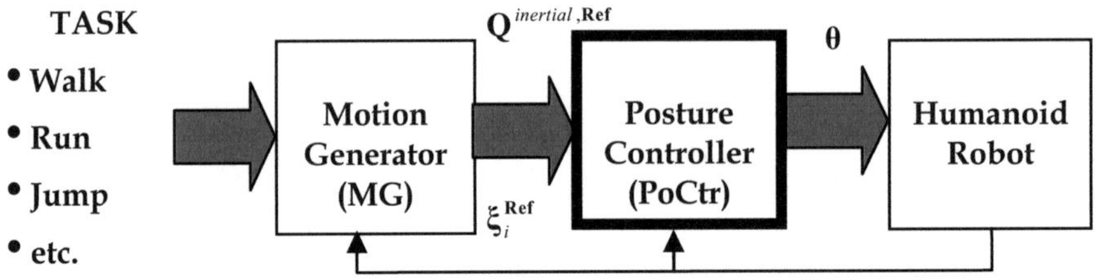

Figure 1. Locomotion control system capable of treating different humanoid locomotion modes

The generalized position (position and orientation) $\xi_i^{\mathbf{Ref}}$ of the extremities of the robot is achieved by an inverse kinematics algorithm. The implementation of a PD controller provides asymptotic reaching of the hands trajectories, which means that an initial error can be compensated by our controller. The inertial forces $\mathbf{Q}^{inertial,\mathbf{Ref}}$ are retrieved by a computed torque like algorithm.

One of the main objectives of our research is to present algorithms that can be implemented in real humanoid robots. In order to prove the effectiveness of our approach, we tested the method on the humanoid robot HRP-2 (Kaneko et al., 2004), shown on Fig. 2. The robot's motion was symmetric (left and right extremities following the same pattern). This allowed us to consider the robot only on the sagital plane, and to apply our general method in reduced order. The inertial forces were selected to obtain an elliptic trajectory of the CoM and zero angular momentum. This motion was chosen because it is dynamic in the sense that the velocity and acceleration of the CoM are not negligible. The hands motion is to asymptotically reach a circular trajectory after being static at the beginning of the motion.

The method presented in this paper for robots commanded in angle (local PID control for each motor) can be also implemented in torque. On (Nunez et al.., 2006) we presented the inertial forces PoCtr considering flight phases and it is applied in torque mode instead of angle mode.

1.2 Related literature

Several works had been presented which covers one or more parts of the block diagram, in Fig. 1, even if this general locomotion paradigm is not often mentioned. Maybe the most representative is the posture controller called Resolved Momentum Control (RMC), presented in (Kajita et al., 2003) and which generalizes the CoM jacobian approach (Sugihara 2003). The inputs for this PoCtr are the desired linear and angular momentum of the whole robot as well as the generalized position of each foot. This posture controller has proven to be useful for treating different locomotion modes like kicking, walking and running (Kajita et al., 2003; Kajita et al., 2005). The advantages of using this kind of methodology are clearly demonstrated on (Sian et al., 2003) where the zmp stability of the motion is considered and the autonomy of the robot is enhanced by the automatic selection of the degrees of freedom (DoF) that should be used for different kind of motions. This approach has been implemented on the humanoid HRP-2 and HRP-2L (Kajita et al., 2005).

The main difference with the work presented here is that RMC is implemented in velocity level (momentum) while the Inertial Forces Posture Controller (IFPC) is implemented in

acceleration level (forces). This opens the possibility of implementing our method in torque and to asymptotically reach the trajectories of the extremities.
The dynamic filters proposed in (Nagasaka et al., 2004) are focused on the motion generation and in particular in the continuity of trajectories of the *Center of Mass* (CoM) for walking, running and jumping motions. This method is implemented on the humanoid QRIO (Nagasaka et al., 2004). Again the locomotion is codified using the generalized motion of the CoM. The posture controller, based on the CoM jacobian, is not detailed on the paper. Regarding the use of whole body dynamics in acceleration level, we can mention the works of (Kondak et al., 2005), based on Tanie's dynamic equations, and (Rok So et al., 2003), based on the Lagrange equations. In both cases the explicit consideration of the zmp constraint instead of a motion generation stage reduces the generality of the approaches.

2. Motion generation

The motion generation stage is supposed to express the desired locomotion (walking, kicking, jumping, etc) using some specific parameters. The generalized inertial forces and position of the extremities are used in our approach, and will be described in this section. We will show how to take into account the zmp stability of the motion using the inertial forces.

2.1 Generalized Position and inertial forces

The generalized position of the extremities (ξ_i) is composed as follows

$$\xi_i = \left(r_i^{\mathrm{T}} \quad \gamma_i^{\mathrm{T}}\right)^{\mathrm{T}} \in \Re^6$$

where $r_i \in \Re^3$ denotes the cartesian position and $\gamma_i \in \Re^3$ the orientation (Euler angles, for instance) of the frame atached to each extremity i which is not in contact with the environment. When the robot is standing on the right foot $i \in \{LF \quad RH \quad LH\}$, which corresponds to the left foot and right and left hands, as shown on Fig. 2. It is clear that i will switch while the robot is walking, running or jumping. In this work we will present in detail the simple suppoer phase. Double support and flight phases are easily obtained implemented based on the same procedure.
The desired generalized position ξ_i^{Ref} of the extremities which are not in contact with the environment should be passed in as inputs to the IFPC presented in next section. The corresponding velocity $\dot{\xi}_i^{\mathrm{Ref}}$ and acceleration $\ddot{\xi}_i^{\mathrm{Ref}}$ will be also needed for implementing a PD controller capable of asymptotically tracking these desired trajectories.
One of the key points of our approach is to consider the ground reaction force and moment as acting at a fixed point on the support polygon[1]. Let us consider the vertical projection of

---

[1] The ground reaction force, distributed along the whole foot surface, can be considered as a force with an associated torque. This torque depends on the considered point (Vernon's Theorem). The zmp is only a special point where this torque is zero, but any other point can be considered (Sardain et al., 2004; Goswami et al., 2004).

the right ankle, the point RH in Fig. 2, for instance. The generalized reaction force acting at this point is denoted by

$$\mathbf{Q}_{RF} = \left(\mathbf{f}_{\mathrm{RF}}^{\mathrm{T}} \quad \tau_{RF}^{\mathrm{T}}\right)^{\mathrm{T}} \in \Re^{6}$$

where $\mathbf{f}_{RF}$ and τ_{RF} are the ground reaction force and torque, respectively.
The Newton-Euler formulation applied to the whole humanoid shown in Fig. 2 states that

$$\begin{pmatrix} \mathbf{f}_{\mathrm{RF}} \\ \tau_{\mathrm{RF}} \end{pmatrix} + \begin{pmatrix} m_T\mathbf{g} \\ \mathbf{r}_{RF,G} \times m_T\mathbf{g} \end{pmatrix} = \begin{pmatrix} m_T\ddot{\mathbf{r}}_{RF,G} \\ \mathbf{r}_{RF,G} \times m_T\ddot{\mathbf{r}}_{RF,G} + \dot{\mathbf{L}}_G \end{pmatrix} \quad (1)$$

or

$$\mathbf{Q}_{RF} = \mathbf{Q}_{RF}^{inertial} - \mathbf{Q}_{RF}^{g} \quad (2)$$

with

$$\mathbf{Q}_{RF}^{inertial} = \begin{pmatrix} m_T\ddot{\mathbf{r}}_{RF,G} \\ \mathbf{r}_{RF,G} \times m_T\ddot{\mathbf{r}}_{RF,G} + \dot{\mathbf{L}}_G \end{pmatrix} \quad (3)$$

$$\mathbf{Q}_{RF}^{g} = \begin{pmatrix} m_T\mathbf{g} \\ \mathbf{r}_{RF,G} \times m_T\mathbf{g} \end{pmatrix}$$

where m_T is the total mass of the robot, $\mathbf{r}_{RF,G}$ is the vector from RF to the CoM of the robot, denoted by G. The angular momentum of the whole robot around G is computed as (Goswami et al., 2004)

$$\mathbf{L}_G = \sum_j \left(\mathbf{r}_{G,Gj} \times m_j\dot{\mathbf{r}}_{G,Gj} + \mathbf{L}_G^j\right)$$

where for each link j, m_j is the mass, $\mathbf{r}_{G,Gj}$ is the position vector from G to the CoM of the link and $\mathbf{L}_G^j = I^j\omega^j$ is the angular momentum around its own CoM.
As well as the trajectories of the hands and swing foot must be specified in velocity and acceleration, the desired inertial forces must be specified also using the integral like terms:

$$i\mathbf{Q}_{RF}^{inertial} = \begin{pmatrix} m_T\dot{\mathbf{r}}_{RF,G} \\ \mathbf{r}_{RF,G} \times m_T\dot{\mathbf{r}}_{RF,G} + \mathbf{L}_G \end{pmatrix} \quad (4)$$

$$ii\mathbf{Q}_{RF}^{inertial} = \begin{pmatrix} m_T\mathbf{r}_{RF,G} \\ \int_0^t \left(\mathbf{r}_{RF,G} \times m_T\dot{\mathbf{r}}_{RF,G} + \mathbf{L}_G\right) d\tau \end{pmatrix} \quad (5)$$

Notice that the generalized inertial forces expressed in (3) and its integral terms (4) and (5) depend not only on the desired integral of position, velocity and acceleration of the CoM of the robot, but also on the angular momentum around G, its derivative and two integral terms. This means that imposing the desired inertial forces is equivalent to specify the linear and angular momentum of the whole robot. The Integral like terms are required in next section to obtain a closed loop PID control on the inertial forces.

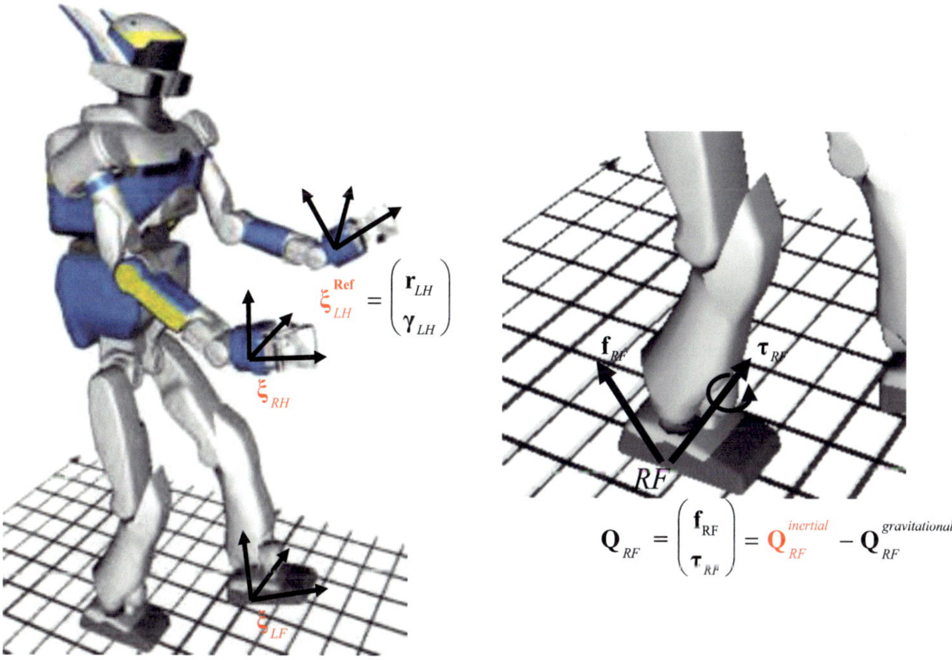

Figure 2. The humanoid robot HRP-2. The position and orientation of the free extremities are denoted by $\mathbf{r}_i$ and γ_i, respectively. The generalized force around a fixed point (RF) on the support foot can be decomposed into gravitational and inertial force

2.2 Stable Inertial forces

Considering the generalized ground reaction force as applied at a fixed point of the support polygon implies that the foot must remain flat on the ground during the whole motion. This is because if the foot rolls over an edge or corner the point where the reaction force is supposed to act will not be in contact with the ground.

The foot not rolling over an edge or corner can be characterized using zmp. The contact with flat foot is characterized by $zmp \in S$ where S is the support polygon. According to Fig. 2, the zmp can be computed using

$$\tau_{RF} = \tau_{zmp} + \mathbf{r}_{RF,ZMP} \times \mathbf{f}_{RF} \tag{6}$$

where

$$\mathbf{r}_{RF,ZMP} = \begin{pmatrix} x_{RF,ZMP} & y_{RF,ZMP} & 0 \end{pmatrix}^{\mathrm{T}}$$

denotes the vector from RF to the zmp as shown in Fig. 3. By definition the torque τ_{zmp} has only a z component, so using eq. (6) we can obtain

$$\tau_{RFx} = y_{RF,zmp} f_z \qquad \tau_{RFy} = -x_{RF,zmp} f_z$$

Considering a square foot, the zmp condition can be stated using the reaction moment as follows

$$-x_{RFB} f_{RF_z} \geq \tau_{RF_y} \geq x_{RFF} f_{RF_z} \tag{7}$$

$$-y_{RFB} f_{RF_z} \leq \tau_{RF_x} \leq y_{RFF} f_{RF_z} \tag{8}$$

where x_{RFB}, x_{RFF}, y_{RFB} and y_{RFF} are used to define the geometry of the foot as shown in Fig. 3.

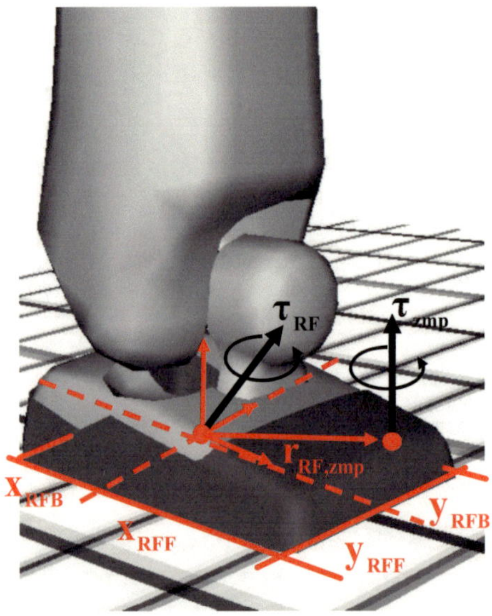

Figure 3. Forces acting at the support foot and its geomety

The second condition for keeping the support foot on the ground is that the reaction force must be positive

$$f_{RFz} \geq 0 \;\; \Rightarrow \;\; \ddot{z}_{RF,G} \geq -g \tag{9}$$

This is because the ground can not pull down the foot. The implication comes from (1) with $g = 9.81 m/s^2$. The last necessary condition is to avoid slippage and depends on the friction constant between the foot and the ground, denoted by μ:

$$\frac{\| f_{RFz} \|}{f_{RFx}} \leq \mu \;\; \Rightarrow \;\; \ddot{x}_{RF,G}\mu \leq \ddot{z}_{RF,G} \tag{10}$$

There are different ways to express the desired motion using the inertial forces. A simple choice is to define a desired trajectory of the center of gravity G to obtain $\mathbf{f}_{RF}^{Ref}$ and their integrals. Then, the only remaining term to complete τ_{RF}^{Ref} would be the desired trajectory of the angular momentum around G. The choice of this trajectory is not evident because it has an influence on the overall motion of the robot. Making $\mathbf{L}_G = 0$ has been proposed as stability index in (Goswami et al. 2004) and is maybe the first choice to be tried. An important fact is that also a desired zmp can be specified by combining (6) with (1). Finding

the adequate angular momentum or zero moment point for a given motion is still, to the authors best knowledge, an open problem.
The inertial force PoCtr will impose the desired motion which must be coded using the trajectories of hands and swing foot and the desired inertial forces. The proposed motion must be feasible, meaning that the hands reachability and singularity must be considered as well as the stability of the desired locomotion, characterized by equations (7), (8), (9) and (10). This must be done in the motion planning stage and some trial and error is up to now necessary to tune the parameters of the motion like the angular momentum and the zmp trajectory.

3. Inertial force position control

In this section the main contribution of this research is presented. The objective of the inertial force PoCtr is to find the desired angles for all the articulations of the humanoid such that the stable inertial forces and positions described in section 2 are obtained.
In order to obtain the Lagrange equations describing the dynamics of the humanoid robot in ground and aerial motions, it is necessary to consider, besides the vector of internal angles of the robot, denoted by $\theta \in R^n$, the position and orientation of a given point of the robot. Usually the hip generalized position ξ_B is used, but any other point can be considered. For simplicity we will consider the coordinate of the support point, i.e. ξ_{RF}. Notice that the model can be modified simply applying the coordinate change $\xi_{RF} = f(\xi_B, \theta)$. The extended Lagrangian model of the robot with extended coordinates

$$\mathbf{q} = \begin{pmatrix} \theta \\ \xi_{RF} \end{pmatrix} \tag{11}$$

can be written as:

$$D(\mathbf{q})\ddot{\mathbf{q}} + \mathbf{vC}(\mathbf{q},\dot{\mathbf{q}}) + \mathbf{g}(\mathbf{q}) = \begin{pmatrix} \tau \\ \mathbf{Q}_{RF} \end{pmatrix} \tag{12}$$

where $\mathbf{vC}$ is the *vector* of centrifugal and coriolis forces, $\mathbf{g}$ the gravity forces, τ are the articulation torques, and $\mathbf{Q}_{RF}$ is the generalized force applied at RF. Considering (11), it is possible to split (12) for the actuated and non actuated coordinates as

$$D_{11}\ddot{\theta} + D_{12}\ddot{\xi}_{RF} + \mathbf{vC}_1 + \mathbf{g}_1 = \tau \tag{13}$$

$$D_{21}\ddot{\theta} + D_{22}\ddot{\xi}_{RF} + \mathbf{vC}_2 + \mathbf{g}_2 = \mathbf{Q}_{RF} \tag{14}$$

In fact, (13) are n equations corresponding to the actuated coordinates while (14) represents 6 equations describing the dynamics of the position and orientation of the support foot. Assuming perfectly inelastic contact, when the robot is standing on its support foot $\dot{\xi}_{RF} = \ddot{\xi}_{RF} = 0$. Substituting this condition in (14) we can get

$$D_{21}\ddot{\theta} + \mathbf{vC}_2 + \mathbf{g}_2 = \mathbf{Q}_{RF} \tag{15}$$

This equation relates all the actuated angles acceleration to the external force, and this will be the base of our control approach. The reason of using (14) instead of (13) for obtaining our control law is because most of humanoid robots, HRP-2 in particular, are controlled in angles position instead of torques. On (Nunez et al., 2006) the use of (13) for a torque based control law is presented.

3.1 Decoupled position control

Let us separate the actuated angles corresponding to each extremity, i.e. legs and arms, of each end effector as

$$\theta = \begin{pmatrix} \theta_{RF}^{\mathrm{T}} & \theta_{LF}^{\mathrm{T}} & \theta_{LH}^{\mathrm{T}} & \theta_{RH}^{\mathrm{T}} \end{pmatrix}^{\mathrm{T}}$$

where θ_{RF} is the vector containing the angles of the right leg, θ_{LF} the left foot, and the last two are the angles of left and right arms. Supposing that the body is not articulated (or that the articulations are not used), the relative generalized position for each extremity *with respect to the hip* is a function of the relative angles of the corresponding extremity, i.e. ${}^{B}\xi_i = \mathbf{f}(\theta_i)$. Deriving twice this expression we obtain

$${}^{B}\ddot{\xi}_i = J_i(\theta_i)\ddot{\theta}_i + \dot{J}(\theta_i, \dot{\theta}_i)\dot{\theta}_i$$

with $i \in \{LF, RH, LH\}$ for the simple support case. If we are dealing with a *standard humanoid robot*, meaning that there are 6 DoF by limb and if furthermore singular configurations are avoided, the jacobian of each extremity J_i is invertible and we can apply the classical PD controller in task space for each extremity as follows

$$\ddot{\theta}^{ref} = J_i^{-1}\left(-\dot{J}\dot{\theta}_i + \mathbf{uP}_i\right) \tag{16}$$

where the position control input is defined as:

$$\mathbf{uP}_i = {}^{B}\ddot{\xi}_i^{ref} + kP_{vi}({}^{B}\dot{\xi}_i^{ref} - {}^{B}\dot{\xi}_i) + kP_{pi}({}^{B}\xi_i^{ref} - {}^{B}\xi_i) \tag{17}$$

where the diagonal matrices kP_{vi} and kP_{pi} must be chosen from stability. The implementation of this control law permits to have initial errors on the desired trajectory of the hands and swing foot, and the rapidity for reaching the target trajectories will depend on kP_{vi} and kP_{pi}.

The over-actuated case can be considered using the pseudo-inverse of the jacobian for optimizing a criteria (manipulability, for instance).

Notice that in order to pass from the absolute position and orientation of the extremities, as mentioned in section 2 to the relatives used in (17) it is necessary to use the hip generalized position ξ_B, because ${}^{B}\xi_i = \xi_i - \xi_B$. In most humanoid robots, including HRP-2, the position velocity and acceleration of the hip can be obtained from sensors (combining gyroscope, accelerometer signal and their numerical derivatives).

3.2 Inertial force position control

Eq. (15) can be separated as

$$\sum_i D_{21i}\ddot{\theta}_i + D_{21RF}\ddot{\theta}_{RF} + \mathbf{vC}_2 + \mathbf{g}_2 = \mathbf{Q}_{RF} \tag{18}$$

with $i \in \{LF, RH, LH\}$, considering the simple support phase.

The basic idea of the generalized force position control is to find the angles of the support leg in order to obtain the desired external force and to compensate the desired motion of the upper body given by the decoupled position control.

This idea can be implemented *using whole body dynamics in acceleration level.* In fact substituting expression (2) on eq. (18) we can obtain

$$D_{21RF}\ddot{\theta}_{RF} = \left(\mathbf{Q}_{RF}^{inertial} - \mathbf{vC}_2 - \sum_i D_{21i}\ddot{\theta}_i\right)$$

The elimination of the gravity term is because the foot coordinates where used as generalized coordinates. It can be shown that $\mathbf{g}_2$ from (2) is equivalent to $-\mathbf{Q}_{RF}^{g}$ in (18).

Assuming D_{21RF} invertible[2] our inertia force control takes the following form

$$\ddot{\theta}_{RF}^{Ref} = D_{21RF}^{-1}\left(\mathbf{uF} - \mathbf{vC}_2 - \sum_i D_{21i}\ddot{\theta}_i^{Ref}\right) \tag{19}$$

with $\mathbf{uF}$ given as a PID control

$$\mathbf{uF} = \mathbf{Q}_{RF}^{iRef} + kF_v i\tilde{\mathbf{Q}}_{RF} + kF_p ii\tilde{\mathbf{Q}}_{RF} + kF_i iii\tilde{\mathbf{Q}}_{RF} \tag{20}$$

The expressions with $\tilde{}$ are the difference between real and desired values of the inertial forces expressed in eq. (3) and the matrices kF_v, kF_p, kF_i are again chosen following stability and performance criteria.

3.3 Double support phase

During the double support phase (d.s.p.) the decoupled position control (16) with (17) should be implemented for both arms, i.e. $i \in \{LF, RF\}$. Then the force position controller (19) with (20) must be implemented for obtaining the angles of one of the support legs, lets suppose the right one.

The angles of the left leg can be then obtained as follows. The generalized position and velocity of the left foot can be obtained as

$${}^{RF}\xi_{LF} = {}^{RF}\xi_B + {}^{B}\xi_{LF}$$

$${}^{RF}\dot{\xi}_{LF} = J_{B,RF}(\theta_{RF})\theta_{RF} + J_{B,LF}(\theta_{LF})\theta_{LF}$$

[2]It is out of the scope of this paper to consider in detail the invertibility of matrix D_{21RF}. For the time being we can only say that this generalized inertia matrix is invertible in all our simulations and experiments.

Because the left foot is in contact with the ground, $\dot{\xi}_{LF} = 0$ and the desired angles of the left foot can be obtained simply as

$$\theta_{LF}^{Ref} = -J_{B,LF}^{-1} J_{B,RF} \theta_{RF}^{Ref}$$

In most humanoid robots, including HRP-2, the motor of each actuated articulation is controlled in angular position, meaning that every sampling period the desired position θ^i must be passed as reference. As shown in next section these values can be obtained from angular acceleration or velocities (for the d.s.p) by simple numerical integration.

4. Experiments

In this section, in order to validate the proposed approach, the experiments on the humanoid robot HRP-2 are presented. The selection motion is simple because the objective of the experiments is only to show the applicability of the proposed inertial force posture control.

The selected movement is symmetric with respect to the sagittal plane, meaning that the left and right parts of the robot move identically. As explained in (Nunez et al. 2005) this kind of motions allows us to consider the robot as constrained to the sagittal plane as shown in Fig. 3. In this case the reference motion can be specified using only

$$\mathbf{Q}_{RF}^{i} = \begin{pmatrix} m_T \ddot{z}_{RF,G} \\ m_T \ddot{x}_{RF,G} \\ m_T (z_{RF,G} \ddot{x}_{RF,G} - x_{RF,G} \ddot{z}_{RF,G}) - \dot{L}_{Gy} \end{pmatrix}$$

$$\xi_{RH} = \begin{pmatrix} X_{RH} \\ Z_{RH} \\ P_{RH} \end{pmatrix} \in R^3$$

where P_{RH} is the pitch of the right hand. The actuated articulations used were only the elbow, shoulder and hip for the upper body and the ankle, knee and hip for the support leg. This means that $\theta_{RF} \in R^3$ and $\theta_{RH} \in R^3$. For the left hand side the same angles were passed to the corresponding articulations.

Once the desired angular acceleration was obtained using (16), (17), (19) and (20) the desired positions to pass to each articulation were obtained by the simple numerical integration algorithm

$$\theta^k = \Delta_t \dot{\theta}^k + \Delta_t \theta^{k-1}$$

$$\theta^k = \Delta_t^2 \ddot{\theta}^k + \Delta_t \dot{\theta}^{k-1} + \theta_{k-1}$$

The same integration method was employed for terms needed in (5) the sample time period being of $\Delta_t = 5ms$. This time was largely enough for making all the computations required by our method.

The desired trajectory of G and the hand position were specified to be ellipsoid as shown in Fig. 4.

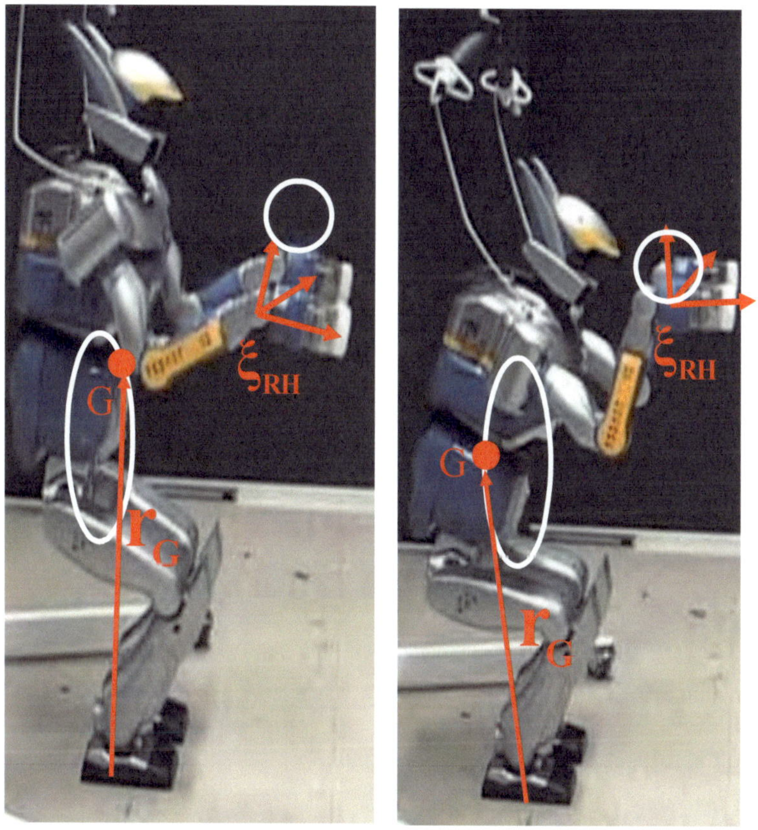

Figure 4. The motion of the robot uses sinus and cosinus functions on the $x-y$ directions in order to obtain the ellipsoid motion shown. During the first 4 s the arms of the robot have not relative motion (left) and after that the arms will move to reach the desired trajectory (right). After $t=12s$ the CoM stay still while the ands continues to move

As mentioned in section 2, the desired trajectory must be specified using first the desired absolute position of the hands ξ^{Ref} and its derivatives. The proposed hand motion was specified using sinus and cosinus functions in the $x-z$ plane with amplitude of $5cm$ and period of $T=1.5s$, this is

$$\xi_{RH}^{Ref} = \begin{pmatrix} x_{RH} \\ z_{RH} \\ \theta \end{pmatrix} = \begin{pmatrix} 0.05\sin\left[2\pi(1/1.5)t\right] \\ 0.05\cos\left[2\pi(1/1.5)t\right] \\ 0 \end{pmatrix}$$

In our experiment in order to verify the closed loop behavior of the controller, we decided to keep the arms fixed during the first 4 seconds of the motion, this means

$$Hands \quad Motion \begin{cases} t \le 4\text{s} & \dot{\theta}_{RH} = \ddot{\theta}_{RH} = 0 \\ t \ge 4\text{s} & \xi_{RH}^{Ref} \end{cases}$$

This means that during this period, only the desired inertial force is specified. We can notice in Fig. 5 and Fig. 6 the transitory after $t=4s$ before reaching the desired trajectory.

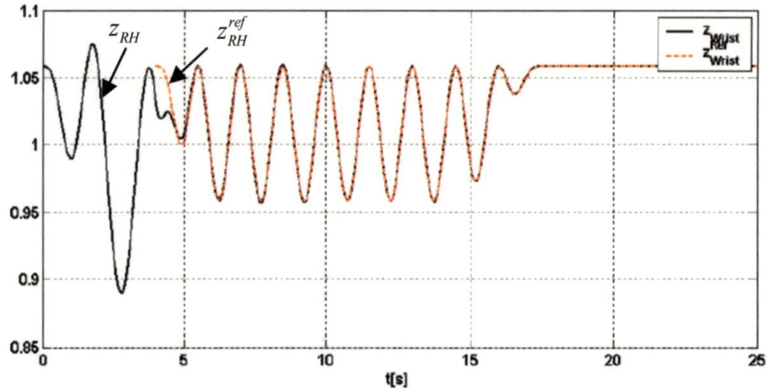

Figure 5. Vertical hands positions. The decoupled position controller is activated at $t = 4[s]$. Because the proposed control is a PD in closed loop, the desired trajectory is reached after a short transition period

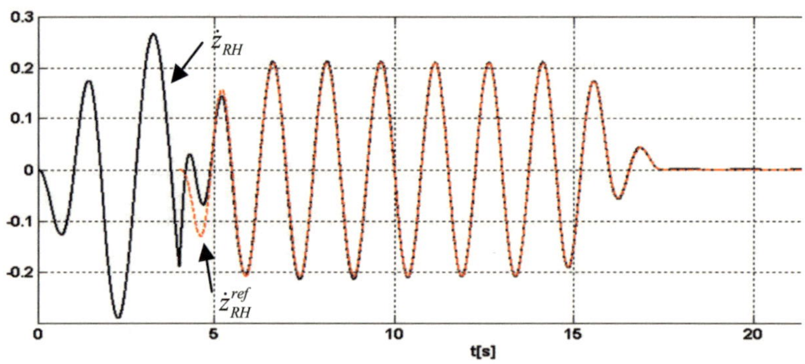

Figure 6 Vertical hands velocity

For the trajectory of the CoM we used cosinus with amplitude of the on the z direction of $5[cm]$ while a sinus of amplitude was $2.5[cm]$ for the x component; both signal with period of $T = 2[s]$. During the first period the signal grows from 0 to the final amplitude, this is in order to respect the initial condition of the robot being static, that is $\dot{\mathbf{r}}_{RF,G}(t) = \dot{\mathbf{r}}_{RF,G}(t) = 0$. This means

$$\begin{pmatrix} r_{Gx} \\ r_{Gz} \\ \int L_{Gy} \end{pmatrix} = \begin{pmatrix} 0.025\sin\left[2\pi(1/2)t\right] \\ 0.05\cos\left[2\pi(1/2)t\right] \\ 0 \end{pmatrix} \tag{21}$$

We can notice that the reference motion of the CoM gradually stops to arrive to zero acceleration at $t = 12[s]$. After this time and before $t = 18s$ the IFPC compensate the hands motion in order to keep G at a fixed position. This is

$$CoM \quad Trajectory \begin{cases} t \leq 12\text{s} & r_G^{\text{Ref}} \\ t \geq 12\text{s} & \dot{r}_G^{\text{Ref}} = \ddot{r}_G^{\text{Ref}} = 0 \end{cases}$$

The consequent ground reaction force and the signals from the force sensor (located at RF in HRP-2) are compared in Fig. 7 and Fig. 8.

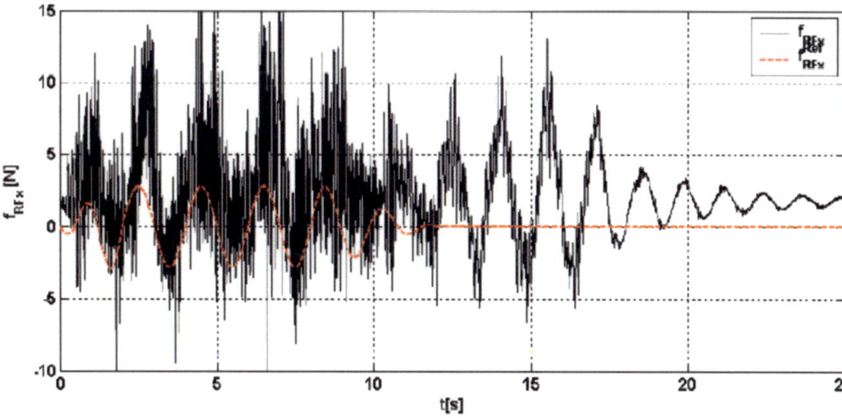

Figure 7. Desired and measured ground reaction force on the x direction

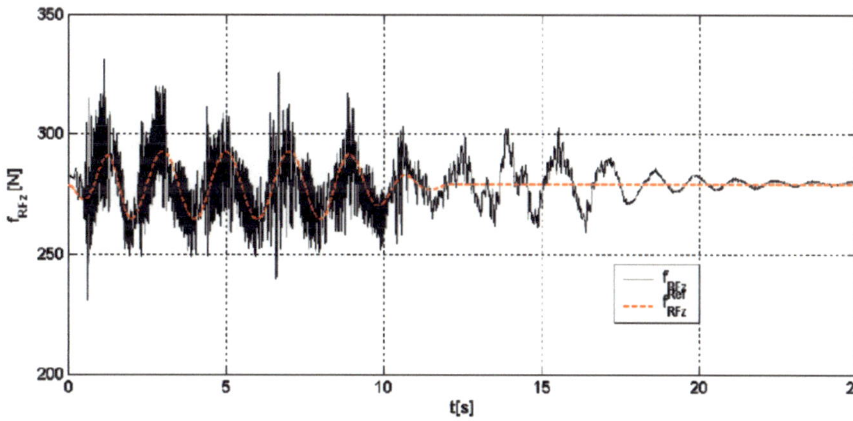

Figure 8. Desired and measured ground reaction force on the z direction

Concerning the angular momentum, as shown on (21) we specified $L_{Gy} = 0$ during the whole motion. The consequent ground reaction torque is shown on next figure.
Finally, the desired zmp point, which is consequence of the desired CoM motion and zero angular momentum as reference is shown in Fig. 10.
On Fig. 7 to Fig. 10 we can see that the following of the inertial forces, related to the CoM trajectory, eq. (3), is better when the CoM of the robot is not static, i.e. before $t = 12s$. The difference between real and reference values in those figures may be explained as the effect of the compliance of the contact between the foot and the ground. When the CoM is stopped after $t = 12s$, this phenomenon becomes more important and the robot bounces a little forward and backwards. Because the presented control method supposes rigid contact

between the foot and the ground, this bouncing can not be taken into account. However we can see that the presented controller, besides the noise on force signals, imposes the desired dynamic motion of the CoM and angular momentum in order to obtain the desired stable motion for a given trajectory of the upper body.

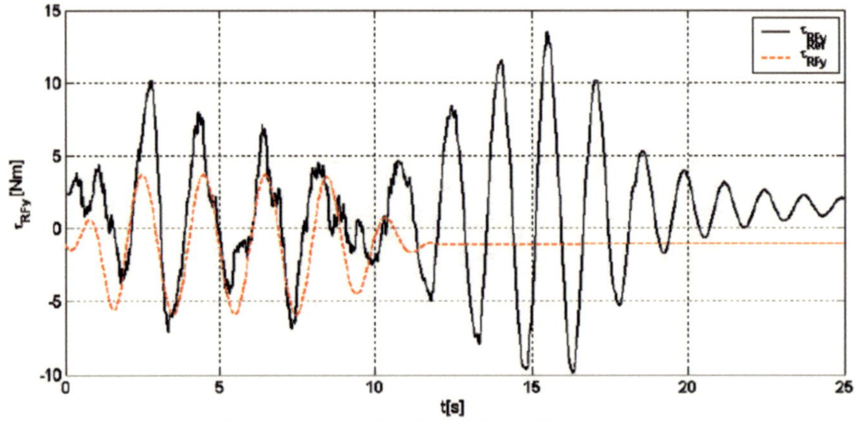

Figure 9. Desired and measured ground reaction torque around y axis

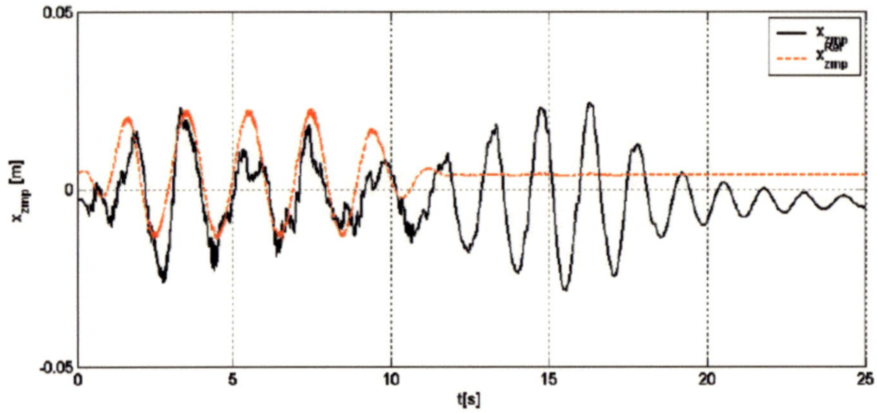

Figure 10. With $L_{Gy} = 0$ the distance from RF to the zmp does not exceed $3.5[cm]$. As a consequence the foot remains flat on the ground

5. Conclusions and perspectives

In order to consider a whole body control approach capable of treating different kinds of locomotion modes, the consideration of a motion planing and a the posture controller stages is important. In this paper this approach is presented for the locomotion of a humanoid robot. With the proposed approach motions including aerial phases, can be considered.

The inertial force posture controller presented here requires the locomotion to be specified using generalized inertial force, besides the trajectory of the extremities not in contact with the ground. This inertial forces can be planed for having zmp stable motion or desired angular momentum.

The angles imposing the robot motion are computed using a decoupled position control and a force controller based on Lagrange equation of motion with external forces applied at one point. The proposed approach was validated on experiments using the humanoid robot HRP-2.
The motion generation for walking and running is our main perspective. Besides, the consideration of the compliance of the ground-foot contact should be considered on future works. Finally the compensation of external forces on the hands (carrying objects) would be an extra for our controller.

6. Acknowledgment

The first author is supported by a scholarship from CONACYT, Mexico. The work presented in this paper was done in the frame of the French-Japanese Joint Robotics Laboratory with support of the Japan Society of Promotion of Science (JSPS).

7. References

Fujiwara, K.; Kanehiro, F. ; Saito, H.; Kajita, S.; Harada, K. & Hirukawa, H. (2004) Falling Motion Control of a Humanoid Robot Trained by Virtual Supplementary Tests, in *Proc. of the IEEE Int. Conf. on Robotics and Automation (ICRA)*, pp.-1077-1082 Vol.2 ISBN:0-7803-8232-3. New Orleans USA. May 2004.

Goswami, A.; Kallem, V. (2004); Rate of change of angular momentum and balance maintenance of biped robots, in *Proc of IEEE Int. Conf on Robotics and Automation (ICRA)* pp.-1057-1064 Vol.2 ISBN:0-7803-8232-3. New Orleans USA. May 2004.

Harada ,K. ; Hirukawa, H. ; Kanehiro, F. ; Fujiwara, K.; Kaneko, K.; Kajita, S. & M. Nakamura (2004). Dynamical Balance of a Humanoid Robot Grasping an Environment, in *Proc. of the IEEE Int. Conf. on Intelligent Robots and System (IROS)*, pp,. 1167-1173, Vol. 2, ISBN.- 0-7803-8463-6, Sendai Japan. October 2004.

Kajita, S.; Kanehiro, F.; Kaneko, K.; Fujiwara, K. ; Harada, K. ; Yokoi, K. ; Hirukawa, H. (2003) Resolved momentum control: humanoid motion planning based on the linear and angular momentum, in *Proc. of the IEEE Int. Conf. on Intelligent Robots and Systems (IROS)*. pp.- 1664-1650, Vol.2, ISBN.- 0-7803-7860-1, Las Vegas, Nevada, October 2003.

Kajita, S.; Nagasaki, T.; Kaneko, K.; Yokoi, K.; Tanie, K. (2005) A running controller of humanoid biped HRP-2L, in *Proc. of IEEE Int. Conf. on Robotics and Automation (ICRA)* pp,. 1167-1173, Vol. 2, ISBN.- 0-7803-8463-6, October 2005.

Kanehiro, F. Hirukawa, H. Kaneko, K. Kajita, S. Fujiwara, K. Harada, K. Yokoi, K. (2004), Locomotion planning of humanoid robots to pass through narrow spaces, in *Proc. Of IEEE International Conference on Robotics and Automation (ICRA)*, 2004. pp 604- 609 Vol.1 ISBN. 0-7803-8232-3, May 2004.

Kaneko, K.; Kanehiro, F.; Kajita, S.; Hirukawa, H.; Kawasaki, T.; Hirata, M.; Akachi, K.; Isozumi, T. (2004); Humanoid robot HRP-2, in *Proc. of IEEE Int. Conf. on Robotics and Automation (ICRA)*, pp. 3785- 3790, Vol. 4, ISBN: 0-7803-8232-3, New Orleans, USA, April 2004.

Kondak, K.; Hommel, G.; Stanczyk, B.; Buss, (2005) M. Robust motion control for robotic systems using sliding mode, *IEEE Int. Conf. on Intelligent Robots and Systems (IROS)*. pp.- 1664-1650, Vol.2, ISBN.- 0-7803-7860-1, Edmonton, Canada, August 2005

Nagasaka, K. ; Kuroki, Y. ; Suzuki, S.; Itoh, Y.; Yamaguchi, J. (2004); Integrated motion control for walking, jumping and running on a small bipedal entertainment robot, in *Proc. of the IEEE Int. Conf. on Robotics and Automation (ICRA)* pp.- 3189-3194, Vol. 4, ISBN.- 0-7803-8232-3, April 2004.

Nunez, V.; Nadjar-Gauthier, N.; (2005) Control Strategy for Vertical Jump of Humanoid Robots, in *Proc. of IEEE Int. Conf. on Intelligent Robots and Systems (IROS)* 2005. pp.- 1564-1550, Vol.2, ISBN.- 0-7803-7860-1, Edmonton, Canada, August 2005

Nunez, V.; Nadjar-Gauthier, N.; Blazevic, P.; (2006) Inertial force position control for jumping of humanoid robots. On *Proc. of the IEEE Int. Symposium on Robotics and Automation (ISRA).* pp.- 3189-3194, Vol. 4, ISBN.- 0-7803-8232-3, San Miguel Regla, Mexico, August 2006.

Rok, B.; Choi, Y. ; Yi, B.; Kim, W (2003); A new ZMP constraint equation with application to motion planning of humanoid using kinematic redundancy, in *Proc. of IEEE Int. Conf. on Intelligent Robots and Systems (IROS),* pp.- 4021-4027, ISBN.- 0-7803-8912-3 ,Las Vegas. Nevada October 2003.

Sakka S.; Yokoi K (2005), Humanoid vertical jump based on force feedback and inertial forces optimization, in *Proc. of the IEEE Int. Conf. on Robotics and Automation (ICRA),* pp,. 1367-1373, Vol. 2, ISBN.- 0-7803-8463-6, October 2005

Sardain, P.; Bessonnet, G.; (2004) Forces acting on a biped robot. Center of pressure-zero moment point, *IEEE Transactions on Systems, Man and Cybernetics,* Part A, Volume 34, Issue 5, Page(s):630 - 637, Sept. 2004.

Sian, N.E.; Yokoi, K.; Kajita, S.; Kanehiro, F.; Tanie, K. (2003); Whole body teleoperation of a humanoid robot - a method of integrating operator's intention and robot's autonomy, in *Proc. of IEEE Int. Conf. on Robotics and Automation (ICRA)* pp. 1651-1656, Vol. 2 ISBN: 0-7803-7860-1. Taipei, Taiwan, September 2003

Sugihara, T. & Nakamura, Y. (2003) Contact phase invariant control for humanoid robot based on variable impedant inverted pendulum model, in Proc. of the *IEEE International Conference on Robotics and Automation (ICRA),* pp.-51-56, Volume: 1, ISBN.- 0-7803-7736-2, Taipei, Taiwan, September 2003.

Tiong, Y. K. & Romyaldy (2004); Conceptual framework for designing humanoid locomotion control system, in *Proc. of IEEE Int. Conf. on Robotics, Automation and Mechatronics,* (RAM) pp. 352 – 258. ISBN 0-7803-8645-0 Singapore, December 2004.

9

Towards Adaptive Control Strategy for Biped Robots

Christophe Sabourin[1], Kurosh Madani[1] and Olivier Bruneau[2]
[1]*Université PARIS-XII, Laboratoire Images, Signaux et Systèmes Intelligents*
[2]*Université Versailles Saint-Quentin-en-Yvelines, Laboratoire d'Ingénierie des Systèmes de Versailles*
France

1. Introduction

The design and the control of humanoid robots are one of the most challenging topics in the field of robotics and were treated by a large number of research works over the past decades (Bekey, 2005) (Vukobratovic, 1990). The potential applications of this field of research are essential in the middle and long term. First, it can lead to a better understanding of the human locomotion mechanisms. Second, humanoid robots are intended to replace humans to work in hostile environments or to help them in their daily tasks. Today, several prototypes, among which the most remarkable are undoubtedly the robots Asimo (Sakagami , 2002) and HRP-2 (Kaneko, 2004), have proved the feasibility of humanoid robots. But, despite efforts of a lot of researchers around the world, the control of the humanoid robots stays a big challenge. Of course, these biped robots are able to walk but their basic locomotion tasks are still far from equalizing the human's dynamic locomotion process. This is due to the fact that the control of biped robot is very hard because of the five following points:

- Biped robots are high-dimensional non-linear systems,
- Contacts between feet and ground are unilateral,
- During walking, biped robots are not statically stable,
- Efficient biped locomotion processes require optimisation and/or learning phases,
- Autonomous robots need to take into account of exteroceptive information.

Because of the difficulty to control the locomotion process, the potential applications of these robots stay still limited. Consequently, it is essential to develop more autonomous biped robots with robust control strategies in order to allow them, on the one hand to adapt their gait to the real environment and, on the other hand, to counteract external perturbations.

In the autonomous biped robots' control framework, our aim is to develop an intelligent control strategy for the under-actuated biped robot RABBIT (figure 1) (RABBIT web) (Chevallereau, 2003). This robot constitutes the central point of a project, within the framework of CNRS ROBEA program (Robea-web), concerning the control of walking and running biped robots. The robot RABBIT is composed of two legs and a trunk and has no foot. Although the mechanical design of RABBIT is uncomplicated compared to other biped robots, its control is a more challenging task, particularly because, in phase of single support, this robot is under-actuated. In fact, this kind of robots allows studying real dynamical walking leading to the design of new control laws in order to improve biped robots' current performances.

Figure 1. RABBIT prototype

In addition to the problems related to control the locomotion process (leg motions, stability), it is important to take into account both proprioceptive and exteroceptive information in order to increase the autonomy of this biped robot. The proprioceptive perception is the ability to feel the position or movements of parts of the body and the exteroceptive perception concerns the capability to feel stimuli from outside of the body. But the both proprioceptive and exteroceptive information are not treated in the same manner. The proprioceptive information, which are for example the relative angles between two limbs and the angular velocity, allow to control the motion of the limbs during one step. The exteroceptive perception must allow to obtain information about the environment around the biped robot. These exteroceptive information allow using predictive strategies in order to adapt the walking gait regarding the environment.
In fact, although the abilities of RABBIT robot are limited in comparison to other humanoid robots, our goal in middle term, is to design a control strategy for all biped robots.
In our previous works, we used CMAC (Cerebellar Model Articulation Controller) neural networks to generate the joint trajectories of the swing leg but, for example, the length of the step could not be changed during the walking (Sabourin, 2005) (Sabourin, 2006). However, one important point in the field of biped locomotion is to develop a control strategy able to modulate the step length at each step. In this manner, in addition to modulate the step length according to the average velocity, like human being, the biped robot can choice at each step the landing point of the swing leg in order to avoid obstacle. But in general, as in the case of human being, the exteroceptive information allowing to give information about obstacles in the near environment of the robot are not precise measures. Consequently, we prefer to use fuzzy information. However this implies to deal with heterogeneous data, which is not a trivial problem. One possible approach consists to use soft-computing techniques and/or pragmatic rules resulting from the expertise of the walking human.

Moreover, this category of techniques takes advantage from learning (off-line and/or on-line learning) capabilities. This last point is very important because generally the learning ability allows increasing the autonomy of the biped robot.
Our control strategy uses a gait pattern based on Fuzzy CMAC neural networks. Inputs of this gait pattern are based on both proprioceptive and exteroceptive information. The Fuzzy CMAC approach requires two stages:

- First, the training of each CMAC neural networks is carried out. During this learning phase, the virtual biped robot is controlled by a set of pragmatic rules (Sabourin, 2005) (Sabourin, 2004). As a result, a stable reference dynamic walking is obtained. The data learnt by CMACs are only the trajectories of the swing leg.
- After this learning phase, we use a merger of the CMAC trajectories in order to generate new gaits.

In addition, a high level control allows us to modify the average velocity of the biped robot. The principle of the control of the average velocity is based on the modification, at each step, of the pitch angle.
The first investigations, only realized in simulation, are very promising and proved that this approach is a good way to improve the control strategy of a biped robot. First, we show that, with only five reference gaits, it is possible to adjust the step of the length as a function of the average velocity. In addition, with a fuzzy evaluation of the distance between feet and an obstacle, our control strategy allows to the biped robot to avoid obstacle using step over strategy.
This paper is organized as follows. After a short description of the real robot RABBIT, section 2 gives the main characteristics of the virtual under-actuated robot used in our simulations. In Section 3, firstly you remind the principles of CMAC neural networks and the Takagi-Sugeno fuzzy inference system, secondly Fuzzy CMAC neural networks are presented. Section 4 describes the control strategy with a gait pattern based on the Fuzzy CMAC structure. The learning phase of each CMAC neural network is presented in section 5. In section 6, we give the main results obtained in simulation. Conclusions and further developments are finally set out.

2. Virtual modelling of the biped robot RABBIT

RABBIT robot has only four joints: one for each knee, one for each hip. Motions are included in the sagittal plane using a radial bar link fixed on a central column that allows to guide robot's advance around a circle. Each joint is actuated by a servo-motor RS420J. Four encoders make it possible to measure the relative angles between the trunk and the thigh for the hip, and between the thigh and the shin for the knee. Another encoder, installed on the bar link, gives the pitch angle of the trunk. Two binary contact sensors detect whether or not the leg is in contact with the ground. Based on the information given by the encoders, it is possible to calculate the length of the step L_{step} when the two legs are in contact with the ground. The duration of the step t_{step} is computed using the contact sensor information (the duration from take-off to landing of the same leg). Furthermore, it is possible to estimate the average velocity V_M using (1).

$$V_M = \frac{L_{step}}{t_{step}} \tag{1}$$

The characteristics (masses and lengths of the limbs) are summarized in table 1.

Limb	Weight(Kg)	Length(m)
Trunk	12	0.20
Thigh	6.8	0.40
Shin	3.2	0.47

Table 1. Robot's limb masses and lengths

Since the contact between the robot and the ground is just one point (passive DOF), the robot is under-actuated during the single support phase: there are only two actuators (at the knee and at the hip of the stance leg) to control three parameters (vertical and horizontal position of the platform and pitch angle). The numerical model of the robot previously described was designed with the software ADAMS[1] (figure 2)

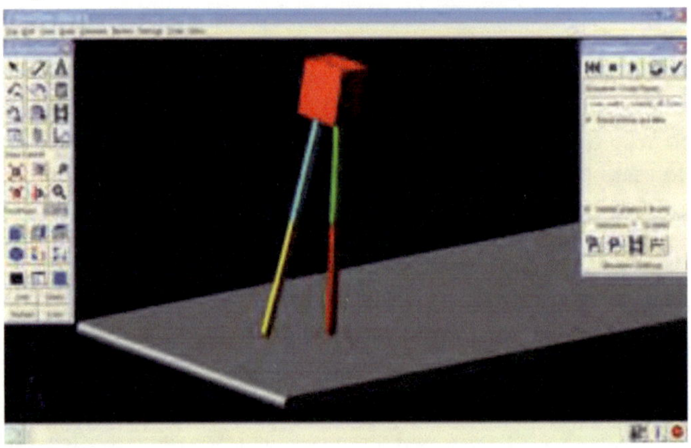

Figure 2. Modelling of the biped robot with ADAMS

This software, from the mechanical system's modelling point of view (masses and geometry of the segments) is able to simulate the dynamic behaviour of such a system and namely to calculate the absolute motions of the platform as well as the limb relative motions when torques are applied on the joints by virtual actuators. Figure 3 shows references for the angles and the torques required for the development of our control strategy.

q_{i1} and q_{i2} are respectively the measured angles at the hip and the knee of the leg i. q_0 corresponds to the pitch angle. T^{sw}_{knee} and T^{sw}_{hip} are the torques applied respectively to the knee and the hip during the swing phase, T^{st}_{knee} and T^{st}_{hip} are the torques applied during the stance phase.

The interaction between feet and ground is based on a spring-damper modelling. This approach allows to simulate more realistic feet-ground interaction namely because the contact between the feet and the ground is compliant. However, in order to take into account the possible phases of sliding, we use a dynamic friction modelling when the tangential contact forces is located outside the cone of friction. The normal contact force F_n is given by equation (2):

1 ADAMS is a product of MSC software.

$$F_n = \begin{cases} 0 & if \quad y > 0 \\ -\lambda_n |y| \dot{y} + k_n |y| & if \quad y \leq 0 \end{cases} \tag{2}$$

y and $\dot{y}$ are respectively the position and the velocity of the foot (limited to a point) with regard to the ground. k_n and λ_n are respectively the generalized stiffness and damping of the normal forces. They are chosen to avoid the bouncing and limit the foot penetration in the ground. Tangential contact force F_t is computed by using equation 3 with F_{t1} and F_{t2} which are respectively the tangential contact force without and with sliding.

$$F_t = \begin{cases} F_{t1} & if \quad \|F_{t1}\| < \mu_s \|F_n\| \\ F_{t2} & if \quad \|F_{t1}\| \geq \mu_s \|F_n\| \end{cases} \tag{3}$$

With:

$$F_{t1} = \begin{cases} 0 & if \quad y > 0 \\ -\lambda_t \dot{x} + k_t (x - x_c) & if \quad y \leq 0 \end{cases} \tag{4}$$

$$F_{t2} = \begin{cases} 0 & if \quad y > 0 \\ -(\text{sgn}(\dot{x}))\lambda_g F_n - \mu_g \dot{x} & if \quad y \leq 0 \end{cases} \tag{5}$$

x and $\dot{x}$ are respectively the foot position and the velocity with regard to the position of the contact point x_c at the instant of impact with the ground. k_t and λ_t are respectively the generalized stiffness and damping of the tangential forces. λ_g is the coefficient of dynamic friction depending on the nature of surfaces coming into contact, μ_g a viscous damping coefficient during sliding, and μ_s is the static friction coefficient.

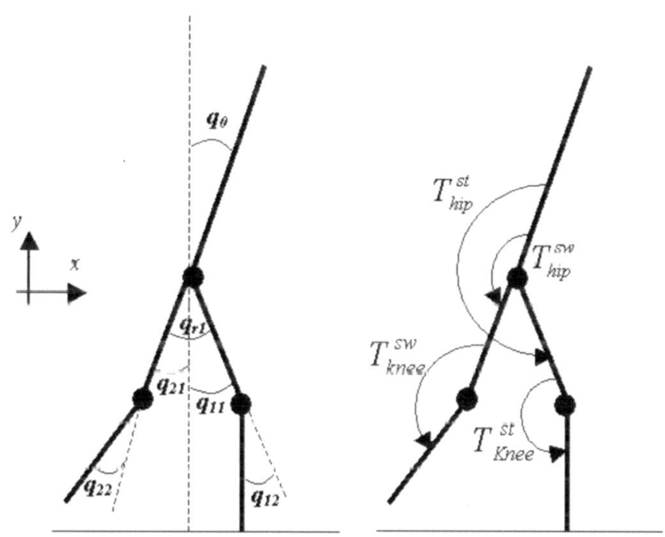

Figure 3. Angle and torque parameters

In the case of the control of a real robot, its morphological description is insufficient. It is thus necessary to take into account the technological limits of the actuators in order to implement the control laws used in simulation on the experimental prototype. From the characteristics of servo-motor RS420J used for RABBIT, we thus choose to apply the following limitations:

- when velocity is included in $[0{,}2000]\,rpm$, the torque applied to each actuator is limited to $1.5\ Nm$ which corresponds to a torque of $75\ Nm$ at the output of the reducer (ration gear is equal to 50),
- when the velocity is included in $[2000{,}4000]\,rpm$ the power of each actuator is limited to $315\,W$,
- when the velocity is bigger than $4000\ rpm$, the imposed torque is equal to zero.

3. Fuzzy-CMAC neural network

The CMAC is a neural network imagined by Albus from the studies on the human cerebellum (Albus, 1975a), (Albus, 1975b). CMAC is a neural network with local generalization abilities. This means that only a small number of weights are necessary to compute the output of this neural network. Consequently, the main interest is the reduction of training and computing times compared with other neural networks (Miller, 1990). This is of course a considerable advantage for real time control. Numerous researchers have investigated CMAC and have applied this approach to the field of control namely for biped robots' control and related applications (kun, 2000), (Brenbrahim, 1997). However, it is pertinent to remind that the memory used by CMAC (e.g. the needed memory size) depends firstly on the input signal quantification step and secondly of the input space size (dimension). For real CMAC based control applications, the CMAC memory size becomes quickly very big. In fact, on the one hand, in order to increase the accuracy of the control the chosen quantification step must be as small as possible; on the other hand, generally in real world applications the input space dimension is greater than two. In order to overcome the problem relating to the size of the memory, a hashing function is used. But in this case, because the size of the memory allowing to store the weights of the neural network is smaller than the size of the virtual addressing memory, some collisions can occur. Another problem occurring in the case of multi-input CMAC is the necessity to set out a learning database covering the whole input space. This is due to the CMAC local generalization abilities and results in yielding enough data (either by performing a large number of simulations available from a significant experimental setup) to wrap all possible states.
We propose a new approach making it possible to take advantage of both local and global generalization capacities with the Fuzzy CMAC neural networks. Our Fuzzy CMAC approach is based on a merger of all the outputs of several Single Input/Single Output (SISO) CMAC neural networks. This merger is carried out using Takagi-Sugeno Fuzzy Inference System. This allows both to decrease the size of the memory and to increase the generalization abilities compared with a multi-input CMAC. In this section, as a first step, we present a short description of SISO CMAC neural network. Sub-section 3.2 describes the Takagi-Sugeno Fuzzy Inference System. Finally, in sub-section 3.3 the proposed Fuzzy-CMAC approach is presented.

3.1 SISO CMAC neural networks

CMAC is an associative memory type neural network. Its structure includes a set of N_d detectors regularly distributed on several N_l layers. The receptive fields of these detectors cover the totality of the input signal but each field corresponds to a limited range of inputs. On each layer, the receptive fields are shifted to a quantification step Δ_q. When the input signal is included in the receptive field of a detector, it is activated. For each value of the input signal, the number of activated detectors is equal to the number of layers N_l (a parameter of generalization). Figure 4 shows a simplified organization of the receptive fields having *14* detectors $(N_d = 14)$ distributed on *3* layers $(N_l = 3)$. Taking into account the receptive fields overlapping, neighbouring inputs will activate common detectors. Consequently, this neural network is able to carry out a generalization of the output calculation for inputs close to those presented during learning (local generalization). The output O of the CMAC is computed using two mappings. The first mapping projects an input space point e into a binary associative vector $D = [d_1, \ldots, d_{Nd}]$. Each element of D is associated with one detector. When one detector is activated, the corresponding element in D of this detector is 1 otherwise it is equal to 0.

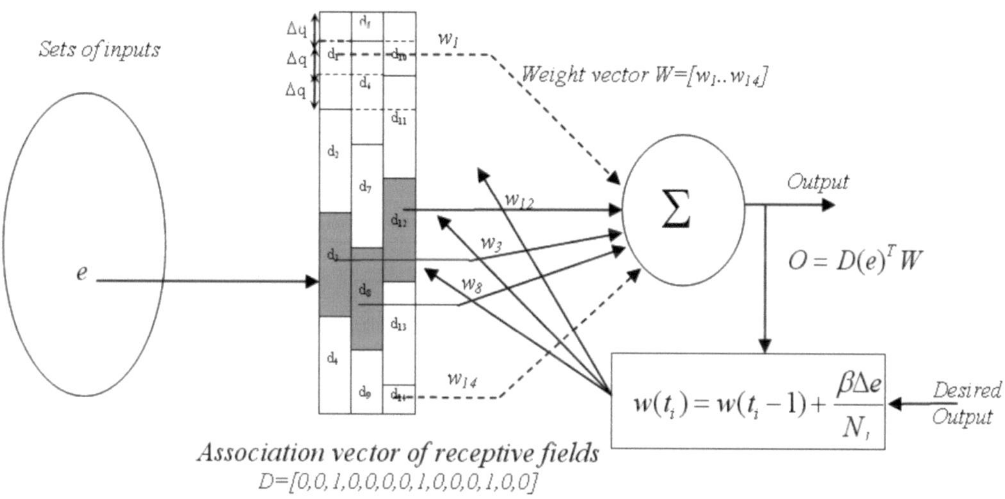

Figure 4. Description of the simplified CMAC with 14 detectors distributed on 3 layers

The second mapping computes the output O of the network as a scalar product of the association vector D and the weight vector $W = [w_1, \ldots, w_{Nd}]$ according to the relation 6, where $(e)^T$ represents the transpose of the input vector.

$$O = D(e)^T W \tag{6}$$

The weights of CMAC are updated by using equation 7:

$$w(t_i) = w(t_{i-1}) + \frac{\beta \, \Delta e}{N_l} \tag{7}$$

$w(t_i)$ and $w(t_{i-1})$ are, respectively, the weights before and after training at each sample time t_i (discrete time). N_l is the generalization number of each CMAC and β is a parameter included in $[0,1]$. Δe is the error between the desired output O^d of the CMAC and the computed output O of the corresponding CMAC.

3.2 Takagi-Sugeno fuzzy inference system

Generally, the Takagi-Sugeno Fuzzy Inference System (TS-FIS) is described by a set of R_k $(k = 1..N_k)$ fuzzy rules such as equation 8:

$$if\ x_1\ is\ A_1^j ...and...x_i\ is\ A_i^j\ then\ y_k = f_k(x_1,...x_{Ni}) \tag{8}$$

x_i $(i = 1..N_i)$ are the inputs of the FIS with N_i the dimension of the input space. A_i^j $(j = 1..N_j)$ are linguistic terms, representative of fuzzy sets, numerically defined by membership functions distributed in the universe of discourse for each input x_i. Each output rule y_k is a linear combination of input variables $y_k = f_k(x_1,...,x_{Ni})$ (f_k is a linear function of x_i). Figure 5 shows the structure of TS-FIS. It should be noted that TS-FIS with Gaussian membership functions is similar to the Radial Basis Function Neural Networks.

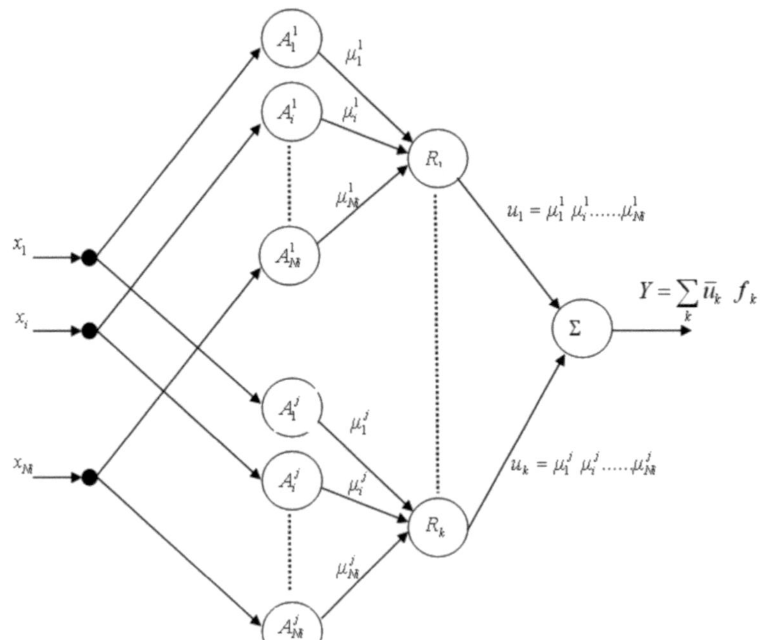

Figure 5. Description of the Takagi-Sugeno Fuzzy Inference System

The calculation of one output of TS-FIS is decomposed into three stages:

- The first stage corresponds to fuzzification. For each condition "x_i *is* A_i^j"; it is necessary to compute μ_i^j which is the numerical value of x_i input signal in the fuzzy set A_i^j.

- In the second stage, the rule base is applied in order to determine each u_k $(k = 1..N_k)$. u_k is computed using equation 9:

$$u_k = \mu_1^j \; \mu_2^j .... \mu_{Ni}^j \tag{9}$$

- The third stage corresponds to the defuzzification phase. But for TS-FIS, the output numerical value Y is carried out using the weighted average of each rule output y_k (equation 10) .

$$Y = \sum_k \overline{u_k} \; y_k \tag{10}$$

With $\overline{u_k}$ is given by equation 11:

$$\overline{u_k} = u_k \,/\, \sum_{k=1}^{N_r} u_k \tag{11}$$

Furthermore, in the case of the zero order Takagi-Sugeno, the rule outputs are a singleton. Consequently, for each k rule, $y_k = f_k(x_1,.....,x_n) = C_k$ where C_k is a constant value independent of the x_i input.

3.3 Fuzzy CMAC

Our Fuzzy CMAC architecture uses a combination of a set of several Single Input/Single Output CMAC neural networks and Takagi-Sugeno Fuzzy Inference System. Figure 6 describes the Fuzzy-CMAC structure with two input signals: e and X. e is the input signal which is applied at all the $CMAC_k$. $X = [x_1,...,x_{Ni}]$ corresponds to the input vector of FIS. Consequently, the output of the Fuzzy CMAC depends on the one hand on TS-FIS and on the other hand on the outputs of a set of SISO CMAC.

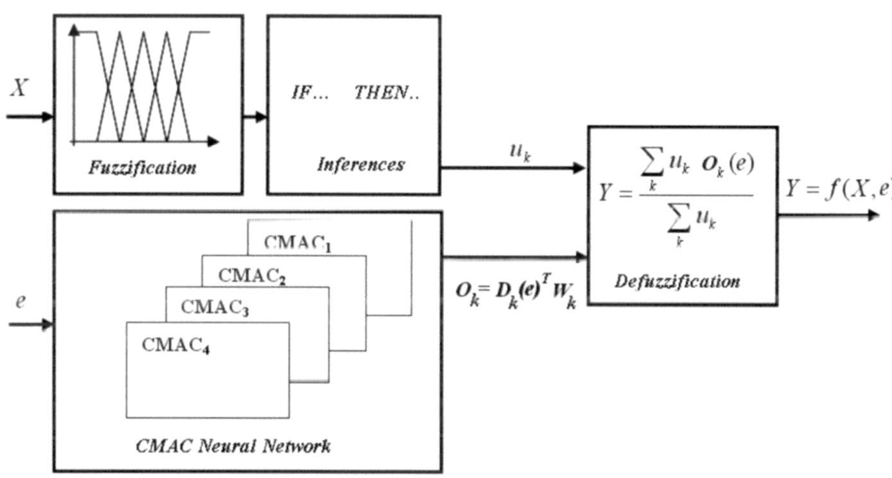

Figure 6. Bloc-diagram of the proposed Fuzzy CMAC structure

The calculation of Y is carried out in two stages:

- First, the output of each $CMAC_k$ is given by equation (12). D_k and W_k are respectively the binary associate vector and the weight vector of each $CMAC_k$ (see section 3.1).

$$O_k(e) = D_k(e)^T W_k \tag{12}$$

- Second, the output Y is carried out using equation (13). In fact, Y is computed using the weighted average of all CMAC outputs.

$$Y = \sum_k \overline{u_k}\, O_k(e) \tag{13}$$

This approach is an alternative solution of the Multi Input/Multi Output CMAC neural networks. The main advantages of the Fuzzy CMAC structure compared to MIMO CMAC are:

- First, the reduction of the size memory because the Fuzzy CMAC uses a small set of SISO CMAC,
- The global generalization capabilities because the Fuzzy CMAC uses a merger of all outputs of CMACs.

In our control strategy, we use Fuzzy CMAC to design a gait pattern for the biped robot. After a training phase of each CMAC, the Fuzzy CMAC allows us to generate the motion of the swing leg. In the next section, we present the principle used to train each CMAC neural network.

4. Training of the CMAC neural networks

During the learning phase, we use an intuitive control, based on five pragmatic rules, allowing us to perform a dynamic walking of our virtual under-actuated robot without reference trajectories. It must be pointed out that during this first stage, we both consider that the robot moves in an ideal environment (without any disturbance) and the frictions are negligible. As frictions are negligible, these fives rules allow us to generate the motions of the legs using a succession of passive and active phases. This intuitive control strategy, directly inspired from human locomotion, allows us to perform a stable dynamic walking using the intrinsic dynamic of the biped robot. It is thus possible to modify the length of the step and the average velocity by an adjustment of several parameters (Sabourin-2004). Consequently, this approach allows us to generate several reference gaits which are learnt by a set of CMAC neural networks.

In the next sub-section, a short description of the pragmatic rules to control the biped robot during the training of the CMAC neural network is presented. In sub-section 4.2, we show how the CMAC neural networks are trained. Finally, we give the main parameters for five walking used during the learning phase (Sub-section 4.3).

4.1. Pragmatic rules

The intuitive control strategy is based on the following five intuitive rules:

- During the swing phase, the torque applied to the hip given by equation (14) is just an impulse with a varying amplitude and a fixed duration equal to $(t_2 - t_1)$.

$$T_1 = \begin{cases} K_{hip}^{pulse} & if\ t_1 < t < t_2 \\ 0 & otherwise \end{cases} \tag{14}$$

Where K_{hip}^{pulse} is the amplitude of the torque applied to the hip at the beginning of the swing phase, and t_1 and t_2 are respectively the beginning and the end of actuation K_{hip}^{pulse}.

- After this impulse, the hip joint is passive until the swing leg is blocked in a desired position using a PD control given by equation (15), which makes it possible to ensure a regular step length.

$$T_2 = K_{hip}^{p}(q_{r1}^{dsw} - q_{r1}) - K_{hip}^{v}\dot{q}_{r1} \quad if\ q_{r1} > q_{r1}^{dsw} \tag{15}$$

q_{r1} and q_{r1}^{dsw} are respectively the measured and desired relative angles between the two thighs, and $\dot{q}_{r1}$ is the relative angular velocity between the two thighs.

- During the stance phase, the torque applied to the hip, given by the equation (16), is used to ensure the stability of the trunk.

$$T_3 = K_{trunk}^{p}(q_0^{d} - q_0) - K_{trunk}^{v}\dot{q}_0 \tag{16}$$

Where q_0 and $\dot{q}_0$ are respectively the angle and the angular velocity of the trunk and q_0^d corresponds to the desired pitch angle of the trunk.

- During the swing phase, the knee joint is free and the torque is equal to zero. At the end of the knee extension, a control torque, given by the equation (17) is applied to lock this joint in a desired position q_{i2}^{dsw}.

$$T_4 = K_{knee}^{p}(q_{i2}^{dsw} - q_{i2}) - K_{knee}^{v}\dot{q}_{i2} \tag{17}$$

q_{i2} and $\dot{q}_{i2}$ are respectively the measured angular position and angular velocity of the knee joint of the leg i.

- During the stance phase, the torque is computed by using equation (18).

$$T_5 = K_{knee}^{p}(q_{i2}^{dst} - q_{i2}) - K_{knee}^{v}\dot{q}_{i2} \tag{18}$$

We choose $q_{i2}^{dst} = 0$ at the impact with the ground in equation (18) which contributes to propel the robot if $q_{i2}^{dst} > q_{i2}^{dsw}$. During the continuation of the stance phase, the same control law is used to lock the knee in the position $q_{i2}^{dst} = 0$.

4.2. Training CMACs

Figure 7 shows the method used to train CMACs neural networks. For each reference gait, four SISO $CMAC_l$ $(l = 1,..,4)$ neural networks learnt the trajectories of the swing leg (in terms of joint positions and velocities). Furthermore, we have considered that the trajectories of each leg in swing phase are identical. This allows to divide by two the number of CMAC and to reduce the training time. Consequently, two SISO CMACs are necessary to memorize the joint angles q_{i1} and q_{i2} and two other SISO CMACs for angular velocities $\dot{q}_{i1}$ and $\dot{q}_{i2}$. q_{i1} and q_{i2} are respectively the measured angles at the hip and the knee of the leg i; $\dot{q}_{i1}$ and $\dot{q}_{i2}$ are respectively the measured angular velocities at the hip and the knee of the leg i (see figure 3).

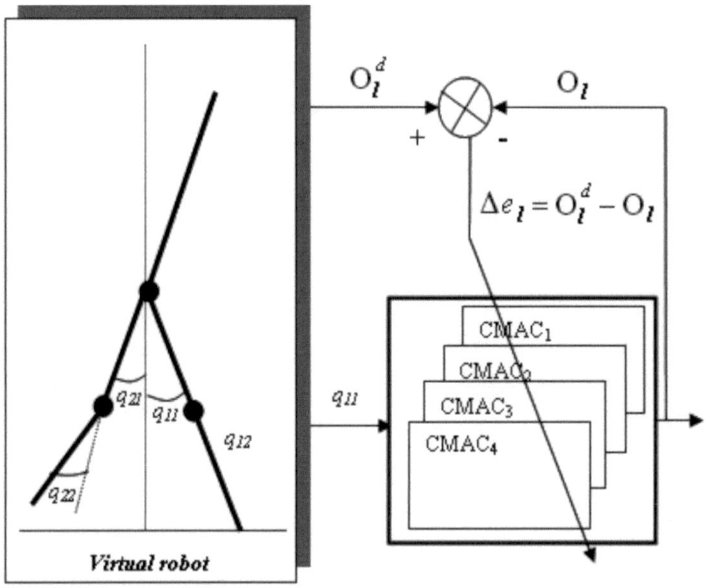

Figure 7. Principle of the learning phase of CMAC neural networks ($e = q_{11}$)

When leg 1 is in support, the angle q_{11} is applied to the input of each $CMAC_l$ ($e = q_{11}$) and when leg 2 is in support, this is the angle q_{21} which is applied to the input of each $CMAC_l$ ($e = q_{21}$). Consequently, the trajectories learnt by the neural networks are a function of the geometrical pattern of the robot. The weights of each $CMAC_l$ are updated by using the error between the desired output O_l^d ($O_1^d = q_{i1}^d, O_2^d = q_{i2}^d, O_3^d = \dot{q}_{i1}^d, O_4^d = \dot{q}_{i2}^d$) of each $CMAC_l$ and the computed output O_l of the corresponding $CMAC_l$. Based on the previous consideration, it is possible to learn N_r different reference walking using $N_r \times 4$ CMACs. In the case of the simulations presented in this section, each CMAC has 6 layers ($N_l = 6$). The width of the receptive fields is equal to $1.5°$ and the quantification step Δ_q is equal to $0.25°$.

4.3. Reference gaits

During the training stage, five reference gaits with an average velocity V_M included in $[0.4..0.8]$ have been learnt by 5×4 single input/single output $CMAC^r$ ($N_r = 5$ and 4 CMACs for one reference walking). Table 2 gives the main parameters which are used during the learning phase according to the average velocity V_M.

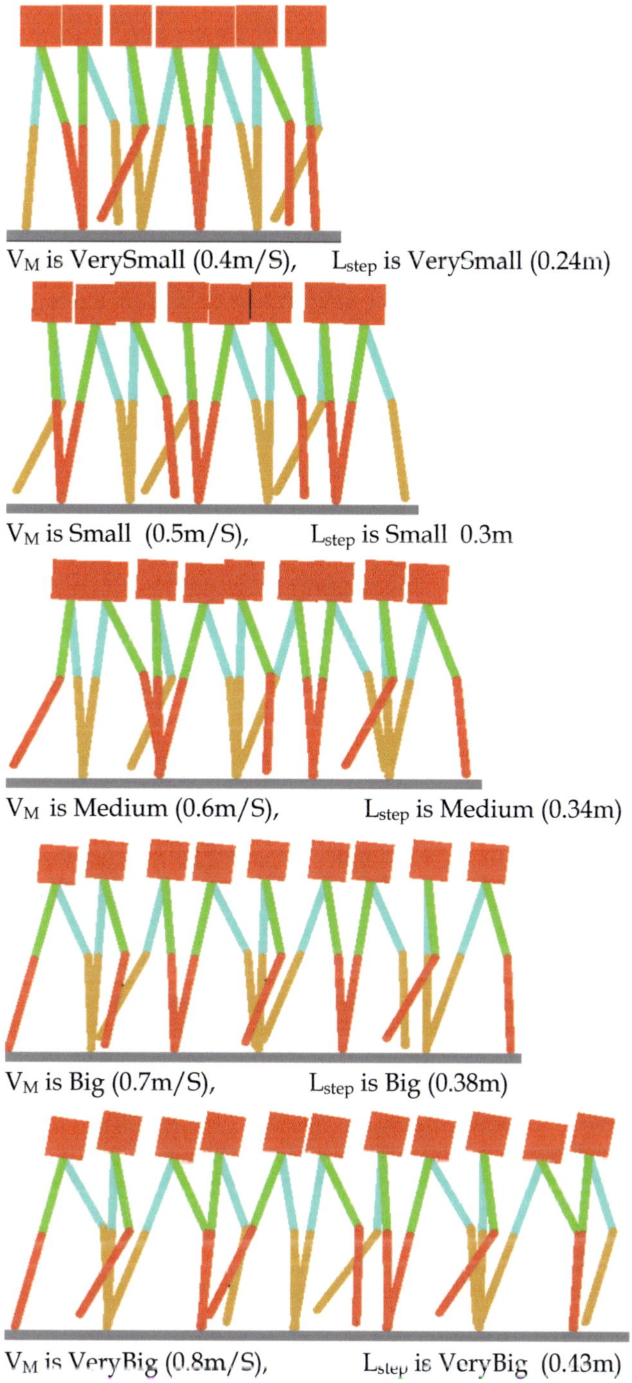

Figure 8. Stick-diagram of the walking robot for five different velocities. V_M and L_{step} are respectively, from the top to the bottom, *VerySmall, Small, Medium, Big and VeryBig*.

	$V_M(m/s)$	$L_{step}(m)$	$q_{r1}^{dsw}(°)$	$q_{i2}^{dsw}(°)$	$q_0^d(°)$
$CMAC^1$	0.4	0.24	20	-7	0
$CMAC^2$	0.5	0.3	25	-10	1.5
$CMAC^3$	0.6	0.34	30	-14	4
$CMAC^4$	0.7	0.38	35	-20	6.5
$CMAC^5$	0.8	0.43	40	-25	10.5

Table 2. Parameters used during the learning stage for five different average velocities

q_{r1}^{dsw} is the desired relative angle between the two thighs (see equation 15), and q_0^d the desired pitch angle of the trunk (see equation 16). q_{i2}^{dsw} corresponds to the desired angle of the knee at the end of the knee extension of the swing leg just before the double contact phase (see equation 17). Each reference walking is characterized by a set of parameters $(q_{r1}^{dsw}, q_{i2}^{dsw}, q_0^d)$ allowing to generate different walking gaits (V_M, L_{step}). Figure 8 shows stick-diagrams representing, for five average velocities V_M and the corresponding step of the length L_{step}, the walking of the biped robot during 2.8s (approximately 6 steps). V_M and L_{step} are respectively, from the top to the bottom, *VerySmall, Small, Medium, Big, VeryBig*. It must be pointed out that each reference gait are really different and the step length L_{step} increases when V_M increases.

Based on the five reference gaits, the goal of our approach is to generate new gaits using a merger of these five learnt gaits. Consequently, after this training phase, we use a mixture between Fuzzy-Logic and the outputs of the CMACs neural network in order to generate the trajectories of the swing leg and consequently to modulate the length of the step.

In the next section, we present the control strategy based on the Fuzzy CMAC neural networks. In addition, we present the control which is used to regulate the average velocity.

5. Control strategy based on both proprioceptive and exteroceptive information

Figure 9 shows the control strategy which is used to control the walking robot. It should be noted that the architecture of this control can be decomposed into three parts:

- The first is used to compute the trajectories of the swing leg from several outputs of the $CMAC_l$ neural networks and a Fuzzy Inference System (Gait pattern). The goal of this part is, on the one hand, to adjust the step length as function of the average velocity, and on the other hand, to adapt step length in order to the robot step over obstacle.
- The second one allows the regulation of the average velocity V_M from a modification of the pitch angle q_0. When the pitch angle increases, the average velocity increases and when the pitch angle decreases, the average velocity decreases. It's in fact a good and easy way to control the average velocity of the biped robot because V_M is function of q_0.
- The third is composed by four PD control in order to ensure the tracking of the reference trajectories at the level of each joint.

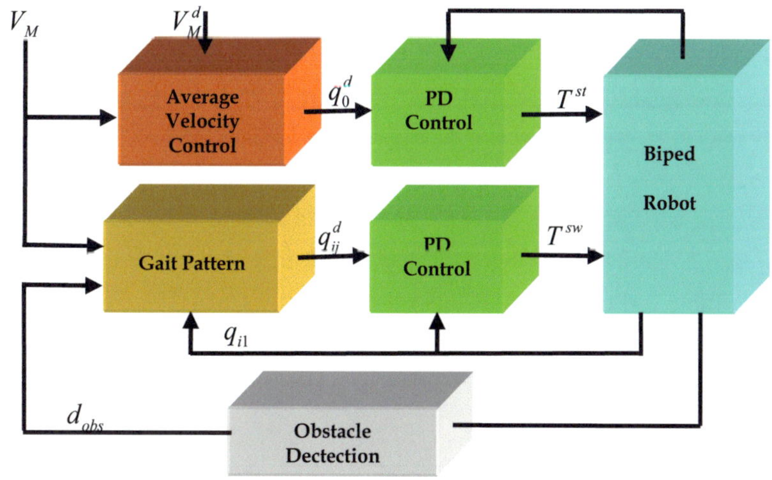

Figure 9. Structure of the control strategy for biped robot

In this section, sub-section 5.1 describes the gait pattern based on the Fuzzy CMAC approach. In sub-section 5.2, the principle of the control of the average velocity is presented. And finally, we give the control laws making it possible the track of the desired trajectories.

5.1 Gait pattern

Our gait pattern is specially designed to adjust the length of the step during walking taking into account of both proprioceptive and exteroceptive information. The inputs of the gait pattern are $e = q_{i1}$ and $X = [V_M, d_{obs}]$ where d_{obs} and V_M represent respectively, the distance between the foot, and the obstacle and the measured average velocity. During the walking, the input e is directly applied at each input of each $CMAC_k^l$. $e = q_{11}$ if leg 1 is in support, and $e = q_{21}$ if leg 2 is in support. But, the measures V_M and d_{obs} are represented using fuzzy sets. Figures 10 and 11 show the membership functions used respectively for V_M and d_{obs}. V_M and d_{obs} are modelled by five fuzzy sets (VerySmall, Small, Medium, Big, VeryBig).

Consequently, the desired angles q_{i1}^d and q_{i2}^d, and the desired angular velocities $\dot{q}_{i1}^d$ and $\dot{q}_{i2}^d$ are carried out by using a merger of the five learnt trajectories. This merger is realized by using TS-FIS. The choice of the fuzzy rules is carried out using pragmatic rules.

Without obstacle ($d_{obs} > 0.5m$), the length of the step is only a function of the average velocity. As human being, more V_M increases and more L_{step} increases. The five following rules allow us to adjust the step of the length as a function of the measured average velocity:

- If d_{obs} *is VeryBig* and V_M *is VerySmall* then L_{step} *is VerySmall*
- If d_{obs} *is VeryBig* and V_M *is Small* then L_{step} *is Small*
- If d_{obs} *is VeryBig* and V_M *is Medium* then L_{step} *is Medium*
- If d_{obs} *is VeryBig* and V_M *is Big* then L_{step} *is Big*
- If d_{obs} *is VeryBig* and V_M *is VeryBig* then L_{step} *is VeryBig*

This implies that when L_{step} *is VerySmall*, L_{step} *is Small*, L_{step} *is Medium*, L_{step} *is Big*, L_{step} *is VeryBig*, the trajectories of the swing leg $[q_{i1}^d, q_{i2}^d, \dot{q}_{i1}^d, \dot{q}_{i2}^d]$ are computed using respectively data held into $CMAC^1$, $CMAC^2$, $CMAC^3$, $CMAC^4$, $CMAC^5$.

When an obstacle is near of the robot ($d_{obs} < 0.5m$), the length of the step depends of the distance between the foot of the robot and this obstacle. Consequently, if d_{obs} *is Big* or d_{obs} *is Medium*, we choice to decrease the length of the step. And, if d_{obs} *is Small* or d_{obs} *is VerySmall*, we prefer to increase step length in order to the robot directly step over the obstacle. Table 3 shows all rules used by the Fuzzy CMAC in the case of the presented gait pattern.

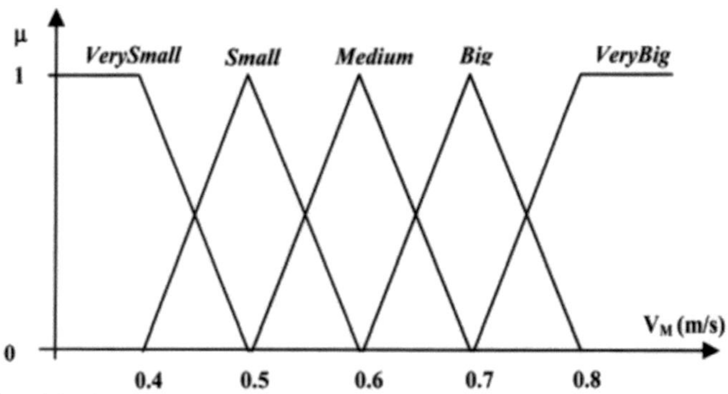

Figure 10. Membership functions used to compute V_M

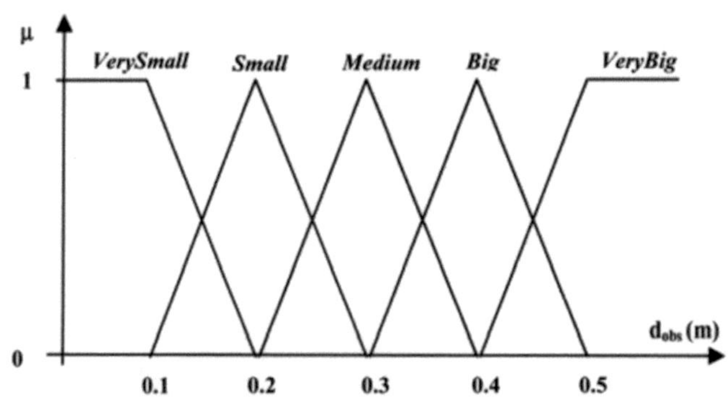

Figure 11. Membership functions used to compute d_{obs}

d_{obs} \ V_M	*VerySmall*	*Small*	*Medium*	*Big*	*VeryBig*
VerySmall	O^4	O^4	O^4	O^4	O^5
Small	O^5	O^5	O^5	O^5	O^5
Medium	O^1	O^1	O^1	O^1	O^1
Big	O^1	O^2	O^3	O^4	O^4
VeryBig	O^1	O^2	O^3	O^4	O^5

Table 3. Fuzzy rules (O^k correspond to the output of the $CMAC^k$)

5.2 Average velocity control

This high level control allows us to regulate the average velocity by adjusting the pitch angle of the trunk at each step using the error between the average velocity V_M and the desired average velocity V_M^d and its derivative. V_M is calculated using equation 1. At each step, Δq_0^d, which is computed using the error between V_M and V_M^d and its derivative (equation 19), is then added to the pitch angle of the previous step $q_0^d(n)$ in order to carry out the new desired pitch angle of the following step $q_0^d(n+1)$ as shown in equation 20.

$$\Delta q_0^d = K^P(V_M^d - V_M) + K^V \frac{d}{dt}(V_M^d - V_M) \tag{19}$$

$$q_0^d(n+1) = q_0^d(n) + \Delta q_0^d \tag{20}$$

5.3 PD control

The third one is composed by four PD control in order to be sure of tracking the reference trajectories on each joint. The torques T_{knee} and T_{hip} applied respectively to the knee and to the hip are computed using the PD control. During the swing stage, the torques are carried out by using equations 21 and 22. q_{ij}^d and $\dot{q}_{ij}^d$ are respectively the reference trajectories (position and velocity) of the swing leg from the output of the Fuzzy-CMAC (j=1 for the hip, j=2 for the knee).

$$T_{hip}^{sw} = K_{hip}^p(q_{i1}^d - q_{i1}) + K_{hip}^v(\dot{q}_{i1}^d - \dot{q}_{i1}) \tag{21}$$

$$T_{knee}^{sw} = K_{knee}^p(q_{i2}^d - q_{i2}) + K_{knee}^v(\dot{q}_{i2}^d - \dot{q}_{i2}) \tag{22}$$

Secondly, the knee of the stance leg is locked, with $q_{i2}^d = 0$ and $\dot{q}_{i2}^d = 0$ (equation 23), and the torque applied to the hip allows to control the pitch angle of the trunk (equation 24). q_0 and $\dot{q}_0$ are respectively the measured absolute angle and angular velocity of the trunk. q_0^d is the desired pitch angle.

$$T_{knee}^{st} = -K_{knee}^p\, q_{i2} - K_{hnee}^v\, \dot{q}_{i2} \tag{23}$$

$$T_{hip}^{st} = K_{trunk}^p(q_0^d - q_0) - K_{trunk}^v\, \dot{q}_0 \tag{24}$$

6. Results

The goal of the two main results presented in this section is to show the interest of the proposed approach. First, we present results about the walking of the biped robot when the average velocity increases. Second, we show that the robot can step over a static obstacle.

6.1. Step length function of average velocity

Figure 12 shows the stick-diagram of the biped robot walking sequence when the desired average velocity increases. It must be noticed that the control strategy, based on the five reference gaits learnt during the training phase of CMAC neural networks (see section 4.3), allows adapting progressively the length of the step as a function of the average velocity.

Figure 13 shows the desired average velocity V_M^d, measured velocity V_M and step length L_{step}. When V_M^d increases from $0.4\,m/s$ to $1\,m/s$, V_M increases gradually and converges towards the new value of V_M^d. L_{step} increases automatically from $0.25\,m$ to $0.43\,m$ from the measured average velocity at each step. The regulation of the average velocity at each step is obtained thanks to an adequate adjustment of the pitch angle (see section 5.2). But, given that the swing leg trajectory depends on the average velocity, the length of the step is automatically adjusted as a function of V_M thanks to the Fuzzy CMAC. It must be pointed out that the average velocity is bigger than $0.8\,m/s$, the length of the step stay constant ($L_{step} = 0.43\,m$).

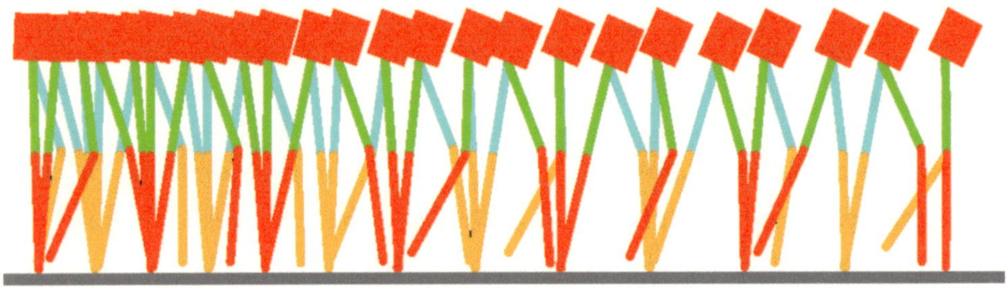

Figure 12. Stick-diagram of the walking robot when the average velocity increases

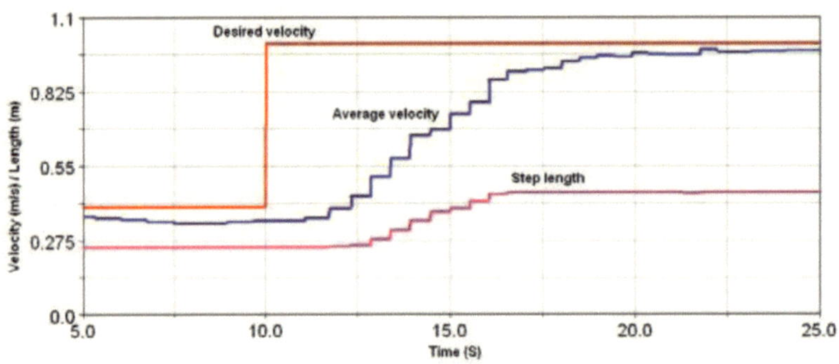

Figure 13. Average velocity and step length when the desired average velocity increases from 0.4m/s to 1m/s

6.2. Avoidance obstacle using step over strategy

The goal of this simulation is to show how the robot can step over an obstacle. In this example, the length and the height of the obstacle are respectively $0.2m$ and $0.05m$. Figures 16 and 17 show respectively stick-diagrams when the biped robot is walking on the floor whit and without obstacle. Without obstacle, the length of the step depends only of the average velocity. Consequently, L_{step} is quasi-constant during the walking. But if an obstacle occurs, our control strategy allows adjusting the step of the length in order to the robot steps over this obstacle. Figure 16 shows the length of the step when the robot is walking on the floor without and with obstacle. In the case of the presented example, the step length is adjusted in order to the landing point of the swing leg is located just before the obstacle. The next step, the step length increases allowing to the robot to step over the obstacle.

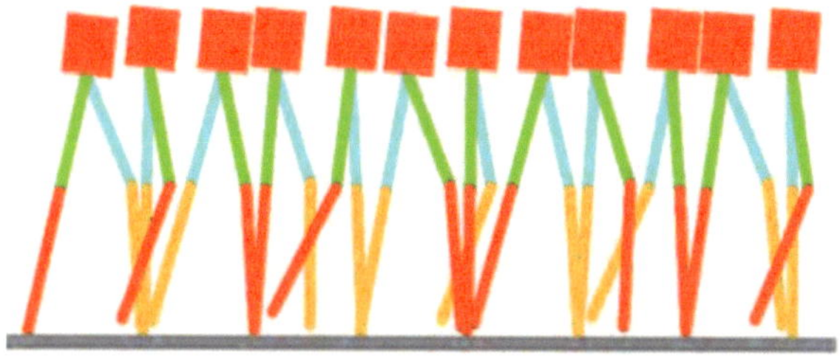

Figure 14. Walking of the biped robot without obstacle on the floor

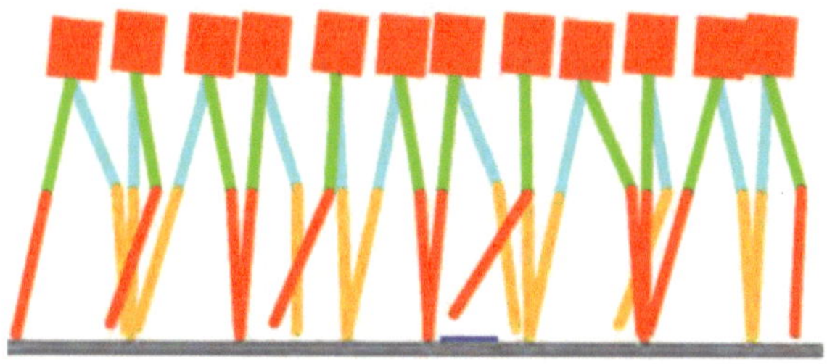

Figure 15. Walking of the robot when it steps over an obstacle

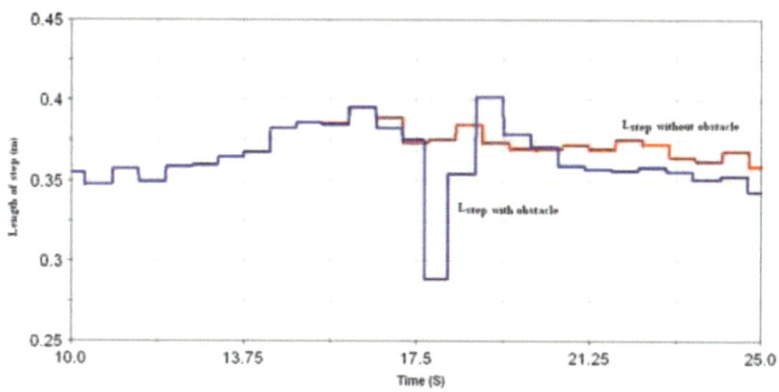

Figure 16. Length of the step when the robot is walking on the floor without and with obstacle

7. Conclusion and further works

In this chapter, we have described a control strategy based on both proprioceptive and exteroceptive information for autonomous biped robots. The first presented results, carried out on the basis of computer based simulation techniques, are very promising and prove that the proposed approach is a good way to improve the control strategy of a biped robot. First, we show that, with only five reference gaits, it is possible to generate other gaits. The adjustment of the step length as a function of the average velocity is due to the gait pattern based on the Fuzzy CMAC structure. Moreover, with a fuzzy evaluation of the distance between the robots' feet and an obstacle, our control strategy allows to the biped robot to avoid an obstacle using step over strategy.
However, it is important to remind that fuzzy rules are based on pragmatic approach and are constructed on the basis of some pre-defined membership functions shapes. For this reason, the presented control strategy may reach some limitation when biped robot comes

across more complex obstacles. Furthermore, in the real word, exteroceptive perception needs to use sensors as camera. Consequently, our further works will focus on two complementary directions: the first one will concern the study of the reinforcement learning strategy in order to increase the abilities of obstacles avoidance; the other one will investigate potentials of the exteroceptive information using vision. Based on these futures works, it will be possible to carry out experimental validations on the real robot RABBIT.

8. References

Albus, J. S. (1975). A new approach to manipulator control: the cerebellar model articulation controller (CMAC). *Journal of Dynamic Systems*, Measurement and Control,pp. 220--227.

Albus, J. S. (1975). Data storage in the cerebellar model articulation controller (CMAC). *Journal of Dynamic Systems, Measurement and Control*, pp. 228--233.

Bekey, G. A. (2005).Autonomous Robots, from *Biological Inspiration to Implementation and Control*. The MIT Press.

Brenbrahim, A.; Franklin, J. (1997). Biped dynamic walking using reinforcement learning. *Robotics and Autonomous Systems*, Vol.22 , pp. 283--302.

Chevallereau, C.; Abba, G., Aoustin, Y.; Plestan, F.; Westervelt, E.R.; Canudas-de-Wit, C.; Grizzle, J.W. (2003). RABBIT: A testbed for advanced control theory. *IEEE Control Systems Magazine*, Vol.23, N°5, pp. 57--79.

Kaneko, K.; Kanehiro, F.; Kajita, S.; Hirukawa, H.; Kawasaki, T.; Hirata, M.; Akachi, K.; Isozumi, T. (2004). Humanoid robot HRP-2. *Proc. IEEE Conf. on Robotics and Automation*, pp. 1083--1090.

Kun, A. L.; Miller, T. (2000). The design process of the unified walking controller for the UNH biped. *Proc. IEEE Conf. on Humanoid Robots.*

Miller, W. T.; Glanz, F. H.; Kraft, L. G. (1990). CMAC: An associative neural network alternative to backpropagation. *Proceedings of the IEEE, Special Issue on Neural Networks*, vol.78, N°10, pp. 1561-1567.

RABBIT-web: http://robot-rabbit.lag.ensieg.inpg.fr/

Robea-web: http://www.laas.fr/robea/

Sabourin, C.; Bruneau, O.; Fontaine, J-G. (2004). Start, stop and transition of velocities on an underactuated bipedal robot without reference trajectories. *Internationnal Journal of Humanoid Robotics*, Vol.1, N°2, pp. 349--374.

Sabourin, C.; Bruneau, O. (2005). Robustness of the dynamic walk of a biped robot subjected to disturbing external forces by using CMAC neural networks. *Robotics and Autonomous Systems*, Vol.23, pp. 81--99.

Sabourin, C.; Bruneau, O.; Buche, G. (2006). Control strategy for the robust dynamic walk of a biped robot. *The International Journal of Robotics Research* (IJRR), Vol.25, N°9, pp. 843--860.

Sakagami, Y.; Watanabe, R.; Aoyama, C.; Matsunaga, S.; Higaki, N.; Fujimura, K. (2002). The intelligent ASIMO: system overview and integration. *Proc. IEEE Conf. on Intelligent Robots and Systems*, pp. 2478--2483.

Vukobratovic, M.; Bocovac, B.; surla, D.; Stokic., D. (1990). Biped locomotion, *Scientific fundamentals of robotics* (vol 7) - Spinger-Verlag.

10

Reinforcement Learning of Stable Trajectory for Quasi-Passive Dynamic Walking of an Unstable Biped Robot

Tomohiro Shibata[1], Kentarou Hitomoi[3], Yutaka Nakamura[2] and Shin Ishii[1]
[1]Nara Institute of Science and Technology, [2]Osaka University, [3]DENSO CORPORATION
Japan

1. Introduction

Biped walking is one of the major research targets in recent humanoid robotics, and many researchers are now interested in Passive Dynamic Walking (PDW) [McGeer (1990)] rather than that by the conventional Zero Moment Point (ZMP) criterion [Vukobratovic (1972)]. The ZMP criterion is usually used for planning a desired trajectory to be tracked by a feedback controller, but the continuous control to maintain the trajectory consumes a large amount of energy [Collins, et al. (2005)]. On the other hand, PDW enables completely unactuated walking on a gentle downslope, but PDW is generally sensitive to the robot's initial posture, speed, and disturbances incurred when a foot touches the ground. To overcome this sensitivity problem, ``Quasi-PDW'' [Wisse & Frankenhuyzen (2003); Sugimoto & Osuka (2003); Takuma, et al. (2004)] methods, in which some actuators are activated supplementarily to handle disturbances, have been proposed. Because Quasi-PDW is a modification of the PDW, this control method consumes much less power than control methods based on the ZMP criterion. In the previous studies of Quasi-PDW, however, parameters of an actuator had to be tuned based on try-and-error by a designer or on *a priori* knowledge of the robot's dynamics. To act in non-stationary and/or unknown environments, it is necessary for robots that such parameters in a Quasi-PDW controller are adjusted autonomously in each environment.

In this article, we propose a reinforcement learning (RL) method to train a controller designed for Quasi-PDW of a biped robot which has knees. It is more difficult for biped robots with knees to walk stably than for ones with no knees. For example, Biped robots with no knee may not fall down when it is in an open stance, while robots with knees can easily fall down without any control on the knee joints.

There are, however, advantages of biped robots with knees. Because it has closer dynamics to humans, it may help to understand human walking, and to incorporate the advantages of human walking into robotic walking. Another advantage is that knees are necessary to prevent a swing leg from colliding with the ground. In addition, the increased degrees of freedom can add robustness given disturbances such as stumbling.

Our computer simulation shows that a good controller which realizes a stable Quasi-PDW by such an unstable biped robot can be obtained with as small as 500 learning episodes, whereas the controller before learning has shown poor performance.
In an existing study [Tedrake, et al. (2004)], a stochastic policy gradient RL was successfully applied to a controller for Quasi-PDW, but their robot was stable and relatively easy to control because it had large feet whose curvature radius was almost the same as the robot height, and had no knees. Their robot seems able to sustain its body even with no control. Furthermore, the reward was set according to the ideal trajectory of the walking motion, which had been recorded when the robot realized a PDW. In contrast, our robot model has closer dynamics to humans in the sense that there are smaller feet whose curvature radius is one-fifth of the robot height, and knees. The reward is simply designed so as to produce a stable walking trajectory, without explicitly specifying a desired trajectory. Furthermore, the controller we employ performs for a short period especially when both feet touch the ground, whereas the existing study above employed continuous feedback control. Since one definition for Quasi-PDW is to emit intermittent control signals as being supplementary to the passivity of the target dynamics, a design of such a controller is important.
The rest of the article is organized as follows. Section 2 outlines our approach. Section 3 introduces the details of the algorithm using policy gradient RL as well as simulation setup. Section 4 describes simulation results. We discuss in section 5 with some directions in future work.

2. Approach Overview

Fig. 1 depicts the biped robot model composed of five links connected by three joints: a hip and two knees. The physical parameters of the biped robot model are shown in Table 1. The motions of these links are restricted in the sagittal plane. The angle between a foot and the corresponding shank is fixed. Because we intend to explore an appropriate control strategy based on the passive dynamics of the robot in this study, its physical parameters are set referring to the existing biped robots that produced Quasi-PDW [Wisse & Frankenhuyzen (2003); Takuma, et al. (2004)]. As described in Fig. 1, θ stands for the absolute angle between the two thighs, θ_{knee1} and θ_{knee2} denote the knee angles, and ω denotes the angular velocity of the body around the point at which the stance leg touches the ground. The motion of each knee is restricted within $[0, \pi/4]$ [rad].

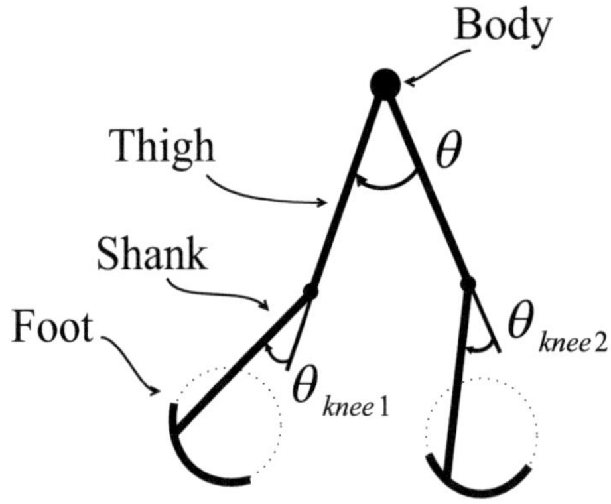

Figure 1. 2D Biped Model

	Length [m]	Mass [kg]
Body	0.0	4.0
Thigh	0.33	0.50
Shank	0.33	0.50
Foot	0.13*	0.020

Table 1. Physical parameters of the robot. * Value of curvature radius

Our approach to achieving adaptive controls consists of the following two stages.
(1) The two knees are locked, and the initial posture which realizes PDW by this restricted system are searched for. The initial posuture is defined by the initial absolute angle between two thighs, θ^s, and the initial angular velocity of the body around the point at which the stance leg touches the ground, ω^s. These values are used for the initial setting of the robot in the next stage.
(2) The two knees are then unlocked, and the robot is controlled by an intermittent controller with adjustable parameters. The parameters are modified by reinforcement learning (RL) so that the robot keeps stable walking.
These two stages are described in detail in the followings.

2.1 Searching for the initial conditions

In the first stage, we searched for an initial posture, denoted by θ^s and ω^s, which realize PDW by the robot with the locked knees, on a downslope with a gradient of ε = 0.03 [rad]. For simplicity, we fixed $\theta^s = \pi/6$ [rad] and searched a region from 0 to π [rad/sec] by $\pi/180$ [rad/sec], for ω^s that maximizes the walking distance. The swing leg of compass-like biped robots which have no knees necessarily collides with the ground, leading to falling down. Thus, in this simulation, the collision between the swing leg and the ground was ignored. We found $\omega^s = 58 \times \pi/180$ [rad/sec] was the best value such to allow the robot to walk for seven steps.

2.2 Design of a Controller

In light of the design of control signals for the existing Quasi-PDW robots, we apply torque inputs of a rectangular shape to each of the three joints (cf. Fig. 2). One rectangular torque input applied during one step is represented by a fourdimensional vector $\tau = \{\tau_{Hip,Amp}, \tau_{Hip,Dur}, \tau_{Kne,Flx}, \tau_{Kne,Ext}\}$. $\tau_{Hip,Amp}$ and $\tau_{Hip,Dur}$ denote the amplitude and the duration of the torque applied to the hip joint, respectively, and $\tau_{Kne,Flx}$ and $\tau_{Kne,Ext}$ are the amplitude of torques that flex and extend the knee joint of the swing leg, respectively. The manipulation of the knees follows the simple scheme described below to avoid the collision of the swinging foot with the ground, so that a swing leg is smoothly changed into a stance leg (cf Fig. 2). First, the knee of the swing leg is flexed with $\tau_{Kne,Flx}$ [Nm] when the foot of the swing leg is off the ground (Fig. 2(b)). This torque is removed when the foot of the swing leg goes ahead of that of the stance leg (Fig. 2(c)), and, in order to make the leg extended, a torque of $-\tau_{Kne,Ext}$ is applied after the swing leg turns into the swing down phase from the swing up phase according to its passive dynamics (Fig. 2(d)). To keep the knee joint of the stance leg being extended, 1 [Nm] is applied to the knee joint. τ is assumed to be distributed as a

Gaussian noise vector, while the mean vector $\bar{\tau}$ is modified by the learning, as described in the next section.

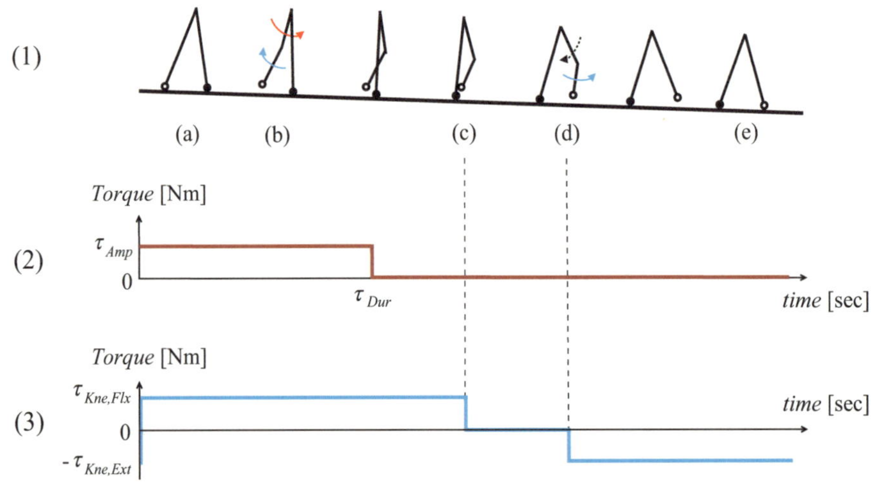

Figure 2. Torque applied to the hip joint and the knee joint. (1) Motions of the swing leg during a single step. (2) Torque applied to the hip joint. (3) Torque applied to the knee joint of the swing leg. (a) A single step starts when both feet touch the ground. (b) After the swing leg is off the ground, the robot begins to bend the knee of the swing leg by applying a torque of $\tau_{Kne,Flx}$ [Nm]. (c) The torque to the knee is removed when the foot of the swing leg goes ahead of that of the stance leg. (d) When the thigh of the swing leg turns into the swing down phase from the swing up phase, a torque of $\tau_{Kne,Ext}$ [Nm] is applied in order to extend the swing leg. (e) The swing leg touches down and becomes the stance leg

3. Learning a Controller

3.1 Policy gradient reinforcement learning

In this study, we employ a stochastic policy gradient method [Kimura & Kobayashi (1998)] in the RL of the controller's parameter $\bar{\tau}$, by considering the requirement that the control policy should output continuous values. The robot is regarded as a discrete dynamical system whose discrete time elapses when either foot touches the ground, i.e., when the robot takes a single step. The state variable of the robot is given by $s_n = (\theta_n, \omega_n)$, where n counts the number of steps, and θ_n and ω_n stand for the absolute angle between two thighs at the n-th step and the angular velocity of the body around the point at which the stance leg touches the ground, respectively.

At the onset of the n-th step, the controller provides a control signal τ_n, which determines the control during the step, according to a probabilistic policy $\pi(\tau \mid \bar{\tau})$. At the end of this step, the controller is assumed to receive a reward signal r_n. Based on these signals, a temporal-difference (TD) error δ is calculated by

$$\delta = \{r_n + \gamma V(s_{n+1})\} - V(s_n), \tag{1}$$

where $\gamma (0 \leq \gamma \leq 1)$ is the discount rate. V denotes the state value function and is trained by the following TD(0)-learning:

$$V(s_n) = V(s_n) + \alpha\delta \tag{2}$$

$$e_n - \frac{\partial}{\partial \overline{\tau}} \ln\left(\pi\left(\tau \middle| \overline{\tau}\right)\right)\Big|_{\tau=\tau_n, \overline{\tau}=\overline{\tau}_n} \tag{3}$$

$$D_n \leftarrow e_n + \beta D_{n-1} \tag{4}$$

$$\overline{\tau}_{n+1} = \overline{\tau}_n + \alpha_p \delta D, \tag{5}$$

where $\boldsymbol{e}$ is the eligibility and $\boldsymbol{D}$ is the eligibility trace. β $(0 \le \beta \le 1)$ is the diffusion rate of the eligibility trace and a_p is the learning rate of the policy parameter. After policy parameter $\overline{\tau}_n$ is updated into $\overline{\tau}_{n+1}$, the controller emits a new control signal according the new policy $\pi(\tau \mid \overline{\tau}_{n+1})$. Such a concurrent on-line learning of the state value function and the policy parameter is executed until the robot tumbles (we call this period an episode), and the RL proceeds by repeating such episodes.

3.2 Simulation setup

In this study, the stochastic policy is defined as a normal distribution:

$$\pi\left(\tau \middle| \overline{\tau}\right) = \frac{1}{\left(2\pi^2\right)\left|\Sigma\right|^{1/2}} \times \exp\left\{-\frac{1}{2}\left(\tau - \overline{\tau}\right)^T \Sigma^{-1}\left(\tau - \overline{\tau}\right)\right\} \tag{6}$$

so that the covariance $\boldsymbol{\Sigma}$ is given by

$$\Sigma = \begin{pmatrix} \sigma_{Hip,Amp^2} & 0 & 0 & 0 \\ 0 & \sigma_{Hip,Dur^2} & 0 & 0 \\ 0 & 0 & \sigma_{Kne,Flx^2} & 0 \\ 0 & 0 & 0 & \sigma_{Kne,Ext^2} \end{pmatrix}, \tag{7}$$

where $\sigma_{Hip,Amp}$, $\sigma_{Hip,Dur}$, $\sigma_{Kne,Flx}$ and $\sigma_{Kne,Ext}$ are constant standard deviations of noise, set at 0.3, 0.05, 0.3 and 0.3, respectively. We assume each component of τ is 0 or positive, and if it takes a negative value accidentally it is calculated again, similarly to the previous study [Kimura, et al. (2003)]. The reward function is set up as follows. If a robot walks stably, ω_n and θ_n should repeat similar values over steps. Furthermore, the robot should take no step in the same place, i.e., θ_{n+1} needs to be large enough. To satisfy these requirements, we define the reward function as

$$r_n = \theta_{n+1} \exp\left(-\left|\theta_{n+1} - \theta_n\right|^2\right) \tag{8}$$

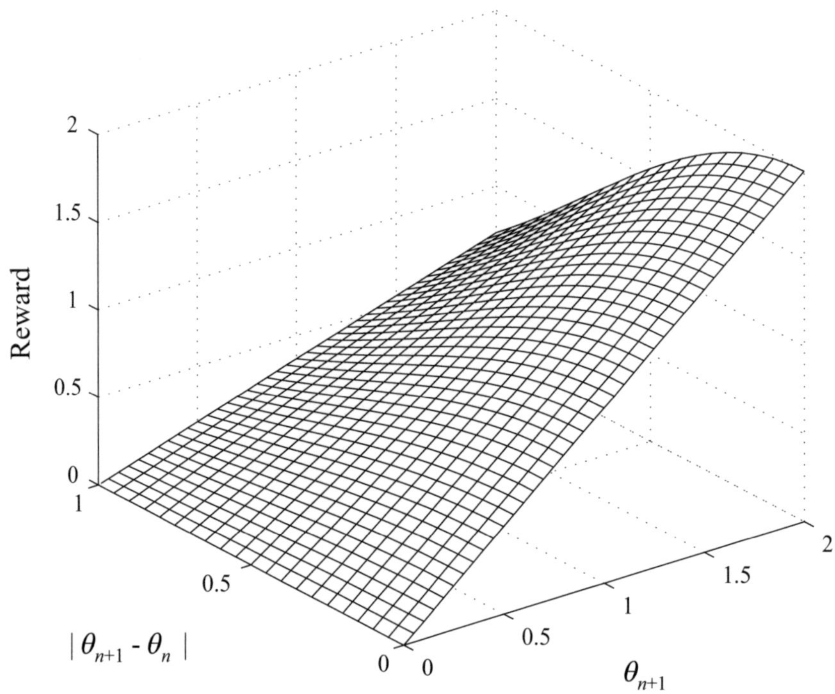

Figure 3. Landscape of the reward function

Figure 3. shows the landscape of this reward function. The value function is represented by a table over grid cells in the state space, and the value for each grid cell is updated by equation (2). In this study, we prepared 10 grid cells; the center of the fifth cell was for θ^s (Fig. 4), and the grid covered the whole state space, by assigning the 0-th cell to the range: $\theta < 0$, and the 9-th cell to the range: $\theta > 2\theta^s$. We used $a = 0.5$, $a_p = 0.01$, and $\beta = \gamma = 0.95$

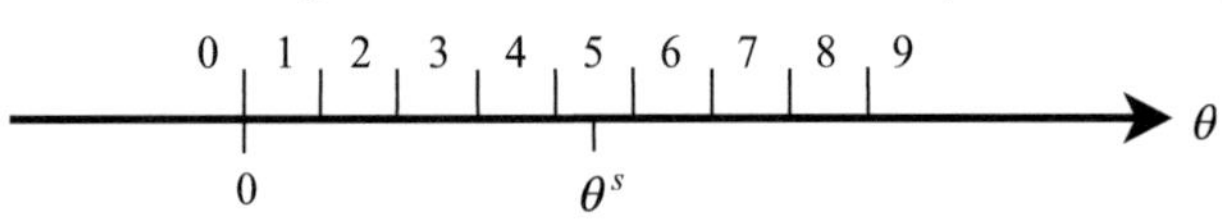

Figure 4. Discretization of the state space

In this study, we used a 3D dynamics simulator, Open Dynamics Engine [ODE]. In simulation experiments, motions of the robot were restricted in the sagittal plane by configuring a symmetric robot model with nine links (Fig. 5). It should be noted this nine-links robot has equivalent dynamics to the five-links model (Fig. 1), under the motion restriction in the sagittal plane; this nine-links model was also adopted by Wisse and Frankenhuyzen (2003) and by Takuma et al. (2004).

4. Simulation Results

Although the physical parameters of our robot were set referring to the existing Quasi-PDW robots, our robot with unlocked knees was not able to produce stable walking by itself. Then, this section describes the way to train the controller according to our RL scheme.

Figure 5. Dynamics simulation of the nine-links model with ODE

4.1 Passive walking without learning

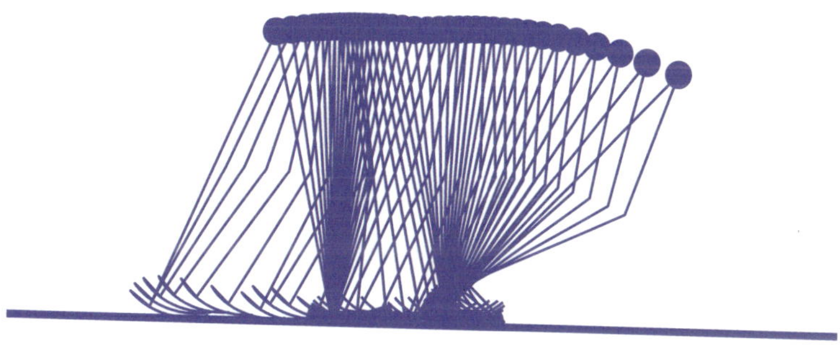

Figure 6. Stick diagram of the passive motion by the robot with knees. Plot intervals are 50 [ms]

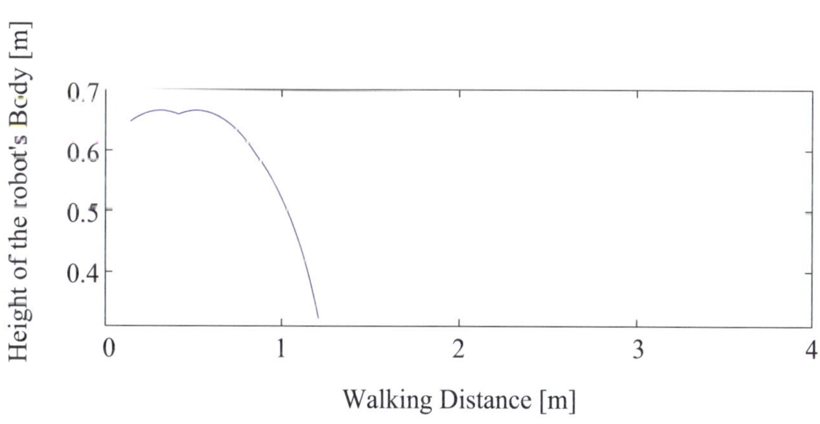

Figure 7. Body's trajectory of the passive robot with knees

First, we examined whether the robot with unlocked knees was able to produce stable walking on a downslope with ε = 0.03 [rad], when it received no controls to the hip joint or the knee joints. The unlocked knees were controlled in the same manner as that described in section 2.2. Initial conditions were set at $\theta_0 = \theta^s$[rad] and $\omega_0 = \omega^s$ [rad/sec], which are the same as those where the knee-locked model performed seven steps walking. As Fig. 7 shows, the robot with unlocked knees walked for 80 cm and then fell down. The robot was not able to walk passively when the knees were unlocked but uncontrolled.

4.2 Learning a controller

The experiment in section 4.1 showed that the robot with unlocked knees was not able to produce stable walking without any control to the hip joint or the knee joints, even when starting from possibly good initial conditions θ^s and ω^s. Then, in this section, we applied on-line RL to the automatic tuning of the parameter $\bar{\tau}$. At the beginning of each episode, the robot's initial conditions were set at $\theta_0 = \theta^s$, $\omega_0 = \omega^s$, and the episode was terminated either when the robot walked for 20 steps or fell down. When the height of the robot's 'Body' became smaller than 80% of its maximum height, it was regarded as a failure episode (falling down). RL was continued by repeating such episodes.

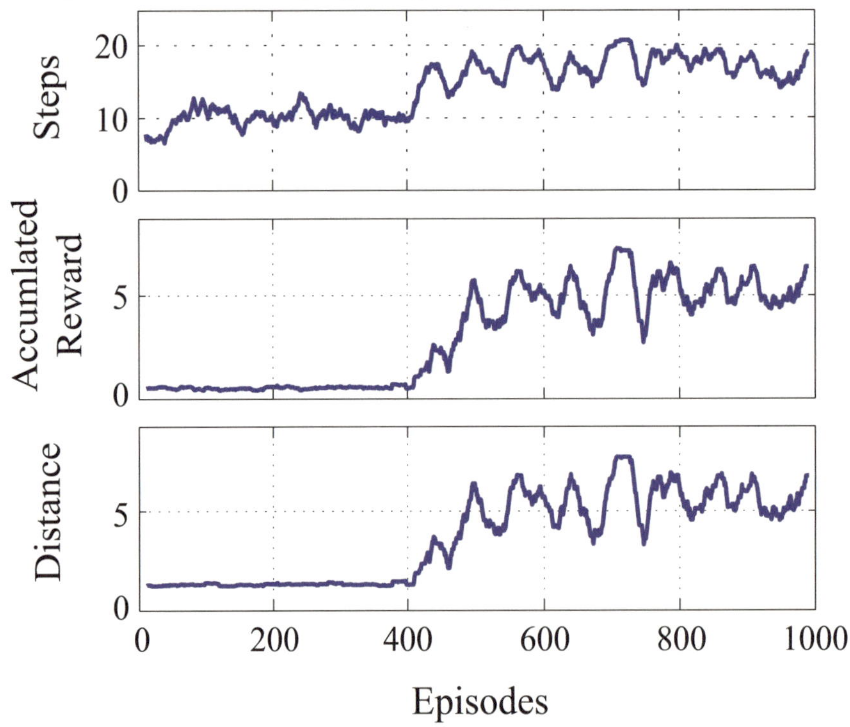

Figure 8. Moving averages of number of steps, cumulative reward, and distance to have walked

Fig. 8 shows the moving averages for ±20 episodes of walking steps (top), cumulative reward (middle), and walking distance (bottom), achieved by the robot. The steps increased after about 400 learning episodes, and went up to nearly 20 steps after about 500 learning episodes. In the early learning stage, the cumulative reward and walked distance were small though the robot walked for more than 10 steps, indicating the robot was walking stumbling with small strides. Using the deterministic controller with the parameter $\overline{\tau}$ after 500 training episodes, the robot was able to walk for more than 20 steps (Fig. 9). The parameter at this time was $\overline{\tau}$ = (0.70, 0.17, 0.93, 0.51).

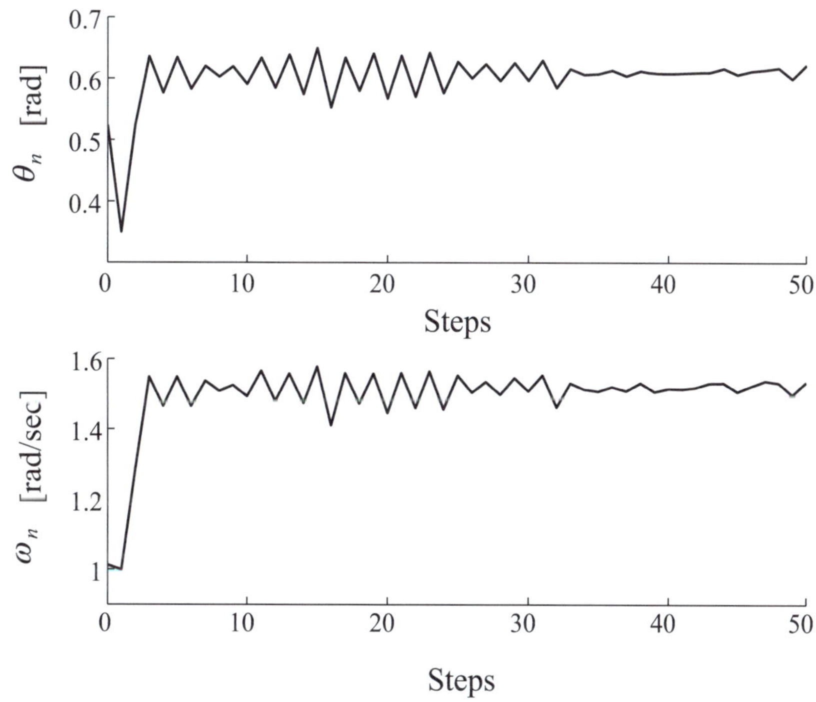

Figure 9. Values of θ_n and ω_n during the walking for 50 steps by the controller after 500 learning episodes

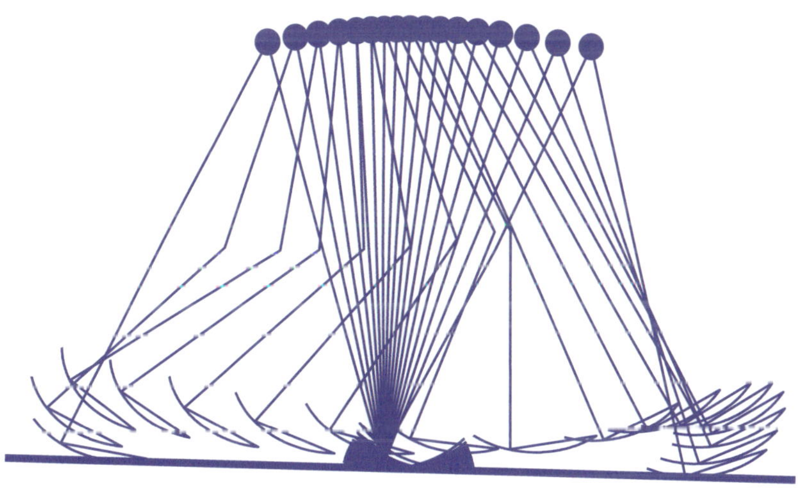

Figure 10. Stick diagram motion by the robot after 500 learning episodes. Plot intervals are 50 [ms]

4.3 Energy efficiency

	This study	Human*	Dynamite*	Asimo*
c_{mt}	0.093	0.05	0.04	1.6

Table 2. Energy efficiency calculated as c_{mt}. *These values are excerpted from literature [12]

Table 2 compares energy efficiency of Quasi-PDW acquired in this study with the others. For this comparison, the dimension-less cost of transport,

$$c_{mt} = (\text{used energy})/(\text{weight} \times \text{distance}) \tag{9}$$

was employed [Collins, et al. (2005)]; c_{mt} is suitable for comparing energy efficiency of simulators with that of real robots, because c_{mt} evaluates the effectiveness of the mechanical design and controller independently of the actuator efficiency. Note that energy from the gravity is included in the calculation of c_{mt} (= 0.093) for our simulated robot. The c_{mt} value achieved by our on-line RL is larger than the one of the PDWcontrolled robot (Dynamite), while it is much smaller than the one of the ZMP criterion (ASIMO).

4.4 Robustness against disturbances

To see the robustness of the acquired Quasi-PDW against possible disturbances from the environment, we conducted two additional experiments.

First, we let the robot with the control parameter after 500 training episodes walk on downslopes with various gradients. Fig. 11 shows the results for ε = 0.02 - 0.05 [rad]. The robot was able to walk for more than 50 steps on downslopes with ε = 0.02 - 0.04 [rad], and 22 steps with 0.05 [rad]; the controller acquired through our on-line RL was robust against the variation (in the gradient) of the environment. Second, we applied impulsive torque inputs to the hip joint during walking. Fig. 12 shows the time-series of θ_n in the same condition as Fig. 9, except that impulsive torque inputs were applied as disturbances at the time points with the arrows. Each disturbance torque was 1 [Nm] and was applied so as to pull the swing leg backward for 0.1 [sec] when 0.4 [sec] elapsed after the swing leg got off the ground. As this figure shows, θ_n recovered to fall into the stable limit cycle within a few steps after disturbances, implying that the attractor of the acquired PDW is fairly robust to noise from the environment.

Additional qualitative analysis by means of return map was performed in order to investigate changes in walking robustness through learning. Fig. 15 plots return maps during walking after disturbances as well as steady state walking at 440 and 500 episodes, respectively. In this figure, the return map is depicted by circles, crosses, and triangles for right after disturbance, next step, and two steps after, respectively (cf. Fig. 14). The maps for steady-state walking (Fig. 15(a),(c)) show the robot was walking stably by keeping θ_n at around 0.6 [rad] in both cases of after 440 and 500 learning episodes. The upper part of Table 3 shows $\langle\theta_n\rangle$, the average θ_n during steady-state walking, after 440, 480, 500, 700, and 900 learning episodes. $\langle\theta_n\rangle$ gets large after 500 episodes, which would be induced by the increase in the accumulated reward (Eq. 8). This reward increase was mainly due to the increase in the step length, even after θ_{n+1}-θ_n became almost zero achieved by making periodic walking.

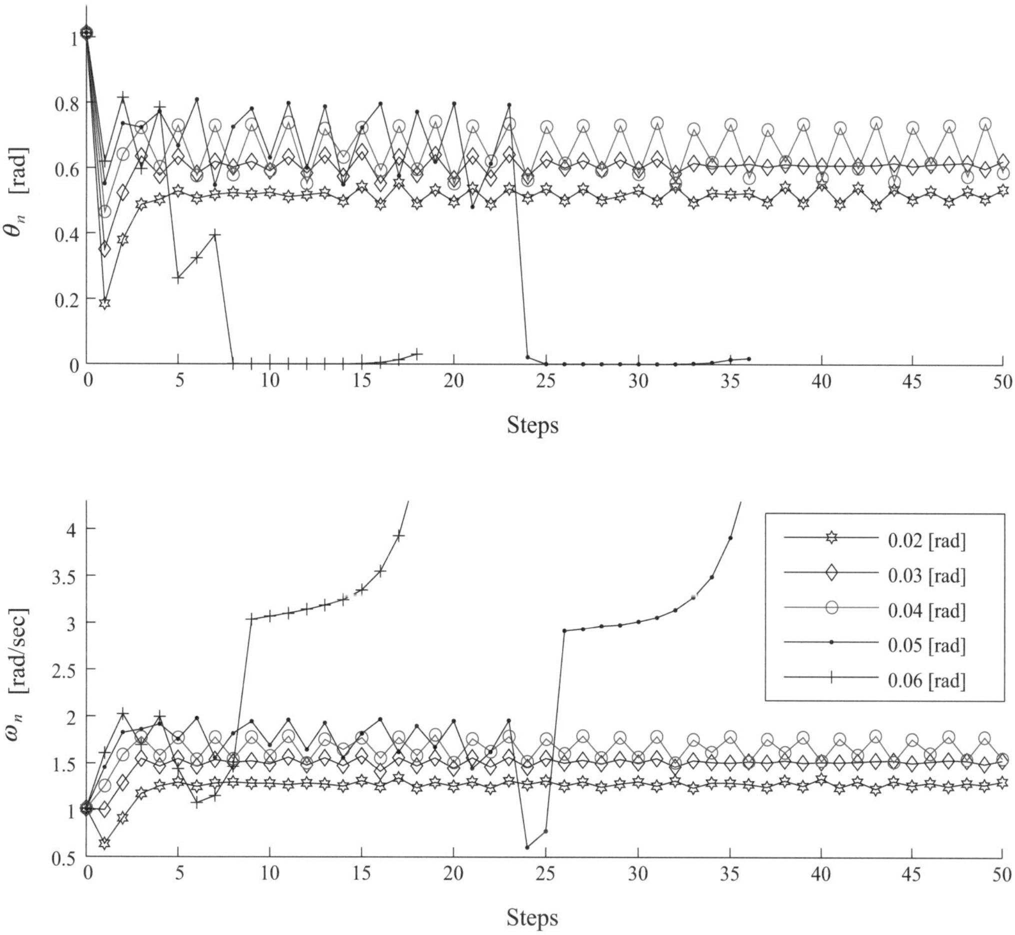

Figure 11. Values of θ_n and ω_n on downslopes with various gradients

The maps after disturbance (Fig. 15(b),(d)) show the disturbed walking recovered quickly the steady-state walking. The lower part of Table 3 shows the average step required for recovery did not decrease in a monotonic fashion, but they are all small enough regardless of the gradual increase in $<\theta_n>$ through learning.

Episodes	440	480	500	700	900
$<\theta_n>$ during steady-state walking	0.5960	0.5957	0.6043	0.6100	0.6120
Average steps for recovery	2.071	2.253	1.182	1.556	2.020

Table 3. Mean values of θ_n during steady walking and of numbers of steps necessary to recover steady walking after the disturbance

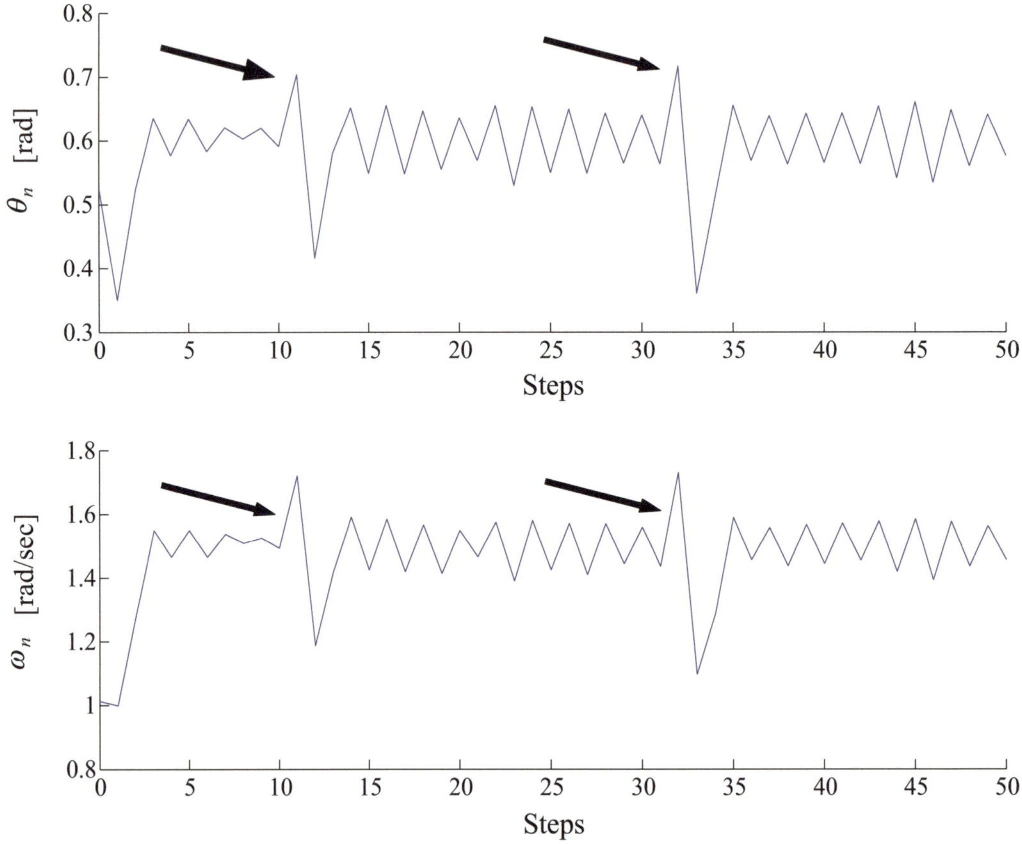

Figure 12. Perturbation of θ against impulsive disturbances (arrowed)

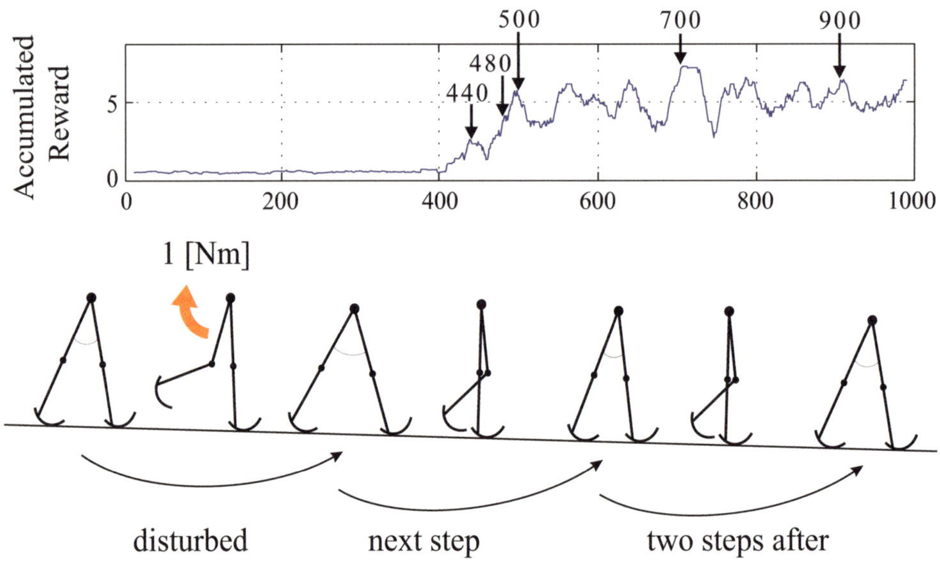

Figure 14. Step counts after disturbance

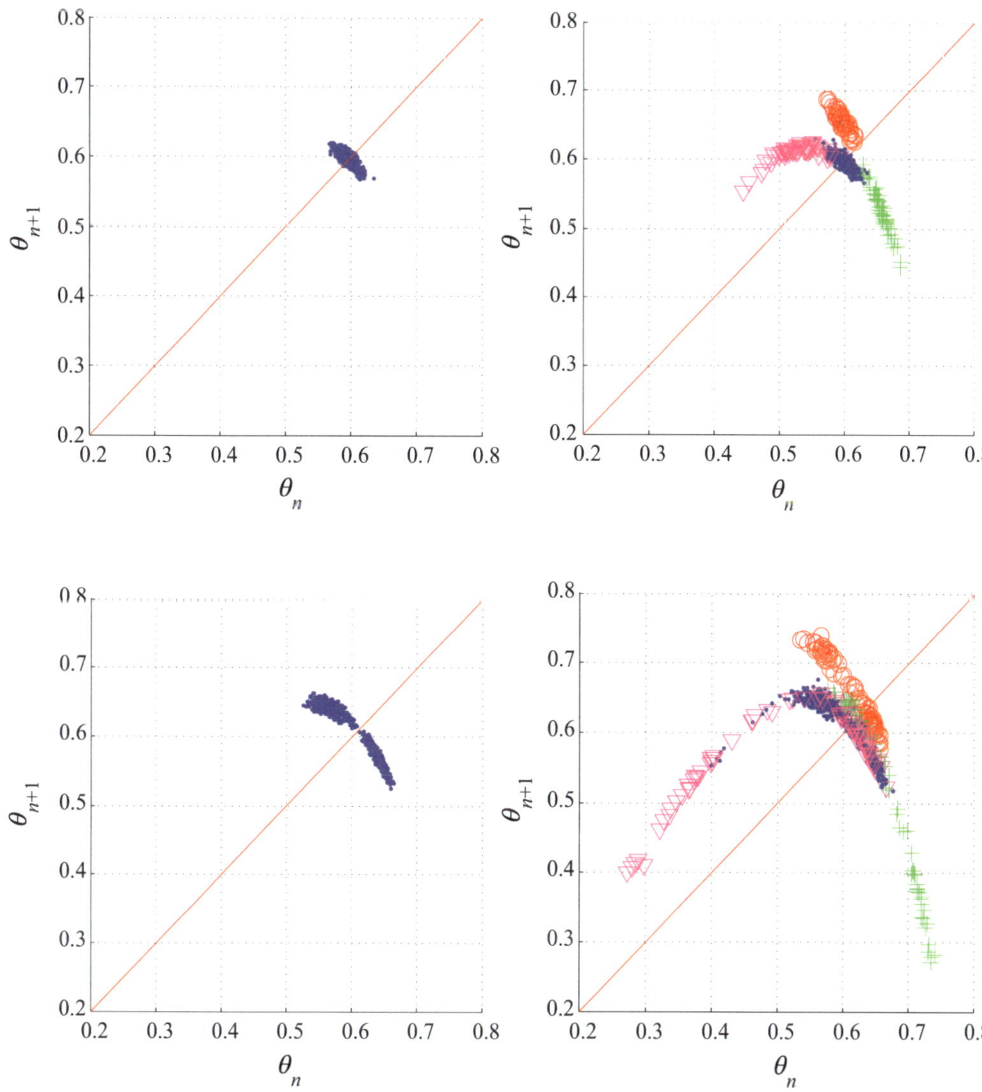

Figure 15. Return maps after 440 or 500 learning episodes

5. Discussion

In this study, we proposed an on-line RL method suitable for Quasi-PDW by a 2D biped robot, whose possession of knees makes the system unstable. Our study is underlaid by the perspective of low energy consumption and good correspondence to human walking. RL was applied not only for the hip joint but also for the knee joints of the robot, and our learning method was successful in making the unstable robot produce stable walking after as small as 500 training episodes, despite of usage of simple intermittent controllers. Although the controller itself was simple, simulation experiments on downslopes with various gradients and through addition of impulsive disturbances have shown that the stochastic policy gradient method with a reward function that encourages continuing rhythmic walking steps have successfully contributed to making the PDW robust against various noise in the environment.

Our learning method consisted of two stages, as described in section 2. After roughly searching in the first stage for an initial angular velocity with which the robot with locked knees walked for several steps, RL was applied to the robot with unlocked knees, starting from the initial condition obtained in the first stage. This two-stages learning reminds us of a developmental progression found at least in humans [Bernstein (1968); Newell & Vaillancourt (2001)] which increases the degree of freedoms as the learning proceeds; after a primitive control is achieved for a restricted system with a low dimensionality, frozen dimensionality is gradually released to realize more complex and smooth movements by the high-dimensional system. Furthermore, animals seem to employ different controllers in the initiation phase and in the maintenance phase for effect e.g., it has been known that three steps in average are required to initiate stationary walking in humans [Miller & Verstraete (1996)]. We consider the first stage of our approach could correspond to the initiation stage above.

As another reason for our successful results, our adaptive controller was trained by RL as to apply intermittent energy for maintaining stable PDW. This intermittent control was inspired by the studies of the measurement of human EMG [Basmajian (1976)] and robot control based on the idea of Quasi-PDW conducted by Collins et al. (2005) or by Takuma et al. (2004). To develop an energy-efficient control method of robots, considerable care about the passivity of the robot should be taken, as Collins suggested. Furthermore, the dynamics of robots with many degrees of freedom generally constitutes a nonlinear continuous system, and hence controlling such a system is usually very difficult. Our approach successfully realized efficient learning, which required as small as 500 learning episodes even with learning for knee joints, by introducing the policy that emits intermittent control signals and a reward function encouraging stable motions, both of which well utilized the passivity of the robot. Our learning method is not restricted to locomotion, since the computational problem and the importance of passivity are both general, although what kind of controllers should be activated or switched when and how are remained as interesting and significant problems for general applicability. From a theoretical point of view, our results indicate that passivity of the robot together with the two-stages scheme effectively restricted a high-dimensional control space of the robot. Nakamura et al. demonstrated that an RL in which a central pattern generator (CPG) was employed succeeded in training a simulated biped robot which had also five links including knees [Nakamura, et al. (2004)]. In their method, the control space was restricted such that the outputs of the controller were likely rhythmic control signals. Combination of such CPG-constrained learning scheme and the passivity constrained learning scheme would be interesting not only for more robust locomotion but also for control of various types of high-dimensional robots. It should also be noted here that CPG seems to be employed for human locomotion [Dietz, et al. (2002)]. Our approach would be plausible in the perspective of energy efficiency and understanding of human walking [Basmajian (1976)]. Along with this issue, how to incorporate the idea of energy efficiency into the reward function is interesting. Another interesting avenue for future work is to devise a method to produce stable walking on a level ground. In addition, we are conducting experiments with a real biped robot [Ueno, et al. (2006)], which would enhance the applicability of the current methodological study.

6. Acknowledgement

We thank Dr. Koh Hosoda and Mr. Takuma at Graduate School of Engineering, Osaka University, for giving us information about Passive DynamicWalking and their biped robot. This study is partly supported by Graint-in-Aid for Scientific Research of Japan Society for the Promotion of Science, No. 16680011 and 18300101.

7. References

Basmajian, J. (1976). *The human bicycle: an ultimate biological convenience*. The Orthopedic clinics of North America, 7, 4, 1027–1029.

Bernstein, N. (1968). *The coordination and regulation of movements*. Pergamon.

Miller, C.A. & Verstraete, M.C. (1996). Determination of the step duration of gait initiation using a mechanical energy analysis. *Journal of Biomechanics*, 29, 9, 1195–1199.

Collins, S.H. & Ruina, A. (2005). A bipedal walking robot with efficient and human-like gait. *Proceedings of IEEE International Conference on Robotics and Automation.*

Collins, S; Ruina, A; Tedrake & M.Wisse. (2005). Efficient bipedal robots based on passivedynamic walkers. *Science*, 307, 1082–1085.

Dietz, V.; Muller, R & Colombo, G. (2002). Locomotor activity in spinal man: significance of afferent input from joint and load receptors, *Brain*, 125, 2626-2634.

Kimura, H. & Kobayashi, S. (1998). Reinforcement learning for continuous action using stochastic gradient ascent. *Intelligent Autonomomus Systems*, 288–295.

Kimura, H. & Kobayashi, S. (1998). An analysis of actor/critic algorithms using eligibility traces: Reinforcement learning with imperfect value function. *Proceedings of 15th International Conference on Machine Learning*, 278–286.

Kimura, H.; Aramaki, T. & Kobayashi, S. (2003). A policy representation using weighted multiple normal distribution. *Journal of the Japanese Society for Artificial Intelligence*, 18, 6, 316–324.

McGeer, T. (1990). Passive dynamics walking. *The International Journal of Robotics Research*, 9, 2, 62–82.

Newell, K.M. & Vaillancourt, D.E. (2001). Dimensional change in motor learning. *Human Movement Science*, 20, 695–715.

Nakamura, Y.; Mori, T. & Ishii, S. (2004). Natural policy gradient reinforcement learning for a CPG control of a biped robot. *Proceedings of International conference on parallel problem solving from nature*, 972–981.

ODE, http://ode.org/.

Sugimoto, Y. & Osuka, K. (2003). Motion generate and control of quasi-passive-dynamic walking based on the concept of delayed feedback control. *Proceedings of 2nd International Symposium on Adaptive Motion of Animals and Machines.*

Takuma, T., Nakajima, S., Hosoda, K. & Asada, M. (2004). Design of self-contained biped walker with pneumatic actuators. *Proceedings of SICE Annual Conference.*

Tedrake, R.; Zhang, T.W.. & Seung,H.S. (2004). Stochastic policy gradient reinforcement learning on a simple 3D biped. *Proceedings of the IEEE International Conference on Intelleigent Robots and Systems.*

Ueno, T.; Nakamura, Y.; Shibata, T.; Hosoda, K. & Ishii, S. (2006). Fast and stable learning of quasi-passive dynamic walking by an unstable biped robot based on off-policy natural actor-critic. *Proceedings of IEEE/RSJ Int Conf Intell Robot Syst*, 5226-5231.

Vukobratovic, M. & Stepanenko, J. (1972).On the stability of anthropomorphic systems. *Mathematical Biosciences*, 15, 1–37.

Wisse, M. & Frankenhuyzen, J. (2003). Design and construction of mike; a 2D autonomous biped based on passive dynamic walking. *Proceedings of 2nd International Symposium on Adaptive Motioon of Animals and Machines.*

11

An Adaptive Biped Gait Generation Scheme Utilizing Characteristics of Various Gaits

Kengo Toda[1] and Ken Tomiyama[2]
[1]*Chiba institute of Technology, Future Robotics Technology Center*
[2]*Chiba institute of Technology, The Dept. of Advanced Robotics*
Japan

1. Introduction

The purpose of this study is to develop a biped locomotion movement generator, named the Sensor-Based Gait Generation system, for humanoid robots that puts biped locomotion to practical use in real environment. The proposed method, in short a gait generator, enables humanoids to use various gaits according to walking surface condition. Policies and structure of the Sensor-Based Gait Generation system are described. A gait generator that is based on the proposed method is designed and implemented onto an original humanoid to demonstrate effectiveness of the proposed method.

Human beings clearly differentiate their walking movements on the downward slope, on the slippery surface and on the flat ground. In other words, humans use a gait that is appropriate to the condition of the walking surface. At present, however, most humanoid robots use only a single gait with possibly adjustable parameters and therefore have clear disadvantage compared with humans. The proposed method imitates gait selection strategy of human and thus eliminates this deficiency.

Many gait generation methods have been proposed already. A gait generated by any one of those methods has good characteristics and shortcomings and therefore has advantages and disadvantages against a given walking surface condition. When humanoids adopt the proposed gait generator, they will be able not only to walk on every road conditions but also to take advantages of good characteristics of each gait.

Especially, we focus on policies of the Sensor-Based Gait Generation system in this paper. One of the two major topics is the explanation of its main components and the other is the configuration of criteria for gait evaluation. In addition, structure of the Sensor-Based Gait Generation system based on this methodology is discussed. After explaining the developed and implemented system that realizes the proposed method, details of environment for experiments are described. Experimental results clearly exhibit practical advantages of the proposed method. Capabilities of the implemental system shown by experimental results are summarized. Conclusions and items for further study are listed at the end.

2. Sensor Based Gait Generation System

The Sensor-Based Gait Generation system realizes continuous gait transition from the current gait to a suitable gait for the condition at the moment. The most important points, therefore, are the gait evaluation and the rule of gait changeover. "What can be the criterion for gait transition?" and "how we evaluate gaits?" are the main subjects of this section. They are described according to the proposed system outline and functions of its components. The policies on those points and the range of application of the proposed method are also described here.

A schematic diagram that shows the processing flow of the proposed system is given in Fig.1. The box with dotted line represents the proposed Sensor-Based Gait Generator and consists of a gait generator and a gait library. The gait generator contains a gait selector and an interpolator. The gait library stores gait modules and a transition module where each gait module is capable of generating a gait according to a conventional method.

When a walk command is given to the system, the gait selector chooses a suitable gait module according to the walking surface condition and robot's own state that are obtained from various sensors. Next, the interpolator generates reference angles for every joint using the selected gait module. The output of the gait generator is distributed to actuators through the gait stabilizer with possible compensations to gaits. Elements and functions of the gait generator and the gait library are explained in detail below.

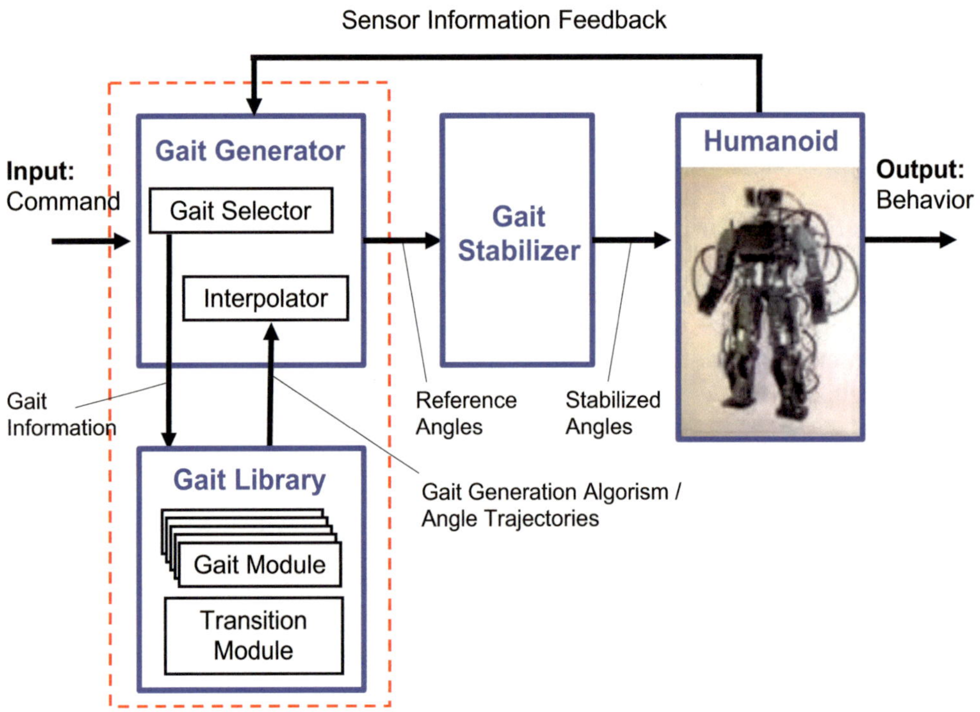

Figure 1. Schematic diagram of the Sensor-Based Gait Generation system

2.1 Gait Selector

Gait Selector is the central component of the proposed system and is responsible for choosing a gait from information on walk surface and on the state of the robot.

When designing the Gait Selector, we need to consider various factors. At least, the following factors must be considered.

Mobility
Mobility is the index of traversability of a given gait. The height that a target humanoid can go over is a good example of traversability. Applicable walking field condition of each gait module is also included. For instance, a gait module for generating constant stride is not applicable to steppingstone condition.
Stability Margin (Falling Risk Management)
To ensure sufficient stability margin is to keep avoiding tumbling. Not only the stability margin but also the risk management of tumbling should be included.
Motion State
Motion state is the index of performance against the desired motion. Gait Selector monitors the gait state to realize desired performance.
Although we have listed mobility, stability margin and motional state as the basic items to be considered, it is obvious that this set is not enough for an all-round gait switching rule for humanoid in real environment. When a humanoid walk on sludge, there are two choices of walking strategy. One way may be to walk faster and to pass the sludge as soon as possible without considering stability. The other way is to walk slowly with sufficient stability without considering the traversal time. The choice is dependent on the assigned task and the risk tolerance of falling. Expected energy efficiency can be important too when the given task requires long time duration. Dynamic priority arrangement of gait selection according to the given task for those situations is a quite interesting problem but is left for the future study. Sensory data are utilized, in this study, only for module switching according to the gait selection rule that is specified by the operator.

2.2 Interpolator
Interpolator is the gait generation engine of the system. It uses the gait module, which Gait Selector has chosen, in composing angle trajectories of joints. There are two procedures to generate gaits depending on the type of the chosen gait module.
Gait modules that can generate joint angles in real time
Gait setting parameters, such as walk period and stride, are given to the gait module chosen by the Gait Selector and reference angles of the links are generated real time basis.
Gait Modules not suited for real time generation
Two angle trajectories, chosen from a set of trajectories that are pre-generated and stored according to an algorithm, are compounded to form the reference angles.

2.3 Gait Modules
A gait module is a self-sustained package of a particular gait. It either contains a gait generating algorithm or a set of pre-generated gaits. Gait modules are identified by gait generation method. Gait modules that are generated by off-line algorithms contain numbers of pre-generated angle trajectories. These gait modules are compiled as a database and stored in Gait Library with required information for each gait module.

2.4 Transition Module
Each gait module simply generates gaits according to the chosen algorithm and therefore, a caution must be exercised when connecting the trajectories from different gait modules. The transition module contains one or more algorithms for generating transitional movements to

maintain dynamical integrity of the walk when a change of gait is selected. In Section 4, the proposed algorithms are explained in detail.

2.5 Gait Evaluation

Numerical index of gait performance is required for evaluation of a gait module or a chain of gait modules. We utilize the following three indices, which can summarize the characteristics of a gait.

Achievement Rate

The rate of success of the desired traversal on conceivable surface environments

Moving Velocity

The traversal velocity on a given surface environment

Energy Efficiency of Locomotion

Traversal distance per unit energy (We use supplemented energy that is defined in the next subsection to calculate this index.)

Characteristics of gaits are converted into numerical values using these indices. In addition, weighted summation of these values is adopted as an integrated evaluation factor of a gait. It is mentioned that other criteria for evaluation can be added to this list when a need arises.

2.6 Supplemented Energy

Supplemented energy per step is defined to be the sum of consumed energy of actuators while a robot walks a step. It is obvious that smaller the supplemented energy less the lost energy. The supplemented energy is derived from the rates of changes of the positional energy and the kinetic energy. In case of a movement that has no lost energy, like a natural response of the inverted pendulum, the potential energy decreases as much as the kinetic energy increases because there is no actuation. Thus, it can be said that a movement is closer to ideal if the sum of the rate of potential energy variation, in short the power, and the rate of kinetic energy variation is closer to zero.

The supplemented energy is computed with the following procedures.

- The powers corresponding to the potential and the kinetic energy

$$P_P = \sum_{i=1}^{n} m_i g \dot{z}_i \tag{1}$$

where

P_P : Power by potential energy

n : Numbers of links

m_i : Mass of link i

g : Acceleration of gravity

z : Height position of CoG of link i

$$P_K = \sum_{i=1}^{n} \left\{ m_i \left(\ddot{x}_i \dot{x}_i + \ddot{y}_i \dot{y}_i + \ddot{z}_i \dot{z}_i \right) + I_X \ddot{\theta}_{Xi} \dot{\theta}_{Xi} + I_Y \ddot{\theta}_{Yi} \dot{\theta}_{Yi} + I_Z \ddot{\theta}_{Zi} \dot{\theta}_{Zi} \right\} \tag{2}$$

where

P_K : Power by kinetic energy

x_i, y_i : Position of CoG of link i

$\theta_{Xi}, \theta_{Yi}, \theta_{Zi}$: Angles of link i about the x, y and z axes

I_{Xi}, I_{Yi}, I_{Zi} : Height of CoG of link i

Next, the sum of the two powers is computed. Note that a positive sum implies that the total torque is applied towards the direction of walk. On the contrary, a negative sum implies that the total torque is acting on the reverse direction of walk. Here, it is assumed that the type of actuators of the robot have no capacity to keep energy. Then, the total 'effort' of actuators can be represented by the absolute value of the sum of powers. Hence, it is used here as the index of the consumed energy.

$$P = \left| \; P_p + P_K \; \right| \tag{3}$$

where

P: Total power

The total supplemented power per step is computed by integration of the total power over the time interval of a step.

$$E = \int_0^T P dt \tag{4}$$

where

E : Supplemented Energy

T : Time interval of a step

3. System Architecture Methodology

The architecture of the Sensor Based Gait Generation system is described in detail. The design procedure of the proposed system is described first. The selection criteria of gait modules are explained afterwards.

3.1 Procedure

The design flow of the Sensor-Based Gait Generation system is as follows:

1. Preparation of gait modules using available gait generation schemes
2. Evaluation of gait modules on each ground condition
3. Designing and development of Gait Selector
4. Installation and architecture optimization

We prepare self-sustained gait modules first. Then, gait modules are categorized according to their mobility and labeled with applicable ground conditions. We evaluate gait modules by rehearsal walking to verify the appropriateness of the relationship between the gait module and ground conditions. Next, Gait Selector is configured by criteria that are based on stability margin and motion state of walking. Finally, we fine-tune Gait Selector by installing the Sensor-Based Gait Generation system onto the target humanoid.

3.2 Selection Criteria of Gait Modules

Among three factors mentioned in Subsection 2.1, the mobility parameter of a gait module is included in the module because it is used only for the test of applicability of the module. Therefore, only gait selection based on stability margin and motion state is explained here. Basically, sensory information is classified roughly into prior information set and posterior one. For example, cameras and laser range finders give prior information of the ground condition. Environment maps that are given by the operator are also included in the prior information set. This information is typically utilized for prediction of ground conditions. Prior information is mostly used in determination of the applicable gait modules for the given ground condition. Preliminary motion for the expected change of ground condition (kajita2003) is a good application example of the prior information. On the other hand, posterior information is utilized to evaluate the stability margin and the motion state. The posterior information is obtained at real-time basis during actual walk. It is very important for the gait selection because disturbances on the balance of gaits can only be detected at real-time basis. Instability that is rooted in ground conditions undetectable by the prior information can, therefore, be absorbed by a gait switching according to the posterior information.

With the above observations, gait modules are selected according to the following policies based on the posterior information.

1. The stability margin must be kept at an appropriate level
2. The current motion state should be made closer to the ideal state

Here, we use the following physical quantities in evaluating the above policies:

- Criterion for stability margin: ZMP (Zero Moment Point)
- Criterion for motion state: Angular Momentum

This set of choices comes from the fact that the most gaits for humanoids are based on the ZMP stability criterion and all of the developed gait modules adopt ZMP criterion. Since ZMP and angular momentum are commonly used, discussions on those criteria are omitted here.

4. Gait Transition Algorithm

Two algorithms that connect joint angle trajectories at the time of gait module changes are described in this section. These algorithms are stored in the transition module.

4.1 Algorithm 1: Transition in Double-Leg Supporting Phase

This transition method is applicable when the switching of gait modules occur during the double-leg supporting phase. It generates motions in this phase for connecting gaits before and after this phase. Two 2-D dynamics models in the sagittal and lateral plane are used to simplify the actual 3-D movement of humanoid. Dynamics model in the sagittal plane is shown in Fig.2 together with the corresponding 3-D model. It is assumed that there is no interference between the sagittal and lateral planes. Trajectories of the waist joint in both sagittal and lateral planes are determined first from positions and speeds of the waist joint at the end of the prior gait and the start of the new gait. It is noted that all other joint angle trajectories of humanoids with geometrical configurations of the 3-D model in Fig. 2 are obtainable from this information.

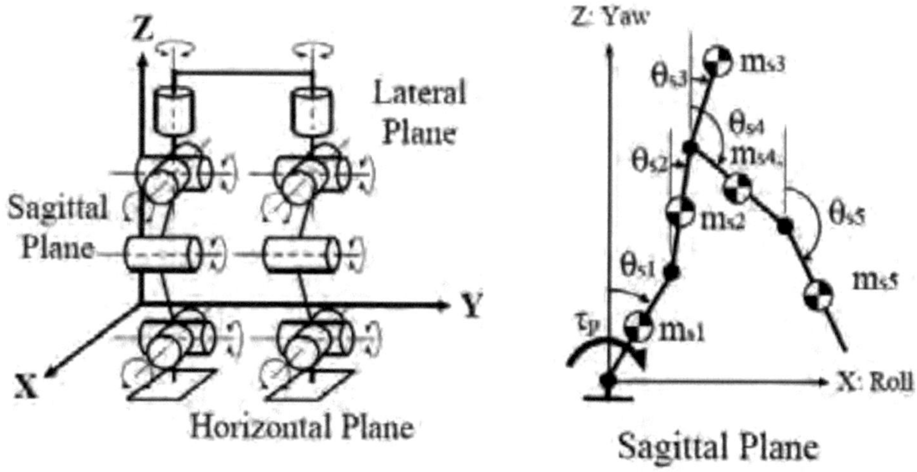

Figure 2. DOF distribution and dynamics model in the sagittal plane

The waist joint trajectory is designed using cubic polynomials as shown in Eq. (5), Eq. (6) and Eq. (7). Note that those functions have enough number of parameters to continuously connect the position and speed trajectories of the waist joint at the start and the end. Both the initial and final conditions of the waist joint trajectory are determined from the supporting leg, which is the hind leg for the initial condition and the fore leg for the final condition. It is also noted that the speed of the waist joint looking from the support-leg expresses the absolute speed of the robot trunk.

$$x_w(t) = \alpha_{x0} + \alpha_{x1}t + \alpha_{x2}t^2 + \alpha_{x3}t^3 \tag{5}$$

$$y_w(t) = \alpha_{y0} + \alpha_{y1}t + \alpha_{y2}t^2 + \alpha_{y3}t^3 \tag{6}$$

$$z_w(t) = \alpha_{z0} + \alpha_{z1}t + \alpha_{z2}t^2 + \alpha_{z3}t^3 \tag{7}$$

where

t : Time

$x_w(t), y_w(t), z_w(t)$: Position of waist at time t

$\alpha_{xn}, \alpha_{yn}, \alpha_{zn}$: Coefficients of cubic polynomial

The waist joint trajectories shown in Eq. (5) and Eq. (7) are used to compute angle trajectories of links in the sagittal plane. Here, the upper body is vertically fixed in order to prevent large movement of the center of gravity. The angle orbit of each link can be determined using Eq. (8) – Eq. (9) from geometrical constraints representing kinematics configuration of the robot. The same procedure is also applicable in the lateral plane.

$$\theta_{s1}(t) = \phi_{wf}(t) + \phi_1 \tag{8}$$

$$\theta_{s2}(t) = \theta_1(t) + \phi_2 - \pi \tag{9}$$

$$\theta_{s3}(t) = 0.00 \tag{10}$$

$$\theta_{s4}(t) = \phi_{wb}(t) - \phi_3 + \pi \tag{11}$$

$$\theta_{s5}(t) = \phi_{wb}(t) + \phi_4 + \pi \tag{12}$$

where

$\theta_{si}(t)$: Angle orbit of link i in sagittal plane

St : Stride

L_{wf} : Length parameter for computation of fore leg

L_{wb} : Length parameter for computation of hind leg

ϕ_{wf} : Angle parameter for computation of fore leg

ϕ_{wb} : Angle parameter for computation of hind leg

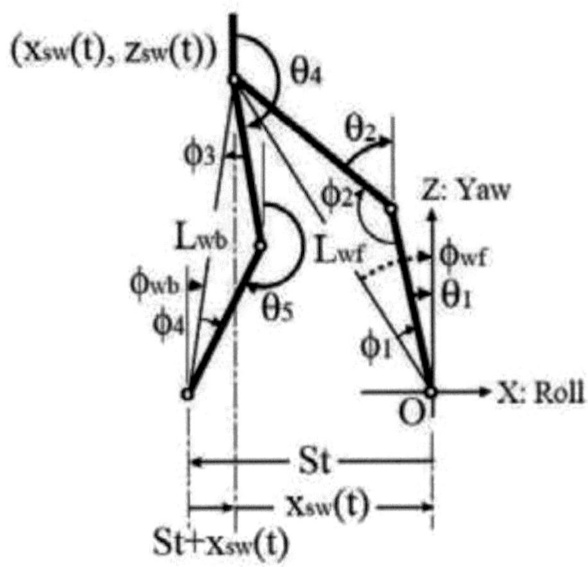

Figure 3. Movement while transition in double-supporting phase

The advantage of this algorithm is that it can easily connect gait modules by the simple geometrical computation with real-time calculation. But, the walk under this algorithm tends to become unstable at the transition of gait module because of discontinuities in acceleration. Nevertheless, this algorithm works most of the time because it takes advantage of the large stability margin resulting from the large supporting polygon of the double-leg supporting phase.

4.2 Algorithm 2: Transition Utilizing Spline Function

The second proposed algorithm utilizes spline functions. This algorithm consists of two processing steps. The first step is for generation of angle trajectories of transitional motion. The second step is for conversion of the generated trajectories into dynamically stable one.

Step 1: Generation of transitional motion

The objective of this step is to generate a set of equations to interpolate trajectories obtained from gait modules. The advantage of this algorithm is to guarantee gait module switching with continuous ZMP transition. This feature is realized by taking second-order derivatives of joint angle trajectories into consideration. We utilize cubic spline functions with four nodes for this purpose.

$$\theta_i = \begin{cases} \alpha_0 + \alpha_1 t + \alpha_2 t^2 + \alpha_3 t^3 & (0 \le t < h) \\ \beta_0 + \beta_1 t + \beta_2 t^2 + \beta_3 t^3 & (h \le t < 2h) \\ \gamma_0 + \gamma_1 t + \gamma_2 t^2 + \gamma_3 t^3 & (2h \le t < 3h) \end{cases} \tag{13}$$

$$h = \frac{1}{3} T_t$$

where

$\alpha_k, \beta_k, \gamma_k$: Coefficients of spline functions

T_t : Transitional period

The following boundary constraints are introduced to keep the continuity of ZMP.

< Boundary Conditions >

- Joint angles at t=0 and t=3h are predetermined from the switching gaits
- Joint angular velocities are continuous at t=0, t=h, t=2h and t=3h
- Joint angular accelerations are also continuous at t=0, t=h, t=2h and t=3h

Step 2: Trajectory stabilization

Transitional motion generated in Step 1 may become unstable dependent on the transition period and boundary conditions. The generated joint angle trajectories are checked for their stability and, if necessary, are modified into stable motion pattern based on the ZMP criterion.

Processing flow of the trajectory stabilization is shown in Fig.4. As described in Fig.4, the motion pattern converter consists of a CoG velocity controller and a referential CoG velocity distributor. The stabilization is processed using these two-step operation.

The transitional angle trajectories from Step 1 and the reference ZMP are supplied to the CoG velocity controller first. CoG of the humanoid is computed by kinematical calculation with the supplied trajectories. In addition, a single-mass model of the humanoid that represents simplified dynamics of the humanoid is applied to obtain the referential CoG velocity. This referential CoG velocity realizes the reference ZMP and stabilizes the transition motion. The referential CoG velocity distributor distributes the CoG velocity to each joint angle by utilizing CoG Velocity Jacobian (Sugihara2002).

This algorithm can realize smooth gait module transition with ZMP continuity. Another advantage of this algorithm is the freedom in the timing of transition. This algorithm can change gait modules in single-supporting phase as well. However, this algorithm requires more calculation effort than algorithm 1.

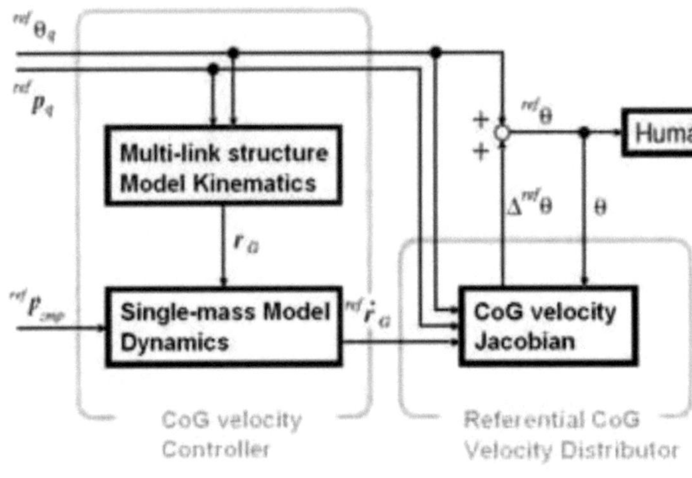

Figure 4. Block diagram of the transitional motion stabilizer

5. Experimental System

The developed and implemented system that realizes the proposed method is explained.

5.1 Hardware Configuration

Hardware configuration of the control system of the robot and a view of the biped walking robot Mk.3 (Furuta2000) used in experiments are shown in Fig.5. Mk.3 was designed for evaluation of gait generating algorithms and walk stability.

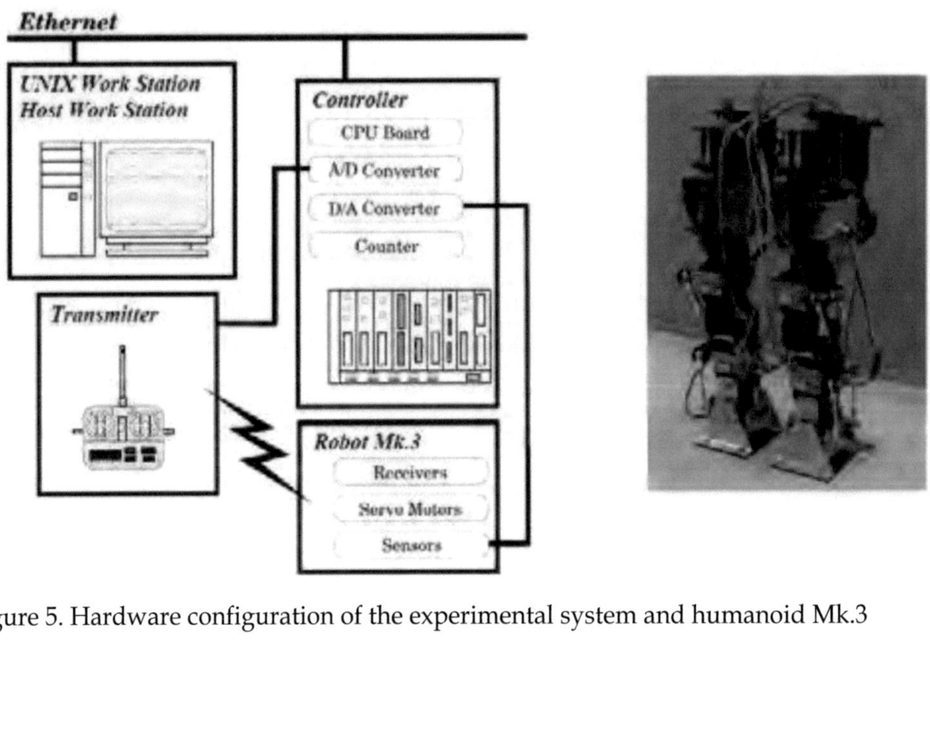

Figure 5. Hardware configuration of the experimental system and humanoid Mk.3

This control system consists of a host computer, a real time controller and a humanoid Mk.3. The reference angle trajectories for links of the robot are distributed wirelessly to motor modules of the robot via a transmitter and receivers. The real time controller uses a commercial real time OS called VxWorks. All sensor values are sent as feedback to the real time controller.

5.2 Developed Gait Modules

Three gait modules based on three kinds of gait generation methods, the "Multi-linked inverted pendulum method (Furuta1997)", the "multi-phase gait (Toda2000)" generating method and the static walk, are constructed and stored in the experimental Gait Library. Although the multi-linked inverted pendulum method has the smallest energy consumption, its movements can easily become unstable since there is no double-leg supporting phase. The stability of this method therefore is established only on level grounds. On the contrary, robots with the multi-phase gait generator can continue walking on rough grounds within limits since certain stabilization of movements during the double-leg supporting phase is possible. However, energy consumption is comparatively large. The static walk has the highest stability margin and can walk through rough grounds within a larger limit than the multi-phase gait. Since the walk cycle is long, however, the walk speed is low and energy consumption is large.
The performance of these gait modules are evaluated in preliminary experiments on even ground, on inclined ground with 5-degree climb and on yielding ground (covered with two sheets of cardboard). Success rate of 10-step walking as the achievement rate, walking speed and the supplemental energy as the energy efficiency for locomotion are measured in the preliminary experiments. These results are summarized in Table 1.

Features \ Gait		Multi-Linked Inverted Pendulum	Multi-Phase Gait	Static Walk
Supplemented Energy [N·m /step]	Sagittal	0.49	1.2	0.84
	Lateral	0.42	0.70	1.0
Walking Speed [m / sec]		0.062	0.071	0.025
Success Rate [%] (ratio)	Even Ground	90 (18/20)	90 (18/20)	90 (18/20)
	Inclined Ground	15 (3/20)	80 (16/20)	80 (16/20)
	Yielding Ground	10 (2/20)	35 (7/20)	80 (16/20)

Table 1. Results of evaluation of gait modules in preliminary experiments

5.3 Experimental Gait Selector

As we have explained in Subsection 3.2, walking state can be judged by monitoring the angular moment of the humanoid because the developed gait modules are based on the ZMP criterion. The flow chart of Gait Selector according to the design policy in Subsection

3.2 is shown in Fig.6. Note that, in this figure, gait modules on the right hand side are more efficient but less stable than those on the left hand side. The right most module, which is for defensive fall, in Fig.6 is selected in the case when stabilization of walk is impossible.

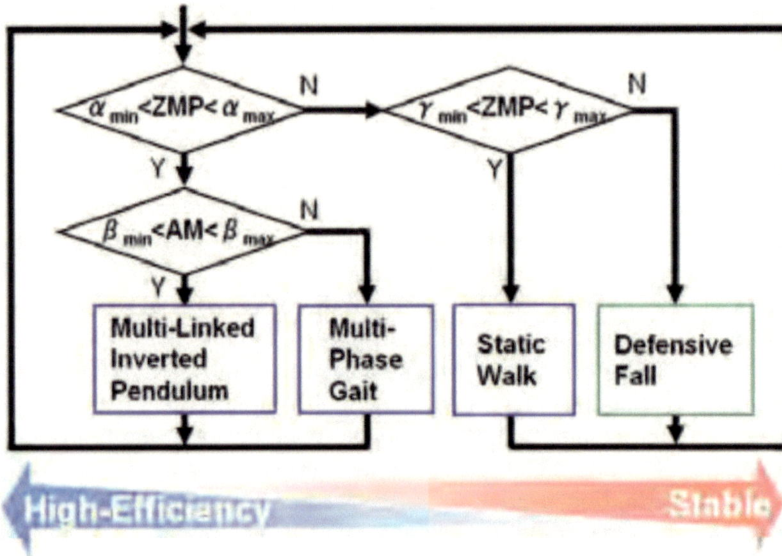

Figure 6. Flow chart of gait selection

At the gait selection, the system first obtains a measured ZMP and determines walk stability margin. If the ZMP deviation is over a threshold determined by α_{min} and α_{max}, imminence of falling is judged. Defensive fall is selected if the stability margin of ZMP equals zero, namely, the outside of thresholds (γ_{min} and γ_{max}). Otherwise, static walk is selected because of the best stability characteristic. If ZMP deviation is within a band defined by the two thresholds α_{min} and α_{max}, then the next gait is selected based on the angular momentum. It is noted that the angular momentum is an index that can express the degree of rotational motion of a robot, just as ZMP is an index that is able to determine the condition of contact between the sole and ground. Therefore, magnitude of the forward motion of a humanoid can best be evaluated by the angular momentum. Since there is an appropriate range of the angular momentum for steady walk, the measured angular momentum is tested if it lies within a set of minimum and maximum thresholds given by β_{min} and β_{max}. If that is the case, then the multi-linked inverted pendulum method is selected as the gait module. If the angular momentum is out of the threshold, multi-phase gait that is more stable than the multi-link inverted pendulum method is selected as the next gait module.

It is known that the evaluation variables used in these criteria are very sensitive and are affected by even microscopic ground conditions. A part of this over sensitivity can be reduced by elimination of high-frequency components of the sensed data. The average of sensor values over 0.080 second interval preceding the gait selection is used for this purpose. A weak point of this operation is the possibility of missing a sharp maximum of ZMP and, as a result, missing the onset of instability. However, this can be overcome by adopting enough stability margins through tactically chosen thresholds.

The following set of threshold values is used:

$$
\begin{aligned}
\alpha_{\min} &= -6.0 \quad [mm] \\
\alpha_{\min} &= \ 10 \quad [mm] \\
\beta_{\min} &= -0.15 \quad [kgm^2/\sec] \\
\beta_{\min} &= -0.030 \ [kgm^2/\sec] \\
\gamma_{\min} &= -40 \quad [mm] \\
\gamma_{\min} &= \ 40 \quad [mm]
\end{aligned}
\tag{14}
$$

Here, the range of α is set at 16 [mm] that is 20% of 80[mm], the actual sole length in traveling direction of Mk.3. In addition, both the thresholds $\alpha_{\min}$ and $\alpha_{\max}$ are shifted forward by 2[mm]. It is because the vertical projection of the center of gravity deviates 2[mm] in the forward direction with our robot. $\gamma_{\min}$ and $\gamma_{\max}$ are set at 40 [mm], sole edge positions, because they represent the limit of stability. For the case of the thresholds of angular momentum, they should be decided based on the desired values derived from the planned motion. Here, the values in the table for the thresholds $\beta_{\min}$ and $\beta_{\max}$ are determined based on the preliminary experiments. The reason for this is a hardware problem. We found that backlashes at gears of the robot have adverse effects on the measured angular momentum through these experiments. It is noted that those thresholds depend only on robot hardware parameters such as the size of the sole, accuracy of sensors, and other physical parameters and not on environmental conditions. Environmental conditions are taken into consideration through real time measurements and gait switchings.

5.4 Installed Gait Transition Algorithm

We have chosen algorithm 1 that was explained in Section 4, namely, transition in double-supporting phase, as the gait transition algorithm. This is because that processing power of the hardware is not enough to execute gait transition with algorithm 2. We have chosen higher priority for real-time operation of gait transition here. It is noted that this transition operation is to be completed within 0.40[sec], which is chosen from the hardware constraint.

6. Experiments

Two purposes of this experiment are the evaluation of the developed experimental system and demonstration of effectiveness of the proposed method.

6.1 Experimental Set-ups

The developed system was implemented onto the control system of the original humanoid robot Mk.3. Gyroscope sensors on each leg link and universal six-axis force sensors installed between the sole and foot were used. Measurement of angular momentum was from gyroscope sensors and measurement of ZMP was from universal force sensors. Measured values were used for judgment of gait module selection at the gait selection brunching points.
The robot is commanded to walk on two kinds of changing road surfaces. In the first case, the surface changes from an upward slope with angle of 5[deg] to an yielding surface (covered with two sheets of cardboard). In the second case, the surface changes from a flat

horizontal ground to an upward slope with angle of 5[deg]. The robot is commanded to walk ten steps in both cases, approximately five steps on each surface.
During the evaluation experiments, ZMP and angular momentum were recorded. At the same time, information on gait selection and overall operation was collected. The obtained data were used for verification of the intended operation of the developed experimental system. Next, success rates of the planned walk, amount of the supplemented energy and traversal time to complete the commanded walk were compared between the proposed method and conventional single gait generation scheme in order to evaluate effectiveness of the proposed method. Major parameter values used for gait generation are listed in Table 2.

Gait	Stride[m]	Period[sec]
Static Walk	0.050	2.0
Multi-Phase Gait		0.70
Multi-Linked IP		0.80

Table 2. Parameter settings of each gait module

Here, selection and change of gait were performed every two steps and at the start of the walk cycle. The reason for every two steps is that gait transition at every step implies that the gait selection of next step must be done while the transient effect of gait change is still prevailing and this will cause errors in selection of gaits.

6.2 Result of the Verification Experiments

Typical trajectories of gait selection, the measured angular momentum and the ZMP from one each of two cases are shown in Result I (Fig.7) and Result II (Fig.8).

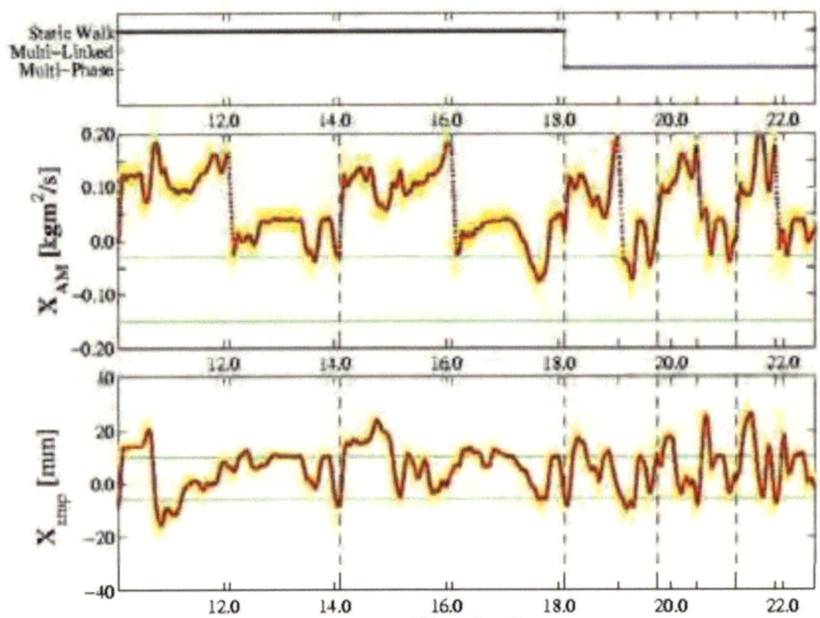

Figure 7. Gait module selection and sensor values I: Walking through an upward slope of 5[deg] and encountering an yielding surface at time 18.1[sec]

Blue lines in the top figures show the selection of gaits. Middle graph is the angular momentum. The lowermost graph is the measured ZMP. Values shown in yellow are the instantaneous measurement of sensors and the red curves are the running average over the 0.080[sec] time duration and are used as indices of gait selection. Green lines show the minimum and maximum thresholds. Vertical dashed lines point the timing of gait selection. In Result I, at the first and second gait selection timings (10.1[sec], the left end of the graph, and 14.1[sec]), static walk was chosen because the averaged ZMP deviated from the range of thresholds. The robot moved onto the yielding surface at the third timing (18.1[sec]) of gait selection. The ZMP came back within the limits of the threshold at this timing but the angular momentum stayed outside of the threshold. It is observed that the selected gait was changed to the multi-phase gait in response to this. In summary, the static walk with the highest stability margin was chosen on the upward slope and the multi-phase gait was chosen on the yielding surface based on a comparatively wider stability margin.

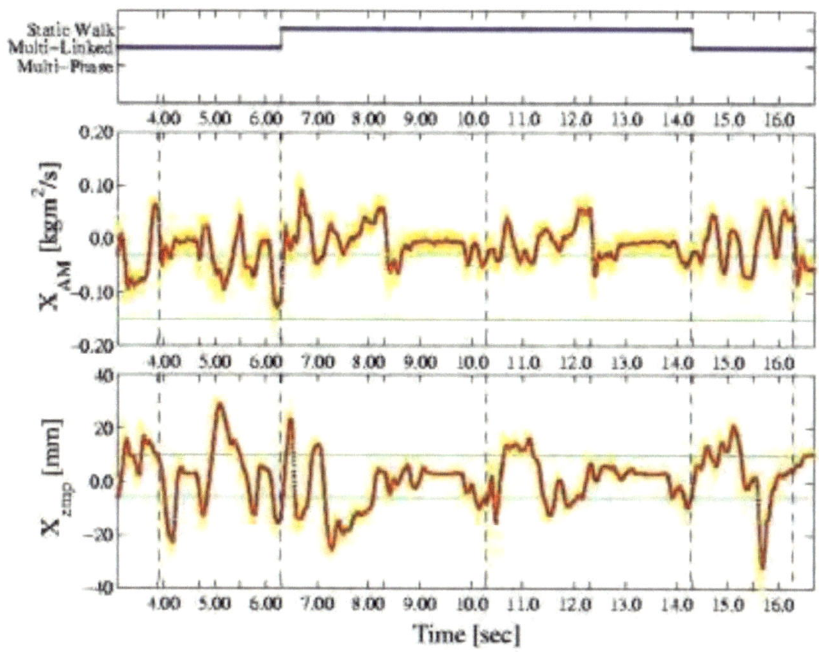

Figure 8. Gait module selection and sensor values II: Walking through a flat horizontal ground and encountering an upward slope at time 6.30[sec]

In Result II, the gait of multi-linked inverted pendulum method was chosen on the initial horizontal ground since the ZMP and the angular momentum were judged to be within the limits of the thresholds. At the second gait selection timing (6.30[sec]) when the robot proceeded to the upward slope, the ZMP deviated out of the threshold. Therefore, static walk with highest stability was chosen. At the fourth timing (14.3[sec]) of gait selection, the gait of multi-linked inverted pendulum method was chosen since the angular momentum returns within the threshold.

The results of these two cases exhibit the gait selection corresponding to the road surface condition is successfully realized using sensor information.

6.3 Effectiveness of the Proposed Method

The sensor-based gait generation and conventional single gait generation are compared in Table 3.

	Success Rate	Traversal Time	Total Energy	Walking Velocity	Energy Efficiency
	[%]	[sec]	[Nm]	[m/sec]	[m/Nm]
Static	16/20	20.0	19.4	0.0250	0.0258
MPG	7/20	7.00	19.0	0.0714	0.0263
MLIP	3/20	8.00	9.10	0.0625	0.0549
SBG-I	10/20	12.4	19.0	0.0403	0.0263
SBG-II	10/20	13.2	15.3	0.0379	0.0327

Table 3. Mobility performance of each gait

The success rates in the upper three lines; static (Static), multi-phased (MPG) and multi-linked (MLIP) gaits, are the averages of success rates on the two experimental walk surfaces discussed in the last subsection over 10 trial walks. All other values; traversal time, total supplemented energy, walking velocity and energy efficiency, are computed based on the reference trajectory generated by those algorithms for the commanded stride and walk period. The lines marked Sensor-Based Gait I and II (SBG-I and SBG-II) in this table correspond to the cases with the proposed Sensor-Based Gait Generation system on the two walk surfaces.

The experimental results in this table show that the walking velocity and the energy efficiency of walk are both enhanced without reducing the success rate of walk in each sensor-based gait. Therefore, it is concluded that the humanoid can acquire sufficient mobility and can make use of the advantages of each gait by adopting the proposed system.

The high success rate of walk comparable to the static gait only case, however, was not obtained in neither of the experiments. The major cause of this is the instability during the transition from a gait to another. This indicates the necessity to improve the transitional motion by utilizing new hardware and/or better algorithm. Installation of higher-end CPU with the transitional algorithm 2 would be a viable approach. By doing this, success rates equivalent to the static walk can be expected in both Sensor-Based Gait I and II cases. It is also noted that it is impossible to increase energy efficiency of the sensor-based gait more than that of the multi-linked inverted pendulum method. This is because the Gait Selector is designed to consider not only the energy efficiency but also the walking stability as the criteria for walk selection.

It is also noted that the success rate is no more than 80% even for the case of static gait in the series of experiments. The reason for this is that no balance control was implemented in the experiments in order to evaluate the effect of the proposed system only.

7. Conclusion and Future Works

A Sensor-Based Gait Generation method was introduced and an experimental system was built. Then, the system was implemented onto an original humanoid robot to evaluate operations and to demonstrate effectiveness of the proposed method. Experimental results exhibited successful gait selection corresponding to the road surface condition obtained from sensor information. Additionally, walking velocity and the energy efficiency are both enhanced without reducing the success rate of walking.

The design approach for Gait Selector based on both ZMP and the angular momentum adopted in this study is a sufficiently general and valid one. The developed Gait Selector should be applicable to many gaits and humanoids. However, more conditional branchings based not only on ZMP and the angular momentum but also on some combinations of them may be necessary depending on such factors as robot hardware, types of gaits and criteria for robot motion evaluation. The fundamental reason for the lack of a fixed design method is that the selection of gait is inherently rooted in factors such as hardware specifications and characteristics of each gait. At present, therefore, we have to redesign the Gait Selector such as that in Fig.6 according to the procedure described in Section 3.

Future studies should be targeted to simplify the design procedure of Gait Selector. The more gait modules and ground conditions are installed into the system, the more complicated parameter tuning must be required. One possibility of avoiding this problem would be to introduce simple learning capability for Gait Selector design. A discrimination method that only utilizes sensor value histories of 3-axis accelerometer to identify several ground conditions (Miyasita2006) was already reported. They employ simple decision tree constructed based on acceleration data that are obtained during several trial motions on each ground condition. There is a possibility of direct acquisition of transition rules by utilizing histories of ZMP and angular momentum with all combinations of a gait module and a ground condition.

Apart from the improvement of the design of Gait Selector, there also is a room for improvements by adding new gait generation modules and improving the success rate of walk through the enhancement of the transition scheme for gait module changes. These are more straightforward tasks if the required additional computational power is available.

8. References

Furuta, T. *et al.* (1999). Biped Walking Using Multiple-Link Virtual Inverted Pendulum Models (in Japanese), *Journal of Robotics and Mechatronics*, Vol.11, No.4 (1999), pp. 304-309, ISSN: 0915-3942

Furuta, T. *et al.* (2001). Design and construction of a series of compact humanoid robots and development of biped walk control strategies, *Robotics and Autonomous Systems*, Vol. 37, No. 2, (November 2001) pp. 81-100(20), ISSN: 0921-8890

Kajita, S. (2002). Zero-Moment Point (ZMP) and Walking Control (in Japanese), *Journal of the Robotics Society of Japan*, Vol. 20, No. 3, (April 2002) pp. 229-232, ISSN: 0289-1824.

Kajita, S. et al. (2003). Biped Walking Pattern Generation by using Preview Control of Zero-Moment Point, *Proceedings of IEEE International Conference on Robotics and Automation (ICRA2003)*, Vol.2, pp. 1620- 1626, ISBN: 0-7803-7736-2, Taipei, September 2003, IEEE

Miyashita, T. and Ishiguro, H. (2006). Behavior Selection and Environment Recognition Methods for Humanoids based on Sensor History (in Japanese), *Proceedings 2006 JSME Conference on Robotics and Mechatronics*, No. 06-4, (CD-ROM) 1P1-E09, Tokyo, May 2006

Nishiwaki, K. *et al.* (2001). Online mixture and connection of basic motions for humanoid walking control by footprint specification, *Proceedings of IEEE International Conference on Robotics and Automation (ICRA2001)*, Vol. 4, pp.4110-4115, ISBN: 0-7803-6576-3, Seoul, May 2001, IEEE

Sugihara,T. et al. (2002). Real-time Humanoid Motion Generation through ZMP Manipulation based on Inverted Pendulum Control, *Proceedings of IEEE International Conference on Robotics and Automation (ICRA2002)*, Vol.2, pp.1404-1409, ISBN: 0-7803-7272-7, Washington D.C., May 2002, IEEE

Toda, K. *et al.* (2004). Sensor-Based Biped Gait Generation Scheme For Humanoid - Implementation and Evaluation -, *Proceedings of 2004 IEEE/RSJ International Conference on Humanoid Robots (Humanoids 2004)*, (CDROM) Paper #61, Santa Monica, November 2004

12

Momentum Compensation for the Fast Dynamic Walk of Humanoids based on the Pelvic Rotation of Contact Sport Athletes

Jun Ueda[1], Kenji Shirae[1], Shingo Oda[2] and Tsukasa Ogasawara[1]
[1]*Nara Institute of Science and Technology,* [2] *Kyoto University*
Japan

1. Introduction

Biped walking for humanoid robots has almost been achieved through ZMP theory (Takanishi, et al., 1985) (Goswami, 1999) (Kajita, et al., 2002). Recently, the research on humanoids has begun to focus on achieving tasks using the arms during walking, in tasks, such as carrying a load (for example, a heavy backpack) or interacting with environments (Harada, et al., 2003). In order to achieve a stable biped-walking, the momentum around the perpendicular axis generated by the swing leg must be counterbalanced. If this momentum exceeds the maximum static friction torque between the floor and the stance foot, the body will begin to slip and rotate around the perpendicular axis. In a normal human walk, the upper body compensates this momentum, i.e., by rotating the thorax (or shoulders) and swinging the arms in an antiphase of the swing leg (van Emmerik & Wagenaar, 1996) (Lamoth, et al., 2002) (LaFiandra, et al., 2003). For humanoid control, research has been presented for momentum compensation using the motion of the entire body including the arms (Yamaguchi, et al., 1993) (Kagami, et al., 2000) (Yamane & Nakamura, 2003) (Kajita, et al., 2003). However, momentum compensation by the upper body is undesirable for a humanoid that uses its arms to achieve a task since this type of compensation limits the degree of freedom (DOF) for the task. In addition, the fluctuation of the upper body has a bad effect not only on the task accomplishment, but also on visual processing since most vision systems are attached to the head part. As a result, it is desirable to preserve as many degrees of freedom of the upper body as possible, and to suppress the fluctuation of the body at the same time. The walking action including momentum compensation should be completed only by the lower body, which leads to a simplification of motion planning.
Improving the performance of humanoids through observations of humans walk seems natural. Recently, however, in the field of exercise and sports science, a clarification of efficient motion in the human has begun, and this clarification has been accompanied by improvements in the measuring equipments used for this endeavour. Many common features can be observed in the motion of contact sport athletes, i.e., they move so as not to twist their trunks as much as possible. This kind of trunk-twistless walk is better understood than before, but is considered inefficient since the walking pattern in the trunk-twistless walk is different from normal one (Steinhaus, 1963) (Ducroquet, et al., 1968). However, a decreased pelvic and thoracic rotation, similar to trunk-twistless walk, has been observed

with a load carriage (LaFiandra, et al., 2003).This decrease in pelvic and thoracic rotation indicates that not twisting the trunk and not swinging the arms, but other momentum compensation, is performed when the intensity of upper-body exercise is high. Therefore, this trunk-twistless walk may be helpful to humanoids for achieving tasks; however, the characteristic of this walk has not been clarified, and its result has not been applied to humanoids.

In this chapter, the characteristics of the trunk-twistless walk are quantitatively investigated from the observation of contact sport athletes. The relative phase of the swing leg and the pelvic rotation appears to be in an antiphase when compared with the normal walk of humans. This leads to the possibility of momentum compensation by pelvic rotation, and this characteristic of the pelvic rotation is implemented to a humanoid in experiments conducted in this chapter. A method of determining the rotation of the humanoid's waist is proposed in conjunction with the pitch angle of the swing legs. In this chapter we confirm that the torque around the perpendicular axis is reduced in the humanoid trunk-twistless walk when compared to a standard humanoid walk without the twisting of the trunk or swinging of the arms. Improvements on the straightness of the walking trajectory and on the reduction in the fluctuation of the upper body during a fast dynamic walk are also confirmed.

2. Walking Measurement

2.1 Methods and Subjects

Three healthy male subjects who are accustomed to the trunk-twistless walk served as subjects. All subjects are contact sport athletes of rugby football, karate, and kendo (the Japanese art of fencing) respectively. All subjects have been coaches. Their mean age, body height, and body weight were 42.6±7.0 years (Mean±S.D.), 171.33±1.52 cm, and 79.3±6.02 kg. Subjects were given several minutes to get used to treadmill walking. The treadmill velocity was set to 1.5 km/h, 3.0 km/h and 4.0 km/h. The normal walk and the trunk-twistless walk were measured for 30 seconds.

A motion capture system with twelve cameras (Vicon Motion Systems Ltd.) was used to measure three dimensional kinematic data (sampling frequency 120Hz) from the reflective markers shown in Fig.1 (a). Two 3-axis accelerometers were attached on both iliac crests to measure the antero-posterior and medio-lateral accelerations of the pelvis. The twisting angle of the trunk was measured using the four markers shown in Fig.1 (b).The thoracic and pelvic rotation around the perpendicular axis, θ_{thorax} and θ_{pelvis} in Fig.2 are measured by the markers on both clavicles and both iliac crests respectively. Both angles are set to 0 when the subject is exactly facing the forward direction. The yaw-axis torque exertion from the stance foot to the floor is defined as τ_{LF} and τ_{RF} for each foot[1]. When τ_{LF} increases to positive and exceeds the maximum static friction, the body begins to rotate clockwise due to the slip that occurs at the stance foot. After walking on the treadmill, the subjects were asked to walk on the pressure distribution sensor (Big-Mat, Nitta Corporation, 440mm×1920mm), as shown in Fig. 3, to measure the trajectory of the COP (Center of Pressure) of the stance foot.

[1] The foot rotation around the perpendicular axis is the main focus of this chapter. Whenever context permits, we use torque/momentum to the torque around the perpendicular/yaw axis.

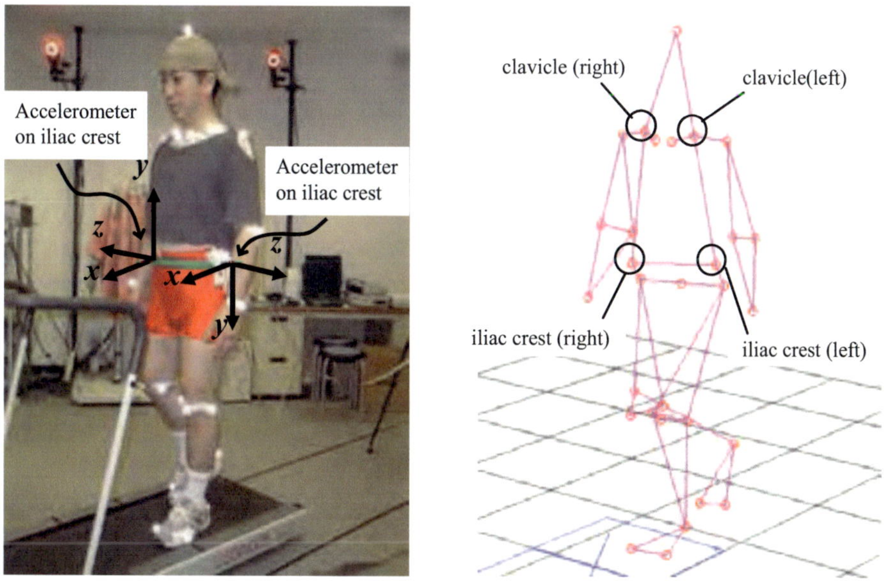

(a) Marker Setup & Acceleration Measurement (b) Captured Human Model
Figure 1. 3-D Motion Capture

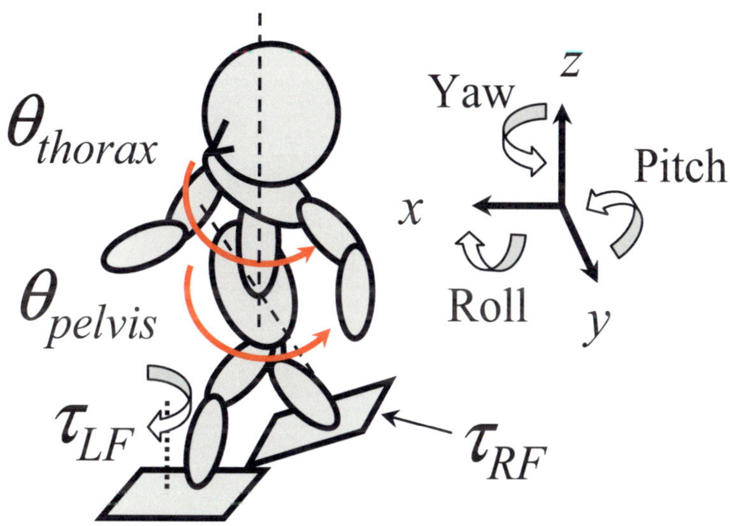

Figure 2. Pelvis-thorax Rotating Angle and Yaw Moment of Stance Foot

2.2 Comparison of Trunk Twisting and Pelvic Rotation

Figure 4 shows typical examples of the captured walking posture in one walking cycle from behind of the subject. In this figure, the postures at LC (Left heel Contact), RO (Right toe OFF), RC (Right heel Contact), LO (Left toe OFF), and next LC are shown.
From Fig.4, it can be observed that the step width of the athlete's walk is wider than the normal walk, and the posture of the stance feet is in external rotation. In addition, the amplitude of pelvic rotation is small, and the relative phase between the swing leg and the pelvis is different compared to the normal walk.

The twisting angle of trunk θ_{twist} is obtained by subtracting θ_{pelvis} from θ_{thorax}:

$$\theta_{twist} = \theta_{pelvis} - \theta_{thorax} \quad (1)$$

Figure 5 shows the typical thorax, pelvis, and twisting angles at 4.0 km/h. The bottom graph shows the stance phase, LC and RC. In the trunk-twistless walk, the relative phase between the pelvic and thoracic rotation is smaller, resulting in a smaller twisting angle of trunk than in the normal walk. In comparison to the stance phase, the relative phase between the leg and the thorax is almost the same for both types of walking, but the difference can be found in the pelvic rotation.

The counterclockwise rotation of the pelvis is observed for the normal walk when the right leg is in the air, whereas in the trunk-twistless walk, the clockwise rotation is observed in the same period. As a result, the relative phase of the swing leg and the pelvic rotation can be said to be in an antiphase for the trunk-twistless walk compared to the normal walk.

Figure 6 shows the walking velocity versus the relative phase of the thoracic and pelvic rotation. For all walking velocities, the relative phase of the trunk-twistless walk is smaller than the normal walk. The relative phase increases when the walking velocity increases as reported in conventional researches (Lamoth, et al., 2002) (LaFiandra, et al., 2003); however the tendency is the same.

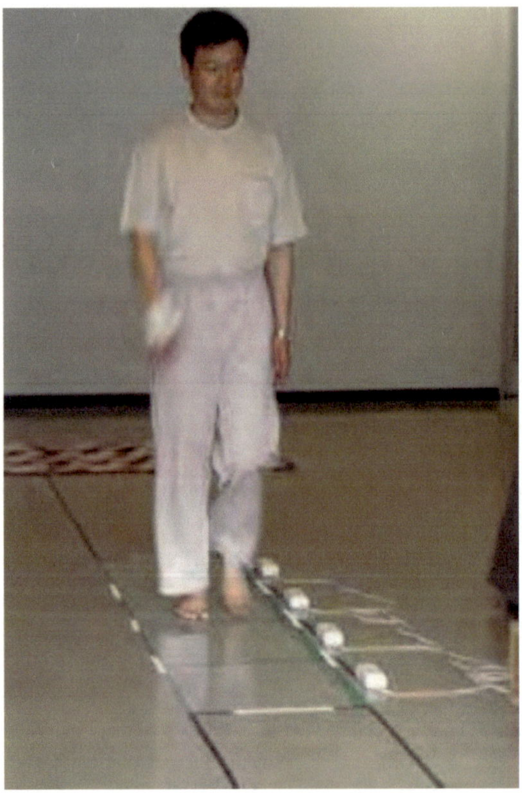

Figure 3. Pressure Distribution Measurement of Stance Foot

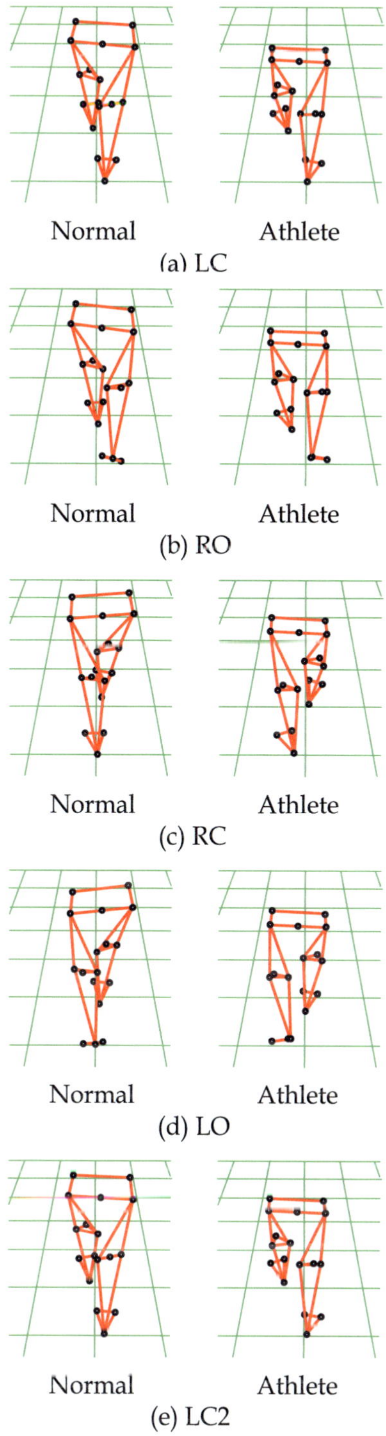

Figure 4. Captured Walking Motion (from behind)

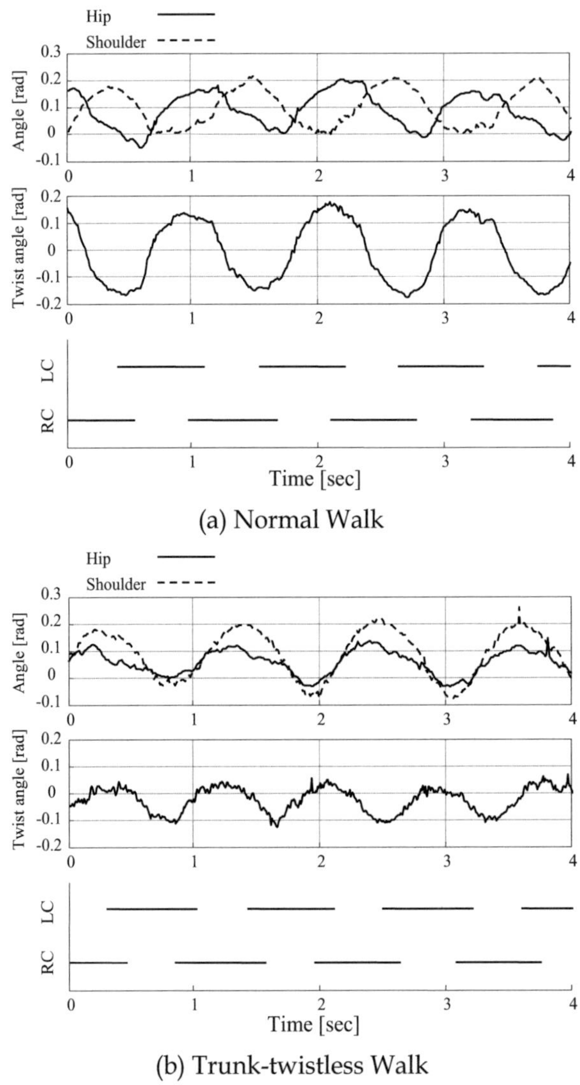

Figure 5. Twisting Angle of Trunk

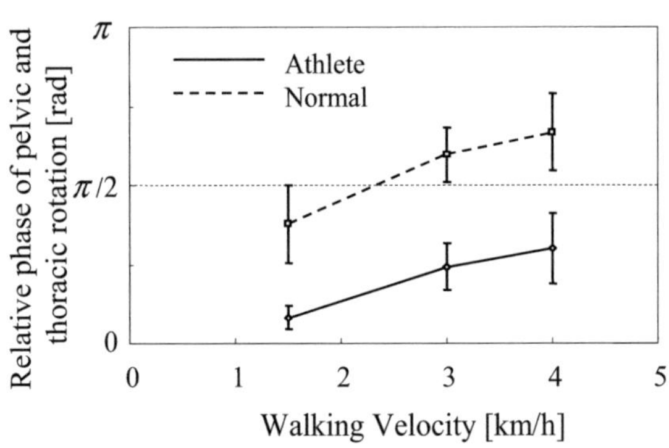

Figure 6. Comparison of Twisting Angle of Trunk (average of 3 subjects)

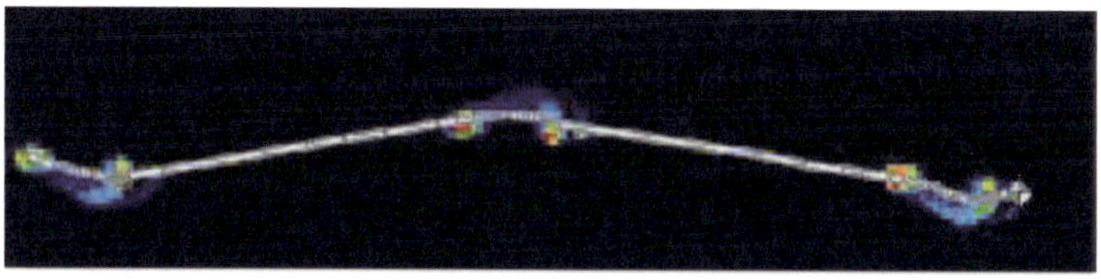

(a) Normal Walk

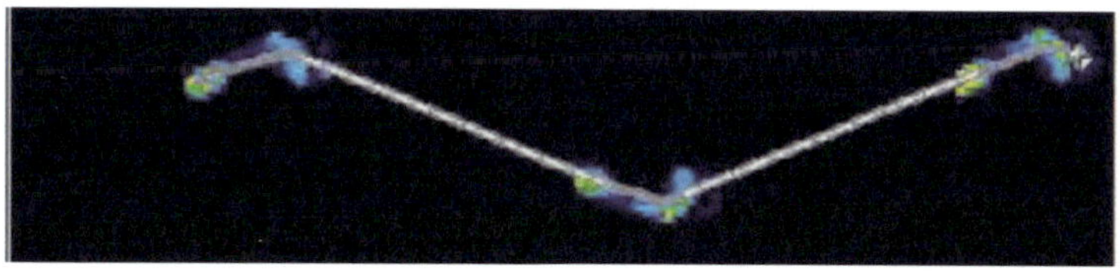

(b) Trunk-twistless Walk

Figure 7. Trajectory of COP

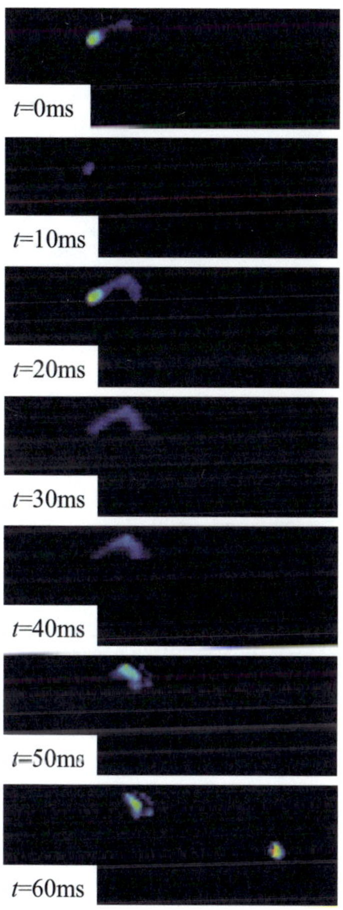

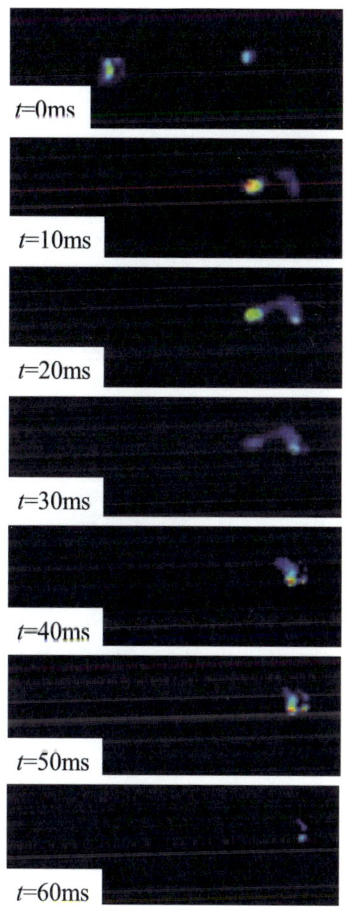

(a) Normal Walk (b) Trunk-twistless Walk

Figure 8. Comparison of Pressure Distribution of Stance Foot

2.3 Yaw-axis Torque between Stance Foot and Floor

Figure 7 shows the trajectories of the COP (Center of Pressure). Similar to Fig.4, the step width of the trunk-twistless walk is wider than the normal walk, and the posture of the stance feet is in external rotation. Figure 8 shows the transition of the pressure distribution of left stance foot for every 10ms from LC to LO. As shown in Fig.8 (a) of the normal walk, the COP moves relatively straight from the heel, thenar eminence, then the first toe. In contrast, the COP moves from the heel, hypothenar eminence, and then the thenar eminence like a curve in the trunk-twistless walk. Note that the first toe is off without being pressured in this case. This difference between the COP trajectories can also be observed in Fig.9 (COP trajectories within the left stance foot).

In order to clarify the reason for this curved COP trajectory, a preliminary experiment was performed. First, a subject stands only on his left foot on the pressure distribution sensor. Next, the subject externally rotates the left hip joint so as to rotate the body clockwise by 90 degrees. At this moment, a counterclockwise torque must be applied to the stance foot due to the body's rotation. Figure 10 shows the measurement result during the process, in which a similarity to Fig.8 (b) is observed. This result indicates that the curved COP trajectory occurs when a torque is applied from the stance foot to the floor.

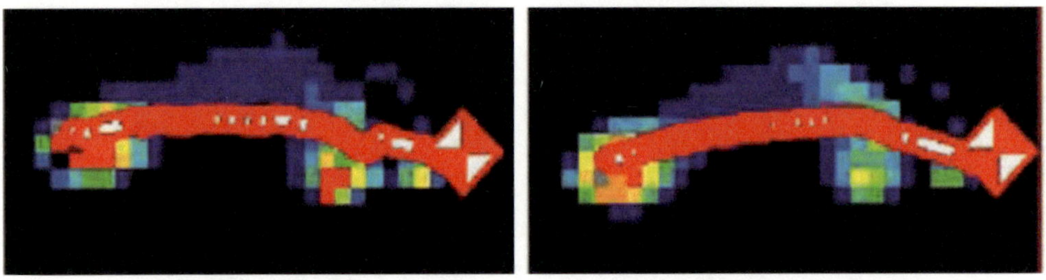

(a) Normal Walk (b) Trunk-twistless Walk

Figure 9. Trajectory of COP of Stance Foot

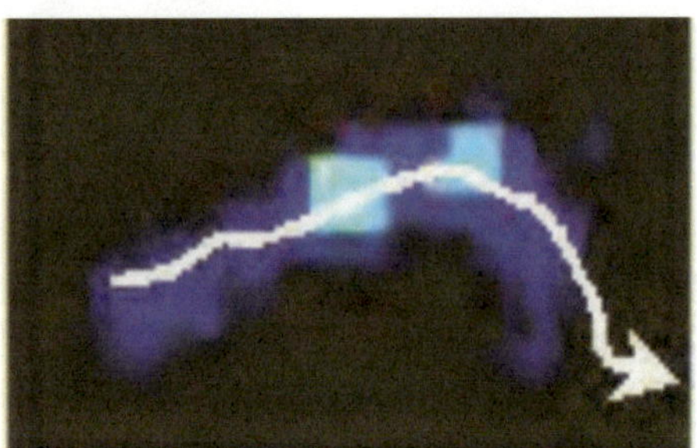

Figure 10. Trajectory of COP when exerting Yaw-axis Moment

2.4 Discussion of Momentum Compensation of the Trunk-twistless Walk

The trunk-twistless walk was quantitatively investigated from the observation of contact sport athletes. The typical characteristics in comparison with the normal walk are:

1. Externally rotated posture of the stance feet,
2. Wide stance width,
3. Small relative phase between the pelvic and thoracic rotation (Small trunk twisting),

4. Curved COP trajectory (Torque transmission at the stance foot),
5. Antiphase of the swing leg and the pelvic rotation.

In a normal human walk, the momentum is compensated by the upper body, i.e., by rotating the thorax (or shoulder) and swinging the arms in antiphase of the swing leg. This motion leads to canceling the torque at the stance foot and avoiding the rotational slip. However, in contact sports, the upper-body posture should be maintained to the front as much as possible in preparation for contacting against the environments. A similar phenomenon can be observed when the contact increases the intensity of the upper-body exercise. It is assumed that the contact sport athletes perform the trunk-twistless motion for the upper-body exercise. A decreased pelvic and thoracic rotation, similar to the above case, has been observed with a load carriage (carrying a heavy backpack) (LaFiandra, et al., 2003). Without twisting the trunk or swinging the arms, the momentum should be compensated by other methods. Our hypothesis is that the momentum is simultaneously compensated passively by the friction at the stance foot and actively by the antiphase pelvic rotation.

First, we discuss the passive compensation caused by the friction. The externally rotated posture of the stance feet and the COP trajectory without passing toes are necessary for the transmission of the torque generated by the hip joints. We have reported that the toes (especially the big and 2nd toes) are important for balance of the upper body (Takemura, et al., 2003). It is assumed the step width becomes wider for the balance. In addition, the translational force at the stance foot can cancel the momentum when the step width becomes wide.

Next, active compensation by the antiphase pelvic rotation is discussed. In Fig.2, τ_{LF} increases when the right leg is accelerated and swung forward in the initial part of left leg stance phase. In the normal walk where the leg and the pelvis are in-phase, the momentum due to the increase of θ_{pelvis} also increases τ_{LF}. In this case, the sum of the momentum should be compensated by trunk twisting and arm swinging. On the other hand, in the trunk-twistless walk where the leg and the pelvis are in an antiphase, the decrease of θ_{pelvis} cancels τ_{LF}. Also, the momentum of inertia of the pelvis is not large when compared to the momentum of inertia of the legs; in other words, the total momentum is not compensated only by this active pelvic rotation. However, the twisting action of the trunk seems unnecessary when combined with passive compensation. In this chapter, we focus on this antiphase pelvic rotation, and apply this antiphase rotation to fast walking by humanoids.

3. Moment Compensation for Humanoids using Waist Rotation

3.1 Humanoid Robot: HRP-2

Figure 11 shows the overview and the actuator configuration of the humanoid HRP-2 (Inoue, 2000) used in this chapter. Its height is 154cm and weight is 58kg, which are close the height and weight of a human. HRP-2 has 30 DOF in total. One characteristic of the HRP-2 is a joint around the perpendicular axis between the chest and the waist. Along with the hip joints, the chest part and the waist part rotate independently around the perpendicular axis as shown in Fig.11 (a). The pelvic rotation discussed in section 2 can be implemented to HRP-2 by correlating the robot's waist to a human's pelvis and its chest to a human's thorax. A simulation software, Open HRP (Hirukawa, et al., 2001), is also available. In this chapter, both experiment and simulation are performed for the investigation, but a very fast walk, which cannot be realized by the real hardware, is examined only in the simulator.

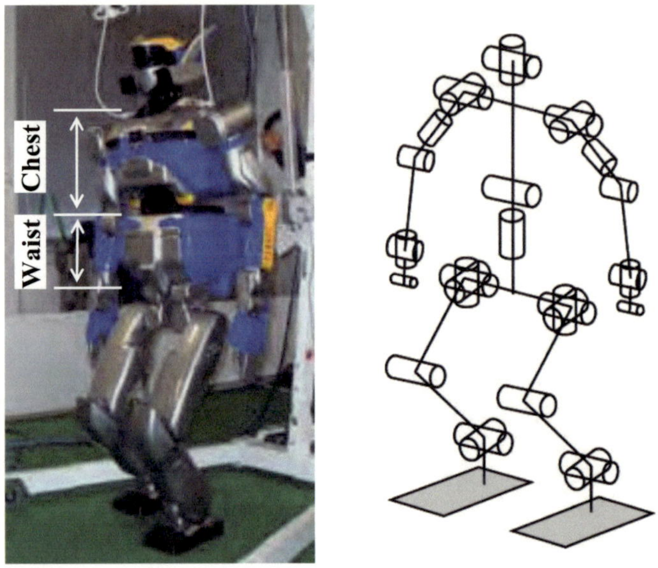

(a) Overview (b) Actuator Configuration

Figure 11. Humanoid Robot HRP-2

	Standard	Waist fixed	Propose
Waist rotation & leg swing	In-phase	0	Antiphase (see Eq.(2))
Arm swing	0	←	←
Chest Rotation	0	←	←
Step width	Shoulder width	←	←

Table 1. Walking Pattern of the Humanoid

3.2 From Athlete Measurements to Humanoids

In Section 2, the antiphase pelvic rotation was observed in the trunk-twistless walk of contact sport athletes. The advantages of the application to humanoids are as follows: One goal of the research on humanoids is to achieve a task where the humanoid carries a load or interacts with environments (Harada, et al., 2003) using its upper body. The use of the upper body for walking itself, however, is undesirable for this purpose. The walking action including momentum compensation should be completed only by the lower body in order to preserve the freedom of the upper body.

In general, humanoids have larger feet for static stability; hence the maximum friction torque of the sole is larger than the human's. The rotational slip was not generated without considering the momentum in a low-speed walking pattern; however, this slip becomes a problem when the walking velocity is increased. In this case, the trunk-twistless walk preserves the upper body's DOF by waist rotation. As a secondary effect, the trunk-twistless walk provides the humanoids with an easy visual processing using a vision system attached to the head part.

As shown in Fig.7 (a) in the walk of normal humans, the feet contact the floor more closely to the center of the body, resulting in a step width narrower than the shoulder. However, in recent version of the humanoid's walk, the feet move in parallel to the walking direction, thus resulting in the same step width of the shoulders (Kajita, et al., 2002). This result merely comes from a simplification of the walking pattern and a collision avoidance pattern between the feet. On the other hand, the expansion of the step length in the trunk-twistless walk is related to balance maintenance and torque transmission. Although humanoids are not alive and athletes are very much so, both can perform the same walk with the same wide step width, which leads to easiness of implementation in the measurement result.

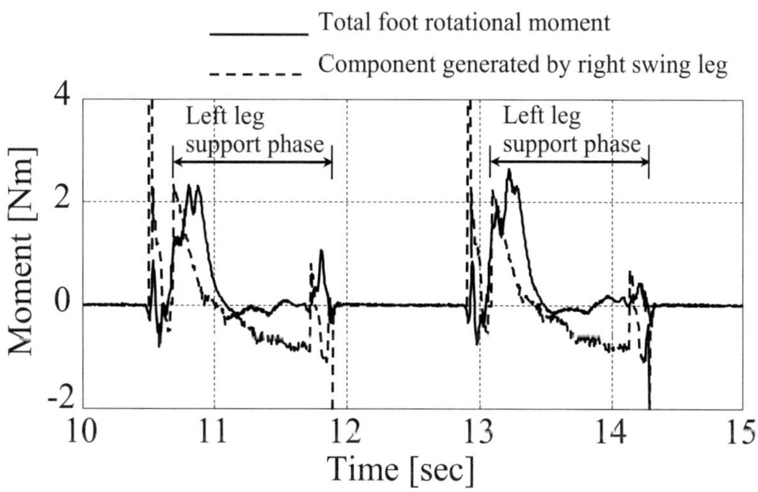

Figure 12. Component of Stance Foot Torque (Simulation, 0.83km/h)

3.3 Evaluation of Humanoid Walk

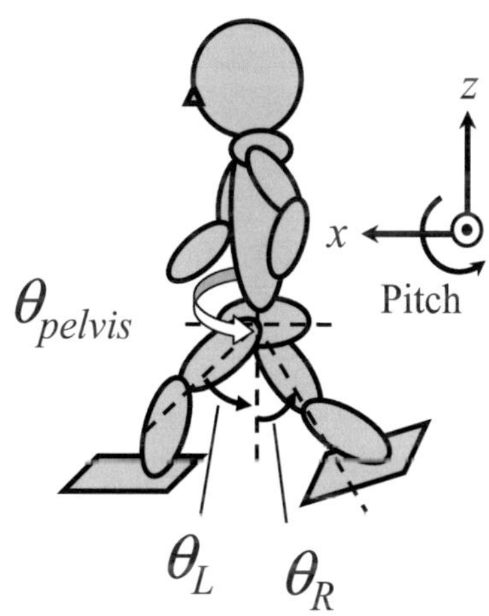

Figure 13. Rotation Angle of Waist and Pitch Angle of Swing Leg

The walking pattern with swinging arms and twisting trunk, which is common to the majority of people, cannot be regarded as a humanoid's standard walk. In this chapter, we apply the antiphase pelvic rotation of athletes to humanoids. First, the standard walk of a humanoid is defined as shown in Table 1 to make clear the effects on momentum compensation without using the upper body. Note that we use `standard' for a humanoid walk to distinguish this walk from the `normal' walk of humans. In the standard walk of a humanoid, the upper body (above the chest) is not twisted and is planned to facing the forward direction. The swinging of arms is not performed; therefore, the walking action is completed only by the lower body.

The waist rotation is set in-phase of the swing leg for the humanoid's standard walk, the same as in a human's normal walk. In contrast, the antiphase rotation of the waist is performed for the proposed walk of the humanoid, and is obtained by (2) presented in the following section. In addition, in the middle of the standard and the proposed walk of the humanoid, the waist-fixed walk is defined without the waist rotation. The step width is set equal to the shoulder's width for all patterns by applying the result of the step width, which allows us to use a standard pattern generator (Kajita, et al., 2002). Note that all three patterns have the same trajectory, velocity, posture, and landing positions for the foot in the air by using the redundancy of the waist and leg part. As a result, the walking velocity, step length, and step width are the same. The only difference is the waist rotation.

3.4 Momentum Compensation by Waist Rotation

Figure 12 shows the total left stance foot torque and the component generated by the right swing leg. As is observed in the graph, the component of the swing leg occupies a large percentage of the total momentum. In order to effectively cancel the stance foot torque by the waist rotation, the waist should be counterbalanced to the largest factor, i.e., the leg motion. The angle of the waist rotation is obtained as follows coupling with the pitch angle of the hip joints:

$$\theta_{pelvis}=k\ [\theta_R(t)-\theta_L(t)]\ /\ \theta_{max} \tag{2}$$

where k=0.175 [rad] (10.0 [deg]) is the maximum amplitude of the waist rotation used to obtain the same amplitude of the standard walk. $\theta_R(t)$ and $\theta_L(t)$ are pitch angles of both swing legs shown in Fig.13, θ_{max} =0.5 [rad] (28.6 [deg]) is the maximum of $\theta_R(t)-\theta_L(t)$ in one walking cycle. By using (2), the antiphase rotation is generated for the proposed walk compared with the standard walk as shown in Fig.14. The stance foot torque due to the acceleration of the swing leg is cancelled by the waist angular momentum.

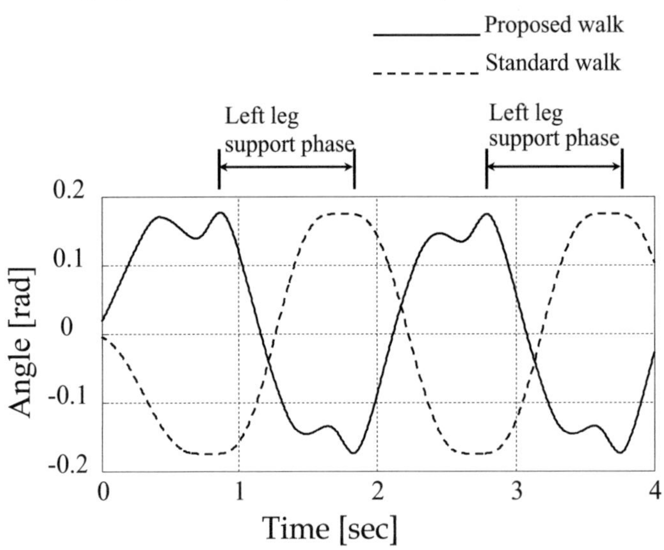

Figure 14. Rotation Angle of Waist (Simulation, 0.83 km/h)

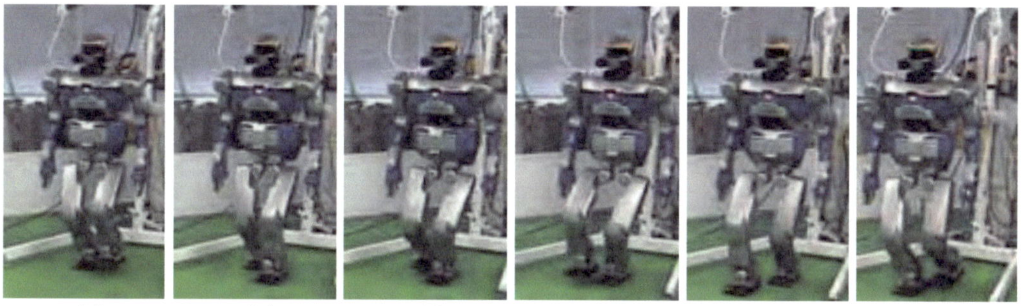

(a) Standard Walk (In-phase waist rotation)

(b) Proposed Walk (Antiphase waist rotation)

Figure 15. Experiments by Humanoid (0.83 km/h)

4. Fast Dynamic Walking of Humanoid

4.1 Comparison of Stance Foot Torque

The effect of the waist rotation for reducing the stance foot torque is confirmed both by simulation and experiment. Figure 15 shows the body posture at 0.83km/h. As shown in the experiment, the body posture for both types of walking is the same except the waist rotation. The antiphase waist rotation is realized in the proposed walk. Figure 16 shows the left stance foot torque by the standard, the waist-fixed, and the proposed walk respectively.

As shown in Fig.16 (a) on the simulation, the torque is decreased by 30% compared with the waist-fixed walk, and by 37% from the standard walk. As shown in Fig.16 (b) on the experiment, the torque is decreased by 30% from the standard one. The validity of the proposed walk can be confirmed in both cases. The difference of the wave form between the simulation and the experiment is due to several unavoidable problems with the experiments, e.g., control delay and modeling errors of the robot itself as well as the impact. Figure 17 shows the RMS (Root Mean Square) of the stance foot torque. The validity can be observed not only in the peak torque but also in the average torque.

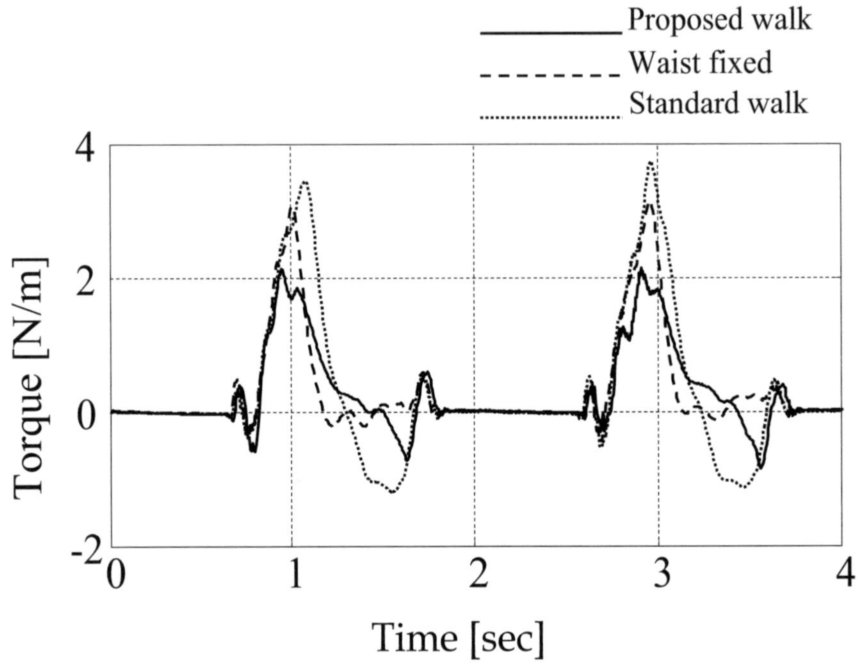

(a) Simulation

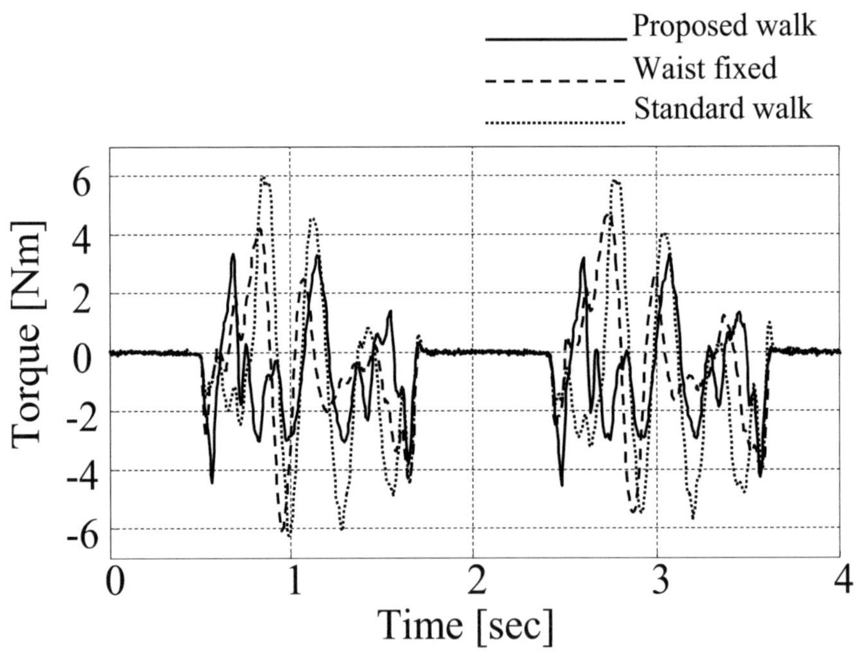

(b) Experiment

Figure 16. Yaw-axis Torque of Stance Foot (0.83 km/h)

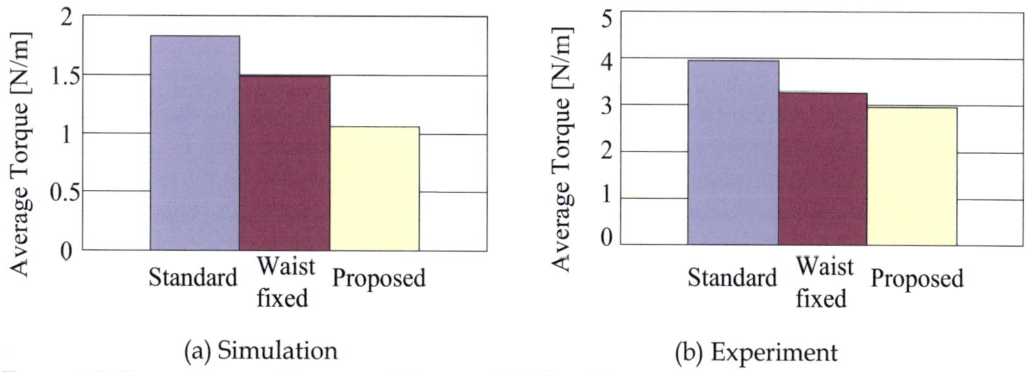

Figure 17. Comparison of Averaged Torque (0.83 km/h)

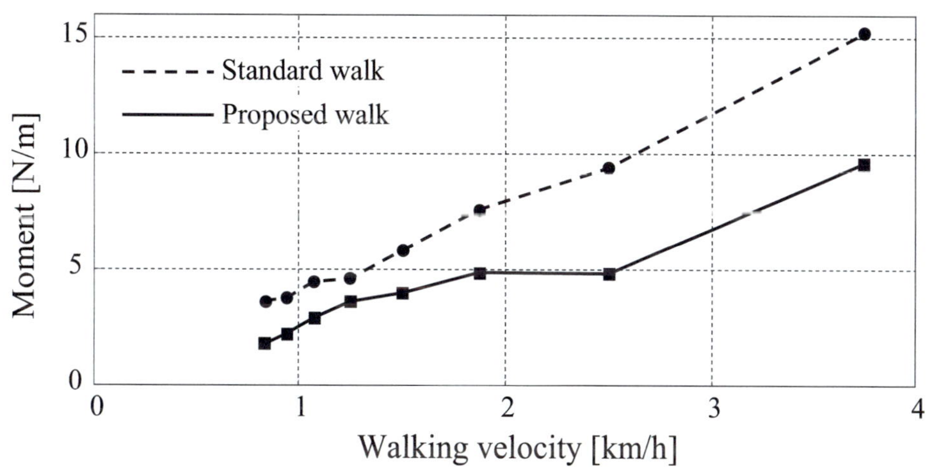

Figure 18. Walking velocity versus Peak Torque of Stance Foot

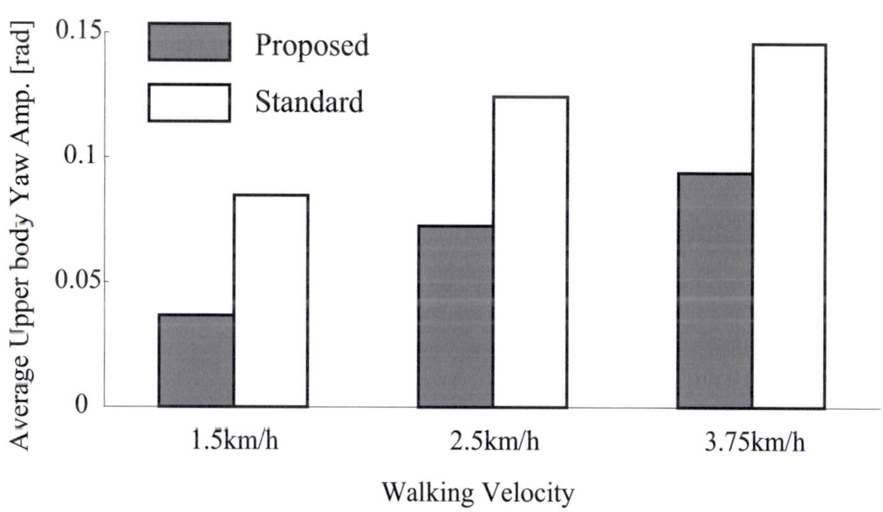

Figure 19. Upper Body Rotation during Fast Walk (Averaged Yaw Amplitude)

4.2 Improvement of Straightness and Upper-body Stability

From a safety perspective, it was difficult for the humanoid to walk fast without slipping at a speed exceeding 1.0 km/h. The following investigations were performed by the simulation: Figure 18 shows the relation between the peak torque and the walking velocity. This relation indicates that a faster walk without slipping can be achieved by the proposed walk for a floor with the same friction coefficient. For example, the peak torque is 9.4Nm at 2.5km/h by the standard walk, whereas it is 9.6 Nm at 3.75 km/h by the proposed walk. In other words, a 1.5 times faster walk can be realized from the view-point of the slip.

Figure 19 shows the average amplitude of the chest's fluctuation at a velocity of 1.5 km/h , 2.5 km/h , 3.75 km/h. In the planned walking pattern, this value should ideally be zero since the upper body is always facing the forward direction. However, the upper body is rotated and fluctuated due to the slip especially in the standard walk. The proposed walk reduces this fluctuation by 40% to 50%.

Figure 20 shows the body posture of every LC at the velocity of 2.5km/h. In the original pattern, the humanoid walks straight in the forward direction. However, the slip changes the walking trajectory, especially in the standard walk. Straightness is improved by the proposed walk by reducing the slip. These results indicate the possibility that walking while using the upper body during a task can be easily accomplished.

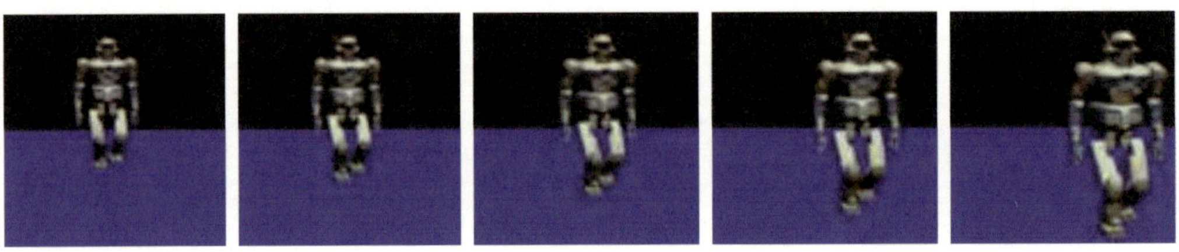

t=0.0s t=0.7s t=1.4s t=2.1s t=2.8s

(a) Standard Walk

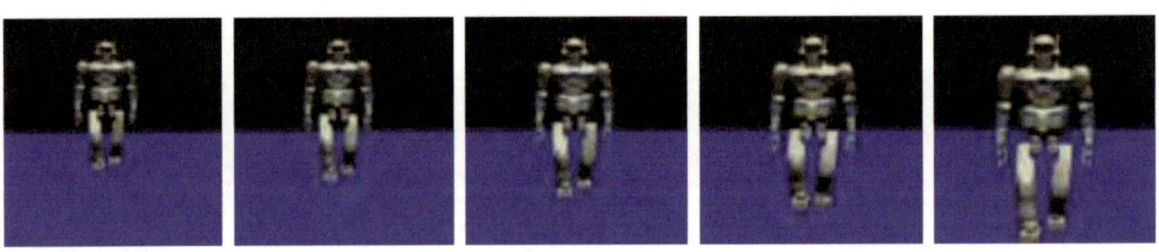

t=0.0s t=0.7s t=1.4s t=2.1s t=2.8s

(b) Proposed Walk

Figure 20. Straightness of Fast Dynamic Walking (2.5km/h)

5. Conclusion

In this chapter, the trunk-twistless walk of contact sport athletes was investigated from a motion measurement. The antiphase pelvic rotation was applied to the fast walk of humanoid HRP-2.The walking action including the momentum compensation was completed only by the lower body, so that the upper body DOF can be used for accomplishing a task. Using the proposed walk, the stance foot torque and the fluctuation of the upper body were both reduced.

The investigation on fast walking in the humanoid was limited to a simulation, and the applied walking velocity was not especially high when compared with human's walking velocity. A more efficient and safer hardware is required for actual experiments. The future work includes an evaluation of the energy efficiency of the trunk-twistless walk, both in humanoids and human athletes. An optimization program for an efficient walking pattern should be investigated. The authors wish to thank Ryoji Oyamada and Eishi Kidera for valuable discussions. The authors also wish to thank Dr. Hiroshi Takemura, Atsutoshi Ikeda, Kazuhiko Hiura, and Takashi Takeuchi of Nara Institute of Science and Technology for the data processing required for this research.

8. References

Takanishi, A.; Ishida, M.; Yamazaki, Y. & Kato, I. (1985). The realization of dynamic walking by biped walking robot WL-10RD, *Proceedings of International Conference on Advanced Robotics*, pp. 459-466, Tokyo, Japan, September, 1985.

Goswami, A. (1999). Postural Stability of Biped Robots and the Foot Rotation Indicator (FRI) Point, *International Journal of Robotics Research*, Vol.18, No.6, pp. 523-533.

Kajita, S.; et al. (2002). A Realtime Pattern Generator for Biped Walking, *Proceedings of 2002 International Conference on Robotics and Automation*, Washington DC, May, 2002, pp. 31-37.

Harada,K. ;Kajita, S.; Kaneko, K. & Hirukawa,H. (2003). ZMP Analysis for Arm/Leg Coordination, *Proceefings of 2003 IEEE/RSJ International Conference on Intelligent Robots and Systems*, Las Vegas, Nevada, October, 2003, pp.75- 81.

van Emmerik,R. E. A.; Wagenaar,R. C. (1996). Effects of walking velocity on relative phase dynamics in the trunk in human walking, *Journal of Biomechanics*, Vol. 29, No. 9, pp. 1175-1184, 1996.

Lamoth, C.J.C.; Beek, P.J. & Meijer, O.G. (2002). Pelvis-thorax coordination in the transverse plane during gait, *Gait & Posture*, Vol. 16, No. 2, pp. 101-114, 2002.

LaFiandra, M; Wagenaar, R.C.; Holt, K.G. & J.P. Obusek. (2003). How do load carriage and walking speed influence trunk coordination and stride parameters?, *Journal of Biomechanics*, Vol. 36, No. 1, pp. 87-95, 2003.

Yamaguchi, J.; Takanishi, A. & Kato, I. (1993). Development of a biped walking robot compensating for three-axis moment by trunk motion, *Proceedings of 1993 International Workshop on Intelligent Robotics and Systems*, Yokohama, Japan, pp. 561-566.

Kagami, S.; Kanehiro, F.; Tamiya, Y.; Inaba, M. & Inoue, H. (2000). AutoBalancer: An Online Dynamic Balance Compensation Scheme for Humanoid Robots, *Proceedings of 4th International Workshop on Algorithmic Foundation on Robotics*, pp. 329-340, 2000.

Yamane K. & Nakamura, Y. (2003). Dynamics Filter-Concept and Implementation of On-line Motion Generator for Human Figures, *IEEE Transactions on Robotics and Automation*, Vol.19, No.3, pp.421-432, 2003.

Kajita, S.; Kanehiro, F.; Kaneko, K.; Fujiwara, K.; Harada, K.; Yokoi, K. & Hirukawa, H. (2003). Resolved Momentum Control: Humanoid Motion Planning based on the Linear and Angular Momentum, *Proceedings of 2003 IEEE/RSJ International Conference on Intelligent Robots and Systems*, pp. 1644-1650.

A. H. Steinhaus. (1963). Toward an understanding of health and physical education, W. C. Brown Co.

Ducroquet, R.; Ducroquet, J. & Ducroquet, P. (1968). Walking and limping, a study of normal and pathological walking, Lippincott, Philadelphia, 1968.

Takemura, H.; Iwama, H.; Khiat, A.; Ueda, J.; Matsumoto, Y. & Ogasawara, T. (2003). Study of the Toes Role in Human Walk by a Toe Elimination and Pressure Measurement System, *Proceedings of 2003 IEEE International Conference on Systems, Man & Cybernetics*, October, Washington, D.C., USA, pp. 2569-2574.

H. Inoue, et al., (2000). HRP, Humanoid Robotics Project of MITI, Proceedings of IEEE/RAS International Conference on Humanoid Robots.

Hirukawa, H.; Kanehiro, F. & Kajita, S. (2001). OpenHRP, Open Architecture Humanoid Robotics Platform, *Proceedings of 2001 International Symposium on Robotics Research.*

13

Vision-based Motion Control of a Biped Robot Using 2 DOF Gaze Control Structure

Shun Ushida and Koichiro Deguchi
Tohoku University
Japan

1. Introduction

Why "*a biped robot*"? Although, for this question, we can give various answers, no doubt we are motivated by existence of powerful controller of human brain. Human can exhibit intelligence and a flexible environmental adaptation ability. The concept of *embodiment of robot,* which is concerned with the human body and the function, allows us to give one of the questions from the point of view of intelligent mechanics. This concept suggests the necessity of the embodiment of humanoid robot in order to obtain an intelligence like human.
Recently, in a community of control engineering and control theory area, a *control biology* and a *bio-mimetic control,* which aim to elucidate and mimic a control mechanism for various organisms respectively, are interested by many researchers. In their investigations, a humanoid robot, in particular a motion control of a biped robot, has attracted attention for not only similarity of appearances between humanoid robots and humans but also the mechanism or intra-dynamics.
For the motion control of biped robots, we can not escape the fact that human motor control depends upon *vision.* Some recent researches on human motion control (Seara et al., 2002; Seara and Schmidt, 2004) have revealed that vision plays a major role in human posture control systems. Also, for the biped robot, vision is the most important sensory system for his posture and motion control.
On the other hand, the humanoid robots nowadays exploit a great variety of sensing technologies such as a angle and torque sensor of each joint, a gyroscope, a pressure sensor in a soul, a CCD camera, a microphone and a range finder. In fact, a humanoid robot shown in Fig. 1, which we use in our experiments, has above sensors in his body. Most of the *sensorium of a robot* is based on the so-called five senses of human. However, for a real humanoid robot, we have to assert that vision is one of underdeveloped sensors. This is mainly caused by two difficulties:

a. *Poor Hardware Resources for vision:* Hardware performance for robot vision on a biped robot is extremely restrictive due to limitations of the available hardware resources. For a small-sized humanoid robot, we can not use large-sized CCD camera, a heavyweight powerful actuator of the vision system.
b. *An enormous amount of information in images:* For image sequences from CCD camera, we have to extract useful information for robot control by using image processing. How ever, there is no systematic methodology of choosing a kind of visual information from an image.

Figure 1. The humanoid robot called HOAP-3 made by Fujitsu Automation Corporation

The realization of vision of biped robots has many difficult problems to be tackled in aspects of both hardware and software (Afzulpurkar et al., 1996; Shibata and Schaal, 2001).
We clarify the structure and the nature of visual information processing system which biped robot should have. For this purpose, human eye structure and the eye movements by brain motor control are considered as an analogy for the humanoid robot system. We shall show that a gaze control mechanism like human is indispensable for vision system of biped robots. Actually, on the humanoid robot, we implemented the 2 DOF gaze control system which consists of the feedback and feedforward factor. Furthermore, we investigate considerable differences between human eyes and the vision of biped robot. This allows us to propose a new, simple and intuitively understandable criterion, called *the range of gaze,* for evaluating gaze control performance.

2. A Vision System of Human

In a human ocular system, we briefly review how one can obtain visual information for motor control efficiently.

2.1 Structure of Ocular

Detail of components of human ocular system has already been investigated from an anatomical point of view. In particular, on robotics, there are two important factors, *eyeball driving system* and *image formation system.*
Human eye movement is driven by 4 rectus muscles and 2 oblique muscles as shown in Fig. 2. 4 rectus muscles control a pan and tilt action of the eyeball. Since 2 muscles of rectus superior and inferior are not parallel on the gaze direction, these muscles cause not only tilt but also pan and slight roll. 2 oblique muscles cancel out this roll action. Hence, precise, quick and smooth eye movements of human can be achieved in superb combination with 6 muscles each other. A remarkable feature of these eye movements is that a motor unit of these muscles is greatly smaller than other muscle. In fact, 1 nerve fiber controls only 6 muscle fibers, whereas for a muscle for a finger movement 1 nerve fiber corresponds to 100-300 muscle ones. A maximum angular speed of eyeball is 300-500 [deg/s] and 1 action finishes within 30-50 [ms]. The eyeball is surrounded with an adipose tissue called corpus adiposum orbitae which plays a role as a rotational bearing.
Another important aspect of human ocular system on robotics is that the image formation system has also individual structure. An incident light from an external source through cornea is projected on retina after refraction by lens. Human eye has *central visual field* and

peripheral visual field. In the former field, we can achieve high-quality and powerful perception of the object, but its scope is extremely narrow. Its angle is about 2.5 [deg]. The central visual field corresponds to a photoreceptor cell on retina called *fovea centralis* which enables us to realize keen vision. On the other hand, the latter is a visual field except for the central one and is used mainly as a motion detector.

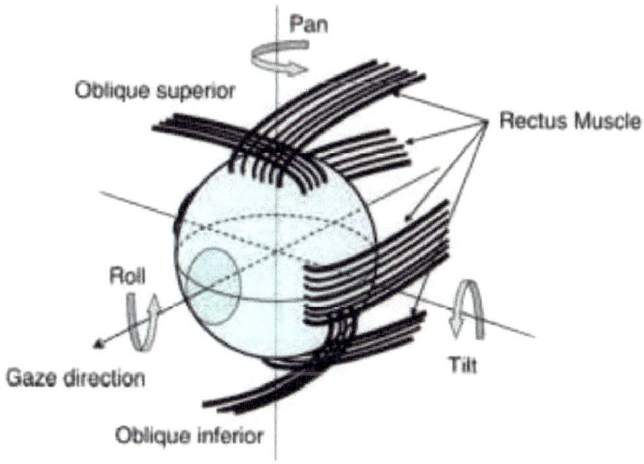

Figure 2. 6 Muscles for eye movement

2.2 Some Ocular Movements

The human eye movements are categorized by its function, a speed of eyeball, an organ caused the movement and so on. Here, we focus on *voluntary eye movement* and *non-voluntary one* due to an intention of human. The voluntary eye movement consists of *saccade* and *smooth pursuit movement,* which are in order of the quickness of the speed. The saccade is stepwise response whose angular velocity is 200-500 [deg/s], whereas the other is smooth tracking movements with 20-120 [deg/s]. On the other hand, the non-voluntary eye movement is generated reflexively. The typical reflex action of an eyeball is called *vestibulo-ocular reflex* (VOR), which is directly due to a stimulus from a vestibular organ for a rotational movement of head. The VOR is related to human gaze control.

2.3 A Constancy of a Position and Hierarchical Structure on Human Motor Control

A model of control mechanism of eye movements have been studied since the middle of the 19th century from the various point of view of experimental psychology, medical science, physiology, engineering, robotics and so on. Due to complexity of human, however, it is difficult to capture the essential nature of eye movements and, therefore, the model in the individual theories, experiments and conjectures tend to be large-scale and complex. In this section, aiming at the applicability to robotics, we concentrate on two simple and intuitively understandable concepts called a *constancy of a position* in medical science, and a *hierarchical structure on human motor control.*

This constancy of a position means a constancy of visual world in our brain. For instance, when we are gazing an external object and moving ourselves by walking, running, bicycle and so on, we can feel the stationary visual world for a stationary scene whereas its image on our retina are moving due to a performance limitation of eye movements. This phenomenon has been conjectured by medical scientists that the transition of the image is

compensated via a cancellation signal from the central nervous system which is stimulated by an other organ except for the retina.
The hierarchical structure on human motor control is related to the achievement of the constancy of a position which is realized by above eye movements such as a saccade and smooth pursuit movement. In this structure, we consider not only eye movements but also the constancy of a position based on eye movements, the perception of an object based on the constancy of a position, and the motor control based on the perception of an object, which form the hierarchical layers. This concept illustrated in Fig. 3, which was inspired by some results of experimental psychology (Recher and Oppenheim, 1955; Kinchla and Wolfe, 1979) and robotics (Westelius, 1995), is used in a subsequent section.

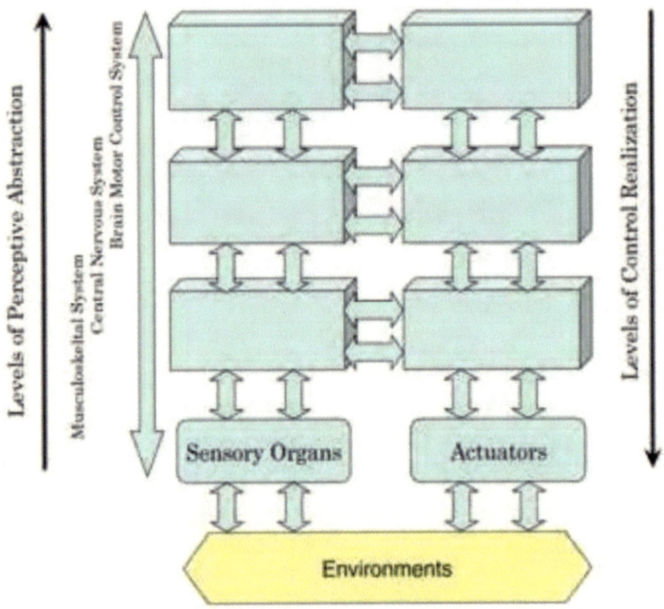

Figure 3. An integrated hierarchical structure on human motor control

3. The Visual Information Processing System in *a* Biped Robot

In this section, we discuss the visual information processing system which a biped robot should quip for vision-based motion control.
A biped robot consists of multiple joints which are precisely controlled by the computer. Since the repeatability for same controlled inputs is very high, the posture control of a biped robot seems to be simple apparently. However, the actual behavior is extremely complex due to a dynamically unstable structure of the biped robot and an interaction between the body and uncertain environments. This implies that the biped robot with a motion can not detect the position and posture himself. The vision enables us to get essential solution for this situation. If a biped robot has some sort of visual information processing system, it can observe an external fixation points as shown in Fig. 4. The vision-based biped robot can control the motion for the external fixation point. This does not mean the robot navigation system for a mobile robot. For unexpected motion error mentioned above, a vision can be used for primitive and basic motion control for autonomous biped robot. A use of the vision

has some advantages. By using the fixed target, it can observe low-frequency postural change, which gyroscope sensors are hard to detect, and it is not affected by drift or accumulation of errors in the integration of sensor data (Seara et al., 2002).

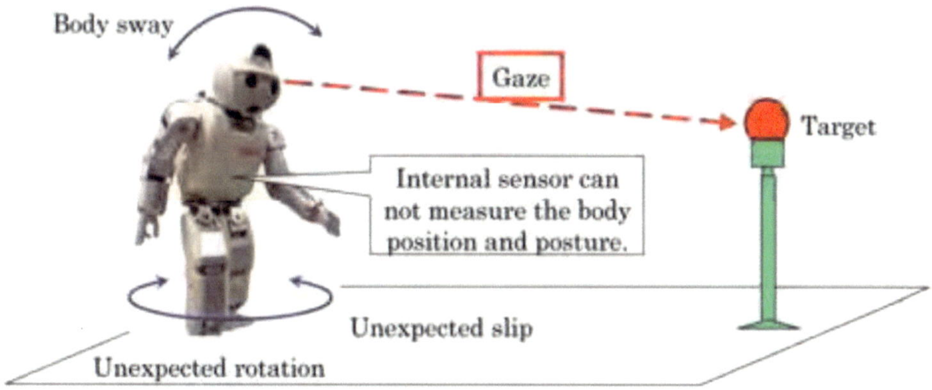

Figure 4. The visual information processing system in the biped robot

3.1 A Necessity of Gaze Control Mechanism

A robot control with a vision, usually called a visual servoing, has studied by many researchers, e.g., (Kuniyoshi et al., 1995; Shibata and Schaal, 2001), who have dealt with a vision-based humanoid or robot arm fixed on a floor, a wheel-type mobile robot on which a CCD camera is installed and so on. However, a vision-based biped robot is absolutely different from others in the following points:

a. A vision is indispensable since the behavior of a biped robot depends on uncertain environment.
b. The vision system is usually located at a robot head, which swings violently.
c. In the motions of a biped robot such as walking and so on, the conflicts between legs and floor are repeatedly included.
d. The heavy visual devises and its processing system decrease the performance of motion significantly.

Under such a situation, the required features of the visual information processing system are the compactness and the enough performances, which, for example, are a wide view without skew, high resolution, high speed, high frame rate, effective real-time image processing algorism. However, it is difficult to satisfy these two features simultaneously. For instance, when the wide view is realized by using all-round camera or fish-eye lens, the skew of the obtained image is not avoidable. The more we require high resolution or frame rate, the more an amount of information of an image and movie increases. This yields a longer processing time. Furthermore, in general, high performance computer causes an increase of the weight of the system (Afzulpurkar et al, 1996; Ishikawa et al, 2002; Kurazume and Hirose, 2000; Shibata and Schaal, 2001).

In order to visual information for motor control on a biped robot without using such an inefficient architecture, introduction of **gaze control mechanism** is indispensable on biped robots. The human vision system in Sec. 2 gives effective suggestions for the situation of (a)-(d). For an actuator which can pan and tilt the CCD camera for *gazing,* a concept of a constancy of a position and the hierarchical structure in Sec. 2.3 are useful.

We can point out the following two important advantages of using a gaze control mechanism for a biped robot, the *decrease of an amount of required information* and the *null method as measurement method.*

For human as shown in Sec. 2.1, our perception with high-resolution is limited within narrow scope called central visual field. In fact, it is known that most of cortices of the visual area in the cerebrum are associated with information processing of the signals from the fovea centralis. In addition to this, an appropriate allocation of roles between the central visual field and the peripheral visual field can achieve a rapid and precise eye movements. In addition, it is known that the frame rate of human eye is at most or less than 30 [fps]. This fact implies that a use of image processing via brain in addition to gaze control can contribute to a decrease of an amount of visual information. On the other hand, in the CCD-image processing system on a biped robot, we can obtain same amount of information from each pixel of which an image consists. Actually, there is no difference of amount of information between a center part and a near part of the edge of an image. This fact can become an advantage and a disadvantage which depend on a required resolution, frame rate, capacity of a memory, a permissible processing time and so on, i.e., we can obtain high resolution image from same amount of information by using an appropriate zoom lens, and however time lag of capturing an image and its processing time can become the bottle neck of a control signal flow. Furthermore, by using appropriate image processing method we can realize a constancy of a position with low frame rate.

Another important nature of gaze control mechanism, the measurement of the posture of biped robot corresponds to a so-called *null method* in which a quantity to be measured is determined by making an adjustable quantity to be zero. When we consider the suitable resolution of an image, the capturing time and the processing time for a operating frequency of gaze control mechanism, this measurement method has some following advantages:

- the achievement and its improvement of a high accuracy of gaze control,
- the possibility of highly sensitive measurement for a motion of the robot, and
- the feasibility of constructing the responsive measurement system by a simple feedback loop structure.

The zero reference point corresponds to target to be gazed.

3.2 A Concept of "the Range of Gaze"

We introduce a notion of *the range of gaze* which is a simple and intuitively understandable criteria for evaluating gaze control performance.

Assume that an external fixation point, we call a target, is located at a center of an image at a time $t = t_0$. If a change of the view field of biped robot due to a motion happens at a time t, the target moves in the image as shown in Fig. 5(a). After a simple image processing, we can center the target in the view field as shown in Fig. 5(b). The image area is characterized by the distances $d_T(t), d_B(t), d_L(t), d_R(t)$ in Fig. 5(b), and we denote this area as $A(d_T(t), d_B(t), d_L(t), d_R(t))$. If the target is not framed in the image, we assume that the area $A(d_T(t), d_B(t), d_L(t), d_R(t))$ vanishes, i.e., all distances are $d_{..}(t) = 0$. Then, we can define a notion of the range of gaze for a motion in a time sequences $t_0, \cdots t_n$ in the following:

$$\bigcap_{t \in \{t_0, \cdots, t_n\}} A(d_T(t), d_B(t), d_L(t), d_R(t)). \tag{1}$$

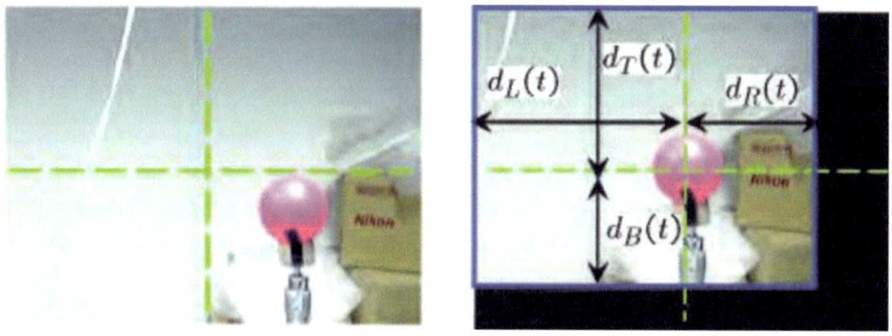

(a) The target position at time t (b) After image shift

Figure 5. Concept of the range of gaze

A meaning of this concept is very clear. This enables us to validate a gaze control performance. For a motion, a wide range of gaze is desirable because this implies that an applicability of various type of image processing methods to this area. We can also achieve a perfect tracking of the target, which is a visual world we, human, are usually seeing. In this sense, obtaining the range of gaze from image sequences corresponds to a part of image processing in our brain.

3.3 A Feedforward and Feedback Control Architecture on a Biped Robot

In this section, we propose a new gaze control mechanism whose structural feature is a feed-forward controller part with a database of robot motions (See also (Takizawa et al., 2004)). Our system as shown in Fig. 6 is the novel motion stabilization system of a biped robot with two degrees of freedom adaptive gaze control system.

This system consists of the fixed PID feedback controller and the adaptive feedforward controller vt which is based on a scheduled motion and drives a pan of CCD camera. It was assumed that the scheduled motion for a walking is a sinusoid. However, this assumption is not necessarily true, which is shown in this paper. According to the control theory, in the 2 DOF control system, each role of the feedback and feedforward controller is mainly to stabilize the closed-loop systems and to improve the time responses, respectively. Therefore, an incorrect pan command generated by the feedforward controller adversely affects the behavior of CCD camera movements.

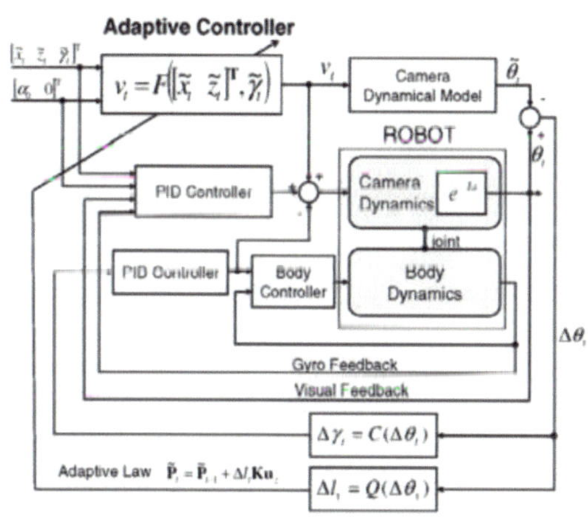

Figure 6. Block diagram of (Takizawa et al., 2004)

To overcome this problem we propose a new feedforward controller which consists of a pan tilt command generator with a database of robot motions and a motion command generator as shown in Fig. 7. The data format in the database is a set of functions f_k, $k = 1, 2, \bullet \bullet \bullet, N$, where N denotes the number of kinds of motions. Each f_k depends on parameters which determine the motion, e.g., a period and a distance of a step and so on. The information of the parameters is provided by the motion command generator which drives all joints of the biped robot, except for the pan-tilt joint, to realize a fe-th scheduled motion. The pan tilt command generator and the motion command generator are synchronized by a reference signal from a timer.

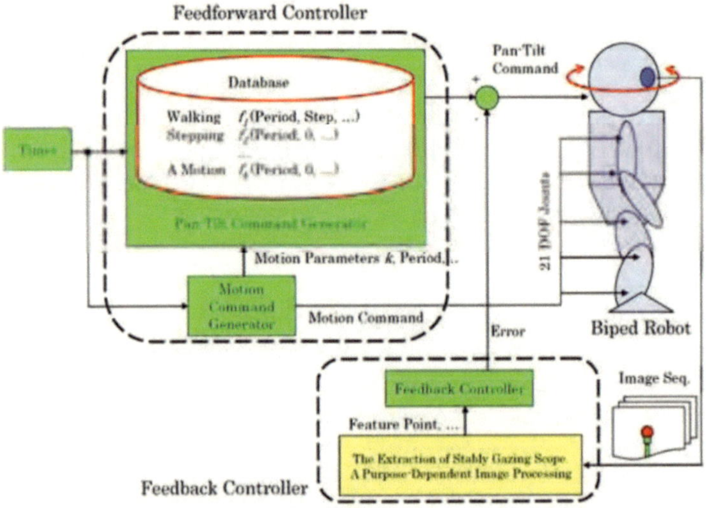

Figure 7. 2 DOF gaze control system with a database structure

As mentioned above, the function f_k directly affects the performance of gaze control. In fact, for human motions which is quick, smooth and precise, it is known that a feedforward factor is predominantly used. Therefore, we can construct f_k by using a precisely motion-dependent signal which are measured by a high-speed external camera as *a priori* information. Hence, we can use a high performance camera system whose frame rate is approximately 20-50 times as a CCD camera of biped robot.

3.4 Experiments

We examined the feedforward control mechanism with a motion database and evaluate it via the range of gaze. To investigate an effect of the database to the gaze control system, we focused on the only feedforward part without any feedback loop. As mentioned in the preceding subsection, the feedforward factor determines predominantly the control performance. This is another reason why we dealt with only the feedforward part. The experimental setting is illustrated in Fig. 8. The biped robot is stepping/walking with the period 0.4 [Hz]. At first, in order to store the feedforward control inputs into the database, we measured these motions of biped robots, HOAP-1 and HOAP-3, by using a high-speed camera, nac Image Technology, Inc., HSV-500c3, 500 [fps]. Both HOAP-1 and HOAP-3 are made by Fujitsu Automation Cor. and HOAP-3, "3" means 3rd generation, is the successor robot of HOAP-1. Therefore, most of specs and performances of HOAP-3 surpass HOAP-1 ones. This setting is illustrated in Fig. 9. By this measurement, we can know how much the head of a biped robot fluctuates during the motions. The results are shown in Fig. 10 and 11. In both case of HOAP-1 and HOAP-3, the difference of the motions appears as the remarkable difference of the fluctuation of the head. This fact supports our proposed gaze control mechanism, which includes motion-dependent pan-tilt command generator in the feedforward controller with database. Furthermore, the differences of HOAP-1 and HOAP-3 for same motions are also conspicuous, which implies that a desirable pan-tilt angle of a camera for gaze control extremely depends on an individual hardware of a biped robot. In our case, in particular, we observed that a structural difference of a pressure sensor in the soul of both robots causes the different behavior for both robots. Hence, as long as the hardware of a biped robot is developing nowadays, we must measure each motion and keep them in a database like our method.

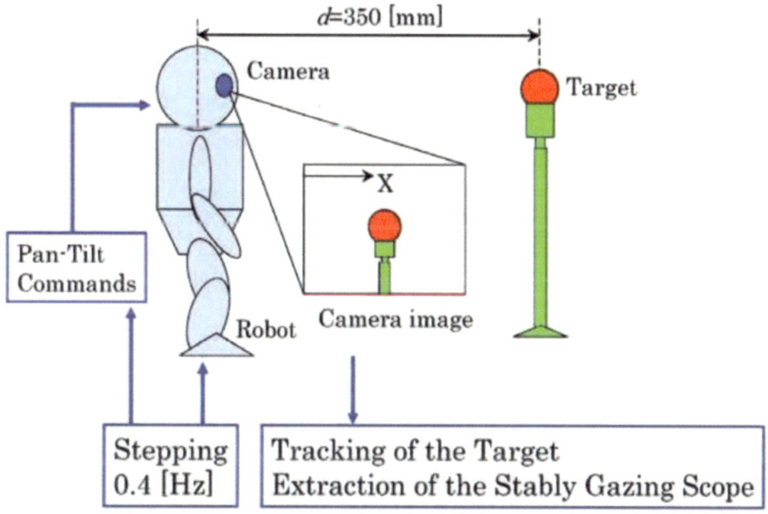

Figure 8. The experimental setting of the gaze control during motions

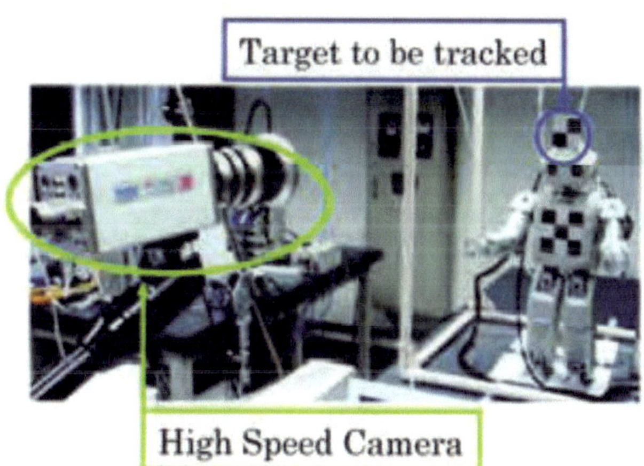

Figure 9. The high-speed camera and the biped robot, HOAP-3

Fig. 12 and 13 show a time response of target displacement in an image and the range of gaze, respectively, during a step motion. The actual range of gaze is shown in Fig. 14. By using gaze control mechanism we can ensure a double and more area of the range of gaze.

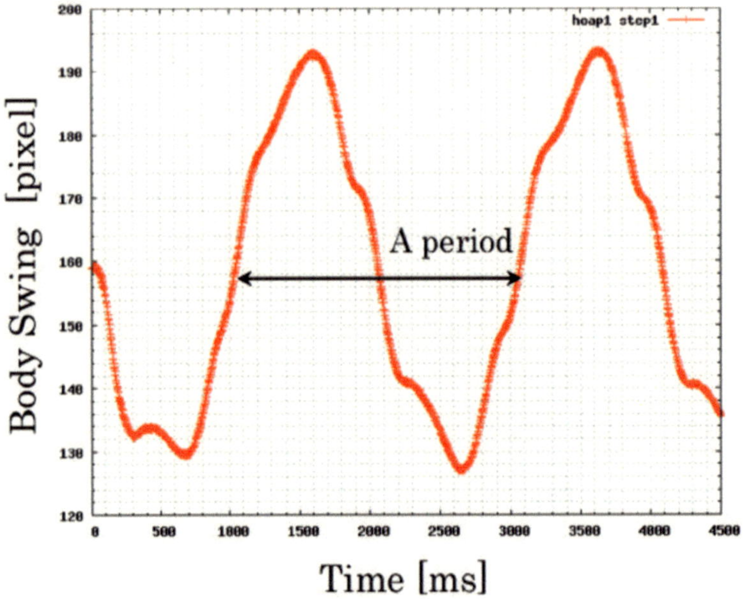

a) Stepping

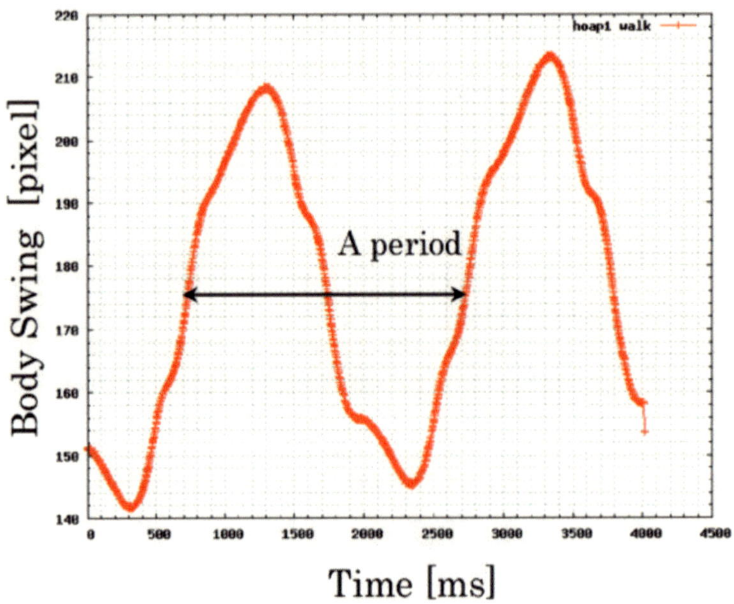

b) Walking

Figure 10. The results of the body swing during stepping/walking of HOAP-1

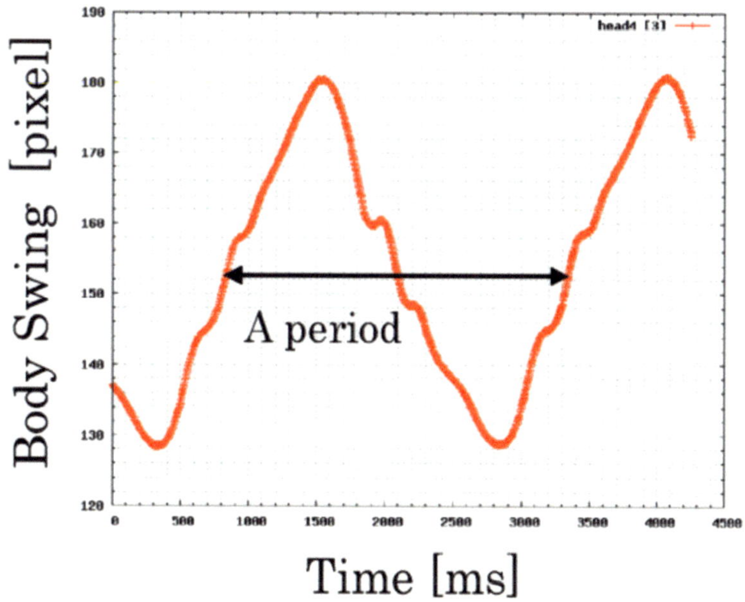

a) Stepping

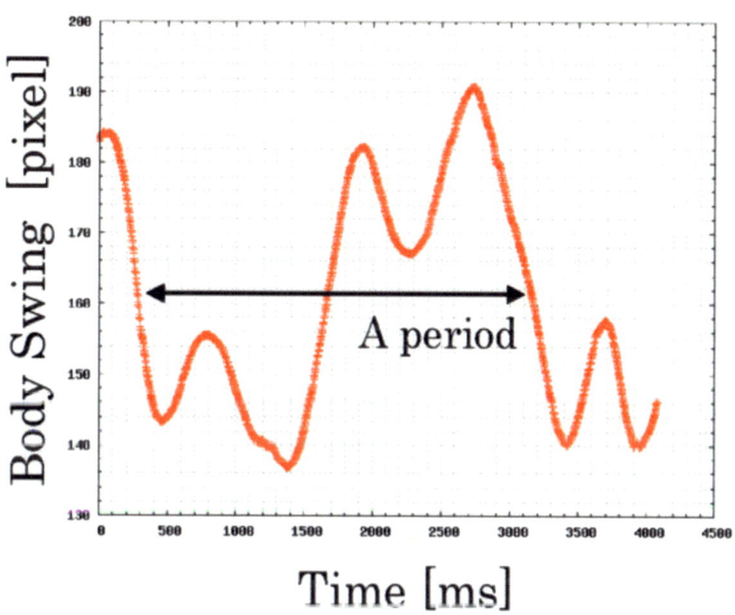

b) Walking

Figure 11. The results of the body swing during stepping/walking of HOAP-3

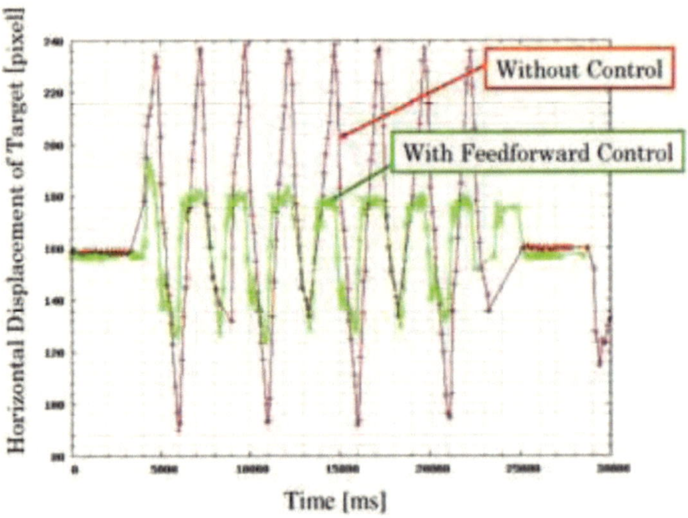

Figure 12. The result of a time response of target displacement for gaze control

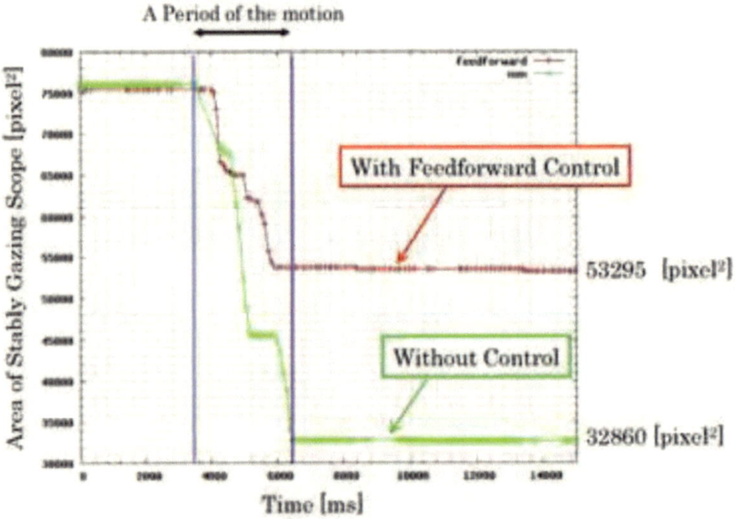

Figure 13. The result of a time response of the range of gaze for gaze control

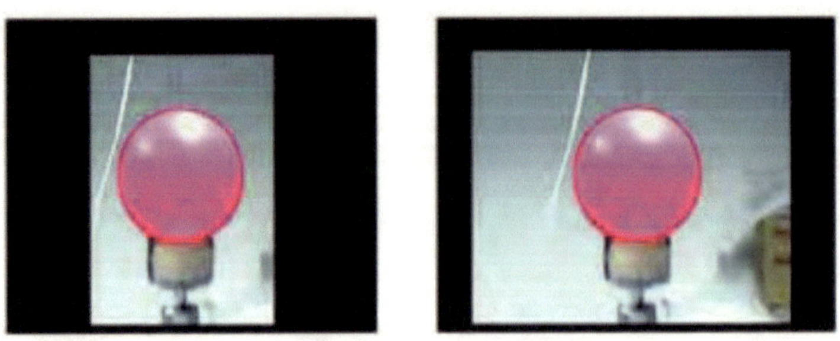

(a) Without any control (b) With gaze control

Figure 14. The actual range of gaze via gaze control

4. Conclusion

In this paper, we discussed both a necessity and importance of gaze control mechanism on vision-based motion control of biped robots. Human eye structure and the eye movements, and brain motor control through a central nervous system were considered as an analogy with robotics. In fact, human has powerful control mechanism with vision in order to realize a various type of motions against uncertain environments. As a result, we showed that gaze control mechanism and integrated hierarchical structure like human motor control are indispensable for a vision system of biped robots.

We can conclude that in order to overcome many difficulties of aspects of both a limitation of available hardware resources for a vision and a real-time image processing for motor control, we need not only gaze control mechanism but also use of the image processing technologies as shown in Fig. 15. We proposed a concept of the range of gaze and gaze control system with database structure. The use of both image processing and control theory allows us to realize an integrated hierarchical structure on the biped robot motion control as shown in Fig. 3. Furthermore, using the range of gaze, we can evaluate the gaze control performance from the point of view of the availability of visual information for robot motion control.

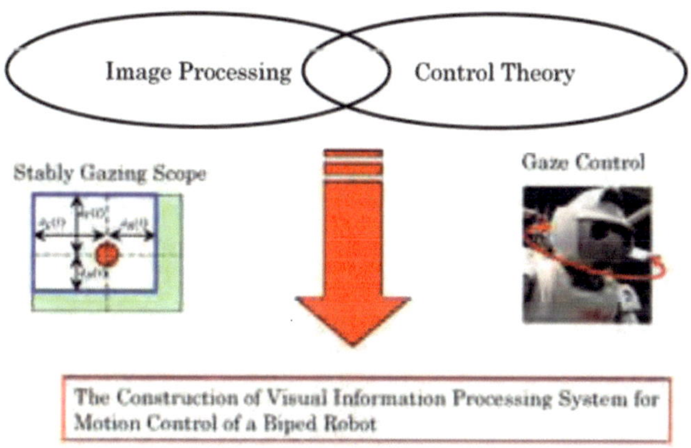

Figure 15. An integration of image processing and control theory on human motor control

5. References

Afzulpurkar, N.; Rougeaux, S.; Berthouze, L. & Kuniyoshi, Y. (1996). A Gaze Based Approach for Visual Servoing, in *Proc. of International Conference on Computational Engineering in System Application,* pp. 570-575

Berthouze, L.; Rougeaux, S.; Chavand, R & Kuniyoshi, Y. (1996). A Learning Stereo-Head Control System, in *Proc. of World Automation Congress, International Symposium on Robotics and Manufacturing, In Robotics and Manufacturing: Recent trends in Research and Applications,* Vol. 6, pp. 43-47

Demiris, J.; Rougeaux, S.; Hayes, G. M.; Berthouze, L. & Kuniyoshi, Y. (1997). Deferred Imitation of Human Head Movements by an Active Stereo Vision Head, in *Proc. of 6th IEEE International Workshop on Robot and Human Communication,* pp. 88-93

Fukuoka, Y; Tanaka, K.; Ishida, A. & Minamitani, H. (1999). Characteristics of Visually Feed-back in Postural Control during Standing, *IEEE Trans. Rehabilitation Engineering,* Vol. 7, No. 4, pp. 427-434

Gutman, J.; Fukuchi, M. & Fujita, M. (2004). Stair Climbing for Humanoid Robots Using Stereo Vision, in the *Proc. of Int. Conf. on Intelligent Robots and Systems,* pp. 1407-1413

Ishikawa, M.; Komuro, T.; Nakabo, Y. & Namiki,A. (2002). The 1ms-Vision System and Its Application Examples, Workshop: Innovative Sensory-Motor Fusion Opens a New Robotic World (Organizers: Masatoshi Ishikawa, Makoto Kaneko), *IEEE International Conference Robotics and Automation (ICRA 2002)*

Jiang, Y.; Nagasaki, S. (2002). Health Evaluation Technology for Elders - Postural Sway Measurement and Postural Stability Evaluation during Upright Stance -, */. of Japanese Society for Non-destructive Inspection,* Vol. 51, No. 9, pp. 567-572

Kinchla, R. A. & Wolfe, J. M. (1979). The Order of Visual Processing: "Top-Down," "Bottom-Up," or "Middle-Out", *Perception and Psychophysics,* Vol. 25, pp. 225-231

Kuniyoshi, Y; Kita, N.; Rougeaux, S. & Suehiro, T. (1995). Active Stereo Vision System with Foveated Wide Angle Lenses, in *Proc. of Second Asian Conference on Computer Vision (ACCV95),* pp. 359-363

Kurazume, R. & Hirose, S. (2000). Development of image stabilization system for a remote operation of walking robots, in the *Proc. IEEE International Conference on Robotics and Automation,* pp. 1856-1861

Nakabo, Y; Fujikawa, N.; Mukai, T; Takeuchi, Y. & Ohnishi, N. (2004). High-Speed and Bio-Mimetic Control of a Stereo Head System, in *Proc. of SICE Annual Conference,* pp. 2371-2376

Navon, D. (1977). Forest before Tree: the Precedence of Global Features in Visual Perception, *Cognitive Psychology,* Vol. 9, pp. 353-383

Rescher, N. & Oppenheim, P. (1955). Logical Analysis of Gestalt Concepts, *British Journal for the Philosophy of Science,* Vol. 6, pp. 89-106

Seara, J.; Strobl, K. & Schmidt, G. (2002). Information Management for Gaze Control in Vision Guided Biped Walking, in *Proc. of Int. Conf. of Intelligence Robots and Systems,* pp. 31-36

Seara, J. & Schmidt, G. (2004). Intelligent Gaze Control for Vision-Guided Humanoid Walking: Methodological Aspects, *Robotics and Autonomous Systems,* Vol. 48, pp. 231-248

Shibata, T. & Schaal, S. (2001). Biomimetic Gaze Stabilization based on Feedback-Error-Learning with Nonparametric Regression Networks, *Neural Networks,* Vol. 14, pp. 201-216

Takizawa, S.; Ushida, S.; Okatani, T. & Deguchi, K. (2001). 2DOF Motion Stabilization of Biped Robot by Gaze Control Strategy, in *Proc. of IEEE/RSJ International Conference on Intelligent Robots and Systems* (IROS2004), pp. 3809-3814

Ushida, S.; Terashita, J. & Kimura, H. (2004). Switching Structural Biomechanical Model of Multisensory Integration during Human Quiet Standing, in *Proc. of 43rd IEEE Conf. on Decision and Control* (CDC2004), pp. 959-965

Westelius, C. J. (1995). Focus of Attention and Gaze Control for Robot Vision, Ph. D Thesis, Linkoping University, Sweden, Dissertation No. 379

14

Limit Cycle Walking

Daan G.E. Hobbelen and Martijn Wisse
Delft University of Technology
The Netherlands

1. Introduction

This chapter introduces the paradigm `Limit Cycle Walking'. This paradigm for the design and control of two-legged walking robots can lead to unprecedented performance in terms of speed, efficiency, disturbance rejection and versatility. This is possible because this paradigm imposes fewer artificial constraints to the robot's walking motion compared to other existing paradigms. The application of artificial constraints is a commonly adopted and successful approach to bipedal robotic gait synthesis. The approach is similar to the successful development of factory robots, which depend on their constrained, structured environment. For robotic walking, the artificial constraints are useful to alleviate the difficult problem of stabilizing the complex dynamic walking motion. Using artificial stability constraints enables the creation of robotic gait, but at the same time inherently limits the performance of the gait that can be obtained. The more restrictive the constraints are, the less freedom is left for optimizing performance.

The oldest and most constrained paradigm for robot walking is that of `static stability', used in the first successful creation of bipedal robots in the early 70's. Static stability means that the vertical projection of the Center of Mass stays within the support polygon formed by the feet. It is straightforward to ensure walking stability this way, but it drastically limits the speed of the walking motions that can be obtained. Therefore, currently most humanoid robots use the more advanced `Zero Moment Point' (ZMP) paradigm (Vukobratovic et al., 1970). The stability is ensured with the ZMP-criterion which constrains the stance foot to remain in flat contact with the floor at all times. This constraint is less restrictive than static walking because the Center of Mass may travel beyond the support polygon. Nevertheless, these robots are still under-achieving in terms of efficiency, disturbance handling, and natural appearance compared to human walking (Collins et al., 2005).

The solution to increase the performance is to release the constraints even more, which will require a new way of measuring and ensuring stability. This is the core of `Limit Cycle Walking'; a new stability paradigm with fewer artificial constraints and thus more freedom for finding more efficient, natural, fast and robust walking motions. Although this is the first time we propose and define the term `Limit Cycle Walking', the method has been in use for a while now. The core of the method is to analyze the walking motion as a limit cycle, as first proposed by Hurmuzlu (Hurmuzlu and Moskowitz, 1986). Most of the research on `Passive Dynamic Walking' initiated by McGeer (McGeer, 1990a) follows this stability method. But also various actuated bipedal robots that have been built around the world fall in the category of `Limit Cycle Walkers'.

This chapter continues as follows. In Section 2 we will give a short discussion on walking stability to create a background for the introduction of Limit Cycle Walking. The exact definition of Limit Cycle Walking follows in Section 3. In Section 4 we show how the stability of Limit Cycle Walking is assessed and clarify this with an exemplary Limit Cycle Walking model. Section 5 gives an overview of the current State of the Art Limit Cycle Walkers. In Sections 6 through 8 we will explain how Limit Cycle Walking is beneficial to the performance of bipedal walkers and substantiate this by showing the State of the Art performance. We will end with a conclusion in Section 9.

2. Bipedal walking stability

Stability in bipedal walking is not a straightforward, well defined concept (Vukobratovic et al., 2006; Goswami, 1999; Hurmuzlu et al., 2004). To create a background for the definition of Limit Cycle Walking, we discuss two extreme classifications of walking stability, the most generic one and an overly restrictive one.

The most generic definition of stability in bipedal walking is `to avoid falling'. This concept is captured with the `viability kernel' by Wieber (Wieber, 2002), the union of all viable states from which a walker is able to avoid a fall. This set of states includes all kinds of possible (periodic) motions or static equilibria and should be established with the presence of possible disturbances. Ultimately, to get optimal performance, bipedal walkers should be designed using this notion of stability, without any more restrictions. However, it turns out that it is not practical for gait synthesis due to its highly nonlinear relation with the state space of a walker. Establishing stability using this definition requires a full forward dynamic simulation or actual experiment starting out at all possible states of the walker, including all possible disturbances, checking whether this results in a fall or not. Given the complex dynamics involved in walking this would be very expensive, numerically as well as experimentally.

The limited practical value of `avoiding to fall' as a stability definition for gait synthesis, has lead a large group of robotic researchers (Sakagami et al., 2002; Ishida, 2004; Hirukawa, 2003) to create bipedal walking based on an overly restrictive classification of stability. We refer to this stability classification as `sustained local stability'. In this case, gait is synthesized as a desired trajectory through state space (usually taken from human gait analysis), which is continuously enforced by applying stabilizing trajectory control. This control aims for sustained local stability, which is obtained if for every point on the nominal trajectory it can be proven that points in its local neighbourhood in state space converge to the trajectory.

The aim for sustained local stability creates two important constraints for bipedal walking: it requires local stabilizability and high control stiffness. Local stabilizability exists when at least one foot is firmly placed on the ground. This constraint is guaranteed by satisfying the Zero Moment Point (ZMP) or Center of Pressure (CoP) criterion (Vukobratovic et al., 1970; Vukobratovic and Borovac, 2004). The constraint of high control stiffness is required to obtain local stability in spite of the presence of the inherently unstable inverted pendulum dynamics of the single stance phase.

In the strive for increasing the performance of bipedal robots, recently a growing group of researchers has decided to let go of the restrictive aim for sustained local stability and adopt a new paradigm for synthesizing bipedal gait, `Limit Cycle Walking'.

3. Definition Limit Cycle Walking

Here we formally define the new (yet popular) paradigm `Limit Cycle Walking':

Limit Cycle Walking is a nominally periodic sequence of steps that is stable as a whole but not locally stable at every instant in time.

With nominally periodic sequence of steps we mean that the intended walking motion (in the ideal case without disturbances) is a series of exact repetitions of a closed trajectory in state space (a limit cycle) putting forward each of the walker's two feet in turn. This trajectory is not locally stable at every instant in time, taking out the necessity of making all points on the trajectory attracting to their local neighbourhood in state space (as it is done in conventional trajectory control). The nominal motion is still stable as a whole because neighbouring trajectories eventually, over the course of multiple steps, approach the nominal trajectory. This type of stability is called `cyclic stability' or `orbital stability' (Strogatz, 2000).

4. Stability analysis

Cyclic stability is the core principle of Limit Cycle Walking. In this section we show how it can be analyzed. This explanation is followed by an example of a Limit Cycle Walking model which shows that it is possible to have cyclic stability without having sustained local stability.

4.1 Method

Cyclic stability of a Limit Cycle Walker is analyzed by observing its motion on a step-to-step basis. One step is considered as a function or `mapping' from the walker's state $\mathbf{v}_n$ at a definite point within the motion of a step (for instance the moment just after heel strike) to the walker's state at the same point in the next step $\mathbf{v}_{n+1}$. This mapping is generally called a Poincaré map in nonlinear dynamics and the definite point within the motion is defined by the intersection of the motion with the Poincaré section (Strogatz, 2000). With regard to walking, the mapping was termed the `stride function' $\mathbf{S}$ by McGeer (McGeer, 1990a):

$$\mathbf{v}_{n+1} = \mathbf{S}(\mathbf{v}_n) \tag{1}$$

This mapping $\mathbf{S}$ is defined by the equations of motion of the walker which are usually solved numerically and integrated over the course of one step.

A periodic motion exists if the mapping of the walker's state gives exactly the same state one step later. This specific state $\mathbf{v}^*$ is called the `fixed point' of the function $\mathbf{S}$:

$$\mathbf{v}^* = \mathbf{S}(\mathbf{v}^*) \tag{2}$$

The cyclic stability of this periodic motion is found by linearizing the function $\mathbf{S}$ at the fixed point $\mathbf{v}^*$, assuming only small deviations $\Delta\mathbf{v}$:

$$\begin{gathered} \mathbf{S}(\mathbf{v}^* + \Delta\mathbf{v}) \approx \mathbf{v}^* + \mathbf{A}\Delta\mathbf{v} \\ \text{with } \mathbf{A} = \frac{\partial \mathbf{S}}{\partial \mathbf{v}} \end{gathered} \tag{3}$$

The matrix **A**, also called the monodromy matrix, is the partial derivative of the function **S** to the state **v**. Stability of the cyclic solution is assured for small state deviations if the eigenvalues λ of the matrix **A** are within the unit circle in the complex plane. In that case (small) deviations from the nominal periodic motion (fixed point) will decrease step after step. The eigenvalues λ are called the Floquet Multipliers and were first used to study the stability of walking by Hurmuzlu (Hurmuzlu and Moskowitz, 1986).

4.2 Example

We will give an example to show what Limit Cycle Walking is and to show that cyclic stability is possible without the constraint of sustained local stability.

Model

The model we will use as an example is the simplest walking model by Garcia et al. (Garcia et al., 1998), shown in Fig. 1. The 2D model consists of two rigid links with unit length l, connected at the hip. There are three point masses in the model, one in the hip with unit mass M and two infinitesimally small masses m in the feet. The model walks down a slope of 0.004 rad in a gravity field with unit magnitude g.

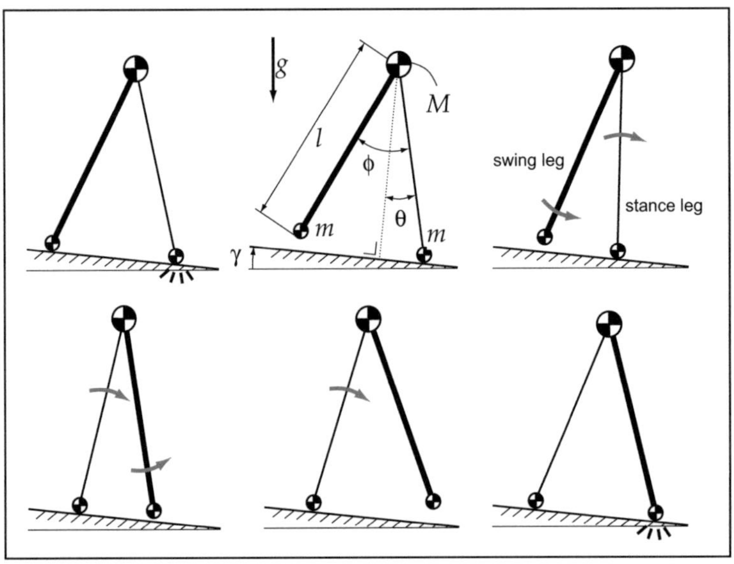

Figure 1. A typical walking step of the simplest walking model. Just after footstrike the swing leg (heavy line) swings forward past the stance leg (thin line) until the swing leg hits the ground and a new step begins. θ is the angle between the stance leg and the slope normal, ϕ is the angle between the two legs, l is the leg length, M is the hip mass, m is the foot mass, g is the gravitational acceleration and γ is the slope angle. Adapted from Garcia et al. (Garcia et al., 1998)

The dynamics of the model consists of two parts. The first part is the continuous dynamics that describes the motion of the stance and swing leg in between footstrikes:

$$\begin{aligned} \ddot{\theta} &= \sin(\theta - \gamma) \\ \ddot{\phi} &= \sin(\phi)\left(\dot{\theta}^2 - \cos(\theta - \gamma)\right) + \sin(\theta - \gamma) \end{aligned} \tag{4}$$

The second part of the dynamics is the discrete impact that describes the footstrike, as this is modeled as a fully inelastic instantaneous collision:

$$\begin{aligned} \theta^+ &= -\theta^- \\ \phi^+ &= -2\theta^- \\ \dot{\theta}^+ &= \cos\left(2\theta^-\right)\dot{\theta}^- \\ \dot{\phi}^+ &= \cos\left(2\theta^-\right)\left(1-\cos\left(2\theta^-\right)\right)\dot{\theta}^- \end{aligned} \tag{5}$$

Note that these equations also incorporate the re-labeling of stance and swing leg angles θ and ϕ.

The nominal cyclic motion that results for these dynamic equations is shown in Fig. 2.

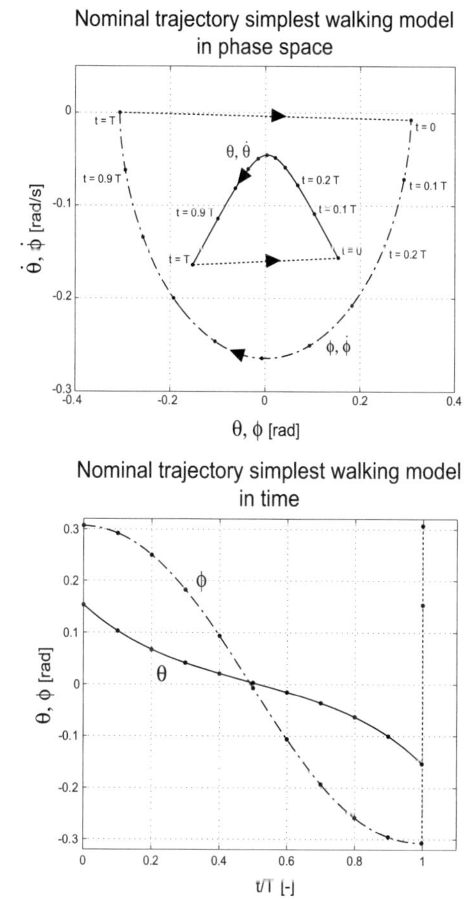

Figure 2. The nominal cyclic motion trajectory of the simplest walking model in phase space (left) and in time (right). In the left plot the solid and dash-dotted lines give the stance leg (θ) and swing leg (ψ) trajectories respectively, resulting from the continuous dynamics of the model. The dotted lines show the jump conditions that correspond to the discrete footstrike. The large dots along the trajectory indicate the amount of time *t* that has elapsed with increments of 1/10 of the nominal step period *T*. The right plot shows the nominal values of θ (solid) and ϕ (dash-dotted) in time

Cyclic stability

To prove the simplest walking model is cyclically stable, we will perform the stability analysis as described in Section 4.1.

The Poincaré section we choose to use for this cyclic stability analysis is defined as the moment just after heelstrike ($2\theta = \phi$). The `stride function' **S** is the mapping from the system states on the Poincaré section of step *n*: $\mathbf{v_n} = \left[\theta_n; \dot{\theta}_n; \dot{\phi}_n\right]$, to the states on the Poincaré section of step *n + 1*: $\mathbf{v_{n+1}} = \left[\theta_{n+1}; \dot{\theta}_{n+1}; \dot{\phi}_{n+1}\right]$. First we find the fixed point v* of the function **S** through a Newton-Raphson search:

$$\mathbf{v}^* = \begin{bmatrix} \theta^* \\ \dot{\theta}^* \\ \dot{\phi}^* \end{bmatrix} = \begin{bmatrix} 0.1534 \\ -0.1561 \\ -0.0073 \end{bmatrix} \tag{6}$$

The monodromy matrix **A** is found by simulating one step for a small perturbation on each of the three states of the fixed point v*. The eigenvalues λ of matrix **A** for the simplest walking model turn out to be:

$$\lambda = \begin{bmatrix} 0.23 + 0.59i \\ 0.23 - 0.59i \\ 0 \end{bmatrix} \tag{7}$$

These eigenvalues are all within the unit circle in the complex plane, rendering the motion of the simplest walking model cyclically stable.

No sustained local stability

Not only is the motion of the simplest walking model cyclically stable, but we can also prove it is not sustained locally stable around its nominal trajectory. This second proof is needed before we can categorize the simplest walking model as a Limit Cycle Walker.

For this proof, we study the behavior of the motion in the neighbourhood of the nominal trajectory. This behavior is described by the dynamics of the deviations from the nominal trajectory $e_\theta = \theta - \theta_{nom}$ and $e_\phi = \phi - \phi_{nom}$:

$$\begin{aligned} \ddot{e}_\theta &= \sin(\theta_{nom} + e_\theta - \gamma) - \sin(\theta_{nom} - \gamma) \\ \ddot{e}_\phi &= \sin(\phi_{nom} + e_\phi)\left((\dot{\theta}_{nom} + \dot{e}_\theta)^2 - \cos(\theta_{nom} + e_\theta - \gamma)\right) \\ &\quad - \sin(\phi_{nom})\left(\dot{\theta}_{nom}^{\,2} - \cos(\theta_{nom} - \gamma)\right) \\ &\quad + \sin(\theta_{nom} + e_\theta - \gamma) - \sin(\theta_{nom} - \gamma) \end{aligned} \tag{8}$$

We linearize these equations by assuming small deviations Δe_θ and Δe_ϕ. The local stability of this linearized continuous system is evaluated by calculating its poles (roots) along the nominal trajectory depicted in Fig. 2. The resulting root-locus diagram and time plot of the real part of the poles are shown in Fig. 3. There is a pole located in the right-half plane (positive real part) along the whole trajectory, indicating that the simplest walking has no sustained local stability; even more it is not locally stable at any point in time. Thus the motion of the simplest walking model is a clear example of Limit Cycle Walking.

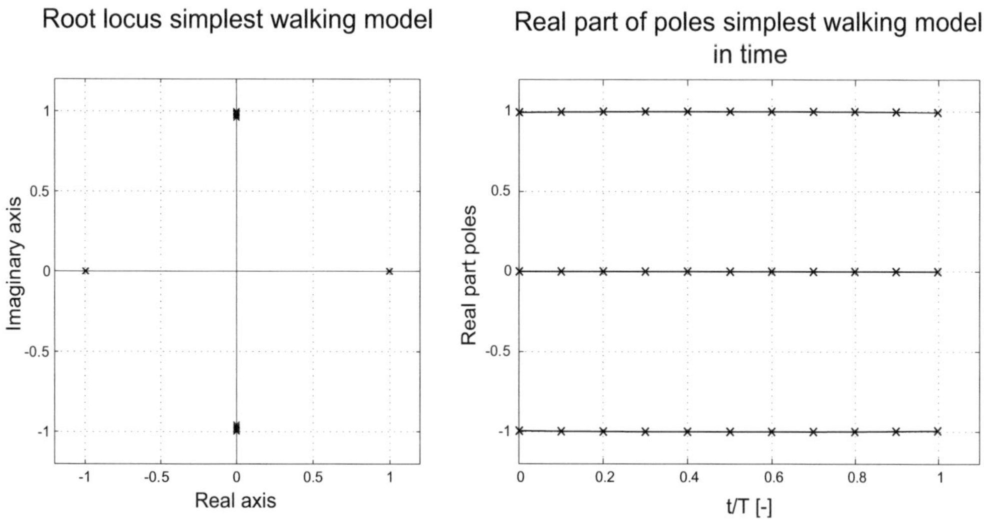

Figure 3. The root locus diagram (left) and time plot of the real part of the poles (right) of the simplest walking model along its nominal trajectory. The pole locations only change slightly throughout the trajectory. The diagrams show that throughout the continuous motion of the simplest walking model there is always a pole in the right-half plane (positive real part); the model is not sustained locally stable

The unnecessary cost of sustained local stability

With this example, we can also show the unnecessary extra constraints that are induced by keeping sustained local stability.

The unstable pole of the simplest walking model corresponds to the inherently unstable motion of the stance leg, being an inverted pendulum. To locally stabilize this motion, the simplest walking model would first need to be able to create a torque τ relative to the ground through a firmly placed foot. Locally stable trajectory tracking (all poles outside the right-hand plane) can be obtained with tracking error feedback with a purely proportional gain K_p only if this feedback gain K_p (normalized for this model) is greater or equal to one:

$$\tau = K_p \cdot e_\theta \quad \text{where } K_p \geq 1 \tag{9}$$

Clearly this shows that the aim for sustained local stability results in unnecessary constraints and unnecessarily high system and control demands. While stable Limit Cycle Walking can be obtained without any effort, sustained local stability requires an extra actuator and tight feedback.

5. State of the Art

There is already a large group of researchers active on Limit Cycle Walking, as we will show in this section. So, one could ask why the concept has not been properly defined earlier. The reason is that many of their robots are derivatives of the so-called Passive Dynamic Walkers (McGeer, 1990a), a subgroup of Limit Cycle Walkers. Passive Dynamic Walking robots will be treated first in this section, followed by actuated Limit Cycle Walkers. Many of these have been referred to as `Passive-based' walking robots, but here we will show that Limit Cycle Walking is a much more accurate label.

5.1 Passive Dynamic Walkers

Passive Dynamic Walking robots are robots that show a perfectly stable gait when walking down a gentle slope without any control or actuation. The simplest walking model that we used in Section 4.2 is a passive dynamic walking model. The stance leg basically is an unstable inverted pendulum and each step is a fall forward. The absence of control clearly places Passive Dynamic Walking as a subgroup of Limit Cycle Walking.
The concept of Passive dynamic walking was introduced in the early 90's by McGeer (McGeer, 1990a), who simulated and built 2D walkers with (McGeer, 1990a) and without knees (McGeer, 1990b) (Fig. 4). As follow-up research, Ruina's group at Cornell University built a 3D "uncontrolled toy that can walk but cannot stand still" (Coleman and Ruina, 1998) (Fig. 5a), the ultimate demonstration of stable Limit Cycle Walking without the existence of any local stability. This group also built a more anthropomorphic 3D prototype with knees and arms (Collins et al., 2001) (Fig. 5b), to show the viability of the concept for the development of humanoid robots. Also in other labs throughout the world passive dynamic walkers have been developed (Mayer et al., 2004; Tedrake et al., 2004a; Ikemata et al., 2006).

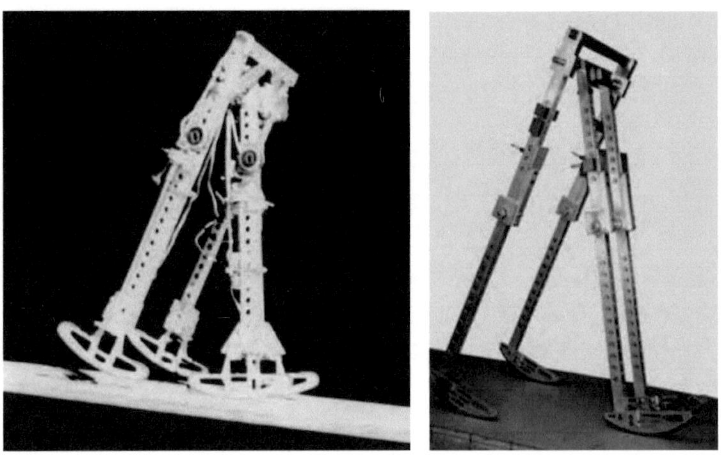

(a) (b)

Figure 4. McGeer's straight legged passive dynamic walker (a) and Cornell University's copy of McGeer's passive dynamic walker with knees (b). Both walkers have 2D dynamics thanks to the double leg pairs (the outer legs form a unit and the inner legs form a second unit)

5.2 Actuated point/arced feet walkers

Perhaps the largest group of typical Limit Cycle Walkers are the actuated robots with point feet or arced feet. They have actuation and control in some of the joints, but the shape of their feet and the resulting point contact or line contact makes these systems underactuated. Although underactuated systems theoretically can be locally stabilized (using the actuated joints, as in the Acrobot (Spong, 1994)), this is not applied in the 2D and 3D robots described here. They depend solely on cyclic stability and thus fall under the category of Limit Cycle Walking.

(a) (b)

Fig. 5: Cornell University's "uncontrolled toy that can walk but cannot stand still" (a) and 3D walker with knees and counter-swinging arms (b)

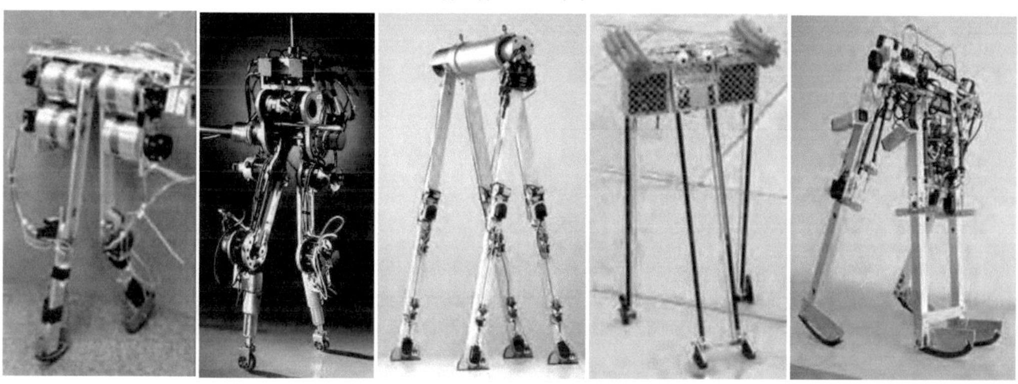

(a) (b) (c) (d) (e)

Figure 6. Two-dimensional actuated point/arced feet walkers: from Carnegie Mellon (a), `Rabbit' from Centre National de Recherche Scientifique (CNRS) and the University of Michigan (b), `Dribbel' from the University of Twente (c), `Cornell Ranger' from Cornell University (d) and `Mike' from Delft University of Technology (e)

The 2D prototypes (Fig. 6) use a four-legged symmetric construction (similar to McGeer's machines) or a guiding boom. Fig. 6a shows a robot with direct drive electric actuators in the hip and knee joints. These weak and highly backdrivable motors make accurate trajectory control practically impossible. Nevertheless, successful walking was obtained with ad-hoc controllers (Anderson et al., 2005), through reinforcement learning (Morimoto et al., 2004) and with CPG control (Endo et al., 2004), demonstrating the potential of the concept of Limit Cycle Walking. At the other end of the spectrum we find the robot `Rabbit' (C.Chevallereau et al., 2003) (Fig. 6b) which has heavily geared electric motors at the hips and knees. The control is based on the concepts of `Hybrid zero dynamics' and `virtual constraints' (Westervelt et al., 2003; de Wit, 2004). It applies tight control on all internal joints, based on the angle of the lower leg with the floor which is left unactuated. As a result, this is a machine with one passive degree of freedom, and so the cyclic stability is calculated with one Floquet multiplier. Between these two extremes, there are currently quite a few 2D Limit Cycle Walking robots (Geng et al., 2006; Dertien, 2006; Wisse and v. Frankenhuyzen, 2003; Wisse et al., 2006), see Fig. 6, including our own pneumatically actuated machines 'Mike' (Wisse and v. Frankenhuyzen, 2003) (Fig. 6e) and 'Max' (Wisse et al., 2006). The great variety of successful controllers and hardware configurations is yet another hint towards the potential of Limit Cycle Walking.

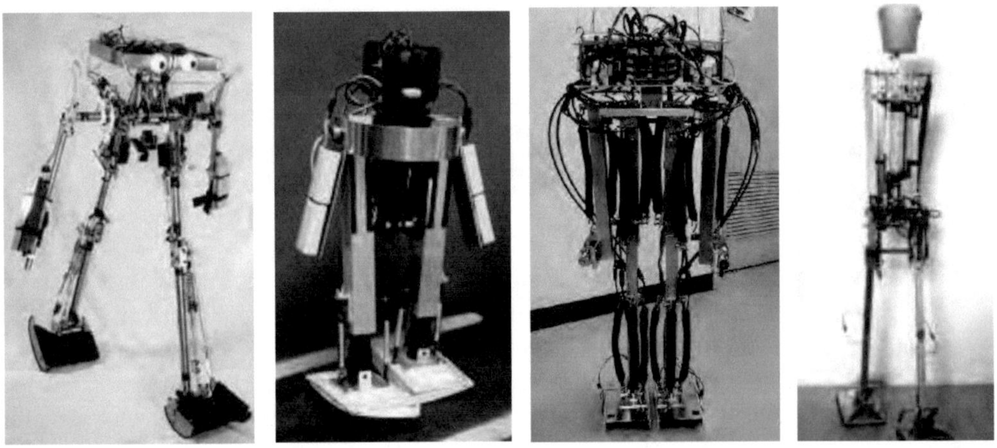

(a) (b) (c) (d)

Figure 7. Three-dimensional actuated arced feet walkers: the `Cornell biped' from Cornell University (a), `Toddler' from Massachusetts Institute of Technology (MIT) (b), `Pneu-man' from Osaka University (c) and `Denise' from Delft University of Technology (d)

There are a few fully three-dimensional (unconstrained) actuated arced feet walkers, as shown in Fig. 7. Cornell University built their `Cornell biped' (Fig. 7a), having an upper body, swinging arms, upper legs, lower legs and arced feet (Collins and Ruina, 2005; Collins et al., 2005). Because of internal mechanical couplings the number of degrees of freedom is only five. The (one DoF) hip joint and knee joints of the `Cornell biped' are fully passive. The only actuation in this machine is a short ankle push-off in the trailing leg just shortly after foot strike at the leading leg. The push-off force is delivered by a spring that is being loaded throughout one step by a small electric motor. `Toddler' (Fig. 7b) is a walker that has been built at MIT (Tedrake et al., 2004a; Collins et al., 2005). It consists of two legs and two arced feet. The (one DoF) hip joint is fully passive, the ankle joint has two degrees of freedom (roll and pitch) which are both activated by position controlled servo motors. Successful gait was obtained with fully feedforward ankle angle trajectories, hand tuned feedback and by applying reinforcement learning (Tedrake et al., 2004b). The `Pneu-man' (Fig. 7c) was developed at Osaka University (Hosoda et al., 2005). It has an upper body, arms, upper legs, lower legs and arced feet. The machine has ten degrees of freedom, two for the arms, two in the hip, two knees and two ankles with two degrees of freedom each. All these joints are actuated by pneumatic McKibben muscles. The `Pneu-man' is controlled through a state machine which is initiated every step at foot strike and subsequently opens and closes pneumatic valves for fixed amounts of time. `Denise' (Fig. 7d) has been developed in our own lab at Delft University of Technology (Wisse, 2005; Collins et al., 2005). It has an upper body, arms, upper legs, lower legs and arced feet. Because of mechanical couplings it only has five internal degrees of freedom. The hip joint is actuated by McKibben muscles while the knee and ankle joints are fully passive.

5.3 Actuated flat feet walkers

We only know of two Limit Cycle Walkers with flat feet, see Fig. 8. `Spring Flamingo' (Fig. 8a) has been built at Massachusetts Institute of Technology and consists of seven links (upper body, upper legs, lower legs and flat feet) (Pratt et al., 2001). All six internal joints are actuated by series elastic actuators, in which springs are intentionally placed in series with geared electric motors (Pratt and Williamson, 1995). This setup allows torque control on all joints. The desired torques for all joints are determined by `Virtual Model Control', an intuitive approach to bipedal walking control, implementing virtual elements such as springs and dampers (Pratt et al., 2001).

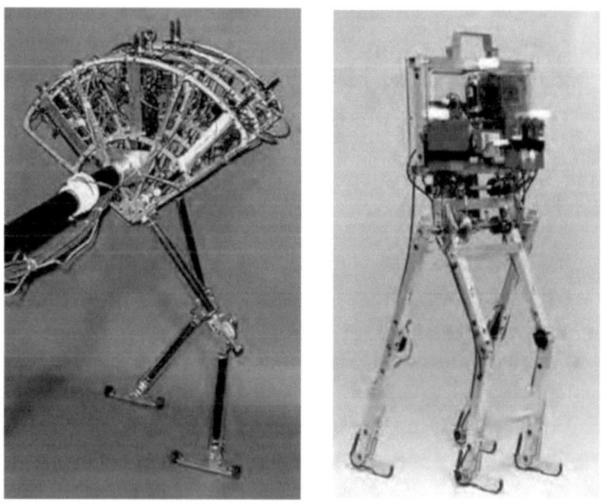

(a) (b)

Figure 8. Actuated flat feet Limit Cycle Walkers: `Spring Flamingo' from Massachusetts Institute of Technology (MIT) (a) and `Meta' from Delft University of Technology (b)

`Meta' (Fig. 8b) is a prototype which is currently being used for research in our lab at Delft University of Technology. It has the same links as `Spring Flamingo', but only applies series elastic actuation at the hip joint and the ankles, the knee joints are passive. Various control strategies are currently being applied to this prototype.

`Spring Flamingo' is constrained to 2D by a boom construction and `Meta' by a symmetric construction of its legs. Currently there are no fully 3D flat feet Limit Cycle Walkers in existence. We have adopted this as a research challenge for our lab for the near future.

6. Energy efficiency

With the introduction of the paradigm Limit Cycle Walking we argue that the reduction of unnecessary constraints can result in increased performance. In this section we discuss the performance of Limit Cycle Walking in terms of energy efficiency. We explain how Limit

Cycle Walking is crucial for increasing energy efficiency and we show that State of the Art Limit Cycle Walkers are indeed highly efficient. First we need to explain how the energy efficiency of bipedal walkers is measured to be able to make a fair comparison.

6.1 Measuring energy efficiency

The energy efficiency of bipedal gait is quantified by the specific cost of transport or specific resistance. This dimensionless number c_t gives the amount of energy that the biped uses per distance traveled per weight of the walker:

$$c_t = \frac{\text{Used energy}}{\text{Weight} \cdot \text{Distance traveled}} \tag{10}$$

To distinguish between the efficiency of the total system design and the efficiency of the mechanical design and the applied forces by the controller, the specific energetic cost of transport c_{et} and the specific mechanical cost of transport c_{mt} are defined. The specific energetic cost of transport c_{et} takes into account the total amount of energy used by the system where the specific mechanical cost of transport c_{mt} only takes into account the amount of mechanical work performed by the actuators.

To give a fair comparison between different walkers, the specific cost of transport should be measured at equal normalized walking speeds. Normalized walking speed is given by the Froude number Fr, which is the actual walking speed divided by the square root of gravity times leg length: $Fr = v/\sqrt{gl}$.

6.2 Limit Cycle Walking improves energy efficiency

Limit Cycle Walking is energy efficient because it allows the use of zero feedback gains or lower feedback gains than are required for sustained local stability.

When disturbances make a walker deviate from its nominal desired trajectory, high feedback gains forcefully redirect the motion to the nominal trajectory. This often results in excessive active braking by the actuators (negative work) and thus increasing energy consumption. The use of high feedback gains actively fights the natural dynamics of the system at the cost of extra energy expenditure. In contrast, Limit Cycle Walking allows the natural dynamics of a walking system to help ensure convergence to the desired motion which takes away (at least part of) the active braking by the actuators.

In case of uncertain or changing system parameters, the use of zero or low feedback gains in Limit Cycle Walking allows a walker to adapt its gait to the changing natural dynamics. This also results in less energy consumption then forcefully maintaining the originally intended motion which constantly requires the walker to fight its natural dynamics.

6.3 State of the Art energy efficiency

Limit Cycle Walkers show high energy efficiency. For instance, McGeer's first passive dynamic machine has a specific cost of transport of approximately $c_{et} = c_{mt} = 0.025$, walking at a Froude number of $Fr = 0.2$. In comparison, the specific energetic cost of transport c_{et} of humans is about 0.2 based on oxygen consumption, humans' specific mechanical cost of transport c_{mt} is estimated at about 0.05 (Collins et al., 2005). Honda's state of the art humanoid Asimo is estimated to have $c_{et} \approx 3.2$ and $c_{mt} \approx 1.6$ (Collins et al., 2005).

The actuated Limit Cycle Walkers show mechanical costs of transport that are similar to the passive dynamic walkers: c_{mt} = 0.06 for `Dribble, $c_{mt}$ = 0.055 for the `Cornell biped', c_{mt} = 0.08 for `Denise' and $c_{mt}$ = 0.07 for `Spring Flamingo' all at a speed of about Fr = 0.15. The specific energetic cost of transport c_{et} for these machines vary significantly, depending on the effort that was put into minimizing the overhead energy consumption of the system. It ranges from c_{et} = 0.2 for the `Cornell biped' to $c_{et} \approx 5.3$ for `Denise'.
Clearly, the performance of the State of the Art Limit Cycle Walkers shows that energy efficiency is improved through the application of Limit Cycle Walking.

7. Disturbance rejection

This section discusses the performance of Limit Cycle Walking in terms of disturbance rejection, the ability to deal with unexpected disturbances. We explain how Limit Cycle Walking is necessary for increasing disturbance rejection and we show the present disturbance rejection in the State of the Art Limit Cycle Walkers. First we need to explain how the disturbance rejection of bipedal walkers is measured to be able to make a fair comparison.

7.1 Measuring disturbance rejection

The way disturbance rejection is measured in Limit Cycle Walkers unfortunately is quite diverse. A range of disturbance rejection measures exists, relating to the rate at which disturbances are rejected (the largest Floquet multiplier (McGeer, 1990a; Tedrake et al., 2004b)), the maximum allowable size of disturbances (the Basin of Attraction (Schwab and Wisse, 2001)) or a combination of the two (the Gait Sensitivity Norm (Hobbelen and Wisse, 2006)).
The currently most commonly used measure is the maximal floor height variation a walker can handle without falling. A Limit Cycle Walker is perturbed by a single floor height difference (step-up/down) of a specific size to observe whether it is able to recover. The size of this perturbation is varied to establish for which maximal size the walker can still consistently prevent falling. The size of this floor height difference is normalized to the walker's leg length for comparison to other walkers.

7.2 Limit Cycle Walking necessary for large disturbance rejection

Obtaining disturbance rejection of large disturbances in walking is not possible with the application of high feedback gains and sustained local stability as it results in inadmissibly high actuation torques. These overly high desired torques will result in saturation of actuators and loss of full contact between the walker's feet and the floor, violating the necessary condition of local stabilizability for obtaining sustained local stability. The solution to this problem is Limit Cycle Walking as it works with lower actuation torques (lower feedback gains) and it does not depend on full contact between the feet and the floor. The crucial notion is that it is allowed to reject a disturbance over an extended amount of time, as long as falling is avoided.
The application of Target ZMP Control (adjustment of the nominal trajectory) by Honda's ASIMO (Website Honda's ASIMO) is a recognition of the fact that Limit Cycle Walking is necessary for increasing disturbance rejection. We feel that the fact that trajectory

adjustment control is applied, makes ASIMO's tight trajectory tracking excessive and unnecessary and thus we would recommend the reduction of this constraint altogether.
For Limit Cycle Walkers there are two ways to increase disturbance rejection: (1) minimizing the divergence of motion throughout a step and (2) increasing the stabilizing effect of step-to-step transitions. An exemplary way to minimize motion divergence is the commonly used application of arced feet instead of point feet in Limit Cycle Walkers. A way to increase the stabilizing effect of step-to-step transitions is foot placement, for instance by applying swing leg retraction (Wisse et al., 2005).

7.3 State of the Art disturbance rejection

The disturbance rejection of passive dynamic walkers is limited. The maximal allowable floor height difference has been established on a passive dynamic walking prototype in our lab. This prototype could handle a step-down of ~1.5% of its leg length (Wisse and v. Frankenhuyzen, 2003).
More information on the maximally allowable size of floor height variations is known from the actuated Limit Cycle Walkers. For two-dimensional Limit Cycle Walkers typical values for a single floor height difference (step-down) that they can handle are: ~1% of its leg length for `Max', ~2% for `Mike' and ~4% for `RunBot'. `Spring Flamingo' was able to handle slope angles of 15°, a disturbance that is comparable to a step-down/up of ~9% of its leg length. For three-dimensional walkers the only data that is available is that `Denise' can handle a single step-down disturbance of ~1% of its leg length.
How this State of the Art disturbance rejection compares to other non-Limit Cycle Walkers is hard to say as there is a limited amount of published data available. From personal communication with other robotic researchers we learned that the amount of unexpected floor variation other robotic walkers can handle is probably in the same order of magnitude. This expectation is supported by the fact that high floor smoothness standards are demanded for bipedal robot demonstrations.
The State of the Art disturbance rejection of bipedal robots certainly is small compared to the human ability to handle unexpected disturbances during walking. We expect that Limit Cycle Walking is an essential paradigm for increasing the disturbance rejection of bipedal robots. Proving this in actual prototypes is a main topic of our lab's future work.

8. Versatility

The last aspect of Limit Cycle Walking performance that we will discuss is gait versatility. The versatility of a Limit Cycle Walker is its ability to purposefully perform different gaits. An important aspect of this versatility is the ability to walk at different speeds. We explain how Limit Cycle Walking creates the opportunity to increase versatility and show the versatility of State of the Art Limit Cycle Walkers. First we need to explain how the versatility of bipedal walkers is measured to be able to make a fair comparison.

8.1 Measuring versatility

The versatility of a bipedal walker is largely described by the range of speeds a walker can obtain in three orthogonal directions. In the sagittal/fore-aft direction the range of walking speeds tells whether a walker is able to walk at high speeds and low speeds, but also if it is able to come to a standstill (zero speed). In the other two directions (lateral/left-right and vertical) it is the ratio between the speed in that direction and the sagittal walking speed that is of importance. In the lateral direction this defines the rate at which a walker can turn (the turn radius it can obtain) or, when the ratio is infinite, whether it can walk sideways. In the vertical direction it defines what types of slopes (or stairs) the walker can handle purposefully. To make walking speeds comparable between different walkers the speed is generally normalized by giving the dimensionless Froude number: $Fr = v/\sqrt{gl}$.

8.2 Limit Cycle Walking increases speed range

Limit Cycle Walking enables walking at high speeds because it uses the more frequent and more violent step-to-step transitions at high speed as a way to create stability instead of having to fight them as disturbances.

In case a bipedal walker applies sustained locally stabilizing control around its nominal trajectory, the step-to-step transitions generally are mainly a source of deviations from the nominal motion which have to be rejected by the trajectory controller. For increasing walking speed, these induced deviations tend to grow in size and occur at a higher frequency, which increases the demands on the trajectory controller. Eventually, the obtainable walking speed is limited by the control bandwidth.

In Limit Cycle Walking this control bandwidth limitation on walking speed does not exist as it does not depend on continuous stabilizing control to reject deviations. Step-to-step transitions are the main source of stabilization and the `bandwidth' of this stabilizing effect automatically increases with increasing walking speed.

8.3 State of the Art versatility

Some of the State of the Art Limit Cycle Walkers have shown to be able to walk at different walking speeds: 'Rabbit' shows a speed range from about Fr = 0.15 to Fr = 0.3, `Toddler' from 0 (it is able to start from standstill) to about $Fr$ = 0.09 and the relatively fast and small `RunBot' from Fr = 0.25 to Fr = 0.5. In comparison Honda's Asimo has a speed range from 0 to Fr = 0.3 and humans from 0 to about Fr = 1.

Research with Limit Cycle Walkers focusing on the ability to purposefully perform turns or walk stairs has gotten little attention so far. Only the prototype `Cornell Ranger' is able to make turns with a minimal radius of about 25 meters. None of the State of the Art Limit Cycle Walkers has shown the ability to purposefully walk up or down stairs.

The State of the Art versatility of Limit Cycle Walkers is quite limited at present in spite of the fact that the paradigm is promising on this aspect. Showing that Limit Cycle Walking prototypes can have a large versatility is another main topic of our lab's future work.

9. Conclusion

In this chapter we have introduced the paradigm `Limit Cycle Walking'. This paradigm has been used for some time by a group of bipedal robotics researchers, but the concept had not been properly defined before:

Limit Cycle Walking is a nominally periodic sequence of steps that is stable as a whole but not locally stable at every instant in time.

Limit Cycle Walking releases the unnecessary stability constraint of sustained local stability that is often applied in bipedal robots. It only requires the much less restrictive cyclic stability. This gives extra freedom for finding better walking performance, it is expected that:

- Limit Cycle Walking increases energy efficiency because it does not apply high feedback gains that fight the natural dynamics of the system in case of disturbances or uncertain parameters.
- Limit Cycle Walking is necessary to increase disturbance rejection because large disturbances tend to violate the constraints that are necessary to enforce local stability.
- Limit Cycle Walking increases versatility by enabling higher walking speeds; the more frequently and violently occurring step-to-step transitions are inherently used for stabilization in contrast to being an increasing source of unwanted disturbances demanding higher control bandwidth.

The increase of energy efficiency has been demonstrated by the State of the Art Limit Cycle Walkers. The expected increase in disturbance rejection and versatility has not yet been shown in existing bipedal walkers. Doing this is the main goal of our lab's future work.

10. Acknowledgements

This research has been partially supported by the European Commission (IST-1-045301-STP) and by the Dutch Technology Fund STW.

11. References

Website Honda's ASIMO. (http://world.honda.com/asimo).

S. O. Anderson, M. Wisse, C. G. Atkeson, J. K. Hodgins, G. J. Zeglin, and B. Moyer (2005). Powered Bipeds Based on Passive Dynamic Principles. *In Proceedings of IEEE/RAS International Conference on Humanoid Robots*, 2005.

C.Chevallereau, G.Abba, Y.Aoustin, F.Plestan, E.R.Westervelt, C.Canudas-De-Wit, and J.W.Grizzle (2003). RABBIT: a testbed for advanced control theory. *IEEE Control Systems Magazine*, 23(5):57-79, 2003.

M. J. Coleman and A. Ruina (1998). An Uncontrolled Walking Toy That Cannot Stand Still. *Physical Review Letters*, 80(16):3658-3661, 1998.

S. H. Collins, M. Wisse, and A. Ruina (2001). A 3-D passive-dynamic walking robot with two legs and knees. *The International Journal of Robotics Research*, 20(7):607-615, 2001.

S. H. Collins and A. Ruina (2005). A Bipedal Walking Robot with Efficient and Human-Like Gait. *In Proceedings of International Conference on Robotics and Automation*, 2005.

S. H. Collins, A. Ruina, R. Tedrake, and M. Wisse (2005). Efficient Bipedal Robots Based on Passive-Dynamic Walkers. *Science*, 307(5712):1082-1085, 2005.

C. Canudas de Wit (2004). On the concept of virtual constraints as a tool for walking robot control and balancing. *Annual Reviews in Control*, 28:157-166, 2004.

E. Dertien (2006). Dynamic walking with Dribbel. *IEEE Robotics and Automation Magazine*, 13(3):118-122, 2006.

G. Endo, J. Morimoto, J. Nakanishi, and G. Cheng (2004). An empirical exploration of a neural oscillator for biped locomotion control. *In Proceedings of IEEE International Conference on Robotics and Automation*, volume 3, 2004.

M. S. Garcia, A. Chatterjee, A. Ruina, and M. J. Coleman (1998). The Simplest Walking Model: Stability, Complexity, and Scaling. *ASME Journal of Biomechanical Engineering*, 120(2):281-288, 1998.

T. Geng, B. Porr, and F. Wörgötter (2006). Fast Biped Walking with a Sensor-driven Neuronal Controller and Real-Time Online Learning. *The International Journal of Robotics Research*, 25(3):243-259, 2006.

A. Goswami (1999). Foot rotation indicator (FRI) point: A new gait planning tool to evaluate postural stability of biped robots. *In Proceedings of International Conference on Robotics and Automation*, 1999.

H. Hirukawa (2003). Humanoid Robotics Platforms Developed in HRP. *In Proceedings of International Conference on Humanoid Robots*, 2003.

D. G. E. Hobbelen and M. Wisse (2006). A disturbance rejection measure for limit cycle walkers: the Gait Sensitivity Norm. *IEEE Transactions on Robotics* (in review), 2006.

K. Hosoda, T. Takuma, and M. Ishikawa (2005). Design and Control of a 3D Biped Robot Actuated by Antagonistic Pairs of Pneumatic Muscles. *In Proceedings of International Symposium on Adaptive Motion in Animals and Machines*, 2005.

Y. Hurmuzlu and G. D. Moskowitz (1986). Role of Impact in the Stability of Bipedal Locomotion. *International Journal of Dynamics and Stability of Systems*, 1(3):217-234, 1986.

Y. Hurmuzlu, F. Gnot, and B. Brogliato (2004). Modeling, stability and control of biped robots - a general framework. *Automatica*, 40:1647-1664, 2004.

Y. Ikemata, A. Sano, and H. Fujimoto (2006). A Physical Principle of Gait Generation and its Stabilization derived from Mechanism of Fixed Point. *In Proceedings of International Conference on Robotics and Automation*, 2006.

T. Ishida (2004). Development of a small biped entertainment robot QRIO. *In Proceedings of International Symposium on Micro-Nanomechatronics and Human Science*, 2004.

N. M. Mayer, A. A. Forough-Nassiraei, Z. Hsu, F. Farkas, and T. Christaller (2004). Stabilizing dynamic walking with physical tricks. *In Proceedings of International Conference on Climbing and Walking Robots*, 2004.

T. McGeer (1990a). Passive Dynamic Walking. *The International Journal of Robotics Research*, 9(2):62-82, 1990.

T. McGeer (1990b). Passive Walking with Knees. *In Proceedings of International Conference on Robotics and Automation*, volume 3, 1990.

J. Morimoto, G. Cheng, C. G. Atkeson, and G. J. Zeglin (2004). A simple reinforcement learning algorithm for biped walking. *In Proceedings of IEEE International Conference on Robotics and Automation*, volume 3, 2004.

G. A. Pratt and M. M. Williamson (1995). Series Elastic Actuators. *In Proceedings of IEEE International Conference on Intelligent Robots and Systems*, 1995.

J. Pratt, C.-M. Chew, A. Torres, P. Dilworth, and G. Pratt (2001). Virtual Model Control: An Intuitive Approach for Bipedal Locomotion. *The International Journal of Robotics Research*, 20(2):129-143, 2001.

Y. Sakagami, R. Watanabe, C. Aoyama, S. Matsunaga, N. Higaki, and M. Fujita (2002). The intelligent ASIMO: System overview and integration. *In Proceedings of International Conference on Intelligent Robots and Systems*, volume 3, 2002.

A. L. Schwab and M. Wisse (2001). Basin of Attraction of the Simplest Walking Model. *In Proceedings of ASME Design Engineering Technical Conferences*, 2001.

M. W. Spong (1994). Swing Up Control of the Acrobot. *In Proceedings of International Conference on Robotics and Automation*, 1994.

S. H. Strogatz (2000). *Nonlinear Dynamics and Chaos.* Westview Press, Cambridge, MA, USA, 2000.

R. Tedrake, T. W. Zhang, M. Fong, and H. S. Seung (2004a). Actuating a Simple 3D Passive Dynamic Walker. *In Proceedings of International Conference on Robotics and Automation*, 2004.

R. Tedrake, T. W. Zhang, and H. S. Seung (2004b). Stochastic Policy Gradient Reinforcement Learning on a Simple 3D Biped. *In Proceedings of International Conference on Robotics and Automation*, 2004.

M. Vukobratovic, A. Frank, and D. Juricic (1970). On the Stability of Biped Locomotion. *IEEE Transactions on Biomedical Engineering*, 17(1), 1970.

M. Vukobratovic and B. Borovac (2004). Zero-Moment Point - thirty years of its life. *International Journal on Humanoid Robotics*, 1(1):157-174, 2004.

M. Vukobratovic, B. Borovac, and V. Potkonjak (2006). Towards a unified understanding of basic notions and terms in humanoid robotics. *Robotica*, 2006.

E. R. Westervelt, J. W. Grizzle, and D. E. Koditschek (2003). Hybrid Zero Dynamics of Planar BipedWalkers. *IEEE Transactions on Automatic Control*, 48(1):42-56, 2003.

P. B. Wieber (2002). On the stability of walking systems. *In Proceedings of International Workshop Humanoid and Human Friendly Robotics*, 2002.

M. Wisse and J. v. Frankenhuyzen (2003). Design and construction of Mike; a 2D autonomous biped based on passive dynamic walking. *In Proceedings of 2nd International Symposium on Adaptive Motion of Animals and Machines*, 2003.

M. Wisse (2005a). Three additions to passive dynamic walking; actuation, an upper body, and 3D stability. *International Journal of Humanoid Robotics*, 2(4):459-478, 2005.

M. Wisse, C. G. Atkeson, and D. K. Kloimwieder (2005b). Swing leg retraction helps biped walking stability. *In Proceedings of IEEE International conference on Humanoid Robots*, 2005.

M. Wisse, D. G. E. Hobbelen, and A. L. Schwab (2006). Adding the upper body to passive dynamic walking robots by means of a bisecting hip mechanism. *IEEE Transactions on Robotics* (in press), 2006.

15

A Human-Like Approach to Footstep Planning

Yasar Ayaz[1,2], Khalid Munawar[1], Mohammad Bilal Malik[1],Atsushi Konno[2] and Masaru Uchiyama[2]
[1]*National University of Sciences and Technology (NUST)*, [2]*Tohoku University*
[1]*Pakistan*, [2]*Japan*

1. Introduction

Terrains in our everyday environment such as homes and offices are custom-designed for biped walkers i.e., human beings. Evolution of humanoid robots was, thus, a natural development in the field of mobile robotics. However, giving legs to a robot instead of wheels gives it a lot more than just resemblance to a human being. Unlike ordinary mobile robots, humanoids have the ability to cross obstacles by stepping over and upon them. This ability is left unexploited if the algorithms used for ordinary mobile robot navigation among obstacles are employed for humanoids too.

Various approaches have been adopted in order to address this problem in the past. (McGhee & Iswandhi, 1979) developed a method that divides the terrain into *'permissible'* and *'impermissible'* stepping positions. Keeping in view the direction in which the ultimate goal position is located, the robot selects the next foothold from amongst the permissible ones in the immediately reachable footholds while taking care of postural stability constraints. While this method targeted a general application to legged robotics, (Yagi & Lumelsky, 1999) presented another one that deals specifically with humanoid robots. Depending upon the distance with the obstacle nearest to the robot in the direction of motion (presumably the direction in which the goal position is located) the robot adjusts the length of its steps until it reaches the obstacle. Now, if the obstacle is small in size, it is overcome by stepping over or upon it whereas if it is too tall to be overcome in this way, the robot starts sidestepping until it moves clear of the obstacle. Obviously, whether to sidestep left or right is also a pre-programmed decision. These and other such *localized* reactive approaches have the tendency to lead the robot into a *local loop* or a *deadlock* in which case the robot would have to be backtracked in order to follow an alternate path.

(Kuffner et al., 2001) argued that since such reactive algorithms failed to consider complexities occurring in the path at a later stage before opting to take it, they ended up stuck in local loops and deadlocks. In order to tackle this problem they presented a footstep planning algorithm based upon game theory that takes into account global positioning of obstacles in the environment. This technique has been tested on H6 (Kuffner et al., 2001), H7 (Kuffner et al., 2003), Asimo Honda (Chestnutt et al., 2005, Michel et al., 2005) and HRP-2 (Michel et al., 2006) humanoid robots with some improvements. The algorithm selects a discrete set of predefined stepping locations in the robot's immediate vicinity while balancing on one leg only. Also predefined are intermediate postures that the robot assumes while moving its feet between any two of these stepping locations. Selecting a set of these possible stepping locations that lead to an equal number of descendants, a tree is spread out from the initial footstep position. Now, after pruning from the tree those branches that do not end up with a step in the destination area, a two-leveled polygon-polygon collision search is conducted in order to identify trajectories that result in collision with the obstacles in the environment. Once these are discarded too, a greedy search is conducted in order to hunt out the best among the paths generated. This strategy cannot distinguish between the paths without obstacles and with some obstacles that can be stepped over due to which the robot always needs to execute high stepping profiles resulting in larger overall energy consumption. In addition, because of generation of a large search tree of possible stepping locations, greedy search has to be applied instead of an exhaustive one. On the other hand, a human in such a situation would start by searching for an obstacle-free path directly connecting initial and goal locations. If found, it would simply be opted. Whereas some options would be considered once an obstacle-free direct path is not found. Again, an obstacle can be dodged from the right, or dodged from the left, or is crossed by stepping over it.

(Ayaz et al., 2006) starts by checking if it is possible for the robot to go directly from the initial to final locations. If so, such a path is adopted. If not, trajectories are formed in order to dodge the hindering obstacle from the right and the left. In addition, keeping in view the dimensional constraints, obstacles are classified into three types. The smallest ones that can be stepped over: type-1. For these an additional trajectory considering stepping over is also formed. The larger ones that cannot be stepped over but the robot can pass right next to them: type-2. And the largest ones that might collide with the robot's arms (sticking out wider than its feet) if the robot passes right next to them: type-3. The robot keeps a larger distance from obstacles of type-3 as compared to those of type-2 in order to avoid collision. In this way, the algorithm starts by considering the shortest path and then expands its alternates on either side spreading out considering the obstacle type and proximity. In addition, by identifying obstacles once encountered, it avoids getting stuck into the local loops and deadlocks. It is noticeable here that branches formed from every node in the search tree can be a maximum of 3, and for several straight line segments each node leads to only one child. This reduces the branching factor making a smallest possible search tree which in turn enables us to apply the exhaustive search for seeking the best possible path. It can be immediately noticed that this search, though exhaustive, is computationally cheap.

This chapter explains our proposed method in detail. In section 2 we analyse the game theory based footstep planning strategy under reference. Section 3 introduces the concept of human-like footstep planning and analyses how it measures up to some existing approaches. Section 4 deals with planning for taking a single step. The footstep planner is explained in detail in section 5. Section 6 presents results of simulation of the proposed algorithm on a model of our humanoid robot 'Saika-3' (Konno et al., 2000) whereas section 7 concludes the chapter while highlighting salient future research prospects.

2. Game Theory Based Footstep Planning

Kuffner's algorithm for footstep planning among obstacles for biped robots (Kuffner et al., 2001; Kuffner et al., 2005) is a global motion planning strategy based on using game theory for finding alternate paths in an obstacle cluttered environment by considering a discrete set of heuristically chosen statically stable footstep locations for a humanoid robot and identifying the better path via cost heuristic implementation using greedy search. The assumptions made by the algorithm are as under:

1. The environment floor is flat and cluttered with non-moving obstacles of known position and height,
2. A discrete set of feasible, statically-stable footstep placement positions and associated stepping trajectories are pre-computed,
3. Only the floor surface is currently allowed for foot placement (not obstacle surfaces).

A static stability analysis is carried out for the stepping motion of the humanoid robot as a result of which a continuous region is identified where the robot can place its lifted foot (while balanced on one leg) without losing stability. Since it is computationally extensive to check every single point in this region for a footstep placement possibility in every step, a discrete set of 'feasible' footstep locations is heuristically identified in this region. Standing balanced on one foot, the robot would alternatively consider placing its foot at each of these locations and from each stepping possibility more nodes are generated in a similar manner. Foot transition from each location to the next is carried out via heuristically identified statically stable intermediate postures Q_{right} and Q_{left} (for standing balanced on right and left foot respectively) in order to limit the number of possible transition trajectories considered in every step. This is followed by better path identification through greedy search by employing a heuristic cost function which attempts to approximate an exhaustive search while applying a greedy one since exhaustive search itself would be very time consumptive given the large search tree of possible stepping locations the algorithm generates (Kuffner et al., 2001; Kuffner et al., 2005).

The algorithm was simulated on a model of H6 humanoid robot (Kuffner et al., 2001; Kuffner et al., 2005). Considering 15 discrete footstep placement options in each step in a room with 20 obstacles, it formed search trees of 6,700 and 830,000 nodes in (Kuffner et al., 2001) and (Kuffner et al., 2005) respectively. Better path was traced out using greedy (Kuffner et al., 2001) and A* (Kuffner et al., 2005) search strategies with total processing time of 12 s on 800 MHz Pentium-II and 4 s on 1.6 GHz Pentium-IV running Linux respectively.

The main drawback of this algorithm is high computational complexity because of generation of a large search tree of possible stepping locations which reflects in terms of time consumption. Moreover, a closer look reveals that since this algorithm is not reactive in nature, it does not distinguish between steps taken to cross over obstacles and those taken ordinarily. Since the same intermediate postures as in ordinary steps (Q_{right} and Q_{left}) have to provide the robot with trajectories capable of stepping over obstacles too, they involve lifting of the foot to a high position as if stepping over obstacles in every step. This, inevitably, is quite consumptive in terms of energy and also tends to make the stepping profiles quite unlike human walk. In addition, when it comes to searching for the best path amongst those traced out by the algorithm, an exhaustive search cannot be applied because of the large tree of possible stepping locations generated (Kuffner et al., 2001; Kuffner et al., 2005). A greedy search, however, might not hunt out the best solution under many circumstances (Cormen et al., 1994).

In order to tackle some of these problems a tiered planning strategy has been introduced that splits the planner into high, middle and low level layers in order to traverse different terrain types (Chestnutt & Kuffner, 2004).

3. Human-Like Footstep Planning – The Concept

3.1 Human-Like Approach

One of the fundamental rationale behind development of humanoid robots is their suitability for navigation in human environments due to these being custom designed for biped walkers. Nevertheless, resemblance in terms of construction alone does not guarantee a superior locomotive ability in such obstacle cluttered terrains unless coupled with a path planner that thinks like a human.

As humans we know that, unlike localized reactive navigation strategies, paths adopted by us to traverse obstacle cluttered terrains are planned taking into account global position of obstacles in the environment. Also, humans do not conduct navigation planning by random spreading of footstep placement possibilities throughout the terrain while ignoring the relative position of obstacles in the environment and later evaluating which of these would result in collision as is done by the game theory based approaches. Quite differently, humans start by first searching for the path directly connecting the start and destination points. If found available, this path is simply taken without considering alternate paths. On the other hand, if the direct path is found unavailable due to presence of an obstacle, paths are considered to dodge it as well as step over it with the better one of these being adopted.

This is precisely the approach we seek to adopt. In our method the planner starts by looking for presence of obstacles in the path directly connecting the initial and goal locations. If available, this path is taken without heed to alternate possibilites. If unavailable, paths are generated from the right and the left of the nearest obstacle hindering the direct passage. In addition, if the obstacle is 'small' enough, a path is also planned by stepping over it. If while planning these alternate steps another obstacle is found hindering an alternate path, paths from right, left and over this obstacle are also analyzed in a similar fashion. In this way the graph starts from the simplest path and moves on to the more complex ones. Being human-like, the method itself is of reactive nature (Ayaz et al., 2006), however, it takes into account global positioning of obstacles in the environment.

3.2 Comparison with Visibility Graph

At a casual glance the algorithm might seem to resemble the well known Visibility Graph Approach (Latombe, 1991) for mobile robot motion planning. However, apart from one obvious difference of planning paths by stepping over obstacles, more fundamental differences of approach reveal themselves upon a keener analysis.

A visibility graph comprises the initial and goal locations and the vertices of all obstacles as nodes in the graph. Contrary to this, our algorithm may not, and in most cases, does not include the vertices of all the obstacles present in the environment in the graph. This is because planning to overcome obstacles not hindering the path of the robot in order to reach the destination intended is unlike the human approach. As explained in the previous subsection, our algorithm expands the graph from simpler to more complex paths. The graph complexity is increased only whenever an obstacle in the environment is found hindering the robot's path towards the destination. Of course, while planning a path to overcome such an obstacle, if another obstacle comes in the way, paths are generated to overcome it as well in a similar fashion and only in such case do its vertices become a part of the graph. In this way, obstacles in the environment not hindering the robot's way are not taken into consideration while expanding the graph.

Another salient difference comes to light when considering graph generation with the same obstacles for two different initial and goal locations. In case of visibility graph approach, the part of the graph formed by linking the obstacle vertices will remain the same in both cases. The only difference would be nodes representing initial and goal positions connected to every node directly connectable to them. However, in case of human-like path planning, nearly the entire graph might change. One reason being that since our intension is to dodge obstacles from right and left, the nodes marking the edges of the near side (and likewise the far side) of an obstacle are not generally interconnected in the graph since this does not represent a meaningful path while moving in this direction. So if the line of motion in case of one set of initial and goal locations is perpendicular to the other, this would result in connections previously absent from the graph even if the same obstacle is dodged in both cases. This is because the sides attributed to as *near* and *far sides* in one case (along which paths do not exist in the graph) might become *right* and *left* in the other (which would represent meaningful paths).

Thus neither the resulting graph nor the way this graph is formed is similar to the visibility graph. In addition, due to the human-like approach towards trajectory generation, the graph formation is efficient and the graph itself is much smaller, taking into account only the obstacles necessarily hindering the robot's way.

4. Taking One Step

While planning footsteps for humanoid robot locomotion in obstacle cluttered environments it is vital to first analyse the kinematics and stability constraints on the robot for taking a single step. In simple words, we have to know the stepping area in which the robot can place one of its feet while balanced on the other without losing stability. Also, there can be numerous trajectories that can be followed by the foot during transition from previous to next stepping locations. Which to adopt out of these is also critical especially when it comes to stepping over obstacles.

4.1 Kinematics and Stability Constraints

We trace forward and inverse kinematics for the humanoid (Saika-3 in our case) by treating both legs as 6-DOF serial manipulator chains originating from the same base link. In our current stepping strategy we only use steps in which the angle of twist in the lifted leg is similar to the angle of extension of the foot. Upper body of the robot is assumed to be static in the body frame at all times (i.e. the robot does not move its arms as it walks).

4.1.1 Stability Criterion

When navigating in indoor human environments, (Kuffner et al., 2001) and (Kuffner et al., 2005) use statically stable footstep locations for the reason that steps in such terrains are to be taken in a careful, and thus slow, manner on account of safety concerns. At the moment we also employ a static stability criterion for our analysis for similar reasons.

For static stability analysis, COG (Centre of Gravity) location for the robot is determined as:

$$m\,{}^{0}\mathbf{P}_{COG} = \sum_{i=1}^{13} m_i\,{}^{0}_{i}\mathbf{T}\,{}^{i}\mathbf{P}_i$$

Where:

m = Total mass of the robot

m_i = Mass of link i of the robot

${}^{0}\mathbf{P}_{COG}$ = Position vector of COG in Σ_0 (i.e. base frame)

${}^{i}\mathbf{P}_{i}$ = Position vector of COG of link i in Σ_i

${}^{0}_{i}\mathbf{T}$ = Matrix for transforming coordinates given in frame Σ_i to ones in Σ_0

$= {}^{0}_{1}\mathbf{T}\,{}^{1}_{2}\mathbf{T}\cdots{}^{i-1}_{i}\mathbf{T}$

As in a static stability analysis, no forces other than gravitation are assumed to be acting on the robot, its ZMP (Vukobratović & Borovac, 2004) must lie directly under the COG during the entire walking motion (Ayaz et al., 2004).

$$ {}^{0}\mathbf{P}_{ZMP} = \begin{bmatrix} {}^{0}x_{ZMP} \\ {}^{0}y_{ZMP} \end{bmatrix} = \begin{bmatrix} {}^{0}x_{COG} \\ {}^{0}y_{COG} \end{bmatrix} \tag{1} $$

Position of the robot's feet at any time can be calculated using forward kinematics as:

$$ {}^{0}\mathbf{P}_{RF} = {}^{0}_{6}\mathbf{T}\,{}^{6}\mathbf{P}_{RF} \tag{2} $$

$$ {}^{0}\mathbf{P}_{LF} = {}^{0}_{12}\mathbf{T}\,{}^{12}\mathbf{P}_{LF} \tag{3} $$

Where:

${}^{i}\mathbf{P}_{RF}$ = Position vector of right foot in Σ_i

${}^{i}\mathbf{P}_{LF}$ = Position vector of left foot in Σ_i

Using above model, if (and only if) the ZMP of the robot is found within the support polygon at all times, the robot is regarded as stable.

4.1.2 Normal Stepping Area

We opt the robot to adopt its normal standing posture such that it stands with its knee bent but hip straight (Fig. 1) to ensure minimum energy consumption while holding its ZMP at the centre of its support polygon. Shifting weight to one foot, ZMP is brought to the foot-centre of one foot so that the other foot can be lifted off the ground (Fig. 1) with maximum manoeuvrability in all directions. Now, using inverse kinematics we find limb angles for placing the lifted foot at a certain point and then using (1) along with (2) and (3) (whichever is applicable keeping in view the current supporting foot), we find whether the robot is statically stable or not. Stability analysis for Saika-3 was conducted for a 0.4 m a side square block (around where its right foot-centre would be in normal posture) with a period of 0.01 m between every two successive stability checks. Fig. 2 displays the results obtained. Areas in Fig. 2 marked by different colour/hatching patterns correspond to placement of the robot's right foot with its foot centre at any point within this area. Each colour indicates the suitability of this point for stepping keeping in view constraints on stability and kinematics.

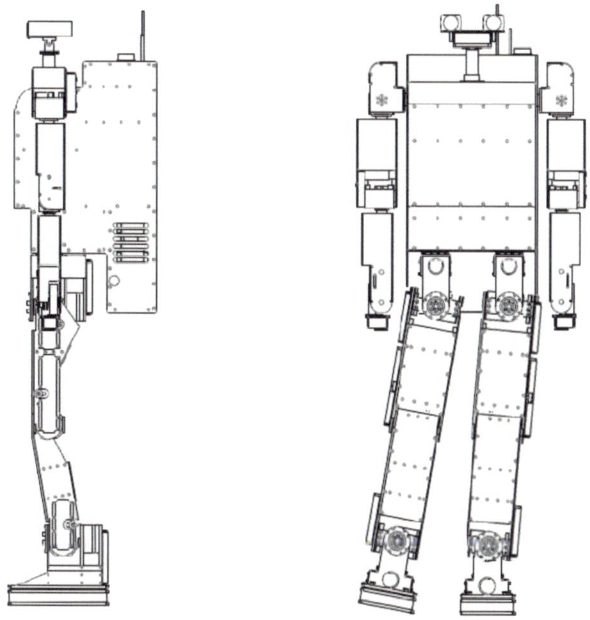

Figure 1. Achieving normal standing posture for taking a step

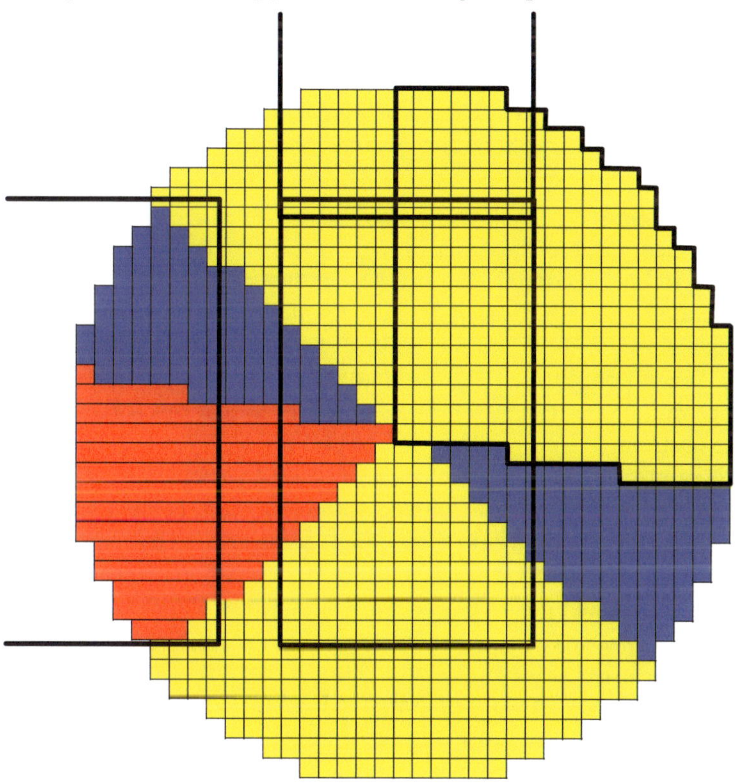

Figure 2. Stepping area for the right foot while balanced on the left with normal step posture

Boxed Pattern : (Yellow)	Robot is stable and the point is included in the workspace of the right leg.
Vertical Lines : (Blue)	Placing the foot here, the robot would have been stable but the point is unavailable for stepping due to constraints on movement of leg joints if the twist in the leg is same as the foot orientation.
Horizontal Lines : (Red)	Satisfies criteria for *Boxed Pattern* but is unavailable for stepping because placing the right foot here would result in stepping over the other foot.

In fact, since in order to stabilize the robot on left foot, the left foot has been moved inwards, we discard the whole left semicircular area for fear of stepping over the other foot. In addition, since we are making a footstep planning algorithm, the area marked by *Boxed Pattern* in the lower half of the circle is not required. This is because this area represents steps taken backwards, a condition which a global footstep planning algorithm is primarily meant to avoid. Thus the area marked by a continuous line in Fig. 2 is the area we shall assume to be available to Saika-3 for stepping while footsteps are planned for its navigation in a human environment.

It is to be noted here that unavailability of the area marked by vertical lines in Fig. 2 is subject to the inverse kinematics used by our step-selection strategy which at the moment only considers those steps in which the angle of twist of the leg is equal to the angle of turning of the foot. For steps in which these two are different, this area could of course be made available for stepping. However this would not change the algorithm's planned footsteps since these areas are to be discarded anyway in order to avoid 'stepping back' and 'stepping on the other foot' conditions as explained above.

4.2 Basic Step Selection

Unlike the game theory based approaches, we do not arbitrarily identify a discrete set of footstep locations to be considered for next step in every step cycle. Instead, we adopt the approach of identifying a direction of motion for the robot; in this direction the step of maximum possible length (0.17 m for Saika-3 using normal posture) is taken during ordinary walk. When approaching the destination or an obstacle (for stepping over), the final step in the walking pattern is shortened in order to just reach the destination or step just short of the obstacle.

Also, while achieving the direction of motion in order to orient the robot to face a certain target location, it has to be borne in mind that humanoids, being nonholonomic robots, cannot turn about their body centre (in most cases for instance in our humanoid robot Saika-3). Thus, if the angle of motion in order to reach a target location is measured with respect to the body centre, the eventual position reached would not be the same as the target position intended (Fig. 3 (a)). Thus we adopt the strategy of measuring the angle of turning of the robot in order to reach the destination according to the initial position of the foot about which the robot would turn. Thus is illustrated in Fig. 3 (b).

The first step in this process is to draw a circle of radius r = *(foot width + distance between two feet)/2* around the destination point. As obvious, both feet of the robot would have their centres somewhere on this circle when the robot reaches the destination. In order to exactly identify the location of feet at the destination, we drop a tangent from the current footstep location to this circle around the destination point. The point at which this tangent reaches the circle is our required destination for the foot from which the tangent was dropped. The turning process begins with the robot first orienting its foot (about which it intends to turn) along the tangent dropped in this way. Following this, the weight of the robot is shifted to the now turned foot and the rest of the body is turned about this foot to bring the robot back to the double support phase with both feet parallel to each other and its orientation towards the required destination.

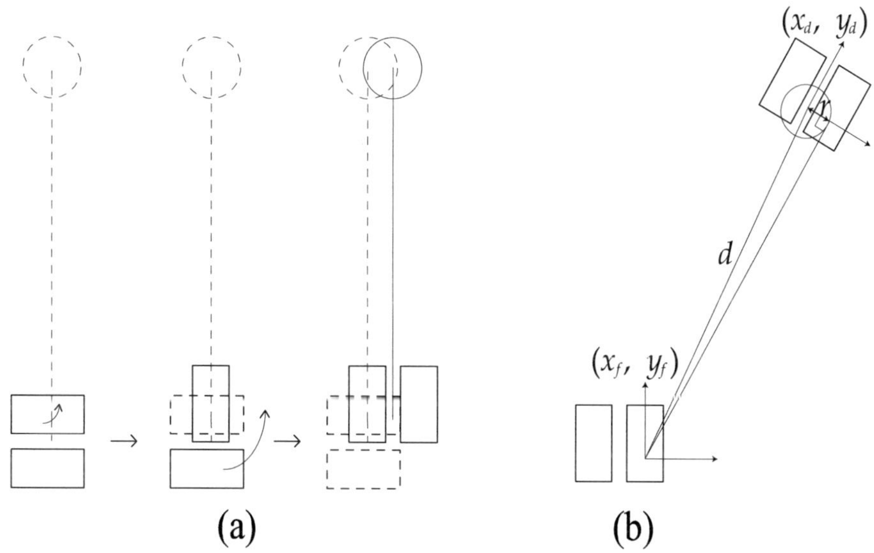

Figure 3. Turning towards a goal location

If the angle of turning is more than the angle of twist achievable by the robot's leg in one step, the turning motion is split into two turnings. Also, apart from simple turning, we also use another turning motion attributed to as 'Extended' or 'On-Walk' turning in which the turning motion is coupled with the stepping motion (i.e. the robot does not place its foot down after twisting the leg only; instead, it twists the leg and then also extends it forwards to take a step before putting it down again and rotating the rest of the body about it to complete the turning motion).

Coming to stepping trajectories for a normal step, any trajectory can be used with minimum foot-lift since no crossing over obstacles is intended (we use foot-lift of 0.05 m for Saika-3).

4.3 Stepping Over Obstacles

4.3.1 Enhancing the Stepping Area

From Fig. 2 it can easily be seen that using this stepping area only, the robot is unable to step over obstacles since even after stretching the right foot forward to its limit, the next placement position overlaps the current one. In order to tackle this constraint we enhance the robot's stepping area by lowering its upper body (and thus the hip joint from where the leg extends forwards). This is done by bending the knee and hip joints of the humanoid robot's supporting leg keeping ZMP of the robot at the centre of the support polygon while lowering its upper body. Repeating the procedure for stability analysis as in section 4.1.2 for this standing posture, we are able to enhance the robot's stepping area by almost three times

(Fig. 4), thus providing a gap between current and next footstep placement locations (0.09 m for Saika-3).
As obvious, enhanced stepping posture is more energy consuming, therefore, it is used only when crossing over obstacles.

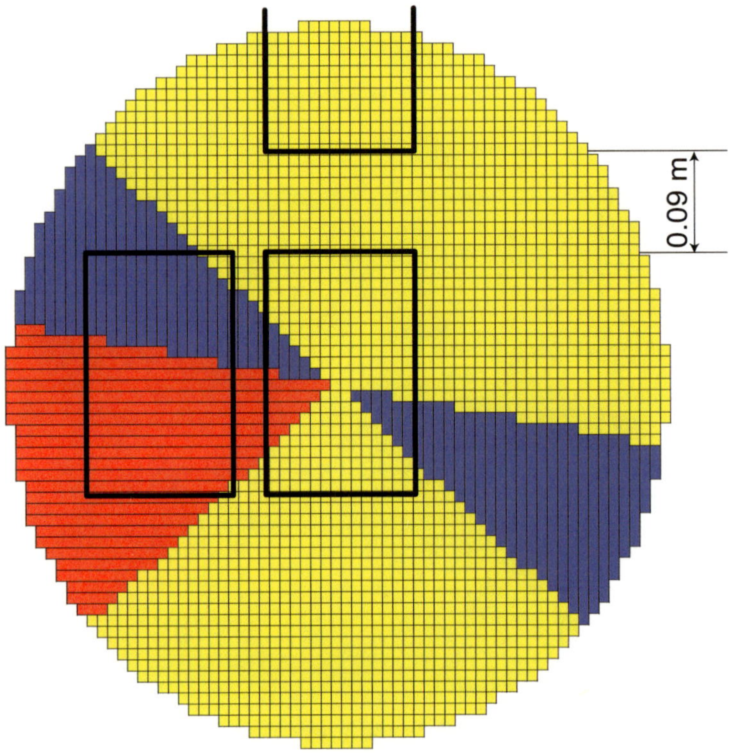

Figure 4. Enhanced stable stepping area. Right foot is extended to its limit

4.3.2 Foot Trajectory

There exist strategies to compute optimum trajectory for the non-supporting foot while stepping over obstacles located perpendicular to the robot's line of motion (Guan et al., 2005; Guan et al., 2006). However, conducting such analysis every time obstacles of different sizes are encountered from different angles would be computationally extensive (Kuffner et al., 2001). On the other hand, if in one step even if the foot is lifted to a height slightly more than minimally required to step over a certain obstacle, it would not amount to much additional energy consumption while making the computation significantly faster. The strategy we adopt, thus, is to fix a trajectory for stepping over obstacles as long as the obstacle satisfies certain dimensional constraints. The foot trajectory we use is shown in Fig. 5.
We assign ankle joints the job of keeping the foot parallel to the ground at all times. Thus, the whole part of the leg below the ankle top remains pointed straight downwards during the stepping motion. Any obstacle with height below the ankle top (0.106 m in case of Saika-3) would thus, not cause a collision while stepping over using this enhanced stepping trajectory if it is small enough to fit into the gap between current and next footstep locations as traced out in the previous subsection. Thus for Saika-3, obstacles fitting within the 0.09 m

x 0.1 m dimensional constraint are marked as *scalable by stepping over*. A point to be noticed here is that the limitation of 0.1 m on the height of the scalable obstacle is there because of the simple strategy employed for stepping over. Surely, for future work, by using more sophisticated stepping over trajectories it is possible to relax this limit.

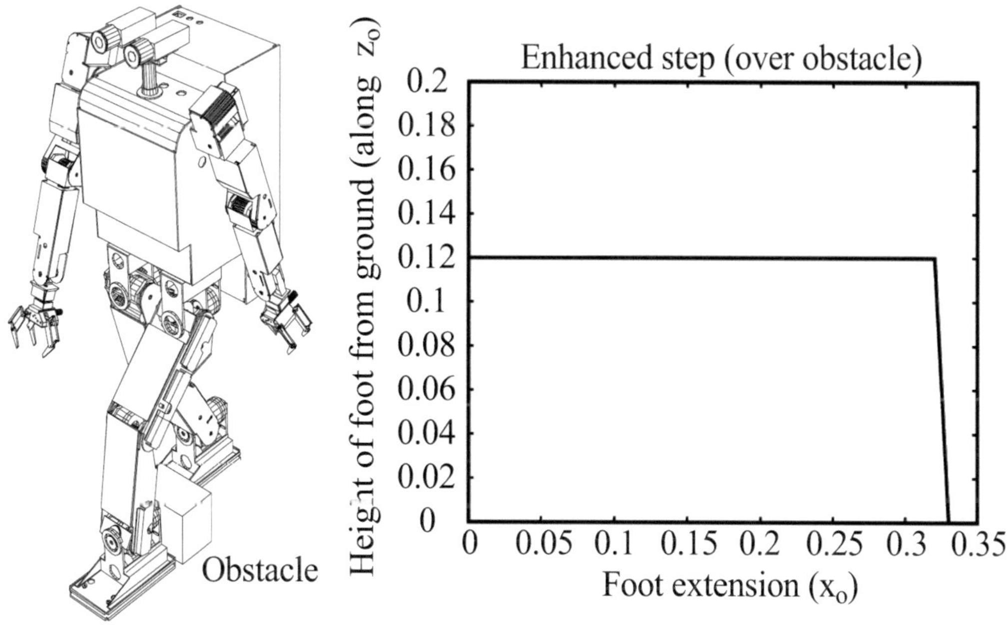

Figure 5. Stepping over an obstacle

4.3.3 Tilted Obstacle

While traversing an obstacle cluttered terrain, the robot might approach obstacles placed at orientations other than directly perpendicular to its line of motion. In this case, where the foot trajectory becomes somewhat complex, the limit on the allowable width of the obstacle possible to be stepped over might be somewhat relaxed (however, we do not make use of this relaxation in our simulations currently). This is achieved by hovering one foot over the obstacle while stepping over instead of placing it down at the mean position (Fig. 6 (c)).

As is evident from Fig. 6, the strategy is for the robot to stop and balance itself in double support state one step short of (i.e. at a distance of *Length of one step +Length of Foot/2* before) reaching the point of intersection between the obstacle's near side and the robot's line of motion (Fig. 6 (a)). Next, if the obstacle's tilt is rightwards compared with the direction of motion, the left foot of the robot is extended one step forward (Fig. 6 (b)) or vice versa. Balanced on this foot now, the robot lifts the right foot to a height of 0.12 m (as for the trajectory in Fig. 5) and then brings it to the mean position. At this point, the foot, while continuously lifted up from the ground, is turned to cross over the obstacle such that it is oriented perpendicular to the obstacle's away side. Extending the foot forward using a similar trajectory as in Fig. 5, only now at the newly acquired orientation, brings the robot to the state shown by Fig. 6 (c). Balancing itself on the right foot now, the left foot is also lifted and then turned to the same orientation as the right, followed by achievement of double support state as shown in Fig. 6 (d) by moving the left foot over the obstacle as well to complete the stepping over process.

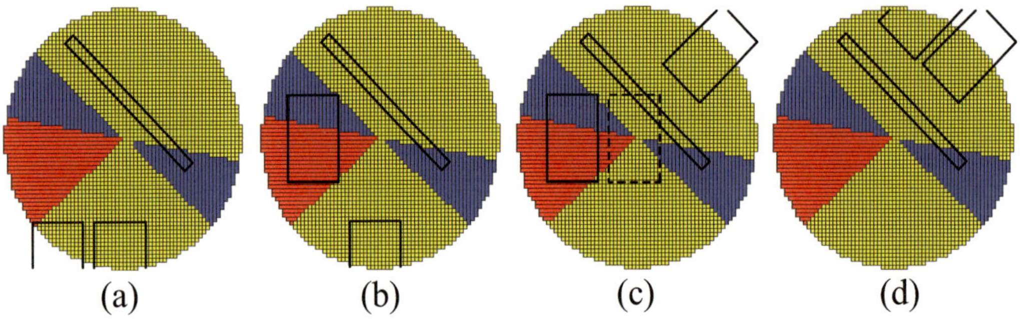

Figure 6. Stepping over an obstacle placed at an angle to the direction of motion.
(a) Stopping one step short of the obstacle to achieve enhanced stepping posture.
(b) Extending foot to balance on while stepping over based upon obstacle's tilt.
(c) Moving one foot over the obstacle. Dotted impression shows foot while lifted.
(d) Moving other foot over the obstacle to complete stepping over.

5. Footstep Planning

Our algorithm inherits assumptions (i) and (iii) from the game theory based footstep planner described in section 2. The aim, as described in section 3, is to follow a human-like strategy for footstep planning moving from simpler to more complex paths while only considering those obstacles that necessarily hinder the robot's way during graph expansion.

5.1 Obstacle Classification

In human environments, obstacles are of a variety of shapes and sizes. Our algorithm requires identification of a box-like boundary around each obstacle such that the obstacle of arbitrary shape is entirely contained inside it. This box-like boundary must be a four cornered shape when viewed from the top but there is no restriction on it being a rectangle or a parallelogram as long as it is a four cornered shape in top view. However, stepping over is only attempted in case the width and height of the boundary that binds the obstacle satisfy the constraints described in Section 4.3.2. Our algorithm at this stage considers the entire area inside this boundary in which the obstacle of arbitrary shape is contained as an obstacle. These obstacles are classified into three groups:

Type-1: Height ≤ 0.1 m	Can be crossed by stepping over if width ≤ 0.09 m.
Type-2: 0.1 m < Height < 0.6 m	Cannot be crossed by stepping over but while dodging it if the foot does not strike it, nothing else will either.
Type-3: Height ≥ 0.6 m	Cannot be crossed by stepping over and while dodging we have to make sure the robot's arms (which stick out wider than its feet) do not collide with it. The 0.6 m height is the distance of the ground surface from the lowest point of Saika-3's arms. This distance will obviously vary from one humanoid robot to another.

5.2 Collision Detection

Extending lines from the current body-corners of the robot to the to-be body-corner locations at the goal position, a gallery is formed. Using two of these boundary-lines and more drawn between corresponding current and destined points, we form a mesh. If any of these line segments is found intersecting a side of an obstacle (marked by a line between two of its neighboring corners), we say that a collision has been detected. Then, based upon distance from intersection point and angles made from current position, near and far sides and left and right corners of the obstacle are identified respectively.

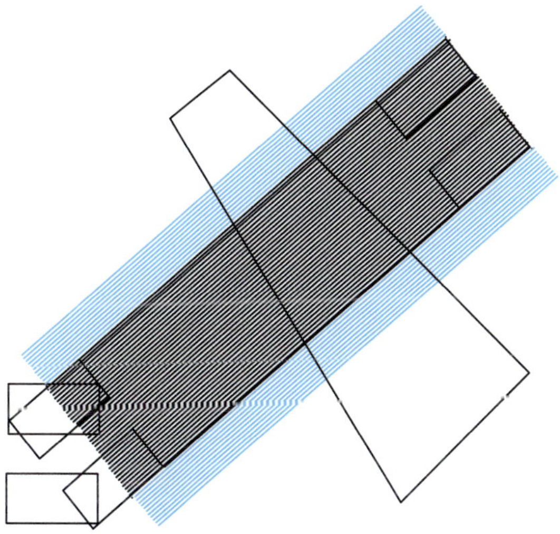

Figure 7. Mesh of lines drawn for collision checking

The collision detection strategy can be understood more easily using Fig. 7. As explained in section 4.2, due to the nonholonomic nature of a humanoid robot, the location of its body center changes while turning. Thus, we have to obtain its intended direction of motion in order to reach the destination using the foot on which it will turn instead of measuring it with respect to the body center. Using this angle of the intended line of motion, an imaginary set of initial footstep locations is generated at the current position which indicates the initial position of the robot after it will orientate itself towards the destination. In this direction of motion, another set of imaginary footstep locations is generated at the destination point, which marks the locations at which the robot will place its feet upon reaching the destination. Joining the outer boundaries of these initial and final imaginary footstep locations, a gallery is formed. Inside this gallery, a mesh of equidistant line segments joining corresponding points at the initial and final locations is generated. Each of these lines is checked for intersection with each of the walls of all obstacles in the environment (which are also marked by line segments when looked at from the top). If an intersection is found, a collision is said to have been detected and the robot seeks to form trajectories in order to overcome this obstacle in the path. This type of collision detection is performed not only between the starting point and the ultimate destination, but between every two points which mark the initial and final locations of every intermediate path the robot considers traversing.

From Fig. 7 it can also be seen that a part of the mesh of lines extends beyond the outer boundary of the footholds (drawn in blue). These are drawn by joining the outer ends of the

arms of the robot (which stick out wider than its feet) at initial and goal locations. Only if an obstacle of type-3 is found to be colliding with a line in this outer part of the mesh (between the feet and the arms), a collision is said to have been detected.
With respect to collision detection, two kinds of simulation results are presented in this chapter. In one, the mesh comprises only 3 lines (two joining body ends and one joining body center at initial and goal locations). In the other, we form a mesh of 27 lines with a distance of less than 2 cm between every two neighboring line segments.

5.3 Overcoming an Obstacle

To dodge an obstacle from a side we choose *pivot points* near the obstacle corners as shown in Fig. 8.

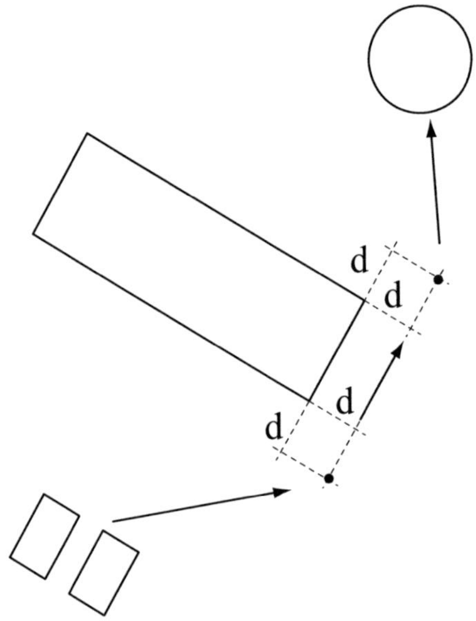

Figure 8. Choosing *pivots* to dodge an obstacle from right

The distance 'd' along the extended side is chosen such that no part of the robot's body collides with the obstacle as the robot stands in double support phase with its body centre at the pivot point. For instance in case of type-1 and type-2 obstacles, this distance is equal to half the length of the diagonal of the rectangular support polygon formed when the robot stands in the double support phase. This is because the outer most edges of the feet are the points closest to the obstacle with which there might be a chance of collision. 'd' can thus be different for each obstacle type but not for obstacles of one group. It is obviously greater for obstacles of type-3 since there is a possibility of collision with the robot's upper body.
As explained in section 4.3, to step over an obstacle, the robot balances itself on both legs one step short of the closest step-location near the obstacle. Next, based on rightward or leftward tilt of the obstacle in relevance with the robot's trajectory, it places forward left or right foot respectively and balances itself on it. Using enhanced stepping motion, the robot now takes a step forward with its movable leg, making sure that the extended foot lands with its back-side parallel to the obstacle's away-side. Fig. 9 displays the trajectories taken to overcome an obstacle of type-1.

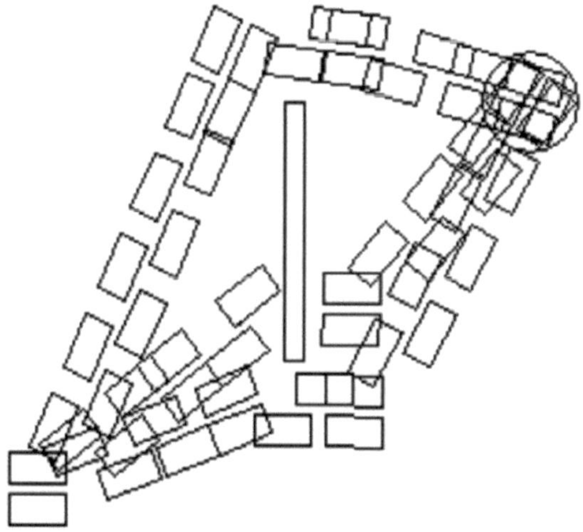

Figure 9. Overcoming an obstacle of type-1

5.4 Local Loops and Deadlocks

Each obstacle in the surroundings is allotted an identification number. The planner keeps a history of the obstacles it has overcome in a path as it plans footsteps. Whenever an obstacle occurs twice, it indicates a local loop or deadlock since attempting to overcome it again is bound to bring the planner back to the same position again and again. Such a trajectory is immediately abandoned and pruned from the search tree.

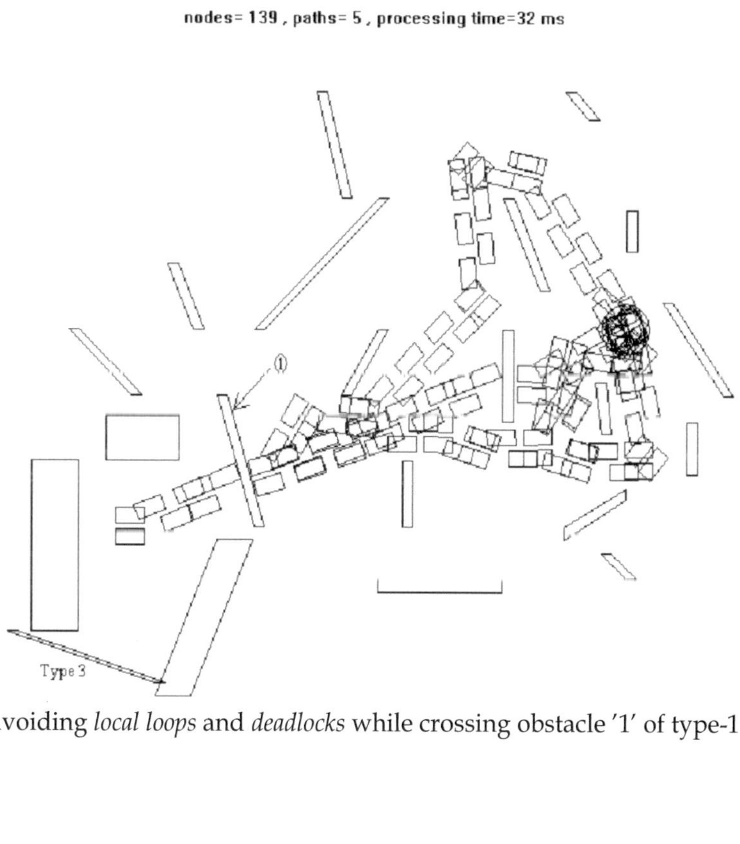

Figure 10. Avoiding *local loops* and *deadlocks* while crossing obstacle '1' of type-1

One such situation where the robot encounters a local loop and deadlocks while trying to dodge the obstacle labelled '1' from both right and left sides is shown in Fig. 10. For instance, when trying to dodge the obstacle labelled '1' in Fig. 10 from the right, the robot chooses pivot points to pass from its right side as elaborated upon in section 5.3. However, this path is obstructed by another obstacle on the robot's right. To dodge this newly encountered obstacle, once again the robot forms trajectories from its right and left. The one attempted to pass from left finds the obstacle labelled '1' in the way again. Since this obstacle is present in the history as one that has already been attempted to be dodged, the trajectory for dodging the newly encountered obstacle from the left is discarded as a situation where a *deadlock* is encountered. The trajectory formed to dodge it from the right, however, finds another obstacle (indicated as a type-3 obstacle in Fig. 10). Attempting to dodge this type-3 obstacle from the left results in a deadlock just as in the previous case whereas the one attempted from its right encounters yet another obstacle. This process is repeated twice more until, in an effort to dodge from the right the obstacle to the left of the obstacle labelled '1', the obstacle labelled '1' is encountered again. This trajectory, therefore, is also abandoned and a *local loop* is said to have been encountered.

A similar situation occurs while trying to dodge the obstacle labelled '1' in Fig. 10 from its left side. Thus the only trajectory possible to overcome this obstacle which is free of local loops and deadlocks is the one formed by stepping over. As we can see, the planner only plans the successful trajectory, avoiding getting stuck in local loops and deadlocks.

5.5 Cost Assignment

A heuristic cost based on the order of complexity of each stepping motion is given at Table 1. These costs are assigned to foot placements of each type as they are planned.

Step type	Cost
Straight	1
Turning	2
Extended	3
Enhanced	4

Table 1. Heuristic costs assigned to steps of different types

5.6 Intermediate Paths

If, in order to proceed towards a pivot point while dodging an obstacle from its side, another obstacle is encountered along the way, alternate paths based upon the obstacle type are formed. All these alternate paths converge at the pivot point, meaning that they will all have similar descendants. Thus, the number of these intermediate alternate paths is multiplied by the number of descendent paths and added to the total number of possible alternate paths. Thus, evaluating cost of each intermediate path and keeping only the best during alternate path creation reduces the number of paths to only useful ones.

5.7 Search for Best Path

In order to identify the best one of the paths formed by our algorithm, we employ a depth first exhaustive search since greedy or A* search would not filter out for us the best path in every scenario (Cormen, 1994).

6. Results

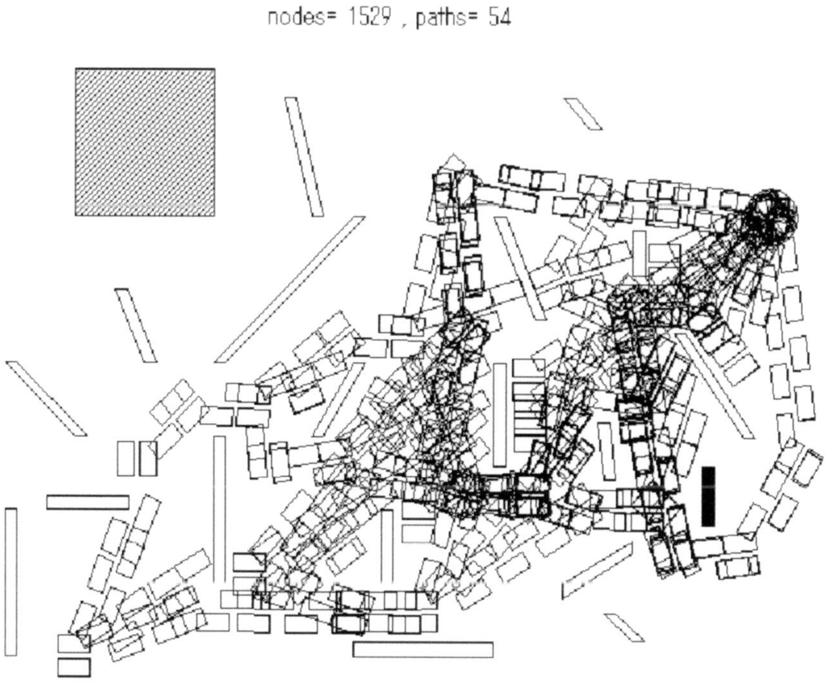

Figure 11. Various paths for reaching the destination in an obstacle cluttered environment

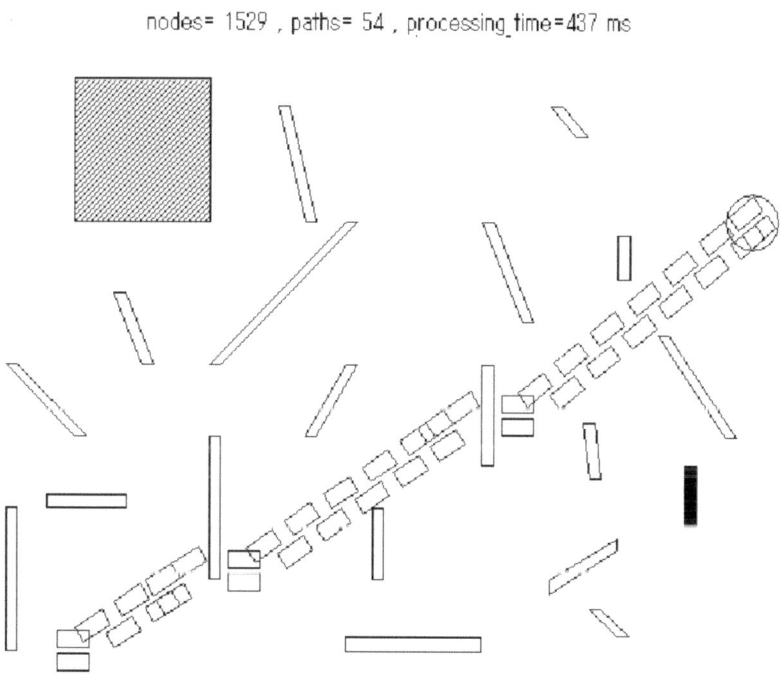

Figure 12. Best path determined using depth first exhaustive search

Figs. 11 and 12 show results of a simulation of Saika-3 planning footsteps in a room with 20 obstacles. We see that a graph of only 1,529 nodes is formed consisting of 54 paths all reaching the destination. The whole process takes only 437 ms using a mesh of 3 lines (for collision detection) and 864 ms using a mesh of 27 lines on a 1.86 GHz Pentium-IV Centrino (1.5 GB RAM) running Windows XP. A comparison with previous techniques reviewed in section 2 shows reduction in the number of nodes as well as in processing time even though the algorithm employs exhaustive search for hunting out best path which guarantees optimality of the chosen path among those traced out by the algorithm.
Some more simulation results are given in Figs. 13-15. We see that the fastest result with the extensive collision checking strategy (27 line mesh) is obtained in Fig. 15. The reason for this is the very small branching that has taken place in the graph. This also reduces the time taken in search for best path. Also, Fig. 14 shows the trajectories formed around the obstacle of type-3 (hatched). It is readily observed that the robot keeps a greater distance with this obstacle than with obstacles of other types.

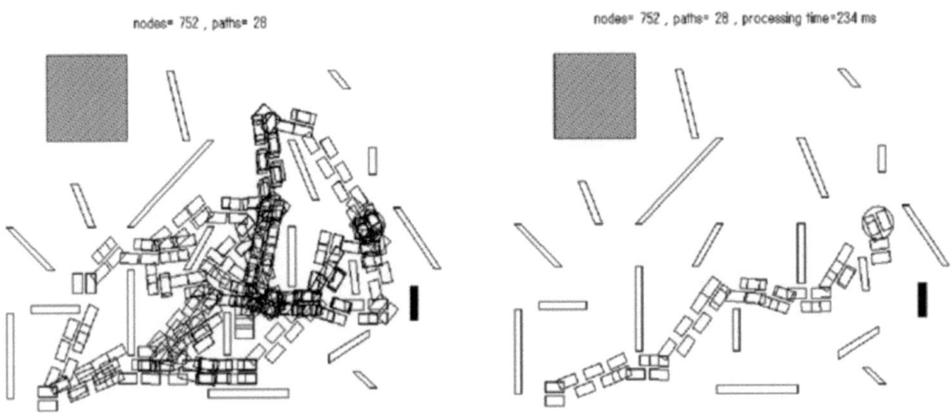

(a) Alternate paths (b) Best path

Figure 13. Result 2: Nodes formed 752, paths 28, time taken 234 ms (3 line mesh)

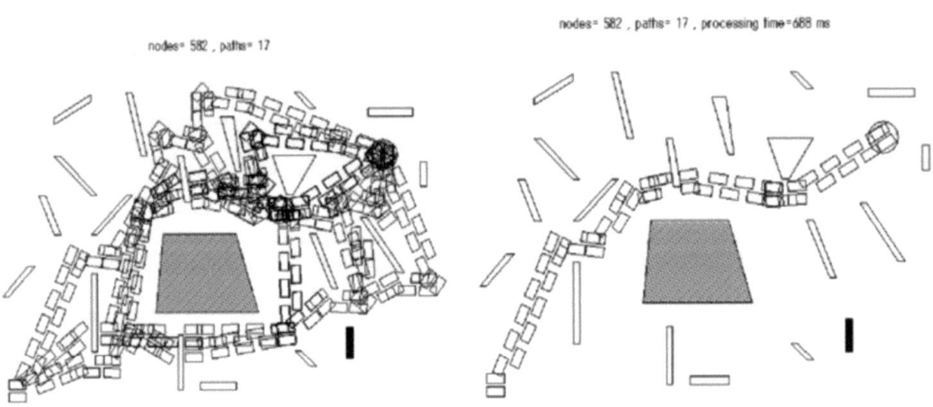

(a) Alternate paths (b) Best path

Figure 14. Result 3: Nodes formed 582, paths 17, time taken 688 ms (27 line mesh)

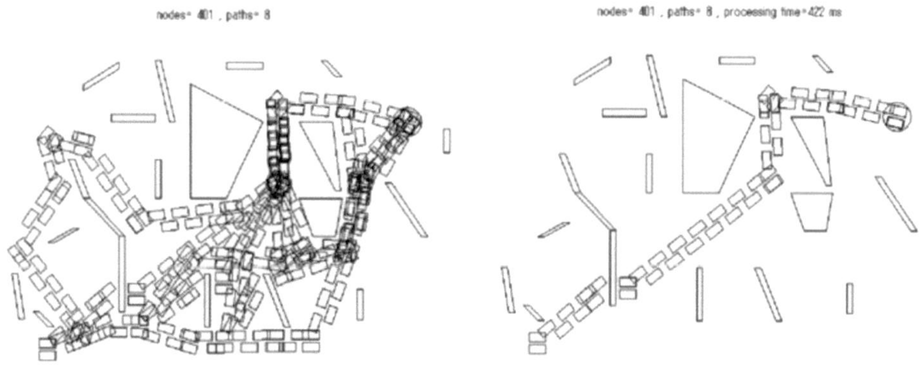

(a) Alternate paths (b) Best path

Figure 15. Result 4: Nodes formed 401, paths 8, time taken 422 ms (27 line mesh)

7. Conclusion

Our algorithm successfully demonstrates a novel global reactive footstep planning strategy with a human-like approach. Incremental graph expansion from simpler to more complex paths ensures formation of a simpler and more useful graph as compared to that formed by approaches such as the visibility graph. The trajectory generated is more energy-efficient since the robot does not have to lift its foot to a high location in every step as in case of game theory based approaches. The algorithm is considerably fast and reduces computational complexity by minimizing the number of alternate steps considered after planning each step. However, basing the cost of each step on energy or time optimisation criteria instead of just the complexity level of the stepping motion can further improve the performance of the algorithm. Possibilities for future work include implementation on real humanoid robot and incorporation of measures for dealing with dynamic obstacles in the environment. In addition, identification of alternate paths by considering postural changes also seems interesting e.g. constraints on many obstacles of type-3 can be reduced to those for type-2 if the robot only lifts up its arms.

8. References

Ayaz, Y.; Afzal, B.; Saeed, M. & Rehman, S.U. (2004). Design, Fabrication and Control of a Two-Legged Walking Robot, *Proceedings of IEEE Int. Workshop. on Robot Motion & Control (RoMoCo)*, pp. 73-78

Ayaz, Y.; Munawar, K.; Malik, M.B. ; Konno, A. & Uchiyama, M. (2006). Human-Like Approach to Footstep Planning Among Obstacles for Humanoid Robots, *Proc. of IEEE/RSJ Int. Conf. on Intelligent Robots and Systems (IROS)*, pp. 5490-5495

Chestnutt, J. & Kuffner, J.J. (2004). A Tiered Planning Strategy for Biped Navigation, *Proceedings of IEEE Int. Conf. on Humanoid Robotics (ICHR)*

Chestnutt, J.; Lau, M.; Cheung, C.; Kuffner, J.J.; Hodgins, J. & Kanade, T. (2005). Footstep Planning for the Honda Asimo Humanoid, *Proceedings of IEEE Int. Conf. on Robotics & Automation (ICRA)*, pp. 629-634

Cormen, T.H.; Leiserson, C.E. & Rivest, R.L. (1994). *Introduction to Algorithms,* McGraw-Hill Book Company

Guan, Y.; Neo, E.S.; Yokoi, K. & Tanie, K. (2005). Motion Planning for Humanoid Robots Stepping over Obstacles, *Proceedings of IEEE/RSJ Int. Conf. on Intelligent Robots and Systems (IROS)*, pp. 364-370

Guan, Y.; Neo, E.S.; Yokoi, K. & Tanie, K. (2006). Stepping Over Obstacles With Humanoid Robots. *IEEE Trans. on Robotics,* Vol. 22, No. 5, pp. 958-973

Konno, A.; Kato, N.; Shirata, S.; Furuta, T. & Uchiyama, M. (2000). Development of a Light-Weight Biped Humanoid Robot, *Proceedings of IEEE/RSJ Int. Conf. on Intelligent Robots and Systems (IROS)*, pp. 1565-1570

Kuffner, J.J.; Nishiwaki, K.; Kagami, S.; Inaba, M. & Inoue, H. (2001). Footstep Planning Among Obstacles for Biped Robots, *Proceedings of IEEE/RSJ Int. Conf. on Intelligent Robots and Systems (IROS)*, pp. 500-505

Kuffner, J.J.; Nishiwaki, K.; Kagami, S.; Inaba, M. & Inoue, H. (2003). Online Footstep Planning for Humanoid Robots, *Proceedings of IEEE/RSJ Int. Conf. on Robotics and Automation (ICRA)*, pp. 932-937

Kuffner, J.J.; Nishiwaki, K.; Kagami, S.; Inaba, M. & Inoue, H. (2005). Motion Planning for Humanoid Robots. *Robotics Research, Paolo Dario and Raja Chatila (Eds.), Springer Tracts in Advanced Robotics,* Vol.15, pp. 365-374

Latombe, J.C. (1991). *Robot Motion Planning,* Kluwer Academic Publishers, Boston, MA

McGhee, R.B. & Iswandhi, G.I. (1979). Adaptive Locomotion of a Multi-legged Robot over Rough Terrain. *IEEE Trans. on Systems, Man and Cybernetics,* Vol. SMC-9, No. 4, pp. 176-182

Michel, P.; Chestnutt, J.; Kuffner, J.J. & Kanade, T. (2005). Vision-Guided Humanoid Footstep Planning for Dynamic Environments, *Proceedings of IEEE/RAS Int. Conf. on Humanoid Robotics (ICHR)*, pp. 13-18

Michel, P.; Chestnutt, J.; Kagami, S.; Nishiwaki, K.; Kuffner, J.J. & Kanade, T. (2006). Online Environment Reconstruction for Biped Navigation, *Proceedings of IEEE Int. Conf. on Robotics & Automation (ICRA)*, pp. 3089-3094

Vukobratović, M. & Borovac, B. (2004). Zero Moment Point---35 Years of Its Life. *International Journal of Humanoid Robotics, World Scientific Publishing Company,* Vol. 1, No. 1, pp. 157-173

Yagi, M. & Lumelsky, V. (1999). Biped Robot Locomotion in Scenes with Unknown Obstacles, *Proceedings of IEEE Int. Conf. on Robotics & Automation (ICRA)*, pp. 375-380

16

Mixed Logic Dynamical Modeling and On Line Optimal Control of Biped Robot

Yingjie Yin and Shigeyuki Hosoe
Bio-mimetic Control Research Center, the Institute of Physical and Chemical Research
Japan

1. Introduction

Walking control of biped robots is classically performed in two steps: (i) the synthesis of walking patterns (Chevallereau et al., 1998) and (ii) the control of the robot to track the prescribed trajectory (Chew & Pratt, 2000; Park & Chung, 2000). The problems of these approaches are that the movements are non-adaptable to environment, and unexpected events are not pre-computable. In this meaning, despite the huge amount of works on the topic, biped control is still inefficient especially in terms of adaptability, robustness and stability. Contrast with this, human being decides his path on line adaptively according to the environment rather than predicting it a very long time ahead. A consideration inspired by human behavior is to introduce the path changes on line while guarantee stability and motion in environment. The automation of biped robots is very difficult but challenging.

The control problem of on line walking adaptation has been studied by some researchers. For instance, the zero moment point (ZMP) was controlled for preserving the dynamic equilibrium (Park & Chung, 1999). However, this adaptation can only involve small changes of the original trajectory. Different with this, a trajectory better adapted to the encountered situation is chosen on line between a set of stored trajectories (Denk & Schmidt, 2001). The correspondent problem is that switching from one trajectory to another may lead to unexpected effects in control. To cope with this problem, a continuous set of parameterized trajectories was used as the candidate of choice (Chevallereau & Sardain, 2000). By it the switches were suppressed, but the set of trajectories has to be defined beforehand.

Different with the previous considerations, a notable approach is to adopt the model predictive control (MPC) for walking adaptation (Azevedo et al., 2004; Kooij et al., 2003). By optimizing on line the joint motion over a receding horizon, biped robots have been controlled without pre-computed reference trajectory and switches at the algorithm level. The walking can auto-adapts to environment changes. However, in (Kooij et al., 2003), accompanying with the approaching of collision, the length of the optimization horizon is shortened step by step within the single support motion phase and finally the landing motion is controlled with only one sampling period optimization. The intendment of the authors is to avoid the difficulty of control when impact phenomenon occurs within the optimization horizon. But consequently, the adaptability to environment is weakened. In (Azevedo et al., 2004), how to predict the occurrence of impact then compensate positively for the effect of impact by MPC is not stated clearly.

A model of biped walker should encapsulate phases of continuous motion, switching between types of motion, and occurrence of impacts. The overall system is naturally a hybrid one. In this research, we provide a unified modeling framework for the biped motion from the hybrid system point of view, model the biped movement as a mixed logic dynamical (MLD) system by emphasizing impact and its effect on the walking dynamics. Based on the MLD model, we adapt MPC approach to the on line walking control of biped robot. The MPC of a MLD system can be solved using powerful mixed integer quadric programming (MIQP) algorithm. Its solution corresponds to the objective-oriented optimization of the gaits. Simulation results validate the effectiveness of our proposal.

2. Modeling of planar biped robots

The robot is comprised of rigid links, connected by revolute joints shown in Fig.1. The model considered is a planar open kinematic chain connected at the hip joint, comprising two identical legs. Torque is applied at hip joint and between the stance leg and the support foot.

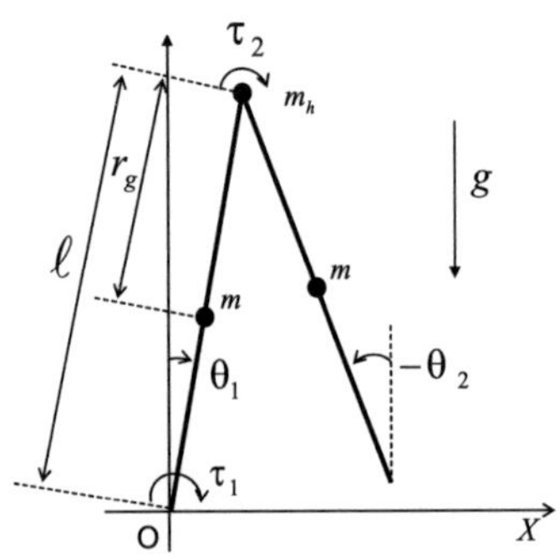

Figure 1. Coordinate of a planar biped robot. X is the progressing direction. The vertical axis is in a parallel but opposite direction with gravitaty g. θ_1, θ_2 is the angle of link 1, link 2 from the vertical axis. τ_1 and τ_2 are the torque for driving joint 1 and joint 2. m_h is the mass of hip, m is the mass of link 1 and 2, ℓ is the length of link 1 and 2, r_g is the distance between hip and the center of mass of each link, respectively

Motions are assumed to take place in the sagittal plane, consist of successive phases of single support and collision event, without slip. The dynamics of feet is neglected. The two phases of the walking cycle naturally lead to a mathematical model of the biped walker consisting of two parts: the differential equation describing the dynamics during the single support phase, and a velocity discontinuity caused of the impact phase. The model equations and its basic features are describing in the following.

2.1 Dynamic equation of swing phase

In the single support phase, one leg is pivoted to the ground while the other is swinging in the forward direction. The dynamic equation of the biped robot during the swing phase can be derived to be

$$M(\theta)\ddot{\theta} + C(\theta,\dot{\theta})\dot{\theta} + g(\theta) = \tau, \tag{1}$$

where $\theta = [\theta_1 \ \theta_2]^T$ is the joint angular vector,

$$M(\theta) = \begin{bmatrix} m_h\ell^2 + m\ell^2 + m(\ell - r_g)^2 & -m\ell\, r_g \cos(\theta_1 - \theta_2) \\ -m\ell\, r_g \cos(\theta_1 - \theta_2) & m\, r_g^2 \end{bmatrix},$$

$$C(\theta, \dot{\theta}) = \begin{bmatrix} 0 & -m\ell\, r_g \sin(\theta_1 - \theta_2)\dot{\theta}_2 \\ m\ell\, r_g \sin(\theta_1 - \theta_2)\dot{\theta}_1 & 0 \end{bmatrix},$$

$$g(\theta) = \begin{bmatrix} -(m_h\ell + m(2\ell - r_g))g_0 \sin\theta_1 \\ m\, r_g\, g_0 \sin\theta_2 \end{bmatrix}.$$

The parameter g_0 is the gravitational acceleration.
In general, the control variable is bounded because of the power limit of actuator,

$$m_\tau \le \tau \le M_\tau, \tag{2}$$

where m_τ, M_τ are the lower bound, the upper bound of τ, respectively.

2.2 Impact model

For simplicity, the collision between the swing leg and the ground is assumed to be inelastic. The contact phenomenon produces two simultaneous events: 1) impact, which causes a discontinuity in the joint angular velocities, with the configuration remaining continuous, and 2) switching due to the transfer of pivot to the point of contact.
Following the law of conservation of angular momentum, we obtain a transition equation of the angular velocity as

$$\dot{\theta}^+ = \lambda(\alpha)\dot{\theta}^-, \tag{3}$$

where $\dot{\theta}^+, \dot{\theta}^-$ denote the angular velocity of post- impact and pre-impact, respectively. α is the half inter-leg angle at the transition instant,

$$\alpha = \frac{\theta_1^- - \theta_2^-}{2} = \frac{\theta_2^+ - \theta_1^+}{2} > 0.$$

The reduction ratio of velocity of the motion transmission, $\lambda(\alpha)$, is

$$\lambda(\alpha) = \left(Q^+(\alpha)\right)^{-1} Q^-(\alpha),$$

$$Q^+(\alpha) = \begin{bmatrix} (m_h + m)\ell^2 + m(\ell - r_g)^2 - m\ell\, r_g \cos(2\alpha) & m\ell^2 - m\ell\, r_g \cos(2\alpha) \\ -m\ell\, r_g \cos(2\alpha) & m\, r_g^2 \end{bmatrix},$$

$$Q^-(\alpha) = \begin{bmatrix} (m_h\ell^2 + 2m\ell(\ell - r_g))\cos(2\alpha) - m\, r_g(\ell - r_g) & -m\, r_g(\ell - r_g) \\ -m\, r_g(\ell - r_g) & 0 \end{bmatrix}.$$

The switching of the pivot to the point of contact is described as

$$\theta^{+} = S\theta^{-}, \qquad S = \begin{bmatrix} 0 & 1 \\ 1 & 0 \end{bmatrix}. \tag{4}$$

The transition from one leg to another is assumed to take place in an infinitesimal length of time. Hence there exists no double support phase.

2.3 A mixed logic dynamical model

The model of a biped walker is hybrid in nature, consisting of continuous dynamics (single support phase) separated by abrupt changes (impact phase) of velocity. How to describe the hybrid system is very important which affects directly the synthesis of biped motion and the control performance.

For a general hybrid dynamical system, one may have several points of view: continuous-time, discrete-event, or mixed. Whether one of these manners applies better than the others depends a lot on the task. For the biped walking problem, we stress the walking motion transmissions, the adaptability to environment, and the robustness to unexpected disturbance. For that, we consider both continuous and discrete dynamics simultaneously including switching between different motion phases, express all the motions in a unified framework, which will allow the synthesis of the walking system in a systematic way.

Define the state variable vector as

$$x = \begin{bmatrix} \theta \\ \dot{\theta} \end{bmatrix}. \tag{5}$$

Then, the dynamical equation of single support phase (1) becomes

$$\dot{x} = A_c x + B_c(t)\tau + b(t) \tag{6}$$

where

$$A_c = \begin{bmatrix} 0 & 1 \\ 0 & 0 \end{bmatrix}, \quad B_c(t) = \begin{bmatrix} 0 \\ M^{-1}(\theta) \end{bmatrix}, \quad b(t) = \begin{bmatrix} 0 \\ -M^{-1}(\theta)(C(\theta,\dot{\theta})\dot{\theta} + g(\theta)) \end{bmatrix}.$$

The transition equations of the angular velocity and joint angle, (3) and (4), become

$$x^{+} = R(\alpha)x^{-} \tag{7}$$

Where x^{-} is the pre-impact state, x^{+} is the post-impact state, and

$$R(\alpha) = \begin{bmatrix} S \\ \lambda(\alpha) \end{bmatrix}.$$

After impact, the biped robot progresses with switched pivot and its state evolves along the dynamical equation (6) again from the reset post-impact state of (7). The right side time derivation of the post-impact state satisfies the following relation.

$$\begin{aligned} \dot{x}^{+} &= A_c x^{+} + B_c(t^{+})\tau + b(t^{+}) \\ &= A_c R(\alpha)x^{-} + B_c(t^{-})\tau + \hat{b}(t^{-}), \end{aligned} \tag{8}$$

where

$$B_c(t^+) = B_c(t^-), \qquad \hat{b}(t^-) = \begin{bmatrix} 0 \\ -M^{-1}(2\alpha)(C(S\theta^-, \lambda\dot{\theta}^-)\lambda\dot{\theta}^- + g(S\theta^-)) \end{bmatrix}.$$

Here we introduce an auxiliary logic variable δ_c to associate the event of collision in order to describe (6) and (8) in a unified equation.

$$\{\text{ impact occurring }\} \leftrightarrow \{\delta_c = 1\}. \tag{9}$$

Then for both the single support phase and the impact phase, the state of the biped robot evolves monotonically along the following unified dynamical equation,

$$\dot{x} = A_c x + B_c(t)\tau + A_c(R(\alpha) - I)z + (\hat{b}(t) - b(t))\delta_c + b(t), \tag{10}$$

where z is the auxiliary variable defined as

$$z = x\delta_c . \tag{11}$$

It is remarkable that (11) can be equivalently expressed by a set of inequalities (Bemporad & Morari, 1999),

$$\begin{aligned} 0_{4\times1} \le z \le \delta_c \max(x), \\ x - (1-\delta_c)\max(x) \le z \le x. \end{aligned} \tag{12}$$

In practice, collision can be measured experimentally or predicted mathematically according to the toe position of swing leg. For a robot with two identical legs walking on smooth plane, a collision occurs if the joint angles satisfy the relation of $\theta_1 = -\theta_2$. If $\theta_1 + \theta_2 > 0$, it is the case of an interpenetration to the ground. For the convenience of numerical calculation, the condition of collision is practically represented as

$$0 \le \theta_1 + \theta_2 \le \varepsilon_2,$$

where ε_2 is arbitrary small positive constant. The right side of the inequality is added to avoid taking deep interpenetration as collision.

In addition, the erected posture of robot, i.e. the case of $\theta_1 = \theta_2 = 0$, has to be distinguished from collisions. Since the point of collision is always in front of the pivot in a nature walking, we set another condition for evaluating the occurrence of collision,

$$\theta_1 > 0.$$

Concluding the previous two conditions, one can see that for a robot with two identical legs walking on smooth plane, the swing leg is in collision with the plane if the following linear inequalities are satisfied,

$$\begin{aligned} \varepsilon_1 \le \theta_1, \\ 0 \le \theta_1 + \theta_2 \le \varepsilon_2. \end{aligned} \tag{13}$$

where $\varepsilon_1 > 0$.

By using (13), we have more concrete expression for (9),

$$\{\ \{\theta_1 \ge \varepsilon_1\} \wedge \{\theta_1 + \theta_2 \le \varepsilon_2\} \wedge \{\theta_1 + \theta_2 \le 0\}\ \} \leftrightarrow \{\delta_c = 1\}. \tag{14}$$

The Boolean relation of (14) can be equivalently expressed by a set of linear inequalities. For that, we induce new logical variables $\delta_1, \delta_2, \delta_3$, which are associated with the following relations,

$$\{\theta_1 \ge \varepsilon_1\} \leftrightarrow \{\delta_1 = 1\}, \tag{15}$$

$$\{\theta_1 + \theta_2 \le \varepsilon_2\} \leftrightarrow \{\delta_2 = 1\}, \tag{16}$$

$$\{\theta_1 + \theta_2 \ge 0\} \leftrightarrow \{\delta_3 = 1\}. \tag{17}$$

Consequently, we obtain

$$\delta_c = \delta_1 \delta_2 \delta_3 \ .$$

According to the known knowledge (Bemporad & Morari, 1999), the equivalent inequalities of (15) are

$$\begin{aligned} \theta_1 &\ge \min(\theta_1) - (\min(\theta_1) - \varepsilon_1)\delta_1 , \\ \theta_1 &\le \varepsilon_1 - \varepsilon_0 + (\max(\theta_1) - \varepsilon_1 + \varepsilon_0)\delta_1 , \end{aligned} \tag{18}$$

where ε_0 is arbitrary small positive constant. Similarly, the equivalent inequalities of (16) are

$$\begin{aligned} \theta_1 + \theta_2 &\le \varepsilon_2 + (\max(\theta_1 + \theta_2) - \varepsilon_2)(1 - \delta_2) , \\ \theta_1 + \theta_2 &\ge \varepsilon_2 + \varepsilon_0 + (\min(\theta_1 + \theta_2) - \varepsilon_2 - \varepsilon_0)\delta_2 . \end{aligned} \tag{19}$$

The equivalent inequalities of (17) are

$$\begin{aligned} \theta_1 + \theta_2 &\ge \min(\theta_1 + \theta_2)(1 - \delta_3) \\ \theta_1 + \theta_2 &\le -\varepsilon_0 + (\max(\theta_1 + \theta_2) + \varepsilon_0)\delta_3 \end{aligned} \tag{20}$$

It means that the joint angles θ_1 and θ_2 determinate the logical values of $\delta_1, \delta_2, \delta_3$ through the linear inequality set (18)-(20), thus the value of the impact flag $\delta_c = \delta_1\delta_2\delta_3$. However, the nonlinear relation $\delta_c = \delta_1\delta_2\delta_3$ requires special solver and complicates calculation. For this problem, we induce one more logical variable δ_4 for variable linearization,

$$\delta_4 = \delta_2 \delta_3. \tag{21}$$

By it the relation $\delta_c = \delta_1\delta_2\delta_3$ finally becomes

$$\delta_c = \delta_1 \delta_4. \tag{22}$$

The Boolean relation (21), (22) can be equivalently expressed by linear inequalities. For (21), the equivalent inequalities are

$$\begin{aligned} -\delta_2 + \delta_4 &\le 0, \\ -\delta_3 + \delta_4 &\le 0, \\ \delta_2 + \delta_3 - \delta_4 &\le 1. \end{aligned} \tag{23}$$

For (23), the equivalent inequalities are

$$\begin{aligned} &-\delta_1 + \delta_c \le 0, \\ &-\delta_4 + \delta_c \le 0, \\ &\delta_1 + \delta_4 - \delta_c \le 1. \end{aligned} \tag{24}$$

Concluding the previous discussion, one can see that we get the unified dynamical equation (10) for the hybrid biped motion, which is subject to constraints of (2), (12), (18)-(20), (23), and (24). In the context, (10) is rewritten as (25). The constraints (2), (12), (18)-(20), (23), and (24) are integrated into (26).

$$\dot{x} = A_c x + B_c(t)\tau + B_z(t)z + B_\delta(t)\delta + b(t), \tag{25}$$

$$E_1 x + E_2 \tau + E_3 z + E_4 \delta \le E_5 , \tag{26}$$

where

$$\delta = [\delta_c \ \delta_1 \ \delta_2 \ \delta_3 \ \delta_4]^T, \quad B_z(t) = A_c(R(\alpha) - I), \quad B_\delta(t) = [\hat{b}(t) - b(t) \ \ 0 \ \ 0 \ \ 0 \ \ 0].$$

The coefficient E_i of (26), $i = 1,\dots,5$, can be known from the context.

For computational convenience, (25) is generally put in discrete-time domain representation.,

$$x(k+1) = Ax(k) + B_1(k)\tau(k) + B_2(k)z(k) + B_3(k)\delta(k) + B_0(k), \tag{27}$$

where with the sampling period T_s,

$$A = e^{A_c T_s}, \qquad B_1(k) = \int_0^{T_s} e^{A_c(T_s - t)} dt \cdot B_c(k), \qquad B_2(k) = \int_0^{T_s} e^{A_c(T_s - t)} dt \cdot B_z(k),$$

$$B_3(k) = \int_0^{T_s} e^{A_s(T_s - t)} dt \cdot B_\delta(k), \quad B_0(k) = \int_0^{T_s} e^{A_s(T_s - t)} dt \cdot b(k) .$$

The MLD model of (25), (26) or (27), (26) describes both the continuous dynamics and the impact event within one framework, which provides the possibility of systematic synthesis and control for the biped motion.

3. Progressing constraint

A successful walking should be stable and successive progress forward. For that, conditions of stable and successive walking have to be taken into account as constraints subject to the optimal gait generation problem.

3.1 Erected body

For supporting the body against gravity, the position of hip should be above a positive level to avoid falling. For example, to achieve

$$\ell^* \le y_{hip} \le \ell,$$

where y_{hip} is the vertical position of hip, we can set

$$-\theta_1^* \le \theta_1 \le \theta_1^* \tag{28}$$

With $\theta^* = \arccos \ell^* / \ell.$.

3.2 Progression

For propel the body in the intended direction with a progression rhythm, horizontal velocity of hip should be kept positive, and strictly monotonically increased. The horizontal position of hip is

$$x_{hip} = \ell \sin \theta_1 .$$

Its time derivation is the horizontal velocity of hip, which should satisfy

$$\dot{x}_{hip} = \dot{\theta}_1 \ell \cos \theta_1 > 0 .$$

For that, we set the following constraint to ensure the forward progression of the body.

$$\dot{\theta}_1 \ge \varepsilon_v \tag{29}$$

where $\varepsilon_v > 0$.

3.3 Environment

For walking at non-smooth ground, we suppose that the environment at the next steps can be detected by sensor such as camera. Then the non-smooth walking surface can be mathematically expressed by a set of equalities, denoted as

$$y = \psi(x).$$

Note that in this case, the occurrence condition of collision, $y_{toe} = \psi(x)$, has to be derived to substitute for (13). Corresponding with this, within single support phase, the toe of the swing leg has to be controlled above the walking surface,

$$y_{toe} > \psi(x). \tag{30}$$

These control constraint (28)-(30) can be integrated together with (26) which is the constraint part of the MLD model of the biped robot. The integrated MLD model of (25), (26), (28)-(30) is the unified description for the biped motion, including both the physical modeling and the motional requirement.

4. Optimal control

The problem is to generate on line the biped motion without pre-defined trajectory, by the simultaneous synthesis and control of walking gait. The optimal gait to current environment consists of the continuous trajectory within each single support phase and the transition time of stance leg.

Basing on our derived MLD model, we handle the control objective by using a systematical method such as the model predictive control (MPC). Note that traditionally, MPC is used to solve the control problem of a constrained but continuous dynamical system. For the biped motion control, because of the existence of logical variables in the MLD model, the MPC of the MLD system is more complex. The idea is to carry out optimization of the biped motion in a receding finite horizon, which can be transformed to solve a mixed integer quadric programming (MIQP) problem (Bemporad and Morari, 1999). Its solution corresponds to the optimal walking gait of the biped robot.
The criterion for minimization is as the following.

$$L = \sum_{i=0}^{N-1} (\|x(k+i+1\mid k) - x_r\|_{Q_1} + \|\tau(k+i)\|_{Q_2} + \|z(k+i) - z_r\|_{Q_3} + \|\delta(k+i) - \delta_r\|_{Q_4}) \tag{31}$$

which is subject to the MLD model of (27), (26), (28)-(30).
In (31), the real number N is the horizon for each minimization. $x(k+i+1\mid k)$ is the state in the $i+1$ steps future of the current state $x(k)$, and can be predicted by the MLD model according to the applied torque sequence $\tau(k),\ldots,\tau(k+i)$. x_r is the desired state at pre-impact point, can be time invariant or time-varying. z_r is the corresponding auxiliary continuous variable of x_r. δ_r is the corresponding auxiliary logical variable of x_r. $Q_1,\ldots,Q_4$ are the weighting matrices.
The first term in (31) is for achieving a progressive walking. In the N steps optimization, the state before impact will be driven to approach the impact point x_r. The state post or after impact will be driven to access the next impact point x_r in order to keep progression. It implies that the criterion (31) and the MLD model can handle the impact effect well even if the impact point appears as an interior point of the optimization, and its solution is optimal for both before impact and after impact state. The second term in (31) is for achieving a minimal torque control purpose.
Note that (27) is a nonlinear MLD model in the sense that it has nonlinear and time-varying coefficients such as $B_2(k)$. However, the optimization of a nonlinear MLD system is difficult to solve because of the system nonlinearity and the curse of dimension. In this study, we linearize the nonlinear MLD system so as to avoid the computational difficulty. The linearization is carried out by freezing the coefficient matrices at the current values $x(k)$ for the next N steps. Basing on the linearized MLD system, we can solve its MPC problem by using a MIQP solver to get the future N steps control inputs $\tau(k),\ldots,\tau(k+N-1)$. Then, we apply only the first control input $\tau(k)$ to the actual nonlinear mechanical system for one sampling period, which results in a new state $x(k+1)$. At this updated working point, the above linearization and optimal calculation are repeated for keeping control.
The control procedure is concluded as follows.

1. Set the sampling period T_s and the horizon N for optimization.
2. Substituting the current state $x(k)$ into the nonlinear MLD model (27), and freezing the coefficient matrices at the current values for the next N steps.
3. Based on the linearized MLD model, solve the MPC problem by MIQP solver.

4. Select only the control input for the next step, and apply it to the nonlinear biped robot. Get the updated state $x(k+1)$.
5. The time is shift from k to $k+1$. Then repeat the step 2-5.

The proposed linearization and the MPC are based on the updating state. This feedback brings robust property to the control system, by which the nonlinear model error and external disturbance can be effectively treated. On the other hand, larger number **N** results in better control performance. But it increases the number of variables and leads the optimal problem more complex. Especially, the number of logical variables determinates the computational time exponentially. The computational burden will obstruct the real time implementation of the control. For this problem, the use of lookup table and the technique of mp-MIQP (Bemporad et al., 2002), which moves the complex computation to off line work, will drastically reduce the on line processing time.

5. Application to a biped walker system

The proposed MLD modeling and MPC approach are applied to a 2 D.O.F. planar biped robot system as shown in Fig.1. The physical parameters of the biped robot are given in Table 1.

m_h	m	ℓ	r_g
10kg	5kg	1m	0.5m

Table 1. Physical parameters of the 2 D.O.F. biped walker

For simplicity, the walking ground is assumed to be smooth plane. The motion of the biped is the automatic transition of single support phase and impact phase. The movement range of link 1 and link 2 is set within $\pm\pi/4$ (rad) which ensures the position of hip above 0.7m to the ground. The torque limitation of link 1 is supposed to be 100(Nm), and that of link 2 is 30(Nm). The minimal progressing velocity of hip, ε_v, is set to be 0.1(rad/s).

¶(6pt)

In our simulations, the biped robot started to walk with an initial state

$$x_0 = [-0.22 \quad 0.22 \quad 0.8 \quad -0.4]^T.$$

To carry out the MPC for generating the walking gait optimally, a reference pre-impact state is required as stated in (31), which is taken as

$$x_r = [0.23 \quad -0.23 \quad 1.30 \quad 1.63]^T.$$

The corresponding auxiliary variables are

$$z_r = [0.23 \quad -0.23 \quad 1.30 \quad 1.63]^T,$$
$$\delta_r = [1 \quad 1 \quad 1 \quad 1]^T.$$

The weighting matrix is chosen to be

$$Q_1 = diag(10^{-4}, 10^{-4}, 10^{-7}, 10^{-7}).$$

The other weighting matrices Q_2, Q_3, Q_4 are appointed to be diagonal matrices, and all of their diagonal elements are equal to 10^{-10}.
All simulations are executed by a computer with a Pentium 3.20GHz CPU and 3G memory. The calculation for the synthesis of walking motion is carried out using MATLAB. The sampling period T_s is 10ms.

The first simulation is for the synthesis of biped motion. The simulation results are shown in Fig.2 and Fig.3. In Fig.2(a), the real line is the trajectory of joint $\theta_1(t)$, and the dot line is the trajectory of joint $\theta_2(t)$. In Fig.2(b), the real line is the trajectory of the angular velocity $\dot{\theta}_1(t)$, and the dot line is the trajectory of the angular velocity $\dot{\theta}_2(t)$. In Fig.3(a), it shows the profile of the applied torque τ_1. In Fig.3(b), it is the profile of the applied torque τ_2.

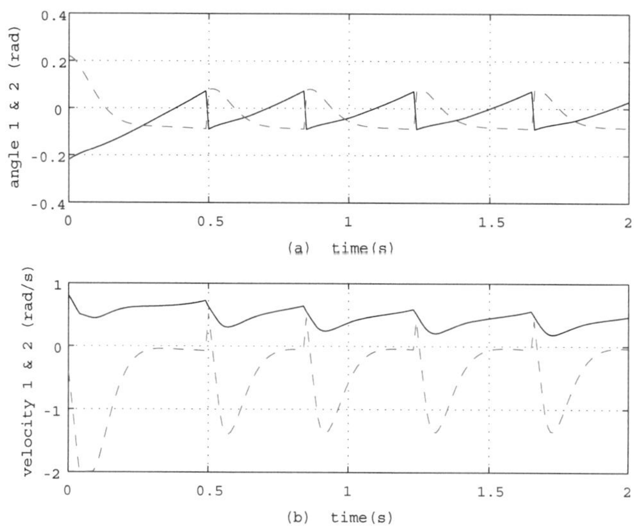

Figure 2. The time histories of joint angle and velocity: (a) the angle of support leg, θ_1 (rad) (the real line) and the angular of swing leg, θ_2 (rad) (the dot line). (b) angular velocity of support leg, $\dot{\theta}_1$ (rad/s) (the real line) and angular velocity of swing leg, $\dot{\theta}_2$ (rad/s) (the dot line)

From these Fig.2 and Fig.3, one can see that the biped robot is controlled to progress 5 steps in 2s. The first step period is 0.5s, slower than the left 4 steps. Leg transition occurred at the position of less than the desired 0.23(rad). Thus the stride is relative small.
The second simulation is to check the robust property of the approach. For that, a pulse type of disturbance vector $w = [4 \quad 10]^T$ is added to the torque at 1s.The profiles of the disturbed torques are shown in Fig.4. From Fig.4 (a) and (b), it is seen that both of the torques converge to their stable trajectories after the disturbance disappeared. The trajectories of angles and angular velocities have not big change from those shown in Fig.2.
By these simulation results, we see that the MLD approach is effective for generating gait pattern, and is robust to external disturbance. However, the computation time for both simulations is about 88s for the 2s walking. Note that these simulations are preliminary studies for validating the effectiveness of MLD modeling and the feasibility of MPC, the computation time is absolutely not optimized and the used MIQP solver is not fast. This

provides us another important research topic of how to decrease the computation time for real time implementation.

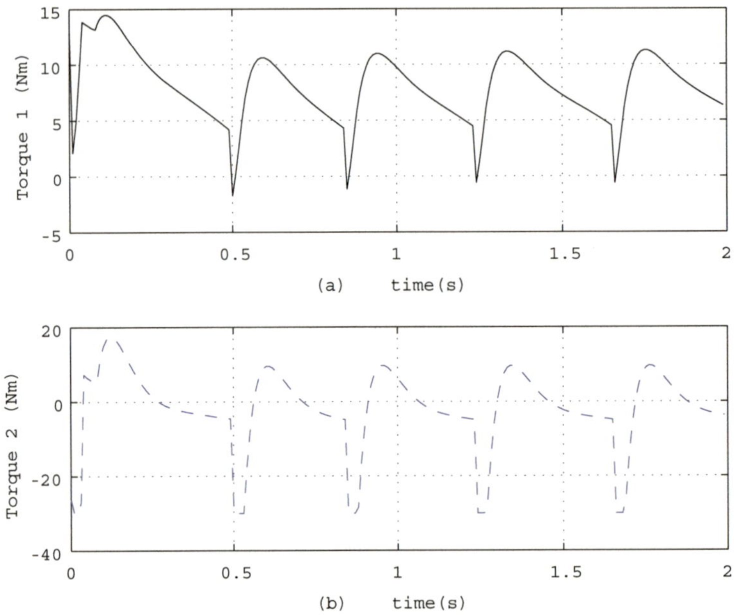

Figure 3. The profiles of torques: (a) torque τ_1 (Nm) of the support leg, (b) torque τ_2 (Nm) of the swing leg

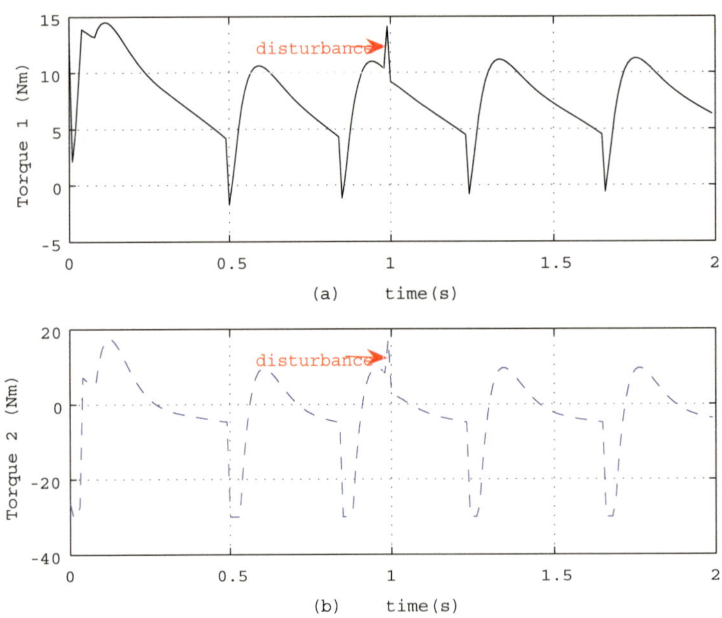

Figure 4. The profiles of the disturbed torques: (a) torque τ_1 (Nm) of the support leg, (b) torque τ_2 (Nm) of the swing leg

6. Conclusion

In this study, we proposed a MLD modeling and MPC approach for the on line optimization of biped motion. Such modeling approach possesses advantage that it describes both the continuous dynamics and the impact event within one framework, consequently it provides a unified approach for mathematical, numerical and control investigations. This MLD model allows model predictive control (MPC) and subsequent stability from the numerical analysis viewpoints, by powerful MIQP solver. Hence the biped robot can be on line controlled without pre-defined trajectory. The optimal solution corresponds to the optimal gait for current environment and control requirement. The feasibility of the MLD model based predictive control is shown by simulations. How to effectively decrease the computation time in order to realize the real time implementation is an important research topic left to future.

Finally, we mention that a human uses his predictive function based on an internal model together with his feedback function for motion, which is considered as a motor control model of a cerebellum (Kawato, 1999). Stimulated by this, a general theoretical study for motion control of hybrid systems is reported in (Yin & Hosoe, 2004) which is based on the MLD model of a hybrid system. We are further developing this theory to help the biped motion synthesis and control. It will be also useful for the realization of complex motion of other bio-mimetic robots.

7. References

Azevedo, C., Poignet, P. & Espiau, B. (2004). Artificial locomotion control: from human to robots, *Robotics and Autonomous Systems,* Vol.47, Issue 4, (July 2004) pp.203-223, ISSN 0921-8890.

Bemporad, A. & Morari, M. (1999). Control of systems integrating logic, dynamics, and constraints, *Automatica,* Vol.35, No.3, (March 1999) pp.407-427, ISSN 0005-1098.

Bemporad, A., Morari, M., Dua, V. & Pistikopoulos, E.N. (2002). The explicit linear quadratic regulator for constrained systems, *Automatica,* Vol.38, No.1, (January 2002) pp.3-20, ISSN 0005-1098.

Chevallereau, C., Formalapos sky, A. & Perrin, B. (1998). Low energy cost reference trajectories for a biped robot, *Proceedings of the IEEE International Conference on Robotics and Automation (ICRA),* pp.1398-1404, ISBN 0-7803-4300-X, Leuven, Belgium, May 1998, IEEE.

Chevallereau, C. & Sardain, P. (2000). Design and actuation optimization of a 4 axes biped robot for walking and running, *Proceedings of the IEEE International Conference on Robotics and Automation (ICRA),* pp.3365-3370, ISBN 0-7803-5886-4, San Francisco, CA, USA, April 2000, IEEE.

Chew, C.M. & Pratt, G.A. (2000). A general control architecture for dynamic bipedal walking, *Proceedings of the IEEE International Conference on Robotics and Automation (ICRA),* pp.3989-3995, ISBN 0-7803-5886-4, San Francisco, CA, USA, April 2000, IEEE.

Denk, J. & Schmidt, G. (2001). Walking primitive synthesis for an anthropomorphic biped using optimal control techniques, *Proceedings of the International Conference on Climbing and Walking Robots (CLAWAR),* pp.819-826, ISBN 1-86058-265-2, Karlsruhe, Germany, September 2001, Professional Engineering Publishing, Bury St Edmunds (u.a.), UK.

Kawato, M. (1999). Internal models for motor control and trajectory planning, *Current Opinion in Neurobiology*, Vol.9, No.6, (December 1999) pp.718-727, ISSN 0959-4388.

Kooij, H. van der, Jacobs, R., Koopman, B. & Helm, F. van der (2003). An alternative approach to synthesizing bipedal walking, *Biological Cybernetics*, Vol.88, No..1, (January 2003) pp.46-59, ISSN 0340-1200.

Park, J.H. & Chung, H. (1999) . Zmp compensation by on-line trajectory generation for biped robots, *Proceedings of the IEEE Conference on Systems, Man and Cybernetics (SMC99)*, pp.960-965, ISBN 0-7803-5731-0, Tokyo, Japan, October 1999, IEEE.

Park, J.H. & Chung, H. (2000). Hybrid control of biped robots to increase stability in locomotion, *Journal of Robotic Systems*, Vol.17, Issue 4, (Published online: March 2000) pp.187-197, ISSN 0741-2223.

Yin, Y.J. & Hosoe, S. (2004). Tracking Control of Mixed Logic Dynamical Systems, *IEICE Trans. on Fundamentals of Electronics, Communications and Computer Sciences*, Vol.E87-A, No.11, (November 2004) pp.2929-2936, ISSN 09168508.

17

Bipedal Walking Pattern Design by Synchronizing the Motions in the Sagittal and Lateral Planes

Chi Zhu[1] and Atsuo Kawamura[2]
[1]Maebashi Institute of Technology, [2]Yokohama National University
Japan

1. Introduction

In these two decades, the technology of bipedal and humanoid robots has been made great progress. Up so far, many bipedal and humanoid robots are successfully developed (Hirai et al., 1998; Takanishi et al., 1985; Ogura et al., 2004; Kaneko et al., 2004; Nakasaka et al., 2004; Löffler et al., 2003; Lohmeier et al., 2004). In these robots, the gait planning and control for bipedal walking are based on ZMP concept (Vukobratovich et al., 1990; Goswami, 1999; Kakami, 2000). There are generally two methods for implementation of the dynamic and stable walking. The first one is based on a multiple rigid body robot model (Takanishi et al., 1985). This model is comparatively precise, but it requires a huge computation cost. The second one is simply based on the assumption that the whole robot mass is concentrated to robot's CoG (Center of Gravity). The typical approach is using a 3D inverted pendulum model (Kajita & Kobayashi, 1987; Kajita et al., 2002; 2003; Zhu et al., 2003, 2004), in which, the robot locomotion in the sagittal and lateral planes are supposed to be independent.

However, up to now, the bipedal walking nature and limitations of walking speed, stride, stride motion time, and so on, have not been completely understood. Partially this is because (1) few attentions are paid on the motion in double support phase and its effect to whole bipedal walking; (2) the investigation is not done intensively even for single support phase; for example, motion in sagittal plane are simply dealt with independent of the motion in lateral plane in most literatures on biped and humanoid robots. As revealed latter, in fact, such an approach is improper since the motions in the two planes are strongly coupled together. (3) ZMP is fixed to a point in robot sole for most of bipedal or humanoid robots.

This paper mainly discusses the above first two problems based on ZMP concept and an inverted pendulum model with the assumption that ZMP is fixed at the center of the robot sole in single support phase. To do this, the relation between the motions in the sagittal and lateral planes is investigated first. By dividing a whole bipedal walking cycle into a double support, a deceleration, and an acceleration phases, and synchronizing the motions in the sagittal and lateral planes, we point out that the motions in these two planes are strongly coupled together, in other words, the motion in the lateral plane greatly restricts the motion in the sagittal plane, vice versa. Then, the role of the double support phase in biped walking is discussed. By changing the start and finish points of the double support phases in the lateral plane, the walking parameters such as walking speed, walking period, phase stride and so forth, are consequently altered. With this, it comes a conclusion that a short double support phase is necessary in the viewpoint of both of a smooth and fast walking after a high enough instantaneous speed in the sagittal plane is obtained at the end of acceleration phase. Consequently, an approach for adjusting the instantaneous speed at the end of the acceleration phase is presented, and a biped walking planning procedure is proposed. Finally a numerical example is given out.

This paper is organized as follows. In Section 2, robot model, trajectories of the robot CoG and ZMP, and the general motion equations of the robot are described. In Section 3, a walking pattern with fixed ZMP is analyzed by being synchronized the motions in the two planes. In Section 4, the effects of the double support phase on bipedal walking are analyzed. Walking speed adjustment, walking planning, and an example are presented in Section 5, and the conclusion is given out in Section 6.

2. Robot Model and Bipedal Walking

2.1 Robot Model and Assumptions

In this study, a bipedal robot is modeled as a 3D inverted pendulum of which the mass of the robot is supposed to concentrate on the point C, the CoG of the robot, as shown in Fig.1. OXYZ is the world coordinate system, where X, Y axes are respectively the walking and swinging directions. XZ plane and YZ plane are respectively called the sagittal plane and lateral plane; and oxyz is the local system fixed to the center of the support sole of the robot. For simplicity, the following assumptions are made.

1. The height of the robot's CoG is a constant, i.e., $\ddot{z} = 0$.
2. The origin o of the local system oxyz is always set at the sole center of support foot.
3. The desired ZMP in single support phase (SSP) is also always set at the sole center of the support foot; hence, the desired ZMP in SSP is identical to o.
4. The equivalent foot length of two feet is b.
5. The robot moves with constant speed in the double support phase.

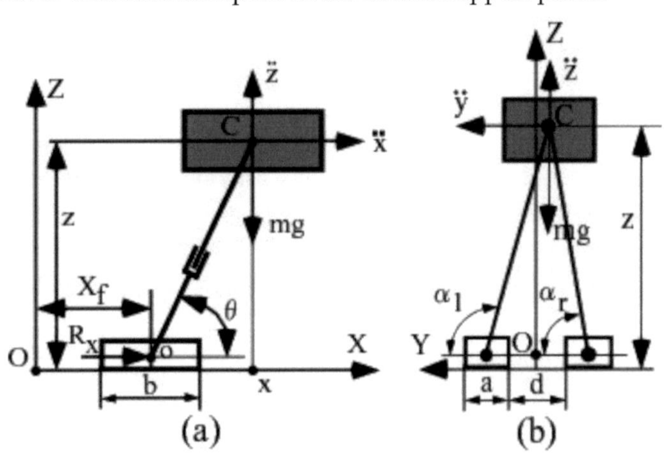

Figure 1. 3D inverted pendulum model of biped robot. (a) motion in sagittal plane. (b) motion in lateral plane

6. The robot just walks forward, i.e., the distance d0 in the lateral plane is a constant (Fig. 2).

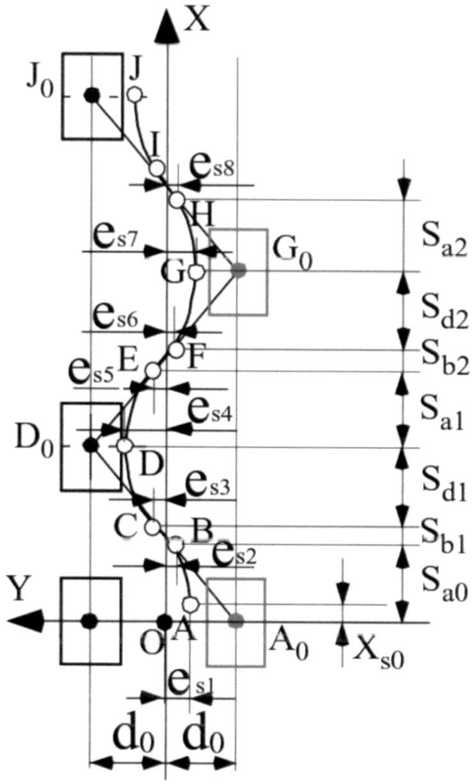

Figure 2. Trajectories of CoG and ZMP in bipedal walking

7. In single support phase, the CoG never passes over the X axis (the midline) of its two feet. This means that the single support phase ends and the double support phase starts before the CoG crosses the X axis.

Here, we make some explanations for the assumptions (2) and (3).

In theory, the origin o of the local system oxyz can be set at any place. Assumption (2) just brings a simple expression of the robot's motion equations (refer to Eqs. (5) and (6)). Hence, it leads great convenience for discussing the robot motion. Otherwise, if the origin o is not at the sole center of support foot, there would be respectively a constant offset in equations (5) and (6).

In SSP, ZMP (Zero moment point) is in fact a center of pressure of the support foot sole, or saying, the acting point of the resultant reaction force of the support floor. Therefore, ZMP is always within the inside of the sole of the support foot in SSP, even when the robot is falling down (at this moment, the ZMP is at one of the edges of the support foot). Consequently, the safest case is that the desired ZMP is at the sole center of the support foot in SSP.

Concretely, with the assumption of the desired ZMP at the sole center of the support foot, the motion equations of the CoG of the robot can be easily derived (as shown by Eqs. (5) and (6)), then this CoG motion is decomposed into each joint's motion (trajectory). The control system is just controlling each joint's motion as planned trajectory. In this way, the control system not only makes the biped robot stable walking, but also guarantees the real ZMP at the desired position (the sole center of the support foot). This approach is being used in most of biped humanoid robots, ex., (Kakami et al., 2000) and (Kajita et al., 2003).

When the real ZMP of the robot deviates from the desired ZMP, an additional on-line ZMP control scheme is necessary by modifying the CoG's motion (in fact, changing some or all joint's motion), for example, as did in (Zhu et al., 2004).
This paper just discusses the CoG's trajectory of the robot with the precondition of the desired ZMP at the sole center of the support foot in SSP, and doesn't consider the detail control problem.

2.2 Trajectories of Robot CoG and ZMP

The robot motion, in which its CoG trajectory is shown in Fig.2, is as follows. Before the robot starts to walk, its CoG is at the midpoint O of the two support feet. When the robot starts to walk, the CoG is first shifted from the point O to A. Here the distances between OA in X and Y direction are respectively X_{s0} and e_{s1}. From the point A, the left leg of the robot (here, assume the left leg of the robot first steps out) steps forward while the right leg supports the robot, and the robot moves forward along the curve AB. When the CoG moves to the point B, the left leg lands to the floor and the robot enters into the double support phase. When the CoG reaches to the point C, the right leg lifts off and the robot is in the single support phase again. During the left leg support phase, the CoG moves along the curve CDE. When the CoG reaches to E, the right leg lands to the floor and the robot is in the double support phase again. In this way, the swing leg and the support leg switches each other in turn at the points B, E, H, the robot moves forward in X direction (sagittal plane) while swings in Y direction (lateral plane). The positions of each point from A to H are shown in Fig.2.
As our assumptions indicate, the ZMP is fixed at the sole center of the support foot in the single support phase, such as points A_0, D_0, G_0, and J_0, and the ZMP moves from A_0 to D_0, from D_0 to G_0 in the double support phase as shown in Fig.2. In the local system *oxyz*, the ZMP can be expressed as

$$x_{ZMP} = x - \frac{\ddot{x}z}{g} \tag{1}$$

$$y_{ZMP} = y - \frac{\ddot{y}z}{g} \tag{2}$$

The above two equations can be rewritten as

$$\ddot{x} = \frac{g}{z}(x - x_{ZMP}) = g \cdot \cot\theta \tag{3}$$

$$\ddot{y} = \frac{g}{z}(y - y_{ZMP}) = g \cdot \cot\alpha \tag{4}$$

Eq.(3) shows that for the motion in the sagittal plane XZ, when the robot is forward leaning (the CoG of the robot is in front of the support foot, $x - x_{ZMP} \geq 0$, the robot has to accelerate forward. Contrarily, when the robot is backward leaning (the CoG is at the rear of the support foot, $x - x_{ZMP} \leq 0$), the robot should decelerate. Therefore, with the switching of the support leg and swinging leg in walking, the robot continually accelerates and decelerates.

In a single support phase, the general motion equations (the solutions of Eqs. (3) and (4) of the CoG of the robot in the two planes are respectively as follows,

$$x(t) = x_{ZMP} + c_1 \cdot e^{\omega t} + c_2 \cdot e^{-\omega t} \tag{5}$$

$$y(t) = y_{ZMP} + c_3 \cdot e^{\omega t} + c_4 \cdot e^{-\omega t} \tag{6}$$

where, x_{ZMP} and y_{ZMP} are respectively constants as our assumption; $\omega = \sqrt{g/z}$; c_1 to c_4 are coefficients determined by initial conditions.

3. Motions in Sagittal and Lateral Planes

In our previous work (Zhu et al., 2003, 2004), a bipedal walking is divided into initial acceleration, double support, deceleration and acceleration phases. Here the motion in each phase is investigated in more detail by combining and synchronizing the motion in the lateral plane. In this paper, the distances e_{sn} (n=0, 1, 2, ···) from the robot's CoG to X axis in Y direction (See Fig. 2) are supposed positive, i.e., $e_{sn} > 0$ (n=0, 1, 2, ···).

3.1 Initial acceleration phase AB

(Motion time: $0 \le t \le t_{a0}$, travelled distance in X and Y direction: $X_{s0} \le x(t) \le S_{a0}$, $-e_{s1} \le y(t) \le -e_{s2}$)

In this phase, the robot starts walking from the standstill and the robot has to accelerate. According to the motion equations (3) and (4), in order to guarantee the ZMP at the sole center of the support foot, the robot has to lean forward in the XZ (sagittal) plane and lean left in the YZ (lateral) plane, respectively. Thus, at the start point A, the offsets X_{s0} and e_{s1} respectively in X and Y axes are necessary as shown in Fig.2. (If X_{s0} = 0, then x_{ZMP} = x. It leads $\ddot{x} = 0$. This means the robot cannot move from the rest ($\dot{x} = 0$). If e_{s1}=0, then the robot will have to move to left and pass over the X axis. This violates the assumption (7)).

This condition implies that the projection of the CoG must be within the support polygon. This is (refer to Fig.1)

$$0 < X_{s0} \le \frac{b}{2} \tag{7}$$

With the initial conditions $x(0)=X_{s0}$, $\dot{x}(0) = 0$, $y(0)=-e_{s1}$, $\dot{y}(0)=0$, $x_{ZMP}=0$, and $y_{ZMP}=-d_0$, the coefficients of Eqs.(5) and (6) are respectively $c_1=c_2=X_{s0}/2$, and $c_3=c_4=(d_0 - e_{s1})/2$. Therefore the two motion equations are respectively

$$x(t) = \frac{1}{2} X_{s0}(e^{\omega t} + e^{-\omega t}) = X_{s0} \cdot \cosh(\omega t) \tag{8}$$

$$\begin{aligned} y(t) &= -d_0 + \frac{1}{2}(d_0 - e_{s1})(e^{\omega t} + e^{-\omega t}) \\ &= -d_0 + (d_0 - e_{s1})\cosh(\omega t) \end{aligned} \tag{9}$$

In the sagittal plane, the terminal conditions at the point B are

$$x(t_{xa0}) = \frac{1}{2} X_{s0}(e^{\omega t_{xa0}} + e^{-\omega t_{xa0}}) = S_{a0} \tag{10}$$

$$\dot{x}(t_{xa0}) = \frac{1}{2} \omega X_{s0}(e^{\omega t_{xa0}} - e^{-\omega t_{xa0}}) = V_{h0} \tag{11}$$

where, t_{xa0} and V_{h0} are respectively the motion time and terminal speed in the sagittal plane in this phase. From (10) and (11), t_{xa0} and V_{h0} can be respectively expressed as

$$e^{2\omega t_{xa0}} = \frac{\omega S_{a0} + V_{h0}}{\omega S_{a0} - V_{h0}} \tag{12}$$

$$V_{h0} = \omega \sqrt{S_{a0}^2 - X_{s0}^2} \tag{13}$$

On the other hand, for the motion in the lateral plane, with the terminal conditions $y(t_{ya0})=-e_{s2}$ and $\dot{y}(t_{ya0}) = V_{ya0}$ (here, t_{ya0} and V_{ya0} are respectively the motion time and the terminal speed of the phase in the lateral plane). With the same way as done for the motion in the sagittal plane, t_{ya0} and V_{ya0} can be respectively expressed as

$$e^{2\omega t_{ya0}} = \frac{\omega(d_0 - e_{s2}) + V_{ya0}}{\omega(d_0 - e_{s2}) - V_{ya0}} \tag{14}$$

$$V_{ya0} = \omega \sqrt{(d_0 - e_{s2})^2 - (d_0 - e_{s1})^2} \tag{15}$$

Obviously, those two motions in the two planes should be synchronized, that is, there should be $t_{xa0}=t_{ya0}$. Thus, substituting Eqs. (13) and (15) into Eqs. (12) and (14) yields

$$\frac{X_{s0}}{S_{a0}} = \frac{d_0 - e_{s1}}{d_0 - e_{s2}} = \rho_{a0} < 1 \tag{16}$$

By substituting (16) into (12) or (14), we can get

$$e^{2\omega t_{a0}} = e^{2\omega t_{xa0}} = e^{2\omega t_{ya0}} = \left(\frac{1 + \sqrt{1 - \rho_{a0}^2}}{\rho_{a0}} \right)^2 \tag{17}$$

Hence, the motion time in this phase is

$$t_{a0} = t_{xa0} = t_{ya0} = \frac{1}{\omega} \ln \frac{1}{\rho_{a0}} \left(1 + \sqrt{1 - \rho_{a0}^2} \right) \tag{18}$$

By substituting (16) into (13) and (15), we can find the following relation

$$\frac{V_{ya0}}{V_{h0}} = \frac{d_0 - e_{s2}}{S_{a0}} \tag{19}$$

Moreover, from (16), there is

$$e_{s2} = \frac{1}{\rho_{a0}} e_{s1} + \left(1 - \frac{1}{\rho_{a0}}\right) d_0 = \frac{S_{a0}}{X_{s0}} e_{s1} + \left(1 - \frac{S_{a0}}{X_{s0}}\right) d_0 \tag{20}$$

Since there should be $e_{s2} \geq 0$ as shown in Fig.2, there is

$$X_{s0} \leq \frac{d_0 - e_{s1}}{d_0} S_{a0} = \left(1 - \frac{e_{s1}}{d_0}\right) S_{a0} \tag{21}$$

If e_{s1} and e_{s2} are determined, then from (7) and (16), S_{a0} should be

$$S_{a0} < \frac{b}{2\rho_{a0}} \tag{22}$$

Eqs.(16), (20), and (21) show that the motions in the sagittal and lateral planes are tightly coupled together. The determination of the initial and terminal positions in one plane automatically and simultaneously decides the initial and terminal positions in another plane, therefore completely decides the motions in the two planes.
Eqs. (16) and (18) show that the motion time is independent of the moving distance, but determined by the ratio of the distance from the CoG to ZMP at the terminal time to the distance at the initial time in either of two planes. Eq. (19) shows that the ratio of the terminal velocities in the two planes is equal to the ratio of two distances from the CoG to ZMP at the terminal time in the two planes. Note that in this phase, an offset x_{s0} is necessary, otherwise the robot cannot start walking.

3.2 Double support phase BC

(Motion time: $0 \leq t \leq t_{ds1}$, travelled distance in X and Y direction: $0 \leq x(t) \leq S_{b1}$, $-e_{s2} \leq y(t) \leq e_{s3}$)
Once the CoG of the robot moves over the point B, the robot enters into the double support phase. We assumed that the robot moves forward with constant velocities V_{h0} and V_{ya0} respectively in X and Y directions. But its ZMP shifts from A_0 to D_0 (Fig. 2) in the double support phase. There are following initial and terminal conditions for ZMP's motion in this phase,

$$\dot{x}_{ZMP}(0) = 0, \quad \dot{y}_{ZMP}(0) = 0 \tag{23}$$

$$x_{ZMP}(0) = 0, \quad y_{ZMP}(0) = -d_0 \tag{24}$$

$$\dot{x}_{ZMP}(t_{ds1}) = 0, \quad \dot{y}_{ZMP}(t_{ds1}) = 0 \tag{25}$$

$$x_{ZMP}(t_{ds1}) = S_{a0} + S_{b1} + S_{d1}, \quad y_{ZMP}(t_{ds1}) = d_0 \tag{26}$$

Therefore, a third order polynomial function of time t is employed to represent ZMP's trajectory.

3.3 Deceleration phase CD

(Motion time: $0 \le t \le t_{d1}$, travelled distance in X and Y direction: $0 \le x(t) \le S_{d1}$, $e_{s3} \le y(t) \le e_{s4}$)

After the robot moves over the point C, its former support leg (here, is the right leg) will lift off and the robot is in the single support. Since the CoG of the robot is at the behind and right of its support foot, the robot must decelerate in both of two directions. This phase CD is called the *deceleration phase.*

With initial conditions $x(0)=-S_{d1}$, $\dot{x}(0)=V_{h0}$, $y(0)=e_{s3}$, $\dot{y}(0)=V_{ya0}$, and $x_{ZMP}=0$, $y_{ZMP}=d_0$, from Eqs.(5) and (6), the motion equations in the sagittal and lateral planes are respectively expressed as follows,

$$x(t)=\frac{V_{h0}}{\omega}\sinh(\omega t)-S_{d1}\cosh(\omega t) \tag{27}$$

$$y(t)=d_0+\frac{V_{ya0}}{\omega}\sinh(\omega t)-(d_0-e_{s3})\cosh(\omega t) \tag{28}$$

The terminal condition in the sagittal plane is $x(t_{d1})=0$, but $\dot{y}(t_{d1})=0$ for the lateral plane. In the same way as shown in subsection 3.1, the terminal speed V_{l1} (the lowest in this phase) in the sagittal plane is expressed as

$$V_{l1}=\sqrt{V_{h0}^{\;2}-(\omega S_{d1})^2} \tag{29}$$

the swinging amplitude e_{s4} in this phase is

$$e_{s4}=d_0-\sqrt{(d_0-e_{s3})^2-(V_{ya0}/\omega)^2} \tag{30}$$

Further, the motion times in the two planes are respectively as follows,

$$e^{2\omega t_{xd1}}=\frac{V_{h0}+\omega S_{d1}}{V_{h0}-\omega S_{d1}} \tag{31}$$

$$e^{2\omega t_{yd1}}=\frac{\omega(d_0-e_{s3})+V_{ya0}}{\omega(d_0-e_{s3})-V_{ya0}} \tag{32}$$

The above two motion times should be the same; therefore there exists the following condition,

$$\frac{V_{h0}}{\omega(d_0-e_{s3})}=\frac{\omega S_{d1}}{V_{ya0}} \tag{33}$$

Consequently, by substituting (13), (15), and (16) into (33), the phase stride S_{d1} can be expressed as

$$S_{d1}=\frac{d_0-e_{s2}}{d_0-e_{s3}}S_{a0}(1-\rho_{a0}^{\;2}) \tag{34}$$

Meanwhile, to guarantee the robot continually moving forward, the terminal speed should be $V_{l1}>0$, that is

$$\omega S_{d1} < V_{h0} \tag{35}$$

With (13) and (14)), the condition (35) implies that

$$0 < e_{s3} < d_0 - (d_0 - e_{s2})\sqrt{1-\rho_{a0}^{\;2}} \tag{36}$$

Further, by substituting Eqs. (29) and (30)) into Eqs. (31)) and (32), we can get

$$e^{2\omega t_{xd1}} = \frac{1+\sqrt{1-\left(\frac{V_{l1}}{V_{h0}}\right)^2}}{1-\sqrt{1-\left(\frac{V_{l1}}{V_{h0}}\right)^2}} = \frac{\left(1+\sqrt{1-\left(\frac{V_{l1}}{V_{h0}}\right)^2}\right)^2}{\left(\frac{V_{l1}}{V_{h0}}\right)^2} \tag{37}$$

$$e^{2\omega t_{yd1}} = \frac{1+\sqrt{1-\left(\frac{d_0-e_{s4}}{d_0-e_{s3}}\right)^2}}{1-\sqrt{1-\left(\frac{d_0-e_{s4}}{d_0-e_{s3}}\right)^2}} = \frac{\left(1+\sqrt{1-\left(\frac{d_0-e_{s4}}{d_0-e_{s3}}\right)^2}\right)^2}{\left(\frac{d_0-e_{s4}}{d_0-e_{s3}}\right)^2} \tag{38}$$

Thus, there are the following relations,

$$\frac{V_{l1}}{V_{h0}} = \frac{d_0-e_{s4}}{d_0-e_{s3}} = \rho_{d1} < 1 \tag{39}$$

$$e_{s4} = (1-\rho_{d1})d_0 + \rho_{d1}e_{s3} \tag{40}$$

and the motion time t_{d1} is

$$t_{d1}=t_{xd1}=t_{yd1}=\frac{1}{\omega}\ln\frac{1}{\rho_{d1}}\left(1+\sqrt{1-\rho_{d1}^{\;2}}\right) \tag{41}$$

Note that when $e_{s3}=e_{s2}$, there is a following relations,

$$e_{s4} = e_{s1} \;\text{ and }\; t_{d1} = t_{a0} \;\; (\text{as } e_{s3} = e_{s2}) \tag{42}$$

3.4 Acceleration phase DE

(Motion time: $0 \le t \le t_{a1}$, travelled distance in X and Y direction: $0 \le x(t) \le S_{a1}$, $e_{s5} \le y(t) \le e_{s4}$)

In the sagittal plane, once passing over the point D, the CoG of the robot will be in front of its support foot. Thus, the robot must accelerate. On the other hand, the robot reaches its

swinging peak at D with zero speed, and its CoG is in the right of its support foot. Therefore, the robot will accelerate in the lateral plane, too. This phase is called the *acceleration phase*.
With the initial conditions x(0)=0, $\dot{x}(0)=V_{l1}$, y(0)=e_{s4}, $\dot{y}(0)=0$, and x_{ZMP}=0, y_{ZMP}=d_0, from Eqs.(5) and (6), the motion equations in the sagittal and lateral planes are respectively,

$$x(t)=\frac{V_{l1}}{\omega}\sinh(\omega t) \tag{43}$$

$$y(t)=d_0-(d_0-e_{s4})\cosh(\omega t) \tag{44}$$

The terminal conditions are x(t_{a1})=S_{a1}, y(t_{a1})=e_{s5}. In the same way as shown in subsection 3.1, the terminal speed V_{h1} and V_{ya1} in the two planes are respectively

$$V_{h1}=\sqrt{V_{l1}^{\;2}+(\omega S_{a1})^2} \tag{45}$$

$$V_{ya1}=\omega\sqrt{(d_0-e_{s5})^2-(d_0-e_{s4})^2} \tag{46}$$

These two velocities are the highest in this phase. Further, the motion time in the two planes can be respectively expressed as follows,

$$e^{2\omega t_{xa1}}=\frac{V_{h1}+\omega S_{a1}}{V_{h1}-\omega S_{a1}} \tag{47}$$

$$e^{2\omega t_{ya1}}=\frac{\omega(d_0-e_{s5})+V_{ya1}}{\omega(d_0-e_{s5})-V_{ya1}} \tag{48}$$

With the condition t_{xa1}=t_{ya1}, by rewriting (45) to express ωS_{a1} in terms of V_{h1} and V_{l1}, and by substituting (46) into (48), then we can get the following condition,

$$\frac{V_{h1}}{V_{l1}}=\frac{d_0-e_{s5}}{d_0-e_{s4}}=\rho_{a1} \tag{49}$$

From (49), e_{s5} can be expressed as

$$e_{s5}=(1-\rho_{a1})d_0+\rho_{a1}e_{s4} \tag{50}$$

Further, from Eqs. (29), (39), (45), and (49), the following three relations can be derived out,

$$S_{a1}=\sqrt{\frac{(d_0-e_{s5})^2-(d_0-e_{s4})^2}{(d_0-e_{s3})^2-(d_0-e_{s4})^2}}\cdot S_{d1} \tag{51}$$

$$V_{h1}^{\;2}-V_{h0}^{\;2}=\omega^2\left(S_{a1}^{\;2}-S_{d1}^{\;2}\right) \tag{52}$$

$$\frac{V_{h1}}{V_{h0}}=\frac{d_0-e_{s5}}{d_0-e_{s3}} \tag{53}$$

These three equations are used to adjust the terminal speed V_{h1} of the acceleration phases. It will be discussed in Section 5. Note that e_{s4} is a function of e_{s3} and determined by eq.(30).

According to our assumptions, there is $e_{s5}>0$, hence from (45) and (49), the phase stride S_{a1} should be determined by

$$S_{a1}<\frac{V_{l1}}{\omega}\sqrt{\left(\frac{d_0}{d_0-e_{s4}}\right)^2-1} \tag{54}$$

Similar to be done in subsection 3.1, the motion time t_{a1} in this phase is

$$t_{a1}=t_{xa1}=t_{ya1}=\frac{1}{\omega}\ln\left(\rho_{a1}+\sqrt{\rho_{a1}^2-1}\right) \tag{55}$$

Note that BCDE is a whole walking cycle which consists of a double support, a deceleration, and a acceleration phases. After the acceleration phase, the robot will enter into the double support phase and start a new walking cycle again. Similar to the deceleration phase CD, for the second deceleration phase FG, we can get

$$S_{d2}=\frac{d_0-e_{s4}}{d_0-e_{s6}}\rho_{a1}S_{a1}=\frac{d_0-e_{s5}}{d_0-e_{s6}}S_{a1} \tag{56}$$

and since $V_{l2}=\sqrt{V_{h1}^2-(\omega S_{d2})^2}>0$, there is

$$\begin{aligned}0<e_{s6}<e_{s6\max}&=d_0-\sqrt{1-\frac{1}{\rho_{a1}^2}}\cdot(d_0-e_{s5})\\&=d_0-\sqrt{(d_0-e_{s5})^2-(d_0-e_{s4})^2}\end{aligned} \tag{57}$$

From the second deceleration phase, the motion is repetitive and the equations are also the same.

4. Double Support Phase and Bipedal Walking

The discussion in Section 3 shows that the motions in the sagittal and lateral planes are strongly coupled together. It also indicates that the initial and terminal points of the double support phases, such as points B, C, E, F, H, I in Fig. 2, greatly affects the robot motion. This section further illustrates the effects of the double support phase to the bipedal walking.

To let the discussion be more general, we discuss the motion in the second walking cycle EFGH in Fig. 2, where, EF is the double support phase, FG and GH are respectively the deceleration and acceleration phases. We assume that when the robot travels to the point E, the parameters before E such as e_{s4}, e_{s3}, S_{d1}, and etc, are determined and known. Here, the parameters used are z=0.50 [m], e_{s1}=0.07 [m], $e_{s2}=e_{s3}=e_{s5}$=0.02 [m].

4.1 Double support phase and deceleration phase

As discussed in Section 3, e_{s5}, or saying, the position of the point E, determines the velocities at the point E that is equal to the velocities at the point F as we assumed. Meanwhile, e_{s6}, the terminal position of the point F of the double support phase EF, affects the motion in phase FG, too. Therefore, both e_{s5} and e_{s6} determine the motion in deceleration phase FG. Note that the selection of e_{s6} should satisfy the condition of ineq.(57), in which, e_{s6max} is the function of e_{s5}, i.e., $e_{s6max}=e_{s6max}(e_{s5})$.

With the variations e_{s5} and e_{s6}, in the phase FG, the motion parameters such as the swinging amplitude e_{s7}, motion time t_{d2}, phase stride S_{d2}, the speed V_{l2} at point G (the lowest one in a whole walking cycle), and the average speed Av_{l2}, will correspondingly change. Their relations with e_{s5} and e_{s6} are respectively shown from Fig. 3 to Fig. 7. In these figures, there are 7 curves corresponding to e_{s5}=0 to e_{s5}=0.07[m] of which the interval of e_{s5} is Δe_{s5}=0.01 [m], and all of these 7 curves start from e_{s6}=0 and end at $e_{s6}=e_{s6max}(e_{s5})$.

Fig. 3 to Fig. 7 respectively show that for

- *e_{s5} fixed but e_{s6} variable case*

  e_{s7}, t_{d2}, and S_{d2} increase, but V_{l2} and Av_{l2} decrease with e_{s6}. Note that (1) a fixed e_{s5} implies that phase stride S_{a1}, the velocities of V_{h1} and V_{ya1} are determined; (2) when $e_{s6}=e_{s5}$, there are $e_{s7}=e_{s4}$, $t_{d2}=t_{a1}$, $S_{d2}=S_{a1}$, and $V_{l2}=V_{l1}$; (3) if $e_{s6}=e_{s6max}(e_{s5})$, then $e_{s7}=d_0$, $t_{d2} \to \infty$, $S_{d2}=S_{d2max}=V_{h1}/\omega$, V_{l2}=0, and Av_{l2}=0 since the robot will stop at the point G in this extreme case.

- *e_{s6} fixed but e_{s5} variable case*

  e_{s7}, t_{d2}, and S_{d2} monotonously decrease with e_{s5}, but V_{l2} and Av_{l2} have not such monotonicity (e_{s5} and e_{s6} should be subject to the condition of ineq.(57). Note that (1) the increase of e_{s5} means the decrease of S_{a1}, V_{h1}, and V_{ya1}; (2) when $e_{s5}=e_{s5max}=e_{s4}$ (=0.07[m], there are $e_{s7}=e_{s6}$, $S_{d2}=S_{a1}$=0, $t_{d2}=t_{a1}$=0, and $V_{l2}=V_{l1}$; in other words, the acceleration phase DE and the deceleration phase FG will disappear; (3) when $e_{s5}=e_{s6}$=0, there is $V_{l2}=V_{l1}$; (4) for a given e_{s6}, there exists a V_{l2max} occuring at a e_{s5}; for example, when e_{s6}=0, V_{l2max}=1.00[m/s]; as e_{s5}=0.0248[m]; (5) similarly, for a given e_{s6}, there exists a Av_{l2max} at a e_{s5}; for instance, when e_{s6}=0, Av_{l2max}=1.16[m/s]; as e_{s5}=0.0167[m] as shown in Fig. 7.

The above phenomena can be explained by the equations discussed in above section just by substituting e_{s4}, e_{s5}, S_{d1}, S_{a1} respectively with e_{s7}, e_{s8}, S_{d2}, S_{a2}, and so forth.

A big V_{l2} (here, V_{h1} is fixed with a fixed e_{s5}) implies a small speed fluctuation (V_{h1} - V_{l2}) and a smooth bipedal walking; meanwhile a big Av_{l2} means the robot walks fast. Therefore, Fig. 6 and Fig. 7 show that small e_{s5} and e_{s6} should be selected in the view of both of smooth walking and fast walking.

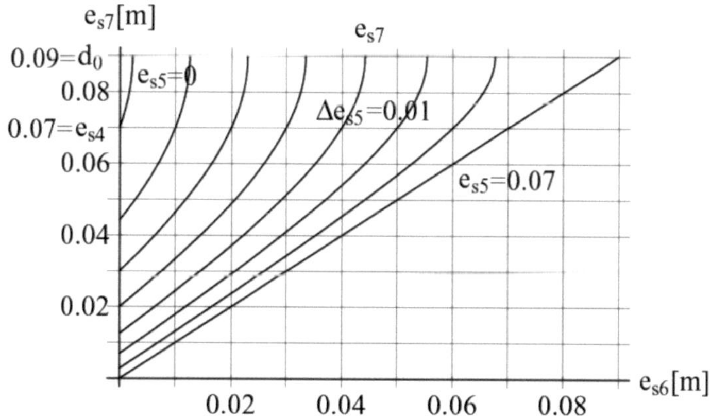

Figure 3. Variation of swinging amplitude e_{s7} with e_{s5} and e_{s6}

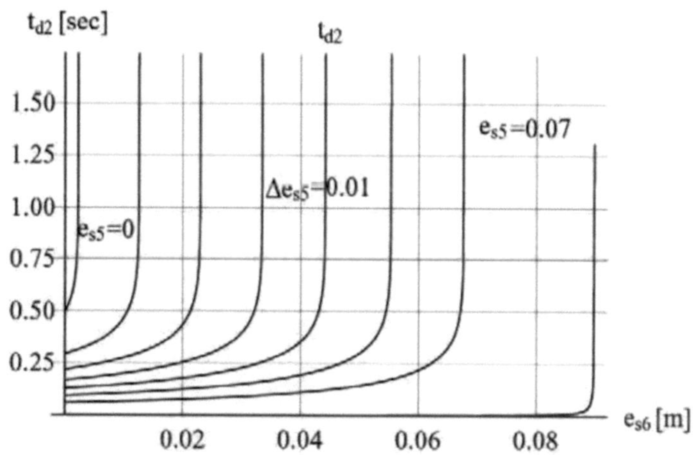

Figure 4. Variation of t_{d2} with e_{s5} and e_{s6}

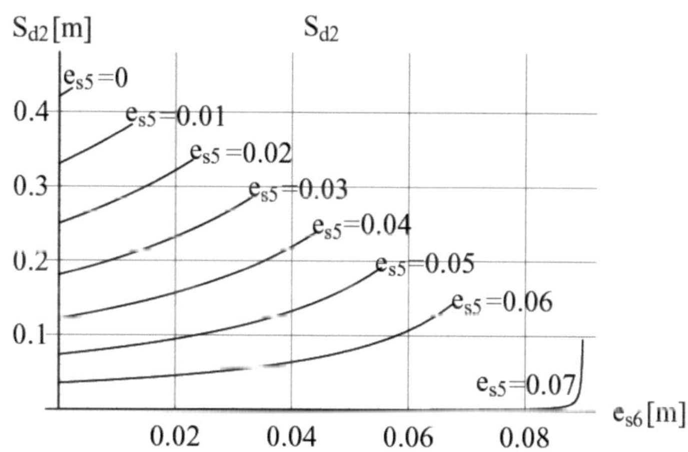

Figure 5. Variation of S_{d2} with e_{s5} and e_{s6}

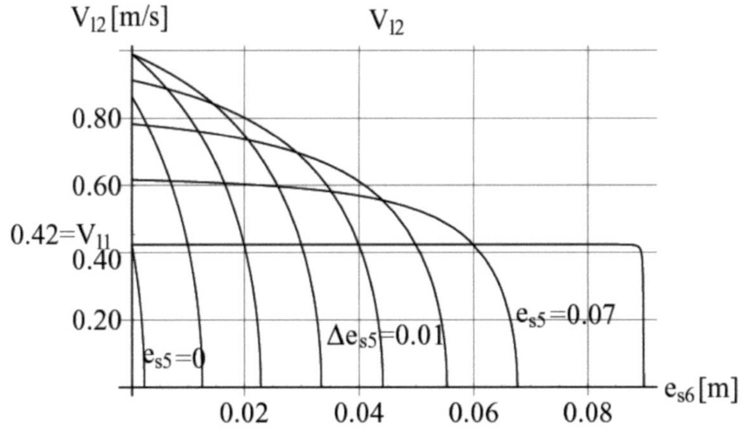

Figure 6. Variation of V_{12} with e_{s5} and e_{s6}

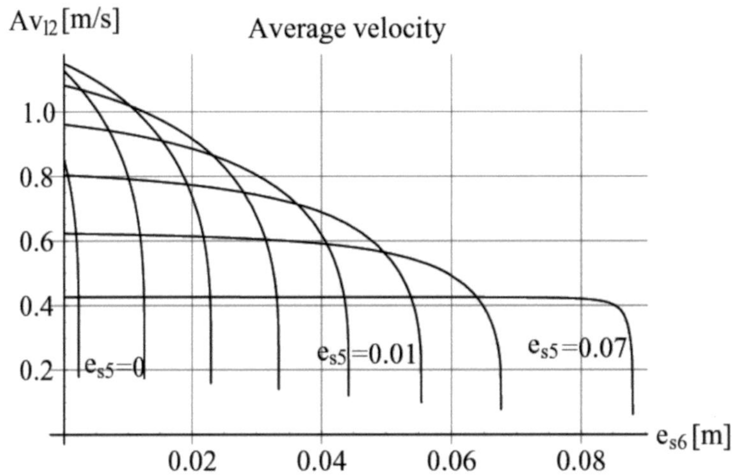

Figure 7. Variation of average speed Av_{12} with e_{s5} and e_{s6}

4.2 Double support phase and acceleration phase

Here, we mainly discuss the effects of e_{s6} and e_{s8} to the motion in acceleration phase GH, because these two parameters determine the motion of the acceleration phase GH. For simplicity, here e_{s5} is set to be equal to e_{s3}, i.e., $e_{s5}=e_{s3}=0.02$[m] (refer to Table. 1. Therefore, the variable ranges of e_{s6} and e_{s8} are respectively $0<e_{s6}<e_{s6max}$ (e_{s5}) (=0.0229[m] and $0<e_{s8}<e_{s7}$ (e_{s6}). As described in subsection 4.1, here e_{s7} is the function of e_{s6}, and e_{s7} (e_{s6max})= d_0.

Similarly to subsection 4.1, with the variations of e_{s6} and e_{s8}, the motion parameters, such as the phase stride S_{a2}, motion time t_{a2}, speed V_{h2} and V_{ya2} at point H, and the average speed Av_{a2} in phase GH, will change. Their change figures are respectively illustrated from Fig. 8 to Fig. 12. In these figures, there are 8 curves corresponding to $e_{s6}=0$ to $e_{s6}=0.021$[m] of which the interval of e_{s6} is $\Delta e_{s5}=0.03$[m], and all of these 8 curves start from $e_{s8}=0$ and end at $e_{s8max}=e_{s7}$ (e_{s6}), where e_{s7} (e_{s6}) means e_{s7} is the function of e_{s6}.

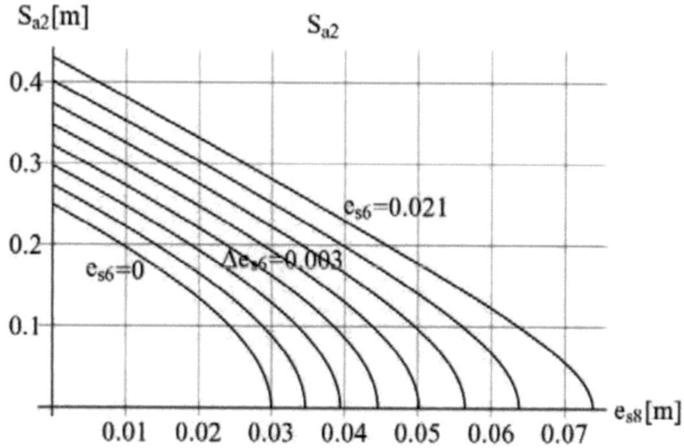

Figure 8. Variation of S_{a2} with e_{s6} and e_{s8}

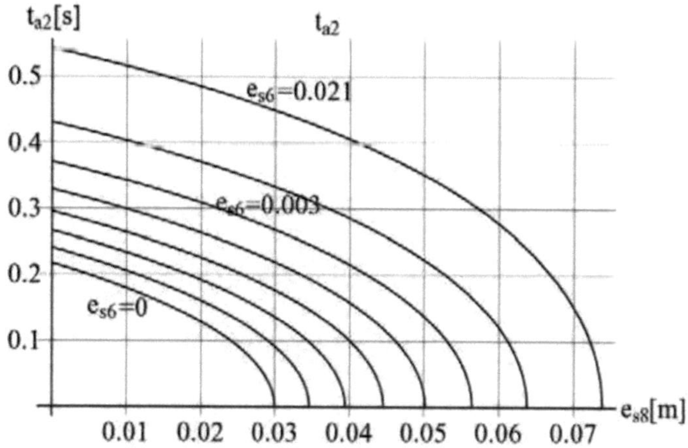

Figure 9. Variation of t_{a2} with e_{s6} and e_{s8}

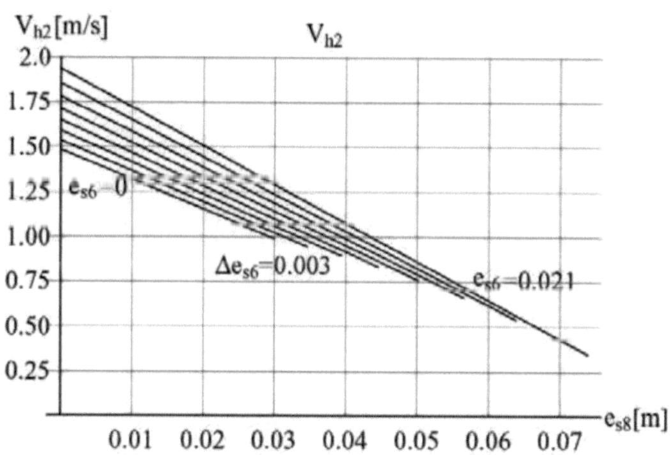

Figure 10. Variation of V_{h2} with e_{s6} and e_{s8}

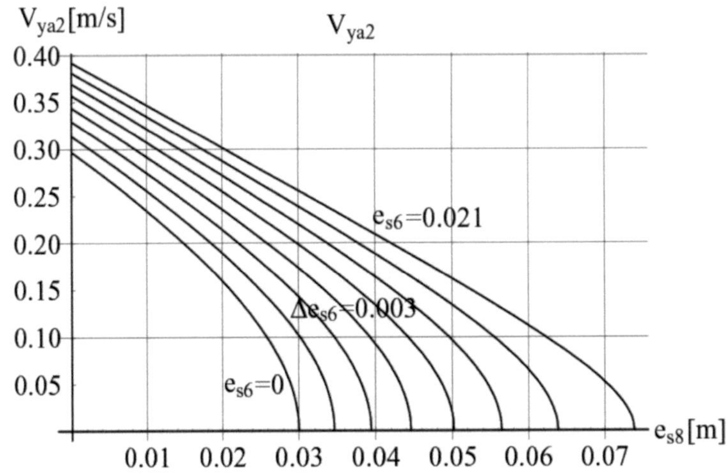

Figure 11. Variation of V_{ya2} with e_{s6} and e_{s8}

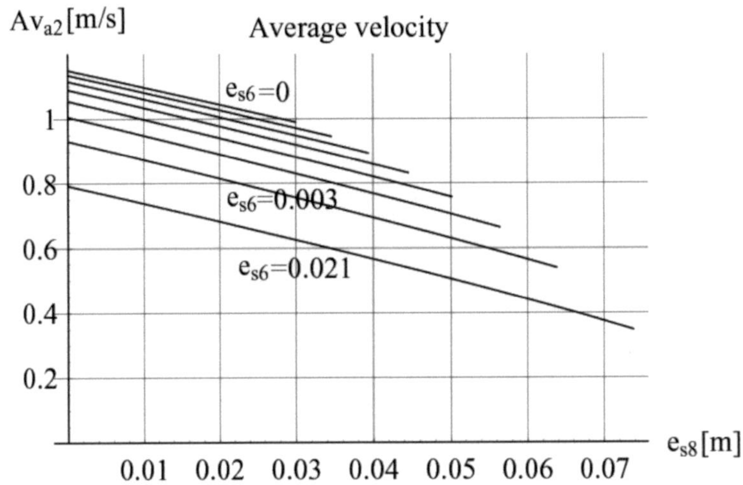

Figure 12. Variation of V_{ya2} with e_{s6} and e_{s8}

Fig. 8 to 12 show that

- *e_{s6} fixed but e_{s8} variable case*
  All of S_{a2}, t_{a2}, V_{ya2}, V_{h2}, and Av_{a2} decrease with e_{s8}. For a fixed e_{s6}, the position of point G (e_{s7}, S_{d2}) and the velocity at G are determined. Thus, the above phenomenon is very straightforward. Note that when $e_{s8}=e_{s7}(e_{s6})$, then there are $S_{a2}=0$, $t_{a2}=0$, $V_{ya2}=0$, $V_{h2}=V_{l2}$, i.e., the double support phase HI will be longest and the acceleration phase GH will disappear.
- *e_{s8} fixed but es6 variable case*
  S_{a2}, t_{a2}, V_{ya2}, and V_{h2} increase, but Av_{a2} decrease with e_{s6}. Note that as aforementioned, when $e_{s6}=e_{s6max}(e_{s5})$, the robot will stop at the point G and cannot enter acceleration phase. Here, this extreme case is omitted.

A high V_{h2} means a high terminal speed, while a low V_{h2} implies a smooth walking. Meanwhile a big Δv_{a2} means a fast walking. Thus, Fig. 10 and 12 show that both of small e_{s6} and e_{s8} are favorite in the view of both of smooth walking and fast walking.

5. Motion Planning and Numerical Example

5.1 Speed Adjustment during Walking

In section 4, we repeatedly point out that a short double support phase (saying, small e_{s5}, e_{s6}, e_{s7}, and etc.) is necessary for a smooth and fast walking. Of course, there is a premise for fast walking, that is, the robot should possess a high enough speed in the sagittal plane before entering a new double support phase. If not, because a short double support phase results in short phase stride and cannot change the robot velocities greatly, therefore the short double support phase cannot produce a fast walking (in this sense, the ability of speed adjustment of a short double support phase is weak). Consequently, we are confronted with a problem: how to get a high speed before entering a new walking cycle.

(The V_{h1} below means the terminal speed at the end of an acceleration phase, while V_{h0} implies the terminal speed at the end of the previous acceleration phase.)

With Eqs. (51) to (53), it is easily concluded that

1. $V_{h1}=V_{h0}$ requires $S_{a1}=S_{d1}$ or $e_{s5}=e_{s3}$. The walking cycle is called *isometric speed cycle.*
2. $V_{h1}>V_{h0}$ implies $S_{a1}>S_{d1}$ or $e_{s5}<e_{s3}$. The walking cycle is *accelerated*. In other words, if the start point (ex., point E) of a double support phase is closer to the midline (X axis) of two feet than the finish point (ex., point C) of the previous double support phase, the robot will be accelerated.
3. Contrarily, $V_{h1}<V_{h0}$ means $S_{a1}<S_{d1}$ or $e_{s5}<e_{s3}$. The walking cycle is *decelerated.*

Note that the above three conditions also imply that if the landing point D_0 is just at the midpoint of CE (Fig. 2), then the robot will just make an isometric speed walking; if the landing point D_0 is at the rear of the midpoint of CE, the robot will accelerate; contrarily, the robot will decelerate if the point D_0 is in front of the midpoint of CE. Thus, by adjusting the phase stride S_a and S_d, or the positions e_{s2}, e_{s3}, e_{s5}, e_{s6}, and e_{s8} of which the start and finish points of double support phases, the walking speed can be changed.

5.2 Motion Planning

To generalize the above discussions, an example of bipedal walking planning for ZMP fixed case is given out, in which, there are two walking cycles BCDE and EFGH. The first one, BCDE, makes an isometric speed walking and the second one, EFGH, is acceleration walking. Here, parameters z or ω, and d_0 are known.

Initial acceleration phase:

1. Determine e_{s1}, e_{s2} and calculate ρ_{a0}.
2. Determine S_{a0} with condition (22).
3. Calculate X_{s0}, V_{ya0}, V_{h0}, and t_{a0}.

Generally, a big e_{s1} results in a big S_{a0} and a high V_{h0}

Double support phase:

1. Determine e_{s3} or e_{s6}.
2. Calculate motion time t_{b1} and t_{b2}.
3. Calculate S_{d1} or S_{d2} with (34) or (56).
4. Determine moving distances of CoG in two directions and determine ZMP function in sagittal plane.

Deceleration phase:
1. Calculate V_{l1} and ρ_{d1} or V_{l2} and ρ_{d2}.
2. Calculate e_{s4} or e_{s7}, and t_{d1}, t_{d2}.

Acceleration phase:
1. Determine e_{s5} or e_{s8}.
2. Calculate S_{a1} or S_{a2}.
3. Calculate ρ_{a1}, V_{ya1}, V_{h1}, t_{a1}, or ρ_{a2}, V_{ya2}, V_{h2}, and, t_{a2}.

5.3 Numerical Example

Parameters used for motion planning are as follows: z=0.50[m], b=0.22[m]. Here, for the first walking cycle BCDE, $e_{s5}=e_{s3}=e_{s2}$, thus this cycle is an isometric speed one. For the second one EFGH, $e_{s8}<e_{s6}<e_{s5}$, it means this cycle is accelerated. In this example, the robot is simply accelerated in the initial acceleration phase AB, since the speed V_{h0} is high enough because of a big S_{a0}. Further, the robot is slightly accelerated in the second walking cycle.
The trajectories and velocities in X and Y directions of the robot's CoG, and ZMP's trajectories in X and Y directions are respectively shown in Fig. 13 to 18. The motion parameters shown in Fig. 13 to Fig. 18 are listed in Table 1. The average speed of the first cycle is about 3.05[km/h], while the one of the second cycle is about 4.04[km/h]. Note that in the second walking cycle, the swinging amplitude e_{s7} gets more narrow, the motion time t_{d2} and t_{a2} are shorter, but the phase strides S_{d2} and S_{a2} only get slightly shorter. Such a walking pattern is very similar to what a walker does in walking race, where the walker's swinging amplitude is small and the walking period is short, but his step length is not changed so much. Resultantly his walking speed is high. (In our example, the speed of the robot is about 4[km/h], but generally a walker's speed in walking race is over 10 [km/h], since the ZMP is variable from the heel to the toe in walking. The bipedal walking based on variable ZMP is beyond the scope of this paper.)

6. Conclusion

In this paper, a new design approach of bipedal walking pattern based on the synchronization of the motions in the sagittal and lateral planes are presented. With the discussion on the motions in these two planes, the fact is clarified that the motions in the sagittal and lateral planes are tightly coupled together. The motion parameters in the sagittal plane such as walking speed, walking time, and phase stride can be easily adjusted by altering the start and finish points of the double support phase in the lateral plane. Therefore, an approach for adjusting the walking speed by controlling the double support phase is naturally developed. Moreover, it is pointed out that a smooth and fast walking can be obtained by shortening the double support phase after a high speed is reached at the end of acceleration phase. Such a walking pattern is very similar to a walker's patter in walking race. The motion planning is also presented and a numerical example is given out. It is expected to apply this theory to a real bipedal robot and extend it to the jumping and running robot.
(This paper is originally published in IEEJ Trans. IA, Vol. 126, No. 8, pp. 1069-1078, 2006)

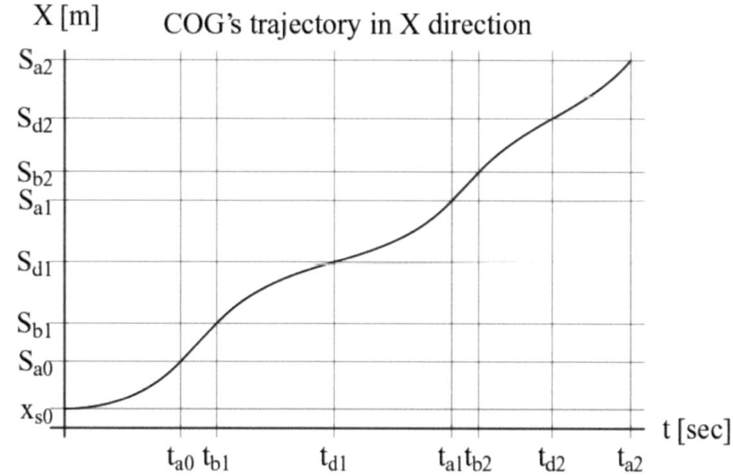

Figure 13. CoG's trajectory in X direction

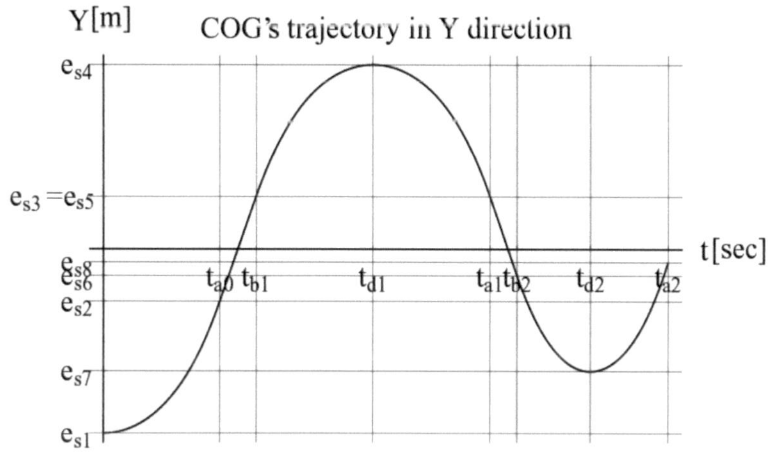

Figure 14. CoG's trajectory in Y direction

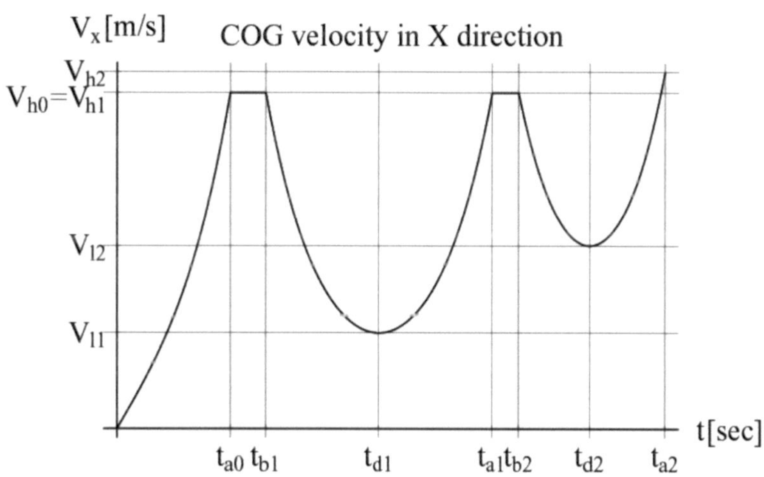

Figure 15. CoG's speed in X direction

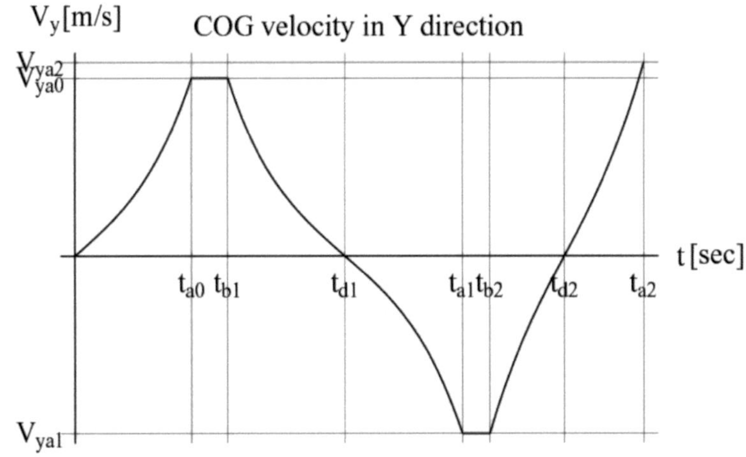

Figure 16. CoG's speed in Y direction

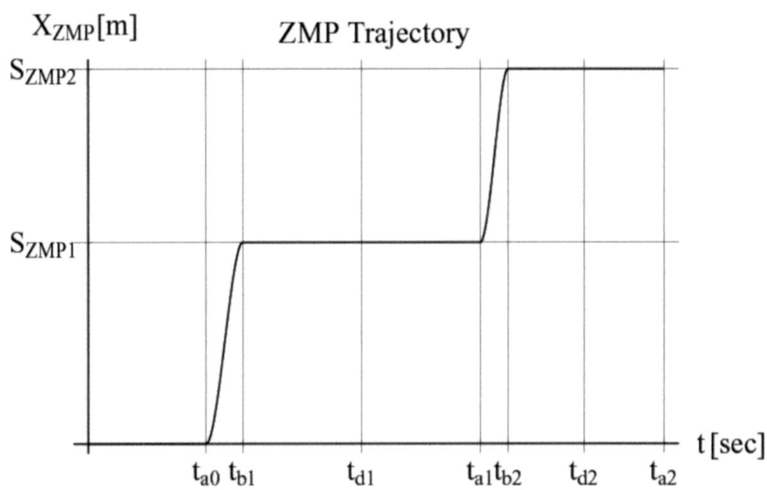

Figure 17. ZMP trajectory in X direction

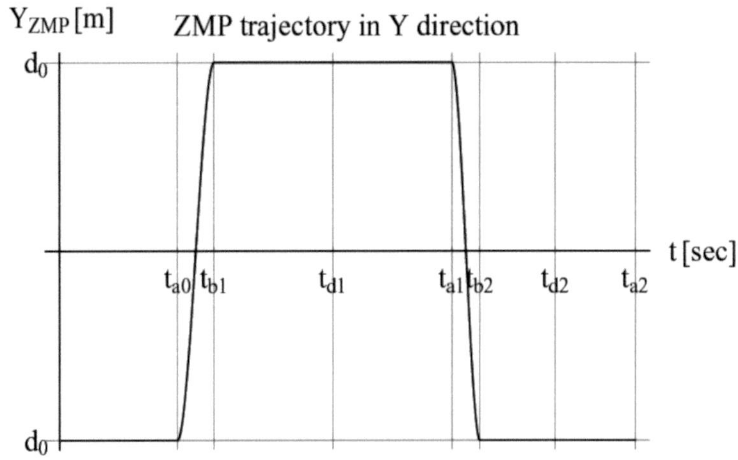

Figure 18. ZMP trajectory in X direction

S_{a0}	S_{b1}	S_{d1}	S_{a1}	S_{b2}	S_{d2}	S_{a2}	S_{ZMP1}	S_{ZMP2}	X_{s0}	d_0	Z_0
0.35	0.20	0.32	0.32	0.15	0.28	0.31	0.87	1.62	0.10	0.09	0.50
t_{a0}	t_{b1}	t_{d1}	t_{a1}	t_{b2}	t_{d2}	t_{a2}	V_{h0}	V_{l1}	V_{h1}	V_{l2}	V_{h2}
0.43	0.13	0.43	0.43	0.10	0.27	0.29	1.48	0.42	1.48	0.81	1.58
d_0	e_{s1}	e_{s2}	e_{s3}	e_{s4}	e_{s5}	e_{s6}	e_{s7}	e_{s8}	V_{ya0}	V_{ya1}	V_{ya2}
0.09	0.07	0.02	0.02	0.07	0.02	0.01	0.046	0.005	0.30	0.30	0.32

Table 1. Motion Parameters for ZMP fixed case

8. References

Hirai, K; Hirose, M.; Haikawa, Y.; & Takenaka, T. (1998), The Development of Honda Humanoid Robot, *Proceedings of the IEEE International Conference on Robotics and Automation,* pp.1321-1326, Leuven, Belgium, 1998.

Takanishi, Ishida, Yamazaki, kato (1985), Realization of Dynamic Walking on Biped Locomotion Robot WL-10RD, *Journal of Robotic Society of Japan*, vol. 3, No. 4, pp. 67-78, 1985. In Japanese.

Ogura, Y.; Aikawa, H.; Lim, H.; & Takanishi, A (2004). Development of a Human-like Walking Robot Having Two 7-DOF Legs and a 2 DOF Waist, *Proceedings of the IEEE International Conference on Robotics and Automation*, pp.134-139, New Orleans, LA, USA, 2004.

Kaneko, K.; Kanehiro, F.; Kajita, S. & et al (2004). Humanoid Robot HRP-2, *Proceedings of the IEEE International Conference on Robotics and Automation*, pp. 1083-1090, New Orleans, LA, USA, 2004.

Nakasaka, K.; & et al (2004). Integrated Motion Control for Walking, Jumping and Running on a Small Bipedal Entertainment Robot, *Proceedings of the IEEE International Conference on Robotics and Automation*, pp.3189-3194, New Orleans, LA, USA, 2004.

Löffler, K.; Gienger, M.; & Pfeiffer, F (2003). Sensor and Control Design of a Dynamically Stable Biped Robot, *Proceedings of the IEEE International Conference on Robotics and Automation*, pp.484-490, Taipei, Taiwan, 2003.

Lohmeier, S.; Löffler, K.; Gienger, M.; Ulbrich, H.; & Pfeiffer, F (2004). Computer System and Control of Biped "Johnnie", *Proceedings of the IEEE International Conference on Robotics and Automation*, pp.4222-4227, New Orleans, LA, USA, 2004.

Vukobratovich, M.; Borovac, B.; Surla, D.; & Stokic, D (1990). Biped Locomotion, Dynamics, Stability, Control and Application, Springer Verlag, 1990.

Goswami, A (1999). Postural Stability of Biped Robots and the FootRotation Indicator (FRI) Point, *International Journal of Robotics Research*, vol.19, no. 6, pp.523-533, 1999.

Kakami, S.; Nishiwaki, K.; Kitakawa, T. & et al (2000). A Fast Generation Method of a Dynamically Stable Humanoid Robot Trajectory with Enhanced ZMP Constraint, *Proceedings of the 2000 IEEE International Conference on Humanoid Robots*, CD-ROM, 2000.

Kajita, S. & Kobayashi, A (1987). Dynamic Walk Control of a Biped Robot with Potential Energy Conserving Orbit, *Transactions of the Society of Instrument and Control Engineers*, vol.23, no.3, pp.281-287, 1987, in Japanese.

Kajita, S. & et al (2002). A Real time Pattern Generator for Biped Walking, *Proceedings of the IEEE International Conference on Robotics and Automation*, Arlington VA (Washington DC), USA, pp.31-37, 2002.

Kajita, S.; & et al (2003). Biped Walking Pattern Generation by Using Preview Control of Zero-Moment Point, *Proceedings of the IEEE/RSJ International Conference on Intelligent Robots and Systems*, pp.1620-1626, Las Vegas, Nevada, USA, October 2003.

Zhu, C.; Okamura, M.; Kawamura, A.; & Tomizawa, Y (2004). Experimental approach for high speed walking of Biped Robot MARI-1, *Proceedings of the 9th International Workshop on Advanced Motion Control*, pp.427-432, 2004.

Zhu, C.; Kawamura, A (2003). Walking Principle Analysis for Biped Robot with ZMP Concept, Friction Constraint, and Inverted Pendulum Model, *Proceedings of the IEEE/RSJ International Conference on Intelligent Robots and Systems*, pp. 364-369, Las Vegas, Nevada, USA, October 2003.

Zhu, C.; Tomizawa, Y.; Luo, X.; & Kawamura, A (2004). Biped Walking with Variable ZMP, Frictional Constraint, and Inverted Pendulum Model, *Proceedings of The 2004 IEEE International Conference on Robotics and Biomimetics (ROBIO2004)*, CD-ROM, Shenyang, China. 2004.

18

Generating Natural Motion in an Android by Mapping Human Motion

Daisuke Matsui[1], Takashi Minato[1], Karl F. MacDorman[2] and Hiroshi Ishiguro[1]
[1]*Osaka University,* [2] *Indiana University*
[1]*Japan,* [2]*USA*

1. Introduction

Much effort in recent years has focused on the development of such mechanical-looking humanoid robots as Honda's Asimo and Sony's Qrio with the goal of partnering them with people in daily situations. Just as an industrial robot's purpose determines its appearance, a partner robot's purpose will also determine its appearance. Partner robots generally adopt a roughly humanoid appearance to facilitate communication with people, because natural interaction is the only task that requires a humanlike appearance. In other words, humanoid robots mainly have significance insofar as they can interact naturally with people. Therefore, it is necessary to discover the principles underlying natural interaction to establish a methodology for designing interactive humanoid robots.

Kanda et al. (Kanda et al., 2002) have tackled this problem by evaluating how the behavior of the humanoid robot "Robovie" affects human-robot interaction. But Robovie's machine-like appearance distorts our interpretation of its behavior because of the way the complex relationship between appearance and behavior influences the interaction. Most research on interactive robots has not evaluated the effect of appearance (for exceptions, see (Goetz et al., 2003; DiSalvo et al., 2002)) — and especially not in a robot that closely resembles a person. Thus, it is not yet clear whether the most comfortable and effective human-robot communication would come from a robot that looks mechanical or human. However, we may infer a humanlike appearance is important from the fact that human beings have developed neural centers specialized for the detection and interpretation of hands and faces (Grill-Spector et al., 2004; Farah et al., 2000; Carmel & Bentin, 2002). A robot that closely resembles humans in both looks and behavior may prove to be the ultimate communication device insofar as it can interact with humans the most naturally.[1] We refer to such a device as an android to distinguish it from mechanical-looking humanoid robots. When we investigate the essence of how we recognize human beings as human, it will become clearer how to produce natural interaction. Our study tackles the appearance and behavior problem with the objective of realizing an android and having it be accepted as a human being (Minato et al., 2006).

---

[1] We use the term *natural* to denote communication that flows without seeming stilted, forced, bizarre, or inhuman.

Ideally, to generate humanlike movement, an android's kinematics should be functionally equivalent to the human musculoskeletal system. Some researchers have developed a joint system that simulates shoulder movement (Okada et al., 2002) and a muscle-tendon system to generate humanlike movement (Yoshikai et al., 2003). However, these systems are too bulky to be embedded in an android without compromising its humanlike appearance. Given current technology, we embed as many actuators as possible to provide many degrees of freedom insofar as this does *not* interfere with making the android look as human as possible (Minato et al., 2006). Under these constraints, the main issue concerns how to move the android in a natural way so that its movement may be perceived as human.

A straightforward way to make a robot's movement more humanlike is to imitate human motion. Kashima and Isurugi (Kashima & Isurugi, 1998) extracted essential properties of human arm trajectories and designed an evaluation function to generate robot arm trajectories accordingly. Another method is to copy human motion as measured by a motion capture system to a humanoid robot. Riley et al. (Riley et al., 2000) and Nakaoka et al. (Nakaoka et al., 2003) calculated a subject's joint trajectories from the measured positions of markers attached to the body and fed them to the joints of a humanoid robot. In these studies the authors assumed the kinematics of the robot to be similar to that of a human body. However, since the actual kinematics and joint structures are different between human and robot bodies, calculating the joint angles from only the human motion data could in some cases result in visibly different motion. This is especially a risk for androids because their humanlike form makes us more sensitive to deviations from human ways of moving. Thus, slight differences could strongly influence whether the android's movement is perceived as natural or human. Furthermore, these studies did not evaluate the naturalness of robot motions.

Hale et al. (Hale et al., 2003) proposed several evaluation functions to generate a joint trajectory (e.g., minimization of jerk) and evaluated the naturalness of generated humanoid robot movements according to how human subjects rated their naturalness. In the computer animation domain, researchers have tackled a motion synthesis with motion capture data (e.g., (Gleicher, 1998)). However, we cannot apply their results directly; we must instead repeat their experiment with an android because the results from an android testbed could be quite different from those of a humanoid testbed. For example, Mori described a phenomenon he termed the "uncanny valley" (Mori, 1970; Fong et al., 2003), which relates to the relationship between how humanlike a robot appears and a subject's perception of familiarity. According to Mori, a robot's familiarity increases with its similarity until a certain point is reached at which slight "nonhuman" imperfections cause the robot to appear repulsive (Fig. 1). This would be an issue if the similarity of androids fell into the chasm. (Mori believes mechanical-looking humanoid robots lie on the left of the first peak.) This nonmonotonic relationship can distort the evaluation proposed in existing studies. Therefore, it is necessary to develop a motion generation method in which the generated "android motion" is perceived as human.

This paper proposes a method to transfer human motion measured by a motion capture system to the android by copying changes in the positions of body surfaces. This method is called for because the android's appearance demands movements that look human, but its kinematics is sufficiently different that copying joint-angle information would not yield good results. Comparing the similarity of the android's visible movement to that of a human being enables us to develop more natural movements for the android.

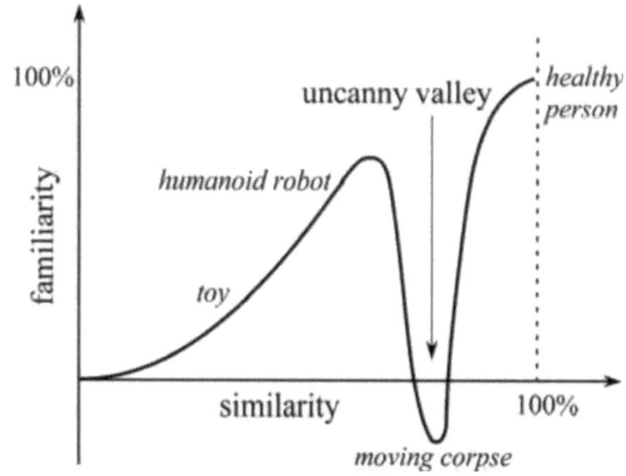

Figure 1. Uncanny valley (Mori, 1970; Fong et al., 2003)

In the following sections, we describe the developed android and mention the problem of motion transfer and our basic idea about the way to solve it. Then we describe the proposed method in detail and show experimental results from applying it to the android.

2. The Developed Android

Fig. 2 shows the developed android called Repliee Q2. The android is modeled after a Japanese woman. The standing height is about 160 cm. The skin is composed of a kind of silicone that feels like human skin. The silicone skin covers the neck, head, and forearms, with clothing covering other body parts. Unlike Repliee R1 (Minato et al., 2004), the silicone skin does not cover the entire body so as to facilitate flexibility and a maximal range of motion. Forty-two highly sensitive tactile sensors composed of PVDF film are mounted under the android's skin and clothes over the entire body, except for the shins, calves, and feet. Since the output value of each sensor corresponds to its deforming rate, the sensors can distinguish different kinds of touch ranging from stroking to hitting. The soft skin and tactile sensors give the android a human appearance and enable natural tactile interaction.

The android is driven by air actuators (air cylinders and air motors) that give it 42 degrees of freedom (DoFs) from the waist up. The legs and feet are not powered; it can neither stand up nor move from a chair. A high power-to-weight ratio is necessary for the air actuator in order to mount multiple actuators in the human-sized body.

The configuration of the DoFs is shown in Table 1. Fig. 4 shows the kinematic structure of the body, excluding the face and fingers. Some joints are driven by the air motors and others adopt a slider-crank mechanism. The DoFs of the shoulders enable them to move up and down and backwards and forwards; this shoulder structure is more complicated than that of most existing humanoid robots. Moreover, parallel link mechanisms adopted in some parts complicate the kinematics of the android, for example in the waist. The android can generate a wide range of motions and gestures as well as various kinds of micro-motions such as the shoulder movements typically caused by human breathing. Furthermore, the android can make some facial expressions and mouth shapes, as shown in Fig. 3. Because the android has servo controllers, it can be controlled by sending data on the desired joint angles (cylinder positions and rotor angles) from a host computer. The compliance of the air actuator makes for safer interaction, with movements that are generally smoother than other systems typically used. Because of the complicated dynamics of the air actuator, executing the trajectory tracking control is difficult.

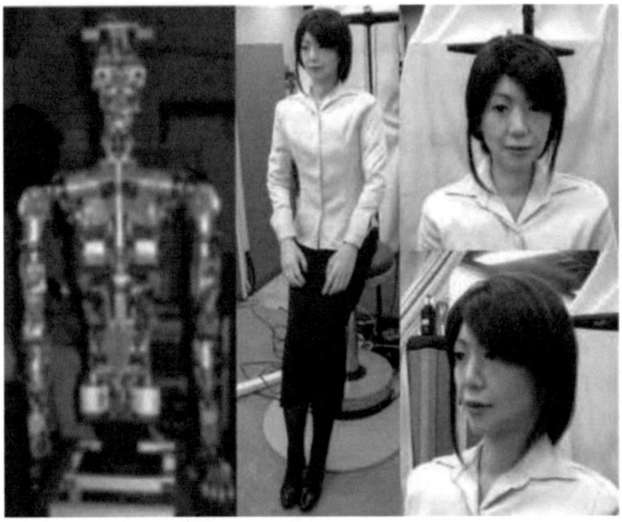

Figure 2. The developed android "Repliee Q2"

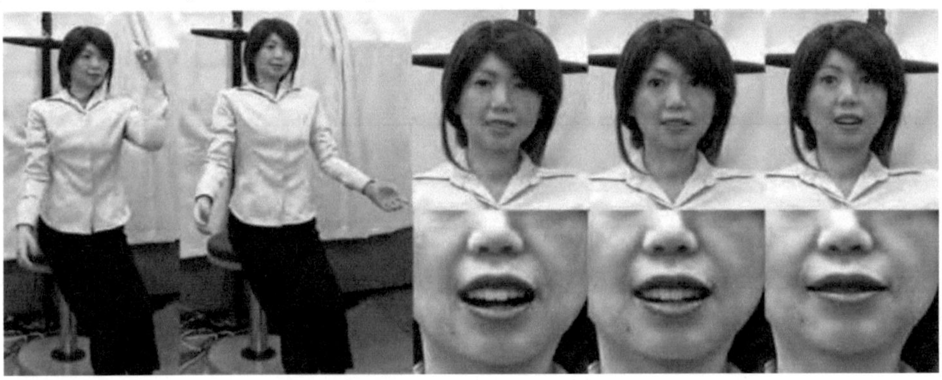

Figure 3. Examples of motion and facial expressions

Degree of freedom	
Eyes	pan×2 + tilt×1
Face	eyebrows×1 + eyelids×1 + cheeks×1
Mouth	7 (including the upper and lower lips)
Neck	3
Shoulder	5×2
Elbow	2×2
Wrist	2×2
Fingers	2×2
Torso	4

Table 1. The DoF configuration of Repliee Q2.

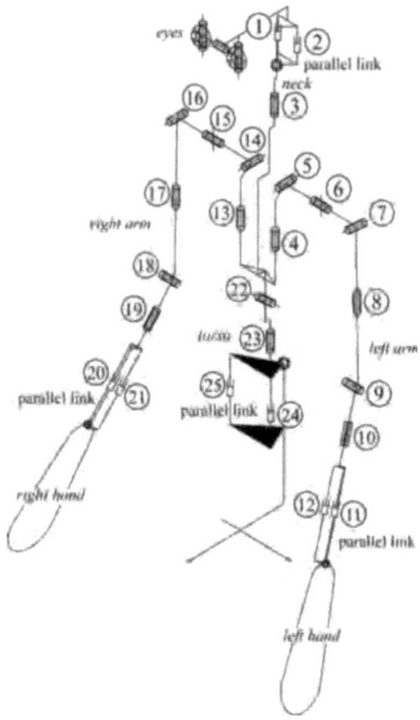

Figure 4. Kinematic structure of the android

3. Transferring Human Motion

3.1 The basic idea

One method to realize humanlike motion in a humanoid robot is through imitation. Thus, we consider how to map human motion to the android. Most previous research assumes the kinematics of the human body is similar to that of the robot except for the scale. Thus, they aim to reproduce human motion by reproducing kinematic relations across time and, in particular, joint angles between links. For example, the three-dimensional locations of markers attached to the skin are measured by a motion capture system, the angles of the body's joints are calculated from these positions, and these angles are transferred to the joints of the humanoid robot. It is assumed that by using a joint angle space (which does not represent link lengths), morphological differences between the human subject and the humanoid robot can be ignored.

However, there is potential for error in calculating a joint angle from motion capture data. The joint positions are assumed to be the same between a humanoid robot and the human subject who serves as a model; however, the kinematics in fact differs. For example, the kinematics of Repliee Q2's shoulder differs significantly from those of human beings. Moreover, as human joints rotate, each joint's center of rotation changes, but joint-based approaches generally assume this is not so. These errors are perhaps more pronounced in Repliee Q2, because the android has many degrees of freedom and the shoulder has a more complex kinematics than existing humanoid robots. These errors are more problematic for an android than a mechanical-looking humanoid robot because we expect natural human motion from something that looks human and are disturbed when the motion instead looks inhuman.

To create movement that appears human, we focus on reproducing positional changes at the body's surface rather than changes in the joint angles. We then measure the postures of a person and the android using a motion capture system and find the control input to the android so that the postures of person and android become similar to each other.

3.2 The method to transfer human motion

We use a motion capture system to measure the postures of a human subject and the android. This system can measure the three-dimensional positions of markers attached to the surface of bodies in a global coordinate space. First, some markers are attached to the android so that all joint motions can be estimated. The reason for this will become clear later. Then the same numbers of markers are attached to corresponding positions on the subject's body. We must assume the android's surface morphology is not too different from the subject's.

We use a three-layer neural network to construct a mapping from the subject's posture x_h to the android's control input q_a, which is the desired joint angle. The reason for the network is that it is difficult to obtain the mapping analytically. To train a neural network to map from x_h to q_a would require thousands of pairs of x_h, q_a as training data, and the subject would need to assume the posture of the android for each pair. We avoid this prohibitively lengthy task in data collection by adopting feedback error learning (FEL) to train the neural network. Kawato et al. (Kawato et al., 1987) proposed feedback error learning as a principle for learning motor control in the brain. This employs an approximate way of mapping sensory errors to motor errors that subsequently can be used to train a neural network (or other method) by supervised learning. Feedback-error learning neither prescribes the type of neural network employed in the control system nor the exact layout of the control circuitry. We use it to estimate the error between the postures of the subject and the android and feed the error back to the network.

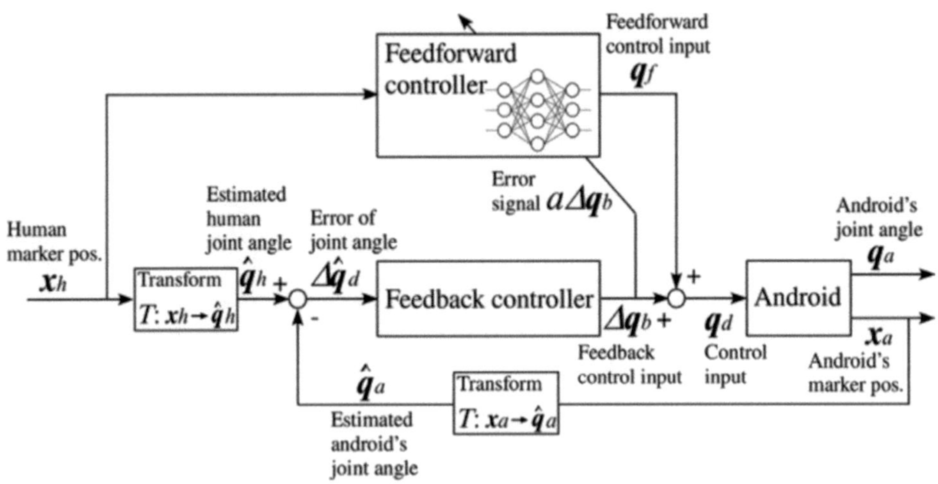

Figure 5. The android control system

Fig. 5 shows the block diagram of the control system, where the network mapping is shown as the feedforward controller. The weights of the feedforward neural network are learned by means of a feedback controller. The method has a two-degrees-of-freedom control architecture. The network tunes the feedforward controller to be the inverse model of the plant. Thus, the feedback error signal is employed as a teaching signal for learning the inverse model. If the inverse model is learned exactly, the output of the plant tracks the reference signal by feedforward control. The subject and android's marker positions are represented in their local coordinates x_h, $x_a \in R^{3m}$; the android's joint angles $q_a \in R^{n}$ can be observed by a motion capture system and a potentiometer, where m is the number of markers and n is the number of DoFs of the android.

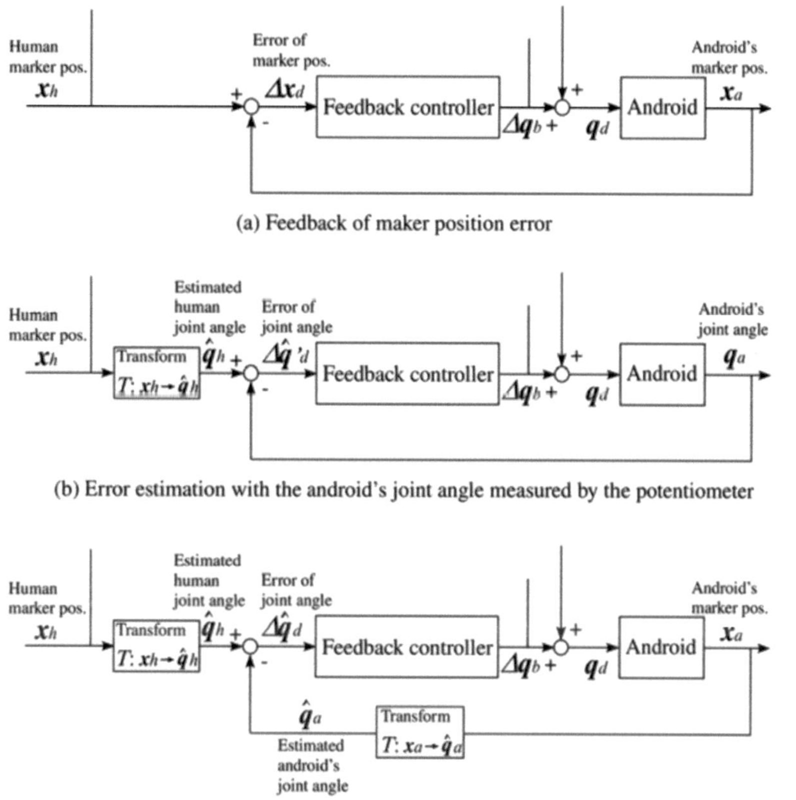

(a) Feedback of maker position error

(b) Error estimation with the android's joint angle measured by the potentiometer

(c) Error estimation with the android's joint angle estimated from the android's marker position

Fig 6. The feedback controller with and without the estimation of the android's joint angle

The feedback controller is required to output the feedback control input Δq_b so that the error in the marker's position $\Delta x_d = x_a - x_h$ converges to zero (Fig 6(a)). However, it is difficult to obtain Δq_b from Δx_d. To overcome this, we assume the subject has roughly the same kinematics as the android and obtain the estimated joint angle $\hat{q}_h$ simply by calculating the Euler angles (hereafter the transformation from marker positions to joint angles is described as T).[2] Converging q_a to $\hat{q}_h$ does not always produce identical postures because $\hat{q}_h$ is an

[2] There are alternatives to using the Euler angles such as angle decomposition (Grood & Suntay, 1983), which has the advantage of providing a sequence independent representation, or least squares, to calculate the helical axis and rotational angle (Challis, 1995; Veldpaus et al., 1988). This last method provides higher accuracy when many markers are used but has an increased risk of marker crossover.

approximate joint angle that may include transformation error (Fig. 6(b)). Then we obtain the estimated joint angle of the android $\hat{q}_a$ using the same transformation T and the feedback control input to converge $\hat{q}_a$ to $\hat{q}_h$ (Fig. 6(c)). This technique enables $\boldsymbol{x}_a$ to approach $\boldsymbol{x}_h$. The feedback control input approaches zero as learning progresses, while the neural network constructs the mapping from $\boldsymbol{x}_h$ to the control input $\boldsymbol{q}_d$. We can evaluate the apparent posture by measuring the android posture.

In this system we could have made another neural network for the mapping from $\boldsymbol{x}_a$ to $\boldsymbol{q}_a$ using only the android. As long as the android's body surfaces are reasonably close to the subject's, we can use the mapping to make the control input from $\boldsymbol{x}_h$. Ideally, the mapping must learn every possible posture, but this is quite difficult. Therefore, it is still necessary for the system to evaluate the error in the apparent posture.

4. Experiment to Transfer Human Motion

4.1 Experimental setting

To verify the proposed method, we conducted an experiment to transfer human motion to the android Repliee Q2. We used 21 of the android's 42 DoFs (n = 21) by excluding the 13 DoFs of the face, the 4 of the wrists (the cylinders 11, 12, 20, and 21 in Fig. 4), and the 4 of the fingers. We used a Hawk Digital System,[3] which can track more than 50 markers in real-time. The system is highly accurate with a measurement error of less than 1 mm. Twenty markers were attached to the subject and another 20 to the android as shown in Fig. 7 (m = 20). Because the android's waist is fixed, the markers on the waist set the frame of reference for an android-centered coordinate space. To facilitate learning, we introduce a representation of the marker position $\boldsymbol{x}_h$, $\boldsymbol{x}_a$ as shown in Fig. 8. The effect of waist motions is removed with respect to the markers on the head. To avoid accumulating the position errors at the end of the arms, vectors connecting neighboring pairs of markers represent the positions of the markers on the arms. We used arc tangents for the transformation T, in which the joint angle is an angle between two neighboring links where a link consists of a straight line between two markers.

The feedback controller outputs $\Delta\boldsymbol{q}_b = K\Delta\hat{\boldsymbol{q}}_d$, where the gain K consists of a diagonal matrix. There are 60 nodes in the input layer (20 markers × x, y, z), 300 in the hidden layer, and 21 in the output layer (for the 21 DoFs). Using 300 units in the hidden layer provided a good balance between computational efficiency and accuracy. Using significantly fewer units resulted in too much error, while using significantly more units provided only marginally higher accuracy but at the cost of slower convergence. The error signal to the network is $t = a\Delta\boldsymbol{q}_b$, where the gain a is a small number. The sampling time for capturing the marker positions and controlling the android is 60 ms. Another neural network which has the same structure previously learned the mapping from $\boldsymbol{x}_a$ to $\boldsymbol{q}_a$ to set the initial values of the weights. We obtained 50,000 samples of training data ($\boldsymbol{x}_a$ and $\boldsymbol{q}_a$) by moving the android randomly. The learned network is used to set the initial weights of the feedforward network.

[3] Motion Analysis Corporation, Santa Rosa, California. http://www.motionanalysis.com/

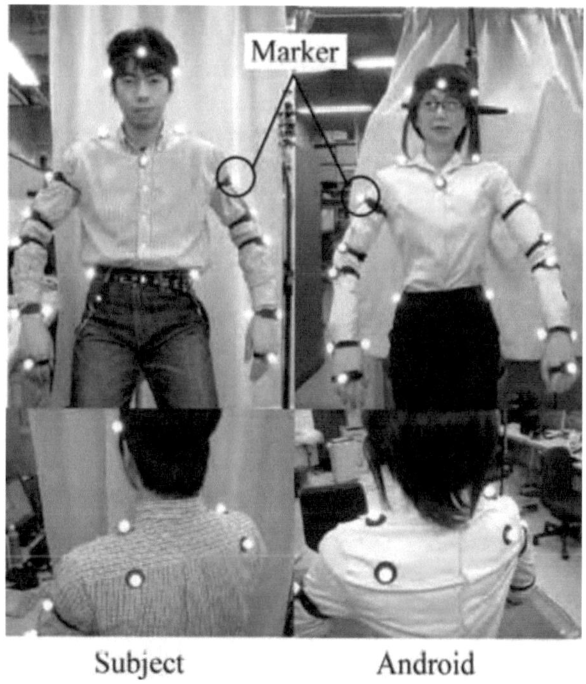

Figure 7. The marker positions corresponding to each other

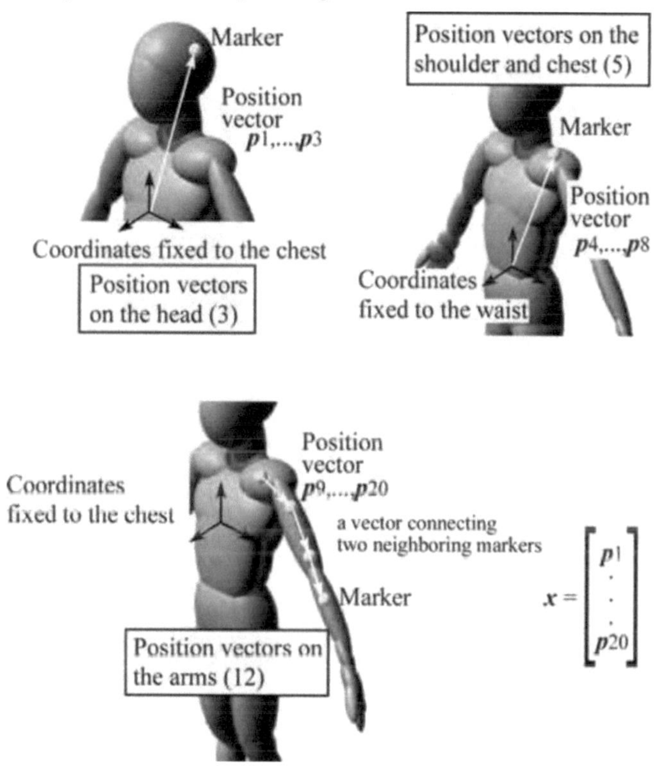

Figure 8. The representation of the marker positions. A marker's diameter is about 18 mm

4.2 Experimental results and analysis

4.2.1 Surface similarity between the android and subject

The proposed method assumes a surface similarity between the android and the subject. However, the male subject whom the android imitates in the experiments was 15 cm taller than the women after whom the android was modeled. To check the similarity, we measured the average distance between corresponding pairs of markers when the android and subject make each of the given postures; the value was 31 mm (see the Fig. 7). The gap is small compared to the size of their bodies, but it is not small enough.

4.2.2 The learning of the feedforward network

To show the effect of the feedforward controller, we plot the feedback control input averaged among the joints while learning from the initial weights in Fig. 9. The abscissa denotes the time step (the sampling time is 60 ms.) Although the value of the ordinate does not have a direct physical interpretation, it corresponds to a particular joint angle. The subject exhibited various fixed postures. When the subject started to make the posture at step 0, error increased rapidly because network learning had not yet converged. The control input decreases as learning progresses. This shows that the feedforward controller learned so that the feedback control input converges to zero.

Fig. 10 shows the average position error of a pair of corresponding markers. The subject also gave an arbitrary fixed posture. The position errors and the feedback control input both decreased as the learning of the feedforward network converged. The result shows the feedforward network learned the mapping from the subject's posture to the android control input, which allows the android to adopt the same posture. The android's posture could not match the subject's posture when the weights of the feedforward network were left at their initial values. This is because the initial network was not given every possible posture in the pre-learning phase. The result shows the effectiveness of the method to evaluate the apparent posture.

4.2.3 Performance of the system at following fast movements

To investigate the performance of the system, we obtained a step response using the feedforward network after it had learned enough. The subject put his right hand on his knee and quickly raised the hand right above his head. Fig. 11 shows the height of the fingers of the subject and android. The subject started to move at step 5 and reached the final position at step 9, approximately 0.24 seconds later. In this case the delay is 26 steps or 1.56 seconds. The arm moved at roughly the maximum speed permitted by the hardware. The android arm cannot quite reach the subject's position because the subject's position was outside of the android's range of motion. Clearly, the speed of the subject's movement exceeds the android's capabilities. This experiment is an extreme case. For less extreme gestures, the delay will be much less. For example, for the sequence in Fig. 12, the delay was on average seven steps or 0.42 seconds.

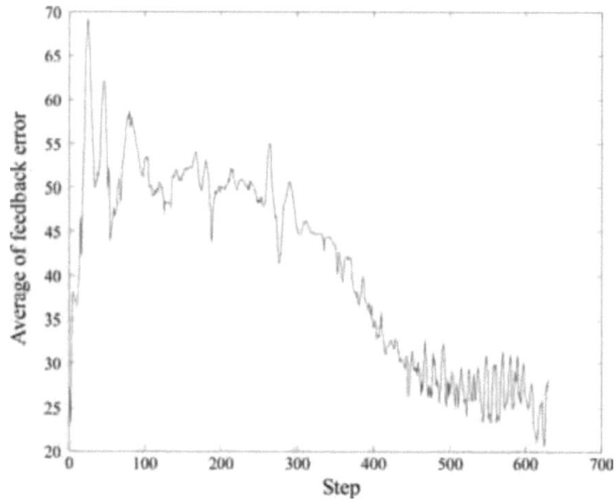

Figure 9. The change of the feedback control input with learning of the network

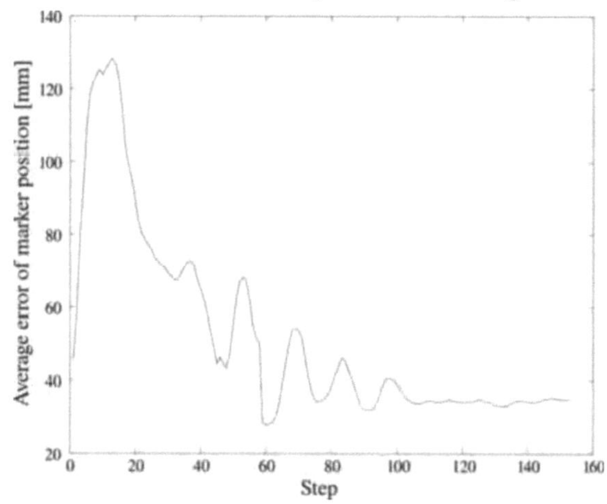

Figure 10. The change of the position error with learning of the network

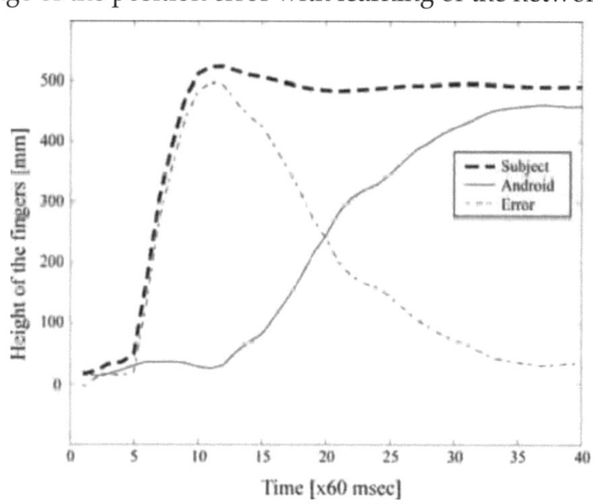

Figure 11. The step response of the android

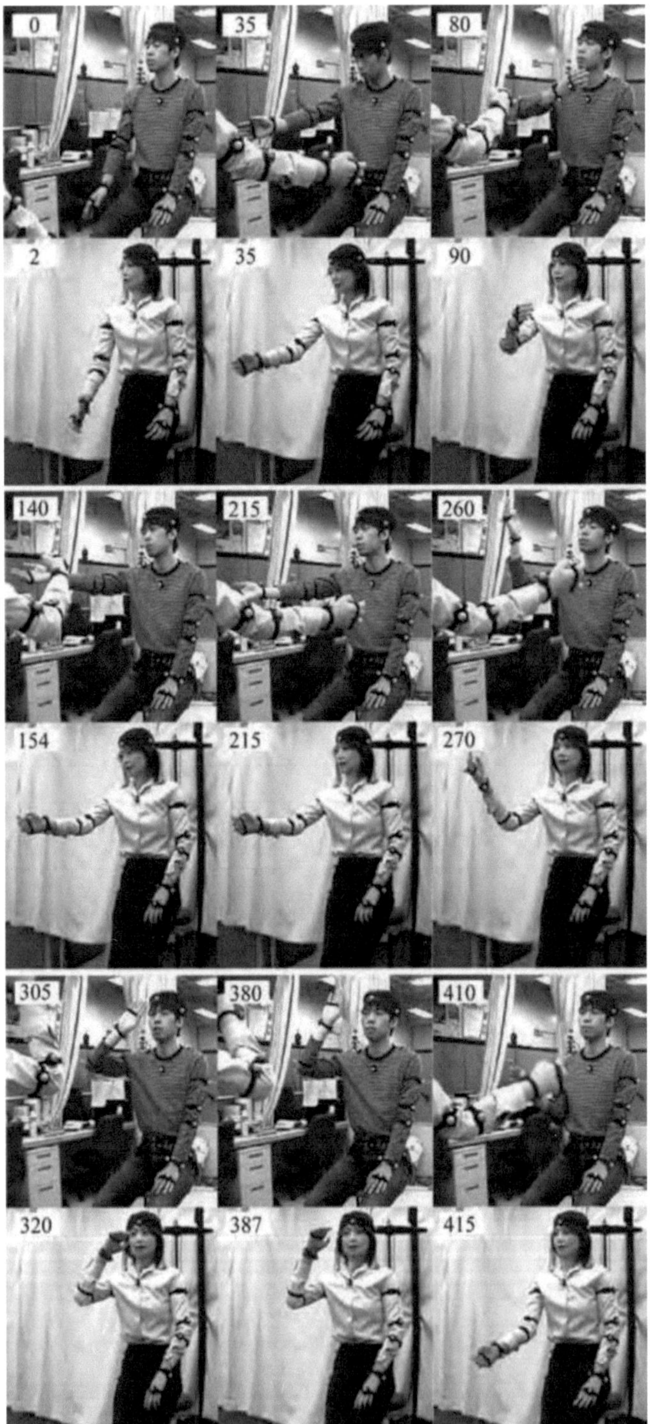

Figure 12. The generated android's motion compared to the subject's motion. The number represents the step.

4.2.4 The generated android motion

Fig. 12 shows the subject's postures during a movement and the corresponding postures of the android. The value denotes the time step. The android followed the subject's movement with some delay (the maximum is 15 steps, that is, 0.9 seconds). The trajectories of the positions of the android's markers are considered to be similar to those of the subject, but errors still remain, and they cannot be ignored. While we can recognize that the android is making the same gesture as the subject, the quality of the movement is not the same. There are a couple of major causes of this:

- The kinematics of the android is too complicated to represent with an ordinary neural network. To avoid this limitation, it is possible to introduce the constraint of the body's branching in the network connections. Another idea is to introduce a hierarchical representation of the mapping. A human motion can be decomposed into a dominant motion that is at least partly driven consciously and secondary motions that are mainly nonconscious (e.g., contingent movements to maintain balance, such autonomic responses as breathing). We are trying to construct a hierarchical representation of motion not only to reduce the computational complexity of learning but to make the movement appear more natural.
- The method deals with a motion as a sequence of postures; it does not precisely reproduce higher order properties of motion such as velocity and acceleration because varying delays can occur between the subject's movement and the android's imitation of it. If the subject moves very quickly, the apparent motion of the android differs. Moreover, a lack of higher order properties prevents the system from adequately compensating for the dynamic characteristics of the android and the delay of the feedforward network.
- The proposed method is limited by the speed of motion. It is necessary to consider the properties to overcome the restriction, although the android has absolute physical limitations such as a fixed compliance and a maximum speed that is less than that of a typical human being.

Although physical limitations cannot be overcome by any control method, there are ways of finessing them to ensure movements still look natural. For example, although the android lacks the opponent musculature of human beings, which affords a variable compliance of the joints, the wobbly appearance of such movements as rapid waving, which are high in both speed and frequency, can be overcome by slowing the movement and removing repeated closed curves in the joint angle space to eliminate lag caused by the slowed movement. If the goal is humanlike movement, one approach may be to query a database of movements that are known to be humanlike to find the one most similar to the movement made by the subject, although this begs the question of where those movements came from in the first place. Another method is to establish criteria for evaluating the naturalness of a movement (Kashima & Isurugi, 1998). This is an area for future study.

4.3 Required improvement and future work

In this paper we focus on reproducing positional changes at the body's surface rather than changes in the joint angles to generate the android's movement. Fig. 6(a) is a straightforward method to implement the idea. This paper has adopted the transformation T from marker positions to estimated joint angles because it is difficult to derive a feedback controller which produces the control input Δq_b only from the error in the marker's positional error Δx_d analytically. We actually do not know which joints should be moved to remove a positional error at the body's surface. This relation must be learned, however, the transformation T could disturb the learing. Hence, it is not generally guaranteed that the feedback controller which converges the estimated joint angle $\hat{q}_a$ to $\hat{q}_h$ enables the marker's position x_a to approach x_h. The assumption that the android's body surfaces are reasonably close to the subject's could avoid this problem, but the feedback controller shown in Fig. 6(a) is essentially necessary for mapping the apparent motion. It is possible to find out how the joint changes relate to the movements of body surfaces by analyzing the weights of the neural network of the feedforward controller. A feedback controller could be designed to output the control input based on the error in the marker's position with the analyzed relation. Concerning the design of the feedback controller, Oyama et al. (Oyama et al., 2001a; Oyama et al., 2001b; Oyama et al., 2001c) proposed several methods for learning both of feedback and feedforward controllers using neural networks. This is one potential method to obtain the feedback controller shown in Fig. 6(a). Assessment of and compensation for deformation and displacement of the human skin, which cause marker movement with respect to the underlying bone (Leardini et al., 2005), are also useful in designing the feedback controller.

We have not dealt with the android's gaze and facial expressions in the experiment; however, if gaze and facial expressions are unrelated to hand gestures and body movements, the appearance is often unnatural, as we have found in our experiments. Therefore, to make the android's movement appear more natural, we have to consider a method to implement the android's eye movements and facial expressions.

5. Conclusion

This paper has proposed a method of implementing humanlike motions by mapping their three-dimensional appearance to the android using a motion capture system. By measuring the android's posture and comparing it to the posture of a human subject, we propose a new method to evaluate motion sequences along bodily surfaces. Unlike other approaches that focus on reducing joint angle errors, we consider how to evaluate differences in the android's apparent motion, that is, motion at its visible surfaces. The experimental results show the effectiveness of the evaluation: the method can transfer human motion. However, the method is restricted by the speed of the motion. We have to introduce a method to deal with the dynamic characteristics (Ben=Amor et al., 2007) and physical limitations of the android. We also have to evaluate the method with different subjects. We would expect to generate the most natural and accurate movements using a female subject who is about the same height as the original woman on which the android is based. Moreover, we have to evaluate the human likeness of the visible motions by the subjective impressions the android gives experimental subjects and the responses it elicits, such as eye contact (Minato et al., 2006; Shimada et al., 2006), autonomic responses, and so on. Research in these areas is in progress.

6. Acknowledgment

We developed the android in collaboration with Kokoro Company, Ltd

7. References

Ben=Amor, H., Ikemoto, S., Minato, T., Jung, B., and Ishiguro, H. (2007). A neural framework for robot motor learning based on memory consolidation. *Proceedings of International Conference on Adaptive and Natural Computing Algorithms*, Warsaw, Poland, 2007.04.

Carmel, D. and Bentin, S. (2002). Domain specificity versus expertise: Factors influencing distinct processing of faces. *Cognition*, Vol. 83, (1–29), ISSN:0010-0277.

Challis, J. H. (1995). A procedure for determining rigid body transformation parameters. *Journal of Biomechanics*, Vol. 28, (733–737), ISSN:0021-9290.

DiSalvo, C. F., Gemperle, F., Forlizzi, J., and Kiesler, S. (2002). All robots are not created equal: The design and perception of humanoid robot heads. *Proceedings of the Symposium on Designing Interactive Systems*, pp. 321–326, ISBN:1-58113-515-7, London, England, 2002.06.

Farah, M. J., Rabinowitz, C., Quinn, G. E., and Liu, G. T. (2000). Early commitment of neural substrates for face recognition. *Cognitive Neuropsychology*, Vol. 17, (117–123), ISSN:0264-3294.

Fong, T., Nourbakhsh, I., and Dautenhahn, K. (2003). A survey of socially interactive robots. *Robotics and Autonomous Systems*, Vol. 42, (143–166), ISSN:0921-8890.

Gleicher, M. (1998). Retargetting motion to new characters. *Proceedings of the International Conference on Computer Graphics and Interactive Techniques*, pp. 33–42, ISBN:0-89791-999-8, Florida, USA, 2003.07.

Goetz, J., Kiesler, S., and Powers, A. (2003). Matching robot appearance and behavior to tasks to improve human-robot cooperation. *Proceedings of the Workshop on Robot and Human Interactive Communication*, pp. 55–60, ISBN:0-7803-8136-X, California, USA, 2003.10.

Grill-Spector, K., Knouf, N., and Kanwisher, N. (2004). The fusiform face area subserves face perception, not generic within-category identification. *Nature Neuroscience*, Vol. 7, No. 5, (555–562), ISSN:1097-06256.

Grood, E. S. and Suntay, W. J. (1983). A joint coordinate system for the clinical description of three-dimensional motions: Application to the knee. *Journal of Biomechanical Engineering*, Vol. 105, (136–144), ISSN:0148-0731.

Hale, J. G., Pollick, F. E., and Tzoneva, M. (2003). The visual categorization of humanoid movement as natural. *Proceedings of the IEEE-RAS/RSJ International Conference on Humanoid Robots*, ISBN:3-00-012047-5, Munich, Germany, 2003.10.

Kanda, T., Ishiguro, H., Ono, T., Imai, M., and Mase, K. (2002). Development and evaluation of an interactive robot "Robovie". *Proceedings of the IEEE International Conference on Robotics and Automation*, pp. 1848–1855, ISBN:0-7803-7272-7, Washington D.C., USA, 2002.05.

Kashima, T. and Isurugi, Y. (1998).Trajectory formation based on physiological characteristics of skeletal muscles. *Biological Cybernetics*, Vol. 78, No. 6, (413–422), ISSN:0340-1220.

Kawato, M., Furukawa, K., and Suzuki, R. (1987). A hierarchical neural network model for control and learning of voluntary movement. *Biological Cybernetics*, Vol. 57, (169–185), ISSN:0340-1220.

Leardini, A., Chiari, L., Croce, U. D., and Cappozzo, A. (2005). Human movement analysis using stereophotogrammetry Part 3. Soft tissue artifact assessment and compensation. *Gait and Posture*, Vol. 21, (212–225), ISSN:0966-6362.

Minato, T., Shimada, M., Ishiguro, H., and Itakura, S. (2004). Development of an android robot for studying human-robot interaction. *Proceedings of the 17th International Conference on Industrial & Engineering Applications of Artificial Intelligence & Expert Systems*, pp. 424–434, ISBN:3-540-22007-0, Ottawa, Canada, 2004.05.

Minato, T., Shimada, M., Itakura, S., Lee, K., and Ishiguro, H. (2006). Evaluating the human likeness of an android by comparing gaze behaviors elicited by the android and a person. *Advanced Robotics*, Vol. 20, No. 10, (1147–1163), ISSN:0169-1864.

Mori, M. (1970). Bukimi no tani [the uncanny valley] (in Japanese). *Energy*, Vol. 7, No. 4, (33–35), ISSN:0013-7464.

Nakaoka, S., Nakazawa, A., Yokoi, K., Hirukawa, H., and Ikeuchi, K. (2003). Generating whole body motions for a biped humanoid robot from captured human dances. *Proceedings of the IEEE International Conference on Robotics and Automation*, ISBN:0-7803-7737-0, Taipei, Taiwan, 2003.09.

Okada, M., Ban, S., and Nakamura, Y. (2002). Skill of compliance with controlled charging/discharging of kinetic energy. *Proceeding of the IEEE International Conference on Robotics and Automation*, pp. 2455–2460, ISBN:0-7803-7272-7, Washington D.C., USA, 2002.05.

Oyama, E., Agah, A., MacDorman, K. F., Maeda, T., and Tachi, S. (2001a). A modular neural network architecture for inverse kinematics model learning. *Neurocomputing*, Vol. 38–40, (797–805), ISSN:0925-2312.

Oyama, E., Chong, N. Y., Agah, A., Maeda, T., Tachi, S., and MacDorman, K. F. (2001b). Learning a coordinate transformation for a human visual feedback controller based on disturbance noise and the feedback error signal. *Proceedings of the IEEE International Conference on Robotics and Automation*, ISBN:0-7803-6576-3, Seoul, Korea, 2001.05.

Oyama, E., MacDorman, K. F., Agah, A., Maeda, T., and Tachi, S. (2001c). Coordinate transformation learning of a hand position feedback controller with time delay. *Neurocomputing*, Vol. 38–40, (1503-1509), ISSN:0925-2312.

Riley, M., Ude, A., and Atkeson, C. G. (2000). Methods for motion generation and interaction with a humanoid robot: Case studies of dancing and catching. *Proceedings of AAAI and CMU Workshop on Interactive Robotics and Entertainment*, pp.35-42, Pennsylvania, USA, 2000.04.

Shimada, M., Minato, T., Itakura, S., and Ishiguro, H. (2006). Evaluation of android using unconscious recognition. *Proceedings of the IEEE-RAS International Conference on Humanoid Robots*, pp. 157–162, ISBN:1-4244-0200-X, Genova, Italy, 2006.12.

Veldpaus, F. E., Woltring, H. J., and Dortmans, L. J. M. G. (1988). A least squares algorithm for the equiform transformation from spatial marker co-ordinates. *Journal of Biomechanics*, Vol. 21, (45–54), ISSN:0021-9290.

Yoshikai, T., Mizuuchi, I., Sato, D., Yoshida, S., Inaba, M., and Inoue, H. (2003).

Behavior system design and implementation in spined muscle-tendon humanoid "Kenta". *Journal of Robotics and Mechatronics*, Vol. 15, No. 2, (143–152), ISSN:0915-3942.

19

Towards an Interactive Humanoid Companion with Visual Tracking Modalities

Paulo Menezes[1], Frédéric Lerasle[2, 3], Jorge Dias[1] and Thierry Germa[2]
[1]*Institute of Systems and Robotics - University of Coimbra,* [2]*LAAS-CNRS,* [3]*Université Paul Sabatier*
[1]*Portugal,* [2, 3]*France*

1. Introduction and framework

The idea of robots acting as human companions is not a particularly new or original one. Since the notion of "robot" was created, the idea of robots replacing humans in dangerous, dirty and dull activities has been inseparably tied with the fantasy of human-like robots being friends and existing side by side with humans. In 1989, Engelberger (Engelberger, 1989) introduced the idea of having robots serving humans in everyday environments. Since then, a considerable number of mature robotic systems have been implemented which claim to be servants, personal assistants (see a survey in Fong et al., 2003). The autonomy of such robots is fully oriented towards navigation in human environments and/or human-robot interaction.

Interaction is facilitated if the robot behaviour is as natural as possible. Two aspects of this are important. The first is to facilitate tasks, which involve direct physical cooperation between humans and robots. The second issue is that robot independent movements must appear familiar and predictable to humans. Furthermore, in order to be more effective towards a seemingly interaction, a similar appearance to humans is an important requirement. These considerations initiated probably the design of humanoid robots. One can mention here commercial robots like QRIO by Sony as well as prototypes like Alpha (Bennewitz et al., 2005), Robox (Siegwart et al., 2003), Minerva (Thrun et al., 2000) or Mobot (Nourbakhsh et al., 2003).

These systems addressed various aspects of human-robot interaction designed by a programmer. This includes all or parts of situation understanding, recognition of the human partner, understanding his intention, and coordination of motion and action and multi-modal communication. Such systems are able to communicate with its a non-expert user in a human-friendly intuitive way by employing the bandwidth of human communication and interaction modalities, typically through H/R interfaces, speech or gestures recognition. It is an evident fact that gestures are natural and rich means, which humans employ to communicate with each other, especially valuable in environments where the speech-based communication may be garbled or drowned out. Communicative gestures can represent either acts or symbols. This includes typically gestures recognition for interaction between humans and robots e.g. waving hands for good-bye, acting hello, and gestures recognition for directions to humanoid e.g. pointing out, stop motion. Unfortunately, a few of the designed robotic systems exhibit elementary capabilities of gesture-based interaction and future developments in the robotic community will be undoubtedly devoted to satisfy this need.

Besides communication process, another and potentially deeper issue is the flexibility as humanoid robots are expected to evolve in dynamic and various environments populated with human beings. Most of the designed robotic systems lack learning representations and the interaction is often restricted to what the designer has programmed.

Unfortunately, it seems impossible to create a humanoid robot with built-in knowledge of all possible states and actions suited to the encountered situations. To face this problem, a promising line of investigation is to conceptualize cognitive robots i.e. permanent learners, which are able to evolve and grow their capacities in close interaction with non-expert users in an open-ended fashion. Some recent platforms e.g. Biron (Maas et al., 2006) or Cog (Fitzpatrick et al., 2003) enjoy these capabilities.

They have no completion and continue to learn as they face new interaction situations both with their environments and the other agents. Basically, they discover a human centred environment and build up an understanding of it. Typically, the robot companion follows a human master in his/her private home so as to familiarise it with its habitat. This human master points out specific locations, objects and artefacts that she/he believes are necessary for the robot to remember. Once such a robot has learnt, all this information, it can start interacting with its environment autonomously, for instance to share/exchange objects with humans.

The robot must also learn new tasks and actions relatively to humans by observing and try to imitate them to execute the same task. Imitation learning (Asfour, 2006), (Shon et al., 2005) addresses both issues of human-like motion and easy teaching of new tasks: it facilitates teaching a robot new tasks by a human master and at the same time makes the robot move like a human. This human instructor must have been logically beforehand identified among all the possible robot tutors, and just then granted the right to teach the robot. Activities and gestures imitation (Asfour, 2006; Nakazawa et al., 2002) is logically an essential important component in these approaches.

These reminders stress that activities/gestures interpretation and imitation, object exchange and person following are essential for a humanoid companion. Recall that two generally sequential tasks are involved in the gestures interpretation, namely the tracking and recognition stages while gestures imitation learning proceeds also through two stages: tracking, and reproduction. All these human-robot interaction modalities require, as expected, advanced tracking functionalities and impose constraints on their accuracies, or on the focus of interest. Thus, person following task requires coarse tracking of the whole human body and image-based trackers is appropriate in such situation. These trackers provide coarse tracking granularity but are generally fast and robust. Tracking hands in image plane is also sufficient to interpret many symbolic gestures e.g. a "hello" sign. On the other side, tracking hands when performing manipulation tasks requires high accuracy and so 3D-based trackers. More globally, many tasks concerning manipulation but also interaction rely on tracking of the whole upper human body limbs, and require inferring 3D information.

From these considerations, the remainder of the paper reports both on 2D and 3D tracking of the upper human body parts or hands from a single camera mounted on mobile robot as most of humanoid robots embed such exteroceptive sensor. This set of trackers is expected to fulfil the requirements of most of the aforementioned human-robot interaction modalities.

Tracking human limbs from a mobile platform has to be able to cope with: (i) automatic initialization and re-initialization after target loss or occlusions, (ii) dynamic and cluttered environments encountered by the robot during its displacements.
The paper is organized as follows. Section 2 gives a brief state-of-art related to human body parts tracking based on one or multiple cameras. This allows to introduce our approach and to highlight particle filters in our context. Our guiding principle to design both 2D and 3D trackers devoted to mobile platform is also introduced. Section 3 sums up the well-known particle filtering formalism and describes some variants which enable data fusion in this framework. The latter involve visual cues which are described in section 4. Sections 5 and 6 detail our strategies dedicated to the 2D and 3D tracking of human hands and their generalization to the whole upper human body parts. Section 7 presents a key-scenario and outlines the visual functions depicted in this paper i.e. trackers of human limbs and face recognition as trackers are classically launched as soon as the current user is identified as human master. These visual functions are expected to endow a universal humanoid robot and to enable it to act as a companion.
Considerations about the overall architecture, implementation and integration in progress on two platforms are also presented. This concerns: (i) person recognition and his/her coarse tracking from a mobile platform equipped with an arm to exchange objects with humans, (ii) fine gestures tracking and imitation by a HRP2 model as a real platform which is recently available at LAAS. Last, section 8 summarizes our contribution and opens the discussion for future extensions.

2. Related works on human body parts tracking

The literature proposes a plethora of approaches dedicated to the tracking of human body parts. Related works can be effectively organized into two broad categories, 2D or image-based tracking, and 3D tracking or motion capture. These categories are outlined in the two next subsections with special emphasis on particle filtering based approaches. Recall that activities/gestures tracking is currently coupled with recognition. Though a state of art related to activities/gestures recognition goes outside the scope of this paper, the interested reader is referred to the comprehensive surveys (Pavlovic, et al., 1997; Wu et al., 1999).

2.1 2D or image-based tracking

Many 2D tracking paradigms of the human body parts have been proposed in the literature which we shall not attempt to review here exhaustively. The reader is referred to (Gavrila, 1999; Eachter et al., 1999) for details. One can mention Kalman filtering (Schwerdt et al., 2000), the mean-shift technique (Comaniciu et al., 2003) or its variant (Chen et al., 2001), tree-based filtering (Thayananthan et al., 2003) among many others. Beside these approaches, one of the most successful paradigms, focused in this paper, undoubtedly concerns sequential Monte Carlo simulation methods, also known as particle filters (Doucet et al., 2000).
Particle filters represent the posterior distribution by a set of samples, or particles, with associated importance weights. This weighted particles set is first drawn from the state vector initial probability distribution, and is then updated over time taking into account the measurements and a prior knowledge on the system dynamics and observation models.
In the Computer Vision community, the formalism has been pioneered in the seminal paper by Isard and Blake (Isard et al., 1998a), which coins the term CONDENSATION for

conditional density propagation. In this scheme, the particles are drawn from the dynamics and weighted by their likelihood w.r.t. the measurement. CONDENSATION is shown to outperform Kalman filter in the presence of background clutter.
Following the CONDENSATION algorithm, various improvements and extensions have been proposed for visual tracking. Isard et al. in (Isard et al., 1998c) introduce a mixed-state CONDENSATION tracker in order to perform multiple model tracking. The same authors propose in (Isard et al., 1998b) another extension, named ICONDENSATION, which has introduced for the first time importance sampling in visual tracking. It constitutes a mathematically principled way of directing search, combining the dynamics and measurements. So, the tracker can take advantage of the distinct qualities of the information sources and re-initialize automatically when temporary failures occur. Particle filtering with history sampling is proposed as a variant in (Torma et al., 2003). Rui and Chen in (Rui et al., 2001) introduce the Unscented Particle Filter (UPF) into audio and visual tracking. The UPF uses the Unscented Kalman filter to generate proposal distributions that seamlessly integrate the current observation. Partitioned sampling, introduced by MacCormick and Isard in (MacCormick et al., 2000a), is another way of applying particle filters to tracking problems with high-dimensional configuration spaces. This algorithm is shown to be well suited to track articulated objets (MacCormick et al., 2000b). The hierarchical strategy (Pérez et al., 2004) constitutes a generalization.

2.2 3D tracking or motion capture

In the recent years, special devices such as data glove (Sturman et al. 1994), immersive environment (Kehl et al., 2004) and marker-based optical motion capturing system (generally Elite or VICON) are commonly used, in the Robotics community, to track the motion of human limbs. Let us mention some developments, which aim at analyzing raw motion data, acquired from the system VICON and reproduct them on a humanoid robot to imitate dance (Nakazawa et al., 2002) or walking gait (Shon et al., 2005). Using such systems is not intuitive and questionable in human-robot interaction session. Firstly, captured motion cannot be directly imported into a robot, as the raw data must be converted to its joint angle trajectories. Secondly, usual motion capture systems are hard to implement while using markers is restrictive.
Like many researchers of the Computer Vision community, we aim at investigating marker-less motion capturing systems, using one or more cameras. Such a system could be run using conventional cameras and without the use of special apparel or other equipment. To date, most of the existing marker-less approaches take advantage of the *a priori* knowledge about the kinematics and shape properties of the human body to make the problem tractable. Tracking is also well supported by the use of 3D articulated models which can be either deformable (Heap et al., 1996; Lerasle et al.,1999; Kakadiaris et al., 2000; Metaxas et al., 2003; Sminchisescu et al., 2003) or rigid (Delamarre et al., 2001; Giebel et al., 2004; Stenger et al., 2003). In fact, there is a trade-off between the modelling error, due to rigid structures, the number of parameters involved in the model, the required precision, and the expected computational cost. In our case, the creation of a simple and light approach that would be adequate to for a quasi-real-time application was one of the ideas that guided the developments. This motivated our choice of using truncated rigid quadrics to represent the limbs' shapes. Quadrics are, indeed, quite popular geometric primitives for use in human body tracking (Deutcher et al., 2000; Delamarre et al., 2001; Stenger et al., 2003). This is due

to the fact that they are easily handled, and can be combined to create complex shapes, and their projections are conic sections that can be obtained in closed form. Our projection method that will be depicted later, although being inspired from (Stenger et al., 2001) has the advantage that it requires less computational power than this one.

Two main classes of 3D model-based trackers can be considered, 3D reconstruction-based approaches (Delamarre et al., 2001; Urtasum et al., 2004) and appearance-based approaches, being both widely investigated. While the former performs a reconstruction of the largest possible number of points of the tracked object or structure and then tries to match them in 3D space, the latter tries to solve the problem of in which configuration should the target be for its representation being the currently observed one. Normally some characteristic features of the object are used to in the construction of a model-to-image fitting process. Our work that is presented in this paper is focused on the use of this kind of approach making no assumptions about clothing and background structure.

To cope with the lack of discriminant visual features, the presence of clutter, and the frequent occurrence of mutual occlusions between limbs, one solution is to base the observation model on multiple views (Delamarre et al., 2001; Deutscher et al., 2000; Gavrila et al., 1996; Lerasle et al., 1999; Stenger et al., 2001; Urtasun et al., 2004). Another solution (Gonçalves et al., 1995; Park et al. 2993; Sidenbladh et al., 2000; Sminchisescu et al., 2003), which is the one we have chosen, is to use a single view and increase the reliability and specificity of the observation model. To do so, a robust and probabilistically motivated integration of multiple measurement modalities is of great help. There are several examples in the literature of such integration like, for example edges and colour cues in (Stenger et al., 2003), edges/silhouette and motion cues in (Sminchisescu et al., 2003) or edges, texture and 3D data cues in (Giebel et al., 2004). In our case, we propose an observation model that combines edges and motion cues for the quadrics limbs, with local colour and texture patches on clothing acting as natural markers. Finally and inspired from (Sminchisescu et al., 2003), we add joints limits and self-body collision removal constraints to the overall model.

Regarding the tracked movements, some approaches rely on simplifications brought in either by using sophisticated learnt motion models, such as walking (Urtasun et al., 2004), or by restricting movements to those contained roughly in a fronto-parallel plane (Sidenbladh et al., 2000). Both simplification choices are well suited to monocular approaches. No specific motion models are used in this work, as we want to be able to track general human motions. In such unconstrained setup, a monocular estimation process suffers necessarily from the inevitable multi-modality of the observation process.

Each of these solutions produces a local minimum in the observation function, by consequence when any single-hypothesis-tracker is started in a position of configuration space too far from the good one, it may simply be trapped in one of the false minima, with the consequent tracking failure target loss.

Reliable tracking requires a powerful multiple hypothesis tracker capable of finding and following a significant number of minima. Local descent search strategies (Delamarre et al., 2001; Lerasle et al., 1999; Kakadiaris et al., 2000; Rehg et al., 1995; Urtasum et al., 2004) do search a local minimum, but with multi-modality there is no guaranty that the globally most representative one is found. Like others (Deutscher et al., 2000; Poon et al., 2002; Wu et al., 2001), we address these problems by employing particle-filtering techniques for the following reasons.

Particle filtering generates random sampling points according to a proposal distribution, which may contain multiple modes encoding ``the good places to look at''. Such probabilistic framework allows the information from different measurements sources to be fused in a principled manner. Although this fact has been acknowledged before, it has not been fully exploited for 3D trackers. Combining a host of cues such as colour, shape, and even motion, may increase the reliability of estimators dedicated to track human limbs.
In what concerns the computational cost, particle filters techniques normally require a substantial computation power, especially in high state-space dimensionality cases, which make the number of required samples to explode. Consequently, large efforts have been devoted to tackle such problem by reducing both the model's dimension through PCA (Wu et al., 2001; Uratasum et al., 2004), and the number of samples by testing stochastic sampling ``variants'' (Deutscher et al., 2000; Sminchisescu et al., 2003).

2.3 Problem statement and guiding principle

2D or 3D human tracking from a mobile platform is a very challenging task, which imposes several requirements. First, the sensor's setup, is naturally embedded on the autonomous robot. By consequence from the camera point of view all scene objects move, this precludes the use of some useful techniques like background subtracting for isolating the target objects. As the robot is expected to evolve in environments that are highly dynamic, cluttered, and frequently subjected to illumination changes, several hypotheses must be handled simultaneously by the trackers. This is due to the multi-modality that appears in the statistical distributions of the measured parameters, as a consequence of the clutter or the changes in the appearance of the target. Consequently, several hypotheses must be handled simultaneously in the developed trackers, and a robust integration of multiple visual cues is required to efficiently localize the good likelihood peaks. Finally, on-board computational power is limited so that only a small percentage of these resources can be allocated to tracking, the remaining part being required to enable the concurrent execution of other functions as well as decisional routines within the robot's architecture. Thus, care must be taken to design efficient algorithms.
The particle-filtering framework is well suited to the above requirements and is widely used in the literature both for 2D or 3D tracking purpose. The popularity of this framework is due to its simplicity, ease of implementation, and modelling flexibility. This framework makes no restrictive assumptions about the probability distributions and enables the fusion of diverse measurements in a simple way. Clearly, combining a host of cues may increase our trackers versatility and reliability. Finally, from the numerous particle-filtering strategies proposed in the literature, one is expected to fit to the requirements of each tracker modality. These considerations lead us to investigate on particle filtering strategies for data fusion. The creation of simple and light monocular-based trackers that would adequate to for a quasi-real time application was another motivation that guided our developments.

3. Particle filtering algorithms for data fusion

3.1 Generic algorithm

Particle filters are sequential Monte Carlo simulation methods for the state vector estimation of any Markovian dynamic system subject to possibly non-Gaussian random inputs (Arulampalam et al., 2002). The aim is to recursively approximate the posterior density function (pdf) of the state vector x_k at time k conditioned on the set of measurements $z_{1:k} = z_1, \ldots, z_k$ through the linear point-mass combination

$$p(x_k|z_{1:k}) \approx \sum_i w_k^{(i)} \delta(x_k - x_k^{(i)}), \quad \sum_{i=1}^{N} w_k^{(i)} = 1. \tag{1}$$

which expresses the selection of a value -or ``particle''- $x_k^{(i)}$ with probability -or ``weight''- $w_k^{(i)}, i = 1, \ldots, N$.

A generic particle filter or SIR is shown on Table 1. The particles $x_k^{(i)}$ evolve stochastically over the time, being sampled from an importance density $q(\cdot)$, which aims at adaptively exploring ``relevant'' areas of the state space. Their weights $\omega_k^{(i)}$ are updated thanks to $p(x_k^{(i)}|x_{k-1}^{(i)})$ and $p(z_k|x_k^{(i)})$, respectively the state dynamics and measurement functions, so as to guarantee the consistency of the approximation (1). In order to limit the degeneracy phenomenon, which says that after few instants the weights of all but one particle tend to zero, step 8 inserts a resampling process. Another solution to limit this effect in addition to re-sampling is the choice of a good importance density.

$[\{x_k^{(i)}, w_k^{(i)}\}]_{i=1}^{N} = \text{SIR\_PF}([\{x_{k-1}^{(i)}, w_{k-1}^{(i)}\}]_{i=1}^{N}, z_k)$

1: **IF** $k = 0$, **THEN** Draw $x_0^{(1)}, \ldots, x_0^{(i)}, \ldots, x_0^{(N)}$ i.i.d. according to $p(x_0)$, and set $w_0^{(i)} = \frac{1}{N}$ **END IF**

2: **IF** $k \geq 1$ **THEN** {—$[\{x_{k-1}^{(i)}, w_{k-1}^{(i)}\}]_{i=1}^{N}$ being a particle description of $p(x_{k-1}|z_{1:k-1})$—}

3: **FOR** $i = 1, \ldots, N$, **DO**

4: "Propagate" the particle $x_{k-1}^{(i)}$ by independently sampling $x_k^{(i)} \sim q(x_k|x_{k-1}^{(i)}, z_k)$

5: Update the weight $w_k^{(i)}$ associated to $x_k^{(i)}$ according to the formula $w_k^{(i)} \propto w_{k-1}^{(i)} \frac{p(z_k|x_k^{(i)}) p(x_k^{(i)}|x_{k-1}^{(i)})}{q(x_k^{(i)}|x_{k-1}^{(i)}, z_k)}$, prior to a normalization step so that $\sum_i w_k^{(i)} = 1$

6: **END FOR**

7: Compute the conditional mean of any function of x_k, e.g. the MMSE estimate $E_{p(x_k|z_{1:k})}[x_k]$, from the approximation $\sum_{i=1}^{N} w_k^{(i)} \delta(x_k - x_k^{(i)})$ of the posterior $p(x_k|z_{1:k})$

8: At any time or depending on an "efficiency" criterion, resample the description $[\{x_k^{(i)}, w_k^{(i)}\}]_{i=1}^{N}$ of $p(x_k|z_{1:k})$ into the equivalent evenly weighted particles set $[\{x_k^{(s^{(i)})}, \frac{1}{N}\}]_{i=1}^{N}$, by sampling in $\{1, \ldots, N\}$ the indexes $s^{(1)}, \ldots, s^{(N)}$ according to $P(s^{(i)} = j) = w_k^{(j)}$; set $x_k^{(i)}$ and $w_k^{(i)}$ with $x_k^{(s^{(i)})}$ and $\frac{1}{N}$

9: **END IF**

Table 1. Generic particle filtering algorithm (SIR)

3.2 Importance sampling from either dynamics or measurements: basic strategies

The CONDENSATION algorithm is instanced from the SIR algorithm as $q(x_k|x_{k-1}^{(i)}, z_k) = p(x_k^{(i)}|x_{k-1}^{(i)})$. A difference relative to the SIR algorithm is that the re-sampling step 8 is applied on every cycle. Resampling by itself cannot efficiently limit the degeneracy phenomenon as the state-space is blindly explored without any knowledge of the observations. On the other side, the ICONDENSATION algorithm (Isard et al., 1998), considers an importance density $q(\cdot)$, which classically relates to the importance function $\pi(x_k^{(i)}|z_k)$ defined from the current image. However, if a particle drawn exclusively from the image is inconsistent with its predecessor in terms of state dynamics, the update formula leads to a small weight. An alternative consists in sampling the particles according the measurements, dynamics and the prior, so that, with $\alpha, \beta \in [0;1]$

$$q\left(x_k^{(i)}|x_{k-1}^{(i)}, z_k\right) = \alpha\pi\left(x_k^{(i)}|z_k\right) + \beta p\left(x_k^{(i)}|x_{k-1}^{(i)}\right) + (1-\alpha-\beta)\,p_0(x_k).$$

3.3 Towards the "optimal" case: the Auxiliary Particle Filter

The Auxiliary Particle Filter (Pitt et al., 1999) noted APF depicted by the algorithm of Table 2 is another variant that aims to overcome some limitations of the ``blind exploration''. This algorithm considers an auxiliary density $p(z_k|\mu_k^{(i)})$, where μ_k^i characterise the density of x_k conditioned on $x_{k-1}^{(i)}$ (step 4). Compared to the CONDENSATION scheme, the advantage of this filter is that it naturally generates points from the sample at k-1 which, conditioned on the current measure, are most likely to be close to the true state and so improve the estimate accuracy. In practice, it runs slightly slower than the CONDENSATION as we need to evaluate the auxiliary weights $\lambda_k^{(i)}$ (step 4) and to perform two weighted bootstraps (steps 4 and 9) rather than one. However, the improvement in sampling will usually dominate these small effects. By making proposals that have high conditional likelihoods, we reduce the cost of sampling many times from particles, which have very low likelihoods and so will not be re-sampled at the second process stage. This improves the statistical efficiency of the sampling procedure and it means that we can reduce substantially the number N of particles.

$[\{x_k^{(i)}, w_k^{(i)}\}]_{i=1}^{N} =$ **AUXILIARY_PF**$([\{x_{k-1}^{(i)}, w_{k-1}^{(i)}\}]_{i=1}^{N}, z_k)$

1: **IF** $k = 0$, **THEN** Draw $x_0^{(1)}, \ldots, x_0^{(i)}, \ldots, x_0^{(N)}$ i.i.d. according to $p(x_0)$, and set $w_0^{(i)} = \frac{1}{N}$ **END IF**

2: **IF** $k \geq 1$ **THEN** $\{-[\{x_{k-1}^{(i)}, w_{k-1}^{(i)}\}]_{i=1}^{N}$ being a particle description of $p(x_{k-1}|z_{1:k-1})-\}$

3: **FOR** $i = 1, \ldots, N$, **DO**

4: From the approximation $\hat{p}(z_k|x_{k-1}^{(i)}) = p(z_k|\mu_k^{(i)})$ – e.g. with $\mu_k^{(i)} \sim p(x_k|x_{k-1}^{(i)})$ or $\mu_k^{(i)} = \mathrm{E}_{p(x_k|x_{k-1}^{(i)})}[x_k]$ –, compute the auxiliary weights $\lambda_k^{(i)} \propto w_{k-1}^{(i)}\hat{p}(z_k|x_{k-1}^{(i)})$, prior to a normalization step so that $\sum_i \lambda_k^{(i)} = 1$

5: **END FOR**

6: Resample $[\{x_{k-1}^{(i)}, \lambda_k^{(i)}\}]_{i=1}^{N}$ –or, equivalently, sample in $\{1, \ldots, N\}$ the indexes $s^{(1)}, \ldots, s^{(N)}$ of the particles at time $k-1$ according to $P(s^{(i)} = j) = \lambda_k^{(j)}$ – in order to get the equivalent evenly weighted particles set $[\{x_{k-1}^{(s^{(i)})}, \frac{1}{N}\}]_{i=1}^{N}$; both $\sum_{i=1}^{N} \lambda_k^{(i)}\delta(x_{k-1} - x_{k-1}^{(i)})$ and $\frac{1}{N}\sum_{i=1}^{N}\delta(x_{k-1} - x_{k-1}^{(s^{(i)})})$ approximate the smoothing pdf $p(x_{k-1}|z_{1:k})$

7: **FOR** $i = 1, \ldots, N$, **DO**

8: "Propagate" the particles by independently drawing $x_k^{(i)} \sim p(x_k|x_{k-1}^{(s^{(i)})})$

9: Update the weights, prior to their normalization, by setting $w_k^{(i)} \propto \frac{p(z_k|x_k^{(i)})}{\hat{p}(z_k|x_{k-1}^{(s^{(i)})})} = \frac{p(z_k|x_k^{(i)})}{p(z_k|\mu_k^{(s^{(i)})})}$

10: Compute the conditional mean of any function of x_k, e.g. the MMSE estimate $\mathrm{E}_{p(x_k|z_{1:k})}[x_k]$, from the approximation $\sum_{i=1}^{N} w_k^{(i)}\delta(x_k - x_k^{(i)})$ of the posterior $p(x_k|z_{1:k})$

11: **END FOR**

12: **END IF**

Table 2. Auxiliary Particle Filter (APF)

4. Importance and measurement functions

Importance functions $\pi(.)$ involve generally discriminant but possibly intermittent visual cues while measurement functions $p(z|x)$ involve cues which must be persistent but are however more prone to ambiguity for cluttered scene (Pérez et al., 2004).

Combining or fusing multiple cues enables the tracker to better benefit from distinct information, and confers robustness w.r.t temporary failures. Measurement and importance functions are depicted hereafter as well as some considerations regarding data fusion.

4.1 Measurement functions

4.1.1 Shape cue

The use of shape cues requires indeed that the class of targets to be tracked is known *a priori* and that contour models can be learnt beforehand, i.e. that coarse 2D ou 3D models of the targeted limbs can be used. For simple view-based shape representation, human limbs are therefore represented by coarse silhouette contours (Figure 1). For 3D tracking, a preliminary 3D model projection and hidden parts removal is required (Delamarre et al, 2001; Deutscher et al., 2001; Menezes et al 2005a, Sminchisescu et al., 2003).

Figure 1. Examples of silhouette templates

The shape-based likelihood is classically computed using the sum of the squared distances between model points $x(j)$ and the nearest closest edges $z(j)$, which lie on the normals that pass through the points $x(j)$. These measurement points are chosen uniformly distributed along the model.

$$p(z^S|x) \propto \exp\left(\lambda_s \frac{D^2}{2\sigma_s^2}\right), \; D = \frac{1}{N_p}\sum_{j=1}^{N_p} |x(j) - z(j)|. \tag{2}$$

where λ_s is a weight dedicated to further 3D tracking purpose (see section 6.2), j indexes the N_p model points, and σ_s a standard deviation being determined a priori.

A variant (Giebel et al., 2004) consists in converting the edge image into a Distance Transform image, noted I_{DT} which is used to peek the distance values. The advantage of matching our model contours against a DT image rather than using directly the edges image is twofold. Firstly, the similarity measure D is a smoother function of the model pose parameters. Secondly, this reduces the involved computations as the DT image can be generated only once, independently of the number N of particles used in the filter. The distance D becomes

$$D = \frac{1}{N_p}\sum_{j=1}^{N_p} I_{DT}(j), \tag{3}$$

where $I_{DT}(j)$ is the associated value in the DT image. Figure 1 (a) and (b) shows two plots of these two likelihoods for an image-based tracker where the target is a 2D elliptical template corresponding coarsely to the head of the right subject in the input image. As expected, the distance (3) appears to be less discriminant to clutter but is shown to enjoy least time consumption for $N \geq 100$.

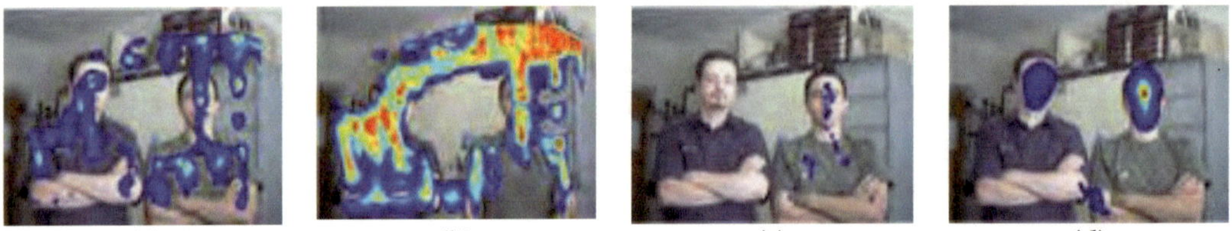

(a) (b) (c) (d)

Figure 2. Likelihoods for 2D tracker for shape (a)-(b), shape and optical flow (c), shape and colour (d)

4.1.2 Shape and motion cues combination

In our context, and contrary to the background, which is assumed to remain static, the human limbs are expected to be moving, possibly intermittently. To cope with cluttered scenes and reject false background attractors, we favour the moving edges, if they exist, as they are expected to correspond to the moving target. As the target can be temporarily stopped, the static edges are not completely rejected, but only made less attractive than the moving ones. The points $z(j)$ in (2) receive the additional constraint that the corresponding optical flow vectors $\vec{f}(z(j))$ must have nonzero norm. The new likelihood $p(z^{MS}|x)$ involves the following similarity measure

$$D = \frac{1}{N_p} \sum_{j=1}^{N_p} |x(j) - z(j)| + \rho\gamma(z(j)), \tag{4}$$

where $\gamma(z(j)) = 0$ (resp. 1) if $\vec{f}(z(j)) \neq 0$ (resp. if $\vec{f}(z(j)) = 0$) and $\rho > 0$ terms a penalty. Figure 2-(c) plots this more discriminant likelihood function for the example seen above. The target is still the subject on the right, but is assumed to be moving.

Regarding the similarity measure (3), shape and motion cues are combined by using two DT images, where the second one $I'_{DT}(j)$ is obtained by filtering out the static edges, based on the local optical flow vector. The distance D becomes

$$D = \frac{1}{N_p} \sum_{j=1}^{N_p} \min\left(I_{DT}(j), K.I'_{DT}(j)\right). \tag{5}$$

where weight values $K \leq 1$ make moving edges more attractive.

4.1.3 Colour cue

Clothes colours create a clear distinction between the observed persons but also the limbs (head, hands and feet, trunk of sleeves) for a given person. Consequently, using clothing patches of characteristic colour distributions, i.e. natural markers, seems very promising. Reference colour models are associated with these targeted ROIs. For a given ROI, we denote h^c_{ref} and h^c_x two N_{bi}-bin normalized histograms in channel c corresponding respectively to the target and a region B_x related to any state x. The colour likelihood model must be defined so as to favour candidate histograms h^c_x close to the reference histogram h^c_{ref}. The likelihood has a form similar to (2), provided that D terms the Bhattacharyya distance (Pérez et al., 2004) between the two histograms. The latter can also depict the similarity of several colour patches related to faces but also clothes, each with its own reference histogram. Let the union $B_c = \bigcup_{p=1}^{N_R} B_{p,c}$ be associated with the set of N_R reference histograms $\{h^c_{p,ref} : c \in \{R,G,B\}, p = 1,\dots,N_R\}$. By assuming conditional independence of the colour measurements, the likelihood $p(z^C|x)$ becomes

$$p(z^C|x) \propto \exp\left(-\sum_{c}\sum_{p=1}^{N_R} \lambda_{p,c} \frac{D^2(h^c_{p,x}, h^c_{p,ref})}{2\sigma_c^2}\right), \tag{6}$$

where $\lambda_{p,c}$ are weighting factors dedicated to further 3D tracking purpose (see section 6.2). This multi-part extension is more accurate thus avoiding the drift, and possible subsequent loss, experienced sometimes by the single-part version (Pérez et al., 2002). Figure 2-(d) plots this likelihood function $p(z^C|x)$ for the above example. Let us note that, from (6), we can also decline a likelihood value $p(z^T|x)$ relative to textured patches based on the intensity component.

4.1.4 Multiple cues fusion

Assuming the measurement models to be mutually independent given the state. Given M measurement sources $(z^1,\dots,z^M)$, the global measurement function can be factorized as

$$p(z^1,\dots,z^M|x) \propto \prod_{m=1}^{M} p(z^m|x). \tag{7}$$

As mentioned before, data fusion is also required for 3D tracking in order to efficiently localize the good likelihood peak in the state space. Figure 3-left shows the plot of the likelihood $p(z^S|x)$ involving the distance (3) and obtained by sweeping a subspace of the configuration space formed by two parameters of a human arm 3D model. Figure 3-middle plots an approximation of the coloured multi-patches likelihood $p(z^C|x)$ entailed in our tracker. The reference colour ROI corresponds to the marked hand. Fusing shape, motion and colour, as plotted in Figure 3-right, is shown to be more discriminant as expected.

Figure 3. Likelihood plots for 3D tracking: shape cue, colour cue, both shape and colour cues

Clearly, mixing all these cues into the measurement function of the underlying estimation scheme helps our 2D or 3D trackers to work under a wide range of conditions encountered by our robot during its displacements.

4.2 Importance function

4.2.1 Shape cue

This importance function $\pi(.)$ considers the outputs from face or hand detectors. Our face detection system is based on the AdaBoost algorithm and uses a boosted cascade of Haar-like features. Each feature is computed by the sum of all pixels in rectangular regions, which can be computed very efficiently using integral images. The idea is to detect the relative darkness between different regions like the region of the eyes and the cheeks (Figure 4-left).

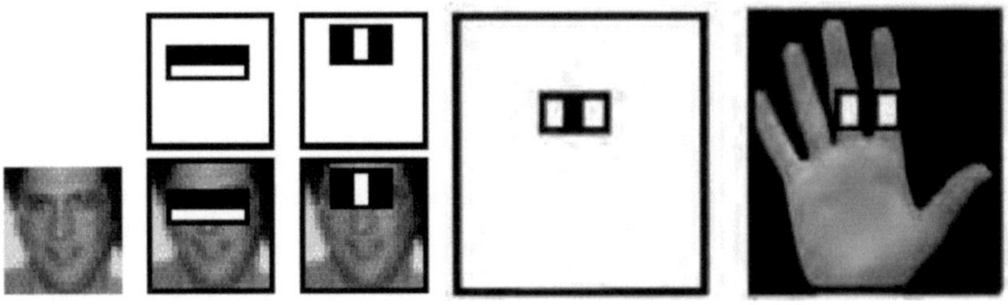

Figure 4. First Haar features for faces (Viola et al., 2001) and for hands

Originally, this idea was developed by Viola et al. in (Viola et al., 2001) to reliably detect faces, in the range of [-45,45] degrees of out-of-plane rotation, without requiring a skin colour model. This widely used detector works quickly and yields high detection rates.
This idea was extended for detecting hands. Our classifier was trained with 2000 images containing upright hands, and 6000 images without hands and used as negative samples. This hand detector exhibits a detection rate slightly smaller than the previous, mainly due to the lack of discriminant contrasts in the hand. Figure 4-right shows example of Haar-like feature used in this context. A video relative to hand detection can be downloaded from the following URL http://www.isr.uc.pt/~paulo/HRI.
Let us characterize the associated importance functions. Given N_B detected faces or hands, and $\mathbf{p}_i = (u_i, v_i), i = 1, \ldots, N_B$ the centroid coordinates of each such region. The associated importance function $\pi(\mathbf{x}|z^S)$ at location $\mathbf{x} = (u_k, v_k)$ follows, as the Gaussian mixture proposal

$$\pi(\mathbf{x}|z^S) \propto \sum_{i=1}^{N_B} \mathcal{N}(\mathbf{x}; \mathbf{p_i}, \mathrm{diag}(\sigma_{u_i}^2, \sigma_{v_i}^2)). \tag{8}$$

where $\mathcal{N}(.;\mu,\Sigma)$ denotes the Gaussian distribution with mean μ and covariance Σ.

4.2.2 Colour cue

Human skin colours have a specific distribution in colour space. Training images from the Compaq database (Jones et al., 1998) enables to construct the associated distributions. The detection of skin-colored blobs is performed by subsampling the input image prior to grouping the classified skin-like pixels. Parts of the segmented regions are filtered regarding their aspect ratio. Then, the importance function $\pi(\mathbf{x}|z^C)$ is defined from the resulting blobs by a Gaussian mixture similar to (8).

4.2.3 Multi-cues mixture

In a mobile robotic context, the efficiency of the above detection modules is influenced by the variability of the environment clutters and the change of viewing conditions. Therefore, the importance function $\pi(.)$ can be extended to consider the outputs from any of the M detectors, i.e.

$$\pi(\mathbf{x}|z^1,\ldots,z^M) = \frac{1}{M}\sum_{j=1}^{M}\pi(\mathbf{x}|z^j). \qquad (9)$$

5. Image-based tracking dedicated to upper human body parts

5.1 Preliminary works for hands tracking

Preliminary investigations (Menezes et al., 2004c) deal with an image-based tracker suitable to estimate fronto-parallel motions of the hand e.g. when performing a ``hello'' or a ``halt'' sign. The aim is to fit the view-based template relative to the targeted hand all along the video stream, through the estimation of its image coordinates (u,v), its scale factor s, and its orientation θ. All these parameters are accounted for in the state vector $\mathbf{x}_k$ related to the k-th frame. With regard to the dynamics model $p(\mathbf{x}_k|\mathbf{x}_{k-1})$, the image motions of observed people are difficult to characterize over time. This weak knowledge is thus formalized by defining the state vector as $\mathbf{x}_k = (u_k, v_k, s_k, \theta_k)'$ and assuming that its entries evolve according to mutually independent random walk models, viz. $p(\mathbf{x}_k|\mathbf{x}_{k-1}) = \mathcal{N}(\mathbf{x}_k|\mathbf{x}_{k-1},\Sigma)$, where $\mathcal{N}(.|\mu,\Sigma)$ terms the Gaussian distribution with mean μ and covariance $\Sigma = diag(\sigma_u^2,\sigma_v^2,\sigma_s^2,\sigma_\theta^2)$

Complex filtering strategies are not necessary in this tracker and we opt logically for the CONDENSATION algorithm as it enjoys the least time consumption. The tracker is launched automatically when detecting hands after agreement between both the Haar-like features based detector and the skin blobs detector outcomes. The particle-weighting step entails the likelihood $p(z^{MS}|x)$ based on the similarity measure (4) and a hand silhouette template (Figure 1). Characteristics and parameter values reported in Table 3 are used in the likelihoods, proposal and state dynamics involved in our hand tracker.

Symbol	Meaning	Value
-	Particle filtering strategy	CONDENSATION
N	Number of particles	100
(nbL, nbC)	Image resolution	(320,240)
σ_s	standard deviation in $p(z^{MS}/x_k)$	36
N_p	Number of model points in similarity measure	30
ρ	Penalty in similarity distance	0.12

Table 3. Characteristics and parameter values for our image-based hand tracker

The running time of this tracker is about 50fps on a PentiumIV-3GHz. Figure 5 shows some image-based tracking snapshots from a sequence involving heavy cluttered background. The entire video can be found at the URL http://www.isr.uc.pt/paulo/HRI. This elementary and specific tracker has not been integrated in the Jido's software architecture (detailed in section 7.1).

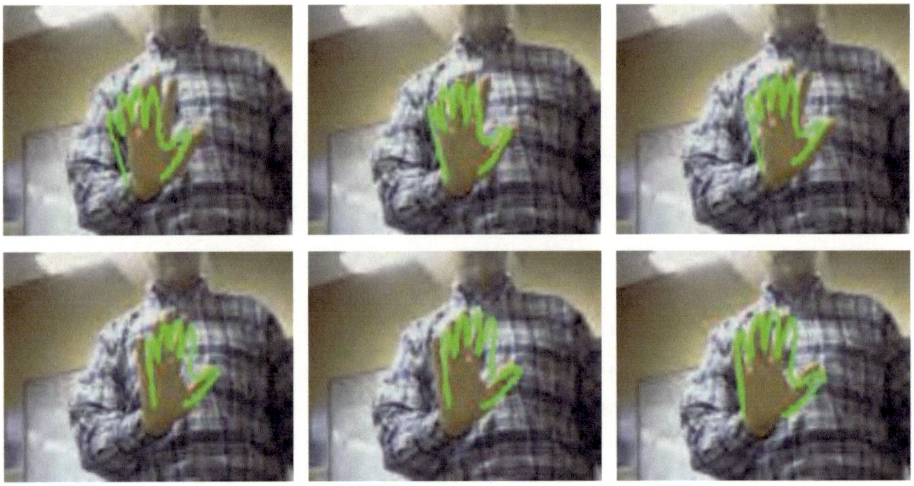

Figure 5. Image-based tracking of hand in heavy cluttered background

5.2 Extension to the upper human body limbs

Same guiding principles, namely data fusion in an appropriate particle filtering strategy, were used to develop an image-based tracker dedicated to the upper human body parts. This tracker can typically be launched for: (i) person following i.e. coordinating the robot's displacements, even if only coarsely, with those of the tracked robot user, (2) people perception in the robot vicinity, for instance to heckle them. This coarse human tracking is used to plan how to position the robot with respect to human beings in a socially acceptable way.

Unfortunately, more than one authorised person can be in robot vicinity, what could make the tracker continuously switch from the targeted person to another. Therefore, for re-identifying individuals information based on face recognition and clothing colour are logically entailed in the characterization of the tracker. These permit to distinguish individuals but also to recover the targeted person after temporary occlusions or out-of-sight. Moreover, any person must be, normally recognized among the potential human masters database before receiving the grant to learn the robot.

5.2.1 Face recognition

This function aims to classify bounding boxes $\mathcal{F}$ of detected faces (see section 4.2) into either one class C_l out of the set $\{C_l\}_{1\leq l\leq M}$ -- corresponding to M users faces presumably learnt offline -- or into the void class $C_\emptyset$. Our approach, clearly inspired by (Turk et al., 1991), consists in performing PCA, and keeps as eigenface bases the first eigenvectors accounting on a certain average of the total class variance. Our evaluations are performed on a face database that is composed of 6000 examples of M=10 individuals acquired by the robot in a wide range of typical conditions: illumination changes, variations in facial orientation and expression, etc. The database is separated into two disjoint sets: (i) the training set (dedicated to PCA) containing 100 images per class, (ii) the test set containing 500 images per class. Each image is cropped to a size of 30x30 pixels. To improve the method, two lines of investigations have been pursued.

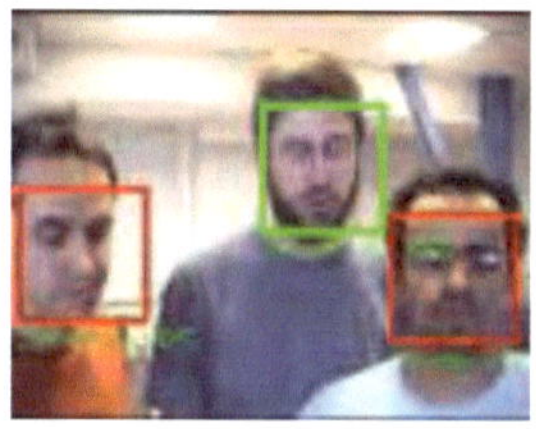

Figure 6. Example of face recognition

Firstly, evaluations highlight that the Distance-From-Face Space (DFFS) error norm leads to the best performances in term of classification rate. For a given face { $\mathcal{F}(j), j \in \{1,...30\times30\}$ }, the DFFS criteria is written as follows

$$\mathscr{D} = \sum_{j=1}^{30\times30} |\mathcal{F}(j) - \mathcal{F}_r(j)| \quad (10)$$

where $\mathcal{F}_r$ is the reconstructed image after projection of $\mathcal{F}$ onto a PCA basis. For a set of M learnt tutors (classes) noted $\{C_l\}_{1\leq l\leq M}$ and a detected face $\mathcal{F}$, we can define for each class C_l, the distance $\mathscr{D}_l = \mathscr{D}(\mathcal{F}, C_l)$ and an a priori probability $P(C_l|\mathcal{F})$ of labelling to C_l

$$\begin{cases} P(C_\emptyset|\mathcal{F}) = 1 \text{ and } \forall l,\ P(C_l|\mathcal{F}) = 0 \text{ when } \forall l,\ \mathscr{D}_l > \tau \\ \forall l,\ P(C_l|\mathcal{F}) = \frac{\mathcal{N}(\mathscr{D}_l|0,\Sigma_\tau)}{\sum_p \mathcal{N}(\mathscr{D}_p|0,\Sigma_\tau)} \text{ otherwise,} \end{cases} \quad (11)$$

where τ is a threshold predefined automatically, $C_\emptyset$ refers the void class and h terms the Heaviside - or ``step'' -- function: h(x)=1 if x>0, 0 otherwise.

Secondly, from the Heseltine et al. investigations in (Heseltine et al., 2002), we evaluate and select the most meaningful image pre-processing in terms of false positives and false negatives. We plot ROC curves based on different image pre-processing techniques for our error norm $\mathscr{D}$. These ROC curves are shown as the sensitivity (the ratio of true positives over total true positives and false negatives) versus the false positive rate. Histogram equalization is shown to outperform the other techniques for our database (Figure 7).

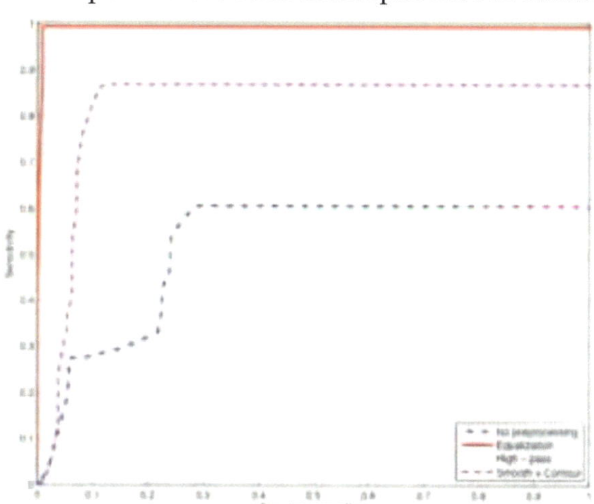

Figure 7. ROC curves image for different pre-processing techniques

Figure 6 shows a snapshot of detected (red)/recognized (green) faces with associated probabilities for a targeted person named Sylvain. More details on the face recognition process can be found in (Germa et al. 2007) and at the URL http://www.laas.fr/~tgerma. This face classifier relates to the module HumRec in the Jido's software architecture (see section 7.1).

5.2.2 Image-based tracking

This tracker is inspired from previously developed ones detailed in (Brèthes et al., 2005). It involves the state vector $\mathbf{x}_k = [u_k, v_k, s_k]'$ - the orientation θ_k being set to a known constant. Regarding the filtering strategy, we opt for the ICONDENSATION scheme, which allows automatic initialization, and aid recovery from transient tracking failures thanks to detection modules. Let us characterize both importance and measurement functions involved in the tracker. The importance function mixes, thanks to (9) the outputs from the colour blobs and face detectors. The importance function (7) becomes

$$\pi(\mathbf{x}_k|z_k^S) \propto \sum_{i=1}^{N_B} P(C_l|\mathcal{F}_i).\mathcal{N}(\mathbf{x};\mathbf{p_i},\mathrm{diag}(\sigma^2_{u_i},\sigma^2_{v_i})) \quad (12)$$

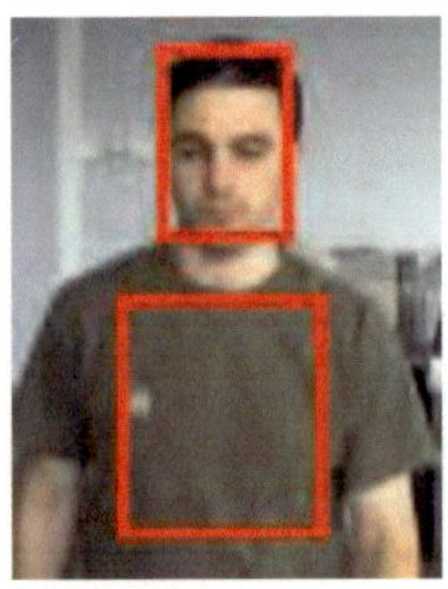

Figure 8. A two-colour patch template template

Two colour models $h^c_{ref_1}$ and $h^c_{ref_2}$ are considered in the colour-based likelihood $p(z_k^C|\mathbf{x}_k)$, respectively for the head and the torso of the guided person (Figure 19 and 10). Their initializations are achieved according to frames, which lead to $P(C_l|\mathcal{F})$ probabilities egal to one. In the tracking loop, the colour model $h^c_{ref_2}$ is re-initialized with the initial values when the user verification is highly confident, typically $P(C_l|\mathcal{F}_i) = 1$. When the appearance of these two ROIs is supposed to change in the video stream, the target reference model is updated from the computed estimates through a first-order filtering process i.e.

$$h^c_{ref,k} = (1-\kappa).h^c_{ref,k-1} + \kappa.h^c_{E[x_k]}, \quad (13)$$

where κ weights the contribution of the mean state histogram $h^c_{E[x_k]}$ to the target model $h^c_{ref,k-1}$, and index p has been omitted for compactness reasons. The models updating can lead to drifts with the consequent loss of the target. To avoid such tracker failures, the global measurement model fuses, thanks to (7), colour but also shape cues. The shape-based likelihood $p(z_k^S|\mathbf{x}_k)$ entails the similarity distance (3) and the head silhouette template (Figure 1). Characteristics and parameter values describing the likelihoods, state dynamics are listed in Table 4.

Due to the efficiency of the face recognition proposal (12), good tracking results are achieved with a reasonably small number of particles i.e. N=150 particles. A PentiumIV-3GHz requires about 40 fps to process the tracking.

Figure 9 and 10 show snapshots of two typical sequences in our context. All regions -- centred on the yellow dots -- close to detected faces with high recognition probabilities corresponding to the person on the background are continually explored. Those – in blue colour -- that do not comply with the targeted person are discarded during the importance-sampling step. Recall that, for large range out-of-plane face rotations ($> |45°|$), the proposal continues to generate pertinent hypotheses from the dynamic and the skin blobs detector. The green (resp. red) rectangles represent the MMSE estimate in step 7 of Table 1 with high (resp. low) confidence in the face recognition process. The proposal generates hypotheses (yellow dots) in regions of significant face recognition probabilities.

Symbol	Meaning	Value
-	Particle filtering strategy	ICONDENSATION
N	Number of particles	150
(nbL, nbC)	Image resolution	(320, 240)
(α, β)	Coeff. in importance function $q(\mathbf{x}_k \mid \mathbf{x}_{k-1}, z_k)$	(0.3, 0.6)
$(\sigma_u, \sigma_v, \sigma_s)$	Standard deviation in random walk models	$(11, 6, \sqrt{0.1})$
$(\sigma_{u_i}, \sigma_{v_i})$	Standard deviation in importance function $\pi(\mathbf{x}_k \mid z^S)$	(6, 6)
$(\sigma_{u_i}, \sigma_{v_i})$	Standard deviation in importance function $\pi(\mathbf{x}_k \mid z^C)$	(6, 6)
N_p	Number of model points in similarity measure	15
σ_s	Standard deviation in shape-based likelihood $p(z_k^S \mid \mathbf{x}_k)$	1.5
N_R	Number of patches in $p(z_k^C \mid \mathbf{x}_k)$	2
σ_C	Standard deviation in color based likelihood $p(z_k^C \mid \mathbf{x}_k)$	0.03
N_{bi}	Number of colour bins per channel involved in $p(z_k^C \mid \mathbf{x}_k)$	32
κ	Coeff. For reference histograms $h_{ref,1}^c, h_{ref,2}^c$ update	(0.1, 0.05)

Table 4. Characteristics and parameter values for our image-based upper human body parts tracker

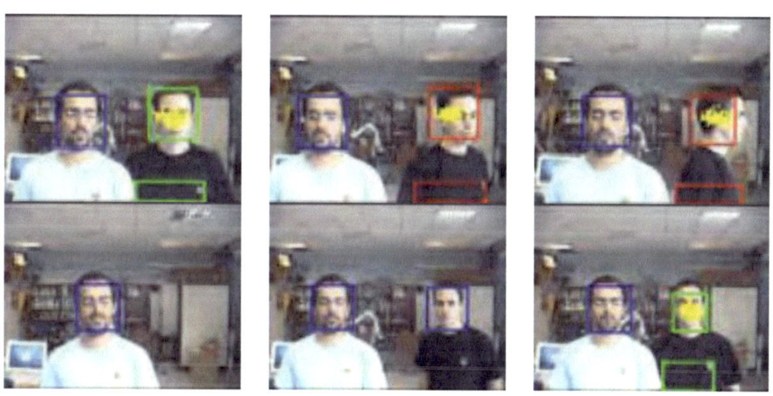

Figure 9. Tracking scenario including two persons with target out-of-sight. Target loss detection and automatic re-initialization

The first scenario (Figure 9), involving sporadic target disappearance, shows that our probabilistic tracker is correctly positioned on the desired person (on the background) throughout the sequence. Although the later disappears, the tracker doesn't lock onto the undesired person thanks to low recognition likelihood. The tracker re-initializes automatically as soon as the target re-appears.

The second scenario (Figure 10) involves occlusion of the target by another person traversing the field of view. The combination of multiple cues based likelihood and face recognition allows keeping track of the region of interest even after the complete occlusion. These videos as well as other sequences are available at the URL http://www.laas.fr/tgerma/videos demonstrate the tracker's performances on a number of challenging scenarios. This tracker relates to the module ICU in the Jido's software architecture (see section 7.1).

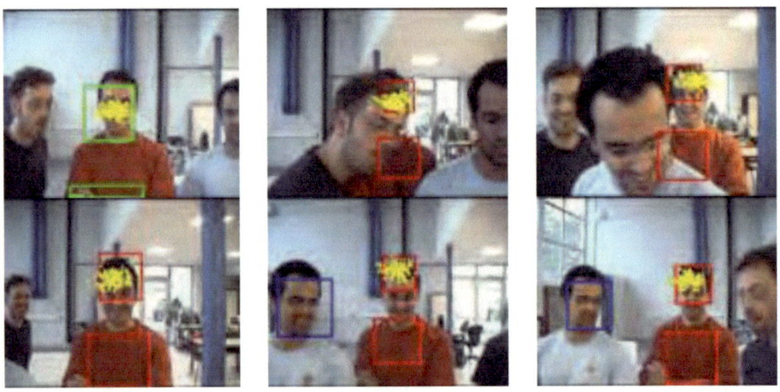

Figure 10. Tracking scenario involving full occlusions between persons

6. 3D tracking dedicated to upper human body parts

6.1 Preliminary works for hands tracking

6.1.1 Basis on quadrics projection

Manipulation tasks, e.g. exchanging object, rely on 3D tracking of the human hands. Therefore, preliminary developments deals with an appearance and 3D model based tracker of hands from a mobile platform. This involves the perspective projection of the associated 3D model. Hands are modelled by deformable quadrics as quadrics and truncated ones can represent approximatively most of the upper human body parts (arm, forearm, torso,...). Let us recall some basis relative to the projection of quadrics. The projection matrix of a quadric in a normalised camera is $\mathbf{P} = [\mathbb{I}_{3\times 2}|\mathbf{0}_{3\times 1}]$. Considering a pinhole camera model, the camera centre and an image point $\mathbf{X}$ define a projective ray, which contains the points given by $\mathbf{X} = [\mathbf{x}\ s]^T$. The depth of a 3D point, situated on this ray, is determined by the scalar. The expression of the quadric $\mathbf{X}^T\mathbf{Q}\mathbf{X} = 0$ can be written as

$$\mathbf{x}^T\mathbf{A}\mathbf{x} + 2s\mathbf{b}^T\mathbf{x} + s^2c = 0, \quad \text{where } \mathbf{Q} = \begin{bmatrix} \mathbf{A} & \mathbf{b} \\ \mathbf{b}^T & c \end{bmatrix}. \tag{14}$$

This expression can be considered as an equation of second degree in s. Then, when the ray represented by $\mathbf{X}(s)$ is tangent to $\mathcal{Q}$, there is a single solution for equation (14) i.e. its discriminant is zero, so:

$$\mathbf{x}^T(\mathbf{b}\mathbf{b}^T - c\mathbf{A})\mathbf{x} = 0. \quad (15)$$

This expression corresponds to a conic $\mathcal{C}$ in the image plane, with $\mathbf{C} = c\mathbf{A} - \mathbf{b}\mathbf{b}^T$, which therefore corresponds to the projection of the quadric $\mathcal{Q}$. For any $\mathbf{x}$ belonging to $\mathcal{C}$, the corresponding 3D point $\mathbf{X}$ is fully defined by finding s_0 which is given by $s_0 = -\mathbf{b}^T\mathbf{x}/c$. This formula can be extended to arbitrary projective camera with $\mathbf{P} = \mathbf{K}[\mathbf{R}|\mathbf{t}]$. Defining a 4x4 matrix $\mathbf{H}$ such that $\mathbf{PH} = [\mathbf{I}|\mathbf{0}]$, and then $\mathbf{x} = [\mathbf{I}|\mathbf{0}]\mathbf{H}^{-1}\mathbf{X}$. Only three types of quadrics are of interest for the next: ellipsoid, cone and cylinder. Hands are here modelled by ellipsoids which image projection for a pinhole camera is an ellipse.

6.1.2 Implementation and results

Our quadric model is deformable to deal with the shape appearance variations depending on the configuration in space. For a targeted hand, the state vector entries relate to the position (t_x, t_y, t_z), orientation (α, β, γ), and axis lengths $(\sigma_x, \sigma_y, \sigma_z)$. As was the case for the above trackers, the state vector entries are assumed to evolve according to mutually independent Gaussian random walk models. From these considerations, the state vector to estimate at time k is

$$\mathbf{x}_k = (t_{x_k}, t_{y_k}, t_{z_k}, \alpha_k, \beta_k, \gamma_k, \sigma_{x_k}, \sigma_{y_k}, \sigma_{z_k})'. \quad (16)$$

We adopt ICONDENSATION as this strategy permits to sample some particles on the target after sporadic disappearances from the field of view. This situation is frequently encountered during manipulation tasks as short human-robot distances ([0.5;1.5] m) are involved. Using the video stream issued from the binocular system embedded on the robot here enlarges the environment perception at such close-range distances. Let us characterize the importance and measurement functions.

A pre-processing step consists in determining correspondences based on heuristics (Schmidt 1996) for skin colour blobs (see 4.2) in the image pair and calculate the 3D position for each match explicitly by triangulation. The face blob is beforehand filtered thanks to the face detector. Given M extracted image blobs at time k, the j-th blob is represented by its centroid $m^L_{j,k}$ (resp. $m^R_{j,k}$) and inertia matrix $\sigma^L_{j,k}$ (resp. $\sigma^R_{j,k}$) in the left and right image. For each triangulated position $P_{j,k} = g(m^L_{j,k}, m^R_{j,k})$, the covariance matrix is estimated as follows

$$\Sigma_{j,k} = G^L_{j,k}.\sigma^L_{j,k}.{G^L_{j,k}}^T + G^R_{j,k}\sigma^R_{j,k}{G^R_{j,k}}^T, \quad (17)$$

where $G^L_{j,k}$ and $G^R_{j,k}$ are the appropriate Jacobian matrices relatively to g(.). Finally, SVD decomposition of matrix $\Sigma_{j,k}$ allows defining a new importance function $\pi(\mathbf{x}_k|z^{3d}_k)$, which related to 3D constraint. Thus, the hands can be understood as three natural markers aiming at positioning the particles during the importance sampling. Note that most blob correspondences between the image planes are correct during this pre-processing step, but that there are also a few outliers (typically the non-targeted hand) that are filtered afterwards in the tracking loop.

Regarding the measurement function, the 3D ellipsoids (corresponding to hypothesis after importance sampling) are projected into a single image plane thanks to (15). The global measurement function fuses shape and colour cues, and can then be formulated as: $p(z^S_k, z^C_k|\mathbf{x}_k) = p(z^S|\mathbf{x}_k).p(z^C_k|\mathbf{x}_k)$. The shape-based likelihood involves the similarity measure (2) while the colour-based likelihood involves histogram updating thanks to (13). Characteristics and parameter values describing the likelihoods state dynamics are listed in Table 5.

Symbol	Meaning	Value
-	Particle filtering strategy	ICONDENSATION
N	Number of particles	100
(nbL, nbC)	Image resolution	(320, 240)
(α, β)	Coeff. In importance function $q(\mathbf{x}_k \vert \mathbf{x}_{k-1}, z_k)$	(0.3, 0.6)
$(\sigma_{t_x}, \sigma_{t_y}, \sigma_{t_z})$	Standard deviation in random walk models	(1, 1, 1) cm
$(\sigma^{\alpha}, \sigma^{\beta}, \sigma^{\gamma})$	Standard deviation in random walk models	(1, 1, 1) degrees
$(\sigma_\alpha, \sigma_\beta, \sigma_\gamma)$	Standard deviation in random walk models	$(10^{-4}, 10^{-4}, 10^{-4})$
N_p	Number of model points in similarity measure	20
σ_s	Standard deviation in shape-based likelihood $p(z_k^S \vert \mathbf{x}_k)$	1.5
N_R	Number of patches in $p(z_k^C \vert \mathbf{x}_k)$	1
σ_c	Standard deviation in colour-based likelihood	0.03
N_{bi}	Number of colour bins per channel involved in $p(z_k^C \vert \mathbf{x}_k)$	32
κ	Coeff. For reference histograms $h^c_{ref,1}$ update	0.1

Table 5. Characteristics and parameter values for our 3D tracker of human hand

The particle filter is run with a set of N=100 particles. The computational time, processed on a 3GHz Pentium IV, is approximatively 25fps. Although tracking a single hand is meaningful when exchanging objects, our approach can be easily extended to track both the two hands.

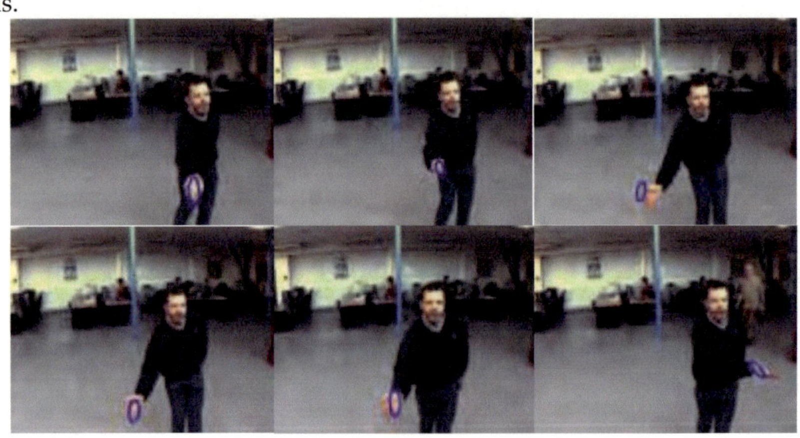

Figure 11. 3D tracking of a hand in heavy clutter. The blue (resp. red) ellipses depict the particles (resp. the MMSE estimate). Automatic re-initialization after target loss

Figure 11 shows the ellipsoid projections on snapshots of a video sequence involving heavy clutter. As one can see, the ellipsoids are deformed, but their positions are still correct. Moreover, the tracker initializes itself automatically and recovers the target after transient failure. This 3D tracker relates to the module GEST in the Jido's software architecture (see section 7.1).

6.2 Extension to the upper human body limbs

This appearance-based tracker is devoted to tasks that rely on understanding/imitation of gestures or activities. We have extended the previous tracker to the whole upper human body parts. The upper human body parts are here modelled by rigid truncated conics. The method to handle model projection and hidden parts removal is inspired from section 6.1.1. Details can be found in (Menezes et al., 2005a). Our emphasis in the paper concerns the characterization of a new measurement model that combines edges and motion cues, with also local color and texture patches on clothing or relative to the hands acting as natural markers. One important point is that our approach could be easily scalable from one single to multiple views as well as to higher number of degrees of freedom (DOF), although it will introduce the consequent increase in computational cost. The following subsections detail some geometric properties the model entailed in the measurement model and the tracker implementation.

6.2.1 Non-observable part stabilisation

Despite the visual cues depicted in section 4, ambiguities arise when certain model parameters cannot be inferred from the current image observations, especially for a monocular system. They include, but are not limited to, kinematical ambiguities. For instance, when one arm is straight and the shape-based likelihood (2) is used, rotation of the upper arm around its axial axis is unobservable, because the model's projected contours remain static under this DOF. Leaving these parameters unconstrained is questionable. Consequently, and like in (Sminchisescu et al, 2003), we control these parameters with a stabiliser cost function that reaches its minimum on a predefined rest configuration x_{def}. This enables the saving of computing efforts that would explore the unobservable regions of the configuration space. In the absence of strong observations, the parameters are constrained to lie near their default values, whereas strong observations unstick the parameters values from these default configurations. The likelihood function for a state x is defined as $p_{st}(x) \propto \exp(-\lambda_{st}\|x_{def} - x\|^2)$. This prior depends on the structure parameters and the factor λ_{st} will be chosen so that the stabilising effect will be negligible for the whole configuration space with the exception of the regions where the other cost terms are constant.

6.2.2 Collision detection

Physical consistency imposes that the different body parts do not interpenetrate. As the estimation is based on a search on the configuration space it would be desirable to a priori remove those regions that correspond to collisions between parts. Unfortunately it is in general not possible to define these forbidden regions in closed form so they could be rejected immediately during the sample phase. The result is that in the particle filter framework, it is possible that configurations proposed by some particles correspond to such impossible configurations, thus exploring regions in the configuration space that are of no interest. To avoid these situations, we use a binary cost function that is not related to observations but only based on a collision detection mechanism. The likelihood function for a state x is $p_{coll}(x) \propto \exp(-\lambda_{co} f_{co})$ with:

$$f_{co}(x) = \begin{cases} 0 & \text{No collision} \\ 1 & \text{In collision} \end{cases}. \tag{18}$$

This function, although being discontinuous for some points of the configuration space and constant for all the remaining, is still usable in a Dirac particle filter context. The advantage of its use is twofold. First it avoids the derivation of the filter to zones of no interest, and second it avoids wasting time in performing the measuring step for unacceptable hypothesis as they can be immediately rejected.

6.3 Observation likelihoods and model priors fusion

Fusing multiple visual cues enables the tracker to better benefit from M distinct measurements $(z^1, \ldots, z^M)$. Assuming that these are mutually independent conditioned on the state, and given L model priors $(p_1(x), \ldots, p_L(x)$, the unified measurement function thus factorizes as

$$p(z^1, \ldots, z^M|x) \propto \prod_{i=1}^{M} p(z^i|x). \prod_{j=1}^{L} p_j(x) \tag{19}$$

6.3.1 Implementation and experiments

In its actual form, our 3D tracker deals two-arms gestures under an 8-DOF model, i.e. four per arm. We assume therefore that the torso is coarsely fronto-parallel with respect to the camera while the position of the shoulders are deduced from the position of the face given by dedicated tracker (Menezes et al., 2004b). All the DOFs are accounted for in the state vector $\mathbf{x}_k$ related to the k-th frame. Kinematical constraints require that the values of these joint angles evolve within anatomically consistent intervals. Samples (issued from the proposal) falling outside the admissible joint parameter range are enforced to the hard limits and not rejected as this could lead to cascades of rejected moves that slow down the sampler.

Recall that pdf $p(\mathbf{x}_k|\mathbf{x}_{k-1})$ encodes information about the dynamics of the targeted human limbs. These are described by an Auto-Regressive model with the following form $\mathbf{x}'_k = A\mathbf{x}'_{k-1} + \mathbf{w}_k$ where $\mathbf{x}'_k = [\mathbf{x}_k, \mathbf{x}_{k-1}]'$, and $\mathbf{w}_k$ defines the process noise. In the current implementation, these dynamics correspond to a constant velocity model. We find this AR model gives empirically better results than usual random walk model, although a more rigorous evaluation would be here desirable. The patches are distributed on the surface model and their possible occlusions are managed during the tracking process. Our approach is different from the traditional marker-based ones because we do not use artificial but natural colour or texture-based markers e.g. the two hands and ROIs on the clothes. We adopt the APF scheme (Table 2), which allows to use some low cost measure or a priori knowledge to guide the particle placement, therefore concentrating them on the regions of interest of the state space. The measurement strategy is as follows: (1) particles are firstly located in good places of the configuration space according to rough correspondences between virtual patches related to the two hands projection and skin-like image ROIs involving likelihood $p(z_k^C|\mathbf{x}_k)$ in step 6, (2) particles' weights are fine-tuned by combining shape and motion cues $(p(z_k^{MS}|\mathbf{x}_k)$ entailing similarity distance (5), three color patches per arm (likelihood (6)) as well as model priors thanks to (19) in step 9. A second and important line of investigation concerns the incorporation of appropriate degrees of adaptability into these multiple cues based likelihoods depending on the target appearance and environmental conditions. Therefore, some heuristics allows weighting the strength of each visual cue in the unified likelihood (7). An a priori confidence criterion of a given coloured or textured patch relative to clothes can be easily derived from the associated likelihood functions where the p-th colour reference histogram $h^c_{ref,p}$ $(h^t_{ref,p}$ for texture one) is uniform and so given by $h^c_{j,ref} = \frac{1}{N_{bi}}, j = 1, \ldots, N_{bi}$ where index p has been omitted for compactness reasons. Typically, uniform coloured patches produce low likelihood values, whereas higher likelihood values characterise confident patches because their associated colour distributions are discriminant and ensure non-ambiguous matching. As stated before, parameters $\lambda_{p,c}$ and $\lambda_{p,t}$ weight the strength of the p-th marker in the likelihood (19). In the same way, the parameter λ_s weights the edges density contribution and is fixed from the first DT image of the sequence.

Symbol	Meaning	Value
-	Particle filtering strategy	APF
N	Number of particles	400
(nbL, nbC)	Image resolution	(320, 240)
λ_{st}	Factor in model prior $p_{st}(x)$	0.5
λ_{co}	Factor in model prior $p_{co}(x)$	0.5
K	Penalty in similarity distance	0.5
σ_s	Standard deviation in similarity distance	1
N_R	Number of patches in color-based likelihood $p(z_k^C \| \mathbf{x}_k)$	6
σ_c	Standard deviation in $p(z_k^C \| \mathbf{x}_k)$	0.3
N_{bi}	Number of color bins per channel involved in $p(z_k^C \| \mathbf{x}_k)$	32

Table 6. Characteristics and parameter values of our upper human body 3D tracker

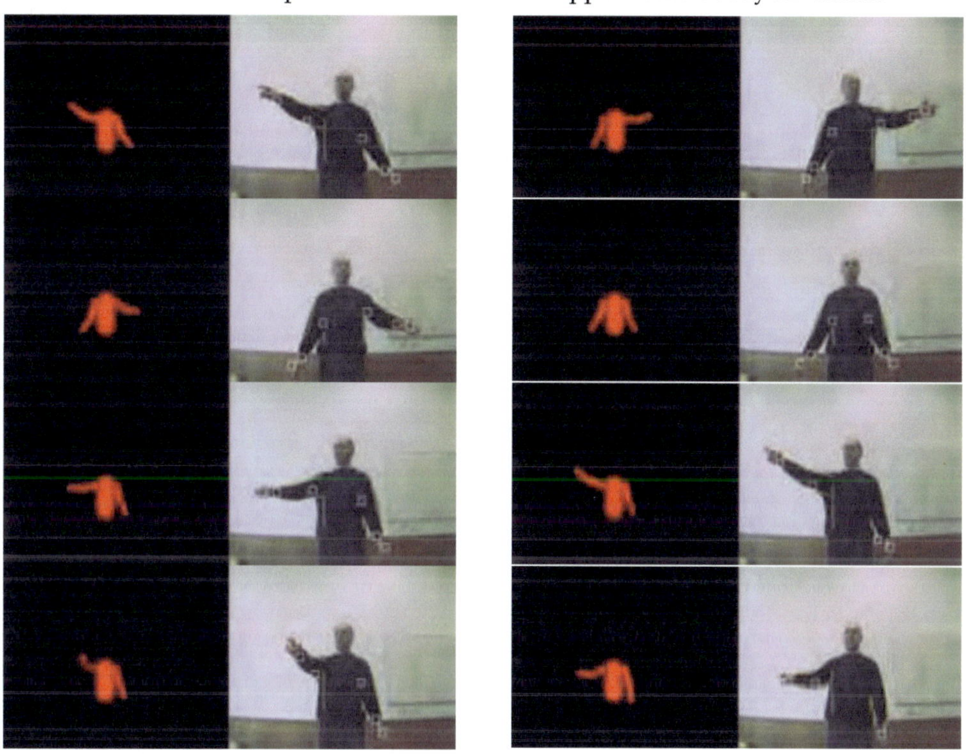

Figure 12. From top-left to bottom right: snapshots of tracking sequence involving deictic gestures

Due to the efficiency of the importance density and the relatively low dimensionality of the state-space, tracking results are achieved with a reasonably small number of particles i.e. N=400 particles. In our non-optimised implementation, a PentiumIV-3GHz requires about 1fps to process the two arm tracking, most of the time being spent in evaluating the observation function. To compare, classic systems take a few seconds per frame to process a single arm tracking. Characteristics and parameter values involved in the likelihoods, proposal and state dynamics of our upper human body 3D tracker are reported in Table 6.
The above-described approach has been implemented and evaluated over monocular images sequences. To illustrate our approach, we show and comment snapshots from two typical sequences acquired from the Jido robot (see section 7) in different situations to highlight our adaptative data fusion strategy. The full images as well as other sequences can be found at the URL http://www.isr.uc.pt/~paulo/HRI . The first sequence (Figure 12) was shot against a white and unevenly illuminated background, but involves loose fitting closing and near fronto-parallel motions. The second sequence (Figure 13) involves coarse fronto-parallel motions over a heavy cluttered background. For each snapshot, the right sub-figures overlay the model projections on to the original images for the MMSE estimate, while the left ones show its corresponding estimated configuration corresponding to the posterior pdf $p(\mathbf{x}_k|z_{1:k})$.

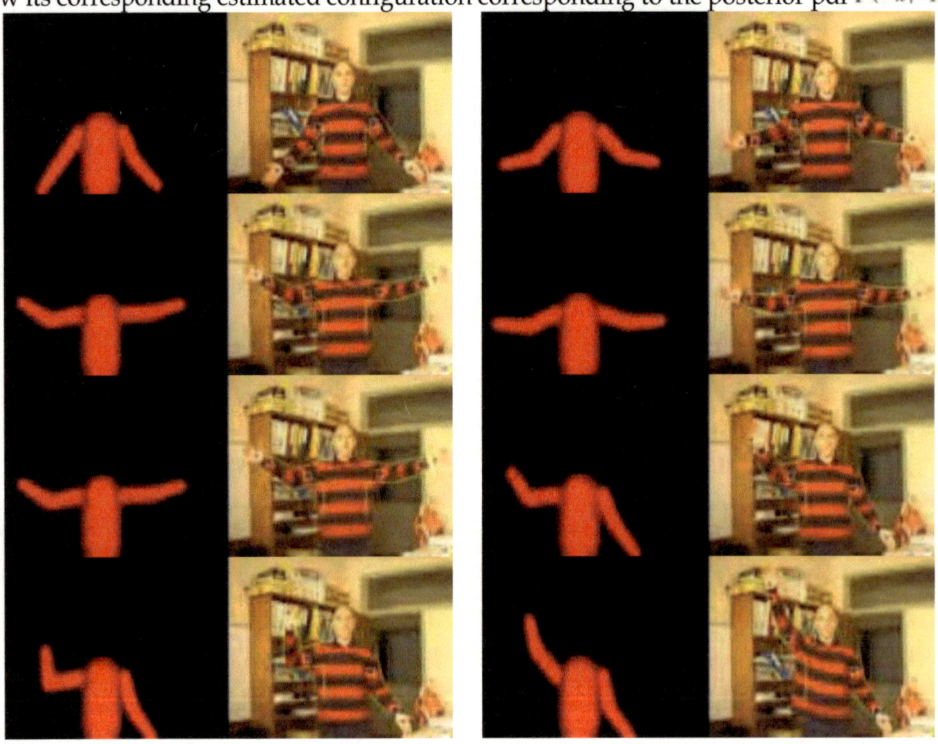

Figure 13. From top-left to bottom-right: snapshots of tracking sequence involving heavy clutter

The first tracking scenario (Figure 12) involves pointing gestures. The target contours are prominent and are weakly disturbed by the background clutter. The high confident contours cue ensures the tracking success. The patches on the uniform sweet are here of little help, as their enclosed colour or texture distributions are quite uniform. The adaptative system allows giving them a weak strength in the unified likelihood cost ($\lambda_{p,c} = 0.1$ against $\lambda_s = 10$). They do not introduce any improvement with respect to their position on the arm, but their benefit comes in the form of inside/outside information, which complements the contours especially when they failed. This permitted the tracking of the arms even when they got out of the fronto-parallel plane thanks to all the patches (Figure 12).

For the second scenario (Figure 13), the tracker deals with significantly more complex scene but tracks also the full sequence without failure. This scenario takes clearly benefit from the introduction of discriminant patches as their colour distributions are far from uniform ones. This leads to higher values of confidence dedicated to the likelihood $p(z_k^c / x_k)$, namely $\lambda_{p,c} = 1$. In these challenging operating conditions, two heuristics allow jointly to release from distracting clutter that might partly resemble human body parts (for instance the cupboard pillar[1]). On the one hand, estimating the edges density in the first frame highlights that shape cue is not a confident one in this context, so its confidence level in the global cost (19) is reduced accordingly during the tracking process i.e. $\lambda_s = 0.1$. On the other hand, optical flow weights the importance relative to the foreground and background contours thanks to the likelihood $p(z_k^{MS}|\mathbf{x}_k)$. If considering only contour cues in the likelihood, the tracker would attach itself to cluttered zones and consequently lose the target. This tracker relates to the module TPB in the Jido's software architecture (see section 7.1).

7. Integration on robotic platforms dedicated to human-robot interaction

7.1 Integration on a robot companion

7.1.1 Outline of the overall software architecture

The above visual functions were embedded on a robot companion called Jido. Jido is equipped with: (i) a 6-DOF arm, (ii) a pan-tilt stereo system at the top of a mast (dedicated to human-robot interaction mechanisms), (iii) a second video system fixed on the arm wrist for object grasping, (iv) two laser scanners, (v) one panelPC with tactile screen for interaction purpose, (vi) one screen to provide feedback to the robot user. Jido has been endowed with functions enabling to act as robot companion and especially to exchange objects with human beings. So, it embeds robust and efficient basic navigation and object recognition abilities. Besides, our efforts focuses in this article concern the design of visual functions in order to recognize individuals and track his/her human body parts during object exchange tasks.

To this aim, Jido is fitted with the "LAAS" layered software architecture thoroughly presented in (Alami et al., 1998). On the top of the hardware (sensors and effectors), the functional level listed in Figure 14, encapsulates all the robot's action and perception capabilities into controllable communicating modules, operating at very strong temporal constraints. The executive level activates these modules, controls the embedded functions, and coordinates the services depending on the task high-level requirements. Finally, the upper decision level copes with task planning and supervision, while remaining reactive to events from the execution control level. The integration of our visual modalities (green boxes) is currently carried out in the architecture, which resides on the Jido robot.

The modules GEST, HumRec, and ICU have been fully integrated in the Jido's software architecture. The module TBP has been devoted preliminary to the HRP2 model (see section

---

[1] with also skin-like color...

7.2). Before integration in Jido, we aim beforehand at extending this tracker to cope with stereoscopic data (Fontmarty et al., 2007).

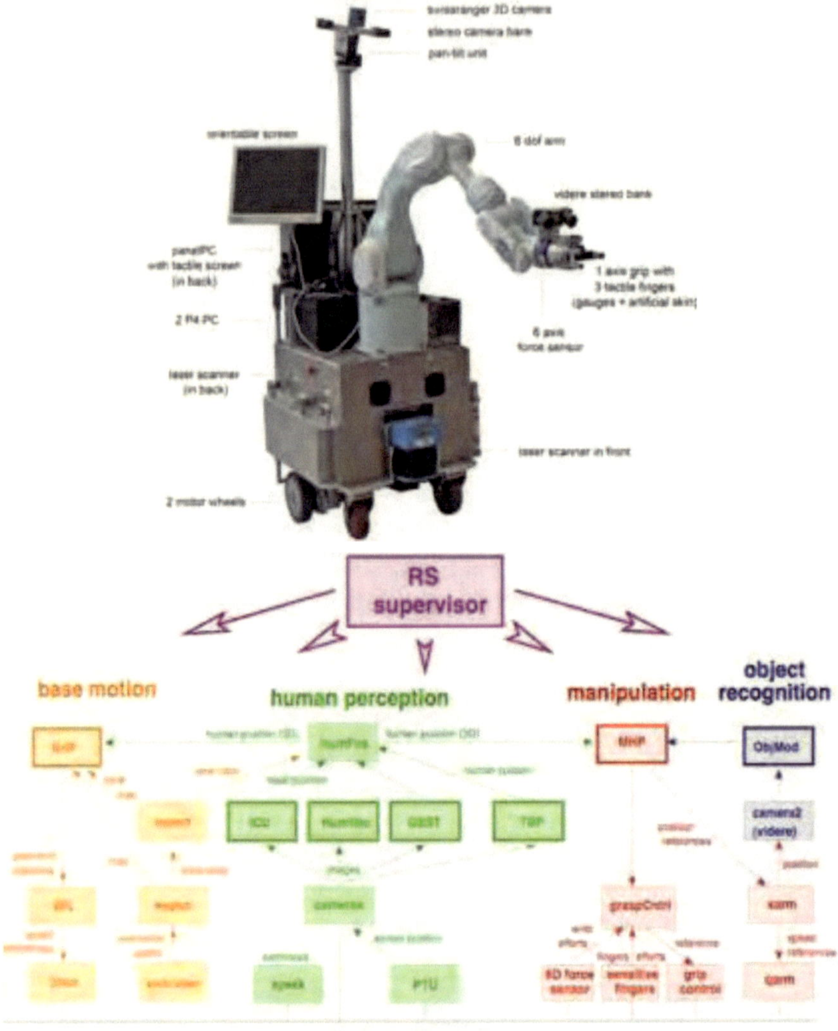

Figure 14. Jido robot and its layered software architecture

7.1.2 Considerations about the visual modalities software architecture

The C++ implementation of the modules are integrated in the ``LAAS'' architecture using a C/C++ interfacing scheme. They enjoy a high modularity thanks to C++ abstract classes and template implementations. This way, virtually any tracker can be implemented by selecting its components from predefined libraries related to particle filtering strategies, state evolution models, and measurement / importance functions. For more flexibility, specific components can be defined and integrated directly. A finite-state automaton can be designed from the vision-based services outlined in section 1. As illustrated in Figure 15, its states are respectively associated to the INIT mode and to the aforementioned vision-based modules while the arrows relate to the transitions between them. Another complementary modalities (blue ellipses), not yet integrated into the robot architecture, have been also added. Heuristics relying on the current human-robot distance, face recognition status, and current executed task (red rectangles) allow to characterize the transitions in the graph. Note that the module ICU can be invoked from all the mentioned human-robot distances ([1;5]m.).

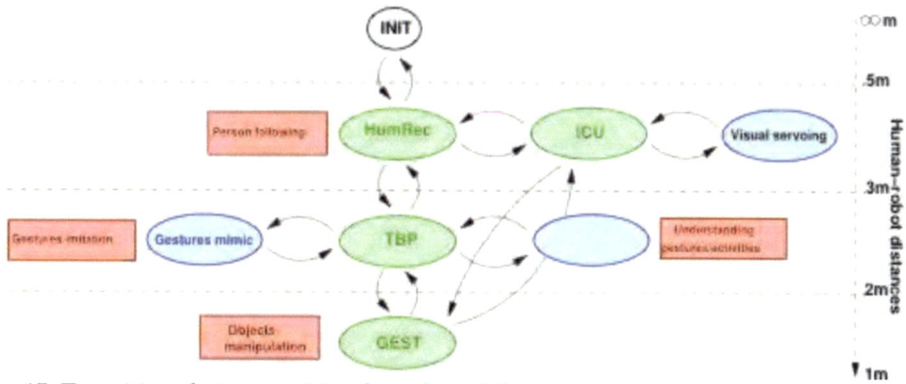

Figure 15. Transitions between vision-based modules

7.2 Integration on a HRP2 model dedicated to gestures imitation

Figure 16. From top-left to bottom-right: snapshots of tracking sequence and animation of HRP2 using the estimated parameters

As mentioned before, a last envisaged application concerns gestures imitation by a humanoid robot (Menezes et al., 2005a). This involves 3D tracking of the upper human body limbs and mapping the joints of our 3D kinematical 3D model to those of the robot. In addition to the previous commented sequences, this scenario (Figure 16) with moderate clutter explores 3D estimation behaviour with respect to problematic motions i.e. non-fronto-parallel ones, elbow end-stops and observation ambiguities. The left column represents the input images and the projection of the model contours superimposed while the right column represents the animation of the HRP2 using the estimated parameters[2]. The first frames involve both elbow end-stops and observation ambiguities. These particular configurations are easily dealt with in our particle-filtering framework. When elbow end-stop occurs, the sampler is able to maintain the elbow angle within its predefined hard limits. Observation ambiguity arises when the arm is straight. The twist parameter is temporary unobservable but remains stable thanks to the likelihood $p_{st}(\mathbf{x}_k)$. As highlighted in (Deutscher et al., 1999), Kalman filtering is quite unable to track through end-stop configurations. Some frames later in Figure 16, the left arm bends slightly towards the camera. Thanks to the patches on the hands, the tracker manages to follow this temporary unobservable motion, although it significantly misestimates the rotation during this motion. The entire video is available at http://www.isr.uc.pt/~paulo/HRI.

8. Conclusion

This article presents the developments of a set of visual trackers dedicated to the upper human body parts. We have outlined visual trackers a universal humanoid companion should deal with in the future. A brief state-of-art related to tracking highlight that particle filtering is widely used in the literature. The popularity of this framework stems, probably, from its simplicity, ease of implementation, and modelling flexibility, for a wide variety of applications.

From these considerations, a first contribution relates to visual data fusion and particle filtering strategies associations with respect to considered interaction modalities. This guiding principle frames all the designed and developed trackers. Practically, the multi-cues associations proved to be more robust than any of the cues individually. All the trackers are applied in quasi-real-time process and have the ability to (re)-initialize automatically.

A second contribution concerns especially the 3D tracker dedicated to the upper human body parts. An efficient method (not detailed here, see (Menezes et al., 2005b) has been proposed in order to handle the projection and hidden removal efficiently. In the vein of the depicted 2D trackers, we propose a new model-image matching cost metric combining visual cues but also geometric constraints. We integrate degrees of adaptability into this likelihood function depending on the human limbs appearance and the environmental conditions. Finally, integration, even if in progress, of the developed trackers on two platforms highlights their relevance and complementarity. To our knowledge, quite few mature robotic systems enjoy such advanced capabilities of human perception.

Several directions are studied regarding our trackers. Firstly, to achieve gestures/activities interpretation, Hidden Markov Models (Fox et al., 2006) and Dynamic Bayesian Network (Pavlovic et al., 1999) are currently under evaluation and preliminary results are actually available. Secondly, we currently study how to extend our monocular-based approaches to account for stereoscopic data as most humanoid robot embed such exteroceptive sensor. Finally, we will integrate all these visual trackers on our new humanoid companion HRP2. The tracking functionalities will be made much more active; zooming will be used to actively adapt the focal lenght with respect to the H/R distance and the current robot status.

---

[2] This animation was performed using the KineoWorks platform and the HRP2 model by courtesy of AIST (GeneralRobotix).

9. Acknowledgements

The work described in this paper has received partial financial support from Fundação para a Ciência e Tecnologia through a scholarship granted to the first author.
Parts of it were conducted within the EU Integrated Project COGNIRON (``The Cognitive Companion'') and funded by the European Commission Division FP6-IST Future and Emerging Technologies under Contract FP6-002020. We want also to thank Brice Burger for implementation and integration involvement regarding the hand 3D tracker.

10. References

Alami, R.; Chatila, R.; Fleury, S. & Ingrand, F. (1998). An architecture for autonomy. *Int. Journal of Robotic Research* (IJRR'98), 17(4):315–337.

Arulampalam, S; Maskell, S.; Gordon, N. & Clapp, T. (2002). A tutorial on particle filters for on-line non-linear/non-gaussian bayesian tracking. *Trans. on Signal Processing*, 2(50):174–188.

Asfour, T.; Gyafas, F.; Azad, P. & Dillman, R. (2006). Imitation learning of dual-arm manipulation tasks in humanoid robot. *Int. Conf. on Humanoid Robots* (HUMANOID'06), pages 40–47, Genoa.

Barreto, J.; Menezes, P. & Dias, J. (2004). Human robot interaction based on haar-like features and eigenfaces. *Int. Conf. on Robotics and Automation* (ICRA'04), New Orleans.

Bennewitz, M.; Faber, F.; Joho, D.; Schreiber, M. & Behnke, S. (2005). Towards a humanoid museum guide robot that interacts with multiple persons. Dans *Int. Conf. on Humanoid Robots* (HUMANOID'05), pages 418–424, Tsukuba.

Brèthes, L.; Lerasle, F. & Danès, P. (2005). Data fusion for visual tracking dedicated to human-robot interaction. Int. Conf. on Robotics and Automation (ICRA'05). pages 2087-2092, Barcelona.

Chen, H. & Liu, T. (2001). Trust-region methods for real-time tracking. Dans *Int. Conf. on Computer Vision* (ICCV'01), volume 2, pages 717–722, Vancouver.

Comaniciu, D.; Ramesh, V. & Meer, P. (2003). Kernel-based object tracking. *Trans. on Pattern Analysis Machine Intelligence* (PAMI'03), volume 25, pages 564–575.

Delamarre, Q.; & Faugeras, O. (2001). 3D articulated models and multi-view tracking with physical forces. *Computer Vision and Image Understanding* (CVIU'01), 81:328–357.

Deutscher, J.; Blake, A. & Reid, I. (2000). Articulated body motion capture by annealed particle filtering. *Int. Conf. on Computer Vision and Pattern Recognition* (CVPR'00), pages 126–133, Hilton Head Island.

Deutscher, J.; Davison, A. & Reid, I. (2001). Automatic partitioning of high dimensional search spaces associated with articulated body motion capture. Dans *Int. Conf. on Computer Vision and Pattern Recognition* (CVPR'01), pages 669–676, Kauai.

Deutscher, J.; North, B.; Bascle, B. & Blake, A. (1999). Tracking trough singularities and discontinuities by random sampling. Dans *Int. Conf. on Computer Vision* (ICCV'99).

Doucet, A.; Godsill, S.J. & Andrieu, C. (2000). On sequential monte carlo sampling methods for bayesian filtering. *Statistics and Computing*, 10(3):197–208.

Engelberger, J.; (1989). *Robotics in Service*, chap 1. Cambridge, MA; MIT Press.

Fitzpatrick, P.; Metta, G.; Natale, L.; Rao, S. & Sandini, G. (2003). Learning about objects through action-initial steps towards artificial cognition. *Int. Conf. on Robotics and Automation* (ICRA'03), pages 3140- 3145, Taipei, Taiwan.

Fong, T.; Nourbakhsh, I. & Dautenhahn, K. (2003). A survey of socially interactive robots. *Robotics and Autonomous Systems* (RAS'03), 42:143–166.

Fontmarty, M.; Lerasle, F. ; Danès, P. & Menezes, P. (2007). Filtrage particulaire pour la capture de mouvement dédiée à l'interaction homme-robot. *Congrès francophone ORASIS*, Obernai, France.

Fox, M.; Ghallab, M. ; Infantes, G. & Long, D. (2006). Robust introspection through learned hidden markov models. *Artificial Intelligence* (AI'06), 170(2):59–113.

Gavrila, D. (1996). 3D model-based tracking of human in actions: A multi-view approach. *Int. Conf. on Computer Vision and Pattern Recognition* (CVPR'96), pages 73-80, San Francisco.

Gavrila, D. (1999). The visual analysis of human movement: a survey. *Computer Vision and Image Understanding* (CVIU'99), 73(1):82–98.

Germa, T.; Brèthes, L.; Lerasle, F. & Simon, T. (2007). Data fusion and eigenface based tracking dedicated to a tour-guide robot. Dans *Association Internationale pour l'Automatisation Industrielle* (AIAI'07), Montréal, Canada. Int. Conf. On Vision Systems (ICVS'07), Bielefeld.

Giebel, J.; Gavrila, D. M. & Schnorr, C. (2004). A bayesian framework for multi-cue 3D object. *European Conf. on Computer Vision* (ECCV'04), Prague.

Goncalves, L.; Bernardo, E. D.; Ursella, E. & Perona, P. (1995). Monocular tracking of the human arm in 3D. *Int. Conf. on Computer Vision* (ICCV'95).

Heap, A. J. & Hogg, D. C. (1996). Towards 3D hand tracking using a deformable model. *Int. Conf. on Face and Gesture Recognition* (FGR'96), pages 140–145, Killington, USA.

Heseltine, T.; Pears, N. & Austin, J. (2002). Evaluation of image pre-processing techniques for eigenface based recognition. *Int. Conf. on Image and Graphics,* SPIE, pages 677–685.

Isard, M. & Blake, A. (1998a). CONDENSATION - conditional density propagation for visual tracking. *Int. Journal on Computer Vision* (IJCV'98), 29(1):5–28.

Isard, M. & Blake, A. (1998b). I-CONDENSATION: Unifying low-level and high-level tracking in a stochastic framework. Dans *European Conf. on Computer Vision* (ECCV'98), pages 893–908.

Isard, M. & Blake, A. (1998c). A mixed-state condensation tracker with automatic model-switching. *International Conference on Computer Vision,* page 107, Washington, DC, USA. IEEE Computer Society.

Jones, M. & Rehg, J. (1998). Color detection. *Rapport technique,* Compaq Cambridge Research Lab.

Kakadiaris, I. & Metaxas, D. (2000). Model-based estimation of 3D human motion. *Trans. on Pattern Analysis and Machine Intelligence* (PAMI'00), 22(12):1453–1459.

Kehl, R. & Gool, L. (2004). Real-time pointing gesture recognition for an immersive environment. *Int. Conf. on Face and Gesture Recognition* (FGR'04), pages 577-582, Seoul.

Lerasle, F.; Rives, G. & Dhome, M. (1999). Tracking of human limbs by multiocular vision. *Computer Vision and Image Understanding* (CVIU'99), 75(3):229–246.

Maas, J. ; Spexard, T. ; Fritsch, J. ; Wrede, B. & Sagerer, G. (2006). BIRON, what's the topic? a multimodal topic tracker for improved human-robot interaction. *Int. Symp. on Robot and Human Interactive Communication* (RO-MAN'06), Hatfield, UK.

MacCormick, J. & Blake, A. (2000a). A probabilistic exclusion principle for tracking multiple objects. *Int. Journal of Computer Vision*, 39(1):57–71.

MacCormick, J. & Isard, M. (2000b). Partitioned sampling, articulated objects, and interface quality hand tracking. *European Conf. on Computer Vision* (ECCV'00), pages 3–19, London. Springer Verlag.

Menezes, P.; Barreto, J. & Dias, J. (2004). Face tracking based on haar-like features and eigenfaces. *IFAC Symp. on Intelligent Autonomous Vehicles*, Lisbon.

Menezes, P.; Lerasle, F.; Dias, J. & Chatila, R (2005a). Appearance-based tracking of 3D articulated structures. *Int. Symp. on Robotics* (ISR'05), Tokyo.

Menezes, P.; Lerasle, F.; Dias, J. & Chatila, R. (2005b). Single camera motions capture system dedicated to gestures imitation. *Int. Conf. on Humanoid Robots* (HUMANOID'05), pages 430–435, Tsukuba.

Metaxas, D. ; Samaras, D. & Oliensis, J. (2003). Using multiple cues for hand tracking and model refinement. *Int. Conf. on Computer Vision and Pattern Recognition* (CVPR'03), pages 443-450, Madison.

Nakazawa, A.; Nakaoka, S.; Kudo, S. & Ikeuchi, K. (2002). Imitating human dance motion through motion structure analysis. *Int. Conf. on Robotics and Automation* (ICRA'02), Washington.

Nourbakhsh, I.; Kunz C. & Willeke, D. (2003). The Mobot museum robot installations: A five year experiment. *Int. Conf. On Intelligent Robots and Systems* (IROS'03), Las Vegas.

Park, J.; Park, S. & Aggarwal, J. (2003). Human motion tracking by combining view-based and model-based methods for monocular vision. Dans *Int. Conf. on Computational Science and its Applications* (ICCSA'03), pages 650–659.

Pavlovic, V.; Rehg, J. & Cham, T. (1999). A dynamic bayesian network approach to tracking using learned switching dynamic models. Dans *Int. Conf. on Computer Vision and Pattern Recognition* (CVPR'99), Ft. Collins.

Pavlovic, V. ; Sharma, R. & Huang, T. S. (1997). Visual interpretation of hand gestures for human-computer interaction : A review. *Trans. On Pattern Analysis and Machine Intelligence* (PAMI'97), 19(7): 677–695.

Pérez, P.; Vermaak, J. & Blake, A. (2004). Data fusion for visual tracking with particles. *Proc. IEEE*, 92(3):495–513.

Pérez, P. ; Vermaak, J. & Gangnet, M. (2002). Color-based probabilistic tracking. Dans *European Conf. on Computer Vision* (ECCV'02), pages 661–675, Berlin.

Pitt, M. & Shephard, N. (1999). Filtering via simulation: Auxiliary particle filters. *Journal of the American Statistical Association*, 94(446).

Poon, E. & Fleet, D. (2002). Hybrid monte carlo filtering: Edge-based tracking. Dans *Workshop on Motion and Video Computing*, pages 151–158, Orlando, USA.

Rehg, J. & Kanade, T. (1995). Model-based tracking of self-occluding articulated objects. *Int. Conf. on Computer Vision* (ICCV'95), pages 612–617, Cambridge.

Rui, Y. & Chen, Y. (2001). Better proposal distributions: Object tracking using unscented particle filter. *Int. Conf. on Computer Vision and Pattern Recognition* (CVPR'01), pages 786–793, Hawai.

Schmid, C. (1996). Appariement d'images par invariants locaux de niveaux de gris. *Thèse de doctorat*, Institut National Polytechnique de Grenoble.

Schwerdt, K. & Crowley, J. L. (2000). Robust face tracking using color. *Int. Conf. on Face and Gesture Recognition* (FGR'00), pages 90–95, Grenoble, France.

Shon, A. ; Grochow, K. & Rao, P. (2005). Imitation learning from human motion capture using gaussian process. *Int. Conf. on Humanoid Robots* (HUMANOID'05), pages 129–134, Tsukuba.

Sidenbladh, H. ; Black, M. & Fleet, D. (2000). Stochastic tracking of 3D human figures using 2D image motion. Dans *European Conf. on Computer Vision* (ECCV'00), pages 702–718, Dublin.

Siegwart, R. ; Arras, O.; Bouabdallah, S.; Burnier, D. ; Froidevaux, G.; Greppin, X.; Jensen, B.; Lorotte, A.; Mayor, L.; Meisser, M.; Philippsen, R.; Piguet, R.; Ramel, G.; Terrien, G. & N. Tomatis (2003). Robox at expo 0.2: a large scale installation of personal robots. *Robotics and Autonomous Systems* (RAS'03), 42:203–222.

Sminchisescu, C. & Triggs, B. (2003). Estimating articulated human motion with covariance scaled sampling. *Int. Journal on Robotic Research* (IJRR'03), 6(22):371–393.

Stenger, B.; Mendonça, P. R. S. & Cipolla, R. (2001). Model-based hand tracking using an unscented Kalman filter. *British Machine Vision Conf.* (BMVC'01), volume 1, pages 63–72, Manchester.

Stenger, B.; Thayananthan, A. ; Torr, P. & Cipolla, R. (2003). Filtering using a tree-based estimator. *Int. Conf. on Computer Vision* (ICCV'03), pages 1063–1070, Nice.

Sturman,D. & Zeltzer, D. (1994). A survey of glove-based input. *Computer Graphics and Applications*, 14(1):30–39.

Thayananthan, A.; Stenger, B.; Torr, P. & R. Cipolla (2003). Learning a kinematic prior for tree-based filtering. Dans *British Machine Vision Conf.* (BMVC'03), volume 2, pages 589–598, Norwick.

Torma, P. & Szepesvari, C. (2003). Sequential importance sampling for visual tracking reconsidered. Dans *AI and Statistics*, pages 198–205.

Thrun, S.; Beetz, M.; Bennewitz, M.; Burgard, W.; Cremers, A.B.; Dellaert, F.; Fox, D.; Halnel, D.; Rosenberg, C.; Roy, N.; Schulte, J. & Schultz, D. (2000). Probabilistic algorithms and the interactive museum tour-guide robot MINERVA. *Int. Journal of Robotics Research* (IJRR'00).

Turk, M. & Pentland, A. (1991). Face recognition using eigenfaces. *Int. Conf. on Computer Vision and Pattern Recognition* (CVPR'91), pages 586–591, Hawai.

Urtasum, R. & Fua, P. (2004). 3D human body tracking using deterministic temporal motion models. *European Conf. on Computer Vision* (ECCV'04), Prague.

Viola, P. & Jones, M. (2001). Rapid Object Detection using a Boosted Cascade of Simple Features. *Int. Conf. on Computer Vision and Pattern Recognition* (CVPR'01), Hawai.

Wachter, S. & Nagel, S. (1999). Tracking persons in monocular image sequences. *Computer Vision and Image Understanding* (CVIU'99), 74(3):174–192.

Wu, Y. & Huang, T. (1999). Vision-based gesture recognition: a review. *International Workshop on Gesture-Based Communication,* pages 103–105, Gif-sur-Yvette.

Wu, Y.; Lin, T. & Huang, T. (2001). Capturing natural hand articulation. I*nt. Conf. on Computer Vision* (ICCV'01), pages 426–432, Vancouver.

20

Methods for Environment Recognition based on Active Behaviour Selection and Simple Sensor History

Takahiro Miyashita[1], Reo Matsumura[2], Kazuhiko Shinozawa[1], Hiroshi Ishiguro[2] and Norihiro Hagita[1]

[1]*ATR Intelligent Robotics and Communication Laboratories,* [2]*Osaka University*
Japan

1. Introduction

The ability to operate in a variety of environments is an important topic in humanoid robotics research. One of the ultimate goals of this research is smooth operation in everyday environments. However, movement in a real-world environment such as a family's house is challenging because the viscous friction and elasticity of each floor, which directly influence the robot's motion and are difficult to immediately measure, differ from place to place. There has been a lot of previous research into ways for the robots to recognize the environment. For instance, Fennema et al. (Fennema et al., 1987) and Yamamoto et al. (Yamamoto et al., 1999) proposed environment recognition methods based on range and visual information for wheeled robot navigation. Regarding humanoid robots, Kagami et al. (Kagami et al., 2003) proposed a method to generate motions for obstacle avoidance based on visual information. They measured features of the environment precisely before moving or fed back sensor information to a robot's controller with a short sampling period. It is still difficult to measure the viscous friction or elasticity of the floor before moving or by using short term sampling data, and they did not deal with such features.

Thus, we propose a method for recognizing the features of environments and selecting appropriate behaviours based on the histories of simple sensor outputs, in order to achieve a humanoid robot able to move around a house. Figure 1 shows how our research differs from previous research according to length of the sensor history and number of types of sensors. The key idea of our method is to use a long sensor history to determine the features of the environment. To measure such features, almost all previous research (Shats et al., 1991; Holweg et al., 1996) proposed methods that used several kinds of sensors with a large amount of calculations to quickly process the sensor outputs. However, such approaches are unreasonable because the robot lacks sufficient space on its body for the attached sensors and processors. Hence we propose using sensor history to measure them because there are close relationships between sensor histories, motions, and environments.

When the robot performs specific motions in specific environments, we can see those features in the sensor history that describe the motion and the environment. Furthermore, such features as viscous friction or floor elasticity do not change quickly. Thus we can use a long history of sensor data to measure them.

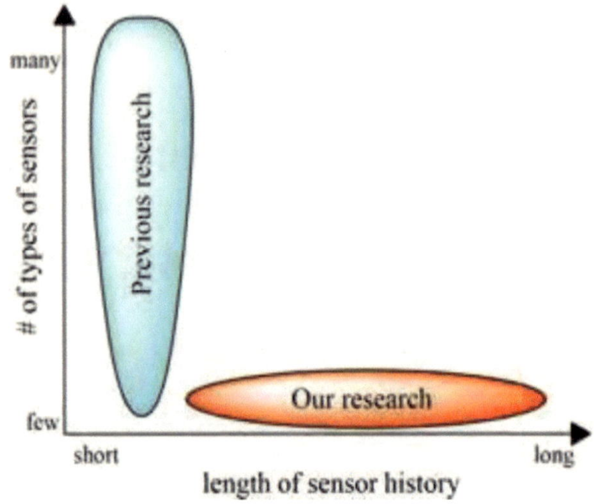

Figure 1. Difference between our research and previous research

In the next section, we describe our method for behaviour selection and environment recognition for humanoid robots. In section 3, we introduce the humanoid robot, named "Robovie-M," that was used for our experiments. We verify the validity of the method and discuss future works in section 4 and 5.

2. Behaviour selection and environment recognition method

2.1 Outline of proposed method

We propose a method for humanoid robots to select behaviours and recognize their environments based on sensor histories. An outline of the method is as follows:

A-1 [preparation 1] In advance, a user of the robot prepares basic motions appropriate to the environment.

A-2 [preparation 2] For each basic motion and environment, the robot records the features of the time series data of its sensors when it follows the motions.

A-3 [preparation 3] For each basic motion, the robot builds decision trees to recognize the environments based on recorded data by using a binary decision tree generating algorithm, named C4.5, proposed by Quinlan (Quinlan, 1993). It calculates recognition rates of decision trees by using cross-validation of the recorded data.

B-1 [recognition 1] The robot selects the motion that corresponds to the decision tree that has the highest recognition rate. It moves along the selected motion and records the features of the time series data of the sensors.

B-2 [recognition 2] The robot calculates reliabilities of the recognition results for each environment based on the decision tree and the recorded data. Then it selects the environments that have reliability greater than a threshold as candidates of the current environment. The threshold is decided by preliminary experiments.

B-3 [recognition 3] The robot again builds decision trees based on the data recorded during the process (**A-2**) that correspond to the selected candidates for the current environment. Go to (**B-1**).

By iterating over these steps, the robot can recognize the current environment and select appropriate motions.

2.2 Robot's motions and features of the environment

Figure 2 shows the motions that the robot has in advance. In our method, there are two kinds of motions: basic and environment-dependent. The basic motions are comprised of a set of motions that can be done in each environment without changing the loci of joints, such as standing up, lying down, etc. All environment-dependent motions are generated in advance by the user. By utilizing our method, once the environment is recognized, the robot can select the suitable motions for it from the environment-dependent motions.

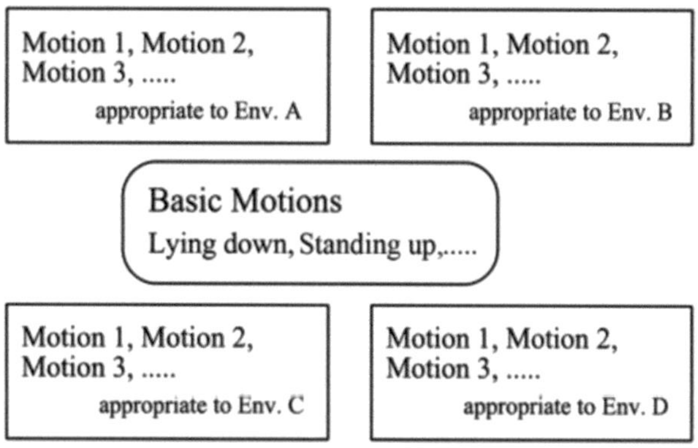

Figure 2. Robots have two kinds of motions: basic and environment-dependent. Both motions are generated by users in advance

In this paper, we use not only averages and standard deviations of the time series data of the sensor outputs, but also averages and standard deviations of the first and second derivatives of those outputs, as the features of the environment. Table 1 shows an example of features of sensor histories by taking different basic motions in a tiled floor environment. We use a set of the features of sensor history $s_n(t)$ as a feature of an environment.

Basic motions	Lying down	Standing up	......
Label of environment	Tiled floor	Tiled floor	......
Ave. of $s_n(t)$	136.19	149.15	......
Std. dev. of $s_n(t)$	21.429	25.64	......
Ave. of $ds_n(t)/dt$	131.13	128.84	......
Std. dev. of $ds_n(t)/dt$	6.1985	6.2903	......
Ave. of $d^2s_n(t)/dt^2$	157.83	132.89	......
Std. dev. of $d^2s_n(t)/dt^2$	11.292	13.554	......

Table 1. Example of features of sensor histories by taking different basic motions in a tiled floor environment. $s_n(t)$ denotes time series data of sensor s_n

2.3 Decision tree based on relationships between basic motions, sensor histories, and environments

A decision tree to recognize the environment is made by C4.5 (Quinlan, 1993), which is a program for inducing classification rules in the form of binary decision trees from a set of given examples. We use the relationships described in Table 1 as examples and make decision trees for each basic motion by using knowledge analysis software WEKA (Witten, 2000) that can deal with C4.5. Figure 3 shows an example of a decision tree for the lying down motion.

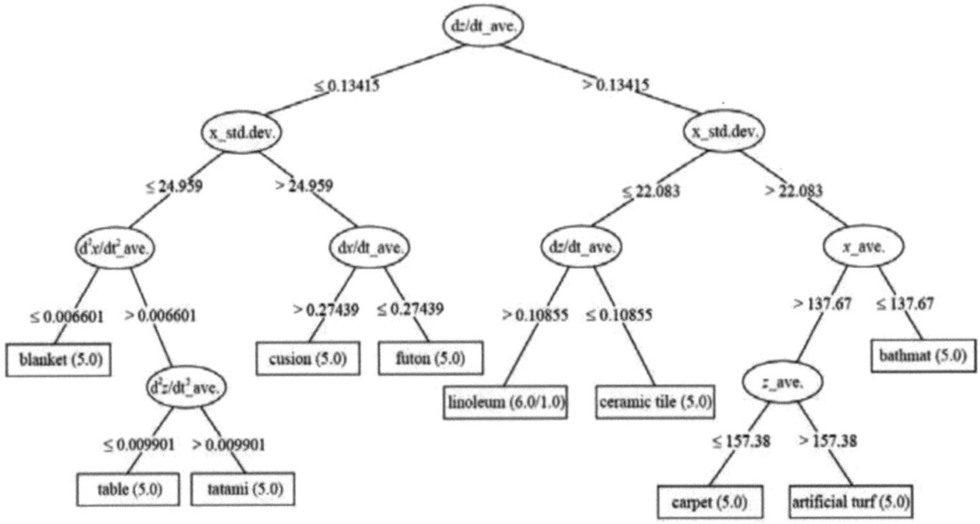

Figure 3. Decision trees recognize environments based on relationships between a motion (lying down), possible environments, and sensor histories. Circles denote features of sensor history. Rectangles denote environments

We can also determine the recognition rate of a decision tree for each basic motion and the reliabilities of the recognition results by cross-validation as follows. The recognition rate of a decision tree for the *k*-th basic motion, r_k, is calculated as follows:

$$r_k = \frac{S_k}{N} , \tag{1}$$

where N and S_k denote the number of all data sets for candidates of the current environment that were obtained in the preparation processes and the number of correctly classified data sets by the decision tree, respectively. After selecting the decision tree that has the highest recognition rate and moving along the *l*-th basic motion that corresponds to the tree, the robot records the data set and obtains a recognition result by using the tree. We calculate following two types of reliabilities from the result. When the environment A is the result of the recognition, the reliability that the result is the environment A, rel_A, is calculated as follows:

$$rel_A = \frac{S_{lAA}}{N_A} , \tag{2}$$

where N_A and S_{lAA} denote the number of all data sets for the environment A that were obtained in the preparation processes and the number of correctly classified data sets by the tree, respectively. The reliability that the result is one of the other environments, for example the environment B, is as follows:

$$rel_B = \frac{S_{lBA}}{N_B} , \tag{3}$$

where N_B and S_{lBA} denote the number of all data sets for the environment B that were obtained in the preparation processes and the number of incorrectly classified data sets that are classified as the environment A by the tree, respectively. This is same as the misrecognition rate that the robot recognizes the environment B as the environment A by the tree.

3. Humanoid robot

In this section, we introduce a small-size humanoid robot, Robovie-M, developed by us. Figure 4 (a) and (b) show an overall view and hardware architecture of Robovie-M. The robot has a head, two arms, a body, a waist, and two legs. Degrees of freedom (DOFs) of the robot are as follows. The robot has 4 DOFs for each arm, 2 DOFs for waist, and 6 DOFs for each leg. The total number of DOFs is 22. As shown in Figure 4 (b), we attached two 2-axial acceleration sensors to the left shoulder of the robot to acquire acceleration values along three orthogonal axes as $s_n(t)$ in Table 1. Table 2 describes specifications of the sensor. Sampling rate of the sensor is 60 [Hz]. The robot can send data of the sensors and get commands of the behaviour from a host PC via a serial cable (RS-232C).

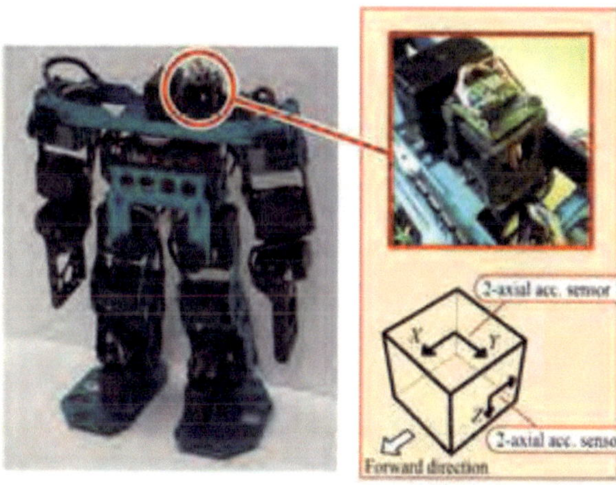

(a) Overall view of Robovie-M and arrangement of sensors

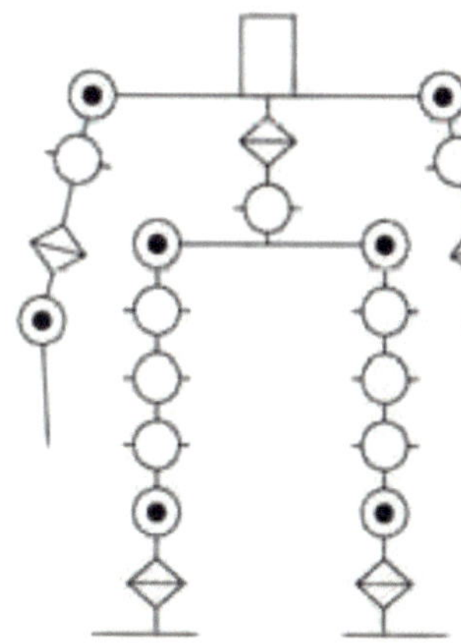

(b) Arrangement of degrees of freedom

Figure 4. Left image shows humanoid robot Robovie-M, and center images indicate sensor arrangement. On the robot's left shoulder, two 2-axial acceleration sensors are attached orthogonally to acquire acceleration values along three axes that describe horizontal and vertical motions. The right image shows an arrangement of robot's degrees of freedom

Model	ADXL202E (ANALOG DEVICES)
# Axis	2
Range	-2 g ~ +2 g
Sensitivity	12.5 % / g
Nonlinearity	0.2 %
Size	5 x 5 x 2 mm

Table 2. Specifications of 2-axial acceleration sensor attached to the robot

4. Experiments

To verify the validity of the proposed method, we conducted a preliminary experiment with our small humanoid robot, Robovie-M. Table 3 and Figure 5 show environments for recognition, the basic motions and time lengths for each motion in the experiments. Figure 6 shows sequences of pictures for each basic motion. We recorded the time series data of the sensor outputs ten times in each environment and for each motion.

Environments		Basic motions	time [s]
Ceramic tiled floor	Linoleum floor	Lying down	4.5
Wooden table	Tatami mat	Standing up	3.0
Cushion	Futon	Tossing and turning	4.0
Carpet	Bathmat	Stepping with one leg	4.5
Blanket	Artificial turf	Stepping with both legs	4.5

Table 3. The left column describes environments used in the experiment. The right column describes the basic motions. Environments are selected from a typical Japanese house

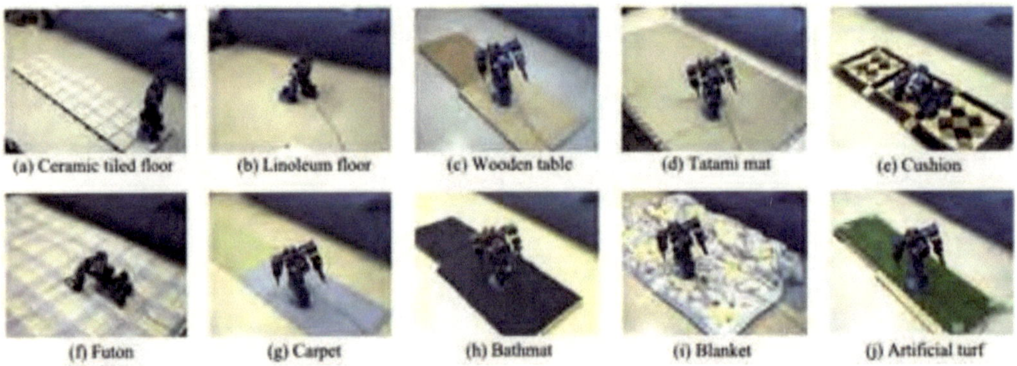

Figure 5. Pictures of environments in experiments

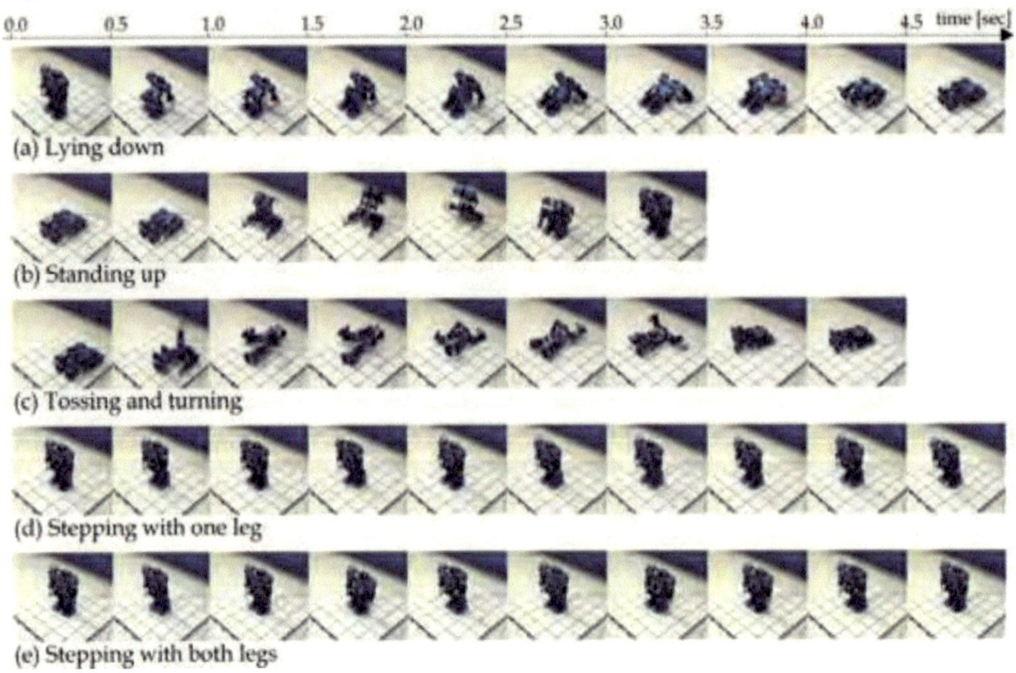

Figure 6. Sequences of pictures for each basic motion

For instance, let us consider the recognition process for the *futon* (a Japanese mattress) environment. First, the robot selected the *stepping with both legs* motion because the motion's decision tree has the highest recognition rate. All recognition rates calculated by equation (1) are described in Figure 7. Second, the robot obtained the sensor history while doing the *stepping with both legs* motion and classified it by using the motion's decision tree. The result of classification was the *blanket* environment. The reliabilities of the result for each environment were obtained, as shown in Table 4. The reliability for the *blanket* environment was calculated by equation (2) and the reliabilities for the others were calculated by equation (3). This time the reliability threshold was 0.2. Then the selected candidates of the current environment were *tatami, futon, artificial turf,* and *blanket*. Next, the robot made decision trees for each basic motion based on the data of the candidates. By calculating their recognition rates, as shown in Figure 8, the robot selected the *stepping with one leg* motion. As a result of performing the selected motion, the robot classified the data as the *futon* environment and obtained *artificial turf* and *futon* as candidates, as shown in Table 5. The robot selected the *lying down* motion from the recognition rates based on the candidate's data shown in Figure 9. Finally, the robot obtained the data while lying down and recognized the current environment as the *futon* environment shown in Table 6. We verified that the robot recognized all environments shown in Table 3 by using our method. The maximum times of the iteration of these processes for the environment recognition was three.

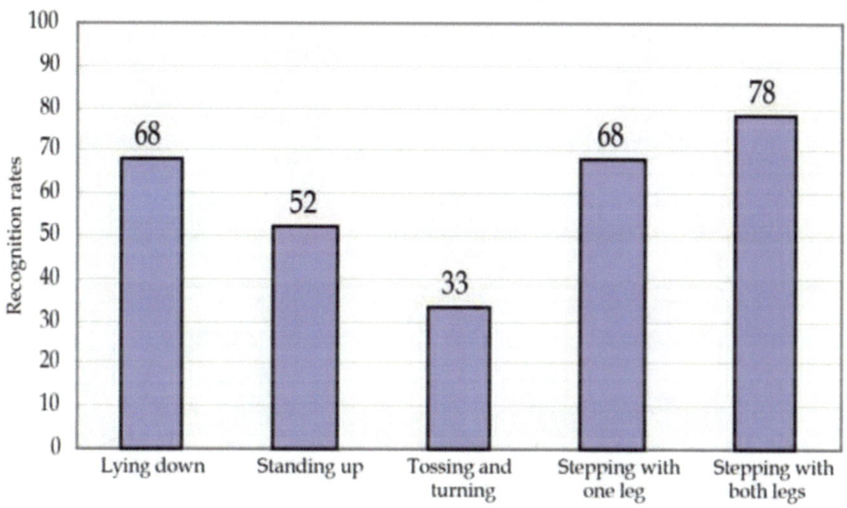

Figure 7. Recognition rates of decision trees for each motion based on all data. The highest rate is obtained by the Stepping on both legs motion

Environment	Reliability	Environment	Reliability
Ceramic tiled floor	0.0	Futon	0.2
Linoleum floor	0.0	Carpet	0.0
Wooden table	0.0	Bathmat	0.0
Tatami mat	0.2	Blanket	0.4
Cushion	0.0	Artificial turn	0.2

Table 4. Reliabilities for each environment when the decision tree of the stepping with both legs motion classifies data to the blanket environment

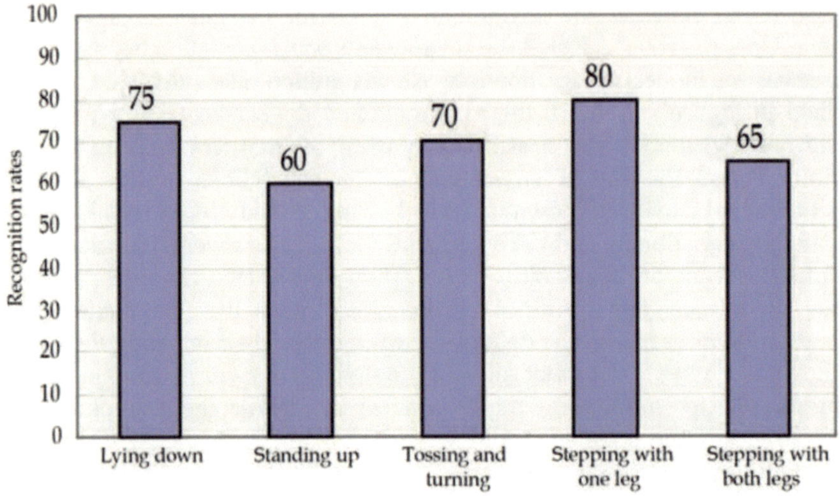

Figure 8. Recognition rates of decision trees for each motion based on data that correspond to tatami, futon, artificial turf, and blanket

Environment	Reliability	Environment	Reliability
Ceramic tiled floor	0.0	Futon	0.8
Linoleum floor	0.0	Carpet	0.0
Wooden table	0.0	Bathmat	0.0
Tatami mat	0.0	Blanket	0.0
Cushion	0.0	Artificial turn	0.2

Table 5. Reliabilities for each environment when the decision tree for the stepping with one leg motion classifies data to the futon environment

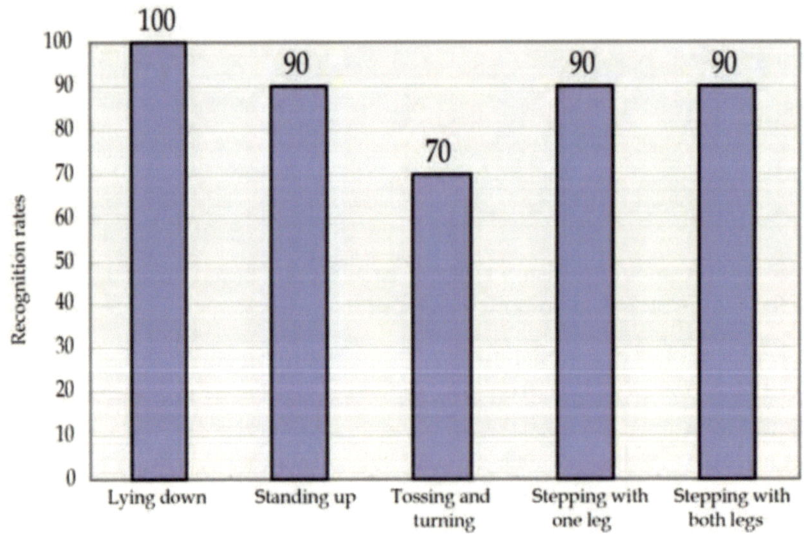

Figure 9. Recognition rates of decision trees for each motion based on data that correspond to futon and artificial turf

Environment	Reliability	Environment	Reliability
Ceramic tiled floor	0.0	Futon	1.0
Linoleum floor	0.0	Carpet	0.0
Wooden table	0.0	Bathmat	0.0
Tatami mat	0.0	Blanket	0.0
Cushion	0.0	Artificial turn	0.0

Table 6. Reliabilities for each environment when the decision tree for the lying down motion classifies data to the futon environment

5. Conclusion

In this paper, we proposed a method for recognizing environment and selecting appropriate behaviours for humanoid robots based on sensor histories. By using the method, the robot could select effective behaviours to recognize current environment.

For ten different environments that are typical in a Japanese family's house, the results of these experiments indicated that the robot successfully recognized them by five basic motions shown in Table 3. However, we should consider the case when number of candidates of current environment does not converge to one. In the case, the robot should acquire new sensor data and rebuild the decision trees, then recognize the environment, again. After these processes, when the number of candidates of the environment becomes one, the robot can decide that the environment is inexperienced. Otherwise, prepared basic motions are not enough for recognizing the environments and an additional basic motion is necessary. In future work, we will clarify dynamical relationships between basic motions and features of environments, and confirm proposed basic motions enough for recognizing the environments. Then, we will extend our method to deal with inexperienced environments.

6. Acknowledgment

This research was supported by the Ministry of Internal Affairs and Communications.

7.References

Fennema, C.; Hanson, A.; Riseman, E.; Beveridge, J.R. & Kumar, R. (1990). Model-Directed Mobile Robot Navigation, *IEEE Transaction on Systems, Man and Cybernetics*, Vol. 20, No. 6, pp. 1352-1369, ISSN: 0018-9472.

Yamamoto, T.; Maeyama, S.; Ohya, A. & Yuta, S. (1999). An implementation of landmark-based position estimation function as an autonomous and distributed system for a mobile robot, *Proceedings of IEEE/RSJ International Conference on Intelligent Robots and Systems (IROS '99)*, Vol. 2, pp. 1141-1148, ISBN: 0 7803 5181 3, Kyongju, South Korea, Oct. 1999.

Kagami, S.; Nishiwaki, K.; Kuffner, J.J.; Okada, K.; Inaba, H. & Inoue, H. (2003). Vision-Based 2.5D Terrain Modelling for Humanoid Locomotion, *Proceedings of IEEE International Conference on Robotics and Automation (ICRA2003)*, Vol. 2, pp. 2141-2146, ISBN: 0-7803-7736-2, Taipei, Taiwan, Sept. 2003.

Shats, S.; Cohen, G. & Cohen, L. (1991). Identification of the linear and nonlinear parameters of an electro-mechanical system, *Proceedings of 17th convention of Electrical and Electronics Engineers in Israel*, pp. 95-97, ISBN: 0-87942-678-0, Tel Aviv, Israel, Mar. 1991.

Holweg, E.G.M.; Hoeve, H.; Jongkind, W.; Marconi, L.; Melchiorri, C. & Bonivento, C. (1996). Slip detection by tactile sensors: algorithms and experimental results, *Proceedings of IEEE International Conference on Robotics and Automation*, Vol. 4, pp. 3234-3239, ISBN: 0-7803-2988-0, Minneapolis, MN, USA, April 1996.

Quinlan, J. R. (1993). *C4.5: Programs for Machine Learning*, Morgan Kaufman, ISBN: 1-55860-238-0, CA, USA.

Witten, I.H. & Frank, E. (2000). *Data Mining - Practical Machine Learning Tools and Techniques with JAVA Implementations* -, Morgan Kaufman, ISBN: 1-55860-552-5, CA, USA.

21

Simulation Study on Acquisition Process of Locomotion by using an Infant Robot

Katsuyoshi Tsujita and Tatsuya Masuda
Dept. of Electrical and Electronic Systems Engineering, Osaka Institute of Technology
Japan

1. Introduction

Locomotion is one of the basic functions of a mobile robot. Using legs is one of the strategies for accomplishing locomotion. The strategy allows a robot to move over rough terrain. Therefore, a considerable amount of research has been conducted on motion control of legged locomotion robots. This chapter treats the motion generation of an infant robot, with emphasis on the emergence of crawling locomotion.

In the future, a walking robot that can carry out various tasks on unstructured terrain will be required. The walking robot is required to achieve real-time adaptability to a changing environment. However, the mechanism from which the adaptive motion pattern emerges is not clear.

Recent biological research and psychological research on acquisition of motion have made great contributions and have given crucial hints as to how to overcome such problems. During spontaneous motion, such as crawling or straight walking, a lot of joints and muscles are organized into a collective unit to be controlled as if this unit had fewer degrees of freedom, but at the same time to retain the necessary flexibility for a changing environment (Bernstein, 1967). Gesell pointed out the principles of motor development in human infants (Gesell, 1946). According to that research, some developmental principles in the acquisition of ontogenetic activities can be observed. One is *directional trends in the acquisition of ontogenetic activities;* others are *functional asymmetry in ontogenetic activities* and *self-regulation in ontogenetic activities.* In addition, various studies have been made on the acquisition of motion especially that of locomotion (Newell, 1990; Thelen et al.,1986,1987; Clark et al.,1988; Burnside, 1927; Adolph, 1997; Savelsbergh, 1993). Moreover, the development of motions has been proposed as being a dynamic interaction between the nervous and musculo-skeletal systems. Rhythmic motion is generated by a central pattern generator (CPG) in the spinal cord (Grillner, 1977,1985). Sensory feedback from the contact sensors or joint angle sensors tunes the oscillation condition of the CPG and makes the locomotion stable in limit cycle (Taga, 1991,1994). Furthermore, biological researches on mode transition of the locomotion according to the situation or variance of the environment are actively going on (Ijspeert, 2001). Based on these biological facts, research has been conducted to clarify the mechanism for humans' acquisition of motion (Yamazaki, 1996; Hase, 2002; Ni et al., 2003; Endo et al., 2004, Kuniyoshi et al., 2004).

The knowledge acquired has inspired robotics researchers, and a considerable amount of research has been done on biologically inspired control systems for walking robots that are based on the CPG controller model, and that will enable them to adapt to variances in the environment (Fukuoka et al., 2003; Tsujita et al., 2001,2005; Lewis et al., 2001).
In this study, we propose a new control system for acquisition of the motion of an infant robot by using oscillators. The proposed control system consists of a spontaneous locomotion controller, reflex controller, and tones controller. The spontaneous locomotion controller is designed as follows. A nonlinear oscillator is assigned to each leg. The periodic trajectory of each leg is calculated as a function of the phase of its oscillator. Touch sensors at the tips of the legs are used to triggers dynamic interactions of the legs. The mutual entrainment of the oscillators generates an appropriate combination of phase differences according to the context of the changing environment, and this leads to a gait pattern. The network connections of the oscillators, specifically, the phase differences of the oscillators, are tuning parameters. These parameters are tuned according to a specific *intension,* i.e., an objective function for learning or optimization, such as stiffening joints or to moving forward at high energy efficiency. The reflex controller generates asymmetrical reflexes on antagonistic pairs of actuators. The idea of the architecture of tones control of the actuators are inspired by the biological studies. According to them, the tones are actively and adaptively controlled by the neuronal system in the 'basal ganglia' (Takakusaki, 2003). And this neuronal system stabilizes the posture, and obtains the stable locomotion by also controlling the oscillation of central pattern generator in the spinal cord. In this study, a type of tones controlling is considered as **PD** feedback control by adaptively changing the gains. The tones controller tunes the stiffness of the joints by changing the feedback gains adaptively according to the situations of the motion pattern of the system. These feedback gains are also refhed through learning or optimization.
We verified the effectiveness of the proposed control system with numerical simulations and hardware experiments.

2. Framework of this study

Fig.l summarizes the developmental acquisition of motion pattern in human infancy.
In this study, developmental acquisition is divided into four stages. In the first stage, the tones controller tunes the stiffness of the joints of the neck, trunk, hips, arms, and legs, in this order. In this stage, the intension of the infant is considered as that of making the controllable body system to perform a motion defhed by a command signal. In the second stage, primitive crawling locomotion emerges from using the alternative motion of hands by applying an asymmetry reflex to the hands' contact with the ground. In this stage, legs are not so skilled at generating propulsion force for locomotion. The intension of this stage is considered that of moving forward. In the third stage, adjustment of the tones of the legs' actuator is completed and the infant begins a perfect crawling locomotion that is fast with high energy-efficiency. The intension of this stage is considered as to move forward faster with less fatigue of actuators. The last stage is bipedal. The intension of this stage is considered as to raise the position of the eye higher.
In this chapter, these intensions are formulated as objective functions that are heuristically and dynamically controlled according to a particular developmental stage of the infant robot. The feedback gains of the actuators that govern tones (stiffness) of the joints and

interaction strength among the oscillators are selected as tuning parameters for learning or optimization.

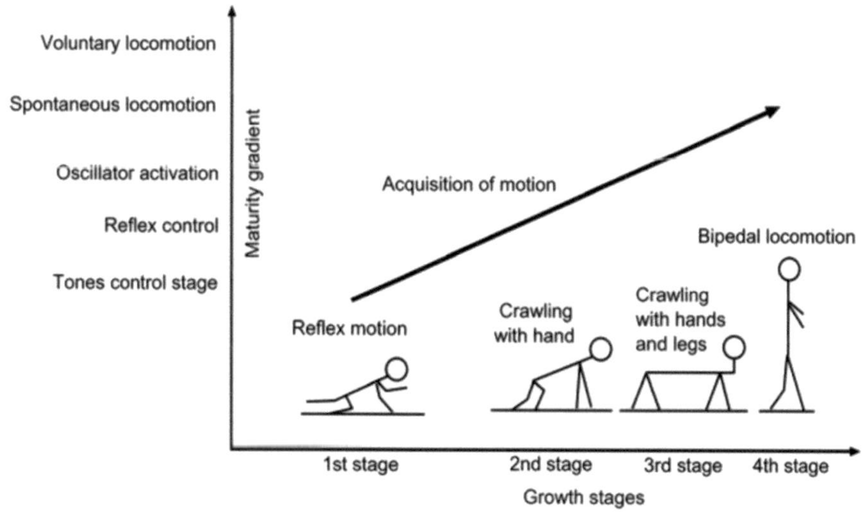

Figure 1. Growth cycle in acquisition of motions

3. Model

Fig. 2 is a schematic model of the infant robot. This robot has 28 DOF (Degrees of Freedom). The robot has a head, a torso, two arms, and two legs. Each arm consists of two links that are connected to each other through three degrees of freedom, (DOF) rotational joints at the shoulders and a one DOF rotational joint at the elbows. Each leg consists of three links that are connected to each other through rotational joints with three degrees of freedom (DOF) at the hip and ankle, and a one DOF rotational joint at the knee. The torso is composed of two parts, upper body and lower body. The upper body and the lower body are connected through a three DOF rotational joint. The arms and legs are connected to the torso with three DOF rotational joints. The head is connected to the torso with a three DOF rotational joint. Each subsystem is enumerated as 0: Upper body, 1:Head, 2:Lower body, 3:Left hand, 4:Right hand, 5:Left leg and 6:Right leg. We define $r_k^{(B)}$ and $\theta_k^{(0)}$ (k = 1, 2, 3) as the components of position vector and Euler angle from inertial space to the coordinate system that is fixed on the upper body, respectively. The joint angle of the lower body to upper body is defined as $\theta^{(2)}$. We also define each joint angle of each corresponding joint, $\theta_j^{(i)}$ (i:subsystem no., j: joint no.).
The state variable is defined as follows,

$$q^T = \begin{bmatrix} r^{(B)T} & \theta^{(0)T} & \theta^{(1)T} & \theta^{(2)T} & \theta_1^{(3)T} & \cdots & \theta_3^{(6)T} \end{bmatrix} \quad (1)$$

Equations of motion for state variable q are derived using Lagrangian formulation as follows;

$$M\ddot{q} + H(q,\ \dot{q}) = G + T + \Lambda \quad (2)$$

where M is the generalized mass matrix and the term $M\ddot{q}$ expresses the inertia. The nonlinear term which includes Coriolis forces and centrifugal forces is $H(q,\ \dot{q})$, and G is the gravity term. The component of T vector $T_j^{(i)}$ and $T^{(B)}$ are the input torque of the actuator at joint j of leg i and at the joint of torso. Λ is the reaction force from the ground at the point where the tip of the leg makes contact. We assume that friction exists between the ends of the legs and the ground. In this study, the candidates of the contact point of the robot body are guessed to be elbows, shoulders, knees, and four points at the rectangular corners of the feet.

The model of friction between the contact points of the ends of hands or legs and the ground is given as follows;

$$\begin{cases} \Lambda_j^i = -K_{E,floor}\Phi_j^i - K_{V,floor}\dot{\Phi}_j^i & |\Lambda_j^i| < \mu\Lambda_3^i \\ \Lambda_j^i = -\mu\Lambda_3^i - K_{V,friction}\dot{\Phi}_j^i & \Lambda_j^i < -\mu\Lambda_3^i \\ \Lambda_j^i = \mu\Lambda_3^i - K_{V,floor}\dot{\Phi}_j^i & \Lambda_j^i > \mu\Lambda_3^i \end{cases} \quad j = 1, 2 \tag{3}$$

where Λ_j^i, j = 1, 2 are horizontal friction force and Λ_3^i is vertical reaction force at contact point i, respectively. $K_{E,floor}$, $K_{V,floor}$, $K_{V,friction}$ and μ are elastic coefficient and viscous coefficient of the visco-elastic model of the contact motion, the coefficient of viscous coefficient of friction of the floor, and the coefficient of stiction, respectively. The position of the end of hand or knee at contact point i is expressed as vector Φ_j^i in the inertial space with the component of $j = 1, 2, 3$.

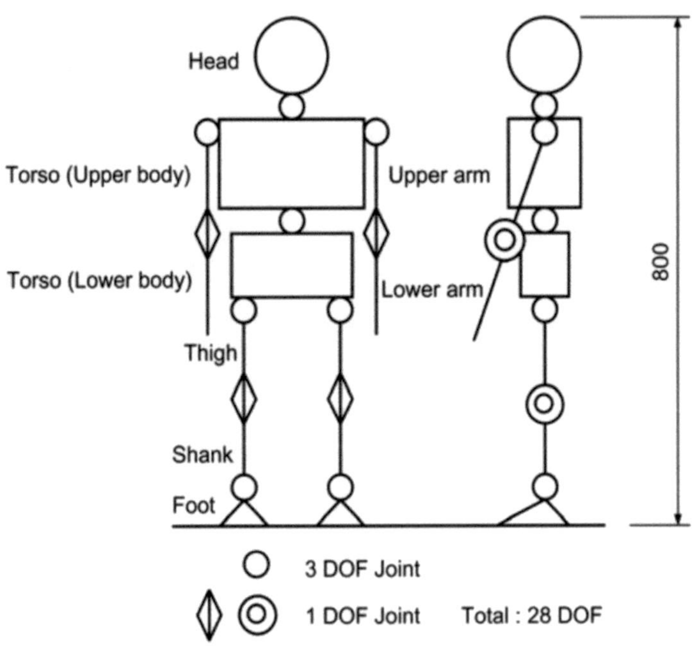

Figure 2. Schematic model of infant robot

4. Control System

The architecture of the proposed control system is diagrammed in Fig. 3. The proposed control system has upward signal-flow and downward signal-flow. An oscillator for determining the rhythm and phase difference of periodic motions during locomotion is assigned for each leg, each hand, the torso, head, and a pace maker. The state of the oscillator for each part is expressed as follows;

$$z^{(i)} = \exp(j\ \phi^{(i)}) \quad (4)$$

$$i = 3, \cdots, 6 \text{ (subsystem no.)} \quad (5)$$

$$z^{(B)} = \exp(j\ \phi^{(B)}) \quad (6)$$

$$z_{pace} = \exp(j\ \phi_{pace}) \quad (7)$$

where $\phi^{(i)}$,$\phi^{(B)}$ and ϕ_{pace} are the phase of oscillator i, torso and that of the pacemaker oscillator.

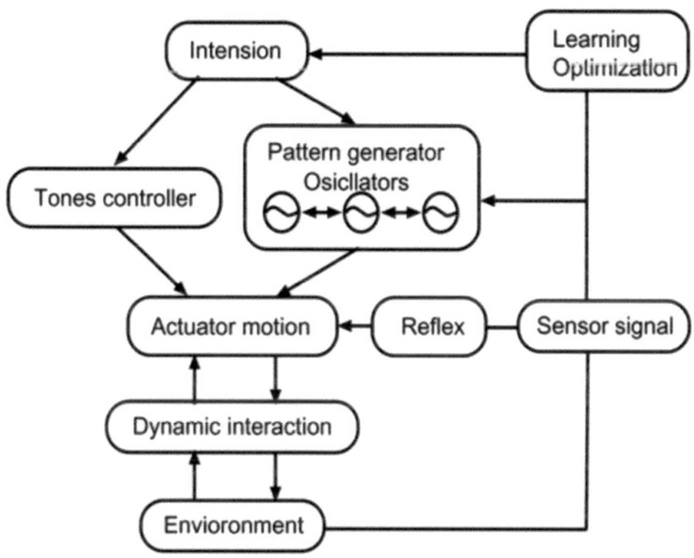

Figure 3. Control system

The pattern generator has the nonlinear oscillator network illustrated in Fig. 4, and it encodes the nominal trajectory of each joint angle of the hand and leg in terms of the phase of the oscillator that is given to the actuator inputs as a command signal. The tones controller tunes and varies the joint stiffness to correspond to the motion.

On the other hand, feedback signals of the contact sensors or attitude sensors are given to the reflex controller, and a pattern generator that is a learning or optimization process for the system. First, the contact sensor signal is input to the reflex controller as feedback signal. The reflex controller makes an immediate reflex to the contact signal or attitude signals to stabilize the posture of the torso. Then, the signal is also input to the pattern generator to reset the oscillator phase from the swinging stage to the supporting stage at the moment of contact, and vice versa. This dynamic interaction loop between pattern generator, environment, and actuator motion creates mutual entrainment and obtains a stable limit cycle for locomotion. Furthermore, the proposed system has a learning or optimization mechanism of using the feedback signal of the sensors. This system-tuning mechanism optimizes the parameters of the phase difference of the oscillators, actuator stiffness, and the amplitude of the synchronizing torso twisting motion.

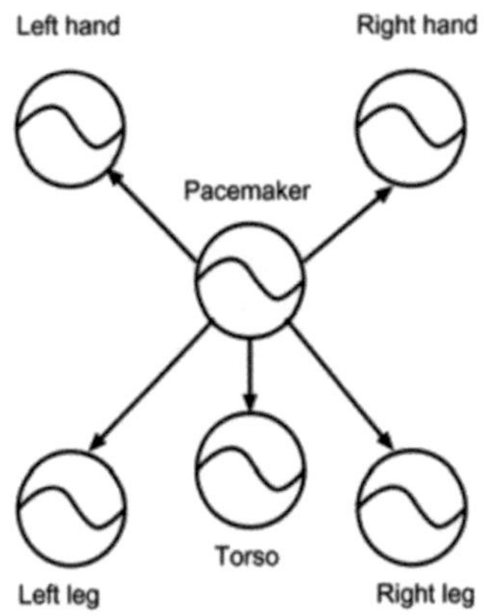

Figure 4. Oscillator network on pattern generator

The oscillator network in the pattern generator is formulated as follows;

$$\phi^{(i)} = \phi_{pace} + \gamma_j^{(i)} \quad (8)$$

$$\phi^{(B)} = \phi_{pace} + \gamma^{(B)} \quad (9)$$

where $\gamma^{(i)}$,$\gamma^{(B)}$ are the phase differences of oscillator i and torso to the pacemaker.

The trajectories of the hands and legs $r_j^{(i)}$ (j: joint no.) are given as function of the oscillator phase as follows;

$$r_j^{(i)} = \begin{cases} f_{sw}(\phi^{(i)}) & \text{Swing stage} \\ f_{sp}(\phi^{(i)}) & \text{Support stage} \end{cases} \quad (10)$$

The motion of the torso is a periodic one as follows;

$$\theta^{(2)} = A_t \sin(\phi^{(B)}) \quad (11)$$

The desired angle of each joint is calculated as a function of oscillator phase as follows;

$$\hat{\theta}_j^{(i)} = \theta_j^{(i)}(r_j^{(i)}(\phi^{(i)})) \quad (12)$$

$$i = 3, \cdots, 6 \text{ subsystem no. } \; j : \text{joint no.} \quad (13)$$

The actuator inputs are designed as follows;

$$T_j^{(i)} = K_{Pj}^{(i)}(\hat{\theta}_j^{(i)} - \theta_j^{(i)}) - K_{Dj}^{(i)}\dot{\theta}_j^{(i)} \quad (14)$$

$$T^{(B)} = K_{PB}(\hat{\theta}^{(2)} - \theta^{(2)}) - K_{DB}\dot{\theta}^{(2)} \tag{15}$$

where $K_{Pj}^{(i)}$,$K_{Dj}^{(i)}$,K_{PB} and K_{DB} are feedback gains and these parameters are controlled as the tones of the actuators and are tuned in the learning process or optimization in the developmental process.

4.1 Optimization

The parameters in the control system are tuned to generate motion patterns such as crawling locomotion. Optimization parameters are as follows;

$$C = [K_{Pj}^{(i)}, K_{Dj}^{(i)}, K_{PB}, K_{DB}, \gamma^{(i)}, \gamma^{(B)}, A_t] \tag{17}$$

The objective function ***U*** is given as follows;

$$U = \alpha \int_0^t \sum_{i,j} (\theta_j^{(i)} - \hat{\theta}_j^{(i)})dt + \beta \int_0^t ||r^{(B)}||dt + wP \tag{18}$$

$$P = \int_0^t \sum_{i,j} T_j^{(i)T} \dot{\theta}_j^{(i)} dt \tag{19}$$

where w is weight parameter, $\boldsymbol{\alpha}$ and $\boldsymbol{\beta}$ are *intension parameters* which makes orientation of the development. These parameters are heuristically controlled as time sequence as follows;

$$\alpha + \beta = w \tag{20}$$

$$\alpha \gg \beta \text{ early stage on development} \tag{21}$$

$$\alpha \sim \beta \text{ middle stage on development} \tag{22}$$

$$\alpha \ll \beta \text{ last stage on development} \tag{23}$$

In this study, for simplicity of the problem, simulated annealing (SA) method is adopted for the optimization method.

5. Numerical Simulation

We implemented numerical simulations to verify the effectiveness of the proposed control system. Table 1 lists the physical parameters of the robot that are used in numerical simulations.

The nominal time period of the swinging stage was choosen as 0.25 [sec]. First, we investigated the tuning properties of the feedback gains. Figures 5 and 6 are optimization profiles (i.e., training profiles) of feedback gains $K_{Pj}^{(i)}$ and $K_{Dj}^{(i)}$, respectively. The feedback gains were well tuned and converged to appropriate values, and these gains made the joints of the body, hands, and legs adequately stiff to control the motions effectively.

Segment length	[m]	Segment mass	[kg]
Head	0.16	Head	1.0
Torso	0.35	Torso	6.2
Upper arm	0.12	Upper arm	0.19
Lower arm	0.14	Lower arm	0.20
Thigh	0.11	Thigh	0.41
Shank	0.12	Shank	0.21
Foot	0.06	Foot	0.12

Table 1. Physical parameters

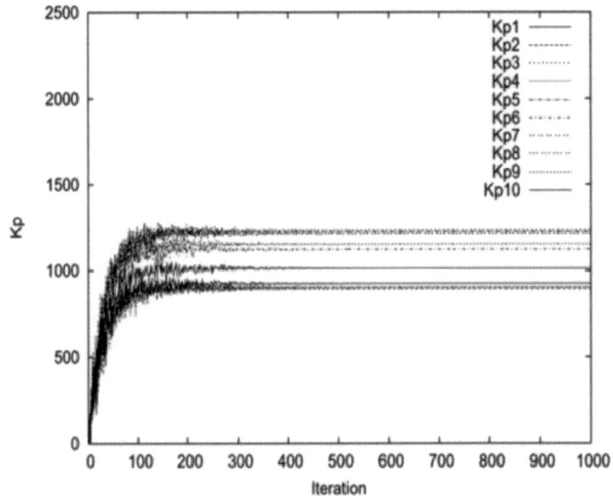

Figure 5. Tuned feedback gains *(K_P)*

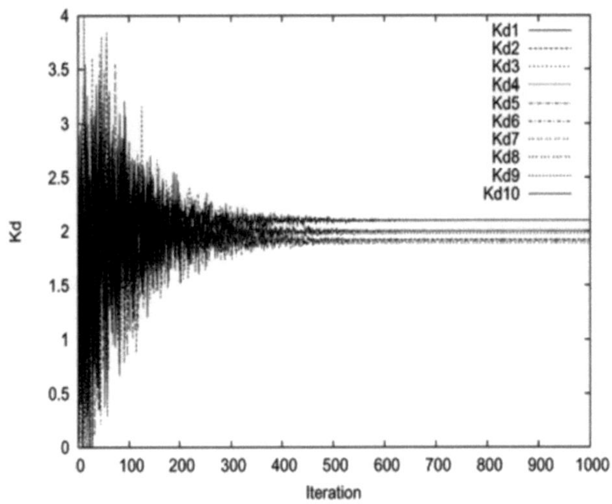

Figure 6. Tuned feedback gains (K_D)

Fig. 7 shows the training properties of the phase differences of the oscillator network in the pattern generator. We found that the converged values indicated the locomotion pattern was a *diagonal sequence,* i.e., left hand and right leg were in phase, right hand and left leg were also in phase, while left and right hands or legs were anti-phase. This phase difference made an effective crawling locomotion and the locomotion was smooth.

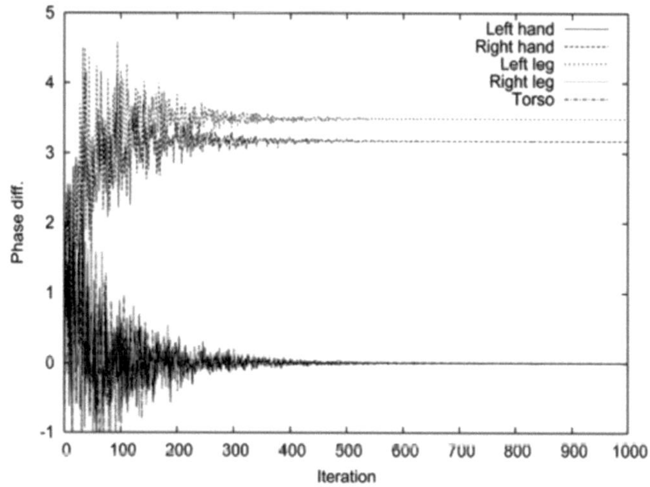

Figure 7. Phase differences of oscillators

Figures 8 and 9 are roll motion of the torso (i.e. roll angle and roll angular velocity in the phase plane). Note that the range of the fgures are not same. The result shows that the well trained system obtained stable and steady locomotion and form a limit cycle. However, in the other case (not trained), there is no limit cycle nor attractor.

Fig. 10 displays the simulation result of the crawling locomotion of the infant robot when it was well trained on its developmental parameters. In this case, the robot used a twisting motion of its torso and achieved a stable and low-energy-cost crawling locomotion. This result implies that the feedback signal of the contact sensor created mutual entrainment of the oscillators, body, and environment, and it achieved a stable limit cycle. Moreover, the appropriate phase difference for the hands, torso, and legs was obtained through the developmental training, i.e., the learning or optimization of control parameters. On the other hand, Fig. 11 shows snapshots of the crawling motion of the untrained robot. We found that in this case the robot lost its stable posture during locomotion. This result was due inappropriate phase differences of the oscillators, and too much actuator stiffness.

Fig. 12 shows the vertical reaction force at the contact point between the leg and the ground. This is for the well-trained robot. We found that the robot had a constant walking cycle and a limited cycle of crawling locomotion. On the other hand, Fig. 13 shows the case of the untrained robot. In this case, the time period from one contact moment to the next contact moment, i.e., the walking cycle, was not constant, but fluctuated. This means the robot had no stable limit cycle in crawling locomotion. Indeed, the robot could not continue stable locomotion and finally fell down.

From these fgures, one can see that the robot with the proposed control system could be well trained, could obtain appropriate actuator stiffness and locomotion conditions, and could autonomously achieve stable crawling locomotion.

The effectiveness of the proposed control system was verified by these results.

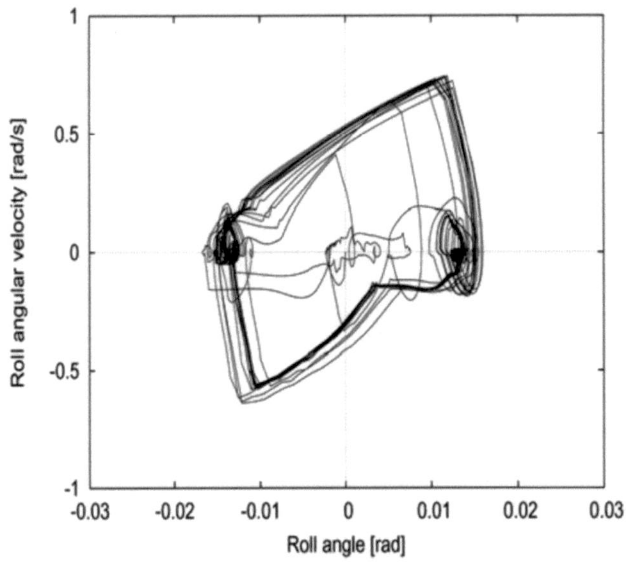

Figure 8. Roll angle and angular velocity (well trained)

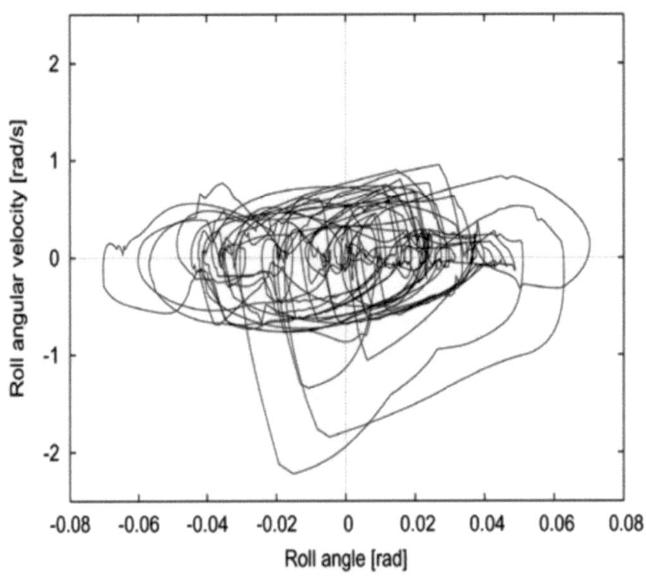

Figure 9. Roll angle and angular velocity (not trained)

Furthermore, we may note that the order of intension parameter dominance, that is, the dynamics of the constraint conditions for learning or optimization, is very important for motion acquisition. Indeed, when the order of intension parameter dominance for tuning the controller was changed from that of Eqs. (20) —(23), we could not obtain the crawling locomotion or appropriate joint stiffness, at least within a practical simulation time.

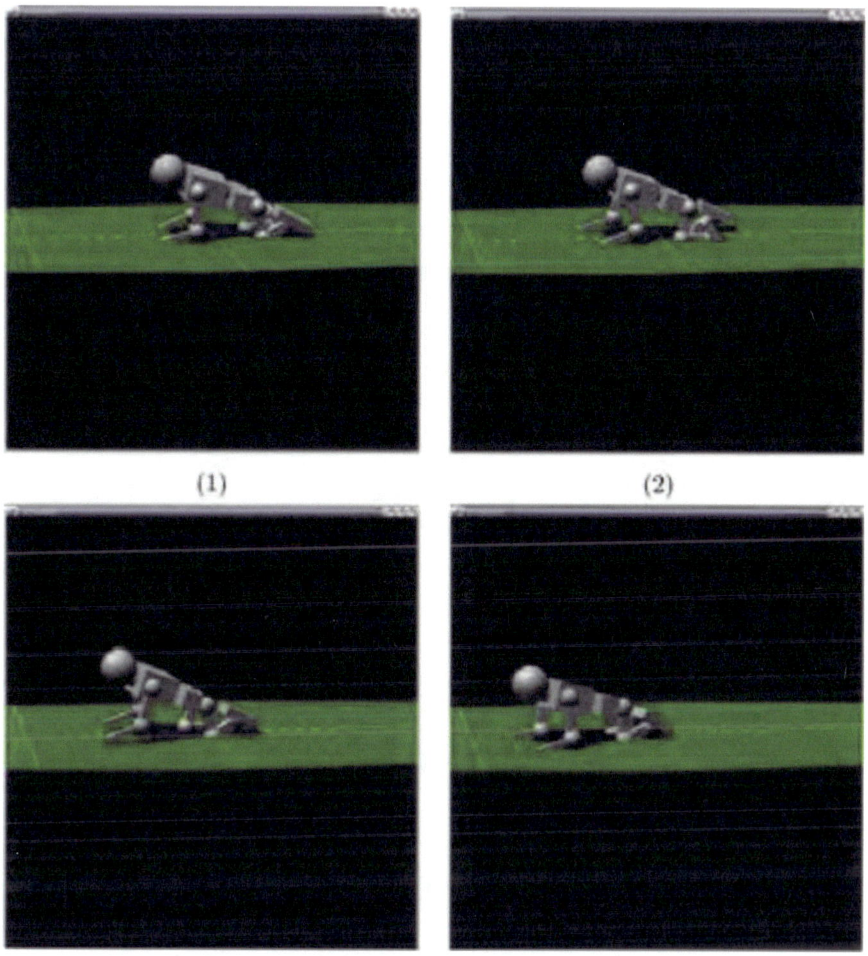

Figure 10. Snapshot taken during locomotion (well trained)

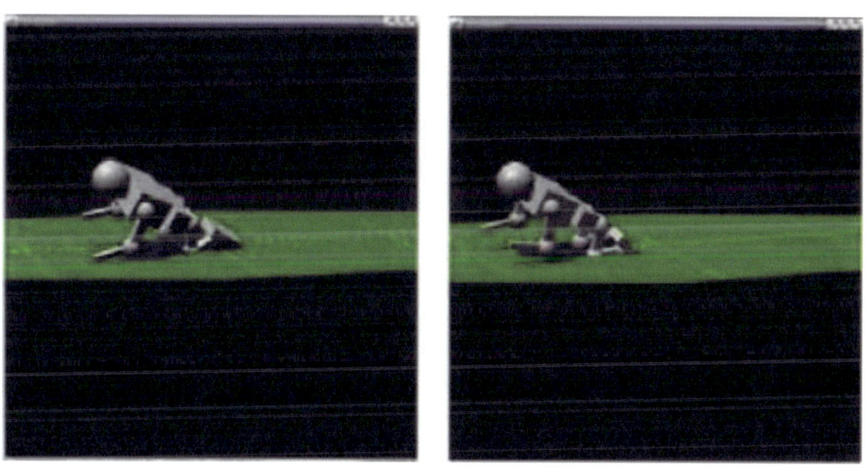

Figure ll. Snapshot taken during locomotion (untrained)

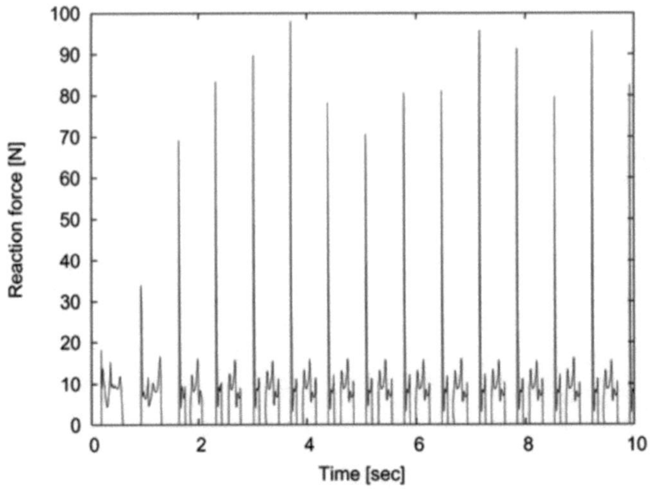

Figure 12. Reaction force on the leg (well trained)

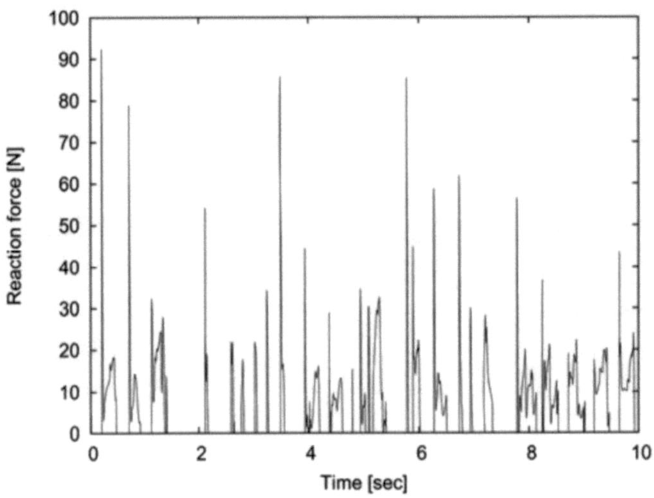

Figure 13. Reaction force on the leg (not trained)

6. Conclusion

This chapter presented a simulation method for simulating, and thereby clarifying, the development process of human locomotion that uses a model of an infant robot. The infant robot has many degrees of freedom. By making constraints and coordination for many DOF of using "intensions" such as "want to stay sitting," or "want to move more forward," etc., various motion patterns emerge according to a specific intension. In this study, this type of intension is given as the objective function and development of motion pattern, and is simulated as an optimization or learning problem. Dynamically changing the intension according to the development process changed the motion pattern of the infant robot, and it developed from sitting to crawling, toward locomotion. We investigated the process of this type of development in the infant robot with numerical simulations.

7. Acknowledgment

This work has been partially supported by a Grant-in-Aid for Scientific Research on Priority Areas "Emergence of Adaptive Motor Function through Interaction between Body, Brain and Environment" from the Japanese Ministry of Education, Culture, Sports, Science and Technology.

8. References

Berstein, N. (1967) *Co-ordination and regulation of movements,* Oxford, Pergamon press, New York Gesell,A. (1946). The ontogenesis of infant behavior, *The ontogenesis of infant behavior,* edited by L. Carmichael, Wiley, New York, pp. 295-331

Newell, K & Emmerik, R. (1990) Are Gesell's developmental principles general principles for the acquisition of coordination? *Advances in motor development research,* Vol.3, pp.143-164

Burnside, L. (1927). Coordination in the locomotion of infants, *Genetic psychology monographs,* Vol.2, No.5, pp.283-341

Thelen, E, Skala,K & Kelso,J. (1987). The Dynamic Nature of Early Coordination: Evidence From Bilateral Leg Movements in Young Infants, *Developmental Psychology,* Vol.23, No.2, pp. 179-186

Thelen, E. (1986). Treadmill-elicited Stepping in Seven-Monthh-Old Infants, *Child Development,* 57, pp. 1498-1506

Clark, J Whitall, J & Phillips, S. (1988). Human interlimb coordination: The first 6 months of independent walking, *Developmental Psychobiology,* Vol.21, 5, pp.445-456

Burnside, L. (1927). Coordination in the Locomotion of Infants, *Genetic Psychology Monographs,* Vol.2, No.5, pp.285-372

Adolph, K. (1997). Learning in the Development of Infant Locomotion, *Monographs of the Society for Research in Child Development,* Vol.62, No.3

Savelsbergh, G. (1993). The Development of Coordination in Infancy, *Advances in Psychology,* No.97

Grillner, S. (1977). Locomotion in vertebrates:Central mechanisms and reflex interaction, *Physiological Review,* No.55, pp.367-371

Grillner, S. (1985). Neurobiological bases of rhythmic motor acts in vertebrates, *Science,* Vol.228, pp.143-149

Taga, G, Yamaguchi, Y & Shimizu, H. (1991). Self-organized control of bipedal locomotion by neural oscillators in unpredictable environment, *Biological Cybernetics,* No.65, pp.147-159

Taga, G. (1994). Emergence of bipedal locomotion through entrainment among the neuro-musculo-skeletal system and the environment, *Physica D,* No.75, pp.190-208

Ijspeert, A. (2001). A connectionist central pattern generator for the atuatic and terrestrial gaits of a simulated salamander, *Biological Cybernetics,*

Yamazaki, N, Hase, K, Ogihara, N & Hayamizu, N. (1996). Biomechanical Analysis of the Development of Human Bipedal Walking by a Neuro-Musculo-Skeletal Model, *Folia Primatologica,* Vol.66, pp.253-271

Hase, K., Yamazaki, N. (2002). Computer simulation study of human locomotion with a three-dimensional entire-body neuro-musculo-skeletal model. I. Acquisition of normal walking, *JSME International Journal,* Series C, 45, 4, pp. 1040-1050

Hase, K., Obinata, G. (2002). Computer simulation study of human locomotion with a three-dimensional entire-body neuro-musculo-skeletal model. II. Biomechanical relationship between walking stability and neuro-musculo-skeletal system, *JSME International Journal,* Series C, 45, 4, pp.1051-1057

Hase, K., Obuchi, S. (2002). Computer simulation study of human locomotion with a three-dimensional entire-body neuro-musculo-skeletal model. III. Simulation of pathological walking and its application to rehabilitation engineering, *JSME International Journal,* Series C, 45, 4, pp.1058-1064

Hase, K., Yokoi, T. (2002). Computer simulation study of human locomotion with a three-dimensional entire-body neuro-musculo-skeletal model. IV. Simulation of running motion and its transition process, *JSME International Journal,* Series C, 45, 4, pp.1065-1072

Ni, J, Hiramatsu,S & Kato,A. (2003). A Model of Neuro-Musculo-Skeletal System for Human Locomotion Under Position Constraint Condition, *Journal of Biomechanical Engineering,* ASME, Vol.125, pp.499-506

G. Endo, G, Morimoto,J, Nakanishi,J & Cheng,G. (2004). An Empirical Exploration of a Neural Oscillator for Bipedal Locomotion Control, *Proc. of the 2004 IEEE Int. Conf. on Robotics and Automation,* New Orleans, pp.3036-3042

Kuniyoshi, et al. (2004). From humanoid embodiment to theory of mind. In: *Embodied artificial intelligence, lecture notes in Computer Science Series, Artificial Intelligence Subseries,* F. lida, R. Pfeifer, L. Steels, & Y. Kuniyoshi (Eds.), Vol. 3139, pp. 202-218, Springer, New York

Fukuoka, Y, Kimura,H, & Cohen,A. (2003). Adaptive Dynamic Walking of a Quadruped Robot on Irregular Terrain Based on Biological Concepts, *The International Journal of Robotics Research,* 22, No. 3, pp.187-202

Tsujita, K, Tsuchiya,K & Onat,A. (2001). Adaptive Gait Pattern Control of a Quadruped Locomotion Robot, *Proc. IEEE/RSJ Int Conf on Intelligent Robots and Systems (IROS 2001),* pp.2318-2325

Tsujita, K, H.Toui & Tsuchiya,K. (2005). Dynamic Turning Control of a Quadruped Locomotion Robot using Oscillators, *Advanced Robotics,* Vol.19, No.10, pp.1115-133

Lewis, M, Hartmann,M, Etienne-cummings,R & Cohen,A. (2001). Control of a robot leg with an adaptive a VLSI CPG chip, *Neurocomp.,*pp.l409-1421

Takakusaki, K, Habaguchi,T, Ohtinata-Sugimoto,J, Saitoh.K & Sakamoto,T. (2003). Basal Ganglia Efferents to the Brainstem Centers Controlling Postural Muscle Tone and Locomotion: A New Concept for Understanding Motor Disorders in Basal Ganglia Dysfunction, *Neuroscience,* Vol.119, pp.293-308

Ogino, M, Katoh,Y, Aono,M, Asada,M, & Hosoda,K. Reinforcement Learing of Humanoid Rhythmic Walking Parameters based on Visual Information,

Morimoto, J, Nakanishi, J, Endo,G, & Cheng,G. (2005). Poincare-Map-Based Reinforcement Learning For Biped Walking, *Proc. of the 2005 IEEE Int. Conf. on Robotics and Automation,* Barcelona

STedrake, R & Seung, H. (2002). Improved Dynamic Stability Using Reinforcement Learning, *Proc. of CLAWAR '02*

Bertenthal, B & Bai, Dina. (1989). Infants' Sensitivity to Optical Flow for Controlling Posture, *Developmental Psychology,* Vol.25, No.6, pp.936-945

22

Visual Attention and Distributed Processing of Visual Information for the Control of Humanoid Robots

Aleš Ude[1,2], Jan Moren[1], and Gordon Cheng[1,3]
[1]*ATR Computational Neuroscience Laboratories, Kyoto,* [2]*Jožef Stefan Institute, Ljubljana,*
[3]*Japan Science and Technology Agency, Saitama*
[1, 3]*Japan,* [2]*Slovenia*

1. Introduction

The function of visual attention is to identify interesting areas in the visual scene so that limited computational resources of a human or an artificial machine can be dedicated to the processing of regions with potentially interesting objects. Already early computational models of visual attention (Koch and Ullmann, 1985) suggested that attention consists of two functionally independent stages:

- in the preattentive stage features are processed rapidly and in parallel over the entire visual field until the focus of attention has been identified, which triggers the eye movement towards the target area;
- in the second phase, the computational resources are dedicated towards the processing of information in the identified area while ignoring the irrelevant or distracting percepts.

Visual attention selectivity can be either overt to drive and guide eye movements or covert, internally shifting the focus of attention from one image region to another without eye movements (Sun and Fisher, 2003). Here we are interested in visual attention that involves eye movements and how to implement it on a humanoid robot. Overt shifts of attention from one selected area to another were demonstrated for example in face recognition experiments (Yarbus, 1967). Although the subjects perceived faces as a whole in these experiments, their eye movements showed that their attention was shifted from one point to another while processing a face. The analysis of fixation points revealed that the subjects performed saccadic eye movements, which are very fast ballistic movements, to acquire data from the most informative areas of the image. Since high velocities disrupt vision and also because the signal that the target had been reached would arrive long after the movement had overshot, saccadic eye movements are not visually guided. The input to the motor system is the desired eye position, which is continuously compared to an efference copy of the internal representation of the eye position.

Many computational models of preattentive processing have been influenced by the feature integration theory (Treisman and Gelade, 1980), which resulted in several technical implementations, e. g. (Itti et al., 1998), including some implementations on humanoid robots (Driscoll et al., 1998; Breazeal and Scasselatti, 1999; Stasse et al., 2000; Vijayakumar et al., 2001). With the exception of (Driscoll et al., 1998), these implementations are mainly concerned with bottom-up, data-driven processing directed towards the generation of saliency maps. However, many theories of visual search, e. g. guided search, suggests that there are several ways for preattentive processing to guide the deployment of attention (Wolfe, 2003). Besides the bottom-up pointers towards salient regions, there is also a top-down guidance based on the needs of the searcher.

Figure 1. Humanoid head used in the experiments. It has a foveated vision system and 7 degrees of freedom (two DOFs in each eye and three DOFs in the neck)

Although bottom-up attention has been studied extensively in the past and is relatively well understood, it is still not easy to implement it in real-time on a technical system if many different feature maps are to be computed as suggested by the feature integration theory. One possible solution is to apply distributed processing to realize the extraction and analysis of feature maps in real-time. We have therefore developed a suitable computer architecture to support parallel, real-time implementation of visual attention on a humanoid robot. The guiding principle for the design of our distributed processing architecture was the existence of separate visual areas in the brain, each specialized for the processing of a particular aspect of a visual scene (Sekuler and Blake, 2002). It is evident from various visual disabilities that the ability of the brain to reassign the processing of visual information to new brain areas is rather limited and that it also takes time. Instead, visual information is transferred along a number of pathways, e. g. magnocellular pathway, parvocellular-blob pathway, and parvocellular-interblob pathway (Rolls and Deco, 2003), and visual processes are executed in well defined areas of the brain. Visual perception results from interconnections between these partly separate and functionally specialized systems. Thus our goal was to design a system that will allow us to transfer information from the source to a number of computers executing specialized vision processes, either sequentially or in parallel, and to provide means to integrate information from various streams coming at different frame rates and with different latencies. The transfer of information in the system can be both feed-forward (bottom-up processing) and feed-backward (top-down effects).

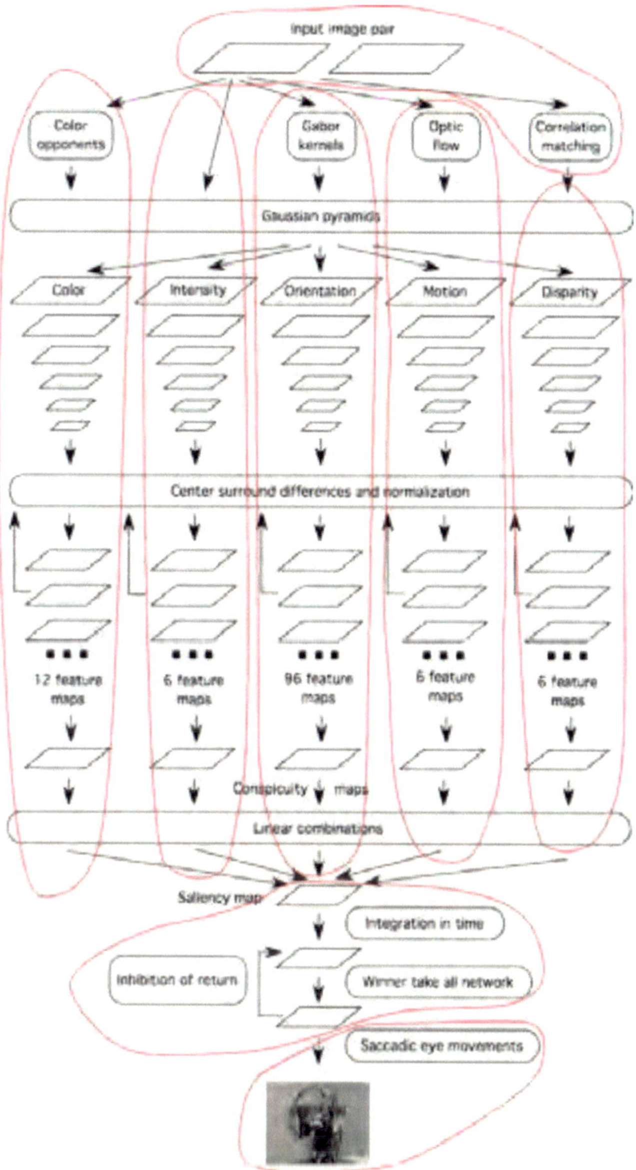

Figure 2. Bottom-up visual attention architecture based on feature integration theory. Compared to the architecture proposed by (Itti et al., 1998), there are two additional streams: motion and disparity. They are both associated with the magnocellular processing pathway in the brain, whereas color, intensity, and orientation are transferred along the parvocellular pathway. Red circles indicate the distribution of visual processes across the computer cluster. Each circle encloses the processes executed by one computer. Our system also includes the control of eye movements

Besides the distributed implementation to realize real-time behavior of the system for the control of a humanoid robot, we also studied the incorporation of top-down information into the bottom-up attention system. Top-down signals can bias the search process towards objects with particular properties, thus enabling the system to find such objects more quickly.

2. Bottom-up preattentive processing

Figure 2 shows our distributed implementation of bottom-up visual attention, which is a modified proposal of (Itti et al., 1998). From the robot's camera, images are distributed across a number of computers and a set of filters is applied to the original stream at each node in the first line of processors. Each of them corresponds to one type of retinal feature maps, which are calculated at different scales. Within each feature processor, maps at different scales are combined to generate a global conspicuity map that emphasizes locations that stand out from their surroundings. The conspicuity maps are combined into a global saliency map, which encodes the saliency of image locations over the entire feature set. The time-integrated global saliency map is supplied as an input to a winner-take-all neural network, which is used to compute the most salient area in the image stream.

2.1 Generation of saliency maps

We have implemented the following feature processors on our system: color, intensity, orientation, motion, and disparity (see also Figure 2). Especially the generation of disparity, motion, and orientation feature maps are time consuming processes and it would be impossible to implement and visualize all of them on one computer and in real-time. The most computationally expensive among them is the generation of orientation feature maps, which are calculated by Gabor filters. They are given by

$$\Phi(\mathbf{x}) = \frac{\|\mathbf{k}_{\mu,v}\|^2}{\sigma^2} \exp\left(-\frac{\|\mathbf{k}_{\mu,v}\|^2 \|\mathbf{x}\|^2}{2\sigma^2}\right)\left(\exp\left(i\mathbf{k}_{\mu,v}^T \mathbf{x}\right) - \exp\left(-\frac{\sigma^2}{2}\right)\right), \tag{1}$$

where $\mathbf{k}_{\mu,v} = k_v[\cos(\phi_\mu), \sin(\phi_\mu)]^T$. Gabor kernels were suggested to model the receptive fields of simple cells in primary visual cortex. In (Itti et al., 1998) a single scale k_v and four orientations ϕ_μ = 0, 45, 90, 135, were used. It has been shown, however, that there exist simple cells sensitive not only to specific positions and orientations, but also to specific scales. We therefore applied Gabor kernels not only at four different orientations but also at four different scales. For the calculation of motion, we used a variant of Lucas-Kanade algorithm. A correlation-based technique was used to generate disparity maps at the available frame rate (30 Hz).

At full resolution (320 x 240), the above feature processors generate 2 feature maps for color (based on double color opponents theory (Sekuler and Blake, 2002)), 1 for intensity, 16 for orientation, 1 for motion and 1 for disparity. Center-surround differences were suggested as a computational tool to detect local spatial discontinuities in feature maps that stand out from their surround. Center-surround differences can be computed by first creating Gaussian pyramids out of the initial feature maps. From the uppermost scale $\mathbf{I}_f(0)$, where f is the corresponding feature, maps at lower scales are calculated by filtering of the map at the previous scale with a Gaussian filter. The resolution of a map at lower scale is half the resolution of the map at the scale above it. Center-surround differences are calculated by subtracting pyramids at coarser scale from the pyramids at finer scale. For this calculation the pyramid maps at coarser scales are up-sampled to finer scales. We calculated the center-surround differences between the pyramids $\mathbf{I}_f(c)$, $c = 2, 3, 4$, and $\mathbf{I}_f(s)$, $s = c + \Delta$, $\Delta = 2, 3$. This results in 6 maps per feature.

The combination of center-surround differences into conspicuity maps for color $\mathbf{J}_c(t)$, intensity $\mathbf{J}_b(t)$, orientation $\mathbf{J}_o(t)$, motion $\mathbf{J}_m(t)$, and disparities $\mathbf{J}_d(t)$ at time t involves normalization to a fixed range and searching for global and local maxima to promote feature maps with strong global maxima. For each modality, center-surround differences are combined into conspicuity maps at the coarsest scale. This process is equivalent to what has been implemented by (Itti et al., 1998) and we omit the details here. The conspicuity maps are finally combined into a global saliency map $\mathbf{S}(t)$

$$\mathbf{S}(t) = w_c \mathbf{J}_c(t) + w_b \mathbf{J}_b(t) + w_o \mathbf{J}_o(t) + w_m \mathbf{J}_m(t) + w_d \mathbf{J}_d(t). \tag{2}$$

The weights w_c, w_b, w_o, w_m, and w_d can be set based on top-down information about the importance of each modality. In the absence of top-down information, they can be set to a fixed value, e. g. 0.2, if all five features are to have the same influence. Finally, to deal with a continuous stream of images, the saliency maps need to be time-integrated

$$\mathbf{S}_{\text{int}}(t) = \gamma^{\delta} \mathbf{S}_{\text{int}}(t-\delta) + \mathbf{G}_{\sigma} * \mathbf{S}(t),\ 0 < \gamma < 1, \tag{3}$$

where $\delta \geq 1$ is the difference in the frame index from the previous saliency map and $\mathbf{G}_{\sigma} * \mathbf{S}(t)$ is the convolution of the current saliency map with the Gaussian filter with standard deviation σ.

2.2 Winner-take-all network

The aim of the preattentive processing is to compute the currently most salient area in the image so that the robot's eye can saccade towards this area and place it into the center of the fovea, thus enabling the robot to dedicate its computational resources to the processing of the foveal image area in the next processing step. Winner-take-all network has been suggested as means to calculate the focus of attention from the saliency map (Koch and Ullmann, 1987). We use the leaky integrate-and-fire model to build a two layer 2-D neural network of first order integrators to integrate the contents of the saliency map and choose a focus of attention over time. It is based on the integration of the following system of differential equations:

$$\begin{aligned} \frac{du_1}{dt}(\mathbf{x},t) + \frac{1}{\tau_1} u_1(\mathbf{x},t) &= \sum_{\mathbf{y}} w_1(\mathbf{x},\mathbf{y}) \mathbf{S}_{\text{int}}(\mathbf{y},t) - \sum_{\mathbf{y}} w_{\text{co}}(\mathbf{x},\mathbf{y}) u_2(\mathbf{y},t) \\ \frac{du_2}{dt}(\mathbf{x},t) + \frac{1}{\tau_2} u_2(\mathbf{x},t) &= \sum_{\mathbf{y}} w_2(\mathbf{x},\mathbf{y}) u_1(\mathbf{y},t) \end{aligned} \tag{4}$$

where $u_i(\mathbf{x}, t)$ is the membrane potential of the neuron of the i-th layer located at $\mathbf{x}$ at time t, τ_i is the time constant of the i-th layer, w_i is the weighting function of the lateral connections of the i-th layer between locations $\mathbf{x}$ and $\mathbf{y}$ and $w_{\text{co}}(\mathbf{x},\mathbf{y})$ is the weighting function of connections between the first and the second layer. Functions $w_i(\mathbf{x},\mathbf{y})$ are given by:

$$w_1(\mathbf{x},\mathbf{y}) = w_2(\mathbf{x},\mathbf{y}) = \frac{1}{2\pi\sigma^2_{\text{inh}}}\exp\left(-\frac{\|\mathbf{x}-\mathbf{y}\|^2}{2\sigma^2_{\text{inh}}}\right). \tag{5}$$

Function $w_{co}(\mathbf{x},\mathbf{y})$ models the coupling effects between the neurons of the network including long-range inhibition and short-range excitation to produce the winning neuron. It is defined as:

$$w_{\text{co}}(\mathbf{x},\mathbf{y}) = \frac{c^2_{\text{in}}}{2\pi\sigma^2_{\text{in}}}\exp\left(-\frac{\|\mathbf{x}-\mathbf{y}\|^2}{2\sigma^2_{\text{in}}}\right) - \frac{c^2_{\text{ex}}}{2\pi\sigma^2_{\text{ex}}}\exp\left(-\frac{\|\mathbf{x}-\mathbf{y}\|^2}{2\sigma^2_{\text{ex}}}\right) \tag{6}$$

We used Euler's method to integrate Equations (4). The integration frequency was set to 100 Hz and is higher than the timing of the vision signal (30 Hz). Hence before updating the integrated saliency map $\mathbf{S}_{\text{int}}$, Equations (4) are integrated a few times as temporal smoothing. When the potential of one of the neurons of the second layer $u_2(\mathbf{x}, t)$ reaches the adaptive firing threshold, the robot eyes move so that the most salient area is placed over the fovea. Vision processing is suppressed during the saccade. Since postattentive processing has not been integrated into the system yet, the robot just waits for 500 ms before moving its eyes back to the original position. At this point the neurons of the second layer are reset to their ground membrane voltage as global lateral inhibition and a local inhibitory signal is smoothly propagated from the first to the second layer at the attended location as inhibition of return. The strength of the inhibitory effect is gradually reduced as the time passes to allow for further exploration of the previously attended regions.

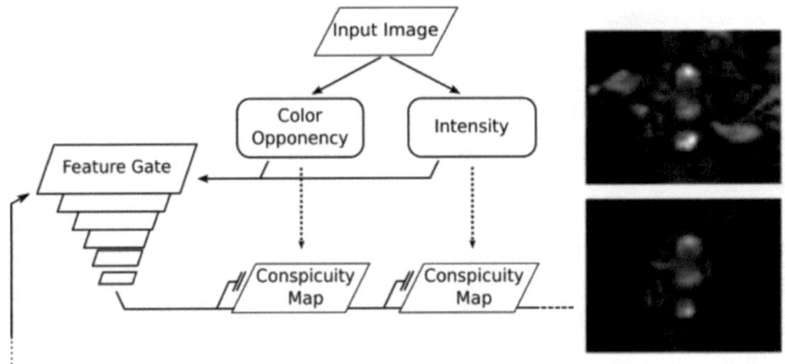

Figure 3. The FeatureGate top-down biasing system added to the simplified architecture from Figure 2. The top-down feature vectors are fed to the FeatureGate system, which finds the inhibitory signals for the conspicuity maps (created in parallel by the bottom-up system). To the right, the resulting saliency map (upper right) and color opponency conspicuity map with inhibition from the FeatureGate subsystem

3. Top-down guidance

As already mentioned in Section 2, feature-wide top-down effects can be introduced into the system by selecting different weights when combining the conspicuity maps into a single saliency map by means of Eq. (2). A recent model by (Navalpakkam and Itti, 2006) computes optimal weights based on the observer's prior beliefs about the scene (target and distractors) to arrive at the linear combination of feature maps that best separates the sought for feature from its expected background. Boosting certain types of features over the others is, however, still a broad mechanism, best suited for biasing higher-level search towards certain kind of data, and not well suited for pinpointing specific features.

Another approach is to introduce context-dependent spatial restrictions on the search, with inhibition on areas not likely to have features the system searches for. (Balkenius et al., 2004) present an adaptive contextual system where the conspicuity map content at the current fixation serves as the contextual cue as to where, in absolute or relative terms, the desired feature is likely or unlikely to be. The saliency map is boosted or inhibited accordingly. This kind of mechanism is more specific, in that it can explicitly focus on, or disregard, areas independently of its bottom-up saliency.

If the goal is to introduce top-down influences looking for specific features, we need a different kind of mechanism. More precisely, we want to be able to give a particular feature vector and bias the saliency towards points that in some way or another match that feature vector. One way to accomplish this has been proposed in FeatureGate model of human visual attention (Cave, 1999). This model introduces top-down effects by lateral inhibition of activation in feature maps. At every given point, the inhibition is a function of this point's nearness to the expected feature vector as compared to the nearness of neighboring points to the same feature vector. The measure of nearness must be defined by a suitable metrics ρ. A point receives inhibition when a neighboring area is closer to the target top-down feature tf than the current location $\mathbf{x}$. The model conversely boosts points proportionally to their distinctiveness at each level (defined as the sum of absolute differences to the neighboring points). Top-down inhibition and local distinctiveness are weighted and combined. The results are gated up from fine to coarse scales, effectively increasing the spatial extent of the inhibition within each level, finally resulting in a pyramid of inhibitory values for different spatial scales.

Let $N_c(\mathbf{x})$ be the neighborhood of location $\mathbf{x}$ at level c in the pyramid and let $S_c(\mathbf{x})$ be all pixels in the neighborhood that are closer to the target than $\mathbf{x}$:

$$S_c(\mathbf{x}) = \left\{ \mathbf{y} \in N_c(\mathbf{x}); \rho\left(\mathbf{I}_{tf}(\mathbf{y};c)\right) < \rho\left(\mathbf{I}_{tf}(\mathbf{x};c)\right) \right\}, \tag{7}$$

Let $\mathbf{I}_{tf}(0)$ be the map generated by processing the image with the top-down target feature processor at full resolution. The top-down inhibition $\mathbf{I}_{tf}^{d}$ is calculated as the value proportional to the difference in the distance from the target feature

$$\mathbf{I}_{tf}^{d}(\mathbf{x};c) = \sum_{\mathbf{y} \in S_c(\mathbf{x})} \left| \rho\left(\mathbf{I}_{tf}(\mathbf{y};c)\right) - \rho\left(\mathbf{I}_{tf}(\mathbf{x};c)\right) \right|, \tag{8}$$

For each j, $j = c, m, i, o, d$, we (optionally) also calculate the distinctiveness $\mathbf{I}_j^d$

$$\mathbf{I}_j^{d}(\mathbf{x};c) = \sum_{\mathbf{y} \in S_c(\mathbf{x})} \left| \mathbf{I}_j(\mathbf{y};c) - \mathbf{I}_j(\mathbf{x};c) \right|. \tag{9}$$

We obtain the signal for inhibition by weighting these two measures

$$\mathbf{I}_j^{\text{inh}}(\mathbf{x};c) = \alpha \mathbf{I}_j^d(\mathbf{x};c) - \beta \mathbf{I}_{tf}^d(\mathbf{x};c). \tag{10}$$

The next, coarser pyramid level is constructed by comparing each point in a small neighborhood $N_{c-1}(\mathbf{x}')$ (2x2 points by default) at the previous level and propagating only the least inhibited point to the point **x** at the next level:

$$\mathbf{I}_{tf}(\mathbf{x};c) = \mathbf{I}_{tf}(\mathbf{y};c-1),\ \mathbf{I}_j(\mathbf{x};c) = \mathbf{I}_j(\mathbf{y};c-1),\ \mathbf{y} = \arg\max_{\mathbf{y}\in N_{c-1}(\mathbf{x}')} \left\{\mathbf{I}_j^{\text{inh}}(\mathbf{y};c-1)\right\}. \tag{11}$$

The process is repeated until we get - at the top-level of the pyramid - a single element representing the globally most salient point with respect to the bottom-up map and, optionally, the distinctiveness. The level below contains the most salient point in each of the four quadrants of the image and so forth. The actual values do not encode how good the matches are as they are relative to other points in the image. With $\alpha = 0$ we get a pure top-down system well adapted for use together with a separate bottom-up system. However, the proposed computational mechanism for the integration of both systems described below is robust enough so that the precise settings are not very important for the overall functionality. It is also possible to set α and β in such a way that the system behaves similarly to the one described by (Cave, 1999) and these were the parameters used in our experiments.

To integrate the result of the above algorithm into the saliency map that can be supplied to the winner-take-all network, we generate a second conspicuity map based on the position (at the appropriate pyramid level) of the most salient top-down point $\mathbf{x}_j$ with respect to the feature map j and the given top-down feature vector. The following formula is used to generate this second map

$$\mathbf{M}_j(\mathbf{x};t) = \begin{cases} 1, & \mathbf{x} = \mathbf{x}_j \\ \mathbf{M}_j(\mathbf{x};t-1) * 0.9, & \mathbf{x} \neq \mathbf{x}_j \end{cases}. \tag{12}$$

Top-down influence	Total fixations	Target fixations	Target fixations percentage
0%	62	23	37%
10%	65	31	48%
20%	53	30	57%
30%	33	19	58%
40%	20	14	70%
50%	19	16	84%
75%	14	13	93%
100%	6	6	100%

Table 1. Fixation data

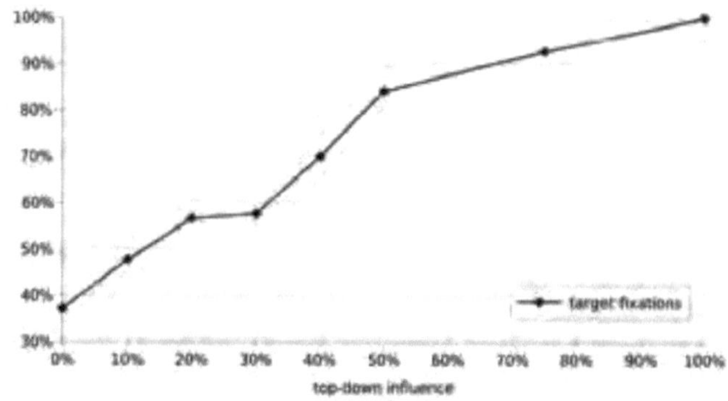

Figure 4. Target fixations as a function of top-down influence. A 30 second image sequence was run through the system with different influence settings. The attended object is fairly salient by itself with 37% of fixations when using the bottom-up saliency system only. The top-down system is able to rapidly boost this ratio, with almost 85% of all fixations when λ is at 0.5

Finally, a new conspicuity map is computed by adding the weighted top-down and bottom-up conspicuity maps $\mathbf{J}_j'(t) = \lambda\mathbf{M}_j(t) + (1-\lambda)\mathbf{J}_j(t)$. Thus the relative importance of bottom-up and top down saliency processing is determined by the parameter λ. In Figure 3, $\lambda = 0.5$ was used and $\mathbf{M}_j$ were initially set to zero, i. e. $\mathbf{M}_j(0) = 0$.

We ran a series of tests to check the effects of top-down biasing. A short image sequence of about 30 seconds depicting an object (teddy bear) being moved around was used as input to the system. In these experiments the system used color opponency and intensity as low-level features and did not generate saccades. The shifts in current region of interest were recorded; note that the saccades that would be performed are selected from a subset of these covert attentional shifts. The top-down system was primed with a vector roughly matching the brightness and color space position of the target object. Given proper weighting factors, the locations selected by FeatureGate are close to the intended target with high probability. On the other hand, by keeping the bottom-up cue in the system we ensure that very salient areas will be attended even if they don't match the feature vector.

Tests were run with all settings equal except for the parameter λ specifying the influence of the top-down system relative to the bottom-up saliency. The data generated is presented in Table 1. We tested the system from 0% influence (only the bottom-up system active) to 100% (only the top-down system used). Fewer saccades are generated overall if there exists a dominant target in the image matching the feature vector and the influence of the top-down cue is high. Since in such cases the behavior of the system changes little as we increase the top-down influences, we tested the system only at two high top-down settings (75% and 100%). Figure 4 demonstrates that the system works much as expected. The target object is fairly salient but it is fixated on less than 40% of the time if only bottom-up saliency is used. With top-down biasing the proportion of fixations spent on the target increases rapidly and with equal influence the target is already fixated 84% of the time. At high levels of top-down influence the target becomes almost totally dominant and the object is fixated 100% of the time when $\lambda = 1$. The rapid dominance of the target as we increase the top-down influence is natural as it is a salient object already. Note that if the top-down selection mechanism has several areas to select from - as it will if there are several objects matching the top-down criteria or if the object has a significant spatial extent in the image - the effect of the top-down system will spread out and weaken somewhat. Also, with two or more similar objects the system will generate saccades that occasionally alternate between them as the inhibition of return makes the current object temporarily less salient overall.

The above experiment was performed with a top-down system closely following the original FeatureGate model in design. Specifically, we still use the distinctiveness estimate at each level. Alternatively, we could apply only the top-down inhibitory mechanism and simply use the map $\mathbf{I}^{d}_{tf}(\mathbf{x};c)$ of Eq. (8) - calculated at the same pyramid level c as the conspicuity maps $\mathbf{J}_j(t)$ - to generate the inhibitory signal. In many practical cases, the behavior of such a system would be very similar to the approach described above, therefore we do not present separate experiments here.

65911 65910 65907 65911 65912	61250 61249 61250 61250 61250	70656 70656 70656 70656 70656
65912 65910 65910 65911 65913	61251 61251 61250 61251 61251	70675 70675 70675 70675 70675
65912 65912 65910 65912 65914	61252 61251 61250 61251 61252	70678 70678 70678 70678 70678
65913 65912 65910 65913 65915	**61253 61253 61250 61253 61253**	70695 70695 70695 70695 70695
65914 65912 65910 65913 65916	61253 61253 61254 61254 61254	70711 70711 70711 70711 70711
65915 65914 65913 65915 65917	61255 61253 61254 61254 61255	70715 70715 70715 70715 70715
65917 65914 65913 65916 65918	61256 61256 61254 61256 61256	**70724 70724 70724 70724 70724**
65918 65916 65913 65916 65919	61257 61256 61257 61257 61257	**70757 70757 70757 70757 70757**
65918 65916 65916 65918 65920	61258 61258 61257 61257 61258	70758 70758 70758 70758 70758
65919 65918 65916 65919 65921	61259 61258 61257 61259 61259	70777 70777 70777 70777 70777
65920 65918 65916 65921 65922	61260 61260 61260 61260 61260	70790 70790 70790 70790 70790
65921 65921 65919 65922 65923	61260 61260 61260 61261 61261	70799 70799 70799 70799 70799
65923 65921 65919 65922 65924	61262 61262 61260 61261 61262	70802 70802 70802 70802 70802
65924 65923 65919 65923 65925	**61263 61262 61260 61263 61263**	70815 70815 70815 70815 70815
65925 65923 65922 65923 65926	61264 61264 61264 61264 61264	70837 70837 70837 70837 70837

Table 2.. Frame indices of simultaneously processed images under different synchronization schemes. In each box, ordered from left to right clumn, the frame indices belong to the disparity, color, orientation, intensity, and motion conspicuity map. See text in Section 4.1 for further explanations

4. Synchronization of processing streams

The distributed processing architecture presented in Figure 2 is essential to achieve real-time operation of the complete visual attention system. In our current implementation, all of the computers are connected to a single switch via a gigabit Ethernet. We use UDP protocol for data transfer. Data that needs to be transferred from the image capture PC includes the rectified color images captured by the left camera, which are broadcast from the frame grabber to all other computers on the network, and the disparity maps, which are sent directly to the PC that takes care of the disparity map processing. Full resolution (320 x 240 to avoid interlacing effects) was used when transferring and processing these images. The five feature processors send the resulting conspicuity maps to the PC that deals with the calculation of the saliency maps, followed by the integration with the winner-take-all network. Finally, the position of the most salient area in the image stream is sent to the PC taking care of motor control. The current setup with all the computers connected to a single gigabit switch proved to be sufficient to transfer the data at full resolutions and frame rates. However, our implementation of the data transfer routines allows us to split the network into a number of separate networks should the data load become too large. This is essential if the system is to scale to a more advanced vision processing such as shape analysis and object recognition.

A heterogeneous cluster in which every computer solves a different problem necessarily results in visual streams progressing through the system at different frame rates and with different latencies. In the following we describe how to ensure smooth operation under such conditions.

4.1 Synchronization

The processor that needs to solve the most difficult synchronization task is the one that integrates the conspicuity maps into a single saliency map. It receives input from five different feature processors. The slowest among them is the orientation processor that could roughly take care of only every third frame. Conversely, the disparity processor works at full frame rate and with lower latency. While it is possible to further distribute the processing load of the orientation processor, we did not follow this approach because our computational resources are not unlimited. We were more interested in designing a general synchronization scheme that allows us to realize real-time processing under such conditions.

The simplest approach to synchronization is to ignore the different frame rates and latencies and to process the data that was last received from each of the feature processors. Some of the resulting frame indices for conspicuity maps that are in this case combined into a single saliency map are shown in the leftmost box of Table 2. Looking at the boldfaced rows of this column, it becomes clear that under this synchronization scheme, the time difference (frame index) between simultaneously processed conspicuity maps is quite large, up to 6 frames (or 200 milliseconds for visual streams at 30 Hz). It does not happen at all that conspicuity maps with the same frame index would be processed simultaneously.

Ideally, we would always process only data captured at the same moment in time. This, however, proves to be impractical when integrating five conspicuity maps. To achieve full synchronization, we associated a buffer with each of the incoming data streams. The integrating process received the requested conspicuity maps only if data from all five streams was simultaneously available. The results are shown in the rightmost box of Table 2. Note that lots of data is lost when using this synchronization scheme (for example 23 frames between the two boldfaced rows) because images from all five processing streams are only rarely simultaneously available.

We have therefore implemented a scheme that represents a compromise between the two approaches. Instead of full synchronization, we monitor the buffer and simultaneously process the data that is as close together in time as possible. This is accomplished by waiting that for each processing stream, there is data available with the time stamp before (or at) the requested time as well as data with the time stamp after the requested time. In this way we can optimally match the available data. The algorithm is given in Figure 5. For this synchronization scheme, the frame indices of simultaneously processed data are shown in the middle box of Table 2. It is evident that all of the available data is processed and that frames would be skipped only if the integrating process is slower than the incoming data streams. The time difference between the simultaneously processed data is cut to half (maximum 3 frames or 100 milliseconds for the boldfaced rows). However, the delayed synchronization scheme does not come for free; since we need to wait that at least two frames from each of the data streams are available, the latency of the system is increased by the latency of the slowest stream. Nevertheless, the delayed synchronization scheme is the method of choice on our humanoid robot.

```
Request for data with frame index n:
  get access to buffers and lock writing
  r = 0
  for i = 1,...,m
    find the smallest b_{i,j} so that n < b_{i,j}
    if such b_{i,j} does not exist
      reply images with frame index n not yet available
      unlock buffers and exit
    if b_{i,(j-1)%M} ≤ n
      j_i = b_{i,(j-1)%M}
    else
      r = max(r, b_{i,j})
  if r > 0
    reply r is the smallest currently available frame index
    unlock buffers and exit
  return { P_{1,j_1} ,..., P_{m,j_m} }
  unlock buffers and exit
```

Figure 5. Pseudo-code for the delayed synchronization algorithm. m denotes the number of incoming data streams, or - in other words - the number of preceding nodes in the network of visual processes. To enable synchronization of data streams coming with variable latencies and frame rates, each data packet (image, disparity map, conspicuity map, joint angle configuration, etc.) is written in the buffer associated with the data stream, which has space for M latest packets. $b_{i,j}$ denotes the frame index of the j-th data packet in the buffer of the i-th processing stream. $P_{i,j}$ are the data packets in the buffers and m is the number of data streams coming from previous processes

We note here that one should be careful when selecting the proper synchronization scheme. For example, nothing less than full synchronization is acceptable if the task is to generate disparity maps from a stereo image pair with the goal of processing scenes that change in time. On the other hand, buffering is not desirable when the processor receives only one stream as input; it would have no effect if the processor is fast enough to process the data at full frame rate, but it would introduce an unnecessary latency in the system if the processor is too slow to interpret the data at full frame rate. The proper synchronization scheme should thus be carefully selected by the designer of the system.

5. Robot eye movements

Directing the spotlight of attention towards interesting areas involves saccadic eye movements. The purpose of saccades is to move the eyes as quickly as possible so that the spotlight of attention will be centered on the fovea. As such they constitute a way to select task-relevant information. It is sufficient to use the eye degrees of freedom for this purpose. Our system is calibrated and we can easily calculate the pan and tilt angle for each eye that are necessary to direct the gaze towards the desired location. Human saccadic eye movements are very fast. The current version of our eye control system therefore simply moves the robot eyes towards the desired configuration as fast as possible.
Note that saccades can be made not only towards visual targets, but also towards auditory or tactile stimuli. We currently work on the introduction of auditory signals into the proposed visual attention system. While it is clear that auditory signals can be used to localize some events in the scene, the degree of cross-modal interactions between auditory and visual stimuli remains an important research issue.

6. Conclusions

The goals of our work were twofold. On the one hand, we studied how to introduce top-down effects into a bottom-up visual attention system. We have extended the classic system proposed by (Itti et al., 1998) with top-down inhibitory signals to drive attention towards the areas with the expected features while still considering other salient areas in the scene in a bottom-up manner. Our experimental results show that the system can select areas of interest using various features and that the selected areas are quite plausible and most of the time contain potential objects of interest. On the other hand, we studied distributed computer architectures, which are necessary to achieve real-time operation of complex processes such as visual attention. Although some of the previous works mention that parallel implementations would be useful and indeed parallel processing was used in at least one of them (Breazeal and Scasselatti, 1999), this is the first study that focuses on issues arising from such a distributed implementation. We developed a computer architecture that allows for proper distribution of visual processes involved in visual attention. We studied various synchronization schemes that enable the integration of different processes in order to compute the final result. The designed architecture can easily scale to accommodate more complex visual processes and we view it as a step towards a more brain-like processing of visual information on humanoid robots.
Our future work will center on the use of visual attention to guide higher-level cognitive tasks. While the possibilities here are practically limitless, we intend to study especially how to guide the focus of attention when learning about various object affordances, such as for example the relationships between the objects and actions that can be applied to objects in different situations.

7. Acknowledgment

Aleš Ude was supported by the EU Cognitive Systems project PACO-PLUS (FP6-2004-IST-4-027657) funded by the European Commission.

8. References

Balkenius, C., Åström, K. & Eriksson, A. P. (2004). Learning in visual attention. *ICPR 2004 Workshop on learning for adaptable visual systems*, Cambridge, UK.

Breazeal, C. & Scasselatti, B. (1999). A context-dependent attention system for a social robot. *Proc. Sixteenth Int. Joint Conf. Artificial Intelligence*, Stockholm, Sweden, pp. 1146-1151.

Cave, K. R. (1999). The FeatureGate model of visual selection. *Psychological Research*, 62:182–194.

Driscoll, J. A.; Peters II, R. A. & Cave, K. R. (1998). A visual attention network for a humanoid robot. *Proc. IEEE/RSJ Int. Conf. Intelligent Robots and Systems*, Victoria, Canada, pp. 1968–1974.

Itti, L.; Koch, C. & Niebur E. (1998). A model of saliency-based visual attention for rapid scene analysis. *IEEE Trans. Pattern Anal. Machine Intell.*, 20(11) :1254–1259.

Koch C. & Ullman S. (1987). Shifts in selective visual attention: towards the underlying neural circuitry. *Matters of Intelligence*, L. M. Vaina, Ed., Dordrecht: D. Reidel Co., pp. 115–141.

Navalpakkam, V. & Itti, L. (2006). An integrated model of top-down and bottom-up attention for optimizing detection speed. *Proc. IEEE Conference on Computer Vision and Pattern Recognition*, New York, pp. 2049-2056.

Rolls, E. T. & Deco, G. (2003). *Computational Neuroscience of Vision*. Oxford, University Press.

Sekuler, R. & Blake, R. (2002). *Perception*, 4th ed. McGraw-Hill.

Stasse, O.; Kuniyoshi Y. & Cheng G. (2000). Development of a biologically inspired real-time visual attention system. *Biologically Motivated Computer Vision: First IEEE International Workshop*, S.-W. Lee, H. H. Bülthoff, and T. Poggio, Eds., Seoul, Korea, pp. 150–159.

Sun, Y. & Fisher, R. (2003). Object-based visual attention for computer vision. *Artificial Intelligence*, 146(1):77-123.

Treisman, A. M. & Gelade, G. (1980). A feature-integration theory of attention. *Cognitive Psychology*, 12(1) :97–136.

Tsotsos, J. K. (2005). The selective tuning model for visual attention. *Neurobiology of Attention*. Academic Press, pp. 562–569.

Vijayakumar, S.; Conradt, J.; Shibata, T. & Schaal, S. (2001). Overt visual attention for a humanoid robot. *Proc. IEEE/RSJ Int. Conf. Intelligent Robots and Systems*, Maui, Hawaii, USA, pp. 2332–2337.

Wolfe, J. M. (2003). Moving towards solutions to some enduring controversies in visual search. *Trends in Cognitive Sciences*, 7(2):70–76.

Yarbus, A. L. (1967) Eye movements during perception of complex objects. In: *Eye Movements and Vision*, Riggs, L. A. (Ed.), pp. 171–196, Plenum Press, New York.

23

Visual Guided Approach-to-grasp for Humanoid Robots

Yang Shen[1], De Xu[1], Min Tan[1] and Ze-Min Jiang[2]
[1]*Laboratory of Complex Systems and Intelligence Science, Institute of Automation, Chinese Academy of Sciences*
[2]*School of Information Science and Technology, Beijing Institute of Technology*
P. R. China

1. Introduction

Vision based control for robots has been an active area of research for more than 30 years and significant progresses in the theory and application have been reported (Hutchinson et al., 1996; Kragic & Christensen, 2002; Chaumette & Hutchinson, 2006). Vision is a very important non-contact measurement method for robots. Especially in the field of humanoid robots, where the robot works in an unstructured and complex environment designed for human, visual control can make the robot more robust and flexible to unknown changes in the environment (Hauck et al., 1999).
Humanoid robot equipped with vision system is a typical hand-eye coordination system. With cameras mounted on the head, the humanoid robot can manipulate objects with his hands. Generally, the most common task for the humanoid robot is the approach-to-grasp task (Horaud et al., 1998). There are many aspects concerned with the visual guidance of a humanoid robot, such as vision system configuration and calibration, visual measurement, and visual control.
One of the important issues in applying vision system is the calibration of the system, including camera calibration and head-eye calibration. Calibration has received wide attentions in the communities of photogrammetry, computer vision, and robotics (Clarke & Fryer, 1998). Many researchers have contributed elegant solutions to this classical problem, such as Faugeras and Toscani, Tsai, Heikkila and Silven, Zhang, Ma, Xu. (Faugeras & Toscani, 1986; Tsai, 1987; Heikkila & Silven, 1997; Zhang, 2000; Ma, 1996; Xu et al., 2006a). Extensive efforts have been made to achieve the automatic or self calibration of the whole vision system with high accuracy (Tsai & Lenz, 1989). Usually, in order to gain a wide field of view, the humanoid robot employs cameras with lens of short focal length, which have a relatively large distortion. This requires a more complex nonlinear model to represent the distortion and makes the accurate calibration more difficult (Ma et al., 2003).
Another difficulty in applying vision system is the estimation of the position and orientation of an object relative to the camera, known as visual measurement. Traditionally, the position of a point can be determined with its projections on two or more cameras based on epipolar geometry (Harley & Zisserman, 2004). Han et al. measured the pose of a door knob relative to the end-effector of the manipulator with a specially designed mark attached on the knob

(Han et al., 2002). Lack of constraints, errors in calibration and noises on feature extraction restrict the accuracy of the measurement. When the structure or the model of the object is prior known, it can be taken to estimate the pose of the object by means of matching. Kragic et al. taken this technique to determine the pose of the workpiece based on its CAD model (Kragic et al., 2001). High accuracy can be obtained with this method for the object of complex shape. But the computational consumption needed for matching prevents its application from real-time measurement. Therefore, accuracy, robustness and performance are still the challenges for visual measurement.

Finally visual control method also plays an important role in the visual guided approach-to-grasp movement of the humanoid robot. Visual control system can be classified into eye-to-hand (ETH) system and eye-in-hand (EIH) system based on the employed camera-robot configuration (Hutchinson et al., 1996). An eye-to-hand system can have a wider field of view since the camera is fixed in the workspace. Hager et al. presented an ETH stereo vision system to position two floppy disks with the accuracy of 2.5mm (Hager et al., 1995). Hauck et al. proposed a system for grasping (Hauck et al., 2000). On the other hand, an eye-in-hand system can possess a higher precision as the camera is mounted on the end-effector of the manipulator and can observe the object more closely. Hashimoto et al. (Hashimoto et al., 1991) gave an EIH system for tracking. According to the ways of using visual information, visual control can also be divided into position-based visual servoing (PBVS), image-based visual servoing (IBVS) and hybrid visual servoing (Hutchinson et al., 1996; Malis et al., 1999; Corke & Hutchinson, 2001). Dodds et al. pointed out that a key to solving robotic hand-eye tasks efficiently and robustly is to identify how precise the control is needed at a particular time during task execution (Dodds et al., 1999). With the hierarchical architecture he proposed, a hand-eye task was decomposed into a sequence of primitive sub tasks. Each sub task had a specific requirement. Various visual control techniques were integrated to achieve the whole task. A similar idea was demonstrated by Kragic and Christensen (Kragic & Christensen, 2003). Flandin et al. combined ETH and EIH together to exploit the advantage of both configurations (Flandin et al., 2000). Hauck et al. integrated look-and-move with position-based visual servoing to achieve 3 degrees of freedom (DOFs) reaching task (Hauck et al., 1999).

In this chapter, issues above are discussed in detail. Firstly, a motion based method is provided to calibrate the head-eye geometry. Secondly, a visual measurement method with shape constraint is presented to determine the pose of a rectangle object. Thirdly, a visual guidance strategy is developed for the approach-to-grasp movement of humanoid robots.

The rest of the chapter is organized as follows. The camera-robot configuration and the assignment of the coordinate frames for the robot are introduced in section 2. The calibration of vision system is investigated in section 3. In this section, the model for cameras with distortion is presented, and the position and orientation of the stereo rig relative to the head can be determined with three motions of the robot head. In section 4, the shape of a rectangle is taken as the constraint to estimate the pose of the object with high accuracy. In section 5, the approach-to-grasp movement of the humanoid robot is divided into five stages, namely searching, approaching, coarse alignment, precise alignment and grasping. Different visual control methods, such as ETH/EIH, PBVS/IBVS, look-then-move/visual servoing, are integrated to accomplish the grasping task. An experiment of valve operation by a humanoid robot is also presented in this section. The chapter is concluded in section 6.

2. Camera-robot configuration and robot frame

A humanoid robot[1] has the typical configuration of vision system as shown in Fig. 1 (Xu et al., 2006b). Two cameras are mounted on the head of the robot, which serve as eyes. The arms of the robot serve as manipulators with grippers attached at the wrist as the hands. An eye-to-hand system is formed with these two cameras and the arms of the robot. If another camera is mounted on the wrist, an eye-in-hand system will be formed.

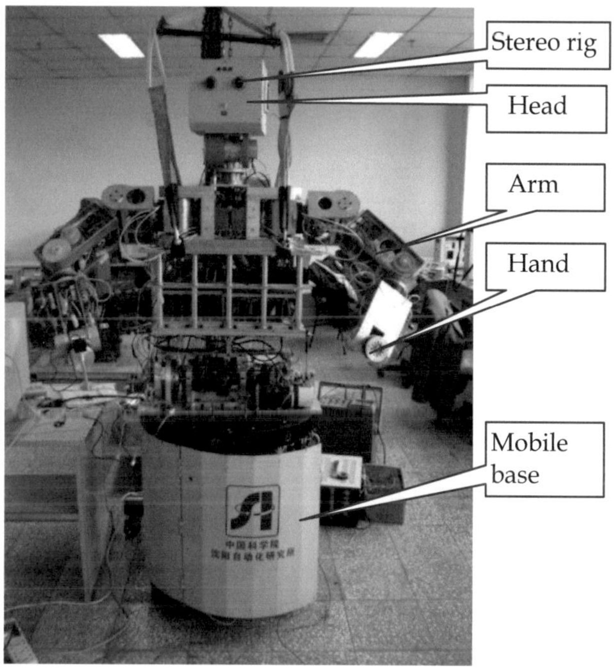

Figure 1. Typical configuration of humanoid robots

Throughout this chapter, lowercase letters (a, b, c) are used to denote scalars, bold-faced ones (**a**, **b**, **c**) denote vectors. Bold-faced uppercase letters (**A, B, C**) stand for matrices and italicized uppercase letters (*A*, *B*, *C*) denote coordinate frames. The homogeneous transformation from coordinate frame *X* to frame *Y* is denoted by ${}^{y}\mathbf{T}_{x}$. It is defined as follows:

$$ {}^{y}T_{x} = \begin{bmatrix} {}^{y}R_{x} & {}^{y}p_{x0} \\ 0 & 1 \end{bmatrix} \tag{1} $$

where ${}^{y}\mathbf{R}_{x}$ is a 3 x 3 rotation matrix, and ${}^{y}\mathbf{p}_{x0}$ is a 3 x 1 translation vector.

Figure 2 demonstrates the coordinate frames assigned for the humanoid robot. The subscript *B*, *N*, *H*, *C*, *G* and *E* represent the base frame of the robot, the neck frame, the head frame, the camera frame, the hand frame, and the target frame respectively. For example, ${}^{n}T_{h}$ represents the pose (position and orientation) of the head relative to the neck.

[1] The robot is developed by Shenyang Institute of Automation, cooperated with Institute of Automation, Chinese Academy of Sciences, P. R. China.

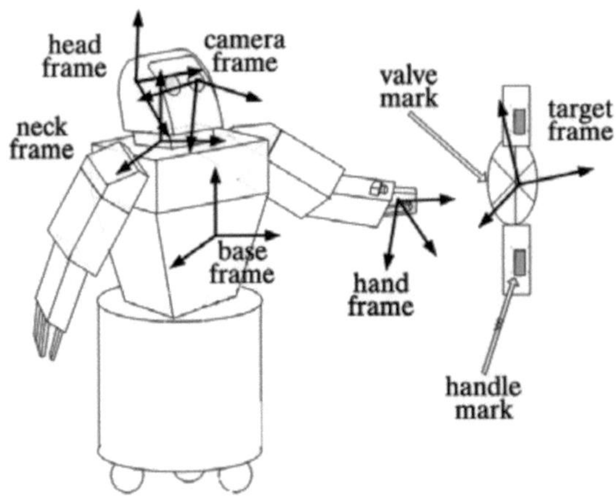

Figure 2. Coordinate frames for the robot

The head has two DOFs such as yawing and pitching. The sketch of the neck and head of a humanoid robot is given in Fig. 3. The first joint is responsible for yawing, and the second one for pitching. The neck frame *N* for the head is assigned at the connection point of the neck and body. The head frame *H* is assigned at the midpoint of the two cameras. The coordinate frame of the stereo rig is set at the optical center of one of the two cameras, e.g. the left camera as shown in Fig. 3.

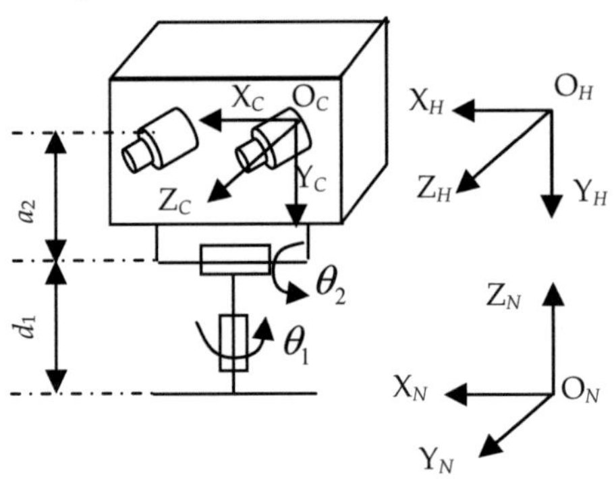

Figure 3. Sketch of the neck and the head

From Fig. 3, the transformation matrix from the head frame *H* to the neck frame *N* is given in (2) according to Denavit-Hartenberg (D-H) parameters (Murray et al., 1993).

$$
{}^{n}T_{c}=\begin{bmatrix} c\theta_1 & -s\theta_1 s\theta_2 & -s\theta_1 c\theta_2 & a_2 s\theta_1 s\theta_2 \\ s\theta_1 & c\theta_1 s\theta_2 & c\theta_1 c\theta_2 & -a_2 c\theta_1 s\theta_2 \\ 0 & -c\theta_2 & s\theta_2 & a_2 c\theta_2 + d_1 \\ 0 & 0 & 0 & 1 \end{bmatrix} \tag{2}
$$

where d_1 and a_2 are the D-H parameters for the two links of the head, θ_1 and θ_2 are the corresponding joint angles, $c\theta$ denotes $\cos\theta$, and $s\theta$ denotes $\sin\theta$.

3. Vision system calibration

3.1 Camera model

The real position of a point on the image plane will deviate from its ideal position as a result of the distortion of the optical lens components. Let (u, v) denote the real pixel coordinates for the projection of a point, (u', v') denote the ideal pixel coordinates without distortion. The distortion is defined as follows:

$$\begin{cases} u = u' + \delta_u(u', v') \\ v = v' + \delta_v(u', v') \end{cases} \tag{3}$$

where δ_u and δ_v represent the distortion in the horizontal and vertical directions respectively.

The distortion can be modeled as a high order polynomial which contains both radial and tangential distortion (Ma et al., 2003). Generally the distortion is formed mainly by the radial component, so the following second order radial distortion model without tangential component is employed for cameras with standard field of view:

$$\begin{cases} u - u_0 = (u' - u_0)(1 + k_u' r'^2) \\ v - v_0 = (v' - v_0)(1 + k_v' r'^2) \end{cases} \tag{4}$$

where (u_0, v_0) are the pixel coordinates of the principle point, (k_u', k_v') are the radial distortion coefficients, and $r' = \sqrt{(u' - u_0)^2 + (v' - v_0)^2}$ is the radius from the ideal point (u', v') to the principle point (u_0, v_0).

When correcting the distortion, the distorted image needs to be corrected to a linear one. So the reverse problem of (4) needs to be solved to obtain the ideal pixel coordinates (u', v') from (u, v). Then the following model is adopted instead of (4):

$$\begin{cases} u'' - u_0 = (u - u_0)(1 + k_u r^2) \\ v'' - v_0 = (v - v_0)(1 + k_v r^2) \end{cases} \tag{5}$$

where (u'', v'') are the pixel coordinates after distortion correlation, k_u, k_v are the correction coefficients, $r = \sqrt{(u - u_0)^2 + (v - v_0)^2}$ is the radius from the point (u, v) to the principle point.

After applying distortion correction, the pixel coordinates (u'', v'') for the projection of a point in the camera frame can be determined with the intrinsic parameters of the camera. Here the four parameters model, which does not consider the skew between the coordinate axes, is employed as follows:

$$\begin{bmatrix} u'' \\ v'' \\ 1 \end{bmatrix} = \begin{bmatrix} k_x & 0 & u_0 \\ 0 & k_y & v_0 \\ 0 & 0 & 1 \end{bmatrix} \begin{bmatrix} x_c / z_c \\ y_c / z_c \\ 1 \end{bmatrix} = M_1 \begin{bmatrix} x_c / z_c \\ y_c / z_c \\ 1 \end{bmatrix} \tag{6}$$

where (x_c, y_c, z_c) are the coordinates of a point in the camera frame, (k_x, k_y) are the focal length in pixel, $\mathbf{M_1}$ is known as the intrinsic parameter matrix of the camera.

Assume the coordinates of a point in the world reference frame W is (x_w, y_w, z_w). Let (x_c, y_c, z_c) be the coordinates of the point in the camera reference frame. Then (x_w, y_w, z_w) and (x_c, y_c, z_c) are related to each other through the following linear equation:

$$\begin{bmatrix} x_c \\ y_c \\ z_c \end{bmatrix} = \begin{bmatrix} n_x & o_x & a_x & p_x \\ n_y & o_y & a_y & p_y \\ n_z & o_z & a_z & p_z \end{bmatrix} \begin{bmatrix} x_w \\ y_w \\ z_w \\ 1 \end{bmatrix} = M_2 \begin{bmatrix} x_w \\ y_w \\ z_w \\ 1 \end{bmatrix} \tag{7}$$

where $\mathbf{n} = (n_x, n_y, n_z)^T$, $\mathbf{o} = (o_x, o_y, o_z)^T$ and $\mathbf{a} = (a_x, a_y, a_z)^T$ are the coordinate vectors for the x-axis, y-axis and z-axis of the world frame in the camera frame, $\mathbf{p} = (p_x, p_y, p_z)^T$ is the coordinate vector of the origin for the world reference frame in the camera frame, $\mathbf{M_2}$ is a 3 x 4 matrix, which is known as the extrinsic parameter matrix of the camera.

3.2 Hand-eye calibration

For a stereo rig, the intrinsic parameters of each camera and the displacement between two cameras can be determined accurately with the method proposed by Xu et al., which is designed for cameras with large lens distortion (Xu et al., 2006a). Then the position of a point in the camera reference frame can be determined with this calibrated stereo rig. The next important step in applying the stereo rig on the humanoid robot is to determine the relative position and orientation between the stereo rig and the head of the robot, which is called head-eye (or hand-eye) calibration.

3.2.1 Calibration algorithm

Refer to Fig. 2, assume that the world coordinate frame is attached on the grid pattern (called calibration reference). The pose of the world frame relative to the camera can be determined with the stereo rig by using the grid pattern. If $\mathbf{T_c}$ represents the transformation from the world reference frame to the camera frame; $\mathbf{T_h}$ is the relative pose of the head with respect to the base of the humanoid robot; $\mathbf{T_m}$ represents the head-eye geometry, which is the pose of the stereo rig relative to the robot head. Then it can be obtained that

$$T_p = T_h T_m T_c \tag{8}$$

where $\mathbf{T_p}$ is the transformation between the grid pattern and the robot base.
With the position and orientation of the grid pattern fixed while the pose of the head varying, it can be obtained that

$$T_{hi} T_m T_{ci} = T_{hi-1} T_m T_{ci-1} \tag{9}$$

where the subscript i represents the i-th motion, i = 1, 2, … , n, $\mathbf{T_{h0}}$ and $\mathbf{T_{c0}}$ represent the initial position of the robot head and the camera.
Left multiplying both sides of (9) by $\mathbf{T_{hi-1}}^{-1}$ and right multiplying by $\mathbf{T_{ci}}^{-1}$ gives:

$$T_{hi-1}^{-1} T_{hi} T_m = T_m T_{ci-1} T_{ci}^{-1} \tag{10}$$

Let $T_{Li} = T_{hi-1}^{-1} T_{hi}$ and $T_{Ri} = T_{ci-1} T_{ci}^{-1}$. $\mathbf{T_{Li}}$ is the transformation between the head reference frames before and after the motion, which can be read from the robot controller. And $\mathbf{T_{Ri}}$ is the transformation between the camera reference frames before and after the movement, which can be determined by means of stereovision method using the grid pattern. Then (10) becomes:

$$T_{Li}T_m = T_mT_{Ri} \tag{11}$$

Solving (11) will give the head-eye geometry $\mathbf{T_m}$. Equation (11) is the basic equation of head-eye calibration, which is called the **AX** = **XB** equation in the literature. Substituting (1) into (11) gives:

$$\begin{bmatrix} R_{Li}R_m & R_{Li}p_m + p_{Li} \\ 0 & 1 \end{bmatrix} = \begin{bmatrix} R_mR_{Ri} & R_mp_{Ri} + p_m \\ 0 & 1 \end{bmatrix} \tag{12}$$

where $\mathbf{R_{Li}}$, R_{Ri} and $\mathbf{R_m}$ are the rotation components of $\mathbf{T_{Li}}$, $\mathbf{T_{Ri}}$ and $\mathbf{T_m}$ respectively, $\mathbf{p_{Li}}$, $\mathbf{p_{Ri}}$ and $\mathbf{p_m}$ are the translation components of $\mathbf{T_{Li}}$, $\mathbf{T_{Ri}}$ and $\mathbf{T_m}$. By (12) it can be obtained that

$$R_{Li}R_m = R_mR_{Ri} \tag{13}$$

$$R_{Li}p_m + p_{Li} = R_mp_{Ri} + p_m \tag{14}$$

Then $\mathbf{R_m}$ and $\mathbf{p_m}$ can be determined with (13) and (14).

3.2.2 Calibration of the rotation component $\mathbf{R_m}$

A rotation **R** can be represented as (Murray et al., 1993):

$$\mathbf{R} = \mathrm{Rot}(\boldsymbol{\omega}, \theta) \tag{15}$$

where Rot(.) is a function representing the rotation about an axis with an angle, $\boldsymbol{\omega}$ is a unit vector which specific the axis of rotation, θ is the angle of rotation. $\boldsymbol{\omega}$ and θ can be uniquely determined from **R** (Murray et al., 1993). The vector $\boldsymbol{\omega}$ is also the only real eigenvector of **R** and its corresponding eigenvalue is 1 (Tsai & Lenz, 1989):

$$\mathbf{R}\boldsymbol{\omega} = \boldsymbol{\omega} \tag{16}$$

Applying (16) to the $\mathbf{R_{Li}}$ and $\mathbf{R_{Ri}}$ in (13) gives (Tsai & Lenz, 1989):

$$\omega_{Li} = \mathbf{R}_m \, \omega_{Ri} \tag{17}$$

where $\boldsymbol{\omega}_{Li}$ and $\boldsymbol{\omega}_{Ri}$ are the rotation axes of $\mathbf{R_{Li}}$ and $\mathbf{R_{Ri}}$ respectively.

Let $\omega_{Li} = (\omega_{Lix}, \omega_{Liy}, \omega_{Liz})^T$, $\omega_{Ri} = (\omega_{Rix}, \omega_{Riy}, \omega_{Riz})^T$ and $R_m = \begin{bmatrix} m_1 & m_2 & m_3 \\ m_4 & m_5 & m_6 \\ m_7 & m_8 & m_9 \end{bmatrix}$.

Then (17) becomes:

$$A_{mi}x_C = b_{mi} \tag{18}$$

where

$$A_{mi} = \begin{bmatrix} \omega_{Rix} & \omega_{Riy} & \omega_{Riz} & 0 & 0 & 0 & 0 & 0 & 0 \\ 0 & 0 & 0 & \omega_{Rix} & \omega_{Riy} & \omega_{Riz} & 0 & 0 & 0 \\ 0 & 0 & 0 & 0 & 0 & 0 & \omega_{Rix} & \omega_{Riy} & \omega_{Riz} \end{bmatrix},$$

$b_{mi} = \omega_{Li} = (\omega_{Lix}, \omega_{Liy}, \omega_{Liz})^T$, $x_C = (m_1 \quad m_2 \quad \cdots \quad m_9)^T$ which is a 9 x 1 vector formed with the columns of the rotation matrix **R**.

Stacking (18) gives the result for the case of n movements of the robot head. A linear equation can be obtained:

$$A_m x_C = b_m \tag{19}$$

where $A_m = \begin{bmatrix} A_{m1} \\ A_{m2} \\ \vdots \\ A_{mn} \end{bmatrix}$ is a $3n$ x 9 matrix, $b_m = \begin{bmatrix} b_{m1} \\ b_{m2} \\ \vdots \\ b_{mn} \end{bmatrix}$ is $3n$ x 1 vector.

Equation (18) indicates that one motion of the head will contribute three equations in (19). Therefore three motions are necessary in order to determine $\mathbf{x_C}$ which has nine independent variables. Providing $n \geq 3$, a least square solution for $\mathbf{x_C}$ is given by

$$x_C = (A_m^T A_m)^{-1} A_m^T b_m \tag{20}$$

Then the rotation component $\mathbf{R_m}$ can be determined from $\mathbf{x_C}$.

3.2.3 Orthogonalization of $\mathbf{R_m}$

The result from previous section gives an estimation of $\mathbf{R_m}$. The deduction of (20) does not consider the orthogonality of $\mathbf{R_m}$. It is necessary to orthogonalize the $\mathbf{R_m}$ obtained from (20). Assume α, β, γ are the Euler angles of the rotation. Then $\mathbf{R_m}$ can be represented as follows (Murray et al., 1993):

$$R_m(\alpha, \beta, \lambda) = Rot(z, \alpha) Rot(y, \beta) Rot(z, \gamma) \tag{21}$$

where Rot(y, .) and Rot(z, .) are functions representing the rotation about y-axis and z-axis. Equation (21) yields that $\mathbf{R_m}$ is a nonlinear function of α, β and γ. Then the problem of solving (17) can be formulated as a nonlinear least squares optimization. The objective function to be minimized, *J*, is a function of the squared error:

$$J(\alpha, \beta, \lambda) = \sum_{i=1}^{n} \left\| \omega_{Li} - R_m(\alpha, \beta, \lambda) \omega_{Ri} \right\|^2 \tag{22}$$

The objective function can be minimized using a standard nonlinear optimization method such as Quasi-Newton method.

$$(\alpha^*, \beta^*, \gamma^*) = \min_{\alpha, \beta, \lambda} J(\alpha, \beta, \lambda) \tag{23}$$

where α^*, β^*, γ^* are the angles where the objective function J reaches its local minimum.

Finally the $\mathbf{R_m}$ is determined by substituting α^*, β^* and γ^* into (21). The orthogonality of $\mathbf{R_m}$ is satisfied since the rotation is represented with the Euler angle as (21). The result from (20) can be taken as the initial value to start the iteration of the optimization method.

3.2.4 Calibration of the translation component p_m

The translation vector $\mathbf{p}_m$ can be determined from (14) once the R_m has been obtained. Rearranging (14) gives:

$$(R_{Li} - I)p_m = R_m p_{Ri} - p_{Li} \tag{24}$$

where **I** stands for a 3 x 3 identity matrix.

It is similar to the derivation from (18) to (19), that a linear equation can be formulated by stacking (24) with the subscript i increasing from 1 to n:

$$C_m p_m = d_m \tag{25}$$

Where $C_m = \begin{bmatrix} R_{L1} - I \\ R_{L2} - I \\ \vdots \\ R_{Ln} - I \end{bmatrix}$ is a $3n$ x 3 matrix, $d_m = \begin{bmatrix} R_m p_{R1} - p_{L1} \\ R_m p_{R2} - p_{L2} \\ \vdots \\ R_m p_{Rn} - p_{Ln} \end{bmatrix}$ is a $3n$ x 1 vector.

Solving (25) gives the translation component $\mathbf{p_m}$ of the head-eye geometry. Giving $n \geq 1$, the least square solution for (25) is as follows:

$$p_m = (C_m^T C_m)^{-1} C_m^T d_m \tag{26}$$

3.3 Experiments and results

The head was fixed at the end of a K-10 manipulator as shown in Fig. 4. A stereo rig was mounted on the head and was faced to the ground. A grid pattern was placed under the head. The world reference frame was attached on the grid pattern with its origin at the center of the pattern. The reference frame of the stereo rig was assigned to the frame of the left camera. The stereo rig was calibrated with the method in (Xu et al., 2006a). The intrinsic parameters of each camera of the stereo rig are shown in Table 1.

Figure 4. Head-eye calibration

	k_u (x 10^{-7})	k_v (x 10^{-7})	u_0	v_0	k_x	k_y
Left	4.2180	3.6959	324.6	248.3	1082.3	1067.8
Right	3.4849	3.6927	335.0	292.0	1252.2	1242.3

Table 1. Parameters of the stereo rig

$\mathbf{T}_{ci}$	$\mathbf{T}_{hi}$	$\mathbf{T}_{pi}$
$\begin{bmatrix} 0.0162 & -0.1186 & 0.9928 & 1100.3 \\ 0.9269 & -0.3706 & -0.0594 & -275.3 \\ 0.3749 & 0.9212 & 0.1040 & 357.7 \end{bmatrix}$	$\begin{bmatrix} 0.9989 & 0.0195 & -0.0418 & 2.0 \\ -0.0362 & 0.8924 & -0.4497 & -9.0 \\ 0.0286 & 0.4508 & 0.8922 & 320.4 \end{bmatrix}$	$\begin{bmatrix} 0.9989 & -0.0437 & 0.0152 & 1148.50 \\ -0.0439 & -0.9990 & 0.0092 & -323.55 \\ 0.0148 & -0.0098 & -0.9998 & 0.02 \end{bmatrix}$
$\begin{bmatrix} 0.9989 & 0.0195 & -0.0418 & 2.0 \\ -0.0362 & 0.8924 & -0.4497 & -9.0 \\ 0.0286 & 0.4508 & 0.8922 & 320.4 \end{bmatrix}$	$\begin{bmatrix} 0.9675 & -0.1499 & 0.2037 & -1.0 \\ 0.2261 & 0.8738 & -0.4306 & -10.1 \\ -0.1134 & 0.4627 & 0.8792 & 313.2 \end{bmatrix}$	$\begin{bmatrix} 0.9992 & -0.0397 & 0.0082 & 1148.60 \\ -0.0398 & -0.9991 & 0.0154 & -322.55 \\ 0.0076 & -0.0158 & -0.9998 & -0.92 \end{bmatrix}$
$\begin{bmatrix} 0.9675 & -0.1499 & 0.2037 & -1.0 \\ 0.2261 & 0.8738 & -0.4306 & -10.1 \\ -0.1134 & 0.4627 & 0.8792 & 313.2 \end{bmatrix}$	$\begin{bmatrix} 0.9675 & -0.1499 & 0.2037 & -1.0 \\ 0.2261 & 0.8738 & -0.4306 & -10.1 \\ -0.1134 & 0.4627 & 0.8792 & 313.2 \end{bmatrix}$	$\begin{bmatrix} 0.9994 & -0.0349 & 0.0085 & 1150.20 \\ -0.0350 & -0.9994 & 0.0045 & -325.44 \\ 0.0083 & -0.0049 & -0.9999 & -0.22 \end{bmatrix}$
$\begin{bmatrix} 0.9675 & -0.1499 & 0.2037 & -1.0 \\ 0.2261 & 0.8738 & -0.4306 & -10.1 \\ -0.1134 & 0.4627 & 0.8792 & 313.2 \end{bmatrix}$	$\begin{bmatrix} 0.9675 & -0.1499 & 0.2037 & -1.0 \\ 0.2261 & 0.8738 & -0.4306 & -10.1 \\ -0.1134 & 0.4627 & 0.8792 & 313.2 \end{bmatrix}$	$\begin{bmatrix} 0.9988 & -0.0489 & 0.0141 & 1150.10 \\ -0.0489 & -0.9988 & 0.0069 & -324.56 \\ 0.0137 & -0.0075 & -0.9999 & -0.66 \end{bmatrix}$

Table 2. The obtained $\mathbf{T}_{ci}$, $\mathbf{T}_{hi}$, and $\mathbf{T}_{pi}$

The displacement between two cameras, which is denoted by ${}^{R}\mathbf{T}_{L}$, is as follows:

$$ {}^{R}T_{L} \begin{bmatrix} 0.9998 & -0.0198 & -0.0063 & -93.4482 \\ 0.0195 & 0.9990 & -0.0403 & 1.4111 \\ 0.0071 & 0.0402 & 0.9991 & 125.7946 \end{bmatrix}. $$

Four pairs of images were acquired by the stereo rig with 3 motions of the head. The relative position and orientation of the grid pattern with respect to the stereo rig, $\mathbf{T}_{ci}$, was measured with the stereovision method. The pose of the head was changed by the movement of the manipulator, while holding the grid pattern in the field of view of the stereo rig. The pose of the end of the manipulator, $\mathbf{T}_{hi}$, was read form the robot controller. Then 3 equations as (11) were obtained. The head-eye geometry, $\mathbf{T}_{m}$, was computed with (20), (21), (23) and (26). The obtained $\mathbf{T}_{ci}$ and $\mathbf{T}_{hi}$ are shown in the first two columns of Table 2, and the calibration result is as follows:

$$ T_m = \begin{bmatrix} -0.0217 & -0.6649 & -0.7466 & 55.1906 \\ -0.0426 & 0.7467 & -0.6638 & -97.8522 \\ 0.9989 & 0.0174 & -0.0446 & 26.0224 \end{bmatrix}. $$

The pose of the grid pattern relative to the robot could be determined by (8) with each group of $\mathbf{T}_{ci}$, $\mathbf{T}_{hi}$ and the obtained $\mathbf{T}_{m}$, as shown in the last column of Table 2. Since the pattern was fixed during the calibration, $\mathbf{T}_{pi}$ should remain constant. From Table 2, the maximum variances of the x, y, z coordinates of the translation vector in $\mathbf{T}_{pi}$ were less than 1.7mm, 2.9mm, and 1.0mm. The results indicated that the head-eye calibration was accurate.

4. Stereo vision measurement for humanoid robots

4.1 Visual positioning with shape constraint

The position and orientation of a object relative to the robot can be measured with the stereo rig on the robot head after the vision system and the head-eye geometry are calibrated. Generally, it is hard to obtain high accuracy with visual measurement, especially the measurement of the orientation, only using individual feature points. Moreover, errors in calibration and feature extraction result in large errors in pose estimation. The estimation performance is expected to be improved if the shape of an object is taken into account in the visual measurement.

Rectangle is a category of shape commonly encountered in everyday life. In this section, the shape of a rectangle is employed as a constraint for visual measurement. A reference frame is attached on the rectangle as shown in Fig. 5. The plane containing the rectangle is taken as the xoy plane. Then the pose of the rectangle with respect to the camera is exactly the extrinsic parameters of the camera if the reference frame on the rectangle is taken as the world reference frame. Assume the rectangle is $2X_w$ in width and $2Y_w$ in height. Obviously, any point on the rectangle plane should satisfy $z_w = 0$.

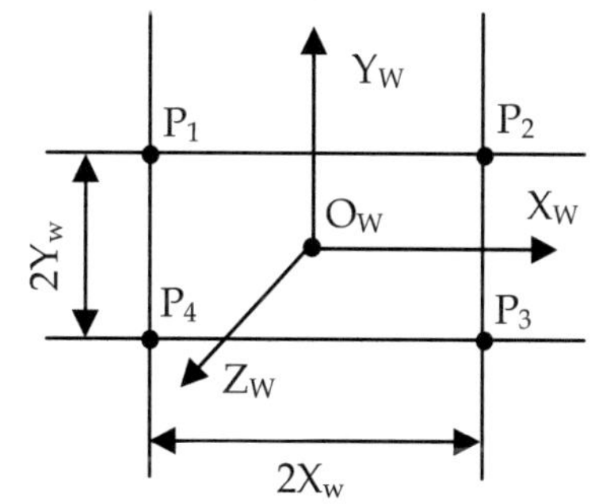

Figure 5. The reference frame and the reference point

4.2 Algorithm for estimating the pose of a rectangle

4.2.1 Derivation of the coordinate vector of x-axis

From (7), and according to the orthogonality of the rotation component of the extrinsic parameter matrix $\mathbf{M}_2$, giving $z_w = 0$, it can be obtained that

$$\begin{cases} o_x x_c + o_y y_c + o_z z_c = y_w + o_x p_x + o_y p_y + o_z p_z \\ a_x x_c + a_y y_c + a_z z_c = a_x p_x + a_y p_y + a_z p_z \end{cases} \tag{27}$$

Assuming that $z_c \neq 0$, $a_x p_x + a_y p_y + a_z p_z \neq 0$, equation (27) becomes:

$$\frac{o_x x'_c + o_y y'_c + o_z}{a_x x'_c + a_y y'_c + a_z} = C_1 \tag{28}$$

where $x'_c = x_c / z_c$, $y'_c = y_c / z_c$ and

$$C_1 = \frac{y_w + o_x p_x + o_y p_y + o_z p_z}{a_x p_x + a_y p_y + a_z p_z}.$$

Points on a line paralleled to the x-axis have the same y coordinate y_w, so C_1 is constant for these points. Taking two points on this line, denoting their coordinates in the camera frame as (x_{ci}, y_{ci}, z_{ci}) and (x_{cj}, y_{cj}, z_{cj}), and applying them to (28) gives:

$$\frac{o_x x'_{ci} + o_y y'_{ci} + o_z}{a_x x'_{ci} + a_y y'_{ci} + a_z} = \frac{o_x x'_{cj} + o_y y'_{cj} + o_z}{a_x x'_{cj} + a_y y'_{cj} + a_z} \tag{29}$$

Simplifying (29) with the orthogonality of the rotation components of $\mathbf{M}_2$ gives:

$$n_x (y'_{ci} - y'_{cj}) + n_y (x'_{cj} - x'_{ci}) + n_z (x'_{ci} y'_{cj} - x'_{cj} y'_{ci}) = 0 \tag{30}$$

Noting that x'_{ci}, y'_{ci}, x'_{cj} and y'_{cj} can be obtained with (5) and (6) if the parameters of the camera have been calibrated, n_x, n_y and n_z are the only unknowns in (30). Two equations as (30) can be obtained with two lines paralleled to the x-axis, and besides, **n** is a unit vector, i.e. $\|\mathbf{n}\|= 1$. Then n_x, n_y and n_z can be determined with these three equations.
It can be divided into two cases to obtain n_x, n_y and n_z. If the optical axis of the camera is not vertical to the rectangle plane, $n_z \neq 0$ is satisfied. Dividing both sides of (30) by n_z gives:

$$n'_x(y'_{ci} - y'_{cj}) + n'_y(x'_{cj} - x'_{ci}) = x'_{cj}y'_{ci} - x'_{ci}y'_{cj} \tag{31}$$

where $n'_x = n_x/n_z$, $n'_y = n_y/n_z$.
Then n'_x and n'_y can be determined with two such equations. The corresponding n_x, n_y, n_z can be computed by normalizing the vector $(n'_x, n'_y, 1)^T$ as follows:

$$\begin{cases} n_x = n'_x \big/ \sqrt{n'^2_x + n'^2_y + 1} \\ n_y = n'_y \big/ \sqrt{n'^2_x + n'^2_y + 1} \\ n_z = 1 \big/ \sqrt{n'^2_x + n'^2_y + 1} \end{cases} \tag{32}$$

If the optical axis is vertical to the rectangle plane, $n_z = 0$ and (30) becomes:

$$n_x(y'_{ci} - y'_{cj}) + n_y(x'_{cj} - x'_{ci}) = 0 \tag{33}$$

Similar to (31), n_x and n_y can be directly computed with two equations as (33), and the n_x, n_y, n_z can be obtained by normalizing the vector $(n_x, n_y, 0)^T$ to satisfy $\|\mathbf{n}\|= 1$.

4.2.2 Derivation of the coordinate vector of z-axis

Similar to (27), by $\mathbf{M_2}$, it can be obtained that:

$$\begin{cases} n_x x_c + n_y y_c + n_z z_c = x_w + n_x p_x + n_y p_y + n_z p_z \\ a_x x_c + a_y y_c + a_z z_c = a_x p_x + a_y p_y + a_z p_z \end{cases} \tag{34}$$

Denote the coordinates of a point in the camera frame as (x_{ci}, y_{ci}, z_{ci}). Assume $z_{ci} \neq 0$. Then (34) becomes:

$$a_x x'_{ci} + a_y y'_{ci} + a_z = C_2(n_x x'_{ci} + n_y y'_{ci} + n_z) \tag{35}$$

where $x'_{ci} = x_{ci}/z_{ci}$, $y'_{ci} = y_{ci}/z_{ci}$, and

$$C_2 = \frac{a_x p_x + a_y p_y + a_z p_z}{x_w + n_x p_x + n_y p_y + n_z p_z}.$$

Since vector **n** and **a** are orthogonal and $a_z \neq 0$, it follows that

$$n_x a'_x + n_y a'_y = -n_z \tag{36}$$

where $a'_x = a_x/a_z$ and $a'_y = a_y/a_z$.
Dividing (35) by a_z and eliminating a'_x from (35) and (36) gives:

$$(n_x y'_{ci} - n_y x'_{ci})a'_y - n_x(n_x x'_{ci} + n_y y'_{ci} + n_z)C'_2 = n_z x'_{ci} - n_x \tag{37}$$

where $C'_2 = C_2/a_z$.
As for the points on a line paralleled to the y-axis, their x coordinate, x_w, are the same, and C'_2 should remain constant. Taking any two points on this line gives two equations as (37). Then a'_y and C'_2 can be obtained with these two equations. Substituting a'_y into (36) gives a'_x. Then a_x, a_y and a_z can be determined by normalizing the vector $(a'_x, a'_y, 1)^T$ as (32).
Finally the vector **o** is determined with **o** = **a** x **n**. The rotation matrix is orthogonal since **n** and **a** are unit orthogonal vectors.

4.2.3 Derivation of the coordinates of the translation vector

Taking one point on the line $y = Y_w$ and the other one on the line $y = -Y_w$, the corresponding constants C_1, which are computed with (28), are denoted as C_{11} and C_{12} respectively. Then it follows that

$$\frac{2(o_x p_x + o_y p_y + o_z p_z)}{a_x p_x + a_y p_y + a_z p_z} = C_{11} + C_{12} \tag{38}$$

$$\frac{2Y_w}{a_x p_x + a_y p_y + a_z p_z} = C_{11} - C_{12} \tag{39}$$

Simplifying (38) and (39) gives:

$$\begin{cases} (2o_x - D_{h1}a_x)p_x + (2o_y - D_{h1}a_y)p_y + (2o_z - D_{h1}a_z)p_z = 0 \\ D_{h2}a_x p_x + D_{h2}a_y p_y + D_{h2}a_z p_z = 2Y_w \end{cases} \tag{40}$$

where $D_{h1} = C_{11} + C_{12}$, $D_{h2} = C_{11} - C_{12}$.
Similarly, the line $x = X_w$ and $x = -X_w$ yield that

$$\begin{cases} (2n_x - D_{v1}a_x)p_x + (2n_y - D_{v1}a_y)p_y + (2n_z - D_{v1}a_z)p_z = 0 \\ D_{v2}a_x p_x + D_{v2}a_y p_y + D_{v2}a_z p_z = 2X_w \end{cases} \tag{41}$$

where $D_{v1} = 1/C_{21} + 1/C_{22}$, $D_{v2} = 1/C_{21} - 1/C_{22}$. C_{21} and C_{22} can be computed with (35).
Then the translation vector **p** = (p_x, p_y, p_z) can be determined by solving (40) and (41).
Xu et al. gave an improved result of the translation vector **p**, where the area of the rectangle was employed to refine the estimation (Xu et al., 2006b).

4.3 Experiments and results

An experiment was conducted to compare the visual measurement method, which considering the shape constraints, with the traditional stereovision method. A colored rectangle mark was place in front of the humanoid robot. The mark had a dimension of 100mm × 100mm. The parameters of the camera are described in section 3.3.
The edges of the rectangle were detected with Hough transformation after distortion corrections. The intersections between the edges of the rectangle and the x-axis and y axis of the reference frame were taken as the feature points for stereovision method. The position and orientation of the rectangle relative to the camera reference frame are computed with the Cartesian coordinates of the feature points.
Three measurements were taken under the same condition. Table 3 shows the results. The first column is the results of the traditional stereovision method, while the 2nd column

shows the results of the algorithm presented in section 4.2. It can be found out that the results of the stereovision method were unstable, while the results of the method with the shape constraints were very stable.

Index	Results with stereovision	Results with the proposed method
1	$\begin{bmatrix} 0.4480 & 0.8524 & 0.2550 & 83.6 \\ -0.8259 & 0.2957 & 0.4850 & 63.9 \\ 0.3423 & -0.4313 & 0.8365 & 921.2 \end{bmatrix}$	$\begin{bmatrix} 0.4501 & 0.8397 & 0.3037 & 91.1 \\ -0.8458 & 0.2917 & 0.4467 & 76.4 \\ 0.2865 & -0.4579 & 0.8415 & 959.6 \end{bmatrix}$
2	$\begin{bmatrix} 0.4420 & 0.8113 & 0.3428 & 83.1 \\ -0.8297 & 0.2642 & 0.5071 & 64.3 \\ 0.3409 & -0.5216 & 0.7899 & 918.4 \end{bmatrix}$	$\begin{bmatrix} 0.4501 & 0.8397 & 0.3037 & 91.1 \\ -0.8458 & 0.2917 & 0.4467 & 76.4 \\ 0.2865 & -0.4579 & 0.8415 & 959.6 \end{bmatrix}$
3	$\begin{bmatrix} 0.4480 & 0.8274 & 0.3053 & 83.4 \\ -0.8259 & 0.2811 & 0.5010 & 63.9 \\ 0.3423 & -0.4861 & 0.8093 & 923.3 \end{bmatrix}$	$\begin{bmatrix} 0.4501 & 0.8397 & 0.3037 & 91.1 \\ -0.8458 & 0.2917 & 0.4467 & 76.4 \\ 0.2865 & -0.4579 & 0.8415 & 959.6 \end{bmatrix}$

Table 3. Measuring results for the position and orientation of an object

More experiments were demonstrated by Xu et al. (Xu et al., 2006b). The results indicate that visual measurement with the shape constraints can give a more robust estimation especially when presented with noises in the feature extraction.

5. Hand eye coordination of humanoids robot for grasping

5.1 Architecture for vision guided approach-to-grasp movements

Differ from industrial manipulators, humanoid robots are mobile platforms and the object for grasping can be placed anywhere in the environment. The robot needs to search and approach the object, and then perform grasping with its hand. In this process, both the manipulator and the robot itself need to be controlled. Obviously the required precision in the approaching process is different from that in the grasping process. The requirement for the control method should also be different. In addition, the noises and errors on the system, including the calibration of the vision system, the calibration of the robot, and the visual measurement, will play an important role in the accuracy of visual control (Gans et al., 2002). The control scheme should be robust to these noises and errors.

The approach-to-grasp task can be divided into five stages: searching, approaching, coarse alignment of the body and hand, precise alignment of the hand, and grasping. At each stage, the requirements for visual control are summarized as follows:

1. Searching: wandering in the workspace to search for the concerned target.
2. Approaching: approaching the target from far distance, only controlling the movement of the robot body.
3. Coarse alignment: aligning the body of the robot with the target to ensure the hand of the robot can reach and manipulate the target without any mechanical constrains; also aligning the hand with the target. Both the body and the hand need to be controlled.
4. Precise alignment: aligning the hand with the target to achieve a desired pose relative to the target at a high accuracy. Only the hand of the robot has to be controlled.
5. Grasping: grasping the target based on the force sensor. The control of the hand is needed.

With the change of the stages, the controlled plant and the control method also change. Figure 6 is the architecture of the control system for visual guided grasping task.

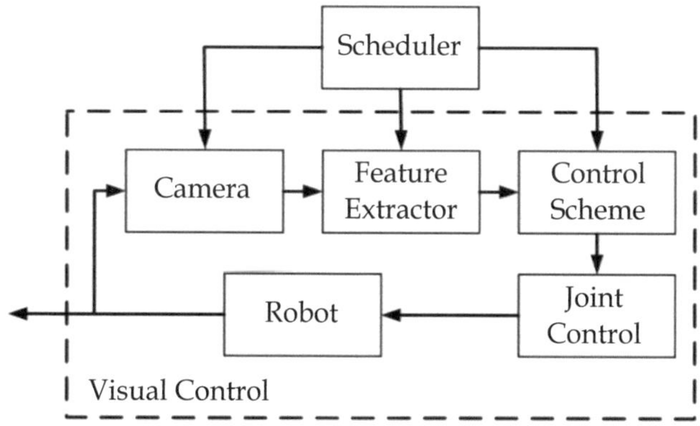

Figure 6. Architecture of the vision control system

In Fig. 6, the visual control block is a basic visual feedback loop. Depending on the used camera-robot configuration and control law, it can be configured as eye-in-hand or eye-to-hand system, with position based or image-based visual control method. The scheduler module monitors the undergoing of the task. It determines which stage should be carried into execution and invokes correspond image processing and visual control module.

5.2 Valve operation by the humanoid robot

An approach-to-grasp task was designed for the humanoid robot, which is shown in Fig. 1. The robot has a head, a body with two arms and a wheeled mobile base. A stereo rig is mounted on the head as the eyes. Two six DOFs manipulators serve as arms. Each arm has a gripper as the hand. The wrist of the hand is equipped with a mini camera and force sensors.

A valve is placed in the workspace of the humanoid robot. A red rectangle mark is attached on the valve for the robot to identify the valve and estimate its pose. Two green marks are attached on the handles of the valve. The robot will search the valve with its stereo rig on the head. Once the robot finds the valve, it moves towards it and operates it with its hands, as shown in Fig. 7. Operations include turning on and turning off the valve.

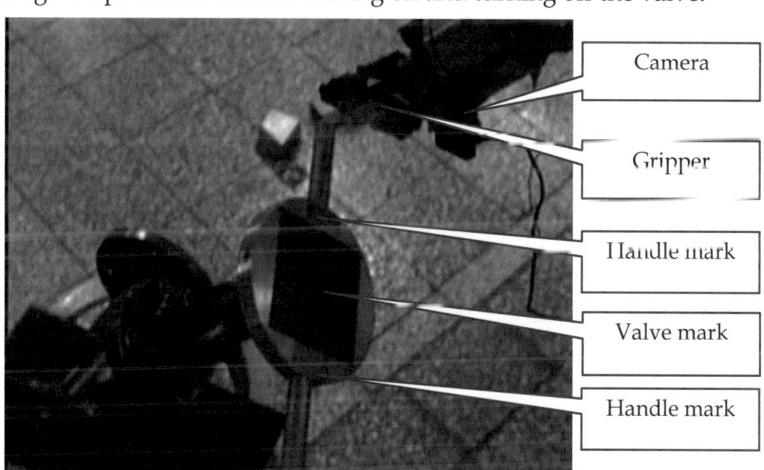

Figure 7. Valve operation with the robot hand

No	Stage	Controlled variables	DOFs	Image features	Camera-robot configuration	Visual control law
1	Searching	${}^{b}\mathbf{T}_{e}$	2 of head 2 of body	Color	EIH	Look-then move
2	Approaching	${}^{b}\mathbf{T}_{e}$	2 of body	Centroid	EIH	Position-based
3	Coarse alignment	${}^{b}\mathbf{T}_{e}$, ${}^{g}\mathbf{T}_{e}$	3 of body 6 of arm	Vertex	EIH, ETH	Position-based
4	Precise alignment	${}^{g}\mathbf{T}_{e}$	4 of arm	Centroid, Area	EIH	Image-based
5	Grasping	${}^{g}\mathbf{T}_{e}$	6 of arm 2 of hand			

Table 4. Summarization of visual control strategy

Table 4 demonstrates the algorithms for each stage of the approach-to-grasp movement. Algorithms for each stage are simple and easy to carry. The methods for calibration and pose estimation discussed in section 3 and section 4 are employed. The complicated approach-to-grasp task such as valve operation can be accomplished with the integration of these methods.

The advantages of both eye-to-hand and eye-in-hand systems are fully exploited in this visual control strategy. The blocking problem for the eye-to-hand system is effectively avoided since cameras on the head are active. The problem of losing targets in the field of view for an eye-in-hand system is resolved, as the hands are only adjusted in a small range.

5.3 Approach and alignment

5.3.1 Approaching

Assume the stereo rig on the head, as well as the head-eye geometry, have been accurately calibrated. After the robot finds the valve based on color with the stereo rig, the center point of the red mark is taken as the image feature. The Cartesian coordinates of the valve can be obtained with the stereovision algorithm using the pixel coordinates of the image feature. A position-based visual control scheme is adopted to control the robot to approach the valve. This is an EIH system if the body of the robot is regarded a "manipulator".

Let the notation ${}^{c}\mathbf{p}_{e}$ represent the position of the mark/valve relative to the camera, ${}^{b}\mathbf{p}_{e}^{*}$ represent the desired position of the valve relative to the robot body, which is about 2m away from the robot. An error function (Hutchinson et al., 1996) can be defined as

$$
{}^{b}\mathbf{e} = ({}^{b}e_{x}, {}^{b}e_{y}, {}^{b}e_{z})^{T} = ({}^{b}\mathbf{P}_{e}^{*} - {}^{b}\mathbf{P}_{e}) = ({}^{b}\mathbf{P}_{e}^{*} - {}^{b}\mathbf{T}_{h} \cdot {}^{h}\mathbf{T}_{c} \cdot {}^{c}\mathbf{p}_{e}) \tag{42}
$$

where ${}^{b}\mathbf{T}_{h}$, ${}^{h}\mathbf{T}_{c}$ can be obtained from robot kinematics and camera calibration.

The robot can only move on the ground, which is the xoy plane of the robot base frame as shown in Fig 2. So the z coordinate of ${}^{b}\mathbf{e}$, can be removed from the error. A proportion control law can be designed to eliminate the error, $({}^{b}e_{x}, {}^{b}e_{y})^{T}$, and make the robot approach the valve. In practice, problems such as route plan and obstacle avoidance should also be considered.

5.3.2 Pose estimation and coarse alignment

When the distance between the robot and the valve is near enough, the approaching stage ends, and the coarse alignment starts. Two types of alignment is required. One is the

alignment of robot body with respect to the valve, and the other one is the alignment of the hand with the valve. For the former, the camera-robot configuration is an eye-in-hand system, which is same to the case in approaching. As for the latter, the cameras on the head and the arm form an eye-to-hand system.
The pose of the valve can be estimated with the red rectangle mark. Four vertexes of the mark are taken as the image features. Then the visual measurement method with shape constraints, described in section 4, is employed to measure the pose of the mark, i.e. the pose of the valve, which is denoted by ${}^{c}\mathbf{T}_{e}$.
First, the alignment of the robot body is carried out. The task function can be defined as:

$$ {}^{b}\mathbf{E} = {}^{b}\mathbf{T}_{b^*} = {}^{b}\mathbf{T}_{h} \cdot {}^{h}\mathbf{T}_{c} \cdot {}^{c}\mathbf{T}_{e} \cdot {}^{e}\mathbf{T}_{b^*} \tag{43} $$

where ${}^{c}\mathbf{T}_{e}$ is the estimated pose, ${}^{e}\mathbf{T}_{b^*}$ is the desired pose of the body relative to the valve.
${}^{b}\mathbf{E}$ is a homogeneous transformation matrix. The translation vector $({}^{b}e_x, {}^{b}e_y, {}^{b}e_z)^T$ and the Euler angle $({}^{b}\theta_x, {}^{b}\theta_y, {}^{b}\theta_z)^T$ of ${}^{b}\mathbf{E}$ is taken as the error. Similar to the case of approaching, only $({}^{b}e_x, {}^{b}e_y, {}^{b}\theta_z)^T$ needs to be controlled. Then a position-based visual servoing law is adopted to align the robot body with the valve. The determination of ${}^{e}\mathbf{T}_{b^*}$ should ensure the robot just stand in front of the valve and the hand can reach the valve. The precision of the visual control for the alignment of the body does not need to be high. On the contrary, the movement of the robot can stop once the arm could reach and manipulate the valve.
The alignment of the arm with the valve is a typical position-based visual servoing process. Let:

$$ {}^{g}\mathbf{E} = {}^{g}\mathbf{T}_{g^*} = {}^{g}\mathbf{T}_{b} \cdot {}^{b}\mathbf{T}_{h} \cdot {}^{h}\mathbf{T}_{c} \cdot {}^{c}\mathbf{T}_{e} \cdot {}^{e}\mathbf{T}_{g^*} \tag{44} $$

where ${}^{e}\mathbf{T}_{g^*}$ is the desired pose of the end-effector relative to the valve, ${}^{c}\mathbf{T}_{e}$ is estimated with the stereo rig, ${}^{h}\mathbf{T}_{c}$ represents the head-eye geometry, ${}^{b}\mathbf{T}_{h}$ is the transformation from the head to the robot base, and ${}^{g}\mathbf{T}_{b}$ is the relative pose between the hand and the robot base frame. ${}^{b}\mathbf{T}_{h}$ and ${}^{g}\mathbf{T}_{b}$ can be obtained with the robot kinematics.
Since ${}^{g}\mathbf{E}$ is relevant to ${}^{c}\mathbf{T}_{e}$, ${}^{g}\mathbf{T}_{b}$, ${}^{b}\mathbf{T}_{h}$ and ${}^{h}\mathbf{T}_{c}$, it would be sensitive to noises and errors in robot kinematics, vision system calibration and pose estimation. It is hard to achieve a high accurate alignment, so this stage is called coarse alignment.
The image obtained from the camera mounted on the wrist is checked while the hand approaching the valve. The coarse alignment will stop if the pixel area of the green mark is large enough or the goal pose ${}^{e}\mathbf{T}_{g^*}$ has been reached.

5.4 Image based visual control for accurate manipulation

In this stage, the arm and the camera on the hand form an eye-in-hand system. an image-based visual servoing method is employed to accurately align the hand with the valve. Figure 8 (a) demonstrates the coordinate frames between the hand and the valve. Assume the valve is in front of the hand, and the orientations of these two frames are the same. The DOFs needed to be controlled for the hand are the translation and the rotation around the z-axis in the hand frame, denoted as $\mathbf{r} = (x, y, z, \theta_z)^T$.
With the green mark on the handle, the pixel coordinates of the center of the mark, the pixel area of the mark, and the angle between the principle axis of the mark and the horizontal direction on the image are taken as the image features, which is denoted as $\mathbf{f} = (u, v, s, \theta)^T$. Figure 8 (b) is the sketch of the image features. The desired image features, $\mathbf{f}^* = (u^*, v^*, s^*, \theta^*)^T$, are obtained off-line by means of teach-by-showing.

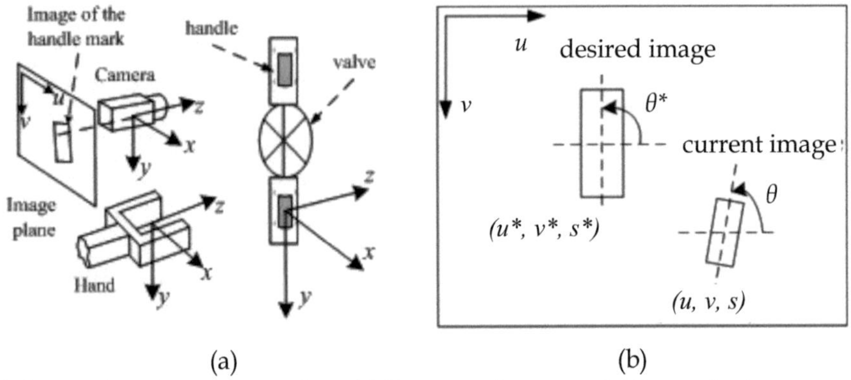

Figure 8. Image-based visual servoing, (a) Frame assignment, (b) Image feature

Assume the camera has been mounted on the wrist in such a manner that the orientation of the camera frame is almost the same as that of the hand frame, as shown in Fig. 8 (a). Then the image will change horizontally when the hand moves along the x-axis of the hand frame, and the image will change vertically when the hand moves along the y-axis. Then the image feature vector, **f**, can be taken as an estimation of **r**. The error is defined as follows:

$$\mathbf{e}(\mathbf{f}) = (u^* - u, v^* - v, (s / s^*) - 1, \theta^* - \theta)^T \tag{45}$$

Then a decoupled proportional control law can be designed:

$$(\Delta x, \Delta y, \Delta z, \Delta\theta_z)^T = -(K_u(u^* - u), K_v(v^* - v), K_z((s/s^*) - 1), K_\theta (\theta^* - \theta))^T \tag{46}$$

where K_u, K_v, K_z and K_θ are the proportional parameters of the controller.

5.5 Experiments and results

Valve operation experiments were conducted with the humanoid robot shown in Fig. 1. The size of the red rectangle mark was 100mm x 100mm. The two handles of the valve were attached with green marks. The head with two MINTRON 8055MK cameras is shown in Fig. 2. Two mini cameras were mounted on the wrists as shown in Fig. 7. The head-eye geometry was well calibrated with the method in section 3.

The size of the image obtained from the stereo rig is 768 x 576 pixels. The parameters of the stereo rig are shown in Table 5 and the displacement between the cameras is given in (47).

Item	Left Camera	Right Camera
K_u	4.2180e-007	3.4849e-007
K_v	3.6959e-007	3.6927e-007
K_x	1.0823e+003	1.2522e+003
K_y	1.0678e+003	1.2423e+003
u	324.6	335.0
v	248.3	292.0

Table 5. Parameters of the stereo rig

$$^{R}T_{L}\begin{bmatrix} 0.9998 & -0.0198 & -0.0063 & -93.4482 \\ 0.0195 & 0.9990 & -0.0403 & 1.4111 \\ 0.0071 & 0.0402 & 0.9991 & 125.7946 \end{bmatrix} \tag{47}$$

Firstly, the robot searched for the valve in the laboratory. When the valve was found, the approaching stage described in section 5.3 started. When the valve was within two meters from the robot, the coarse alignment began. The position of the mobile base of the robot was adjusted according to the pose of the valve until the hand could reach the valve. When the robot stopped moving, the position and orientation of the valve was measured by the stereo rig on the robot head. Table 6 shows the pose of the valve relative to the base frame of the robot, which is attached at the chest. In the process of approaching the valve, two arms were positioned so that they did not block the stereo rig to observe the valve. A pair of images obtained from the stereo rig at the end of the coarse alignment is shown in Fig. 9. It is shown that the hand was at the place near to the handle with an appropriate pose.
The hands of two arms would move to the two handles of the valve respectively. At the same time the cameras on the head were inactive, while the camera at each hand was in operation to observe the green mark on the valve handle. Then accurate alignment of the hand was started. The hands of robot were guided to the handle stably and accurately. Finally, a hybrid control method using force and position was employed to rotate the valve after the hands reached the handle and grasped it successfully.
In a series of experiments, the humanoid robot was able to autonomously find, approach and operate the valve successfully. The advantages of combining both eye-to-hand and eye-in-hand systems are clearly demonstrated.

n	o	a	p(mm)
-0.3723	-0.6062	-0.6795	-905.8
0.0279	0.7225	-0.6862	-36.7
0.9277	-0.3326	-0.2522	124.5

Table 6. Position and orientation of the valve

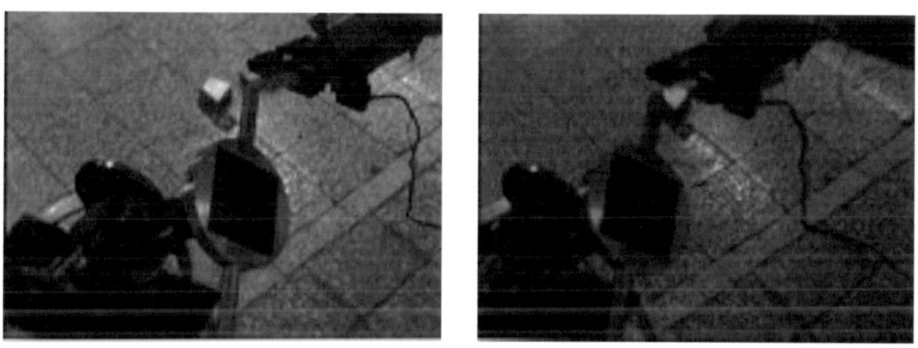

(a) (b)

Figure 9. Images obtained with the stereo rig on the robot head, (a) Left, (b) Right

6. Conclusion

Issues concerning with the approach-to-grasp movement of the humanoid robot are investigated in this chapter, including the calibration of the vision system, the visual measurement of rectangle objects and the visual control strategy for grasping.

A motion based method for head-eye calibration is proposed. The head-eye geometry is determined with a linear equation and then refined by a nonlinear least squares optimization to ensure the orthogonality of the rotation matrix.
A visual measurement algorithm is provided for the pose estimation of a rectangle object. Both the accuracy of the measurement and the robustness to noises in feature extraction are improved with the shape constraints of the rectangle object
A visual control strategy is presented in this chapter, which integrates different visual control method to fulfil the complicated approach-to-grasp task. The whole process of the grasping task is divided into several stages. The precision requirement of each stage is matched with an appropriate visual control method. Eye-to-hand and eye-in-hand architectures are integrated, at the same time, position-based and image-based visual control methods are combined to achieve the grasping.
A valve operating experiment with a humanoid robot was conducted to verify these methods. The results show that the robot can approach and grasp the handle of the valve automatically with the guidance of the vision system.
Vision is very important for humanoid robots. The approach-to-grasp movement is a basic but complex task for humanoid robots. With the guidance of the visual information, the grasping task can be accomplished. The errors on calibration of the vision system and the robot system will affect the accuracy of the visual measurement and the visual control. Improving the robustness of algorithms to these errors and noises should be the efforts in the future work. In addition, methods for stably identifying and tracking objects in an unstructured environment also need to be studied.

7. Acknowledgement

The authors would like to acknowledge the National High Technology Research and Development Program of China (grant No. 2006AA04Z213), the National Key Fundamental Research and Development Project of China (grant No. 2002CB312200) and the National Natural Science Foundation of China (grant No. 60672039) for the support to this work. The authors would also like to acknowledge the United Laboratory of Institute of Automation, Chinese Academy of Sciences (CASIA) and University of Science and Technology of China (USTC) for the support to this work.

8. References

Chaumette, F. & Hutchinson, S. (2006). Visual servo control Part I: Basic approaches. *IEEE Robotics and Automation Magazine*, Vol. 13, No. 4, pp. 82-90, ISSN: 1070-9932

Clarke, T. A. & Fryer, J. G. (1998). The development of camera calibration methods and models. *Photogrammetric Record*, Vol. 16, No. 91, pp. 51-66, ISSN: 0031-868X

Corke, P. I. & Hutchinson, S. A. (2001). A new partitioned approach to image-based visual servo control. *IEEE Transactions on Robotics and Automation*, Vol. 17, No. 4, pp. 507-515, ISSN: 1042-296X

Dodds, Z.; Jagersand, M; Hager, G. D & Toyama, K. (1999). A hierarchical vision architecture for robotic manipulation tasks, *Proceedings of First International Conference on Computer Vision System*, pp. 312-330, ISSN: 0302-9743, Las Palmas, Gran Canaria, Spain, Jan. 1999, Springer, Berlin, German

Faugeras, O. D. & Toscani, G. (1986). The calibration problem for stereo. *IEEE Computer Society Conference on Computer Vision and Pattern Recognition*, pp. 15-20, ISBN: 0-8186-0721-1, Miami Beach, USA, 1986, IEEE Press, New York, USA

Flandin, G.; Chaumette, F. & Marchand, E. (2000). Eye-in-hand/Eye-to-hand cooperation for visual servoing. *Proceedings of 2000 IEEE International Conference on Robotics and Automation,* pp. 2741-2746, ISBN: 0-7803-5886-4, San Francisco, California, USA, Apr. 2000, IEEE Press, New York, USA

Gans, N. R.; Corke, P. I & Hutchinson, S. A. (2002). Performance tests of partitioned approaches to visual servo control, *Proceedings of 2002 IEEE International Conference on Robotics and Automation*, pp. 1616-1623, ISBN: 0-7803-7272-7, Washington, DC, USA, May 2002, IEEE Press, New York, USA

Hager, G. D.; Chang, W. C. & Morse, A. S. (1995). Robot hand-eye coordination based on stereo vision. *IEEE Control Systems Magazine*, Vol. 15, No. 1, pp. 30-39, ISSN: 0272-1708

Han, M.; Lee, S.; Park, S. K. & Kim, M. (2002). A new landmark-based visual servoing with stereo camera for door opening, *Proceedings of International Conference on Control, Automation and Systems*, pp. 1892-1896, Muju Resort, Jeonbuk, Koreaa, Oct. 2002

Hartley, R. & Zisserman, A. (2004). *Multiple view geometry in computer vision (Second Edition)*, Cambridge University Press, ISBN: 0521540518, London, UK

Hashimoto, K.; Kimoto, T; Ebine, T. & Kimura, H. (1991). Manipulator control with image-based visual servo, *Proceedings of 1991 IEEE International Conference on Robotics and Automation,* pp. 2267-2272, ISBN: CH2969-4, Sacramento, California, USA, Apr. 1991, IEEE Press, New York, USA

Hauck, A.; Sorg, M.; Farber, G. & Schenk, T. (1999). What can be learned from human reach-to-grasp movements for the design of robotic hand-eye system?. *Proceedings of 1999 IEEE International Conference on Robotics and Automation,* pp. 2521-2526, ISBN: 0-7803-5180-0-5, Detroit, Michigan, USA, May 1999, IEEE Press, New York, USA

Hauck, A.; Passig, G. Schenk, T. Sorg, M. & Farber, G. (2000). On the performance of a biologically motivated visual control strategy for robotic hand-eye coordination, *Proceedings of 2000 IEEE/RSJ International Conference on Intelligent Robotics and Systems*, pp. 1626-1632, ISBN: 0-7803-6348-5, Takamatsu, Japan, Oct. 2000, IEEE Press, New York, USA

Heikkila, J. & Silven, O. (1997). A four-step camera calibration procedure with implicit image correction. *Proceedings of 1997 IEEE Computer Society Conference on Computer Vision and Pattern Recognition,* pp. 1106-1112, ISBN: 1063-6919, San Juan, Puerto Rico, Jun. 1997, IEEE Press, New York, USA

Horaud, R.; Dornaika, F. & Espiau, B. (1998). Visually guided object grasping. *IEEE Transactions on Robotics and Automation*, Vol. 14, No. 4, pp. 525-532, ISSN: 1042-296X

Hutchinson, S.; Hager, G. D. & Corke, P. I. (1996). A tutorial on visual servo control. *IEEE Transactions on Robotics and Automation*, Vol. 12, No. 5, pp. 651-670, ISSN: 1042-296X

Kragic, D.; Miller, A. T. & Allen, P. K. (2001). Real-time tracking meets online grasp planning, *Proceedings of IEEE International Conference on Robotics and Automation*, pp. 2460-2465, ISSN: 1050-4729, Seoul, Korea, May 2001, IEEE Press, New York, USA

Kragic, D. & Christensen, H. I. (2002). Survey on visual servoing for manipulation. *Technical report, Computational Vision and Active Perception Laboratory(CVAP),* Jan. 2002, CVAP259, ISRN KTH/NA/P--02/01--SE

Kragic, D. & Christensen, H. I. (2003). A framework for visual servoing. *Proceedings of International Conference on Computer Vision Systems*, pp. 345-354, ISSN: 0302-9743, Graz, Austria, Apr. 2003, Springer, Berlin, German

Ma, L.; Chen Y. Q. & Moore, K. L. (2003). Flexible camera calibration using a new analytical radial undistortion formula with application to mobile robot localization. *Proceedings of 20003 IEEE International Symposium on Intelligent Control*, pp. 799-804, ISBN: 0-7803-7891-1, Houston, Texas, USA, Oct. 2003, IEEE Press, New York, USA

Ma, S. D. (1996). A self-calibration technique for active vision system. *IEEE Transactions on Robotics and Automation*, Vol. 12, No. 1, pp. 114-120, ISSN: 1042-296X

Malis, E.; Chaumette, F. & Boudet, S. (1999). 2½D visual servoing. *IEEE Transactions on Robotics and Automation*, Vol, 15, No. 2, pp. 238-250, ISSN: 1042-296X

Murray, R. M.; Li, Z. X. & Sastry, S. S. (1993). *A mathematical introduction to robotic manipulation*, CRC Press, ISBN: 0-8493-7981-4, Boca Raton, Florida, USA

Tsai, R. Y. (1987). A versatile camera calibration technique for high-accuracy 3D machine vision metrology using off-the-shelf TV cameras and lenses. *IEEE Transactions on Robotics and Automation*, Vol. 3, No. 4, pp. 323-344, ISSN: 0882-4967

Tsai, R. Y. & Lenz, R. K. (1989). A new technique for fully autonomous and efficient 3D robotics hand/eye calibration. *IEEE Transactions on Robotics and Automation*, Vol. 5, No. 3, pp. 345-358, ISSN: 1042-296X

Xu, D.; Li, Y. F. & Tan, M. (2006a). Method for calibrating cameras with large lens distortion. *Optical Engineering*, Vol. 45, No. 4, pp. 043602-1-043602-8 , ISSN: 0091-3286

Xu, D.; Tan, M.; Jiang Z. M. & Hu, H. S. (2006b). Use of colour and shape constraints in vision-based valve operation by robot. *International Journal of Advanced Robotic Systems*, Vol. 3, No. 3, pp. 267-274, ISSN: 1729-8806

Zhang, Z. (2000). A flexible new technique for camera calibration. *IEEE Transactions on Pattern Analysis and Machine Intelligence*, Vol. 22, No. 11, pp. 1330-1334, ISSN: 0162-8828

24

Dexterous Humanoid Whole-Body Manipulation by Pivoting

Eiichi Yoshida[1], Vincent Hugel[2], Pierre Blazevic[2], Kazuhito Yokoi[1] and Kensuke Harada[3]
[1]*AIST/ISRI-CNRS/STIC Joint French-Japanese Robotics Laboratory (JRL), (AIST),*
[2]*Laboratoire d'Ingenierie des Systemes de Versailles,*
[3] *Humanoid Research Group, Intelligent Systems Research Institute, (AIST)*
[1, 3]*Japan,* [2]*France*

1. Introduction

Recent progress in research on humanoid robots is making them to complicated tasks, such as manipulation, navigation in dynamic environments, or serving tasks. One of promising application areas for humanoid robots include the manipulation task thanks to their high potential ability of executing of a variety of tasks by fully exploiting their high mobility and adaptability coming from its large number of degrees of freedom. Especially, manipulating bulky objects through a whole-body motion is suitable for them, which has been difficult for other types of robots.

This paper focuses on whole-body manipulation of large objects by a humanoid robot using a method called "pivoting" (Y. Aiyama, et al, 1993). This manipulation method has several advantages such as dexterity and stability over other methods like pushing or lifting. Moreover, the requirement of maintaining the equilibrium of the robot during the manipulation cannot be managed in the same way as in the case of the robot simply walking. To cope with those problems, an impedance control framework is first introduced to hold and manipulate the object, together with a whole-body balancing control. Next, a control framework called resolved momentum control (RMC) (S. Kajita et al., 2003) is adopted to allow the robot to step forward after manipulation by keeping the hand position with the object.

The next section presents overview of the subject of manipulation tasks. Section 3 addresses an algorithm to deal with the manipulation, followed by the description of control techniques included in the algorithm in Section 4. Section 5 gives simulation results using the dynamic simulator OpenHRP. In Section 6 the simulated results of manipulation are verified by hardware experiments using HRP-2 humanoid platform described before concluding the paper.

2. Pivoting as dexterous manipulation method

For manipulation of large objects that cannot be lifted we can make use of "non-prehensile manipulation" methods such as pushing (M. Mason, 1986, K. Lynch, 2002) or tumbling (A.

Bicchi, 2004). Aiyama et al. proposed a so-called "graspless manipulation" using pivoting motion (Y. Aiyama, et al, 1993), and also the analysis and planning have been reported by Maeda and Arai (Y. Maeda and T. Arai, 2003).
In another research field, whole-body manipulation tasks by humanoid robot have been investigated recently. The main manipulation methods developed so far are pushing (H. Yoshida et al., 2002, Y. Hwang et al., 2003, K. Harada et al., 2004, T. Takubo et al., 2004) and lifting (K. Harada et al, 2005).
We believe pivoting manipulation (Fig. 1), which we humans also often employ for example to move an oil drum or a barrel, is considered to be advantageous over those methods in for the following three reasons.

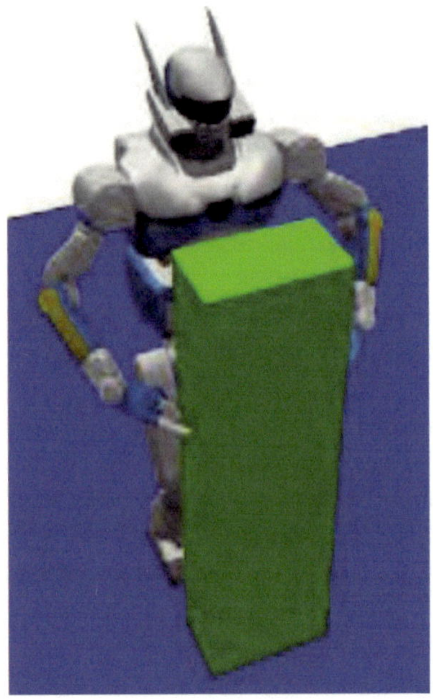

Figure 1. Pivoting manipulation by a whole-body motion of a humanoid robot

	Precise positioning	Adaptability	Stability
Pushing	×	×	○
Lifting	○	▲	×
Tumbling	▲	▲	○
Pivoting	○	○	○

○ : Suitable, ▲: Average, × : Not suitable

Table 1. Various manipulation methods compared for moving a large object by a humanoid

First, pivoting is expected to be precise in positioning. Since the object is not held in pushing, uncertainty is more significant than pivoting and tumbling has also limitation in reachability. Although some of those shortcomings can be overcome using sophisticated control methods for pushing (K. Lynch, 1992, A. Bicchi, 2004) pivoting allows the robot to move the object to the desired position in a more straightforward manner.

The second advantage is the adaptability to various terrain conditions. Using pushing, it is difficult to be used in rough terrains whereas pivoting can be easily adapted. Lifting and tumbling can be used, but pivoting is more advantageous in terms of flexibility of manipulation and variety in shape and weight of manipulated objects.

Lastly, we can expect more stable manipulation through pivoting. It has advantage of lower risk of falling over lifting when manipulating large or heavy objects. Moreover, the manipulated object can help widening stability margin in some cases.

This comparison is summarized in Table 1. However, so far pivoting manipulation has not fully been exploited for humanoid robots due to the difficulty in coordinating manipulating motion with body balancing at whole-body level.

Therefore, the goal of this paper is to establish a control method for pivoting manipulation by a whole-body motion of humanoid robot. The contribution of this paper is two-fold: the algorithm for executing pivoting manipulation for a humanoid robot and integration of control techniques for this manipulation. For the first point, since pivoting manipulation that has not been implemented using humanoid platform so far, the challenge in this paper provides new potential application fields that need dexterous manipulation by humanoid robots. The second contribution addresses integration of such control techniques as impedance control for contact force, body balancing during manipulation and RMC for stepping motion for the purpose of whole-body manipulation. In the following sections, pivoting manipulation by a humanoid robot is presented based on those techniques.

3. Pivoting Manipulation Algorithm

The manipulation is executed by repeating the following two phases, the manipulation and robot motion phases.

In the former phase, the manipulation is done in a quasi-static way by repeating rotation of the object about an axis. The contact force to hold the object is controlled using impedance control. The manipulation is performed through whole-body motion coordination to achieve both manipulation and stable body balancing. The detailed descriptions are given in sections 4.1, 4.2 and 4.3.

In the latter phase, the robot itself moves to advance with the hands at the same position to continue manipulation. The body motion is planned through resolved momentum control (RMC) as detailed in section 4.4.

The algorithm of the pivoting manipulation task is summarized in Fig. 2. The detailed manipulation sequence is illustrated in Fig. 3 where the dotted and solid lines in each figure denote the states before and after the manipulation respectively.

a. The object is inclined by angle α around an axis a that includes a vertex v on the plane so that the object has a point contact with the plane at v (Fig. 3a).
b. The object is rotated by angle β around the vertical axis z on vertex v to move to the desired direction (Fig. 3b). As a result, the axis a is rotated to another axis $\boldsymbol{a}'$ on the plane.
c. The object is rotated by $-\alpha$ around the rotated axis $\boldsymbol{a}'$ to reposition the front bottom edge of the object touches ground. The displacement of the object is defined as the distance between the projected points of its center of mass before and after the manipulation. (Fig. 3c).

1. If the displacement of the object exceeds a value *D*, the robot steps toward the desired direction.
2. Otherwise continue to step 1 or terminate the motion.

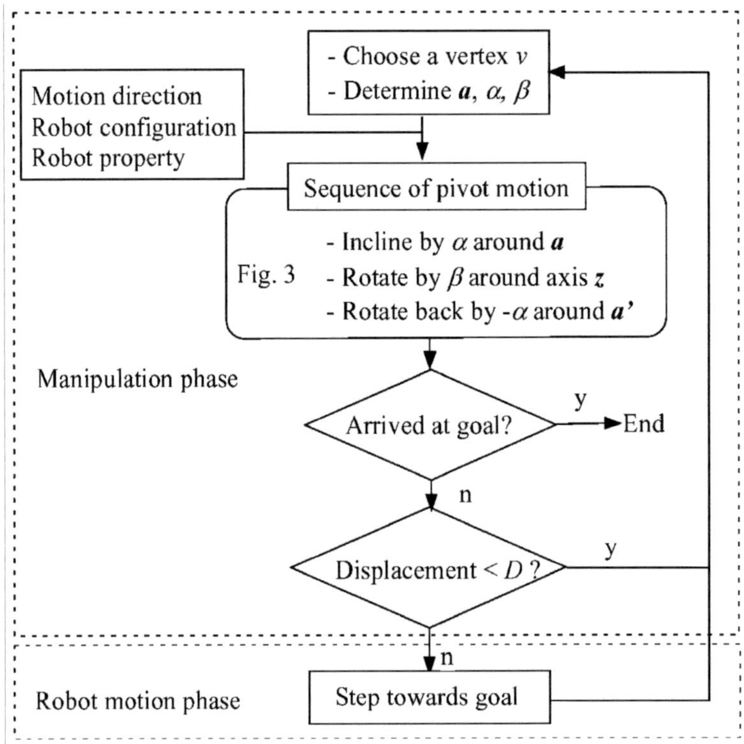

Figure 2. Flow of pivoting manipulation

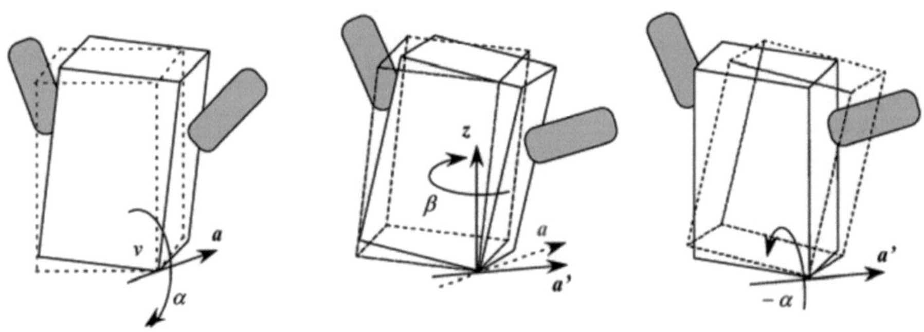

(a) Inclining (b) Rotating around z-axis (c) To initial contact

Figure 3. Sequence of pivot motion of a box object composed of three phases, (a) inclination, (b) horizontal rotation and (c) repositioning

According to the desired trajectory, the parameters *a*, *β* and *D* must be designed so that the manipulated object can follow the desired trajectory (Fig. 4). The working area and physical properties of the robot body must also be taken into account. The *a* axis and the vertex around which to incline the object are selected not to lose the stability of the robot and the object. In our case, the closer vertex to the robot is selected since smaller torque is required

for the movement, and the direction of axis a is chosen so that the robot does not make a large forward motion.
A vertical axis to the horizon is chosen as axis z since no work is required for quasi-static motion around this axis. The angle β depends on the constraint of robot motion capability as well as the desired trajectory of the robot. The hand trajectories can be calculated geometrically from the pivoting motion and contact points of the hand on the object. To execute more complicated trajectory than that of Fig. 4, we will further need a planning strategy to determine those parameters, taking account of combination of stepping motion. The next section details the manipulation control and robot motion for the pivoting task.

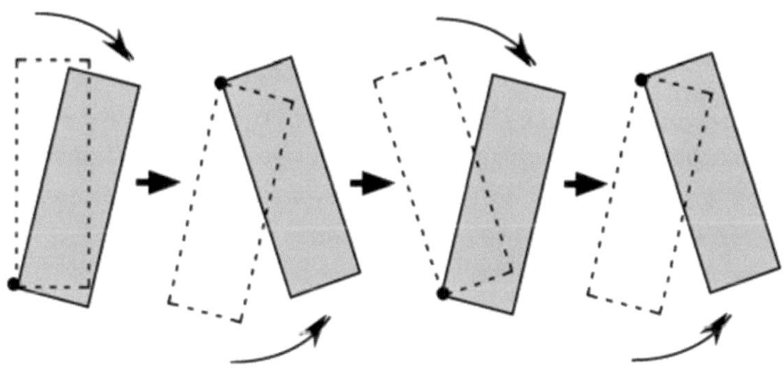

Figure 4. Sequence of pivoting manipulation to transport an object

4. Control Techniques for Manipulation and Robot Motion

In this section we describe how the manipulation task is controlled by a humanoid robot. In the manipulation control phase, the robot is supposed to grasp the object by two hands firmly (Fig. 1) without slipping. First, we introduce impedance control to control the contact force to hold the object. Then, a method of body balancing is adopted to maintain the center of mass (CoM) in the supporting phase during manipulation. We assume that the manipulation is done for a rectangular box-shaped object in a quasi-static way by repeating the rotations on a plane as described in Fig. 3.

4.1 Object holding

Since we assume quasi-static motion, we adopt position control for robot hands to achieve the trajectory for the desired motion. For position-controlled robot as HRP-2 used for simulations and experiments, the output of the following impedance control is added to the position command of manipulation to regulate the contact force (K. Harada et al, 2004). The robot hands are placed on the two sides of the object so that they can apply forces in the opposite directions. Let x_H be the hand position which is controlled in the direction perpendicular to the object face, and let f_x be the force exerted to the hand. The impedance control law to lead the force f_x to the desired force f_{xd} is given as

$$m\ddot{x}_H + c\dot{x}_H = f_{xd} - f_x \tag{1}$$

where m and c are the mass and damper parameters. This control scheme is illustrated in Fig. 5. For grasping, one of the hands is controlled by this law and the other is controlled in position to avoid unnecessary oscillation.

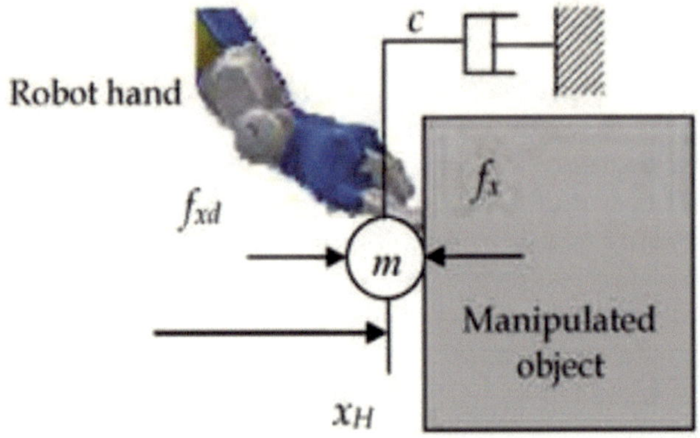

Figure 5. Impedance control for grasping based on force. To achieve the desired force f_{xd}, the output force f_x is computed by using impedance using virtual mass and damper parameters m and c

Since the humanoid robot HRP-2 we are using is based on position control, the control law is discretized in the simulation and experiment implementation as follows.

$$x_H(t+\Delta t) - x_H(t) = \frac{\Delta t^2}{m}\left(f_{xd} - f_x(t) - c\frac{x_H(t) - x_H(t-\Delta t)}{\Delta t} \right) + x_H(t) - x_H(t-\Delta t) \tag{2}$$

where Δt denotes the sampling period of robot control.

4.2 Pivoting Manipulation

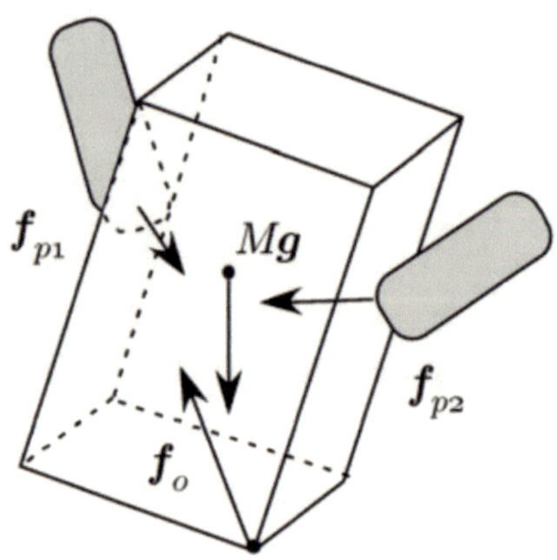

Figure 6. Forces for pivot motion using two contact points

We assume that the robot holds the object rigidly by two contact points of its hands as illustrated in Fig. 6. Let p_{pi}, $(i=1, 2)$ be the vector of position of each hand, p be the position of the center of gravity and f_{pi} be the force applied to each grasping point. Also, let Mg and f_0 denote the gravity vector and resistance force at the contact point at the ground respectively. Then the object is in static equilibrium when the following conditions are satisfied.

$$f_o + \sum_{i=2}^{2} f_{pi} + Mg = o \tag{3}$$

$$\sum_{i=2}^{2} p_{pi} \times f_{pi} + p_g \times Mg = o \tag{4}$$

We assume that the manipulation is executed quasi-statically and that no slipping occurs at the contact points. We also assume that the geometry and location of the object are known. The object is manipulated by moving each hand under the above condition. The contact forces are controlled by impedance control (1) by giving the desired contact force f_{xd}.

4.3 Whole-body Balancing

The humanoid robot should not only apply the necessary force to the hand for manipulation but also keep the balance of the whole body throughout the manipulation task.
We adopt the method developed by Harada et al. (K. Harada et al, 2005). The key idea is to control the waist position in order that the "static balancing point" is on the center of the foot supporting polygon. The static balancing point $p_s=[x_s, y_s, 0]^T$ is the point on the floor to which all the resistance force from both hands and gravity are applied, which is equivalent to ZMP when only quasi-static motion is generated. The relationship between the contact force and static balancing point is described in the following equations.

$$x_s - \bar{x} = -\sum_{i=1}^{2} \frac{z_{pi} f_{pi}^{x} + (x_{pi} - x_s) f_{pi}^{z}}{M_R g} \tag{5}$$

$$y_s - \bar{y} = -\sum_{i=1}^{2} \frac{z_{pi} f_{pi}^{y} + (y_{pi} - y_s) f_{pi}^{z}}{M_R g} \tag{6}$$

where $f_{pi} = [f_{pi}^{x}, \; f_{pi}^{y} \; f_{pi}^{z}]^T (I = 1,2)$, $p_{pi} = [x_{pi}, y_{pi}, z_{pi}]^T$ and $g = [0,0,-g]^T$. The position of the CoM without external force is denoted by $\bar{p} = [\bar{x}, \bar{y}, \bar{z}]^T$ here and M_R is the total mass of the robot. Note that the sign in right-hand side of (5) and (6) is negative since the reaction force at each hand is $-f_{pi}$.

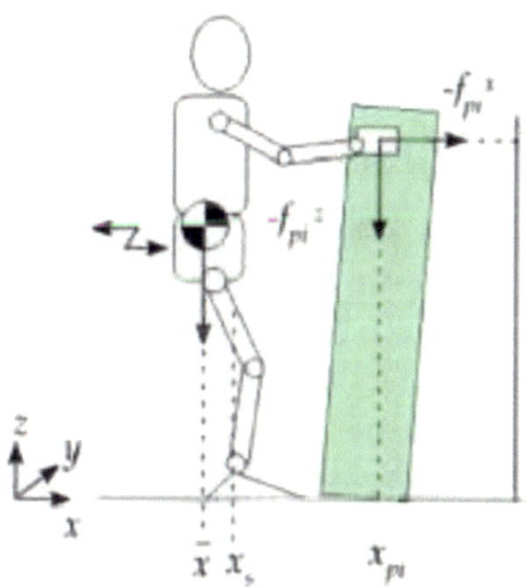

Figure 7. Body-balancing by moving waist position. According to the contact forces at the hands, the humanoid robot adjusts its waist position to maintain the "static balancing point" at the center of the support polygon

The balancing control is performed by moving the position of the waist so that the position of the static balancing point $\boldsymbol{p}_s$ is at a desired position inside the convex hull of supporting area(s) (Fig. 7). To ensure maximum stability margin, we usually employ the center of the support polygon as desired position of $\bar{\boldsymbol{p}}$. The movement of the waist position can be approximated by the CoM position !!!p from the desired position $\boldsymbol{p}_d$ of $\boldsymbol{p}_s$ (K. Harada et al, 2005).
In this way, the force of pivoting manipulation is generated as a resulting compensation of disturbance due to the contact with the object.

4.4 Stepping motion

Now the robot itself needs to move in the desired direction of object transportation after a sequence of manipulation that displaces the object to a more distant position from the robot. For this robot motion, it is preferable that robot keep the hands on the object to easily go on the manipulation.
We here introduce RMC that is a control framework where a humanoid robot is controlled such that the resulted total linear/angular momenta reach specified values (S. Kajita, 2003). It is well known that, for a humanoid robot to walk in a stable manner, the ZMP (Zero Moment Point) must be within the convex hull of the supporting area(s). By using RMC, we can control the CoM of the robot so that it is always within the convex hull of the supporting area(s) to maintain robot balance. Since the linear momentum $\boldsymbol{P}$ depends on the time derivative of CoM position $\boldsymbol{p}_g$ through the total mass M_R as $\boldsymbol{P} = M_R\dot{\boldsymbol{p}}_g$, the position of CoM can be controlled by manipulating the linear momentum as $\boldsymbol{P} = kM_R(\tilde{\boldsymbol{p}}_g - \boldsymbol{p}_g)$, where the tilde denotes the reference value, and k is the gain of the control scheme. Using this equation we are can calculate the desired linear momentum P to control the robot CoM.
The hand position can be controlled based on an extended method of RMC developed by Neo et al. (N. E. Sian, 2003) by keeping the position of CoM inside the supporting area. In this way, the robot can step towards the object with its both hands at the position keeping the contact, while the CoM is controlled always inside the convex hull of supporting area(s). Moreover, keeping the hand position on the object may help to maintain the equilibrium of the robot body.

5. Simulation results

To validate the proposed pivoting manipulation method, simulations using the humanoid robot HRP-2 (K. Kaneko et al., 2004) have been conducted using the humanoid robot simulator OpenHRP (F. Kanehiro et al., 2001). The humanoid robot HRP-2 has 30 degrees of freedom with 1.54 [m] in height and 58 [kg] in weight, with wrists and ankles equipped with force sensors. Since the input of force sensor is noisy, we use an average value of the last 20 samplings measured every 5 [ms]. The width, depth, and height of the object are 0.35[m], 0.2[m], and 1.2[m] respectively, with the mass of 3[kg].
In Table 2, the parameters of the pivoting sequence are shown. The directions of axes x, y and z are given in Fig. 8(a). Here parameter D of object displacement is set to 50[mm] to prevent the arms from going into singular stretched configuration and also from losing stability by extending arm too much. In the impedance control we adopt m = 10[kg], c = 300[N m^{-1} s] and f_{xd}= 25[N] respectively. The contact force f_{xd} is determined for the robot to hold firmly enough to prevent the object from slipping because of gravity. The z position of

contact point is set to 0.62[m] height and at the midpoint of the depth of the object in x direction.

In the simulation, first the robot grasps the object rigidly at the tip of the hands and keeps the hands at the original position until the contact force by impedance control stabilizes. The hand trajectories during the manipulation are computed geometrically by using the contact points of both hands and the motions described in Table 2. Position control is performed to follow the calculated hand trajectories and the impedance control law (1) is applied to one of the hands, in our case the right hand, to maintain grasping during the manipulation while as the other hand is position-controlled.

Fig. 8 shows the snapshots of the simulation. As can be seen, the humanoid robot successfully displaces the object by pivoting.

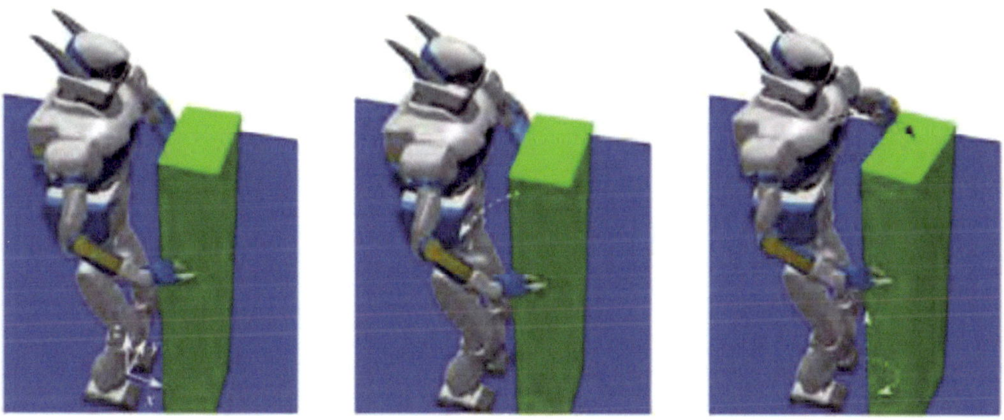

(a) Initial state (b) Step 1: inclining rightward (c) Step 3: rotating CW

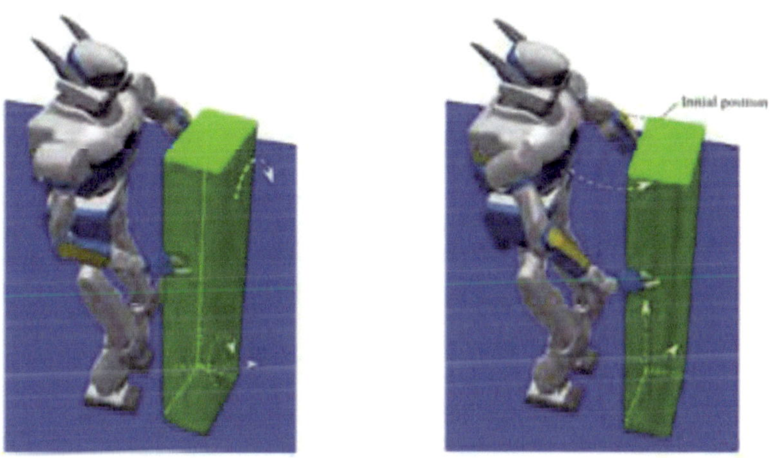

(d) Step 4: inclining leftward (e) Step 5: rotating CCW

Figure 8. Simulation results of pivoting manipulation. Starting from the initial state (a), first the robot inclines the object (b) to support it on a vertex and then turns it horizontally clockwise (c). After repositioning in a horizontal position, the object is inclined on the support of another vertex (d) and turned counter-clockwise to advance the object in a parallel position with the initial one

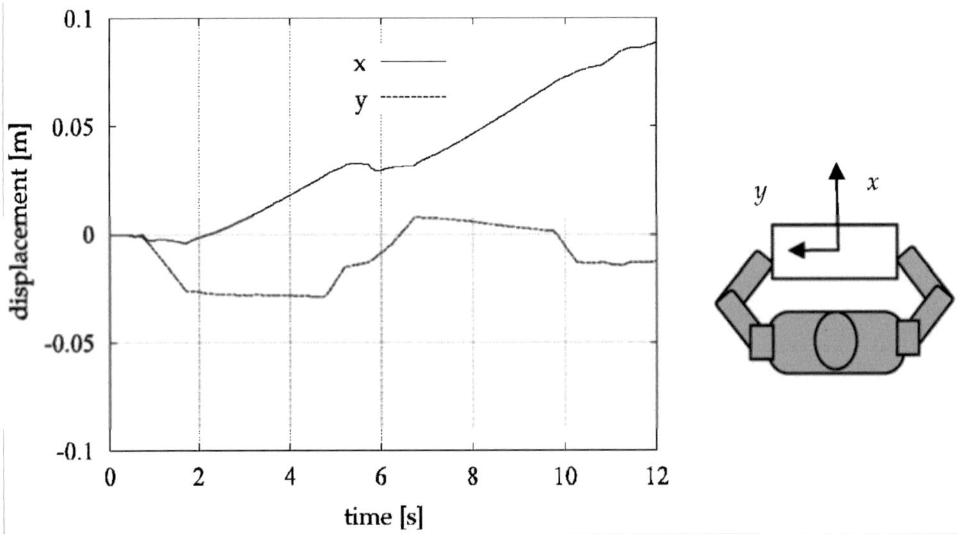

Figure 9. Simulation result of displacement of the center of mass of the manipulated object in x and y directions shown at right-hand side

Step	Type (in Fig. 3)	Axis*	Angle [deg]	Time [s]	Contact vertex **	Description**
1	(a)	$\boldsymbol{a}_1$= (3, -1, 0)	α = 5.0	1.0	NR	Inclining rightward
2	(b)	z= (0, 0, 1)	β =-15.0	3.0	NR	Rotating CW
3	(c)	$\boldsymbol{a}'_1$= (3, -1, 0)***	α = -5.0	1.0	NR	Inclining back
4	(a)	$\boldsymbol{a}_2$= (3, 1, 0)	α = -5.0	1.0	NL	Inclining leftward
5	(b)	z= (0, 0, 1)	β = 15.0	3.0	NL	Rotating CCW
6	(c)	$\boldsymbol{a}'_2$= (3, 1, 0)***	α = 5.0	1.0	NL	Inclining back

* See Fig. 8a for the reference frame. The angle between the axis x and $\boldsymbol{a}_1$, $\boldsymbol{a}_2$ is 18.4 [deg] and -18.4[deg] respectively.
** NR: Near-Right, NL: Near-Left, CW: Clockwise, CCW: Counter-clockwise
*** Axis $\boldsymbol{a}'$ is rotated vector a around z by β.

Table 2. Parameters of pivot motion

In the simulation, we measure the displacement of the manipulated object, the contact forces, and center of the mass of the robot respectively to evaluate the performance of the manipulation itself, impedance control, and the balance control.
The displacement of the center of mass of the manipulated object projected on the ground is shown in Fig. 9. Since the object is first inclined around the axis a before rotating around z

axis, the x displacement reduces first and then increases. This occurs twice and we can see the object is displaced by 0.073 [m] in x direction.

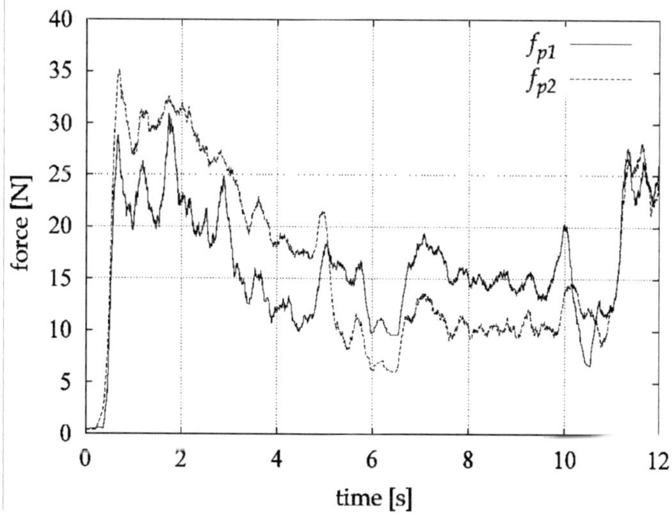

Figure 10 Contact forces of each hand measured from the force sensors during simulation

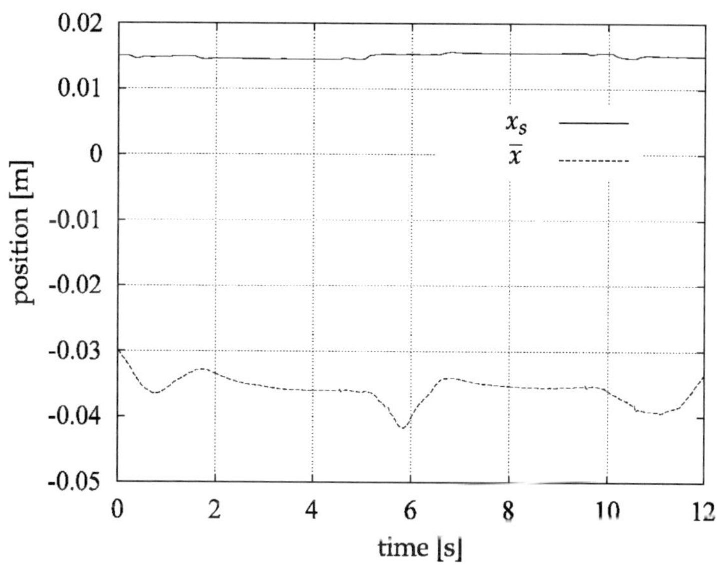

Figure 11. Simulation result of the x component of static balancing point (p_s) and CoM position (p) during the manipulation. The static balancing point is maintained during simulation

Fig. 10 shows the simulation results of the contact forces f_{p1}, f_{p2} for right and left hands respectively. The robot hands are approaching to the object at the dose positions at t=0 [sec]. As shown in the figure, the contact occurs at around f=0.8 [sec]. Then both the contact forces stabilize around the desired force f_{xd}=25[N] by t=1.0 [sec]. The manipulation starts at this moment and lasts for 10[sec] as indicated in Table 1. We can observe that the contact force decreased during the manipulation. It may be because the arms go to a stretched configuration where it is difficult to generate the force in the desired direction during manipulation especially for the position controlled arm. However, the forces regain the desired value after the manipulation is finished at t=11 [sec]. From this simulation result of measured contact force, it is verified that the usage of impedance control for contact force is sufficiently effective.

The x coordinates of static balancing point $\boldsymbol{p}_s$ and CoM position $\overline{\boldsymbol{p}}$ without external force, x_s and $\overline{x}$, plotted in Fig. 11. Here $\overline{\boldsymbol{p}}$ represents an approximated movement of the waist position. Without balance controlling, x_s would keep increasing as the robot reaches out the arms and force is applied to the hand during the manipulation until the robot falls down. In the graph, the static balancing point is regulated by moving the waist position so that it coincides with the center of the feet (shift along x-axis of 0.015[m]) according to the balancing control.

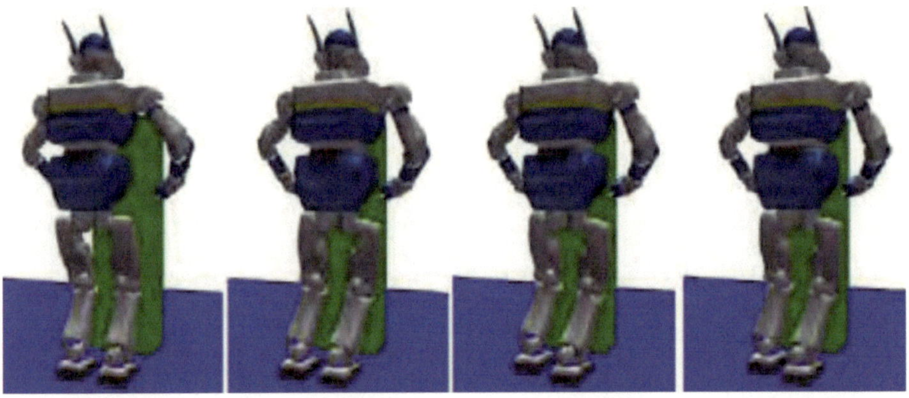

(a) Initial position | (b) Moving CoM rightward | (c) Stepping left leg | (d) Left leg landed

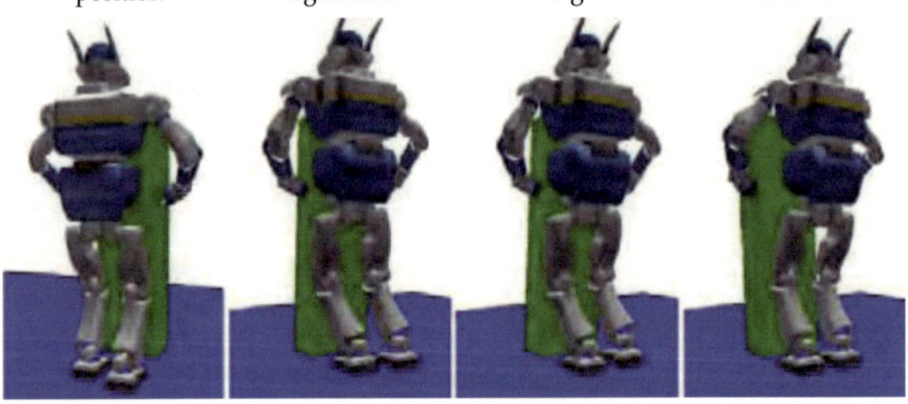

(e) Moving CoM center | (f) Moving CoM leftward | (g) Stepping right leg | (h) Stepping finished

Figure 12. Stepping forward keeping the hands on the object using RMC. First making a step by supporting on left foot (a) - (d), then another step on right leg (e) - (h) by keeping the CoM inside the support polygon

The robot motion phase is simulated based on the proposed method. Fig. 12 shows the sequence of stepping for forward motion. After the manipulation, the robot steps forward by 50[mm] by moving its feet alternatively with the hands fixed on the object, using RMC to maintain the whole body balance. As shown in Fig. 12, first robot moves its CoM on the right foot and then moves the left foot forward. The same sequence is repeated for the right foot. The simulation shows that robot can effectively moves towards the desired direction of manipulation.

6. Experimental Results

We have conducted an experiment of the manipulation control part in the same condition as in simulation using HRP-2 humanoid hardware platform. Thanks to the architecture of OpenHRP with binary compatibility with the robot hardware, the developed simulation software can be directly utilized in hardware without modification.

Fig. 13 shows snapshots of the experiments using a box of the same size and weight as in simulation. As can be seen, the pivoting manipulation has been executed appropriately and the displacement in x direction was around 0.06[m] as expected from simulation.

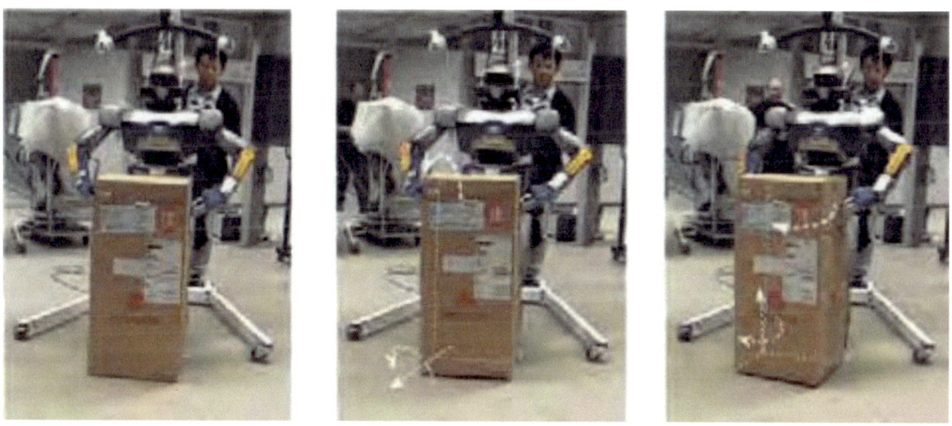

(a) Initial state (b) Step 1: inclining rightward (c) Step 3: rotating CW

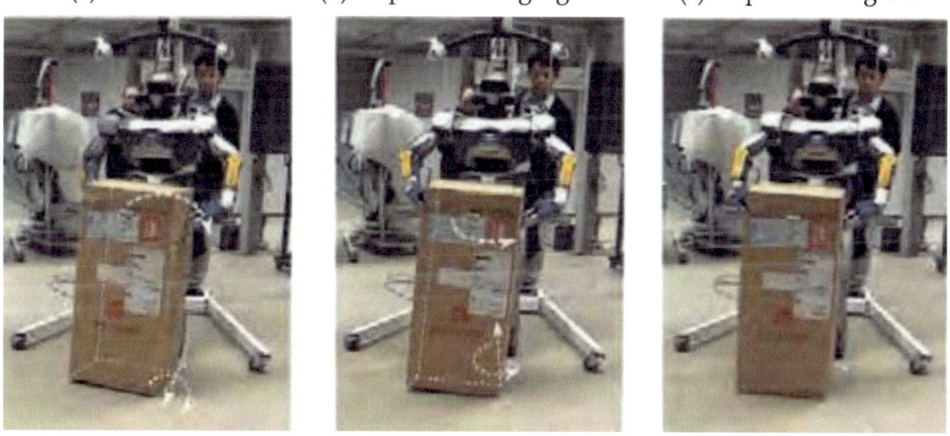

(d) Step 4: inclining leftward (e) Step 5: rotating CCW (f) Final state

Figure 13. Experiment of pivoting motion. Starting from the initial position (a), first the object is inclined (b) to rotate clockwise horizontally (c). Then the humanoid robot inclines the object on the other vertex (d) to rotate counter-clockwise (e) to finish the manipulation (f)

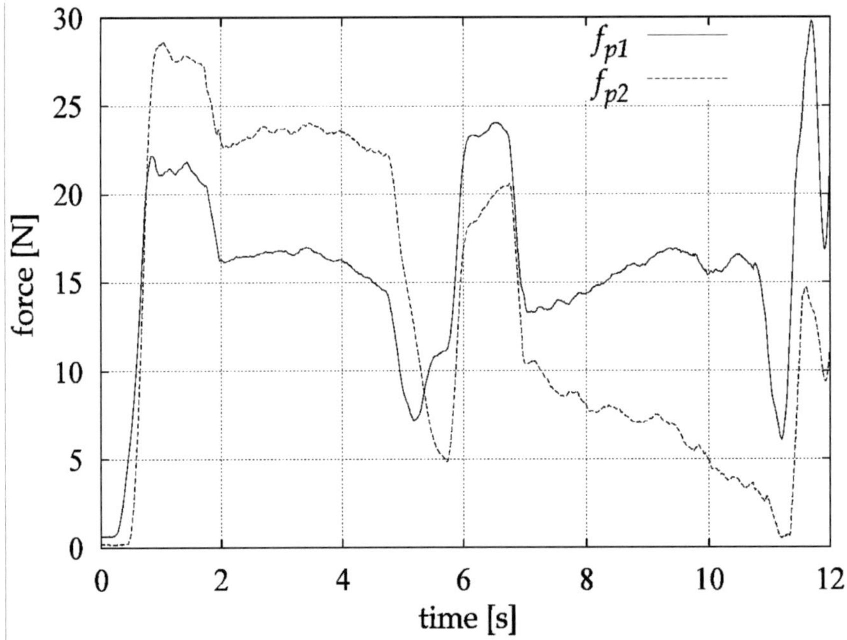

Figure 14. Experimental result of contact forces of each hand. The grasping start at t=1[sec] and finish in 10 seconds. Enough force for the manipulation is maintained but

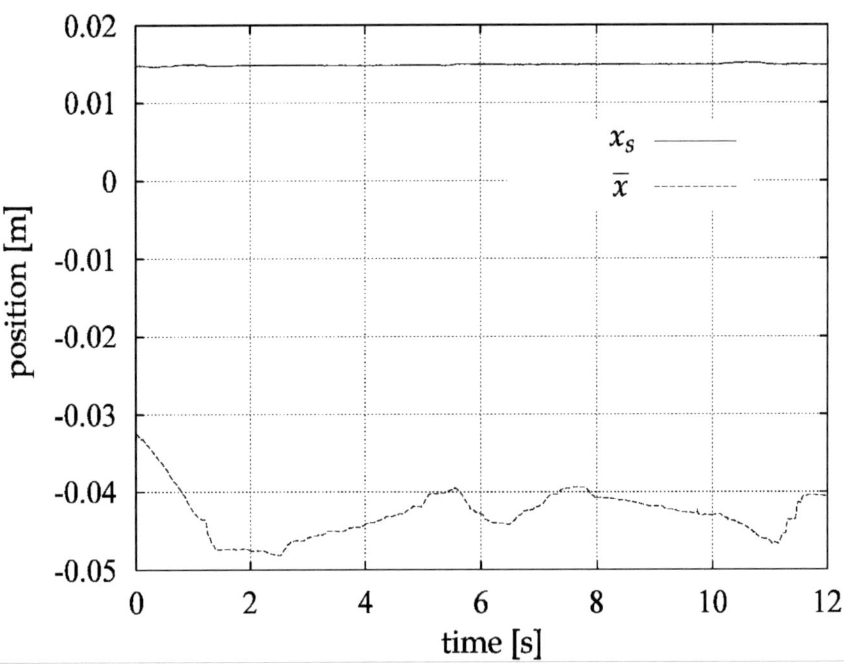

Figure 15. Experimental result of static balancing point (x_s) and CoM position !!!(x). The static balancing point is maintained near the center of the support polygon (x=0) by changing the waist position

The experimental result of contact forces measured from wrist force sensors is shown in Fig. 14. Although the measured force shows similar characteristics with the simulation, one of the forces drops drastically from desired force of 25 [N] at the end of manipulation even though it was enough to execute the manipulation task. This is considered to come from the stretched configuration of arm that makes it difficult to generate desired force to hold the object. The manipulation experiment was successful; however, the control of arm configuration and grasping position need to be investigated for more reliable manipulation. The experimental result of the static balancing point and CoM position plotted in Fig. 15 shows the effectiveness of balance control to keep the static balancing point in stable position during the manipulation.

To conclude the experimental results, we could verify the validity of proposed pivoting manipulation based on whole-body motion.

7. Conclusions

In this paper a pivoting manipulation method has been presented to realize dexterous manipulation that enables precise displacement of heavy or bulky objects. Through this application, we believe that the application area of humanoid robot can be significantly extended.

A sequence of pivoting motion composed of two phases has been proposed, manipulation control and robot stepping motion. In the former phase, an impedance control and balancing control framework was introduced to control the required contact force for grasping and to maintain stability during manipulation respectively. Resolved momentum control is adopted for stepping motion in the latter phase.

We then showed a sequence of pivoting motion to transport the objects towards the desired direction. We have shown that the proposed pivoting manipulation can be effectively performed by computer simulation and experiments using a humanoid robot platform HRP-2.

As a future work, the method will be improved to adapt to various object shapes of transportation in pursuit of wide application in future developments. One of other extension is the manipulation planning for more general trajectories with experimentation of both manipulation and stepping phases. Integration of identification of physical properties of the objects or environments (Yu et al., 1999, Debus et al., 2000) is also an important issue to improve the robot's dexterity.

8. References

Y. Aiyama, M. Inaba, and H. Inoue (1993). Pivoting: A new method of graspless manipulation of object by robot fingers, *Proc. of the IEEE/RSJ Int. Conf. on Intelligent Robots and Systems*, 136 -143,1993.

S. Kajita, F. Kanchiro, K. Kaneko, K. Fujiwara, K. Harada, K. Yokoi, and H. Hirukawa (2003). Resolved momentum control: Humanoid motion planning based on the linear and angular momentum, *Proc. of IEEE/RSJ Int. Conf. on Intelligent Robots and Systems*, 1644-1650, 2003.

M. Mason (1986). Mechanics and Planning of Manipulator Pushing Operation, *Int. J. Robotics Research*, 5-3,53-71,1986.

K. Lynch (1992). The Mechanics of Fine Manipulation by Pushing, *Proc. IEEE Int. Conf. on Robotics and Automation*, 2269-2276,1992.

A. Bicchi, Y. Chitour, and A. Marigo (2004). Reachability and steering of rolling polyhedra: a case study in discrete nonholonomy, *IEEE Trans, on Automatic Control*, 49-5, 710-726,2004.

Y. Maeda and T. Arai (2003). Automatic Determination of Finger Control Modes for *Graspless Manipulation, Proc. of IEEE/RSJ Int. Conf. on Intelligent Robots and Systems*, 2660-2665,2003.

H. Yoshida, K. Inoue, T. Arai, and Y. Mae (2002). Mobile manipulation of humanoid robots-optimal posture for generating large force based on statics, *Proc. of IEEE Int. Conf. on Robotics and Automation*, 2271 -2276,2002.

Y. Hwang, A. Konno and M. Uchiyama (2003). Whole body cooperative tasks and static stability evaluations for a humanoid robot, *Proc. of IEEE/RSJ Int. Conf. on Intelligent Robots and Systems*, 1901 -1906,2003.

K. Harada, S. Kajita, F. Kanehiro, K. Fujiwara, K. Kaneko, K. Yokoi, and H. Hirukawa (2004). Real-Time Planning of Humanoid Robot's Gait for Force Controlled Manipulation, *Proc. of IEEE Int. Conf. on Robotics and Automation*, 616-622,2004.

T. Takubo, K. Inoue, K. Sakata, Y. Mae, and T. Arai. Mobile Manipulation of Humanoid Robots Control Method for CoM Position with External Force -, *Proc. of IEEE/RSJ Int. Conf. on Intelligent Robots and Systems*, 1180-1185,2004.

K. Harada, S. Kajita, H. Saito, M. Morisawa, F. Kanehiro, K. Fujiwara, K. Kaneko, and H. Hirukawa (2005). A Humanoid robot carrying a heavy object, *Proc. of IEEE Int. Conf. on Robotics and Automation*, 1724-1729,2005.

N. E. Sian, K. Yokoi, S. Kajita, and K. Tanie (2003). Whole Body Teleoperation of a Humanoid Robot Integrating Operator's Intention and Robot's Autonomy - An Experimental Verification -, *Proc. of IEEE/RSJ Int. Conf. on Intelligent Robots and Systems*, 1651-1656,2003.

F. Kanehiro, N. Miyata, S. Kajita, K. Fujiwara, H. Hirukawa, Y. Nakamura, K. Yamane, I. Kohara, Y. Kawamura and Y. Sankai (2001). Virtual humanoid robot platform to develop, *Proc. of IEEE/RSJ Int. Conf. on Intelligent Robots and Systems*, 1093 -1099, 2001.

K. Kaneko, F. Kanehiro, S. Kajita, H. Hirukawa, T. Kawasaki, M. Hirata, K. Akachi and T. Isozumi (2004). The Humanoid Robot HRP-2, *Proc. of IEEE/RSJ Int. Conf. on Robotics and Automation*, 10831090, 2004.

Y. Yu, K. Fukuda, and S. Tsujio (1999). Estimation of Mass and Center of Mass of Graspless and Shape-Unknown Object, *Proc. of IEEE Int. Conf. on Robotics and Automation*, 2893-2898, 1999.

T. Debus, P. Dupont and R. Howe: Automatic Identification of Local Geometric Properties During Teleoperation, *Proc. of IEEE International Conference on Robotics and Automation*, 3428 -3434, 2000.

25

Imitation Learning Based Talking Heads in Humanoid Robotics

Enzo Mumolo and Massimiliano Nolich
DEEI, Universitá degli Studi di Trieste
Italy

1. Introduction

The main goal of this Chapter is to describe a novel approach for the control of Talking Heads in Humanoid Robotics.
In a preliminary section we will discuss the state of the art of the research in this area. In the following sections we will describe our research results while in the final part some experimental results of our approach are reported. With the goal of controlling talking heads in mind, we have developed an algorithm which extracts articulatory features from human voice. In fact, there is a strong structural linkage between articulators and facial movements during human vocalization; for a robotic talking head to have human-like behavior, this linkage should be emulated. Exploiting the structural linkage, we used the estimated articulatory features to control the facial movements of a talking head. Moreover, the articulatory estimate is used to generate artificial speech which is - by construction - synchronized with the facial movements.
Hence, the algorithm we describe aims at estimating the articulatory features from a spoken sentence using a novel computational model of human vocalization. Our articulatory features estimator uses a set of fuzzy rules and genetic optimization. That is, the places of articulation are considered as fuzzy sets whose degrees of membership are the values of the articulatory features. The fuzzy rules represent the relationships between places of articulation and speech acoustic parameters, and the genetic algorithm estimates the degrees of membership of the places of articulation according to an optimization criteria. Through the analysis of large amounts of natural speech, the algorithm has been used to learn the average places of articulation of all phonemes of a given speaker.
This Chapter is based upon the work described in [1]. Instead of using known HMM based algorithms for extracting articulatory features, we developed a novel algorithm as an attempt to implement a model of human language acquisition in a robotic brain. Human infants, in fact, acquire language by imitation from their care-givers. Our algorithm is based on imitation learning as well.
Nowadays, there is an increasing interest in service robotics. A service robot is a complex system which performs useful services with a certain degree of autonomy. Its intelligence emerges from the interaction between data gathered from the sensors and the management algorithms. The sensorial subsystem furnishes environment information useful for motion tasks (dead reckoning), auto-localization and obstacle avoidance in order to introduce

reactiveness and autonomy. Humanoid robotics has been introduced for enabling a robot to give better services. A humanoid, in fact, is a robot designed to work with humans as well as for them. It would be easier for a humanoid robot to interact with human beings because it is designed for that. Inevitably, humanoid robots tend to imitate somehow the form and the mechanical functions of the human body in order to emulate some simple aspects of the physical (i.e. movement), cognitive (i.e. understanding) and social (i.e. communication, language production) capabilities of the human beings. A very important area in humanoid robotics is the interaction with human beings, as reported in [2]. Reference [2] describes the Cog project at MIT and the related Kismet project which have been developed under the hypothesis that *humanoid intelligence requires humanoid interactions with the world.* In this chapter we deal with human-humanoid interaction by spoken language and visual cues, i.e. with talking heads in humanoid robotics. In fact, human-like artificial talking heads can increase a person's willingness to collaborate with a robot and helps create the social aspects of the human-humanoid relationship. The long term goal of the research in talking heads for a humanoid is to develop an artificial device which mechanically emulates the human phonatory organs (i.e. tongue, glottis, jaw) such that unrestricted natural sounding speech is generated. The device will be eventually contained in an elastic envelop which should resemble and move as a human face. Several problems have to be addressed towards this goal. First of all the complex phenomena in the human vocal organs should be mechanically emulated to produce a good artificial speech. Second, the control of the mechanical organs must be temporally congruent with human vocalization and this can be very complex to manage. The result is that at the state of the art the quality obtained with mechanical devices is only preliminar, yet interesting. For these reasons, and waiting that the mechanical talking heads reach a sufficient quality, we just emulate a talking head in a graphical way while the artificial speech is algorithmically generated.

It is worth emphasizing now the objective of this Chapter, which is the description of a novel algorithm to the control of a humanoid talking head and to show some related experimental results. This means that we estimate a given set of articulatory features to control the articulatory organs of a humanoid head, either virtual or mechanical. Two applications are briefly described: first, a system which mimicry human voice and, second, a system that produces robotic voice from unrestricted text, both of them with the corresponding facial movements.

Although almost all the animals have voices, only human beings are able to use words as mean of verbal communication. As a matter of fact, voice and the related facial movements are the most important and effective method of communication in our society. Human beings acquire control methods of their vocal organs with an auditory feedback mechanism by repeating trials and errors of hearing and uttering sounds. Humans easily communicate each other using vocal languages. Robotic language production for humanoids is much more difficult. At least three main problems must be solved. First, concepts must be transformed into written phrases. Second, the written text must be turned into a phonemic representation and, third, an artificial utterance must be obtained from the phonemic representation. The former point requires that the robot is aware of its situational context. The second point means that graphemic to phonemic transformation is made while the latter point is related to actual synthesis of the artificial speech.

Some researchers are attempting to reproduce vocal messages using mechanical devices. For instance, at Waseda University researchers are developing mechanical speech production systems for talking robots called WT-1 to WT-5, as reported in [3, 4, 5, 6, 7, 8, 9, 10, 11]. The authors reported that they can generate Japanese vowels and consonants (stops, fricatives and nasal sounds) reasonably clearly, although not all the utterances sound natural yet. On the other hand, the researchers of the robot Kismet [12] are expanding their research efforts on naturalness and perception of humanness in robots. An important step toward talking heads development is to estimate accurate vocal tract dynamic parameters during phonation. It is known, in fact, that there is a very high correlation between the vocal tract dynamic and the facial motion behavior, as pointed out by Yehia et al. [13]. For a mechanical talking robot, the artificial head should have human like movements during spoken language production by the robot, provided that the artificial head is tied to the vocal tract by means of some sort of elastic joint. In any case, the mechanical vocal tract should be dynamically controlled to produce spoken language. This requires enough knowledge of the complex relations governing the human vocalization. Until now, however, there has been no comprehensive research on the speech control system in the brain, and thus, speech production is still not clearly understood. This type of knowledge is pertaining to articulatory synthesis, which includes the methods to generate speech from dynamic configuration of the vocal tract (articulatory trajectory).

Our algorithm is based on imitation learning, i.e. it acquires a vocalization capability in a way similar to human development; in fact, human infants learn to speak through interaction by imitation with their care-givers. In other words, the algorithm tries to mimic some input speech according to a distance measure and, in this way, the articulatory characteristics of the speaker who trained the system are learned. From this time on, the system can synthesize unrestricted text using the articulatory characteristics estimated from a human speaker. The same articulatory characteristics are used to control facial movements using the correlation between them. When implemented on a robot, the audio-synchronized virtual talking head give people the sense that the robot is talking to them. As compared to other studies, our system is more versatile, as it can be easily adapted to different languages provided that some phonetic knowledge of that language is available. Moreover, our system uses an analysis-by-synthesis parameter estimation and it therefore makes available an artificial replica of the input speech which can be useful in some circumstances.

The rest of this chapter is organized as follows. In Section 2 some previous work in graphical and mechanical talking heads is briefly discussed. In Section 3 the imitation learning algorithm based on fuzzy model of speech is presented, and the genetic optimization of articulatory parameters is discussed. In Section 4 some experimental results are presented; convergence issues, acoustical and articulatory results are reported. In this Section also some results in talking head animation are reported. Finally, in Section 5 some final remarks are reported.

2. Previous work on talking heads

The development of facial models and of virtual talking heads has a quite long history. The first facial model was created by F.Parke in 1972 [14]. The same author in 1974 [15] produced an animation demonstrating that a single model would allow representation of many expressions through interpolated transitions between them. After this pioneer work, facial models evolved rapidly into talking heads, where artificial speech is generated in synchrony with animated faces. Such developments were pertaining to the human-computer interaction field, where the possibility to have an intelligent desktop agent to interact with, a virtual friend or a virtual character for interacting with the web attracted some attention. As regards these last points, Lundeberg and Beskow in [16] describe the creation of a talking head for the purpose of acting as an interactive agent in their dialogue system. The purpose of their dialogue system is to answer questions on chosen topics using a rich repertoire of gestures and expressions, including emotional cues, turntalking signals and prosodic cues such as punctuators and emphasisers. Studies of user reactions indicated that people had a positive attitude towards the agent. The FAQBot describes in [17] a talking head which answers questions based on the topics of FAQs. The user types in a question, the FAQBot's AI matches an answer to the question, and the talking head speaks, providing the answer to the user.

Other applications of talking head have been envisaged in many other field, such as the improvement of language skills, education and entertainment as reported in [18]. As regards entertainment, interactive input devices (e.g. Facial Animation, instrumented body suits, data gloves and videobased motion-tracking systems) are often used to drive the animation. In [19, 20] approaches for acquiring the expressions of the face of a live actor, and to use that information to control facial animation are described. Also the MIT Media Laboratory Perceptual Computing Section has developed systems that allow realtime tracking and recognition of facial expressions as reported in [21, 22].

The field of assistive technology has been also explored: in [23] a set of tools and technologies built around an animated talking head to be used in daily classroom activities with profoundly deaf children has been described. The students enters commands using speech, keyboard and mouse while the talking head responds using animated face and speech synthesis. On the other hand, if accurate face movements are produced from an acoustic vocal message uttered by a human, important possibilities of improving a telephone conversation with added visual information for people with impaired hearing conversation are introduced [24].

2.1 Social implication of a talking head in humanoid robotics

A brief description of social implication of talking heads is worth of because many current research activities are dealing with that. Socially-situated learning tutors with robot-directed speech is discussed in [25]. The robot's affective state and its behavior are influenced by means of verbal communication with a human care-giver via the extraction of particular cues typical of infant-directed speech as described in [26]. Varshavskaya in [27] dealt with the problem of early concept and vocal label acquisition in a sociable robot. The goal of its system was to generate "the kind of vocal output that a prelinguistic infant may produce in the age range between 10 and 12 months, namely emotive grunts, canonical babblings, and a formulaic proto-language". The synthesis of a robotic proto-language through interaction of a robot either with human or a robotic teacher was also investigated in [28].

Other authors (for example [29, 30, 31, 32]) have investigated the underlying mechanisms of social intelligence that will allow it to communicate with human beings and participate in human social activities. In [33] it was described the development of an infant-like humanoid robot (Infanoid) for situating a robot in an environment equivalent to that experienced by a human infant. This robot has a human-like sensori-motor systems, to interacts with its environment in the same way as humans do, implicitly sharing its experience with human interlocutors, sharing with humans the same environment [32]. Of course, talking heads have a very important role in this developments.

2.2 Graphical talking heads

In achieving the above goals, facial animation synthesis often takes two approaches: 3D mesh based geometry deformations and 2D image manipulations. In a typical 3D mesh approach, a mesh model is prepared which contains all the parameters necessary for the subsequent animations. Noh and Neumann describe in [34] several issues of graphical talking heads. The model is animated by mesh node displacements based on motion rules specified by deformation engine such as vector muscles [35, 36], spring muscles [37, 38], free form deformations [39], volume morphing [40], or simple interpolation [41]. If only the frontal movements are required, like in application based on mouth animation only, a 2D image-based approach is sufficient. 2D based approaches are also attractive for lip reading.
Ezzat et al. described in [42] a text to audiovisual translator using image warping and morphing between two viseme images. Gao et al. described in [43] new mouth shapes by linear combinations of several base images. Koufakis et al. describe in [44] how to use three basis images captured from different views and synthesize slightly rotated views of the face by linear combination of these basis images. Cosatto et al. in [45] describe their algorithm based on collecting various image samples of a segmented face and parameterize them to synthesize a talking face. By modeling different parts of the face from different sample segments, synthesized talking faces also exhibit emotions from eye and eyebrow movements and forehead wrinkles. Methods that exploit a collection of existing sample images must search their database for the most appropriate segments to produce a needed animation.
Other work, in particular that described in [43, 44, 46] used Mesh based texture mapping techniques. Such techniques are advantageous because warping is computed for relatively few control points.
Finally, there have been attempts to apply Radial Basis Functions (RBF) to create facial expressions. In [47] one of these approaches is described. Most approaches warped a single image to deform the face. However, the quality obtained from a single image deformation drops as more and more distortions are required. Also, single images lack information exposed during animation, e.g., mouth opening. Approaches without RBF using only single images have similar pitfalls.

2.3 Mechanical talking heads in humanoid robotic

When applied to a robot, mechanical talking heads give people a compelling sense that the robot is talking to them at a higher level as compared to virtual ones. At Waseda University the talking robots WT-1 to WT-5 [3, 4, 5, 6, 7, 8, 9, 10, 11] have been reported to the scientific community starting from 2000. The WT1 to 5 talking heads have been developed for generated human vocal movements and some human-like natural voices. For emulating the human vocalization capability, these robots share human-like organs as lungs, vocal cords, tongue, lips, teeth, nasal cavity and soft palate. The robots have increasing features, as an increasing number of degree of freedom (DOF) and the ability to produce some human-like natural voices. The anthropomorphic features were further improved in WT-4 and WT-5. WT-4 had a human-like body to make the communication with a human more easily, and has an increased number of DOF. This robot aimed to mimic continuous human speech sounds by auditory feedback by controlling the trajectory and timing. The mechanical lips and vocal cords of WT-5 have similar size and biomechanical structure as humans. As a result, WT-5 could produce Japanese vowels and consonant sounds (stops, fricatives and nasals) of 50 Japanese sounds for human-like speech production. Also at Kagawa University researchers dealt with talking heads from about the same years [48, 49, 50, 51]. They developed and improved mechanical devices for the construction of advanced human vocal systems with the goals to mimicry human vocalization and for singing voice production. They also developed systems for open and closed loop auditory control.

3. An algorithm for the control of a talking head using imitation learning

The block diagram of the algorithm described in this paper is reported in Fig. 1. According to the block diagram, we now summarize the actions of the algorithm.

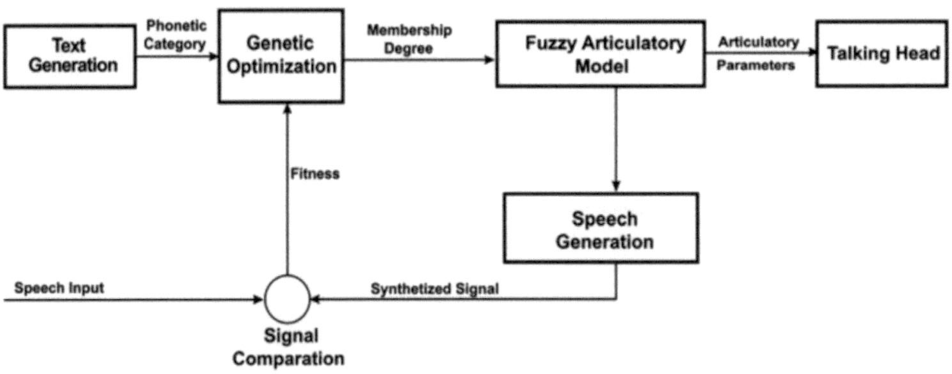

Figure 1. Block diagram of the genetic-fuzzy imitation learning algorithm

First, the operator is asked to pronounce a given word; the word is automatically selected from a vocabulary defined to cover all the phonemes of the considered language. Phonemes are described through the 'Locus Theory' [52].
In particular, the transition between two phonemes is described using only the target one. For example we do not consider that in the transition, say, 'no', the phoneme /o/ comes from the phoneme /n/ but only an average target configuration of phoneme /o/ is considered.

	Round	Open	Anterior	Sonorant	Bilaial	Labiodental	Alveolar	Prepalatal	Palatal	Vibrant	Dental	Velar
/sil/												
/a/												
⋮						⋮						
/n/	0	0	0	0.8	0.6	0	0	0	0.8	0	0.4	0
/o/	0.6	0.8	0.5	0.9	0	0	0	0	0	0	0	0
⋮						⋮						
/z/												

Figure 2. Membership degrees of phoneme transitions coming from 'any' phoneme. The membership degrees for the utterance 'no' are shown.

Each phoneme is therefore described in terms of articulatory as described in Fig. 2. The number corresponding to the articulatory feature is the degree of membership of that feature. These degrees of membership are obtained through genetic optimization, as described shortly. For example, in Fig. 3 a string of membership values for the utterance /no/ obtained with genetic optimization is reported.

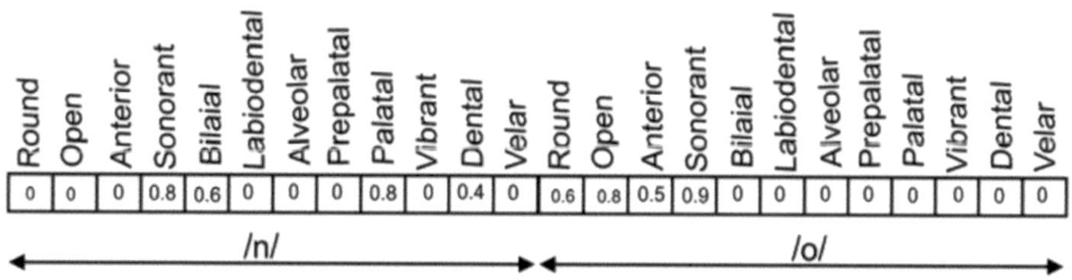

Figure 3. String of membership degrees for the utterance 'no'

To summarize, the learning mechanism of the articulatory parameters works as follows: the operator, acting as care-giver, pronounces a word and the robot generates an artificial replica of the word based on the articulatory and acoustic estimation. This process iterates until the artificial word matches the original one according to the operator's judgement. At this point the robot has learnt the articulatory movements of the phoneme contained in the word. The operator must repeat this process for a number of words. After these phases, the speech learning process is completed.
The synthesis of the output speech is performed using a reduced Klatt formant synthesizer [53], whose block diagram is represented in Fig. 4.

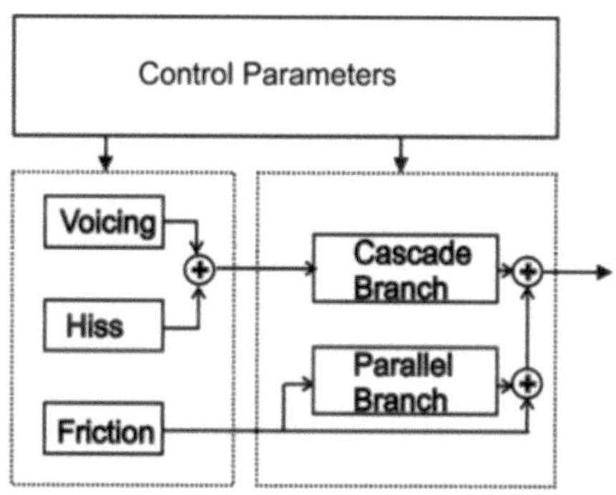

Figure 4. Simplified Klatt model used in this work

time	AV	AH	AF	F1	B1	F2	B2	F3	B3	A2F	A3F	A4F	A5F	A6F	AB
0	0	0	0	500	60	1500	90	2500	150	0	0	0	0	0	0
5	0	0	0	505	60	1441	90	2440	150	0	0	0	0	0	0
10	0	0	0	510	60	1381	91	2379	150	0	0	0	0	0	0
15	0	0	0	516	60	1322	91	2319	150	0	0	0	0	0	0
20	0	0	0	521	60	1262	92	2258	151	0	0	0	0	0	0

Figure 5. Acoustic parameters of a vowel sound for the first 20 ms

This system is basically composed by a parallel filter bank for the vocal tract modeling for unvoiced sounds and a cascade of filters for the vocal tract modeling for voiced sounds. It is controlled by fifteen parameters, namely the first three formants and bandwidths, the bypass AB, the amplitude AV for voiced sounds and the amplitudes AF, AH and A2F-A6F

for the fricative noise generator, updated every 5 ms. For instance, in Fig. 5 we report the parameters of a vowel sound in the very first interval, 20 ms long.
Since the fuzzy rules, however, describe the locus of the acoustical parameters, a model of the parameters profiles has been introduced. The profile of each synthesis parameter 'p' is described with four control features, namely the initial and final intervals I(p) and F(p), the duration D(p) and the locus L(p), as reported Fig. 6.

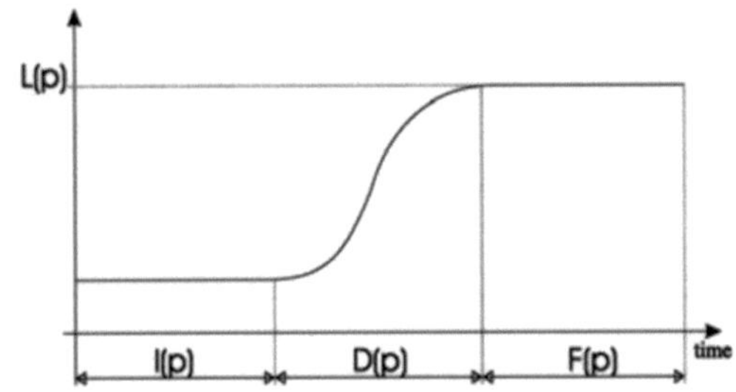

Figure 6. Synthesis parameters profiling in terms of Initial, Final, Duration and Locus fuzzy variables, I(p), F(p), D(p) and L(p), respectively

The I(p) control feature determines the length of the starting section of the transition, whose slope and target values are given by the D(p) and L(p) features. The parameter holds the value specified by their locus for an interval equal to F(p) ms; however, if other parameters have not completed their dynamic, the final interval F(p) is prolonged. The I(p), F(p), and D(p) parameters are expressed in milliseconds, while the target depends on what synthesis control parameter is involved; for example, for frequencies and bandwidths the locus is expressed in Hz, while for amplitudes in dB. It is worth noting that the initial values of the transition depend on the previous target values.

3.1 Phoneme and Control Parameters Fuzzification

As mentioned above, the phonemes are classified into broad classes by means of the manner of articulation; then, the place of articulation is estimated by genetic optimization. Therefore, each phoneme is described by an array of nineteen articulatory features, six of them are boolean variables and represent the manner of articulation and the remaining thirteen are fuzzy and represent the place of articulation.
Representing the array of features as (vowel, plosive, fricative, affricate, liquid, nasal | any, rounded, open, anterior, voiced, bilabial, labiodental, alveolar, prepalatal, palatal, vibrant, dental, velar), the /a/ phoneme, for example, can be represented by the array:

[1,0, 0, 0,0,0 | 1, 0.32,0.9, 0.12,1,0,0,0, 0, 0,0,0,0]

indicating that /a/ is a vowel, with a degree of opening of 0.9, of rounding of 0.32, and it is anterior at a 0.12 degree. The /b/ phoneme, on the other hand, can be considered a plosive sonorant phoneme, bilabial and slightly velar, and therefore it can be represented by the following array:

[0,1, 0, 0,0,0 | 1, 0,0,0,0.8, 0.9,0, 0, 0,0,0,0, 0.2].

The arrays reported as an example have been partitioned for indicating the boolean and the fuzzy fields respectively. Such arrays, defined for each phoneme, are the membership values of the fuzzy places of articulation of the phonemes.

On the other hand, the I, D, F and L fuzzy variables, defined in a continuous universe of discourse, can take any value in their interval of definition. The fuzzy sets for these variables have been defined as follows:

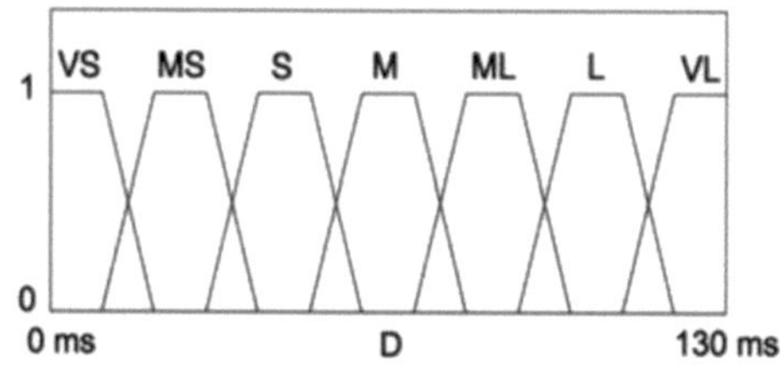

Figure 7. Fuzzy sets of the D(p) fuzzy variable

- Duration D(p). The global range of this fuzzy variable is 0-130 ms, with trapezoidal membership functions as shown in Fig. 7. In Fig. 7 such values are indicated as follows:

$$\text{Very Short, Medium Short, Short, Medium, Medium Long, Long, Very Long} \tag{1}$$

- Initial Interval I(p). As D(p), this fuzzy variable is divided into trapezoidal membership functions in a 0-130 ms interval. The fuzzy values are indicated, in this case:

$$\text{Instantaneous, Immediate, Quick, Medium, Medium Delayed, Delayed, Very Much Delayed} \tag{2}$$

- Final Interval F(p). The numeric range is $0-130$ ms and the fuzzy values are the same as indicated for the Initial Interval I(p).
- Locus L(p). The fuzzy values of this variable depend on the actual parameter to be controlled. For AV, AH and AF the fuzzy values are:

$$\text{Zero, Very Low, Low, Medium Low, Medium, Medium High, High, Very High} \tag{3}$$

and their membership functions are equally distributed between 12 and 80 dB with the trapezoidal shape shown in Fig. 7. The other gain factors, namely A2F-A6F and AB, take one of the following values:

$$\text{Very Low, Low, Medium Low, Medium, Medium High, High, Very High} \tag{4}$$

in the range 0-80 dB with the same trapezoidal shape as before. The values of L(F1), L(F2) and L(F3) are named as in (4), with trapezoidal membership functions uniformly distributed from 180 to 1300 Hz, 550 to 3000 Hz and 1200 to 4800 Hz for the first, second and third formant respectively. For example, the fuzzy sets of L(F1) are shown in Fig. 8.

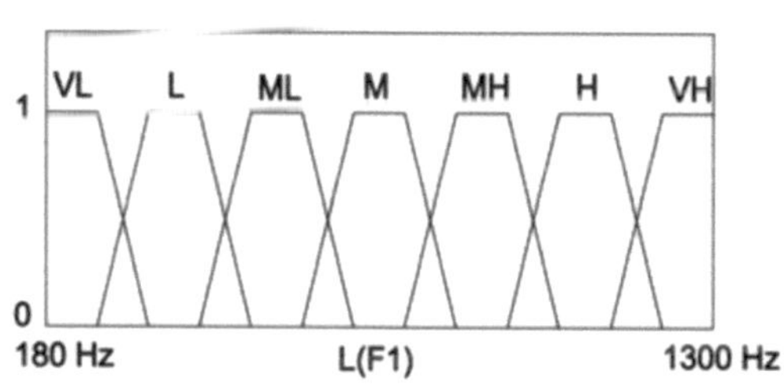

Figure 8. Fuzzy sets of the F1 locus

Finally, the loci of the bandwidths Bl, B2 and B3 take one of the fuzzy values described in (4), and their trapezoidal membership functions are regularly distributed in the intervals 30-1000 Hz for Bl, 40-1000 Hz for B2 and 60-1000 Hz for B3.

3.2 Fuzzy Rules and Defuzzification

By using linguistic expressions which combine the above linguistic variables with fuzzy operators, it is possible to formalize the relationship between articulatory and acoustic features.

We report as an exemplification, the semplified fuzzy rule of the transitions Vowel-Fricative.

```
IF PO IS ANY AND PI IS SONORANT THEN {
                                        B(AV) IS MEDIUM-HIGH
                                              }
IF PO IS ANY AND PI IS ^SONORANT THEN {
                                        B(AV) IS ZERO
                                              }
IF PO IS ^LAB AND PI IS ANY THEN {      D(F1) IS MEDIUM-LONG ;
                                        D(F2) IS MEDIUM-LONG ;
                                        D(F3) IS MEDIUM-LONG
                                              }
IF PO IS LAB*SON AND PI IS ANY THEN { I(AV) IS IMMEDIATE ;
                                        D(F1) IS MEDIUM ;
                                        D(F2) IS MEDIUM ;
                                        D(F3) IS MEDIUM
                                                }
IF PO IS LAB*^SON AND PI IS ANY THEN { D(F1) IS MEDIUM-SHORT ;
                                        D(F2) IS MEDIUM-SHORT ;
                                        D(F3) IS MEDIUM-SHORT
                                                }
IF PO IS ANY AND PI IS OPEN THEN { B(F1) IS MEDIUM
                                        }
IF PO IS ANY AND PI IS ^OPEN THEN { B(F1) IS VERY-SHORT
                                        }
IF PO IS ANY AND PI IS ANTERIOR THEN { B(F2) IS MEDIUM-HIGH
                                            }
IF PO IS ANY AND PI IS ^ANTERIOR THEN { B(F2) IS LOW
                                            }
IF PO IS ANY AND PI IS ROUND THEN { B(F3) IS LOW
                                        }
IF PO IS ANY AND PI IS ^ROUND THEN { B(F3) IS MEDIUM
                                        }
```

Since the manner of articulation well partitions the phonemes in separated regions, the rules have been organized in banks, one for each manner.

That is, calling PO and PI the actual and the future phonemes respectively, the set of rules is summarized in Fig. 9. The rule decoding process is completed by the defuzzification operation, which is performed with the fuzzy centroid approach.

Concluding, as shown in Fig. 9, there are several transitions which are performed with the same set of rules. For example, all the transition toward fricatives and liquid phonemes are realized with the same bank of rules. This is because the related transitions can be approximated with a strong discontinuity, and thus they can be considered independent from the starting phonemes; the symbol 'CO' used in these banks stands, in fact, for a generic consonant sounds. Other banks are missing; this is because they are concerned with transitions which occur very rarely in Italian language.

3.3 Genetic optimization of articulatory and acoustic parameters

Let us take a look at Fig. 1. Genetic optimization estimates the optimum values of the degrees of membership for the articulatory features used to generate an artificial replica of the input signal by comparing the artificial with the real signal. The optimal membership degrees of the articulatory places minimize the distance from the uttered signal; the inputs are the number of phonemes of the signal and their classification in terms of manner of articulation.

One of the most important issues of the genetic algorithm is chromosome coding. The chromosome used for the genetic optimization of a sequence of three phonemes is shown in Fig. 10.

P0 \ P1	Vowel	Plosive	Fricative	Affricate	Liquid	Nasal
Vowel	VO->VO 9 rules	VO->PL 13 rules	VO->FR 13 rules	VO->AF 4 rules	VO->LI 5 rules	VO->NA 8 rules
Plosive	PL->VO 15 rules		CO->FR 13 rules		CO->LI 12 rules	PL->NA 15 rules
Fricative	FR->VO 12 rules		CO->FR 13 rules		CO->LI 12 rules	
Affricate	AF->VO 11 rules		CO->FR 13 rules		CO->LI 12 rules	
Liquid	LI->VO 14 rules	CO->PL 15 rules	CO->FR 13 rules	CO->AF 4 rules	CO->LI 12 rules	
Nasal	NA->VO 11 rules	CO->PL 15 rules	CO->FR 13 rules	CO->AF 4 rules	CO->LI 12 rules	NA->NA 9 rules

Figure 9. Outline of the bank of fuzzy rules. P0 and P1 represent the actual and target phonetic categories. CO denotes a generic consonant

It represents the binary coding of the degrees of membership. Typical values of mutation and crossover probability are around 0.1 and 0.7 respectively.

An important aspect of this algorithm is the fitness computation, which is represented by the big circle symbol in Fig. 1. For the sake of clarity of the Section, we now briefly summarize the mel-cepstrum distance measure.

3.3.1 Mel-Cepstrum distance measure

Our distance measure uses the well known band-pass Mel-scale distributed filter bank approach, where the output power of each filter is considered. We can interpret the output of a single band-pass filter as the k-th component of the DFT of the input sequence x(n):

$$X(k) = \sum_{n=0}^{N-1} x(n) e^{-j\frac{2\pi}{N}nk} \tag{5}$$

Since we are interested in the center frequencies of the band-pass filters, the k-th frequency is moved, with respect to eq. (5), by *ir/N*. Therefore:

$$\hat{X}(k) = \sum_{n=0}^{N-1} x(n) e^{-j\left(\frac{2\pi}{N}k + \frac{\pi}{N}\right)n} = \sum_{n=0}^{N-1} x(n) e^{-j\frac{2\pi}{N}n(k-0.5)} \quad (6)$$

Since *x(n)* is real, the sequence *X(n)* is symmetric; hence only the first N/2 points are sufficient. The inverse transform of the last relation is thus:

$$x(k) = \frac{2}{N} \sum_{n=0}^{N/2-1} \hat{X}(k) \cos\left(\frac{2\pi}{N}(k-0.5)n\right) \quad (7)$$

Figure 10. The binary chromosome obtained by coding

In conclusion, setting $N = N/2$, the Mel-cepstrum coefficients can be computed with the following relation:

$$c_i = \sum_{n=0}^{\overline{N}-1} X(k) \cos\left(\frac{2\pi}{N}(k-0.5)i\right) \quad i = 0 \ldots M-1 \quad (8)$$

where *X(k)* is the logarithm of the output energy for the k-th triangular filter, N is the number of filters and M is the number of Mel-cepstrum coefficients. As reported above, the computation of the cepstral coefficients is performed using a band-pass filter bank distributed on a Mel frequency scale. The Mel scale is a non linear scale, motivated by perceptual studies of the frequency response characteristics of the human auditory system. Within the audible frequency range, a pure tone or the fundamental frequency of a complex sound produces a psychological effect on the listener. The psychological effect of a given frequency is measured in [Mel] according to the following definition: A 40 dB tone at 1000Hz produces a 1000 Mel effect. Different frequencies produce different effects. It was experimentally evaluated that a 2000 Mel response is obtained for a 3100 Hz tone, while a 500 Mel response is obtained at 400 Hz. The relation between frequency in Hz and Mel it was found to be well described by a linear scale in Hz below 1000 Hz and a logarithmic scale above 1000 Hz: $Mel = 1000 \log_2\left(1 + \frac{Hz}{1000}\right)$.

The mel-cepstral distance measure is then defined by:

$$D = \sum_{i=0}^{M-1} (c_i^a - c_i^o)^2 \quad (9)$$

where c_i^a and c_i^o are the *i*-th mel cepstral coefficients of the artificial and original phrase.

3.3.2 Fitness computation and articulatory constraints

The fitness, which is the distance measure between original and artificial utterances and is optimized by the genetic algorithm, is an objective measure that reflects the subjective quality of the artificially generated signal. The mel-cepstrum measure is used to compare the artificial signal generated by the fuzzy module and the speech generation module against the original input signal. The original and the artificial utterances are first aligned and then divided into frames and the average squared Euclidean distance between spectral vectors obtained via critical band filters is computed. The alignment between the original and artificial utterances is performed by using dynamic programming [54], with slope weighting as described in [55] and shown in Fig. 11.

Therefore, using the mapping curve between the two signals obtained with dynamic programming, the mel-cepstral distance D between original and artificial utterances represented respectively with X and Y is computed as follows:

$$D(X,Y) = \frac{1}{L_\Phi} \sum_{k=0}^{T} \left[\sum_{j=0}^{K} [c_x(\Phi_x(k),j) - c_y(\Phi_y(k),j)]^2 \, m(k) \right]$$

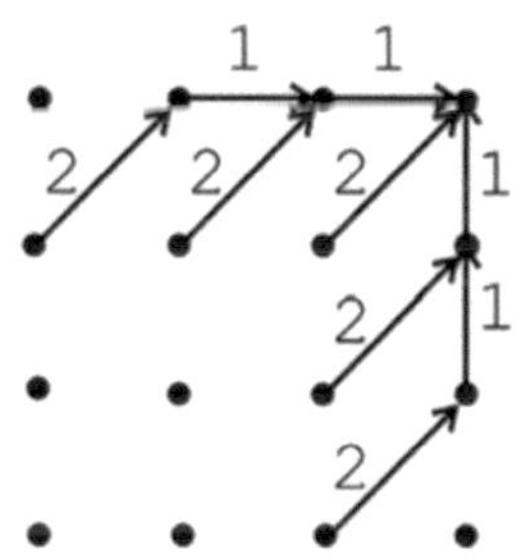

Figure 11. Slope weighting performed by dynamic programming

where T is the number of frames, K is the number of cepstrum coefficients, $\Phi = (\Phi_x, \Phi_y)$ is the non-linear mapping obtained with dynamic programming, L_Φ is the length of the map, $c_x(i,j)$ is the j-th Mel-cepstrum of the i-th frame of the original utterance, $c_y(i,j)$ is the j-th Mel-cepstrum of the i-th frame of the artificial utterance, and $m(k)$ are the weights as shown in Fig. 11.

The fitness function of the Place of Articulation (PA), i.e. the measure to be maximized by the genetic algorithm, is then computed as:

$$Fitness\,(PA) = \frac{1}{D(X,Y)}$$

Therefore, the goal of the genetic optimization is to find the membership values that lead to a maximization of the fitness, i.e. the minimization of the distance $D(X, Y)$, namely $PA = argmax\ \{Fitness\ (PA)\}$, $PA = \bigcup PA_i$, $i = 1, \dots, N$, where PA_i is the degree of membership of the i-th place of articulation, N is the number of phonemes of the input signal. However, in order to correctly solve the inverse articulatory problem, the following constraints, due to the physiology of the articulations, have been added to the fitness:

- it is avoided that a plosive phoneme is completely dental and velar simultaneously;
- it is avoided that a nasal phoneme is completely voiced;

- it is avoided that all the membership degrees are simultaneously less than a given threshold;
- it is avoided that two or more degrees of membership are simultaneously greater than another threshold.

The fitness is therefore given by:

$$Fitness\,(PA) = \frac{1}{D(X,Y) + \sum_{j=1}^{N_c} P_j}$$

where Pj is the j-th penalty function and N_c is the number of constraints.
In conclusion, the optimization of places of articulation (PA) can be expressed as follows:

$$PA = argmax\,\{Fitness\,(PA)\}$$

4. Experimental results

Our imitation learning algorithm has been tested considering several different aspects. Initially, we have considered the convergence issue. In Fig. 12 a typical convergence behavior is represented, where 1/*Fitness (PA)* against number of generation is shown. Basing on the results shown in Fig. 12, the experimental results presented in the following are obtained with a population size of 500 elements, mutation rate equal to 0.1 and crossover probability of 0.75.

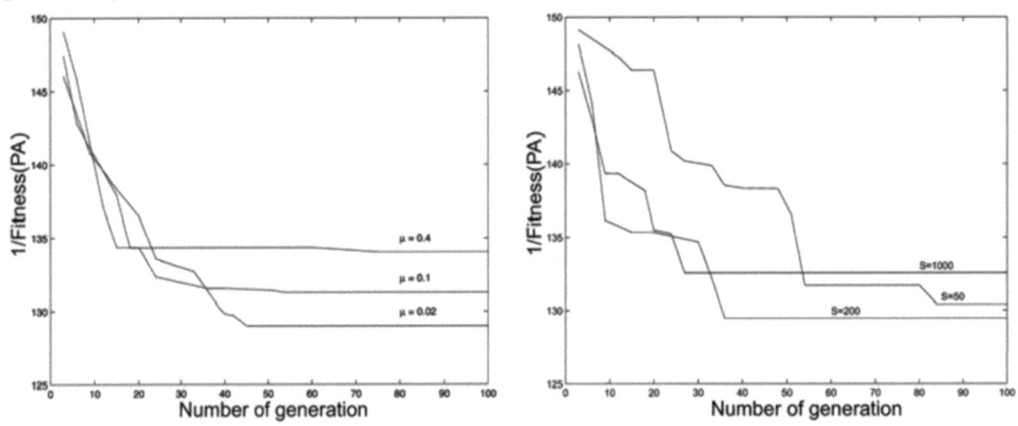

Figure 12. Convergence diagram, i.e. 1/*Fitness (PA)* versus number of generation for six different learning experiments. In the left panel the mutation rate μ is varied and the population size is maintained constant to 200 elements. In the right panel the population size S is varied and the mutation rate is maintained constant to 0.02

As outlined in Section 3, the speech learning mechanism works as follows: the operator pronounces one word and the talking robot generates an artificial replica of the word based on the articulatory and acoustic estimation. This process iterates until the artificial word matches the original one according to the operator judgement. More precisely, the robot learns how to pronounce a word in terms of articulatory movements using several utterances from the same talker. In Fig. 13 is shown a typical fitness error behavior considering different utterances of the same word and both non-iterative and iterative learning: in non-iterative learning (left panel), the optimization process starts randomly in each optimization process; in iterative learning (right panel) the optimum obtained using the previous utterance of the same word is used as a starting point for the new optimization process using the new utterance of the word, obtaining usually a better fitness error. Iterative learning has another important feature: it allows us to obtain a mean behavior of the articulatory parameters for a given word and a given talker. The operator must repeat this iterative learning process for a number of words. After these phases, the speech learning process is completed.

Coming back to Fig. 13, it is worth noting that the three descending curves are related to three subsequent different pronunciations of the same word.

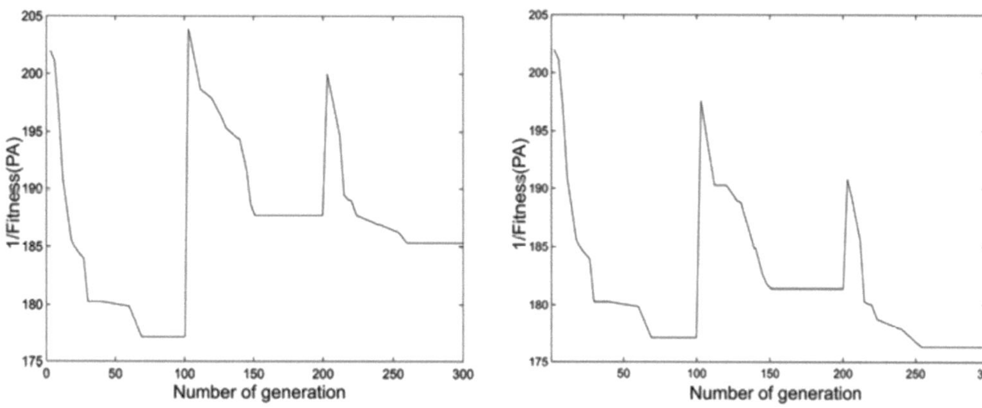

Figure 13. Convergence diagram for non-iterative (left panel) and iterative (right panel) learning algorithm. Using non-iterative learning, each learning process starts from random initial conditions. Using iterative learning, the new learning process starts from the optimal parameters obtained from the learning of another utterance of the same word. Each curve of this figure is related to the Italian word 'nove' pronounced three times; averaged values are depicted

In Fig. 14 and in Fig. 15 are reported some experimental results related to the analysis of the Italian word 'gentile' ('kind'). In the upper panel of Fig. 14 the dynamic behavior of the first three formant frequencies is reported. The vertical lines denote the temporal instants of the stationary part of each phoneme. It is worth noting that this segmentation is done on the synthetic signal but it can be related to the original signal using the non—linear mapping between the original and synthetic word obtained by dynamic programming. In the lower panel of Fig. 14 the behavior of low and high frequencies amplitudes are shown.

The trajectories of the places of articulation, estimated with the algorithm, and reported as an example in Fig. 15, can be used to shape the various organs of a mechanical vocal tract, and consequently, by means of a mechanical linkage, of the mechanical talking head.

Since the facial motion can be determined from vocal tract motion by means of simple linear estimators as shown by Yehia et al. in [13], we used the same parameters to control a graphical talking head. Yehia et al. in [13] built an estimator to map vocal-tract positions to facial positions. Given a vector y of vocal-tract marker positions to a vector x of facial positions, an affine transformation is defined by

$$\tilde{x} - \mu_x = T_{yx}(X - \mu_y) \tag{10}$$

with $\mu_x = E[x]$ and $\mu_y = E[y]$ the expected values of x and y. Arranging all the frames of vocal-tract and facial data training sets in singles matrices Y and X, where M_{tr} is the number of vectors contained in the training set, an estimation of T_{yx} is given by

$$T_{yx} \approx X_0 Y_0^T (Y_0 Y_0^T)^{-1} \tag{11}$$

where Y_0 and X_0 are given by Y and X subtracting the corresponding expected value from each row respectively. Using this linear estimation and the articulatory parameters of our algorithm, we animated a graphical talking head.

To generate an artificial face, a synthetic 3D model was used and visualized in OpenGL under controlled illumination as shown in Fig. 16. The Geoface 3-D facial mesh [56] was used for the experiment and tesselated to a high degree for utmost smoothness in the surface normal computation. Some muscles around the mouth have been added in order to obtain the correct facial movements during articulation of vowels and consonants.

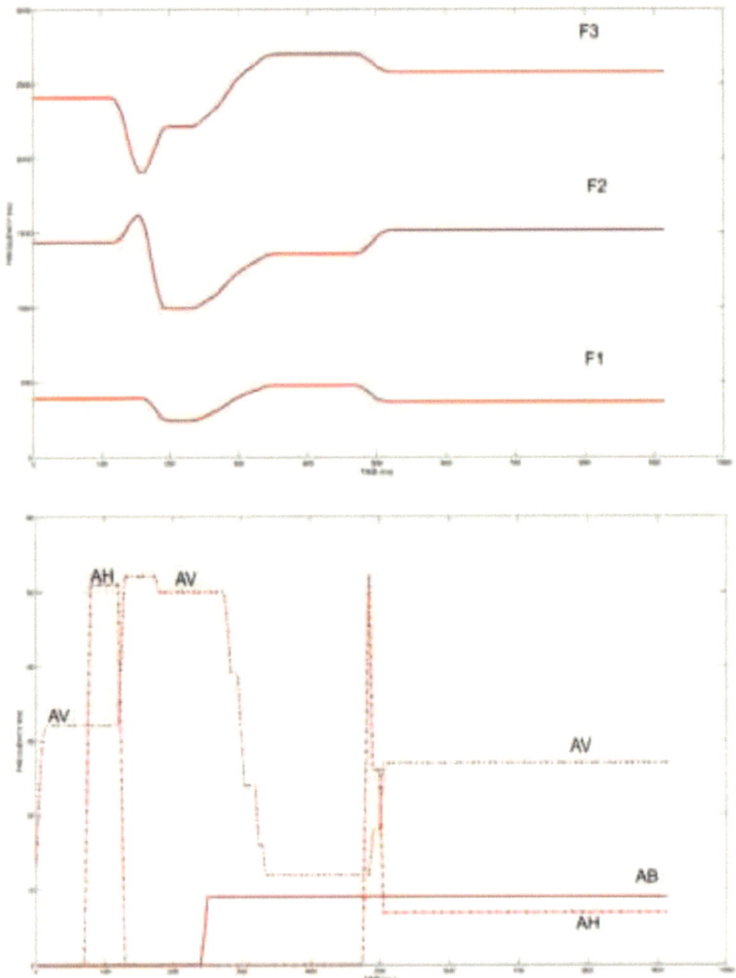

Figure 14. Acoustic analysis of the Italian word 'gentile' obtained with the genetic-fuzzy imitation learning algorithm

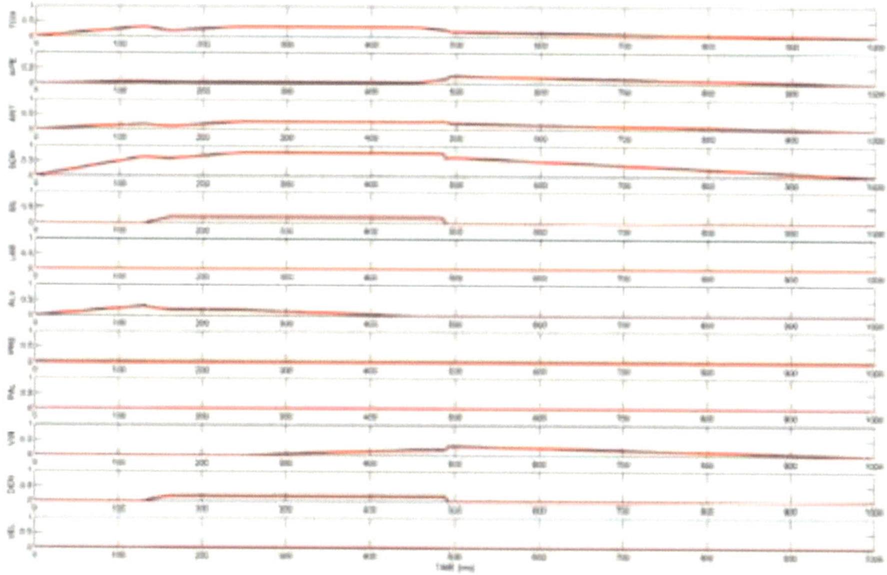

Figure 15. Places of articulation of the Italian word 'gentile' estimated with the genetic-fuzzy imitation learning algorithm

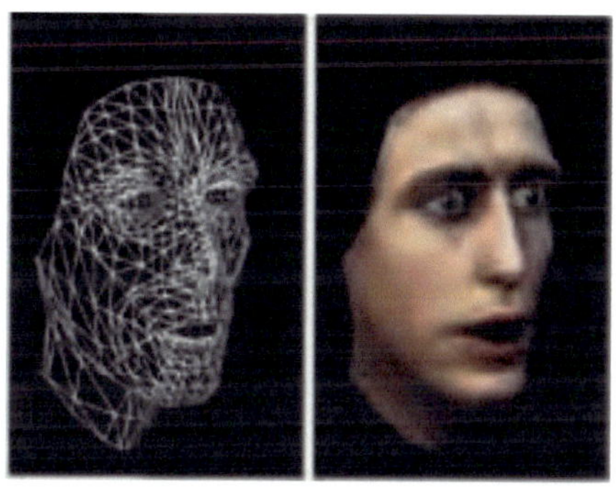

Figure 16. In the left panel is shown the mesh model of the talking head. In the right panel, a skin has been added to the model

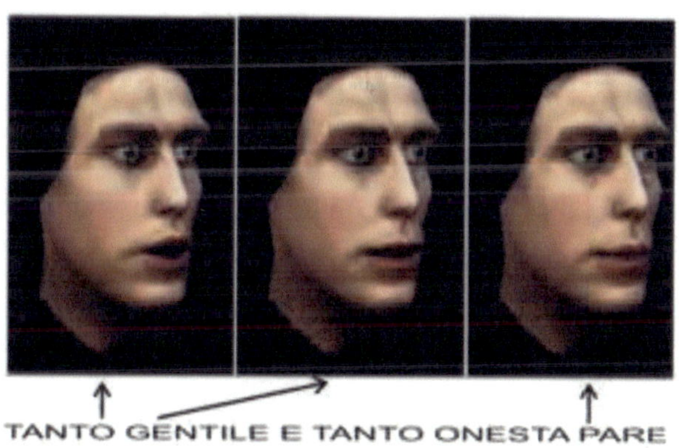

Figure 17. Some frames of the Italian utterance "tanto gentile e tanto onesta pare" as pronounced by the talking head

The linear model between the articulatory movements and the facial movements, as described by eq. (11), has been estimated on the basis of the MOCHA-TIMIT data base. MOCHA-TIMIT [57] is a database of articulatory features that considers a set of 460 sentences designed to include the main connected speech processes in English, pronounced by two speakers, a male and a female. This database includes articulatory and acoustic data recorded in studies on speech production. Some instruments, namely EMA and EPG, have been used for the production of MOCHA-TIMIT. EMA (electromagnetic articulography) is a technique which allows articulatory movements to be monitored by means of small electromagnetic coils attached to vocal-tract structures in the mid-sagittal plane. Possible measurement points are situated on the tongue, the upper and lower lips, the mandible, and the velum. In addition, coils are generally attached to fixed structures such as the bridge of the nose and the upper central incisors to provide a maxillary frame of reference. Alternating magnetic fields generated by transmitter coils make it possible to measure the position of each receiver coil relative to two orthogonal axes in the midsagittal plane, with a measurement bandwidth typically ranging from DC up to about 250 Hz [58]. EPG (electropalatography) is a technique for recording the timing and location of tongue-palate contact patterns during speech [59]. It involves the subject wearing an artificial plate moulded to fit the upper palate and containing a number of electrodes mounted on the surface to detect lingual contacts.
A set of common phonemes between English and Italian language pronounced by the male speaker has been extracted from MOCHA-TIMIT database, and the movements of the lips related to the articulatory features has been recorded in the form of eq. (11).
In Fig. 17 are shown three picture of the talking head pronouncing the Italian utterance: "Tanto gentile e tanto onesta pare" . Informal audio-visual tests show that the algorithm is able to produce correct results.

5. Discussion and Final remarks

The estimation of articulatory static and dynamic configuration is one of the most difficult problems in voice technology. Several approaches have been attemped during the past years, and most of the difficulties are due to the fact that the articulatory-acoustic relation is not unique: different articulatory configurations can produce the same signal. The problem can be faced with suitable constraints which are needed to avoid the unrealistic configurations. One on the last work in this area has been reported in [60], which estimated articulatory parameters by finding the maximum a posteriori estimate of articulatory parameters for a given speech spectrum and the state sequence of a HMM—based speech production model.
This model consists of HMMs of articulatory parameters for each phoneme and an articulatory-to-acoustic mapping that transforms the articulatory parameters into the speech spectrum for each HMM state. The authors constructed the model by using simultaneously observed articulatory and acoustic data for sentence utterances, which was collected by using an electro-magnetic articulo-graphic (EMA) system.
Our system does not use EMA at all, trying to get the right articulations with several penalty factors. For this reason, the audio-video result is not always correct and it needs a judgment by the operator.
We emphasize the following final remarks.

- Our algorithm generates allophonic variations of the phonemes. In other words, since the described procedure performs a sort of interpolation among the acoustical parameters, the actual phonemic realization depends on the phonetic context.
- Our fuzzy model can be easily modified and tuned because the fuzzy rules are expressed in a linguistic manner.
- Many further optimizations are possible, in terms of genetic algorithm and in terms of the fuzzy rules which could be automatically estimated instead of being defined from phonetic knowledge on its own, as we did.
- Our algorithm could be used as a design tool of mechanical talking heads because the artic ulatory parameters can be easily added or deleted from the fuzzy rules and their effects of the modification can be immediately verified.
- Finally, is has to be noted that in this paper we implemented only the rules pertaining to the Italian language. This does not limit the generality of our method: if a different language has to be considered, new banks of fuzzy rules could be added and the previous banks could be modified.

6. Conclusions

In this Chapter we have dealt with the articulatory control of talking heads, which can be either graphical or mechanical. A novel approach for the estimation of articulatory features from an input speech signal is described. The approach uses a set of fuzzy rules and a genetic algorithm for the optimization of the degrees of membership of the places of articulation. The membership values of the places of articulation of the spoken phonemes have been computed by means of genetic optimization. Many sentences have been generated on the basis of this articulatory estimation and their subjective evaluations show that the quality of the artificially generated speech is quite good. As compared with other works in acoustic to articulatory mapping, which generally compute the vocal tract area functions from actual speech measurements, our work present a method to estimate the place of articulation of input speech through the development of a novel computational model of human vocalization.

7. References

E. Mumolo, M. Nolich, and E. Menegatti. A genetic-fuzzy algorithm for the articulatory imitation of facial movements during vocalization of a humanoid robot. In *Proceeding of the 2005 IEEE Int. Conf. on Humanoid Robotics*, pages 436-441, 2005. [1]

R. Brooks, C. Breazeal, M. Marjanovic, B. Scassellati, and M. Wiliamson. The Cog Project: Building a Humanoid Robot. *Lecture Notes in Artificial Intelligence, Springer-Verlag*, 1998. [2]

K. Nishikawa, K. Asama, K. Hayashi, H. Takanobu, and A. Takanishi. Development of a Talking Robot. In *Proceedings of the 2000 IEEE/RSJ International Conference on Intelligent Robots and Systems*, pages 1760—1765, 2000. [3]

K. Nishikawa, K. Asama, K. Hayashi, H. Takanobu, and A. Takanishi. Mechanical Design of a Talking Robot for Natural Vowels and Consonant Sounds. In *International Conference on Robotics and Automation*, pages 2424-2430, May 2001. [4]

K. Nishikawa, A. Imai, T. Ogawara, H. Takanobut, T. Mochidal, and A. Talcanishi. Speech Planning of an Anthropomorphic Talking Robot for Consonant Sounds Production. In *Proceedings of the 2002 IEEE International Conference on Robotics and Automation,* pages 363-368, 2002. [5]

K. Nishikawa, H. Takanobu, T. Mochida, M. Honda, and A. Takanishi. Speech Production of an Advanced Talking Robot based on Human Acoustic Theory. In *Proceadlngr dthe 2004 IEEE International Conference on Robotics and Automation,* pages 363-368, 2004. [6]

K. Nishikawa, T. Kuwae, H. Takanobu, T. Mochida, M. Honda, and A. Takanishi. Mimicry of Human Speech Sounds using an Anthropomorphic Talking Robot by Auditory Feedback. In *Proceeding of the 2004 IEEE International Conference on Robotics and Automation,* pages 272-278, 2004. [7]

K. Fukui, K. Nishikawa, S. Ikeo, E. Shintaku, K. Takada, H. Takanobu, M. Honda, and A. Takanishi. Development of a Talking Robot with Vocal Cords and Lips Having Humanlike Biological Structures. In *Proceedings of the 2005 IEEE/RSJ International Conference on Intelligent Robots and Systems,* pages 2023-2028, 2005. [8]

K. Fukui, K. Nishikawa, T. Kuwae, H. Takanobu, T. Mochida, M. Honda, and A. Takanishi. Development of a New Human-like Talking Robot for Human Vocal Mimicry. In *Proceedings of the 2005 IEEE International Conference on Robotics and Automation,* pages 1437-1442, 2005. [9]

K. Fukui, K. Nishikawa, S. Ikeo, E. Shintaku, K. Takada, A. Takanishi, and M. Honda. New Anthropomorphic Talking Robot having Sensory Feedback Mechanism and Vocal Cords based on Human Biomechanical Structure. In *The First IEEE/RAS-EMBS International Conference on Biomedical Robotics and Biomechatronics,* BioRob 2006, pages 1095—1100, 2006. [10]

K. Fukui, K. Nishikawa, S. Ikeo, M. Honda, and A. Takanishi. Development of a Human-like Sensory Feedback Mechanism for an Anthropomorphic Talking Robot. In *Proceedings of the 2006 IEEE International Conference on Robotics and Automation,* pages 101—106, 2006. [11]

C. Breazeal. *Designing Sociable Robots.* MIT Press, 2004. [12]

H. Yehia, P. Rubin, and E. Vatikiotis-Bateson. Quantitative association of vocal-tract and facial behavior. *Speech Communication,* V.26:23-43, 1998. [13]

F. I. Parke. Computer generated animation effaces. Master's thesis, University of Utah, 1972. [14]

F. I. Parke. A Parame.te.ric Model for Human Faces, *PhD thesis,* University of Utah, 1974. [15]

M. Lundeberg and J. Beskow. Developing a SDagent for the august dialogue system. In *Proceedings of the Auditory Visual Speech Processing '99 Conference (AVSP'99),* 1999. [16]

Grossman B. Cechner P. Beard, S. and A. Marriott. Faqbot. In *Sydney Area Workshop on Visual Information Processing.,* 1999. [17]

Cohen M. A. Massaro, D. W. and M. A. Berger. Creating talking faces: Applying talking faces. In *In Perceiving Talking Faces: From Speech Perception to a Behavioral Principle,* 1998. [18]

L. Williams. Performance-driven facial animation. In *Computer Graphics,* 1990. [19]

D. Terzopoulos and K. Waters. Analysis and synthesis of facial image sequences using physical and anatomical models. In *IEEE Transactions on Pattern Analysis and Machine Intelligence,* 1993. [20]

I. A. Essa and A. Pentland. A vision system for observing and extracting facial action parameters. In *Proceedings of IEEE Computer Vision Pattern Recognition Conference,* 1994. [21]

I. A. Essa and A. P. Pentland. Facial expression recognition using a dynamic model and motion energy. In *International Conference on Computer Vision,* 1995. [22]

Cole R. et al. New tools for interactive speech and language training: Using animated conversational agents in the classrooms of profoundly deaf children. In *Proceedings of ESCA/SOCRATES Workshop on Methods and Tool Innovations for Speech Science Education.* London, UK., 1999. [23]

J.J. Williams and A.K. Katsaggelos. An HMM-Based Speech-to-Video Svnthesizer. *IEEE Trans, on Neural Networks,* V.13(N.4):900-915, 2002. [24]

C. Breazeal and L. Aryananda. Recognition of affective communicative intent in robot-directed speech. In *Autonomous Robots,* 2002. [25]

A. Fernald. Four-month-old infants prefer to listen to motherese. In *Infant Behavior and Development,* 1985. [26]

P. Varshavskaya. Behavior-based early language development on a humanoid robot. In *2nd Int. Conf. on Epigenetics Robotics,* 2002. [27]

K. Dautenhahn and A. Billard. Studying robot social cognition within a developmental psychology framework. In *3rd Int. Workshop on Advanced Mobile Robots,* 1999. [28]

C. Breazeal and B. Scassellati. Infant-like social interactions between a robot and a human caretaker. In *Adaptive Behavior,* 2000. [29]

K. Dautenhahn. I could be you: The phenomenological dimension of social understanding. In *Cybernetics and Systems Journal,* pages 417—453, 1997. [30]

B. Scassellati. Theory of mind for a humanoid robot. In *Proceedings of the 2000 IEEE/RSJ International Conference on Humanoid Robotics,* 2000. [31]

J. Zlatev. The epigenesis of meaning in human beings, and possibly in robots. In *Lund University Cognitive Studies,* 1999. [32]

H. Kozima and J. Zlatev. An epigenetic approach to human-robot communication. In *International Workshop on Robot and Human Interactive Communication ROMAN-2000,* 2000. [33]

Ulrich Neumann Jun-yong Noh. Talking Faces. In *IEEE International Conference on Multimedia and* Expo (II), 2000. [34]

K. Waters. A muscle model for animating threedimensional facial expression. In *Computer Graphics,* 1987. [35]

J. Frisbie K. Waters. A Coordinated Muscle Model for Speech Animation. In *Graphics Interface,* 1995. [36]

K. Waters Y. C. Lee, D. Terzopoulos. Realistic face modeling for animation. In *Siggraph proceedings,* 1995. [37]

N. Badler S. Platt. Animating facial expression. In *Computer Graphics,* 1981. [38]

N. M. Thalmann D. Thalmann P. Kalra, A. Mangili. Simulation of Facial Muscle Actions Based on Rational Free From Deformations. In *Eurographics,* 1992. [39]

T. Kim R. Enciso U. Neumann D. Fidaleo, J-Y. Noh. Classification and Volume Morphing for Performance-Driven Facial Animation. In *Digital and Computational Video,* 1999. [40]

D. Lischinski R. Szeliski D. H.Salesin F. Pighin, J. Hecker. Synthesizing Realistic Facial Expressions from Photographs. In *Siggraph proceedings,* 1998. [41]

T. Poggio T. Ezzat. Mike Talk: A Talking Facial Display Based on Morphing Visemes. In *Computer Animation 1998,* 1998. [42]

Y. Ohta L. Gao, Y. Mukaigawa. Synthesis of Facial Images with Lip Motion from Several Real Views. In *Automatic Face and Gesture Recognition,* 1998. [43]

B.F. Buxton I. Koufakis. Very low bit rate face video compression using linear combination of 2D face view and principal components analysis. In *Image and Vision Computing,* 1999. [44]

H. P. Graf E. Cosatto. Sample-Based Synthesis of Photo Realistic Talking Heads. In *Computer Animation 1998,* 1998. [45]

M. Ouhyoung W. Perng, Y. Wu. Image Talk: A Real Time Synthetic Talking Head Using One Single Image with Chinese Text-To-Speech Capability. In *IEEE,* 1998. [46]

D. Reisfeld Y Yeshurun N. Arad, N. Dyn. Image Warping by Radial Basis Functions: Application to Facial Expressions. In *Graphical Models and Image Processing,* March 1994. [47]

T. Higashimoto and H. Sawada. Speech Production by a Mechanical Model Construction of a Vocal Tract and its Control by Neural Network. In *Proceedings of the 2002 IEEE International Conference on Robotics and Automation* (ICRA 2002), pages 3858-3863, 2002. [48]

H. Sawada, M. Nakamura, and T. Higashimoto. Mechanical Voice System and Its Singing Performance. In *Proceedings of 2004 1EEE/RSJ International Conference on Intelligent Robots and Systems,* October 2004. [49]

H. Sawada, M. Nakamura, and T. Higashimoto. Mechanical voice system and its singing performance. In *Proceedings of IEEE/RSJ International Conference on Intelligent Robots and Systems,* 2004 (IROS 2004), pages 1920-1925, October 2004. [50]

M. Nakamura and H. Sawada. Talking Robot and the Analysis of Autonomous Voice Acquisition. In *Proceedings of the 2006 IEEE/RSJ International Conference on Intelligent Robots and Systems,* pages 4684-4689, October 2006. [51]

P.C. Delattre, A.M. Liberman, and F.S. Cooper. Acoustic Loci and Transitional Cues for Consonants. *Journal of the Acoustical Society of America,* V.27(N.4), 1955. [52]

D. H. Klatt. Review of Text-to-Speech Conversion in English. *J. Acous. Soc. Am.,* pages 737-793, 1987. [53]

H. Sakoe and S. Chiba. Dynamic Programming Algorithm Optimization for Spoken Word Recognition. *IEEE Trans. Acoust. Speech Signal Processing,* V.26:43-49, 1978. [54]

L. Rabiner and B.-H. Juang. *Fundamentals of Speech Recognition.* Prentice—Hall, 1993.[56]

F. Parke and K. Waters. *Computer Facial Animation.* A.K. Peters, 1996. [55]

A. Wrench. MOCHA-TIMIT http://www.cstr.ed.ac.uk/research/projects/artic/mocha.html, November 1999. [57]

J.S. Perkell, M.H. Cohen, M.A. Svirsky, M.L. Matthies, I. Garabieta, and M.T.T. Jackson. Electromagnetic midsagittal articulometer systems for transducing speech articulatory movements. *Journal of the Acoustical Society of America,* V.92:3078-3096, 1992. [58]

W.J. Hardcastle, W. Jones, C. Knight, A. Trudgeon, and G. Calder. New developments in electropalatography: a state—of—the—art report. *Clinical Linguistic and Phonetics,* V.3:1—38, 1989. [59]

S. Hiroya and M. Honda. Estimation of Articulatory Movements From Speech Acoustics Using an HMM-Based Speech Production Model. *IEEE Transactions on Speech and Audio Processing,* Vol. 12(N. 2):175, March 2004. [60]

26

Bilinear Time Delay Neural Network System for Humanoid Robot Software

Fumio Nagashima
Fujitsu Laboratories Ltd.
Japan

1. Introduction

Recently there are many humanoid robot research projects for motion generation and control, planning, vision processing, sound analysis, voice recognition and others. There are many useful results for these elementary technologies by the effort of researchers. The humanoid robot HOAP (Fig.1, Fujitsu Automation) contributed these researchers and I also received a benefit from this robot for motion generation and control investigation. And in parallel, several researchers are working on formulation as a system based on these elementary technologies. The researchers in this area think how to integrate the system using many technologies and how to realize a suitable system for a certain purpose.

Figure 1. Humanoid robot, HOAP

There are two ways of thought to build the system, "top-down" and "bottom-up" (Minsky, 1990). Standing to "top-down" approach, they analyze the system regarding the total system balance and determine the interface between the elements from the total system demand. In this case, they must consider the efficiency to build the system and investigate the re-use

technology. The typical example using this way is Humanoid Robot Project (HPR, Nakamura et al., 2001) in Japan. They analyze the tasks for each purpose, search common technologies and integrate these modules.
On another hand, "bottom-up" approach researcher try to design the small units and to make some rules for building the system using these units. In this area, there are some important results (Hawking, 2004). But there are very few results far less than ones based on "top-down" approach, because it's difficult to make a small unit model and to make rules for all purposes needed for humanoid robots.
I show the design of a unit for ``bottom-up'' approach in this chapter. This design is based on an artificial neural network model. The concept of this design differs from ordinary neural networks. It's for building the total software system of humanoid robot.

2. Neural Network

Because an artificial neural network is a mathematical model of biological neural network, it is expected to be a suitable model for building humanoid robot software system. But several researches on this area do not stand to this purpose. Many of them are related to new type mathematical problems, non-linear equation system. Such non-linearity is a part of reason for difficulty to build the large scale software system. The non-linearity must be in the connections, must not be in the neuron model itself. I give a reason and explain in last part of this section.
In this section, I outline the suitable model for building the large scale software system like humanoid robot one. The non-linearity scrambled out of neuron. First, I describe ordinary non-linear neural network shortly. It's as a non-linear investigation tools.

2.1 Tool for Non-linear Research

The research of artificial neural network has a long history (Mcculloch, 1943). Some people believe this research area is a new activity for strong non-linearity and chaos related phenomena. Usually, the mathematical neuron models have the non-linearity, for example, sigmoid function. And the very useful learning method, back propagation (BP) is developed based on such a non-linearity. But it's important to remind that the non-linear phenomena research and the neural network one are two different things.
The neural network using strong non-linearity has ups and downs. In the case of high dimensional problem, the strong non-linear network has an upper hand over linear ones (Barron, 1993). The invention of back propagation method may be a landmark event.
This very useful method is applied to various kinds of problem. Some researchers use the threshold for a layered neural network (Nakano, 1972) and developed the method for association. It might be a promising step to brain function. In some cases, the non-linear factor has a superiority than the combination of linear factors (Barron, 1993). Some researchers also use the threshold for a neural oscillator (Kimura, 2003) and are interested in a phenomenon of limit cycle and entrainment between input signal and internal state. There researches demonstrate that the very small scale non-linear oscillator has functions of motion generation and adjusting itself to the variety of environment. It's amazing result. But this group suffers from the disadvantage of scale problem.
A particular kind of non-linearity can produce multiple jobs. On the other hand, a linear system cannot perform like this. But it's impossible to combine a number of non-linear

networks because there are strong interferences among networks even if each network is well constructed.
The artificial neural network is the mathematical model of biological neural network. Fundamentally speaking, an artificial neural network is a word categorized in a biologically inspired engineering and the non-linearity is a mathematical term. These are the words that are quite different research regions. All mathematical model based on biological neural network is an artificial neural network.
The research of engineering using non-linear effect is interesting and might be useful and proper biological model. But, we must give more attention to an alternative roll of artificial neural network, common notation for a large scale system.
Human's or animals' brain consist of biologically neural network. And these are very large scale systems. Therefore it is easy to get the idea that the artificial neural network can be applied to a large scale system.

2.2 Tool for Building the robot system

Generally a notation of neural network is simple. It consists of neurons and wirings as shown in Fig. 2. Such a uniform notation has a possibility for a common language of a huge variety of systems. For whole humanoid robot software system, we need many categories of systems, motion generation and control, vision processing, sound analysis and etc. I discuss the neural network model that is suitable for that sake of common language for all purpose of humanoid robot software system.

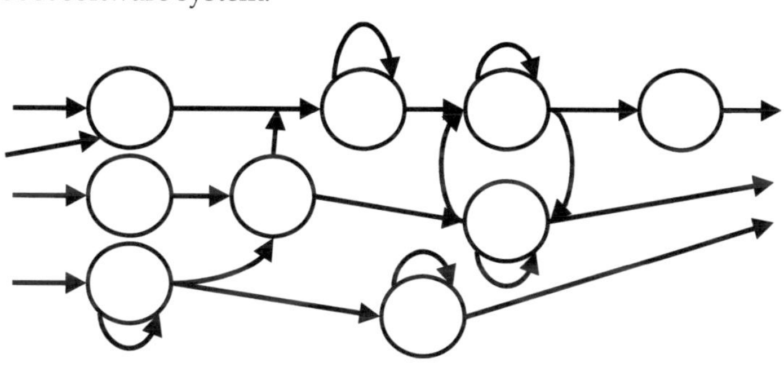

Figure 2. Neural Network

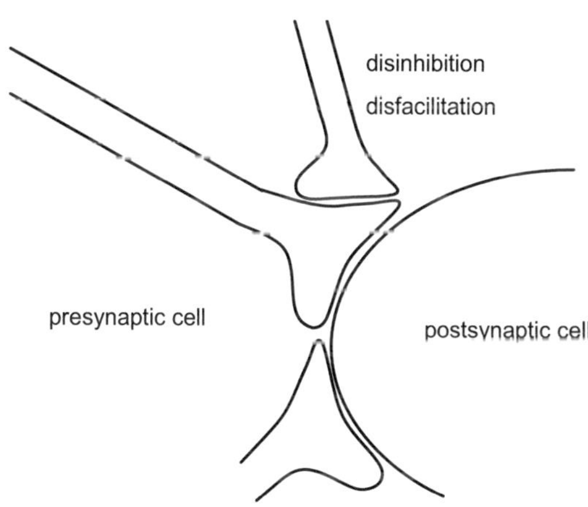

Figure 3. Disinhibition, disafacilitation

Because the many engineering systems draw on linear operations, the goal system must be able to express the linear system for both algebraic and differential equation. There are many examples indicating this fact. PD control, the fundamental feedback control, outputs the value proportional to the input of position and velocity. Correlation calculation is for recognition of many things and it needs inner product between input and template. There operations are based on linear algebra. Moreover, PID control needs the integrator. The linear differential equation can output the some special functions, that is, triangle functions, polynomials and other eigen functions. In addition, the logical calculation is important for conditional process. As a result, the system must describe the following:

- linear algebra
- differential equation
- logical calculation

There are several formulations satisfying the above conditions. I use the following simple formula in the system.

$$\varepsilon_i \frac{dy_i}{dt} + y_i = \sum_j C_{ij} y_j + \sum_{j,k} C_{ijk} y_j y_k + \sum_{j,k} C'_{ijk} H(y_j) y_k \tag{1}$$

The left hand side represents the "delay". The parameter ε_i is the "delay" scale. y_i, y_j, y_k are state quantity of neurons. $C_{ij}, C_{ijk}, C'_{ijk}$ are connection weight, constant. $H()$ is a step function. Because all information transmission has delay in natural world, this formula is reasonable. The first term of right hand side is a linear connection, the second term is a bilinear connection and the third term is a digital bilinear connection. It becomes the linear equation putting $\varepsilon_i, C_{ijk}, C'_{ijk} \to 0$ as follows:

$$y_i = \sum_j C_{ij} y_j \tag{2}$$

Adding y_i to right hand side as follows:

$$\varepsilon_i \frac{dy_i}{dt} + y_i = y_i + \sum_{j,j\neq i} C_{ij} y_j + \sum_{j,k} C_{ijk} y_j y_k + \sum_{j,k} C'_{ijk} H(y_j) y_k \tag{3}$$

It becomes an integrator for solving the differential equations.

It becomes a neuron keeping constant value, putting $\varepsilon_i \to \infty$ as follows:

$$\lim_{\varepsilon_i \to \infty} \left\{ \frac{dy_i}{dt} + \frac{1}{\varepsilon_i} y_i = \frac{1}{\varepsilon_i} \left(\sum_j C_{ij} y_j + \sum_{j,k} C_{ijk} y_j y_k + \sum_{j,k} C'_{ijk} H(y_j) y_k \right) \right\}$$
$$\longrightarrow \frac{dy_i}{dt} = 0 \tag{4}$$

Bilinear connection is for the important operation, that is, measurement, higher order polynomial and triangle functions, inner product and etc. This mechanism is the model of biological fact, disinhibition or disfacilitation (Fig.3).

I should notice that this model has the similarities with fluid mechanics or something like that. Because Newtonian equation show the acceleration is exactly proportional to the external force, the motion equation of a infinitely small fluid particle is linear. But, The Navier-Stokes equation is a famous as non-linear equation system. In this case, the non-linearity also come from the interaction among particles. Most non-linearity is come from the interactions among a number of linear systems in natural world.

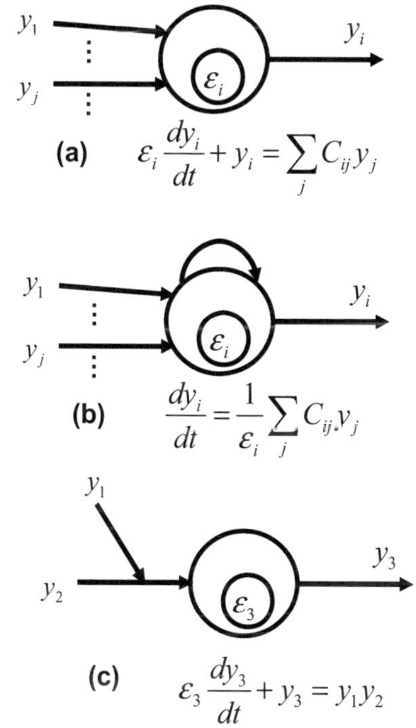

Figure 4. Neuron model

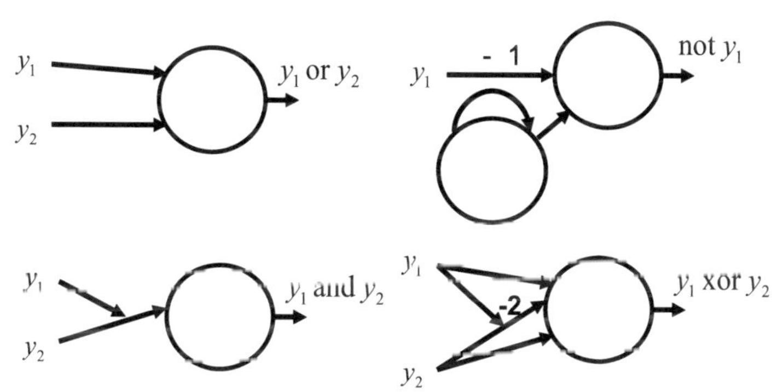

Figure 5. Logical operation

I show the outline of the model in Fig.4. Using this neural network, the logical operations (or, and, not, xor) can be expressed as Fig.5. I developed a language for the above concepts. I call this RNN language. NueROMA (Nagashima, 2004), Fujitsu Automation's Product, includes RNN language simulator, RNN language executer, dynamics simulator of rigid body system, RNN graphic editor as shown in Fig.6. You can download the trial NueROMA software including language specification using BNF (Dick, 1991).Table 1 show the neuron notation in this system. Table 2 show the connection notation. Table3 show the simple example of network. The digital switch is a digital bilinear connection. The threshold is a special case of digital switch. Table 4 show the practical example.

Network notation	RNN language notation	Remarks
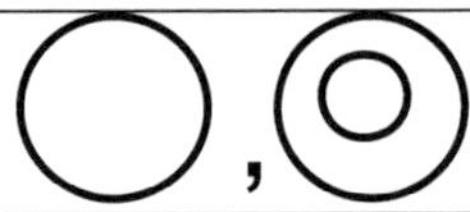	var a;	Neuron
	const V=1; var a=V;	Neuron with initial value
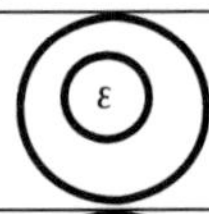	const eps=0.1; var a(eps);	Neuron with delay parameter

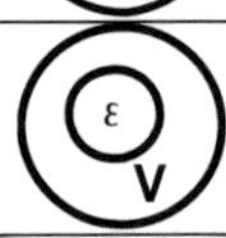	const V=1;
	const eps=0.1;
	var a(eps)=V;
		Neuron with delay parameter and initial value

	var a(0.0);	Terminal neuron

Table 1. Neuron notation

Network notation	RNN language notation	Remarks

	const C=1.2;
	var n1, n2;
	...
	n1 := C*n2;
		Connection

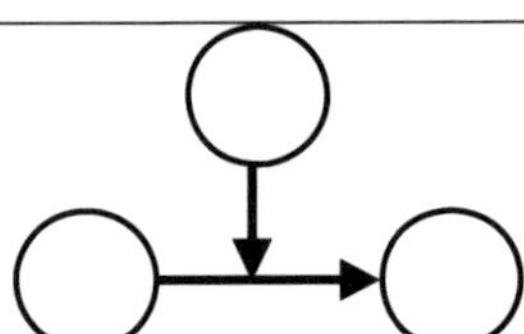	var n1, n2, n3; ... n1 := n2 * n3;	Bilinear connection
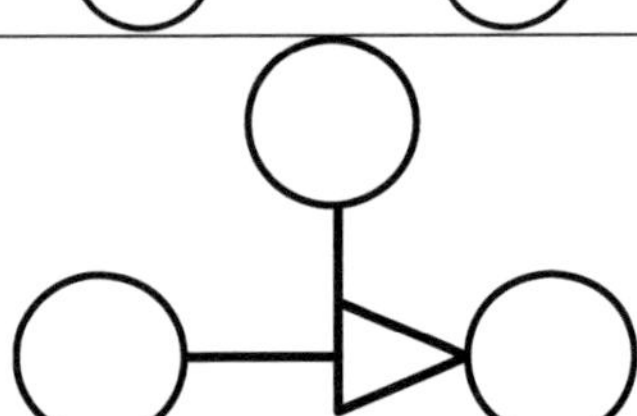	var n1, n2, n3; ... n1 := if(0<n2) 1 * n3;	Digital switch
	var n1, n2; ... n1 := 1 * (0<)n2;	Threshold

	var(eps00) = V0;
	...
	var n1(eps);
		Variable delay parameter

Table 2. Connection Notation

Network notation	RNN language notation	Mathematical notation

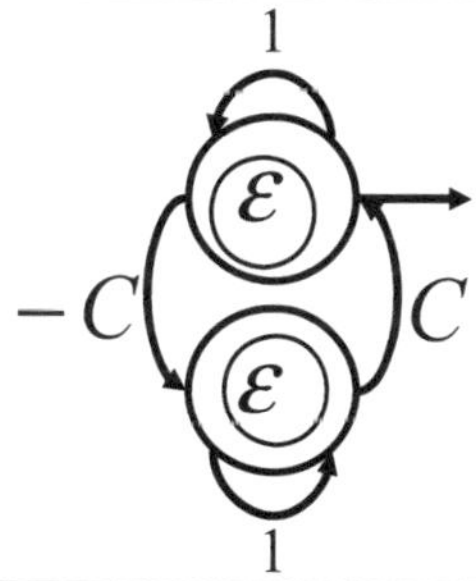	circuit cir1 {
	const C = 1.0;
	const eps = 0.1;
	var y1(eps) = 0.0;
	var y2(eps) = 0.0;
	y1 := 1.0 * y1 + C * y2;
	y2 := 1.0 * y2 - C * y1;
	}
		$\varepsilon \frac{dy_1}{dt} + y_1 = y_1 + Cy_2$
		$\varepsilon \frac{dy_2}{dt} + y_2 = y_2 - Cy_1$

Table 3. RNN Example

Network	Remarks
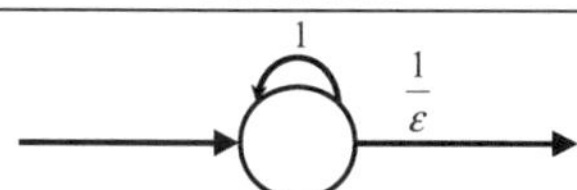	Integrator
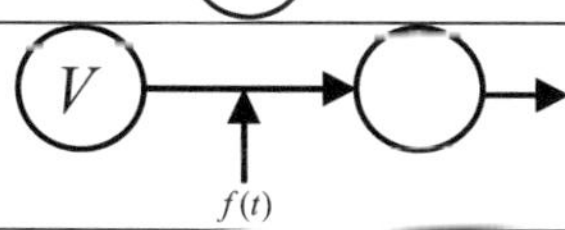	Inner Product

Table 4. Practical RNN Example

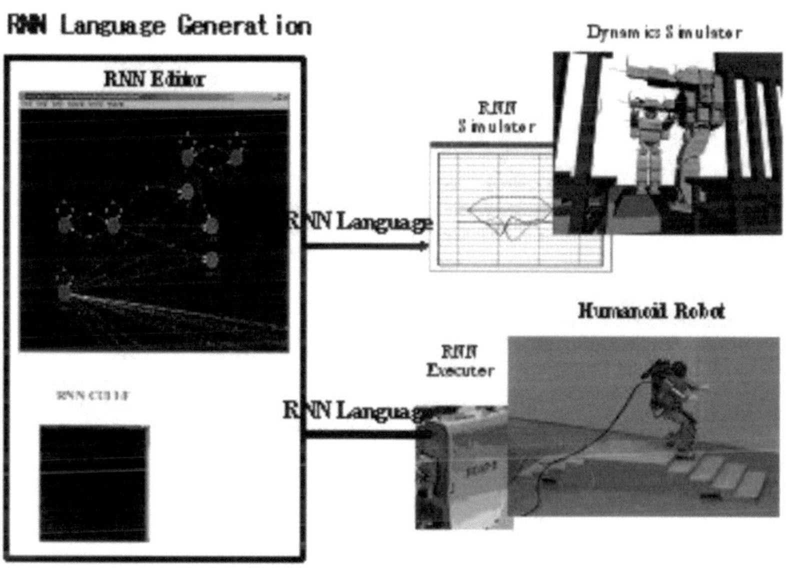

Figure 6. Nueroma, the neural network editing and simulation tool

3. Basic Element Structure

There are many kinds of neural network until now. But the relationship among them is not discussed circumstantially. Especially, the researchers did not give much thought to the transformation between these. It's not a problem if neural network is for a part of software system. But it's not a good approach for building the whole software system.

The purpose of the discussion in this chapter is to propose the neural network system for total robot software system. The goal of this discussion is the notation for dynamically growing neural network. For this goal, it is necessary to discuss the relationship between some models and the transformation from a typical neural network model to another. In this section, I discuss the important sub-networks and their relationship.

3.1 Linear Mapping, Correlation Calculation, Associative Memory

Linear transformation is a basic operation for proposed model. Using this operation I can build the correlation calculation network for estimate the similarity between an input and the template. Adding the threshold mechanism to this, the associative memory can be gotten (Nakano, 1972).

3.2 CPG

CPG is investigated by biological researcher (Grillner, 1985). It becomes known that a biological neural network can generate the non-zero output without input. This fact can be modelled using a successive linear transformation. I show the modification process using simple example. First, think the simple linear transformation, the above chart of Fig.7 can be describe by following equation:

$$\begin{bmatrix} y_1 \\ y_2 \end{bmatrix} = \begin{bmatrix} 1 & -1 \\ 1 & 1 \end{bmatrix} \begin{bmatrix} x_1 \\ x_2 \end{bmatrix} \tag{4}$$

This equation is a kind of associative memory. Next, think the successive transformation of this equation as follows:

$$\begin{aligned} y_1^{(0)} &= 1 \\ y_2^{(0)} &= 0 \\ y_1^{(i+1)} &= y_1^{(i)} - y_2^{(i)} \\ y_2^{(i+1)} &= y_1^{(i+1)} + y_2^{(i)} \end{aligned} \tag{5}$$

This numerical sequence becomes as middle below chart of Fig.7. The cycle is just 6. Lastly, think the limitation of infinitely small time i and the following equation can be gotten:

$$\frac{dy_1}{dt} = -y_2, \frac{dy^2}{dt} = y_1 \tag{6}$$

Then,

$$\frac{dy_1}{dt} + y_1 = 0 \tag{7}$$

This equation has a triangle function with the cycle, $2\pi(\approx 6)$. I call the network like this CPG. CPG is considered as a successive associative memory. Fig.7 and Fig.8 show the process diagram of the transition from the simple linear transformation to CPG. CPG in Fig.8 output a linear function.

If it is assumed that this network is an associative memory, output of the network called a recall. Second network shows a successive recall using the same network. A broken line in a bottom graph shows this network output through time. Third network shows a continuous recall. A solid line in a bottom graph shows the continuous recall output. It is a CPG.

3.3 Bilinear, Parametric Resonance

Bilinear connection is a variable coefficient equation from the view of linear mathematics. Using the mechanism, the inner product, parametric resonance and some important calculation are realized.

3.4 Chaos

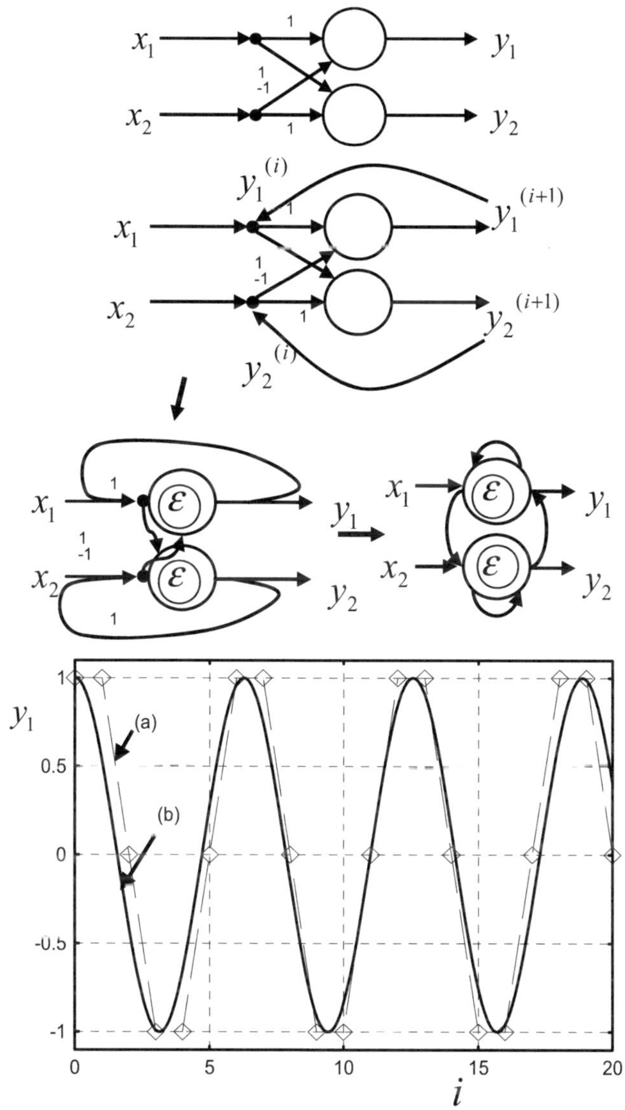

Figure 6. Associative memory and cyclic CPG

The low order chaos can be realized by a proposed system. But to keep the network understandable is very important to realize the large scale system. It is strongly recommended to use such a network carefully. Fig. 9 shows the network representing a Lorentz equation (Lorenz, 1963) as follows:

$$\begin{aligned} \frac{dx}{dt} &= -\sigma x + \sigma y \\ \frac{dy}{dt} &= -y + rx - xz \\ \frac{dz}{dt} &= -bz + xy. \end{aligned} \tag{8}$$

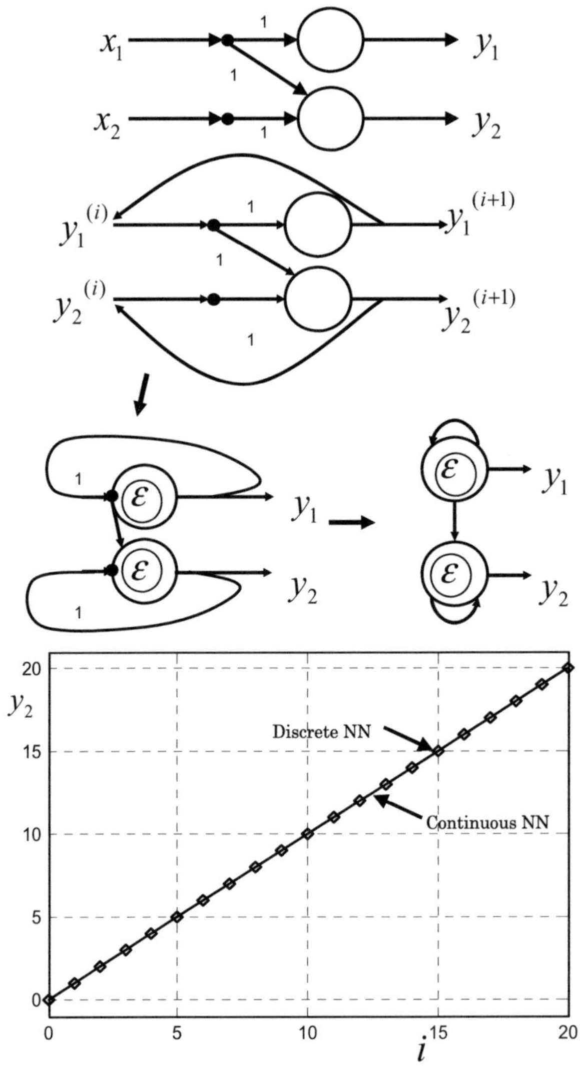

Figure 7. Associative memory and non-cycle CPG

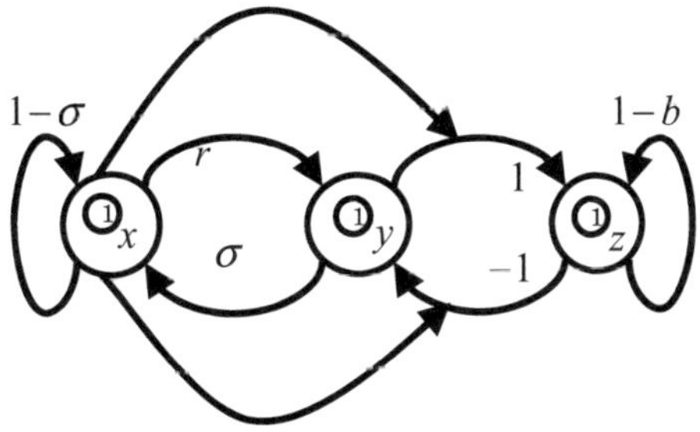

Figure 8. Neural network expression of Lorenz equation

4. Guideline for System Building

In this section, I discuss about the guideline for building the system. Usually the environments around a robot are very complicated. It is expected that a robot adjust itself for such environments. Mathematically these complexities are represented by non-linear equation systems. We can formulate an equation system in a few cases and cannot formulate that in many cases. Hence we need an algorithm to solve the non-linear equation system in the both given and unknown cases systematically.

4.1 Perturbation Method

In old days, astronomical scientist wanted to determine the orbit of planets. But it was hard because the governing equation system has non-linearity (Poincare, 1908) and they did not have a good calculation machine for numerical solution. Therefore, they developed perturbation method for getting the approximate solution of such a non-linear equation system.

If we can set up the equation system to determine the coefficient of neural network, we can use the perturbation techniques even if the system has non-linearity. A lot of techniques are developed (Hinch, 1991, Bellman, 2003). In this section, I show you a basic example using following equation(Hinch, 1991):

$$x^2 + \varepsilon x - 1 = 0 \tag{9}$$

This equation is analytical solved and has the solution as follows:

$$x = -\frac{1}{2} \pm \sqrt{1 + \frac{1}{4}\varepsilon^2} \tag{10}$$

For the explanation of perturbation technique, I assume ε is a small parameter and can be ignored in a first step as follows:

$$x^2 - 1 = 0 \tag{11}$$

This equation is easily solved and has solution, $x = \pm 1$. This solution is a first approximation of the solution of Eq. (11) in a sense of perturbation technique. ε is small but not equal to zero, we can assuming x as follows:

$$x(\varepsilon) = x_0 + x_1\varepsilon + x_2\varepsilon^2 + \cdots \tag{12}$$

Substituting this equation into Eq.(11), we get the following sequential equations and solutions:

$$
\begin{aligned}
O(\varepsilon^0) &: x_0^{\ 2} - 1 = 0 \longrightarrow x_0 = 1 \\
O(\varepsilon^1) &: 2x_1 = -x_0 \longrightarrow x_1 = -\frac{1}{2} \\
O(\varepsilon^2) &: 2x_2 = -x_1^{\ 2} + x_1 \longrightarrow x_2 = -\frac{1}{8}
\end{aligned}
\tag{13}
$$

I choose the positive solution in the above. I show these approximate solutions and exact solution in Fig.10. It is noteworthy that it is good approximation even in the region, $\varepsilon \approx 1$.

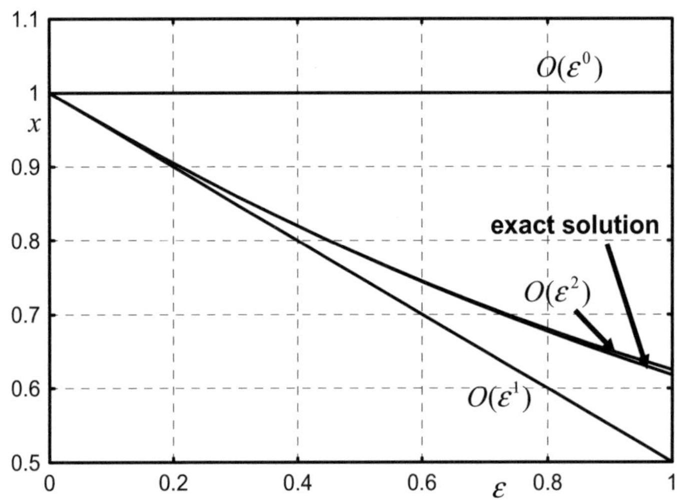

Figure 9. Simple example of perturbation

In the case of differential equation system, the same concept is applied. For example, the following solution is a one of typical expansion for the equation having cyclic solution.

$$x(t) = c_0 + c_1 \cos t + s_1 \sin t + c_2 \cos 2t + \cdots \tag{12}$$

For the equation having non-cyclic solution, the expansion $t, t^2, t^3, \cdots$ is suitable assuming t is small. There are many types of expansion. See the references (Hinch, 1991, Bellman, 2003), for the variety of examples.

4.2 Numerical Perturbation Method

Unluckily if we cannot get the governing equation system, the same concept can be applied to solving the problem. I call this process "numerical perturbation method" and discuss in the section "Motion Generation and Control" and its references.

5. Applications

In this section, I show some applications based on this method concretely.

5.1 Forward Kinematics

First of all, I show the basic problem in robotics, forward kinematics. The forward kinematics is the combination of triangle functions. Then, I start the discussion with the triangle function. To realize the triangle function in this system, Taylor-Maclaurin expansion is used as follows:

$$y(\theta) = c_0 + c_1\theta + c_2\theta^2 \cdots \tag{19}$$

where y is a arbitrary function of θ and c_0, c_1, c_2 are constants. I show the general neural network for such a expansion in Fig.11(a), and the concrete example, $\sin\theta$ (Eq.(20)) in Fig.11(b).

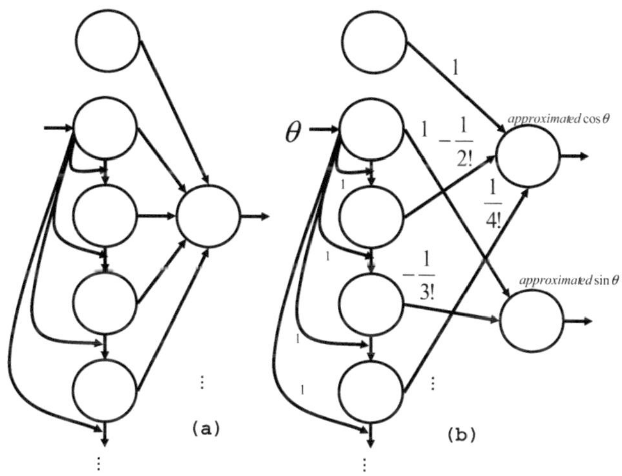

Figure 11. Taylor expansion neural network

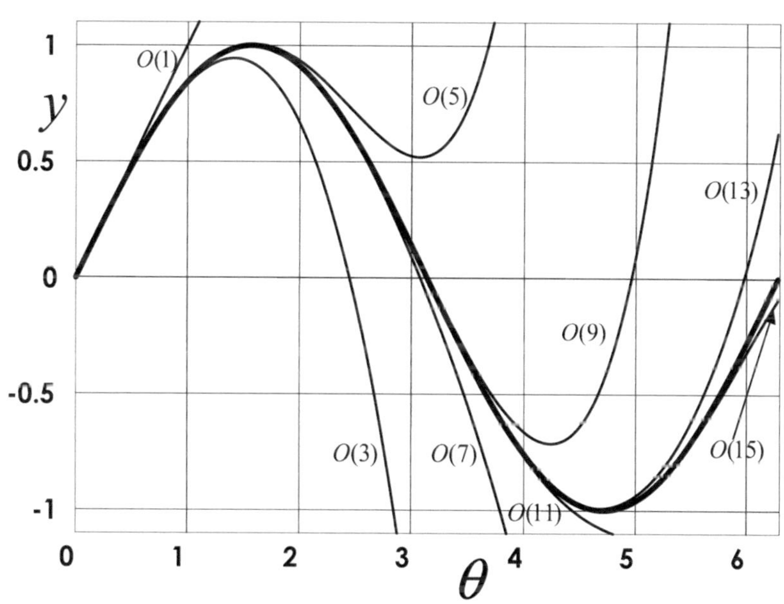

Figure 12. Approximate triangle function by neural network

$$y(\theta) = \sin\theta = \theta - \frac{1}{3!}\theta^3 + \frac{1}{5!}\theta^5 - \frac{1}{7!}\theta^7 + \frac{1}{9!}\theta^9 - \frac{1}{11!}\theta^{11} + \cdots \tag{20}$$

Fig.12 shows the result of such an approximation. In this figure, the heavy line is $\sin\theta$, the other lines are approximate solution, $O(n)$ represents $O(\theta^n)$ solutions. In this case, $O(17)$ solution is drawn, but it's just on the exact solution's line.

5.2.1 Analytical Method

In this sub-section, I show illustrative case. The simple arm (Fig.13) kinematics solution can be solved analytically as follows:

$$\begin{aligned} x &= \sin\theta_1 (l_1 \sin\theta_2 + l_2 \sin(\theta_2 + \theta_3)) \\ y &= \sin\theta_1 (l_1 \sin\theta_2 + l_2 \sin(\theta_2 + \theta_3)) \\ z &= l_1 \sin\theta_2 + l_2 \sin(\theta_2 + \theta_3) \end{aligned} \tag{21}$$

The neural network representing this equation is as Fig. 14

5.2.2 Numerical Method

The kinematics problem is a fitting problem for multivariable polynomial. Then this problem is determining coefficients problem assuming the position is as follows:

$$\mathbf{X}(\boldsymbol{\theta}) = \mathbf{X}(0) + \left(\delta\boldsymbol{\theta} \bullet \frac{d}{d\boldsymbol{\theta}}\right)\mathbf{X}(0) + \frac{1}{2!}\left(\delta\boldsymbol{\theta} \bullet \frac{d}{d\boldsymbol{\theta}}\right)\mathbf{X}(0) + \cdots \tag{22}$$

To determine coefficients problem is calculation follwing values from the actual measurements.

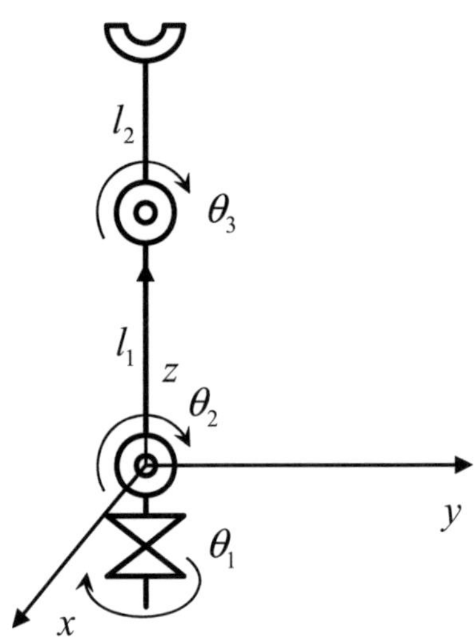

Figure 13. Simple Arm

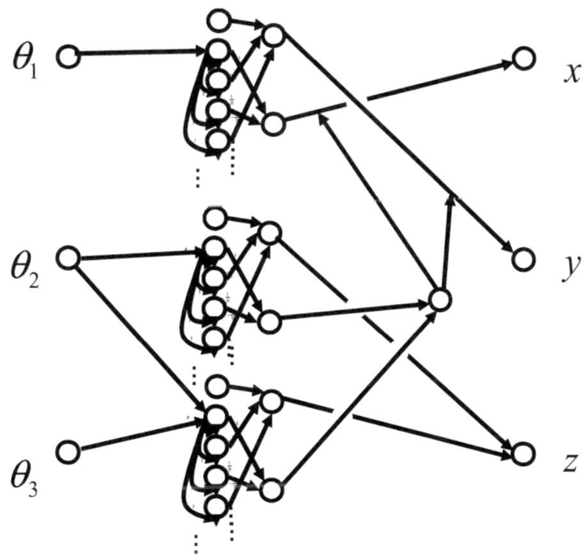

Figure 14. Kinematics neural network

$$x_i, \frac{dx_i}{d\theta_j}, \frac{d^2x_i}{d\theta_j d\theta_k}, \frac{d^2x_i}{d\theta_j d\theta_k d\theta_l}, \cdots \tag{23}$$

Fig. 15 show the growing neural network in this process.

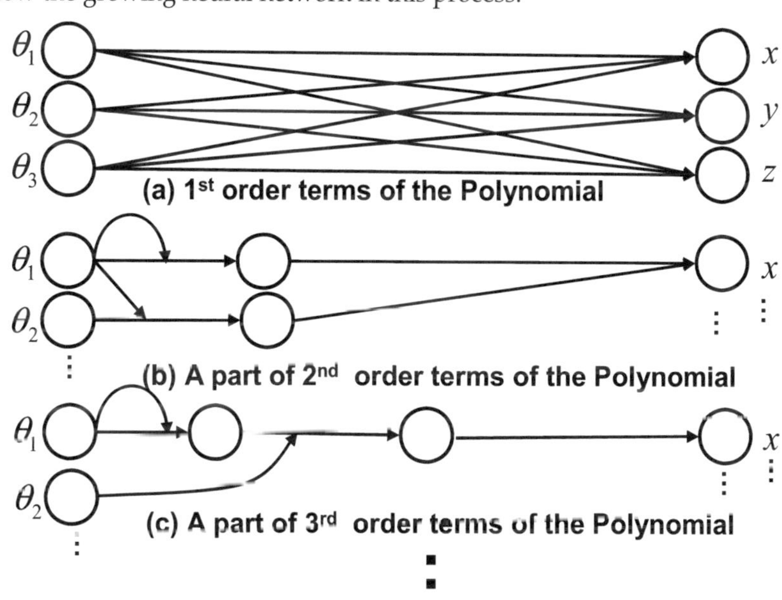

Figure 15. Kinematics neural network by learning process

5.2 Inverse Kinematics

In this sub-section, I discuss the inverse kinematics problem. In this problem, the solution has an inverse trigonometric function.

5.2.1 Analytical Method

The inverse kinematics of the arm (Fig.13) has 4 type solutions. In this sub-section, I only discuss following solution. The other solutions can be considered in the similar way.

$$\begin{aligned}
\theta_1 &= \tan^{-1}\frac{x}{y} \\
\theta_3 &= \cos^{-1}\left\{\frac{1}{2l_1 l_2}\left[\left(x^2+y^2+z^2-\left(l_1^{\,2}+l_2^{\,2}\right)\right)\right]\right\} \\
\theta_2 &= \cos^{-1}\frac{z}{r}+\alpha
\end{aligned} \tag{24}$$

In the similar fashion as the forward kinematics, there is a convergence radius problem of the Taylor expansion of the term, $1/y, \tan^{-1}(x/y), \cos^{-1}(z/r)$. It can be avoidable using the concept of analytical continuation. For example, because the function, $\tan^{-1}(x/y)$ has the singular points at $\pm i$, the expansion near $x/y=0$ breaks down near ± 1. Fig.16 show the example of region splitting for $\tan^{-1}(x/y)$. Such techniques keep a high accuracy for wide range. Fig. 17 shows a part of the neural network using such techniques. It includes the digital switch neuron.

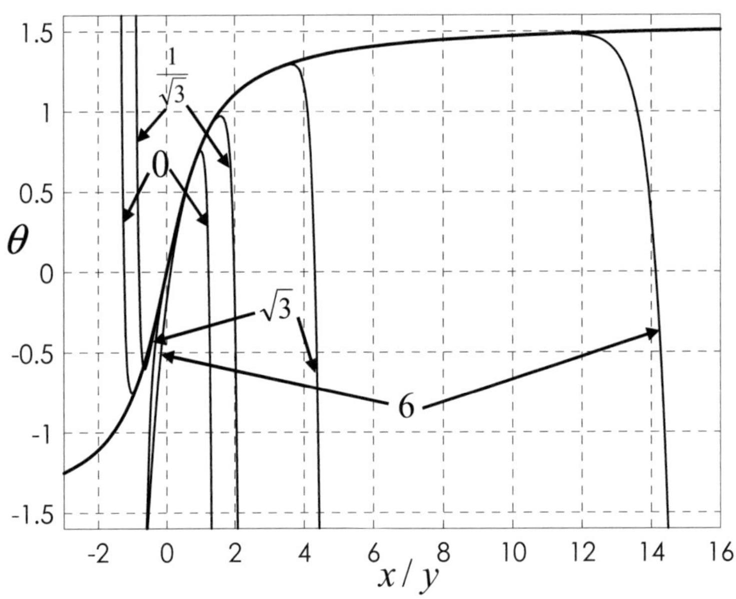

Figure 16. Approximate solution of $\theta = \tan^{-1}\frac{x}{y}$

5.2.2 Numerical Method

If we do not have the information of arm, the above technique can be applied numerically. In this case, the neural network has the some polynomials with digital switch (Fig.17).

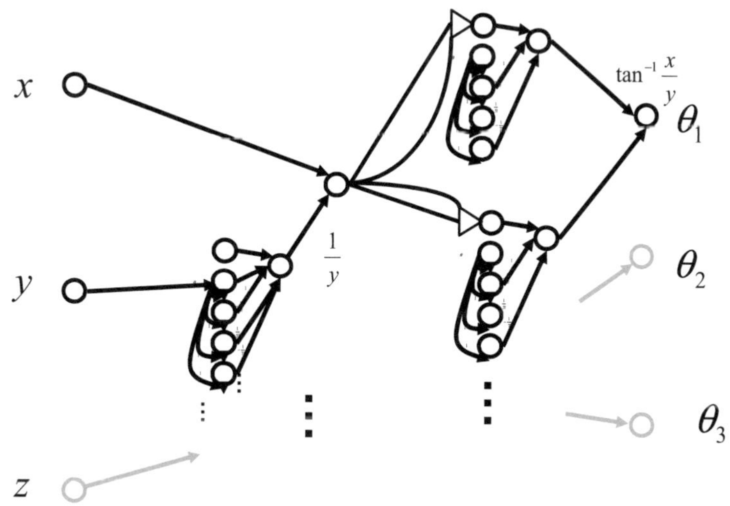

Figure 17. A part of inverse kinematics neural network

5.3 Motion Generation and Control

There are many references about the motion generation and control using this system. See references(Nagashima, 2003). Fig. 18 show the example of growing neural network for motion.

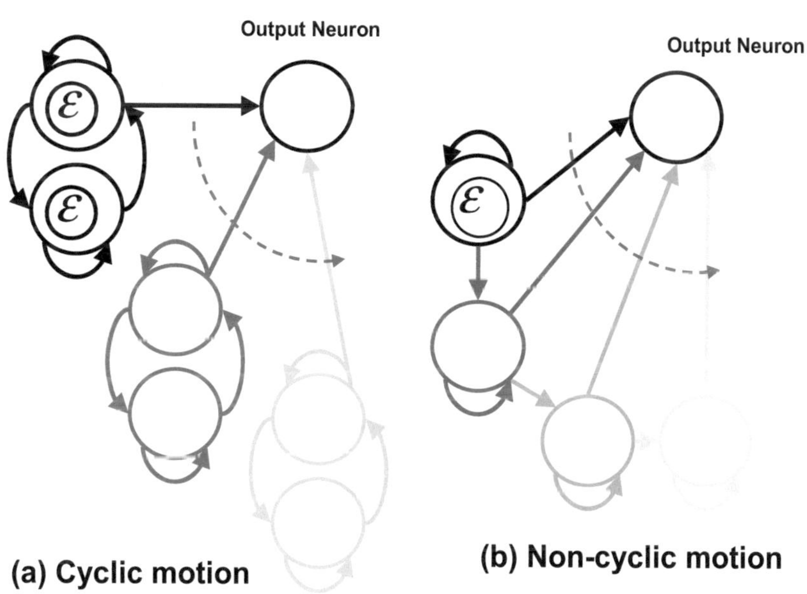

Figure 18. Typical perturbation process for motion neural network

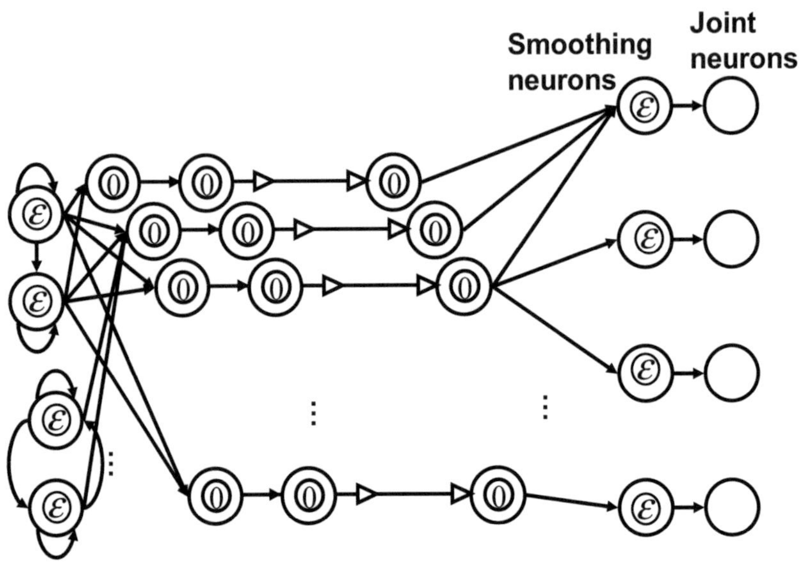

Figure 19. An example of motion neural network

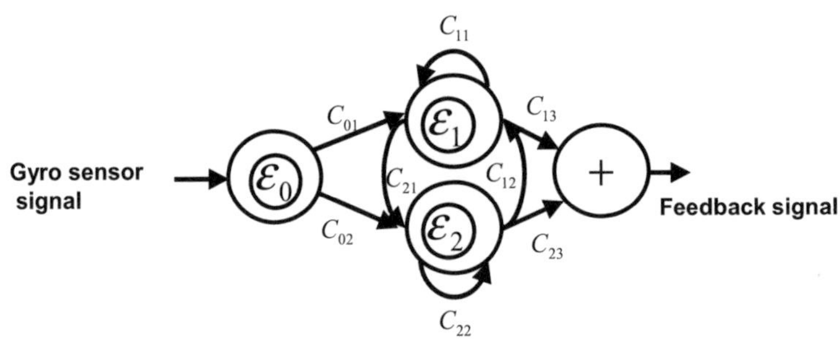

Figure 20. An example of feedback neural network

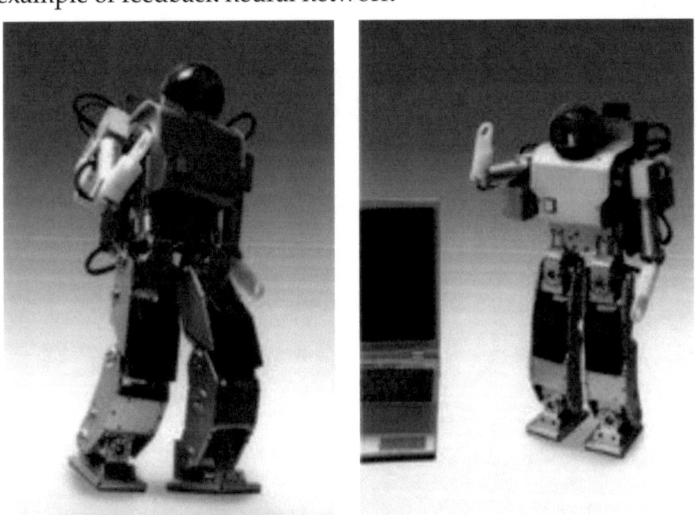

Figure 21. Experiments using HOAP

Fig. 19 shows the neural network example for real motion. Fig. 20 shows the feedback neural network for stabilize the upper body of humanoid robot. Fig. 21 shows the HOAP, humanoid robot for experiment of walk.

5.4 Sound Analysis

It is popular that the Fourier transformation (FT) is applied to the pre-processing of a sound analysis. Usually neural network for sound analysis uses the result of this FT. But it's unnatural and wrong in a sense of neural network as a total system basis. In this section, I discuss the problem of the transformation of signal.

Proposed model can compose the differential equation for triangle functions as shown in previous section. I call this network CPG. Putting the signal to the neuron and wire of this CPG, this becomes the resonator governed by the following equation,

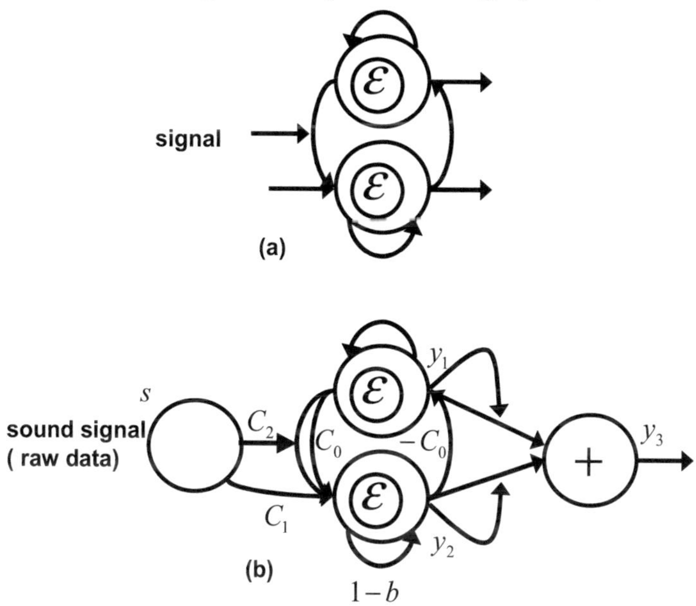

Figure 22. Central Pattern Recognizer (CPR)

$$\frac{d^2 y_1}{dt^2} + \beta \frac{dy_1}{dt} + \omega^2 (1 + \delta_m s) y_1 = \delta_f s \tag{25}$$

$$y_2 = -\omega \frac{dy_1}{dt} \tag{26}$$

$$y_3 = y_1^2 + y_2^2 \tag{27}$$

where s is input signal, $\beta, \omega, \delta_m, \delta_f$ are constants determined by connection weights, y_1, y_2 are neuron value. Fig. 22 shows the neural network for this. It vibrates sympathetically with the input signal and this can recognize a specific frequency signal. I call this network Central Patten Recognizer (CPR). Using a number of this network, it is possible to create the network which function is very similar to FT. Fig.23 shows the elements of cognitive network, CPR. Fig.24 shows the output of this network and it is a two-dimensional pattern. The sound analysis problem becomes the pattern recognition of this output. The pattern recognition problem is solved by the fitting function problem similar to kinematics problem.

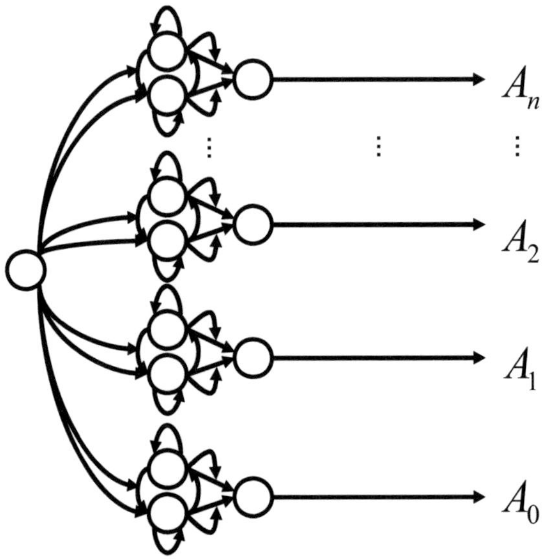

Figure 23. Sound analyzing neural network

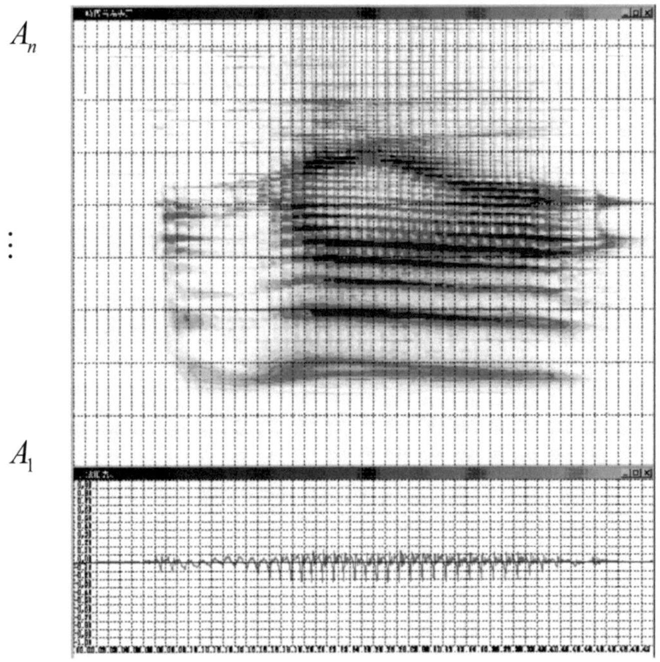

Figure 24. Example of sound analyzing results using CPR

5.5 Logical Calculation

Logical calculation is a basic problem for neural network in old days. Especially, exclusive-OR problem offer the proof that perceptron cannot solve the non-linear problem. In this section, I show examples which can solve the nonlinear logic problem using proposed system. Fig.5 shows the basic logical calculations, (or, and, not, xor). Fig.25 shows the simple application network. It shows the half adder which is the lowest bit adder circuit in an accumulator.

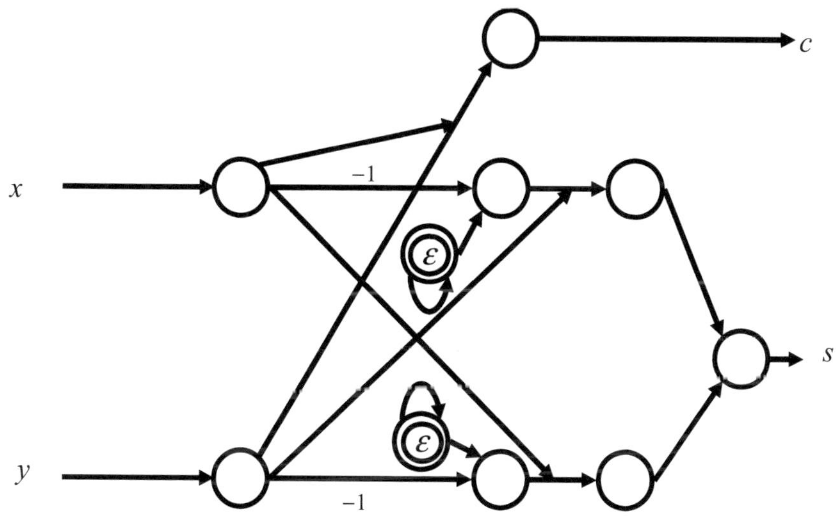

Figure 25. Half adder

5.6 Sensor Integration

The goal of this method is an integration of software system. The fusion of sensor problem is a well-suited application of this model. The concept of an associative memory is known as the introduction to a higher cerebral function. Especially, autocorrelation associative memory is important (Nakano, 1972). This concept can restore the noisy and muddy information to original one. Proposed model can work out this concept through the multiple sensing information. This fact is very important. Proposed model can treat the all sensing data evenly. In a similar fashion, the sensing result information at any levels can be treated evenly.

6. Conclusion

In this chapter, I describe the neural network suitable for building the total humanoid robot software system and show some applications. This method is characterized by

- uniform implementation for wide variety of applications
- simplicity for dynamically structure modification

The software system becomes flexible by these characteristics. Now, I'm working on the general learning technique for this neural network. There is a possibility free from NFL problem (Wolper, 1997).

This chapter is originally written for the RSJ paper in Japanese (Nagashima, 2006).

7. References

Barron, A,(1993). Universal Approximation Bounds for Superpositions of a Sigmoidal Function, *IEEE Trans. on Information Theory*, IT-39, pp.930-944.

Bellman, R,(2003). Perturbation Techniques in Mathematics, *Engineering and Physics*, Dover Publications, Reprint Edition, June 27.

Fujitsu Automation. http://jp.fujitsu.com/group/automation/en/.

Grillner, S, Neurobiological Bases of Rhythmic Motor Acts in Vertebrate, *Science 228*, 143-149

Grune, D and Ceriel J.H.Jacobs, Parsing Techniques,(1991). *A Practical Guide*, Ellis Horwood Ltd.

Hawking, J and Blakeslee, S, (2004). *On Intelligence*, Times Books

McCulloch, W. and Pitts, W. ,(1943). A Logical Calculus of the Ideas Immanent in Nervous Activity, *Bulletin of Mathematical Biophysics*, 5, pp. 115-133

Hinch, E.J., etc., (1991). *Perturbation Methods*, Cambridge University Press, October 25.

Kimura, H, Fukuoka, Y. and Cohen A.H.,(2003). Biologically Inspired Adaptive Dynamic Walking of a Quadruped Eobot, in *8th International Conference on the Simulation of Adaptive Behavior*, 2003, pp. 201-210

Lorenz, E.N., (1963). Deterministic nonperiodic flow, J. Atmos. Sci., 20, 130

Minsky, M. (1990). Logical vs. Analogical or Symbolic vs. Connectionist or Neat vs. Scruffy, *Artificial Intelligence at MIT*, Vol.1, Expanding Frontiers, MIT Press

Nagashima, F, (2003). A Motion Learning Method using CGP/NP, *Proceedings of the 2nd International Symposium on Adaptive Motion of Animals and Machines*, Kyoto, March, 4-8

Nagashima, F, (2004). - NueROMA -, Homanoid Robot Motion Generation System, *Journal of the Robtics Society of Japan*, Vol.22, No.2, pp.34-37 (in Japanese)

Nagashima, F ,(2006). A Bilinear Time Delay Neural Network Model for a Robot Software System, *Journal of the Robotics Society of Japan*, Vol.24, No.6,pp.53-64 (in Japanese)

Nakamura, Y. et al. (2001). Humanoid Robot Simulator fot the METI HRP Project, *Robotics and Autonomous Systems*, Vol.37, pp.101-114

Nakano, K., (1972). Associatron – A Model of Associative Memory, *IEEE Trans. on Systems Man and Cybernetics*, SMC 2, pp. 381-388

Poincare,(1908). *Science et Methode*

Bellman, R, (2003). Perturbation Techniques in Mathematics, Engineering and Physics, Dover Publications; Reprint Edition, June 27

Rumelhart, D. E. , Hinton, G. E. and McCelland, J. L. (1986). A General Framework for Parallel Distributed Processing, In D.E.Rumelhart and J.L.McClelland(Eds.),: *Parallel Distributed Processing: Explorations in the Microstructure of Cognition*, Cambridge, Ma, MIT Press, 1, pp. 45 - 76

Shan, J, Nagashima, F. (2002). Neural Locomotion Controller Design and Implementation for Humanoid Robot HOAP-1, *RSJ conference* , 1C34

Wolper, D.H. and Macready, W.G. (1997). No Free Lunch Theorems for Optimazation, *IEEE Transaction on Evolusionary Computation*, 1, 1, pp. 67-82

Robot Learning by Active Imitation

Juan Pedro Bandera, Rebeca Marfil, Luis Molina-Tanco, Juan Antonio Rodríguez, Antonio Bandera and Francisco Sandoval
Grupo de Ingeniería de Sistemas Integrados, Universidad de Málaga
Spain

1. Introduction

A key area of robotics research is concerned with developing social robots for assisting humans in everyday tasks. Many of the motion skills required by the robot to perform such tasks can be pre-programmed. However, it is normally agreed that a truly useful robotic companion should be equipped with some learning capabilities, in order to adapt to unknown environments, or, what is most difficult, learn to perform new tasks.

Many learning algorithms have been proposed for robotics applications. However, these learning algorithms are often task specific, and only work if the learning task is predefined in a delicate representation, and a set of pre-collected training samples is available. Besides, the distributions of training and test samples have to be identical and the world model is totally or partially given (Tan et al., 2005). In a human world, these conditions are commonly impossible to achieve. Therefore, these learning algorithms involve a process of optimization in a large search space in order to find the best behaviour fitting the observed samples, as well as some prior knowledge. If the task becomes more complicated or multiple tasks are involved, the search process is often incapable of satisfying real-time responses.

Learning by observation and imitation constitute two important mechanisms for learning behaviours socially in humans and other animal species, e.g. dolphins, chimpanzees and other apes (Dautenhahn & Nehaniv, 2002). Inspired by nature, and in order to speed up the learning process in complex motor systems, Stefan Schaal appealed for a pragmatic view of imitation (Schaal, 1999) as a tool to improve the learning process. Current work has demonstrated that learning by observation and imitation is a powerful tool to acquire new abilities, which encourages social interaction and cultural transfer. It permits robots to quickly learn new skills and tasks from natural human instructions and few demonstrations (Alissandrakis et al., 2002, Breazeal et al., 2005, Demiris & Hayes, 2002, Sauser & Billard, 2005).

In robotics, the ability to imitate relies upon the robot having many perceptual, cognitive and motor capabilities. The impressive advance of research and development in robotics over the past few years has led to the development of this type of robots, e.g. Sarcos (Ijspeert et al., 2002) or Kenta (Inaba et al., 2003). However, even if a robot has the necessary skills to imitate the human movement, most published work focus on specific components of an imitation system (Lopes & Santos-Victor, 2005). The development of a complete imitation architecture is difficult. Some of the main challenges are: how to identify which features of an action are important; how to reproduce such action; and how to evaluate the performance of the imitation process (Breazeal & Scassellati, 2002).

In order to understand and model imitation ability, psychology and brain science can provide important items and perspectives. Thus, the theory of the development of imitation in infants, starting from reflexes and sensory-motor learning, and leading to purposive and symbolic levels was proposed by Piaget (Piaget, 1945). This theory has been employed by several authors (Kuniyoshi et al., 2003, Lopes & Santos-Victor, 2005) to build robots that exhibit abilities for imitation as a way to bootstrap a learning process. Particularly, Lopes and Santos-Victor follow a previous work of Byrne and Russon (Byrne & Russon, 1998) to establish two modes of imitation defined in terms of what is shared between the model and the imitator (Lopes & Santos-Victor, 2005):

- Action level: The robot replicates the behaviours of a demonstrator, without seeking to understand them. The robot does not relate the observed behaviour with previously memorized ones. This mode is also called 'mimicking' by the authors.
- Program level: The robot recognizes the performed behaviour so it can produce its own interpretation of the action effect.

These modes can be simultaneously active, allowing for an integrated effect.

This chapter is focused on the development of a complete architecture for human upper-body behaviour imitation that integrates these two first modes of imitation (action and program levels). However, in order to simplify the described imitation architecture, and, in particular, to simplify the perception system, manipulated tools will not be taken into account.

Two main hypothesis guide the proposed work. The first is the existence of an innate mechanism which represents the gestural postures of body parts in supra-model terms, i.e. representations integrating visual and motor domains (Meltzoff & Moore, 1977). This mechanism provides the action level ability and its psychological basis will be briefly described in Section 2. The second hypothesis is that imitation and learning by imitation must be achieved by the robot itself, i.e. without employing external sensors. Thus, invasive items are not used to obtain information about the demonstrator's behaviour. This approach is exclusively based on the information obtained from the stereo vision system of a HOAP-I humanoid robot. Its motor systems will be also actively involved during the perception and recognition processes. Therefore, in the program level, the imitator generates and internally performs candidate behaviours while the demonstrator's behaviour is unfolding, rather than attempting to classify it after it is completed. Demiris and Hayes call this "active imitation", to distinguish it from passive imitation which follows a one-way perceive - recognize - act sequence (Demiris & Hayes, 2002).

The remainder of this chapter is organized as follows: Section 2 briefly discusses several related work. Section 3 presents an overview of the proposed architecture. Sections 4 and 5 describe the proposed visual perception and active imitation modules. Section 6 shows several example results. Finally, conclusions and future work are presented in Section 7.

2. Related work

2.1 Action level imitation

Action level imitation or mimicking consists of replicating the postures and movements of a demonstrator, without seeking to understand these behaviours or the action's goal (Lopes & Santos-Victor, 2005). This mode of imitation can be shared with the appearance and action levels of imitation proposed in (Kuniyoshi et al., 2003).

Psychology can help to develop the action level imitation mode in a robot. Thus, different theories have been proposed to justify the mimicking abilities presenting very early neonatal children. The innate release mechanism (IRM) model (Lorenz, 1966) can be briefly stated as the mechanism which predisposes an individual organism to respond to specific patterns of stimulation from its external environment. Thus, this model postulates that the behaviour of the teacher simply triggers and releases equivalent fixed-action-patterns (FAPs) by the imitator. Although IRM can be used to model the action level imitation, there is an important reason that makes it a bad candidate to inspire the general approach to this imitation mode on an autonomous robot. IRM denies any ontogenetic value to immediate imitation and emphasizes instead the developmental role of deferred imitation (Piaget, 1945). This implies the complete knowledge of the set of FAPs. The precise specification of this set is always complex and, at present, it has not been provided. In any case, it is clear that the range of imitated actions is wide and difficult to define. This claim has been also discussed from the psychology point of view. Research has shown that it is very probable that humans present some primitive capacity for behavioral matching at birth (Meltzoff & Moore, 1989). It is difficult to explain the imitation ability of a very early neonatal child based on its knowledge of a complete and previously established set of FAPs. Meltzoff and Moore pose two alternative explanations to the early presence of this mimicking ability in neonatal children (Meltzoff & Moore, 1977): i) the existence of an innate mechanism which represents the postures of body parts in terms integrating visual and motor domains; and ii) the possibility of creating such supra-modal representations through self exploratory "body babbling" during the fetus period. Although this self-learning stage will be really performed, it would not permit to imitate behaviours like facial expressions that the neonatal child has never seen before. Therefore, these authors theorize that neonatal imitation is mediated by a process of active intermodal mapping (AIM) (Meltzoff & Moore, 1989). AIM hypothesis postulates that imitation is a matching-to-target process. Infants' self-produced movements provide proprioceptive feedback that can be compared to the visually-perceived target. AIM proposes that such comparison is possible because the perception and generation of human movements are registered within a common supra-modal representational system. Thus, although infants cannot see their own bodies, these are perceived by them. They can monitor their own movements through proprioception and compare this felt activity to what they see. A similar hypothesis has been formulated by Maurer and Mondloch (Maurer & Mondloch, 2005), but while Meltzoff's AIM hypothesis appears to be activated as a choice made by the infant, they argue that, largely because of an immature cortex, the neonatal child does not differentiate stimuli from different modalities, but rather responds to the total amount of energy, summed across all modalities. The child is aware of changes in the pattern of energy and recognizes some patterns that were experienced before, but is unaware of which modality produced the pattern. As a result, the neonatal child will appear to detect cross-modal correspondences when stimuli from different modalities produce common patterns of energy change. Thus, the response of an infant is a by-product of what is termed neonatal synesthesia, i.e., the infant confuses input from the different senses.

Many mobile robot imitation approaches are closer to these hypothesis models, especially when the goal is not to recognize the behaviour performed by the demonstrator, but imitate it directly. Besides, research on imitation in robotics usually takes the approach of studying learning by imitation, assuming that the robot already possesses the skill to imitate successfully and in turn exploits this ability as a means to acquire knowledge. That is, it is typically assumed the innate presence of an imitation ability in the robot. Thus, the robot in (Hayes & Demiris, 1994) tries to negotiate a maze by imitating the motion of another robot, and it only maintains the distance between itself and the demonstrator constant. The humanoid robot Leonardo imitates facial expressions and behaviours, in order to learn new skills and also to bootstrap his social understanding of others, by for example inferring the intention of an observable action (Breazeal et al., 2005). A computational architecture that follows more closely the AIM model was proposed in (Demiris et al., 1997). Experiments performed on a robot head in the context of imitation of human head movements show the ability of this approach to imitate any observed behaviour that the hardware of the robot system can afford.

In the AIM model, children may use imitation for subsequent learning; but they do not have to learn to imitate in the first place. Other authors support the hypothesis that the supra-modal representations that integrate visual and motor domains can be created by the robot through self-exploration. The biological basis of this approach can be found in the Asymmetric Tonic Neck reflex (Metta et al., 2000) which forces neonatal children to look at their hands, allowing them to learn the relationships between visual stimuli and the corresponding motor action. In the action-level imitation models described in (Lopes & Santos-Victor, 2005, Kuniyoshi et al., 2003), the robot learns the supra-modal representations during an initial period of self-exploration while performing movements as both visual and proprioceptive data are available. These representations can be learnt sequentially, resembling human development stages (Metta et al., 2000). Although this theory can satisfactorily explain the development of arm/hand imitation abilities, it is difficult to justify the neonatal children ability to imitate face expressions present at birth. The body babbling is therefore considered as a pre-imitation stage in which random experimentation with body movements is involved in order to learn a set of motor primitives that allow the neonatal child to achieve elementary body configurations (Breazeal et al., 2005).

2.2 Program level imitation

Robotics researchers have recognized the potential of imitation to ease the robot programming procedure. Thus, they realized that instead of going through complex programming, robots could learn how to perform new assembly tasks by imitating a human demonstrator. It must be noted that program level imitation is not always achieved from visual observation. Thus, Ogata and Takahashi (Ogata & Takahashi, 1994) use a virtual reality environment as a robot teaching interface. The movement of the demonstrator in the virtual reality space is interpreted as a series of robot task-level operations using a finite automaton model. In contrast with virtual reality, (Tung & Kak, 1995) presents a method in which a robot can learn new assembly tasks by monitoring the motion of a human hand in the real world. Their work relies on invasive sensing and can not be used easily to get accurate and complete data about assembly tasks. A more accurate method to track human hand motion is presented in (Kang & Ikeuchi, 1993). Although their work also employs a glove wired to the computer to take input from the demonstrator's hand, it uses stereo vision to improve results. One of the first examples of non-invasive teaching method is the work of Inaba and Inoue (Inaba & Inoue, 1989). This paper describes a vision-based robot programming system via a computer vision interface. Kuniyoshi et al. develop a system which can be taught reusable task plans by watching a human performing assembly tasks via a real-time stereo vision system (Kuniyoshi et al., 1994). The human instructor only needs to perform a task in front of the system while a robot extracts task description automatically without disturbing it.

In all previously described work, the same strategy has been successfully used to allow robots to perform complex assembly tasks. This strategy can be resumed in the plan from observation (APO) paradigm proposed in (Ikeuchi & Suehiro, 1992). This passive paradigm postulates that the imitation process proceeds serially through the three stages of perception, recognition and reproduction. In a passive scheme, there is not substantial interaction between all stages, nor any relation of the perception and recognition stages to the motor systems. The motor systems are only involved in the final reproduction stage (Demiris & Hayes, 2002). Therefore, a passive paradigm implies that program level imitation should require at least an elementary level of representation, which allows for recognition of the perceived actions. The psychology basis of this passive approach can be found in the IRM model described in the previous subsection. As a learning mechanism, the IRM presents a new problem that complicates its application out of industrial assembly tasks. IRM determines that true imitation have to be novel and not already in the repertoire. Therefore, imitation is a special case of observational learning occurring without incentives, without trial and error, and requiring no reinforcement (Andry et al., 2001). Then, imitation only can provide new behaviours to the repertoire and it is not employed to improve the quality of imitated tasks or recognize known behaviours.

These claims have been discussed by neuroscientists and psychologists. While there is still some debate to define what behaviours the term imitation is exactly refering to, it is assumed that imitation is the ability to replicate and learn new skills by the simple observation of those performed by others (Billard, 2001). Thus, imitation (or program level imitation) is contracted to mimicking (or action level imitation), where imitation relies on the ability to recognize observed behaviours and not only to reproduce them by transforming sensed patterns into motor commands. In this context, experiments show that repeated imitative sessions improve imitation or recognition of being imitated (Nadel, 2004). The implication of the motor system in the imitation process defines the so-called active imitation which is biologically supported by the mirror neural system. The mirror neurons were first detected in the macaque monkey pre-motor cortex (PM), posterior parietal cortex (PPC) and superior temporal sulcus (STS) (Rizzolatti et al., 1996). Later, brain imaging studies of the human brain highlighted numerous areas, such as STS, PM and Broca (Decety et al., 2002). While the discovery of this system is certainly an important step toward a better understanding of the brain mechanisms underlying the capability of the primates to imitate, the role of the mirror neuron system as part of the general neural processes for imitation is still not completely explained.

Sauser and Billard present a model of a neural mechanism by which an imitator agent can map movements of the end effector performed by other agents onto its own frame of reference (Sauser & Billard, 2005). The model mechanism is validated in simulation and in a humanoid robot to perform a simple task, in which the robot imitates movements performed by a human demonstrator. However, this work only relies on the action level imitation (mimicking). It does not distinguish between known and novel movements, i.e. all movements are processed and imitated through the same mechanism. Therefore, there is no mechanism to improve the quality of the imitated behaviour. The passive and active paradigms are combined into a dual-route architecture in (Demiris & Hayes, 2002): known behaviours are imitated through the active route; if the behaviour is novel, evident from the fact that all internal behaviours have failed to predict adequately well, control is passed to the passive route which is able to imitate and acquire the observed behaviour. Lopes and Santos-Victor (Lopes & Santos-Victor, 2005) propose a general architecture for action and program level visual imitation. Action level imitation involves two modules. A view-point transformation module solves the correspondence problem (Alissandrakis et al., 2002) and a visuo-motor map module maps this visual information to motor data. For program level imitation an additional module that allows the system to recognize and generate its own interpretation of observed behaviours to produce similar behaviours at a later stage is provided.

3. Overview of the Proposed Architecture

Fig. 1 shows an overview of the proposed architecture. The whole architecture is divided into two major modules related to visual perception and active imitation. The goal of the proposed visual perception system is the detection and tracking of the demonstrator's upper-body movements. In this work, it is assumed that in order to track the global human body motion, it is not necessary to capture with precision the motion of all its joints. Particularly, in the case of upper-body movements, it is assumed that the robot only needs to track the movement of the head and hands of the human, because they are the most significant items involved in the human-to-human interaction processes. This system works without special devices or markers, using an attention mechanism to provide the visual information. Since such system is unstable and can only acquire partial information because of self-occlusions and depth ambiguity, a model-based pose estimation method based on inverse kinematics has been employed. This method can filter noisy upper-body human postures. Running on a 850MHz PC, the visual perception system captures the human motion at 10 to 15 Hz. Finally, a retargeting process maps the observed movements of the hands onto the robot's own frame of reference. Section 4 will describe the different modules of the proposed visual perception system.

The active imitation module performs the action level and program level imitation modes. To achieve the mimicking ability, it only needs to solve the visuo-motor mapping. This mapping defines a correspondence between perception and action which is used to obtain the angle joints which move the robot's head and hands to the visually observed positions. Elbows are left free to reach different configurations. Angle joints are extracted through the use of a kinematic model of the robot body. This model includes an important set of constraints that limit the robot's movements and avoid collisions between its different body parts (these constrains are necessary, as the robot has no sensors to help in preventing collisions). The body model also determines the space that the robot's end-effectors can span. This space will be quantified to ease the memorization of behaviours. Thus, each behaviour is coded as a sequence of items of this reduced set of possible postures. In order to recognize previously memorized behaviours, the active imitation system includes a behaviour comparison module that uses a dynamic programming technique to make this comparison. Section 5 describes the proposed active imitation system.

The proposed work is inspired by the possible role that mirror neurons play in imitative behaviour. Particularly, it is related to the recent work of Demiris et al. (Demiris & Hayes, 2002, Demiris & Khadhouri, 2005), Breazeal et al. (Breazeal et al., 2005) and Lopes and Santos-Victor (Lopes & Santos-Victor, 2005). Thus, the proposed system emphasizes the bidirectional interaction between perception and action and employs a dual mechanism based on mimicking and behaviour recognition modules, as proposed in (Demiris & Hayes, 2002). However, the proposed approach does not use the adapted notion of mirror neurons to predictive forward models which match a visually perceived behaviour with the equivalent motor one. Therefore, the mirror neuron-inspired mechanism is achieved by a process where the imitator behaviours are represented as sequences of poses., Behaviours are used in the imitator's joint space as its intermodal representation (Breazeal et al., 2004). Thus, the visually perceived behaviours must be mapped from the set of three-dimensional absolute coordinates provided by the visual perception module onto the imitator's joint space. In Breazeal's proposal (Breazeal et al., 2005) this process is complicated by the fact that there is not a one-to-one correspondence between the tracked features and the imitator's joints. To solve this problem, it is proposed that the robot learns the intermodal representation from experience. In the proposed system, the imitator robot establishes this one-to-one correspondence by mapping movements of the end effectors performed by the demonstrator onto its own frame of reference (Sauser & Billard, 2005).

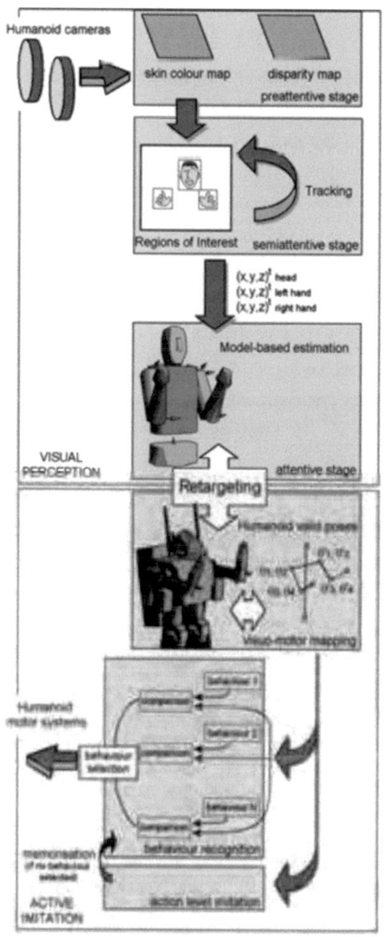

Figure 1. Overview of the proposed architecture

This transformation is achieved by a grid-based retargeting algorithm (Molina-Tanco et al., 2006). The importance given to this algorithm is influenced by the work of Lopes and Santos-Victor. These authors define an architecture based on three main modules: viewpoint transformation, visuo-motor mapping and behaviour recognition (Lopes & Santos-Victor, 2005). However, their work only address postures as behaviours to imitate. In the proposed work, more complex behaviours, where the temporal chaining of elementary postures is taken into account, are addressed.

4. Visual Perception System

To interact meaningfully with humans, it is interesting that robots will be able to sense and interpret the same phenomena that humans observe (Dautenhahn & Nehaniv, 2002). This means that, in addition to the perception required for conventional functions (localization, navigation or obstacle avoidance), a robot that interacts with humans must posses perceptual capabilities similar to humans.

Biological-plausible attention mechanisms are general approaches to imitate the human attention system and its facility to extract only relevant information from the huge amount of input data. In this work, an attention mechanism based on the Feature Integration Theory (Treisman & Gelade, 1980) is proposed. The aim of this attention mechanism is to extract the human head and hands from the scene. The proposed system integrates bottom-up (data-driven) and top-down (model-driven) processing. The bottom-up component determines and selects salient image regions by computing a number of different features (preattentive stage). In order to select the demonstrator's head and hands as relevant objects, skin colour has been included as input feature. Disparity has been also employed as input feature. It permits to take into account the relative depth of the objects from the observer. Similar features has been used in (Breazeal et al., 2003). The top-down component uses object templates to filter out data and track only relevant objects (semiattentive stage). The tracking algorithm can handle moving hands and head in changing environments, where occlusions can occur. To support the tracking process, the model includes weighted templates associated to the appearance and motion of head and hands. Then, the proposed system has three steps: parallel computation of feature maps, feature integration and simultaneous tracking of the most salient regions. The motivation of integrating an attention mechanism in this architecture to reduce the amount of input data is twofold: i) the computational load of the whole system is reduced, and ii) distracting information is suppressed. Besides, although in the current version of the proposed architecture, the attention mechanism only employs skin colour and depth information to extract the relevant objects from the scene, new features like colour and intensity contrasts could be easily included in subsequent versions. Thus, the mechanism could be used to determine where the attention of the observer should be focused when a demonstrator performs an action (Demiris & Khadhouri, 2005).

The outputs of the semiattentive stage are the inputs of a third module that performs the attention stage. In this work, a model of human appearance is used in the attention stage with the main purpose of filtering fast, non-rigid motion of head and hands. Besides, it can provide the whole range of motion information required for the robot to achieve the transformation from human to robot motion. To estimate articulated motion, the human model includes a 3D geometric structure composed of rigid body parts.

4.1 Preattentive stage

In this work, the visual perception system is applied to track simultaneously the movements of the hands and the head of a human demonstrator in a stereo sequence. The depth of the tracked objects is calculated in each frame by taking into account the position differences between the left and right images. The preattentive stage employs skin colour and disparity information computed from the available input image in order to determine how interesting a region is in relation to others. Attractivity maps are computed from these features, containing high values for interesting regions and lower values for other regions. The integration of these feature maps into a single saliency map provides to the semiattentive stage the interesting regions of the input video sequence.

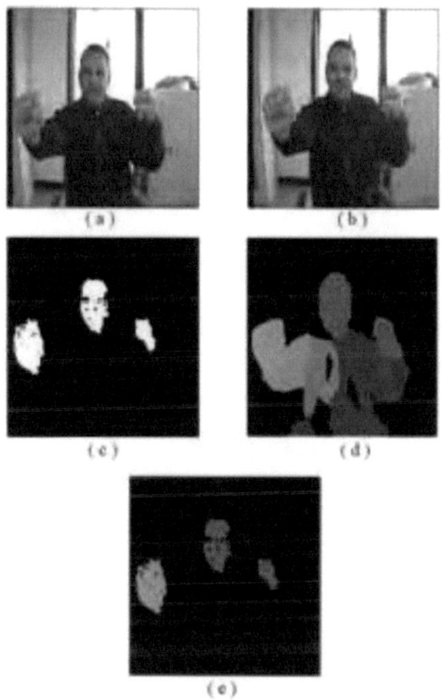

Figure 2. a-b) Input stereo pair; c) skin colour; d) disparity map; and e) saliency map

Fig. 2 shows an example of saliency map obtained from a stereo pair. In order to extract skin colour regions from the input image, an accurate skin chrominance model using a colour space can be computed and then, the Mahalanobis distance from each pixel to the mean vector is obtained. If this distance is less than a given threshold T_s then the pixel of the skin feature map is set to 255. In any other case, it is set to 0. The skin chrominance model used in the proposed work has been built over the TSL colour space (Terrillon & Akamatsu, 1999). Fig. 2b shows the skin colour regions obtained from the left image of the stereo pair (Fig. 2a). On the other hand, the system obtains the relative depth information from a dense disparity map. Closed regions, with high disparity values associated, are considered more important. The zero-mean normalized cross-correlation measure is employed as disparity descriptor. It is implemented using the box filtering technique that allows to achieve fast computation speed (Sun, 2002). Thus, the stereo correlation engine compares the two images for stereo correspondence, computing the disparity map at about 15 frames per second. Fig. 2c shows the disparity map associated to the stereo pair at Fig. 2a. Finally, and similarly to other models (Itti & Koch, 2001), the saliency map is computed by combining the feature maps into a single representation (Fig. 2d). The disparity map and the skin probability map are then filtered and combined. A simple normalized summation has been used as feature combination strategy, which is sufficient for systems with a small number of feature maps.

4.2 Semiattentive stage

The semiattentive stage tracks the head and hands of the human demonstrator, which are selected from the input saliency map. Firstly, a Viola-Jones face detector (Viola & Jones, 2001) runs on each significant region to determine whether it corresponds to a face. The closest face to the vision system is considered as the demonstrator's face. Connected components on the disparity map are examined to match the hands which correspond with this selected face (Breazeal et al., 2003). Finally, a binary image including the head and hands of the demonstrator is built. It must be noted that this process is run only as an initialization step, i.e. to search for a human demonstrator. Once the demonstrator has been found, hands and head are tracked over time.
The proposed method uses a weighted template for each object to track which follows its viewpoint and appearance changes. These weighted templates and the way they are updated allow the algorithm to successfully handle partial occlusions. To reduce the computational cost, templates and targets are hierarchically modeled using Bounded Irregular Pyramids (BIP) that have been modified to deal with binary images (Marfil et al., 2004, Molina-Tanco et al., 2005).

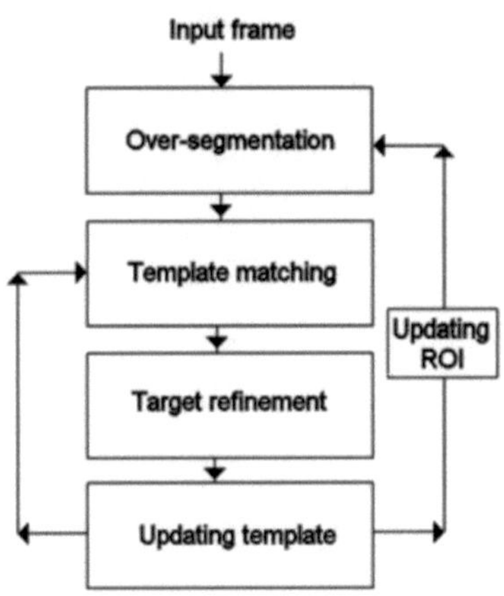

Figure 3. Data flow of the tracking algorithm

The tracking process is initialized as follows: once the demonstrator's head and hands are found, the algorithm builds their hierarchical representations using binary BIPs. These hierarchical structures are the first templates and their spatial positions are the first regions of interest (ROIs), i.e. the portions of the current frame where each target is more likely located. Once initialized, the proposed tracking algorithm follows the data flow shown in Fig. 3. It consists of four main steps which are briefly described bellow (see Appendix A for further details):

- Over-segmentation: in the first step of the tracking process a BIP representation is obtained for each ROI.
- Template matching and target refinement: once the hierarchical representation of the ROIs are obtained, each target is searched using a hierarchical template matching process. Then, the appearance of each target is refined following a top-down scheme.

- Template updating: as targets can represent severe viewpoint changes over a sequence, templates must be updated constantly to follow up varying appearances. Therefore, each template node includes a weight which places more importance to more recent data and allows to forget older data smoothly.
- Region of interest updating: once the targets have been found in the current frame, the new ROIs for the next frame are obtained.

4.3 Attentive stage

In the proposed system, our robot performs imitation and learning by imitation by itself, i.e. without employing external sensors. No markers or other invasive elements are used to obtain information about the demonstrator's behaviour. Therefore, this approach is exclusively based on the information obtained from the stereo vision system of the imitator. In order to filter the movements of all tracked items, the attentive stage employs an internal model of the human.

This work is restricted to upper-body movements. Therefore, the geometric model contains parts that represent hips, head, torso, arms and forearms of the human to be tracked. Each of these parts is represented by a fixed mesh of few triangles, as depicted in Fig. 4. This representation has the advantage of allowing fast computation of collisions between parts of the model, which helps in preventing the model from adopting erroneous poses due to tracking errors.

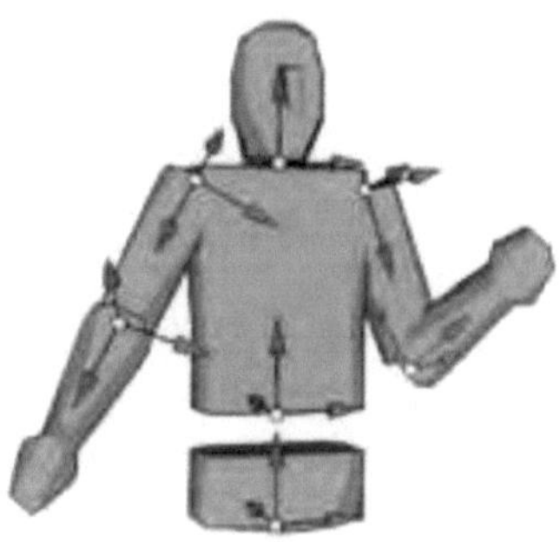

Figure 4. Illustration of the human upper-body kinematic model

Each mesh is rigidly attached to a coordinate frame, and the set of coordinate frames is organized hierarchically in a tree. The root of the tree is the coordinate frame attached to the hips, and represents the global translation and orientation of the model. Each subsequent vertex in the tree represents the three-dimensional rigid transformation between the vertex and its parent. This representation is normally called a skeleton or kinematic chain (Nakamura & Yamane, 2000) (Fig. 4). Each vertex, together with its corresponding body part attached is called a bone. Each bone is allowed to rotate --but not translate-- with respect to its parent around one or more axes. Thus, at a particular time instant t, the pose of the skeleton can be described by $\Phi^{(t)} = (R^{(t)}, \vec{s}^{(t)}, \phi^{(t)})$ where $R^{(t)}$ and $\vec{s}^{(t)}$ are the global orientation and translation of the root vertex, and $\phi^{(t)}$ is the set of relative rotations between successive children. For upper-body motion tracking, it is assumed that only φ needs to be updated -this can be seen intuitively as assuming that the tracked human is seated on a chair.

The special kinematic structure of the model can be exploited to apply a simple and fast analytic inverse kinematics method which will provide the required joint angles from the Cartesian coordinates of the tracked end-points (see Appendix B).

4.4 Retargeting

In any form of imitation, a correspondence has to be established between demonstrator and imitator. When the imitator body is very similar to that of the demonstrator, this correspondence can be achieved by mapping the corresponding body parts (Nehaniv & Dautenhahn, 2005). Thus, Lopes and Santos-Victor propose two different view-point transformation algorithms to solve this problem when the imitator can visually perceive both the demonstrator's and its own behaviour (Lopes & Santos-Victor, 2005). However, the similarity between the two bodies is not always sufficient to adopt this approach. Often the imitator's body will be similar to the demonstrator's, but the number of degrees of freedom (DOFs) will be very different. In these cases, it is not possible to establish a simple one-to-one correspondence between the coordinates of their corresponding body parts. Thus, more complex relations and many-to-one correspondences are needed to perform imitation correctly (Molina-Tanco et al., 2006). Sauser and Billard (Sauser & Billard, 2005) describe a model of a neural mechanism by which an imitator can map movements of the end-effector performed by other agents onto its own frame of reference. Their work is based on the mapping between observed and achieved subgoals, where a subgoal is defined as to reach a similar relative position of the arm end-effectors or hands. Our work is based on the same assumption.

In this work, the mapping between observed and achieved subgoals is defined by using three-dimensional grids, associated to each demonstrator and imitator hand. Fig. 5b shows the grid associated to the left hand of the demonstrator. This grid is internally stored by the robot and can be autonomously generated from the human body model. It provides a quantization of the demonstrator's reachable space. The demonstrator's reachable space cells can be related to the cells of the imitator's grids (Fig. 5c). This allows defining a behaviour as a sequence of imitator´s grid elements. This relation is not a one-to-one mapping because the robot's end-effector is not able to reach to all the positions that the human's hand can reach. Thus, the proposed retargeting process involves a many-to-one correspondence that has to solve two main problems: i) how to perform re-scaling to the space reachable by the imitator's end-effectors and ii) how to obtain the function that determines the egocentric (imitator) cell associated to an observed allocentric (demonstrator) cell. The presented system solves these problems using look-up tables that establish a suitable many-to-one correspondence. Briefly, two techniques can be used to relate demonstrator and imitator cells. See (Molina-Tanco et al., 2006) for further details:

- Uniform scale mapping (USM). The length of a stretched arm for both the demonstrator and the imitator gives the maximum diameters of the corresponding grids. The relation between these lengths provides a re-scaling factor applied to demonstrator's end-effector position to obtain a point in the imitator grid. The nearest cell to this point is selected as imitator's end-effector position. Although it is a valid option, USM may distort quality of the imitated behaviour if a large part of the motion is performed in an area that the imitator cannot reach.
- Non-uniform scale mapping (NUSM). This approach is applied when it is possible to set a relation between the shapes of demonstrator and imitator grids. This relation

allows to apply a geometric transformation to all demonstrator poses that roughly translate the movements to the imitator's reachable area. NUSM improves USM results but it needs a better *a priori* knowledge about both demonstrator's and imitator's bodies. As in this case we are able to determine both human and HOAP-I reachable spaces, and it is possible to set a rough relation between them (Molina-Tanco et al., 2006), the NUSM approach is chosen.

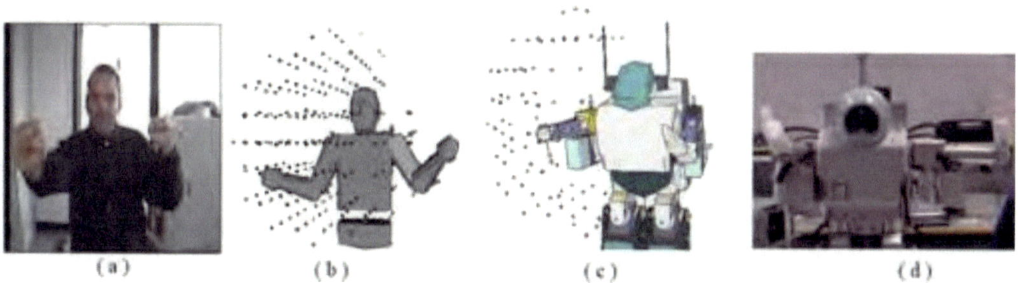

Figure 5. Overview of the retargeting process: a) human demonstrator; b) internal human model (only a sub-region of the grid is shown for clarity); c) internal humanoid model (only a sub-region of the grid is shown for clarity); and d) real humanoid imitator

5. Active Imitation Module

The main characteristic of an active imitation architecture is that the same motor representations that are responsible for the generation of a movement are recruited in order to perform behaviour recognition (Demiris & Hayes, 2002). Thus, the active imitator does not execute the passive perception -- recognition -- action cycle, but actively generates possible behaviours concurrently with the perception of the demonstrator's behaviour. The most similar behaviour to the perceived one is selected. If the observed behaviour is not recognized, it is added to the memorized repertoire. The original idea of active imitation was proposed by Demiris and Hayes (Demiris & Hayes, 2002). The same concept has been subsequently employed by other authors, e.g. (Breazeal et al., 2005) or (Lopes & Santos-Victor, 2005). The proposed approach is similar to the one presented in (Demiris & Hayes, 2002). The following section introduces the three components of the active imitation module: visuo-motor mapping, action level imitation and behaviour recognition (Fig. 1).

5.1 Visuo-motor mapping

The visuo-motor mapping (VMM) defines a correspondence between an observed behaviour and executed action. In the proposed system, the demonstrator hands are tracked and their coordinates are used to extract a coherent upper-body human pose using a human model. The hand poses provided by this model are translated to an egocentric image by the retargeting method described in Section 4.4. The VMM will relate the egocentric image coordinates of these hands to the actual joint angles, in terms of forward/inverse kinematics. Then, the VMM can be used to infer the motor commands used to achieve the perceived pose. This ability is subsequently used to make recognition in motor space and to imitate (Demiris & Hayes, 2002, Lopes & Santos-Victor, 2005).

Several authors propose a VMM algorithm that defines a direct translation from the imitator's end-effector coordinates to the imitator's joint angles which must be sent to the robot motors to achieve a pose similar to the observed one (Lopes & Santos-Victor, 2005, Molina-Tanco et al., 2005). In the proposed approach, from the coordinates of the hands on the imitator's frame of reference, the imitator obtains the rest of joint coordinates using a model of its own body (Figs. 5b-d). The mechanism to obtain the joint angles from the hand positions is explained in Appendix B. The robot model (Fig. 5c) is used not only to extract the pose of the joints, but also to impose a set of constraints that limit the robot's movements and avoid collisions between its different body parts. The joint limits can be taken into account by the motor control software, but there is no internal mechanism to avoid the collision between two robot parts. Although the implementations are very different, the proposed approach can be related to the free-elbow method proposed in (Lopes & Santos-Victor, 2005).

5.2 Action level imitation

The action level imitation or mimicking can be directly achieved by sending to the robot motor controllers the joint angles provided by the VMM. However, in the proposed imitation architecture, the action level imitation module is also the responsible of generating the representation of every observed behaviour. This representation will be memorized if it is not present in the actual repertoire of behaviours.
The behaviour is not exactly memorized as it was observed. All the different modules of the architecture introduce some errors, resulting in noisy observed behaviour. Therefore, the action level imitation module performs outlier removal during the acquisition process. Outliers are removed using local windows centered at each element of the observed sequence. They are detected as small size clusters on the sequence. Fig. 6a shows the trajectory for an observed behaviour when outlier removal has not been performed. It can be noted that the movement trajectory is noisy. On the contrary, Fig. 6b illustrates the same behaviour after outlier removal. This filtered representation will be memorized and included into the behaviour repertoire.
Once outliers have been removed, the behaviour is stored as a sequence of transitions between different elements of the grid. Self-transitions are discarded.

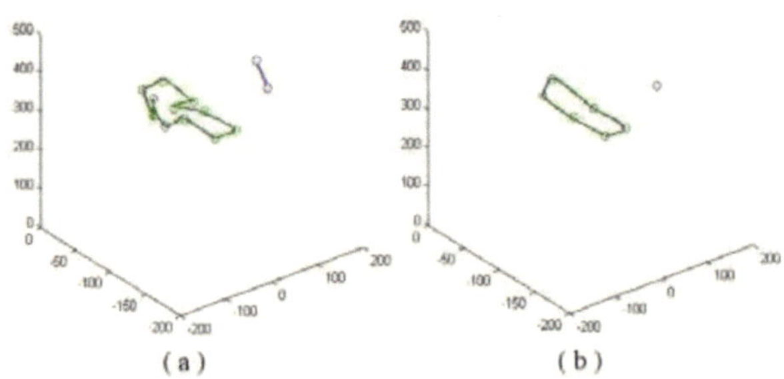

Figure 6. a)Trajectories of the end-effectors for an observed behaviour; and b) trajectories for the same behaviour in a) after outlier removal.

5.3 Behaviour recognition

When comparing an observed behaviour of arbitrary duration to the repertoire of memorized behaviours, time must be normalized out. In this work, an online version of a classical time normalization algorithm based on dynamic programming, which is commonly

known as Dynamic Time Warping (DTW) (Sakoe & Chiba, 1978), is employed. DTW simultaneously calculates the optimal time warping function and the resulting minimum distance between two patterns.
Demiris and Hayes introduced the concept of confidence value, as an indicator of how confident an imitator's behaviour is that it can match an observed behaviour (Demiris & Hayes, 2002). In the proposed system, the confidence value for an observed behaviour is inversely proportional to the distance between the observed and memorized behaviours.

5.4 Behaviour learning

When the human demonstrator performs a behaviour that the imitator does not know, none of the memorized behaviours' confidence value stands out. This behaviour is therefore considered as novel, and subsequently added to the behaviour repertoire, i.e. learnt. In this work, it is assumed that learning is the process of acquiring a novel behaviour, either its trajectory specifications or the motor commands needed to achieve it (Demiris & Hayes, 2002). Learning as used here does not imply generalization or adaptation to different circumstances or any other processes as used in the field of machine learning. Once the novel behaviour is learnt, it can be recognized in subsequent performances of the future demonstrators.

6. Experimental Results

The first experiments reported here show that the proposed architecture can provide mimicking abilities to our HOAP-I humanoid robot. HOAP-I is a mini-humanoid robot built by Fujitsu, provided with 20 degrees of freedom, including 4 on each arm. Fig. 7 illustrates the trajectories of the demonstrator's and imitator's hands for several examples. It can be observed that the imitation is qualitatively good although it suffers a non-uniform scale mapping to bring observed motion into the reachable space of the robot. Using this mimicking ability, we have generated a set of behaviours which will constitute the behavioural repertoire of the robot. Behaviours in Fig. 7 constitute the set of memorized trajectories. For this experiment we have chosen the trajectories to correspond to different diving signals. Each signal conveys a message to fellow divers: `(It is) Cold', `(I am) OK', etc. The meaning of each signal is illustrated Fig. 7. In the rest of this Section, the abbreviated terms are used: [Cold], [OK], etc.
In the second experiment, different demonstrators perform behaviours of a reduced repertoire that the robot has memorized. Table 1 shows the confidence values associated to these demonstrations. Note that this is a much more challenging task for the robot, and thus we chose a reduced signal repertoire. No feedback was provided to the demonstrators as to how 'well' the diving signals were performed. The movements were described verbally to the performers, and they had not seen the motion of previous demonstrators nor had information about their results during the experiment. The main consequence of these restrictions is that some movements were performed in quite different ways by different demonstrators. Particularly, the [Help] behaviour in which the extended right arm moves up and down generated some confusion, as users did not know relative position of the hand nor amplitude of movements. Table 1 shows that this situation makes the system consider the unrecognized motion as a new behaviour, and consequently it is stored again in the data base. The rest of the movements were correctly recognized as Table 1 depicts.

		1. Performer #1				Performer #2				Performer #3			
		2. [Cold]	3. [OK]	4. [Help]	5. [Well]	[Cold]	[OK]	[Help]	[Well]	[Cold]	[OK]	[Help]	[Well]
Stored behaviours	[Cold]	**1.00**	*0.38*	*0.16*	*0.26*	**0.69**	*0.55*	*0.31*	*0.24*	**0.89**	*0.56*	*0.35*	*0.26*
	[OK]	*0.39*	**1.00**	*0.20*	*0.47*	*0.44*	**0.75**	*0.46*	*0.39*	*0.52*	**0.79**	*0.51*	*0.39*
	[Help]	*0.16*	*0.20*	**1.00**	*0.53*	*0.17*	*0.19*	*0.11*	*0.38*	*0.17*	*0.22*	*0.24*	*0.40*
	[Well]	*0.26*	*0.45*	*0.54*	**1.00**	*0.31*	*0.32*	*0.23*	**0.98**	*0.27*	*0.36*	*0.33*	**0.97**
		6. Training data				7. Test data							

Table 1. Confidence values

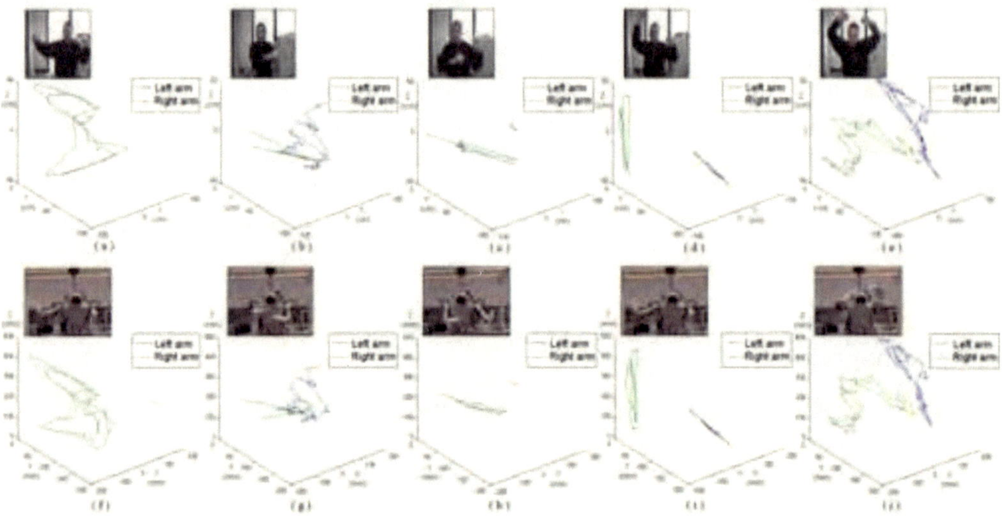

Figure 7. a-e) Trajectories followed by the demonstrator's hands; and f-j) trajectories followed by the imitator's hands. These behaviours also constitute the memorized repertoire: a) [Help]; b) [Cold]; c) [NoAir]; d) [GoUp] and e) [Well]

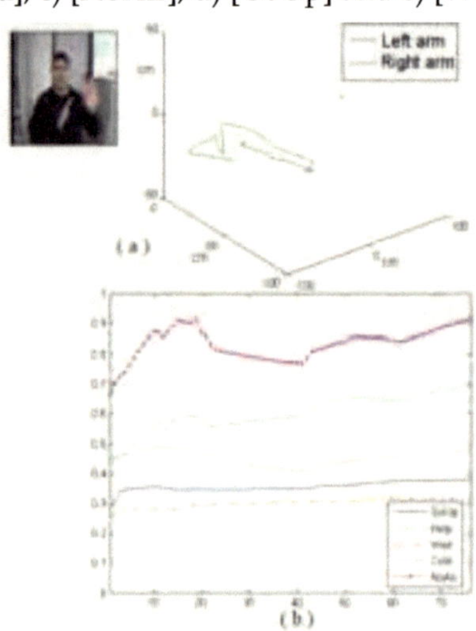

Figure 8. a) Trajectory followed by the demonstrator's hands –behaviour [NoAir]–; and b) confidence values of memorized behaviours {[Help], [Cold], [NoAir], [GoUp], [Well]} when demonstrator executes behaviour [NoAir]

Recognition can be achieved even when only a part of the stored motion is detected by the system, as the online version of the DTW algorithm updates confidence values for each received frame. Fig. 8a shows the evolution of these confidence values while a demonstrator is executing the behaviour [NoAir]. It can be noted that the memorized behaviour [NoAir] not only consistently gets the highest confidence value at the end of the motion, but it is also the most similar behaviour during all the sequence. Thus, it could have been correctly recognized even if the communication would have been interrupted during the test. Although a result based on partial observation should be always considered with caution, it can be very useful in real interaction scenarios where the data flow is usually disturbed by external factors.

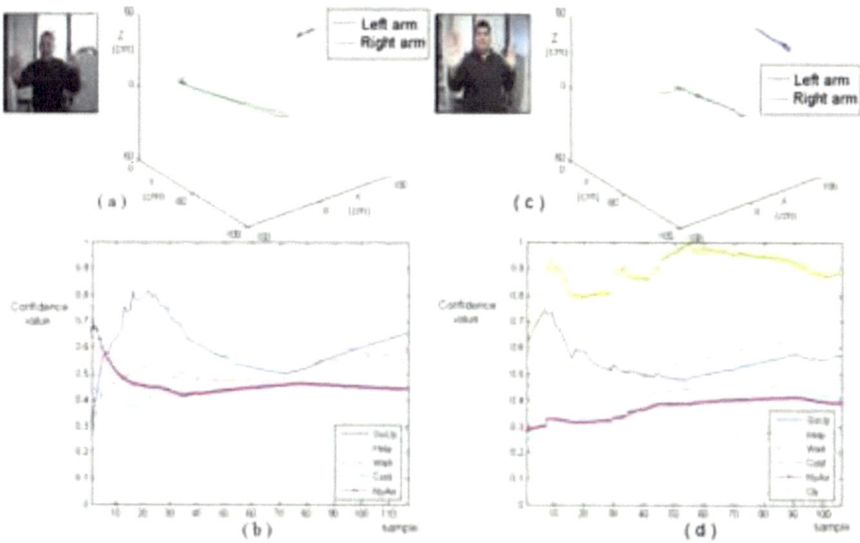

Figure 9. a) Trajectory followed by the demonstrator's hands -behaviour [OK]-; b) confidence values of memorized behaviours {[Help], [Cold], [NoAir], [GoUp], [Well]} when demonstrator executes behaviour [OK]; c) trajectory followed by the demonstrator's hands -behaviour [OK]-; and d) confidence values of memorized behaviours {[Help], [Cold], [NoAir], [GoUp], [Well], [OK]} when demonstrator executes behaviour [OK]

Finally, Fig. 9a illustrates an observed behaviour which is not in the memorized repertoire. Fig. 9b shows the confidence values for this behaviour. It can be appreciated that none of the behaviours' confidence value stands out. This triggers the learning module, which adds the novel behaviour which to the imitator's repertoire (in the experiment illustrated by Fig. 9a-b, the novel behaviour corresponds to the signal [Well]). If a different demonstrator performs the behaviour [Well] (Fig. 9c), the imitator has now this behaviour into the memorized set. Fig. 9d illustrates the confidence values. It can be appreciated that, in this occasion, the confidence value of the recently learn behaviour stands out, resulting in the behaviour being correctly recognized.

7. Conclusions and Future Work

In this chapter, an architecture that endows a robot with the ability to imitate has been described. This architecture has modules that provide action level and program level

capabilities. The program level imitation is achieved by a behaviour recognition module which compares previously memorized and observed behaviours. If there are no behaviours that can match sufficiently well, the action level imitation module provides a representation acquired from the observed behaviour. This representation is added to the repertoire of memorized behaviours. Experiments have been performed using a real humanoid robot HOAP-I and different human demonstrators. These experiments have showed that the architecture is able to imitate known behaviours, as well as acquiring new ones which are successfully employed later.
As behaviour complexity increases, other communication channels between demonstrator and imitator such as verbal instruction or attentional cues are required (Nicolescu & Mataric, 2005). In the proposed work, imitation learning is not augmented by allowing the demonstrator to employ additional instructive activities. Neither the learned representations are refined through generalization from multiple learning experiences, nor through direct feedback from the teacher. These items will be taken into account in future work. Current efforts focus on improving the proposed attention mechanism and including verbal communication modules in the architecture.

8. Appendix A. Tracking using BIPs

In the binary BIP, each node n is identified by (i,j,l) where l represents the level and (i,j) are the co-ordinates within the level. To build the different levels of the pyramid, each node has two associated parameters:

- Homogeneity, $Hom(i,j,l)$. This is set to 1 if the four nodes immediately underneath have homogeneity values equal to 1. Otherwise, it is set to 0. It must be noted that in the base of the structure (level 0) only nodes which correspond to image pixels of the interest regions have homogeneity values equal to 1.
- Parent link, $(X,Y)_{(i,j,l)}$. If $Hom(i,j,l)$ is equal to 1, the values of the parent link of the four nodes immediately underneath are set to (i,j). Otherwise, these four parent links are set to a null value.

Each template M and target T is represented by using binary BIP structures:

$$M^{(t)}(l) = \bigcup_{ij} m^{(t)}(i,j,l) \quad T^{(t)}(l) = \bigcup_{ij} q^{(t)}(i,j,l) \tag{1}$$

$M^{(t)}(l)$ and $T^{(t)}(l)$ being the level l of the pyramid structure corresponding to the template and target in frame t, respectively. Each level of the template and the target are made up of a set of homogeneous nodes.

A.1 Oversegmentation

The hierarchical representation of a region of interest $ROI^{(t)}$ in the current frame t depends on the target position in the previous frame, and is updated as described in Subsection 8.4. The hierarchical structure can be represented in each level as:

$$ROI^{(t)}(l) = \bigcup_{ij} p^{(t)}(i,j,l) \tag{2}$$

being p a node of the bounded irregular pyramid built over the ROI.

A.2 Template matching and target refinement

The process to localize the target in the current frame t is a top-down process which starts at a working level $l^{(t)}_w$ and stops in the level where the target is found. In each level l, the template $M_i^{(t)}(l)$ is placed and shifted in $ROI^{(t)}(l)$ until the target is found or until $ROI^{(t)}(l)$ is completely covered. If $ROI^{(t)}(l)$ was completely covered and the target was not found, the target localization would continue in the level below. The displacement of the template can be represented as $d_k^{(t)}=(d_k^{(t)}(i),d_k^{(t)}(j))$, being $d_0^{(t)}$ the first displacement and $d_f^{(t)}$ the final displacement. $d_f^{(t)}$ is the displacement that situates the template in the position where the target is placed in the current frame. The algorithm chooses as initial displacement in the current frame $d_0^{(t)} = d_f^{(t-1)}$. In order to localize the target and obtain $d_f^{(t)}$, the overlap $O^{(t)}_{d_k^{(t)}}$ between $M^{(t)}(l)$ and $ROI^{(t)}(l)$ in each template displacement k is calculated:

$$O^{(t)}_{d_k^{(t)}} = \sum_{ij\in\xi} w^{(t)}(m(i,j,l_w^{(t)})) \tag{3}$$

being $w^{(t)}(m(i,j,l))$ a weight associated to $m^{(t)}(i,j,l)$ in the current frame t, as explained in Section 8.3. ξ is the subset of nodes that satisfy the following two conditions:

$$\begin{aligned} &Hom(f(m^{(t)}(i,j,l_w^{(t)}),a(t))) = 1 \\ &Hom(p^{(t)}(i+d_k^{(t)}(i),j+d_k^{(t)}(j),l_w^{(t)})) = 1 \end{aligned} \tag{4}$$

being $f(m^{(t)}(i,j,l_w^{(t)}),a(t))$ a coordinate transformation of $m^{(t)}(i,j,l_w^{(t)})$ that establishes the right correspondence between $m^{(t)}(i,j,l_w^{(t)})$ and $p^{(t)}(i+d_k^{(t)}(i),j+d_k^{(t)}(j),l_w^{(t)})$. $a(t)$ denotes the parameter vector of the transformation, which is specific for the current frame. Equation 4 is satisfied when a match occurs. All the ROI nodes that match with nodes of the template are marked as nodes of the target. Thus, the hierarchical representation of the target $T^{(t)}$ is obtained.
In order to refine the target appearance, its hierarchical representation is rearranged level by level following a top-down scheme (Marfil et al., 2004). This process is applied to all homogeneous nodes of *ROI* which have not been marked as target nodes in the template matching process. If one of these nodes has a homogeneous neighbor node that belongs to the target, it is also marked as a target node.

A.3 Template updating

In order to update the template, a new parameter is included in the template model:

- $w^{(t)}(m(i,j,l))$. It is a weight associated to each node $m^{(t)}(i,j,l)$ of the template $M^{(t)}$ in the current frame t.

The whole template is updated at each sequence frame:

$$m^{(t+1)}(i,j,l) = \begin{cases} m^{(t)}(i,j,l) & \text{if no match} \\ f^{-1}(q^{(t)}(i,j,l),a^{(t)}) & \text{if match} \end{cases} \tag{5}$$

$$w^{(t+1)}(m(i,j,l)) = \begin{cases} w^{(t)}(m(i,j,l)) - \alpha & \text{if no match} \\ 1 & \text{if match} \end{cases} \tag{6}$$

where the superscript *(t)* denotes the current frame and the forgetting constant, *a*, is a predefined coefficient that belongs to the interval *[0,1]*. This constant dictates how fast the forgetting action will be.

A.4 Region of interest updating

This process has two main steps:

1. *$ROI^{(t+1)}(0)$ selection:* Level 0 of the new region of interest is obtained by taking into account the position where the target is placed in the original image of the frame *t*. Firstly, the bounding-box of $T^{(t)}(0)$ ($BB(T^{(t)}(0))$) is computed. Then, $ROI^{(t+1)}(0)$ will be made up of the pixels of the next frame $p^{(t+1)}(i,j,l)$ which are included in the bounding box plus the pixels included in an extra border ε of the bounding box. This extra border tries that the target in the next frame will be placed in the new ROI.

$$ROI^{(t+1)}(0) = \bigcup_{ij \in \{BB(T^{(t)}(0)) + \varepsilon\}} p^{(t+1)}(i,j,0) \tag{7}$$

2. *Over-segmentation of $ROI^{(t+1)}(0)$:* The hierarchical structure $ROI^{(t+1)}$ is built. This step is performed at the beginning of the tracking process *t+1* (subsection 8.1).

9. Appendix B. Model pose estimation

As shown in Fig. 11, each arm is modelled with a two-bone kinematic chain. The parent bone corresponds to the upper arm and is allowed to rotate around three perpendicular axes. This provides a simplified model of the shoulder joint. $T({}^{w}_{1}R)$ is the local transformation between the upper-arm reference frame O_1 and a coordinate frame attached to the torso and centered at the shoulder joint w. The bone representing the lower arm is allowed to rotate around a single axis, corresponding to the elbow joint. $T({}^{2}_{1}R, {}^{1}\vec{l}_1)$ denotes the local transformation between the upper-arm reference frame O_1 and the lower-arm reference frame O_2, where ${}^{1}\vec{l}_1 = (0,0,l_1)^T$, being l_1 the length of the upper-arm, and ${}^{2}_{1}R$ corresponds to the rotation θ_ε about the elbow axis.

Given a desired position for the end-point of the arm at time instant *t+1*, ${}^{w}\vec{p}_d^{(t+1)}$, and given the rotation matrices ${}^{w}_{1}R^{(t)}$ and ${}^{1}_{2}R^{(t)}$ at the previous time instant *t*, the problem is then to find the updated matrices ${}^{w}_{1}R^{(t+1)}$ and ${}^{1}_{2}R^{(t+1)}$. A simple geometric method is summarized here that can solve such problem. See (Mitchelson, 2003) for further details.

1. Bring ${}^{w}\vec{p}_d^{(t+1)}$ within reach of the arm:

$$\text{if } \left|{}^{w}\vec{p}_d^{(t+1)}\right| > (l_1 + l_2) \text{ then } {}^{w}\vec{p}_d^{(t+1)} \leftarrow {}^{w}\vec{p}_d^{(t+1)} \frac{l_1 + l_2}{\left|{}^{w}\vec{p}_d^{(t+1)}\right|}$$

2. *Compute elbow circle:* Posing the model arms is an under-constrained problem, as four degrees of freedom must be specified from only three constraints, corresponding to the

co-ordinates of the desired end-point position ${}^{w}\vec{p}_d^{(t+1)}$. The elbow circle is defined as the set of positions that the elbow is free to adopt when the end-point of the arm reaches ${}^{w}\vec{p}_d^{(t+1)}$. It has a radius r and it is contained in a plane perpendicular to the vector ${}^{w}\vec{p}_d^{(t+1)}$ at a distance b to the shoulder joint.

$$r^2 = \frac{(d+l_1+l_2)(-d+l_1+l_2)(d-l_1+l_2)(d+l_1-l_2)}{2d}$$

$$b = \sqrt{l_1^2 - r^2}$$

where $d = \left|{}^{w}\vec{p}_d^{(t+1)}\right|$

3. *Choose updated elbow axis* ${}^{w}\vec{x}_2^{(t+1)}$ *and location* ${}^{w}\vec{l}_2^{(t+1)}$: We chose the elbow axis at time instant $t+1$ to be the closest to the one at the previous time instant, ${}^{w}\vec{x}_2^{(t)}$:

$${}^{w}\vec{x}_2^{(t+1)} = ({}^{w}\vec{p}_d^{(t+1)} \wedge {}^{w}\vec{x}_2^{(t)}) \wedge {}^{w}\vec{p}_d^{(t+1)}$$

$${}^{w}\vec{l}_1 = b\frac{{}^{w}\vec{p}_d^{(t+1)}}{|{}^{w}\vec{p}_d^{(t+1)}|} + r\frac{{}^{w}\vec{x}_2^{(t+1)} \wedge {}^{w}\vec{p}_d^{(t+1)}}{|{}^{w}\vec{x}_2^{(t+1)} \wedge {}^{w}\vec{p}_d^{(t+1)}|}$$

4. *Fill updated rotation matrices* ${}^{w}_{1}R^{(t+1)} = ({}^{w}\vec{x}_1\, {}^{w}\vec{y}_1\, {}^{w}\vec{z}_1)$ *and* ${}^{1}_{2}R^{(t+1)} = ({}^{1}\vec{x}_2\, {}^{1}\vec{y}_2\, {}^{1}\vec{z}_2)$ *with:*

$${}^{w}\vec{x}_1 = {}^{w}\vec{x}_2 \qquad {}^{1}\vec{x}_2 = (1,0,0)$$

$${}^{w}\vec{z}_1 = {}^{w}\vec{l}_1 / |{}^{w}\vec{l}_1| \qquad {}^{1}\vec{z}_2 = {}^{w}R_1({}^{w}\vec{p}_d - {}^{w}\vec{l}_1)$$

$${}^{w}\vec{y}_1 = {}^{w}\vec{z}_1 \wedge {}^{w}\vec{x}_1 \qquad {}^{1}\vec{y}_2 = {}^{1}\vec{z}_2 \wedge {}^{1}\vec{x}_2$$

The proposed inverse kinematics method can obtain an arm pose that will put the hand of the model in the required position. But this pose must be analyzed in order to determine if it respects model joint limits and does not produce a collision between different links.
The detection of limits violation and collisions is merely used to correct tracking errors and produce a more natural motion in the human model. But for HOAP-1 model, these two features become a crucial part of the motion generation, as an incorrect pose could damage the real robot if it is not previously detected and avoided.

- *Joint limits detection.* Given the updated shoulder and elbow rotation matrices, it is necessary to extract joint angles from these matrices that correspond to the real DOFs of the model.
  This process is made by applying a parameterization change to rotation matrices. There is a direct correspondence between Denavith-Hartenberg (DH) (Craig, 1986) parameters and model joint angles, so the local axes referred angles are converted to DH parameters. The shoulder conversion can be done applying the following parameterization to the rotation matrix ${}^{w}_{1}R$:

$$ {}_{1}^{w}R = \begin{pmatrix} c\theta_2 c\theta_3 & -c\theta_2 s\theta_3 & s\theta_2 \\ s\theta_1 s\theta_2 c\theta_3 + c\theta_1 s\theta_3 & -s\theta_1 s\theta_2 s\theta_3 + c\theta_1 c\theta_3 & -s\theta_1 c\theta_2 \\ -c\theta_1 s\theta_2 c\theta_3 + s\theta_1 s\theta_3 & c\theta_1 s\theta_2 s\theta_3 + s\theta_1 c\theta_3 & c\theta_1 c\theta_2 \end{pmatrix} \quad (8) $$

where θ_1, θ_2 and θ_3 are the real DOFs of the model arm, $c\theta_i$ is $\cos(\theta_i)$ and $s\theta_i$ is $\sin(\theta_i)$.
The elbow angle is much easier to obtain: as there is only one DOF in the elbow, the local rotation angle is equal to model θ_4 angle.
Both human and humanoid robot models distribute the DOFs in the same way: two located in the shoulder, and two in the elbow. But the inverse kinematics method provides a solution in which the shoulder contains three DOFs. It is required to move this DOF from shoulder to elbow. This operation can be easily done, given the chosen parameterization, as the third DOF of the shoulder corresponds with the rotation along the segment axis and so it can be directly translated as the first elbow DOF.
Once the model DOFs are computed, the system can directly check if any of them lies beyond its limits.

- *Collision detection.* This process is much more complex than previous one. Our method uses RAPID (Gottschalk et al., 1996) as the base of the collision detection module. This library provides functions that can quickly and efficiently check collisions between meshes composed by triangles, and attached to model links.

Once the system detects an incorrect position (i.e. joint limit or collision), it follows these steps:

1. The system looks for alternative poses (i.e. different arm configurations). Imitation requires to place hands in certain coordinates, but the elbow is free to move in the circle presented in Fig. 10a. Thus, alternative poses will preserve hand positions, but will move the elbow in this circle.
2. The motion of the arm should be as smooth as possible. Thus, alternatives should be more densely searched near the current elbow location. This is implemented by exponentially distributing the alternatives around the initial incorrect elbow position, as shown below:

$$ \theta_{2i} = \pi \frac{1}{100^{\frac{(n-i)}{n}}} \quad (9) $$

   where θ_{2i} and $\theta_{(2i+1)}$ correspond to two symmetric alternatives on the elbow circle with respect to the current pose, and $n = N/2$, being N the number of alternative poses that are tested when the current pose is erroneous.
   Fig. 10b shows alternatives given a certain pose. As required, alternative poses are placed on the elbow circle (Fig. 10a) and are more deeply distributed near the current elbow position.
3. The system chooses the nearest valid alternative.
4. If there is no valid alternative, the arm remains in the last valid position.

The speed of the process depends on the number of alternatives it needs to check. A system using a correct number of alternatives should produce smooth movements and work in real-time even if all of them are to be tested.
The alternative evaluation module has been also used when the system is in a valid pose: in these cases, the two nearest alternatives to current pose are checked. If one of them locates

the elbow in a lower vertical position, and do not produce limits violation nor collisions, then the elbow is moved to that position. This allows the model to adopt more natural poses when possible.

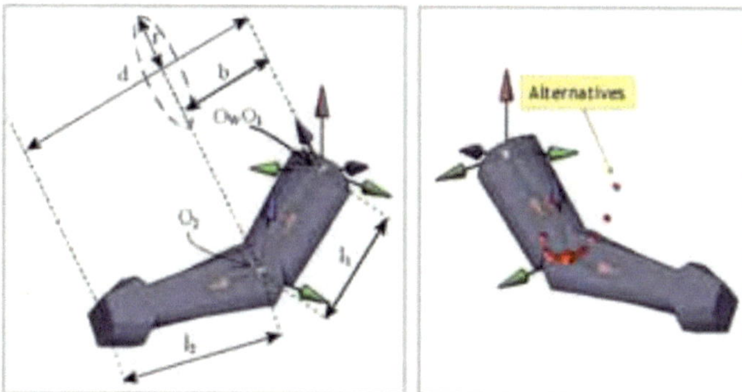

(a) (b)

Figure 10. a) Kinematic model of the human arm showing local coordinate frames and elbow circle; and b) alternative poses (red spheres) for a given elbow position

10. Acknowledgement

This work has been partially granted by the Spanish Ministerio de Educación y Ciencia (MEC) and FEDER funds, Project n. TIN2005-01359, and by the European Robotics Research Network (EURON), Project VISOR.

11. References

Alissandrakis, A.; Nehaniv, C. & Dautenhahn, K. (2002). Imitation With ALICE : Learning to Imitate Corresponding Actions Across Dissimular Embodiments, *IEEE Trans. on Systems, Man and Cybernetics – Part A: Systems and Humans*, Vol. 32, No. 4, July 2002, pp. 482-496.

Andry, P.; Gaussier, P.; Moga, S.; Banquet, J. P. & Nadel, J. (2001). Learning and communication via imitation: an autonomous robot perspective, *IEEE Trans. on Systems, Man and Cybernetics – Part A: Systems and Humans*, Vol. 31, No. 5, September 2001, pp. 431-442.

Billard, A. (2001). Learning motor skills by imitation: a biologically inspired robotic model, *Cybernetics and Systems Journal*, Vol. 32, 2001, pp. 155-193.

Breazeal, C. & Scassellati, B. (2002). Challenges in building robots that imitate people. In: *Imitation in Animals and Artifacts*, Dautenhahn, K. & Nehaniv, C. (Ed.), pp. 363-390, MIT Press, Cambridge, MA, USA.

Breazeal, C.; Brooks, A.; Gray, J.; Hancher, M.; McBean, J.; Stichl, D. & Strickon, J. (2003). Interactive robot theatre, *Communications of the ACM*, Vol. 46, No. 7, July 2003, pp. 76-84.

Breazeal, C.; Buchsbaum, D.; Gray, J.; Gatenby, D. & Blumberg, B. (2005). Learning From and About Others: Towards Using Imitation to Bootstrap the Social Understanding of Others, *Artificial Life*, Vol. 11, No. (1-2), January 2005, pp. 31-62.

Byrne, R. & Russon, A. (1998). Learning by imitation: a hierarchical approach, *Behavioral and Brain Sciences*, Vol. 21, No. 5, October 1998, pp. 667-721.

Craig, J. J. (1986). *Introduction to Robotics*, Addison-Wesley, Boston, MA, USA.

Dautenhahn, K. & Nehaniv, C. (2002). *Imitation in animals and artifacts*, MIT Press, Cambridge, MA, USA.

Decety, J.; Chaminade, T.; Grezes, J. & Meltzoff, A. (2002). A PET exploration of the neural mechanisms involved in reciprocal imitation, *Neuroimage*, Vol. 15, January 2002, pp. 265–272.

Demiris, J.; Rougeaux, S.; Hayes, G. M.; Berthouze, L. & Kuniyoshi, Y. (1997). Deferred imitation of human head movements by an active stereo vision head, *Proceedings of the 6th IEEE Int. Workshop on Robot Human Communication*, pp. 88-93, Sendai-Japan, September-October 1997.

Demiris, J. & Hayes, G. M. (2002). Imitation as a dual-route process featuring predictive and learning components: A biologically plausible computational model, In: *Imitation in Animals and Artifacts*, Dautenhahn, K. & Nehaniv, C. (Ed.), pp. 327-361, MIT Press, Cambridge, MA, USA.

Demiris, Y. & Khadhouri, B. (2005). Hierarchical, Attentive Multiple Models for Execution and Recognition (HAMMER), *Proceedings of the Int. Conf. on Robotics and Automation, Workshop on Robot Programming by Demonstration*, pp. 38-41, Barcelona-Spain, April 2005.

Gottschalk, S.; Lin, M. C. & Manocha, D. (1996). OBB-Tree: A Hierarchical Structure for Rapid Interference Detection, *Computer Science*, Vol. 30, August 1996, pp. 171-180.

Hayes, G. M. & Demiris, J. (1994). A robot controller using learning by imitation, *Proceedings of the 2nd International Symposium on Intelligent Robotic Systems*, pp. 198-204, Grenoble-France, July 1994.

Ijspeert, A.; Nakanishi, J. & Schaal, S. (2002). Movement imitation with nonlinear dynamical systems in humanoid robots, *Proceedings of the IEEE Int. Conf. on Robotics and Automation*, pp. 1398-1403, Washington-USA, May 2002.

Ikeuchi, K. & Suehiro, T. (1992). Towards an assembly plan from observation - part I: assembly task recognition using face-contact relations (polyhedral objects), *Proceedings of the IEEE Int. Conf. on Robotics and Automation*, pp. 2171-2177, Nice-France, May 1992.

Inaba, M. & Inoue, H. (1989). Visual-based robot programming, *Proceedings of the 5th Int. Symp. on Robotics Research*, pp.129-136, Tokio, August 1989.

Inaba, M.; Mizuuchi, I.; Tajima, R.; Yoshikai, T.; Sato, D.; Nagashima, K. & Inoue, H. (2003). Building spined muscle-tendon humanoid, In: *Robotics Research: The Tenth International Symposium*, Jarvis, R. A. & Zelinsky, A. (Ed.), pp. 113-130, Springer-Verlag, Berlin.

Itti, L. & Koch, C. (2001). Feature combination strategies for saliency-based visual attention systems, *Journal of Electronic Imaging*, Vol. 10, No. 1, January 2001, pp. 161-169.

Kang, S. B. & Ikeuchi, K. (1993). Toward automatic robot instruction from perception - Recognizing a grasp from observation, *IEEE Trans. on Robotics Robots and Automation*, Vol. 9, No. 4, 1993, pp. 432-443.

Kuniyoshi, Y.; Inaba, M. & Inoue, H. (1994). Learning by watching: extracting reusable task knowledge from visual observation of human performance, *IEEE Trans. on Robotics and Automation*, Vol. 10, No. 6, December 1994, pp.799-822.

Kuniyoshi, Y.; Yorozu, Y.; Inaba, M. & Inoue, H. (2003). From visual-motor self learning to early imitation - a neural architecture for humanoid learning, *Proceedings of the 2003 IEEE Int. Conf. on Robotics and Automation*, pp. 3132-3139, September 2003.

Lopes, M. & Santos-Victor, J. (2005). Visual learning by imitation with motor representations, *IEEE Trans. on Systems, Man and Cybernetics - Part B: Cybernetics*, Vol. 35, No. 3, June 2005, pp. 438-449.

Lorenz, K. (1966). *On aggression*, Harcout, Brace and World, Orlando.

Marfil, R.; Bandera, A.; Rodríguez, J. A. & Sandoval, F. (2004). Real-time Template-based Tracking of Non-rigid Objects using Bounded Irregular Pyramids, *Proceedings of the IEEE/RSJ Int. Conf. on Intelligent Robotics and Systems*, Vol. 1, pp. 301-306, Sendai-Japan, September-October 2004.

Maurer, D. & Mondloch, C. (2005). The infant as synaesthete, *In: Attention and Performance XXI: Processes of Change in Brain and Cognitive Development*, Munakata, Y. & Johnson, M. (Ed.).

Meltzoff, A. & Moore, M. (1977). Imitation of facial and manual gestures by human neonates, *Science*, Vol. 198, October 1977, pp. 75-78.

Meltzoff, A. & Moore, M. (1989). Imitation in newborn infants: Exploring the range of gestures imitated and the underlying mechanisms, *Developmental Psychology*, Vol. 25, 1989, pp. 954-962.

Metta, G.; Manzotti, R.; Panerai, F. & Sandini, G. (2000). Development: Is it the right way toward humanoid robotics?, *Proceedings of the IAS*, pp. 249-253, Venice-Italy, July 2000.

Mitchelson, J. R. (2003). *Multiple-Camera Studio Methods for Automated Measurement of Human Motion*, Ph.D. dissertation, CVSSP, School of Electronics and Physical Sciences, Univ. of Surrey, UK.

Molina-Tanco, L.; Bandera, J. P.; Marfil, R. & Sandoval, F. (2005). Real-time human motion analysis for human-robot interaction, *Proceedings of the IEEE/RSJ Int. Conf. on Robotics and Intell. Systems*, pp. 1808-1813, Alberta-Canada, August 2005.

Molina-Tanco, L.; Bandera, J.P.; Rodríguez, J.A.; Marfil, R.; Bandera, A. & Sandoval, F. (2006). A Grid-based Approach to the Body Correspondence Problem in Robot Learning by Imitation, *Proceedings of the European Robotic Symposium (EUROS 2006), Workshop on Vision Based Human-Robot Interaction*, Palermo-Italy, March 2006.

Nadel, J. (2004). Early imitation and the emergence of a sense of agency, *Proceedings of the 4th Int. Workshop on Epigenetic Robotics*, pp. 15-16, Genoa-Italy, August 2004.

Nakamura, Y. & Yamane, K. (2000). Dynamics computation of structure-varying kinematic chains and its application to human figures, *IEEE Trans. on Robotics and Automation*, Vol. 16, No. 2, April 2000, pp. 124-134.

Nehaniv, C. & Dautenhahn, K. (2005). The correspondence problem in social learning: What does it mean for behaviors to "match" anyway?, In: *Perspectives on Imitation: From Cognitive Neuroscience to social science*, Hurley, S. (Ed.), MIT Press, Cambridge, MA, USA.

Nicolescu, M. & Mataric, M. (2005). Task learning through imitation and human-robot interaction, In: *Models and mechanisms of imitation and social learning in robots, humans and animals*, Dautenhahn, K. & Nehaniv, C. (Ed.), Cambridge University Press, Cambridge, UK.

Ogata, H. & Takahashi, T. (1994). Robotic assembly operation teaching in a virtual environment, *IEEE Trans. on Robotics and Automation*, Vol. 10, No. 3, 1994, pp. 391-399.

Piaget, J. (1945). *La formation du symbole chez l'enfant*, Delachaux et Niestle, Paris.

Rizzolatti, G.; Fadiga, L.; Gallese, V. & Fogassi, L. (1996). Premotor cortex and the recognition of motor actions, *Cognitive Brain Research*, Vol. 3, No. 2, March 1996, pp. 131–141.

Sakoe, H. & Chiba, S. (1978). Dynamic Programming Algorithm Optimization for Spoken Word Recognition, *IEEE Trans. on Acoustics, Speech, and Signal Processing*, Vol. ASSP-26, No. 1, February 1978, pp. 43-49.

Sauser, E. & Billard, A. (2005). View sensitive cells as a neural basis for the representation of others in a self-centered frame of reference, *Proceedings of the Third International Symposium on Imitation in Animals and Artifacts*, pp. 119-127, Hatfield-UK, April 2005.

Schaal, S. (1999). Is imitation learning the route to humanoid robots?, *Trends in Cognitive Sciences*, Vol. 3, No. 6, June 1999, pp. 233-242.

Sun, C. (2002). Fast stereo matching using rectangular subregioning and 3D maximum-surface techniques, *Int. Journal of Computer Vision*, Vol. 47, No. 1/2/3, April-June 2002, pp. 99-117.

Tan, K. C.; Chen, Y. J.; Tan, K. K. & Lee, T. H. (2005). Task-Oriented Developmental Learning for Humanoid Robots, *IEEE Trans. on Industrial Electronics*, Vol. 52, No. 3, June 2005, pp. 906-914.

Terrillon, J. C. & Akamatsu, S. (1999). Comparative performance of different chrominance spaces for color segmentation and detection of human faces in complex scene images, *Proceedings of the 12th Conf. on Vision Interface*, Vol. 2, pp. 180-187, May 1999.

Treisman, A. & Gelade, G. (1980). A feature integration theory of attention, *Cognitive Psychology*, Vol. 12, No. 1, January 1980, pp. 97-136.

Tung, C. P. & Kak, A. C. (1995). Automatic learning of assembly tasks using a dataglove system, *Proceedings of the Int. Conf. on Intelligent Robots and Systems*, pp. 1-8, 1995.

Viola, P. & Jones, M. (2001). Rapid object detection using a boosted cascade of simple features, *Proceedings of the IEEE Conf. on Computer Vision and Pattern Recognition*, pp. 511-518, Kauai-USA, December 2001.

28

Affective Communication Model with Multimodality for Humanoids

Hyun Seung Yang, Yong-Ho Seo, Il-Woong Jeong and Ju-Ho Lee
AIM lab, EECS, KAIST (Korea Advanced Institute of Science and Technology)
Republic of Korea

1. Introduction

A humanoid robot that can naturally interact with humans must be capable of high-level interaction skills; that is, it must be able to communicate with humans in the form of human-like skills such as gestures, dialogue communication with mutual sympathy.

In terms of a human symbiotic situation, it is necessary for a robot to be sociable in terms of thinking and feeling. In this aspect, social interaction and communication that share emotions are important issues in the field of a humanoid robot research.

We propose an affective communication model for humanoid robots, so that these robots achieve high level communication through the framework of the affective communication model with multimodality.

The proposed model comprises five subsystems: namely, perception, motivation, memory, behavior, and expression. In the perception system, we implemented a bimodal emotion recognizer. To ensure a humanoid robot can respond appropriately to the emotional status of users and itself, we designed subsystems that use their own drive, emotions, and memory.

The major improvement in our framework is the construction of a memory system that stores explicit emotional memories of past events. The literature from cognitive science and neuroscience suggests that emotional memories are vital when the human brain synthesizes emotions (Ledoux, J., 1996). While previous research on sociable robots either ignores the emotional memory or maintains the emotional memory implicitly in high-level interaction, we need to establish explicit memories of emotions, events, and concepts. Therefore we have adopted the concept of emotional memory for our humanoid robot. Our memory system maintains explicit memories of previous emotional events. Thus, the emotional system can synthesize emotions on the basis of emotional memories.

Since 1999, AIM lab directed by Hyun S. yang in KAIST has focused on building new humanoid robots with a self-contained physical body, perception to a degree which allows the robot to be autonomous, and social interaction capabilities of an actual human symbiotic robot. This study was built on previous researches about the social interactions and the developments of the first generation humanoid robots, AMI, AMIET coupled with a human-size biped humanoid robot, AMIO in the AIM Lab (Yong-Ho Seo, Hyun S. Yang et al., 2003, 2004, 2006).

The proposed model was used to enhance the interaction between human and humanoid robot. Accordingly, we designed and implemented the methods which are mentioned above and successfully verified the feasibility through the demonstrations of human and humanoid robot interactions with AIM Lab's humanoids.

2. Previous Sociable Robots and Affective Communication Model

In recent times, the concept of emotion has increasingly been used in interface and robot design. Moreover, numerous studies have been performed to integrate emotions into products including electronic games, toys, and software agents (C. Bartneck et al., 2001). Furthermore, Affective social interaction between human and robot is hot issues in robotics research area currently.

Many researchers in robotics have been exploring this issues such as a sociable robot, 'Kismet' which conveys intention through facial expression and engages in infant like interaction with a human caregiver (Breazeal, C., 1999). An AIBO, an entertainment robot, behaves like a friendly and life like dog which interact with human by touch and sound (Arkin and Fujita et al., 2003). Mel, a conversational robot, introduced concepts that a robot leads the interactions by explaining its knowledge (Sidner, C.L. et al., 2005). Cat Robot was developed to investigate the emotional behavior of physical interaction between a cat and a human (Shibata, T. at al., 2000).

Tosa & Nakastu have researched on emotion recognition system through speech and its applications. Their initial work, MUSE and MIC recognized human emotions involved in speech and expressed emotions through computer graphics. Since then, they developed some application systems utilizing their initial concept (Nakastsu, Tosa et al., 1996,1999).

To perform high-level tasks while contending with various situations in actual, uncertain environments, a robot needs various abilities such as the ability to detect human faces and objects using a camera in addition to the speech processing, the ability to sense an obstacle using several sensors, and the capability of manipulation and bipedal navigation.

In addition, it is necessary to integrate these software functions efficiently and reliably. Therefore, to operate the robot, control architecture or framework was usually planned based on behavior architecture. The affective communication model which we designed is an unified control architecture to perform complex tasks successfully using a set of coordinated behaviors. A high-level task is driven from a motivation system. The motivation system activates set of coordinated behaviors. Finally, an expression system activates multimodal human interfaces such as a voice, gestures and 3D facial expressions.

3. Framework of Affective Communication Model

Motivated the human brain structure discovered by cognitive scientist (Ledoux, J., 1996), we have designed the framework for our sociable humanoid robots. We designed the affective communication framework to include the five subsystems shown in Fig. 1. Our framework is similar to the creature kernel framework for synthetic characters (Yoon, S.Y., Burke, R.C. et al., 2000). The similar framework was also applied to the software architecture of Kismet (Breazeal, C., 1999). However, since our goal is to enable dialogue interactions, we improved the framework so that our robot can preserve explicit memories. Thus, our system has two major benefits over previous systems.

The fisrt is a memory system. We added a memory system to the referred framework. The memory system enables a robot to represent, and reflect upon, itself and its human partners. It also enhances a robot's social skills and fosters communication with humans. To enable

affective interaction between a robot and humans, we enabled a robot to preserve its emotional memory of users and topics.
The second is a dialogue based human robot interaction. Other affective robots, such as Kismet, were based on an infant-caretaker interaction, but our system is based mainly on a dialogue interaction. Accordingly, our internal design and implementation differ from other robots because of our distinct goal of multimodal affective communication.
The main functions of each subsystem are summarized as follows. The perception system, which mainly extracts information from the outside world, comprises the subsystems of face detection, face recognition, emotion recognition, and motion and color detection. The motivation system is composed of a drive and an emotional sys-tem. Drives are motivators; they include endogenous drives and externally induced desires. The emotional system synthesizes a robot's artificial emotions. The memory system, as mentioned above, preserves the emotional memories of users and topics. We improved the subsystems of our previous humanoid robots (Yong-Ho Seo, Hyun S. Yang et al., 2003, 2004).

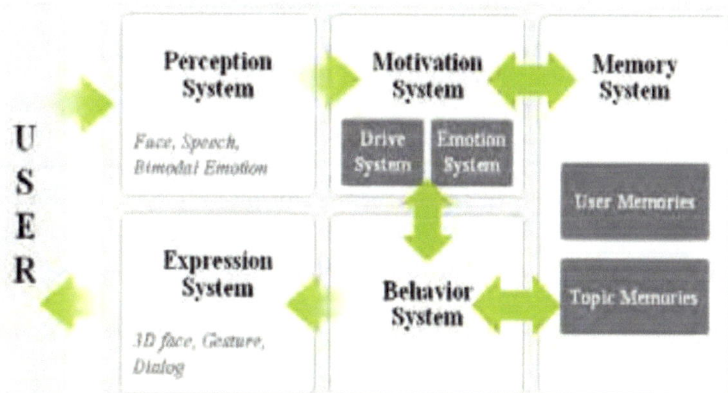

Figure 1. Framework of Affective communication model

4. Perception System

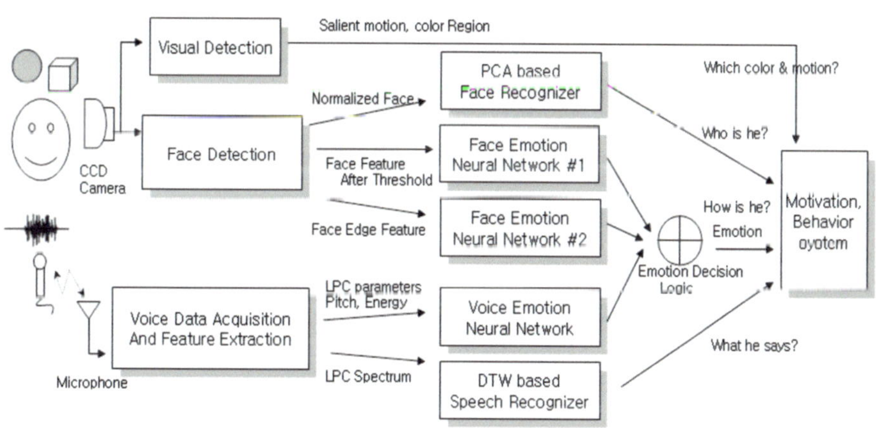

Figure 2. Structure of Perception system

The perception system comprises face detection, face recognition, speech recognition, emotion recognition, and visual attention detection that enables a robot to detect objects and color. Accordingly, through the perception system, the robot can learn basic knowledge such as the identity of the human user, the user's emotional state, and what the user is saying and doing. The overall process of the perception system is shown Fig. 2.

4.1 Face Detection and Recognition

The face detection system finds human faces in an image from CCD cameras. For robust and efficient face detection, the face detection system used a bottom-up, feature-based approach. The system searches the image for a set of facial features such as color and shape, and groups them into face candidates based on the geometric relationship of the facial features. Finally, the system decides whether the candidate region is a face by locating eyes in the eye region of a candidate's face. The detected facial image is sent to the face recognizer and to the emotion recognizer. The face recognizer determines the user's identity from the face database. To implement the system, we used an unsupervised PCA-based face classifier commonly used in face recognition.

4.2 Bimodal Emotion Recognition

This section describes each emotion recognition and then bimodal emotion recognition and discusses the emotion recognition performance. We estimated emotion through facial expression and speech, and then integrate them to enable bimodal emotion recognition. Emotion recognition plays an important role in the perception system because our system enables a robot to recognize human partner's emotional status, and behave appropriately considering the recognized emotion status.

For emotion recognition through facial expression, we normalized the image captured. We then extracted the following two features, which are based on Ekman's facial expression features (Ekman, P. et al., 1978). The first feature is a facial image of lips, brow and forehead. After applying histogram equalization and the threshold of the standard distribution of bright of normalized face image, we extracted the parts of lips, brow and forehead from the entire image.

The second feature is an edge image of lips, brow and forehead. After applying histogram equalization, we extracted the edges around the regions of the lips, brow and forehead.

Each of the extracted features is then trained using two neural networks for five emotions (neutral, happy, sad, angry, and surprise). These emotions are chosen among the basic six emotions because people generally have difficulty in making facial expression especially for the emotions, fear and disgust, which show lower recognition rate than other four emotions. Each neural network is composed of 1610 input nodes, 6 hidden nodes and 5 output nodes. The 1610 input nodes receive 1610 pixels of the input image and the output nodes represent 5 emotions: neutral, happy, sad, angry, and surprise. The number of hidden nodes is determined by experiment.

Finally, the decision logic determines the final emotion result from two neural network results. The face emotion decision logic utilizes the weighted sum of two neural network results and voting method of result transitions over time domain.

The overall process of the emotion recognition through facial expression is shown in Fig. 3.

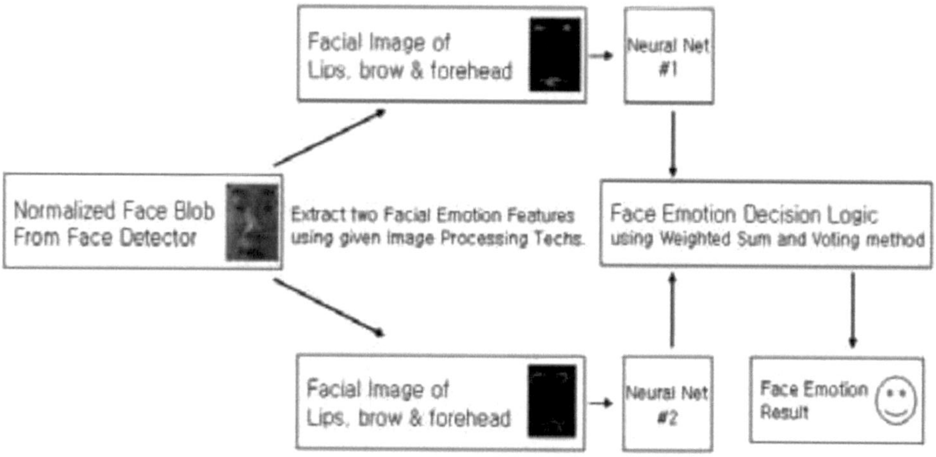

Figure 3. Process of Emotion Recognition through Facial Expression

The followings are about emotion recognition through speech. For emotion recognition through speech, we adopted a recognition method similar to the one used in the life-like communication agents MUSE and MIC (Nakastsu, Tosa et al., 1996, 1999). The speech features influenced by emotions are speech-influencing pitch, energy, timing, and articulation (Cahn, J., 1990). For feature extraction, we extracted phonetic and prosodic features in accordance with the state of the art research (Nakatsu, Tosa et al., 1999). Neural network is used to train each feature vector and recognize the emotions.

Two kinds of features are used in emotion recognition. One is a phonetic feature and the other is a prosodic feature. LPC (linear predictive coding) parameters, which are typical speech feature parameters often used for speech recognition, are adopted for the phonetic feature. The prosodic feature, on the other hand, consists of two factors: Speech power and pitch structure. Speech power and pitch parameters can be obtained in the LPC analysis. In addition, a delta LPC parameter is adopted, which is calculated from LPC parameters and expresses a time variable feature of the speech spectrum, since this parameter corresponds to a temporal structure.

The speech feature calculation is carried out in the following way. Analog speech is first transformed into digital speech by passing it through a sound card with an 8 KHz sampling rate and 16 bits accuracy. The digitized speech is then arranged into a series of frames, where each frame is a set of 256 consecutive sampled data points. LPC analysis is carried out in real time and the following feature parameters are obtained for each of these frames. Thus, for the t-th frame, the obtained feature parameters can be expressed by

$$Ft = (Pt, pt, dt, c1t, c2t, \ldots, c12t) \quad (1)$$

(where speech power is P; pitch is p, Delta LPC parameter is d; LPC parameters are c1t, c2t, …, c12t.)

The sequence of this feature vector is fed into the speech period extraction stage.

For extraction of speech period, we used the information of speech power. Speech power is compared with a predetermined threshold value; if the input speech power exceeds this threshold value for a few consecutive frames, the speech is decided to be uttered. After the beginning of the speech period, the input speech power is also compared with the threshold value; if the speech power is continuously below the threshold for another few consecutive frames, the speech is decided not to exist. The speech period is extracted from the whole data input through this process. Ten frames are extracted for the extracted speech period where each is situated periodically in the whole speech period and kept the same distance from adjacent frames. Let these ten frames be expressed as f1, f2, ..., f10. The feature parameters of these ten frames are collected and the output speech features are determined as a 150 (15x10) dimensional feature vector. This feature vector is expressed as

$$FV = (F1, F2, \ldots\ldots., F10) \tag{2}$$

(where Fi is a vector of the fifteen feature parameters corresponding to frame fi.)
This feature vector (FV) is then used as input for the emotion recognition stage.
For emotion recognition, each feature vector was trained using a neural network for five emotions, neutral, happy, sad, and angry. The reason the four emotions was taken as classifiers is that these emotions have higher recognition rate than other emotions such as surprise, fear and disgust from the experimental results of previous systems to recognize emotion through speech.
Accordingly, the neural network was composed of 150 (15x10) input nodes corresponding to the dimension of speech features, 4 hidden node and 5 output nodes (neutral, sad, happy, angry and surprise). The neural network configuration is similar with the previous work of Nakatsu, R., Nicholson, J. and Tosa, N. (Nakatsu, Tosa et al., 1999), however internal lerning weights of neural network and nomarlization algorithm of speech parameters are somwhat different.
For bimodal emotion recognition, we used decision logic to integrate the two training results. The final result vector of the decision logic (Rfinal) is as follows:

$$Rfinal = (RfaceWf + RspeechWs) + Rfinal\text{-}1 - \delta t \tag{3}$$

(where Rface and Rspeech are the results vector of the emotion recognition through facial expression and speech. Wf and Ws are the weights of the two modalities. δ is a decay term that restores the emotional status to neutral.)

4.3 Performance Evaluation of Bimodal Emotion Recognition

The overall correctness of the implemented bimodal emotion system recognition was about 80 percent for each of the five testers. By resolving confusion, the bimodal emotion system performed better than facial-only and speech-only systems.
We found that the two modalities have complementary property for each emotion from the experimental results shown in Table 1 and Table 2, as well as the previous research of bimodal emotion recognition (Huang, T.S., et al., 1998). The complementary property shows that for happy emotion, facial expression has higher recognition rates than speech. For sad emotion, voice has higher recognition rates. For angry emotion, facial expression and speech have similar recognition performance. Therefore, we made the final decision logic to conduct weighted summation. Accordingly, we appropriately determined the weight valuables, Wf and Ws in the reference of the experimental results.
Actually, it is difficult to extract hidden emotional features from natural facial expression and speech. Therefore, we used intentionally exaggerated facial expression and speech for each emotion to achieve high recognition rate compared to the previous research experiments.

For testing facial-only emotion recognition, we conducted experiments with four people. For training, we used five images per each emotion of each person. We set aside one from each category as test data, use the rest of samples as training data. The recognition result is shown in Table 1. Facial expression-only emotion recognition yield performance of 76.5% and 77.1% for the two neural networks. Therefore, we conducted weighted-summation to select the best result for each emotion from two neural networks and then achieved higher recognition rate of 79.5%.

Facial Expression - Neural Net. #1		Facial Expression - Neural Net. #2	
Happy	85.00 %	Happy	84.00 %
Sad	77.50 %	Sad	80.00 %
Neutral	77.50 %	Neutral	65.00 %
Angry	65.00 %	Angry	81.50 %
Surprise	77.50 %	Surprise	75.00 %
Total	76.5 %	Total	77.1 %

Table 1. Performance of Emotion from Facial Expression

In emotion recognition through facial expression, there is a little variation between people according to the Ekman's facial expression features [35]. On the other hand, there is a big difference in emotion recognition through speech because people have distinct and different voice. Especially, the speech features of men and women are largely different. Accordingly, we divided experiments into the two groups of men and women. In addition, we selected 4 emotions except surprise, because it is hard to recognize surprise from speech sentences.
For four people (two men and two women), we trained 15 sentences frequently used in communication with the robot. The testers repeated one sentence for each emotion five times. We set aside one from each category as test data, used the rest of samples as training data. The average recognition rate of men and women is shown in Table 2.

Speech Expression – NN for Men		Speech Expression – NN for Women	
Happy	72.00 %	Happy	77.50 %
Sad	82.50 %	Sad	85.50 %
Neutral	75.50 %	Neutral	70.00 %
Angry	84.00 %	Angry	75.00 %
Total	78.5 %	Total	77 %

Table 2. Performance of Emotion from Speech Expression

The bimodal emotion system integrated facial and speech systems with one decision logic. We evaluated the bimodal system for four people in real-time environment with varying scales and orientations using a variety of complex backgrounds.
The participants were asked to make emotional facial expressions while speaking out the sentence emotionally for each emotion at five times during a specified period to time. The overall bimodal emotion system yielded approximately 80 % for each of four testers. It achieved higher performance results than facial-only and speech-only by resolving some confusion. The higher result of this emotion recognition system compared to the other systems is caused by the limited number of users. Therefore, if the more users are

participated in this recognition system, the lower recognition result is expected. It's the limitation of these emotion recognition systems.

5. Motivation System

The motivation system sets up the robot's nature by defining its "needs" and influencing how and when it acts to satisfy them. The nature of the robot is to affectively communicate with humans and ultimately to ingratiate itself with them. The motivation system consists of two related subsystems, one that implements drives and a second that implements emotions. Each subsystem serves as a regulatory function for the robot to maintain its "well-being"

5.1 Drive System

The motivation system defines the robot's nature by defining its "needs" and influencing how and when it acts to satisfy them. The nature of the proposed humanoid robot is to socially interact with humans and ultimately to ingratiate itself with them. The motivation system consists of two related subsystems, one that implements drives and a second that implements emotions. Each subsystem serves as a regulatory function for the robot to maintain its "well-being"

In our previous research, three basic drives were defined for a robot's affective communication with humans (Yong-Ho Seo, Hyun S. Yang et al., 2004). In the new drive system for a humanoid robot operating and engaging interactions with human, four basic drives were defined for the robot's objectives as they related to social interaction with a human: a drive to obey a human's commands; a drive to interact with a human; a drive to ingratiate itself with humans and a drive to maintain its own well-being.

The first drive motivates a robot to perform a number of predefined services according to a human's commands. The second drive activates the robot to approach and greet humans. The third drive prompts the robot to try to improve a human's feelings. When the robot interacts with humans, it tries to ingratiate itself while considering the human's emotional state. The forth drive is related to robot's maintenance of its own well-being. When the robot's sensors tell it that extreme anger or sadness is appropriate, or when its battery is too low, it stops interacting with humans

5.2 Emotion System

Emotions are significant in human behavior, communication and interaction (Armon-Jones, C., 1985). A synthesized emotion influences the behavior system and the drive system as a control mechanism. To enable a robot to synthesize emotions, we used a model that comprises the three dimensions of emotion (Schlossberg, H., 1954). This model characterizes emotions in terms of stance (open/close), valence (negative/positive) and arousal (low/high). Our system always assumes the stance to be open, because a robot is always openly involved in interactions. Therefore, we only consider valence and arousal, implying that only three emotions are possible for our robots: happiness, sadness, and anger.

The arousal factor (Arousal about current user) is determined by factors such as whether a robot finds the human, and whether the human responds. Low arousal increases the emotion of sadness.

The valence factor (Response about current user) is determined by whether the human responds appropriately to robot's requests. A negative response increases the emotion of anger; a positive response increases the emotion of happiness.
The synthesized emotion is also influenced by the drive and the memory system. The robot's emotional status is computed by the following equation.

$$\text{If } t = 0,\ E_i(t) = M_i \quad (t = 0 \text{ when new face appears})$$

$$\text{If } t \neq 0,\ E_i(t) = A_i(t) + E_i(t-1) + \sum D_i(t) + M_i - \delta t. \tag{4}$$

Where $E_i(t)$ is the robot's emotional status, t is time(when new face appears), i = {joy, sorrow, anger}. $A_i(t)$ is the emotional status calculated by the mapping function of [A, V, S] from the current activated behavior. D_i is the emotional status defined by the activation and the intensity of unsatisfied drives in the drive system. M_i is the emotional status of the human recorded in the memory system. Finally, δt is a decay term that eventually restores the emotional status to neutral.

6. Memory System

Topic memories contain conversational sentences that a robot has learned from users. The topic memories are first created when the perception system recognizes that the frequency of a keyword has exceeded a threshold; that is, when the user has mentioned the same keyword several times. After the behavior system confirms that the current user is talking about a particular keyword, the memory system makes a new topic memory cell for that keyword. In the memory cell, the sentences of the user are stored and an emotional tag is attached with respect to robot's current emotion.

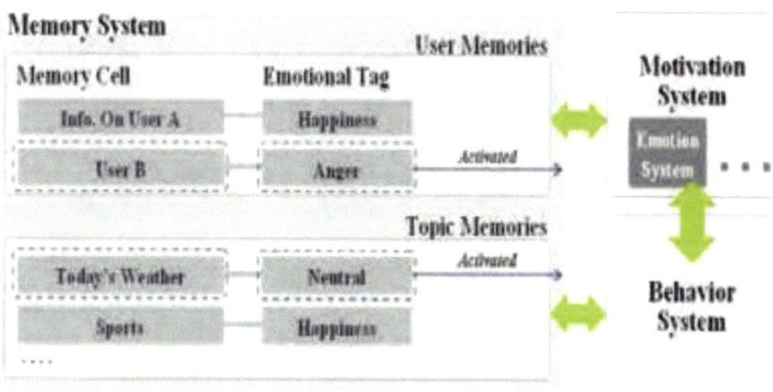

Figure 4. Activation of Memory cells in the Memory System

Of all the topic memories, only the one with the highest activation value is selected at time t. We calculated the activation values of the topic memories, Ti (t), as follows:

$$\text{If COMM} = 0,\ T_i(t) = W_{mt} \sum E_k(t)\, ET_i(t)$$

$$\text{If COMM} = i,\ T_i(t) = 1 \tag{5}$$

COMM represents the user's command to retrieve specific topic memory, t is time, $E_k(t)$ is AMI's current emotion, and $ET_i(t)$ is the emotional tag of the topic. Thus, $\sum E_k(t)\, ET_i(t)$

indicates the extent of the match between robot's current emotion and the emotion of the memory of the topic. Finally, W_{mt} is a weight factor. The activation of the memory system is shown in following Fig. 4.

7. Behavior and Expression System

We designed the structure of the behavior system that has three levels, which address the three drives of the motivation system as mentioned above. As the system moves down a level, more specific behavior is determined according to the affective relationship between the robot and human.

The first level of the behavior system is called drive selection. The behavior group of this level communicates with the motivation system and determines which of the three basic drives should be addressed. The second level, called high-level behavior selection, decides which high-level behavior should be adopted in relation to the perception and internal information in the determined drive. In the third level, called low-level behavior selection, each low-level type of behavior is composed of dialogue and gestures, and is executed in the expression system. A low-level type of behavior is therefore selected after considering the emotion and memory from other systems. The Fig. 5 shows the hierarchy of the behavior system and its details.

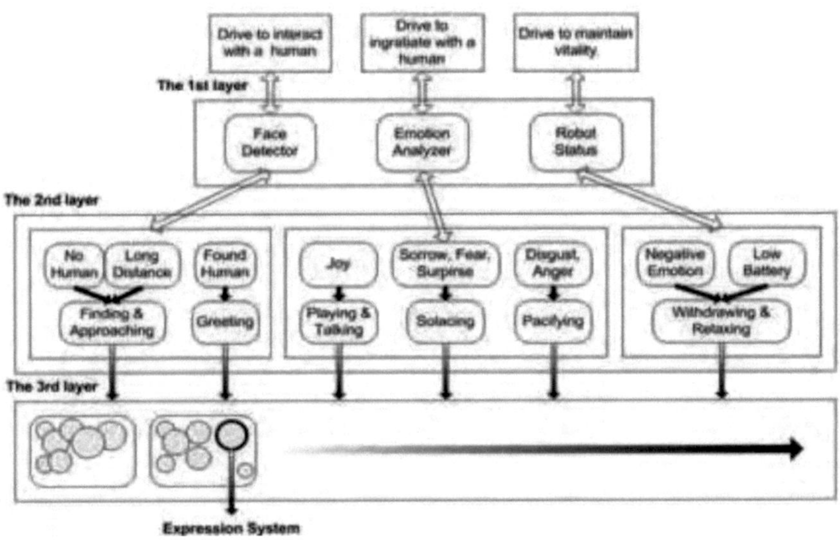

Figure 5. Hierarchy of the Behavior System

The expression system is the intermediate interface between the behavior system and robot hardware. The expression system comprises three subsystems: a dialogue expression system, a 3D facial emotion expression system and a gesture expression system.

The expression system plays two important functions. The first function is to execute the behavior received from the behavior system. Each type of behavior consists of a dialogue between the robot and the human. Sometimes the robot uses interesting gestures to control the dialogue's flow and to foster interaction with the human. The second function is to express robot's emotion. The robot expresses its own emotions through facial expressions but it sometimes uses gestures to covey its intentions and emotions.

7.1 Dialogue Expression

Dialogue is a joint process of communication sharing of information (data, symbols, context) between two or more parties. In addition, humans employ a variety of paralinguistic social cues (facial displays, gestures, etc.) to regulate the flow of dialogue (M. Lansdale, T. Ormerod, 1994). We consider there to be three primary types of dialogue: low level (prelinguistic), non verbal, and verbal language. Among them, the robot communicates with a human through daily verbal language with appropriate gestures.

However, it is difficult to enable a robot to engage in natural dialogue with a human because of the limitation of the current technique on speech recognition, natural language processing, etc. Accordingly, we predefined dialogue flow and topics. In order to make natural dialog possible in the limit that the robot could recognize only the limited number of speech, we constructed dialog as follows: First, the robot lead actively dialog by asking user's intention ahead of him to avoid the possibility that the robot cannot understand human speech. Second, the robot answers the most frequently used responses when they cannot understand to avoid unnatural dialog.

The dialogue expressions comprise the most commonly used speeches according to the selected behavior of Finding & Approaching, Greeting, Talking, Playing, Consoling, Pacifying, Withdrawing & Resting behavior group. In Finding & Approaching, the robot mainly finds a human by calling a human. In greeting, the robot says hello to a human and asks his name, etc. In Talking, dialog consists of various common topics such as hobby, weather, movies, etc. In playing, the robot plays with a human through a kind of jokes, OX quiz and Nonsense Quiz. In Consoling and Pacifying, the robot asks what the human is angry about and then makes a joke to console him or give pleasure to him. Further, the robot asks his worries by recognizing his sad emotion and listens to his sayings as his friend and counselor. The followings are the examples of the conversations in human robot interaction using the proposed model.

Sample conversation between AMI and users

Conversation 1

(GREETING)
AMI: Is there anyone who wants to talk to me?
User 1: Hello, AMI.
User 1 approaches. AMI finds a human face.
AMI: I heard someone calling me.
AMI: Ah, you are there. You're (User 1's name).
AMI: Let's shake hands.
User 1: OK.
User 1 and AMI shake hands.
AMI: How are you today?
User 1: Not very good. I had a pretty bad day.
AMI: I'm sorry to hear that.
(TALKING)
AMI: Is it cold today?
User 1: Yes.
AMI: I think so. I like warm and sunny days. How about you?
User 1: Well, I also prefer warm weather.

AMI: I thought so.
(CONSONLING)
AMI: Anyway, you look somewhat sad. It there anything wrong with you?
User 1: Yes.
AMI: Yeah, you look pretty unhappy. Can you tell me what happened to you today?
User 1: OK.
AMI: Tell me then.
User 1: It was a terrible day for me. I had a lot of homework to submit and I also had an exam today. I didn't have enough time to study for the exam because of my homework. So my exam was terrible, and that's why I'm not feeling good now.
AMI: I really fell sorry about that. Why don't we listen to some music?
User 1: Good idea.
AMI plays some music.
... (The conversation continues until the user says, "AMI, let's stop talking.")

Conversation 2

(GREETING)
AMI: Hello? Is anyone there?
User 2: Hi.
User 2 starts speaking.
AMI: Ah. I head a voice. Where are you?
User 2: Turn left.
AMI: I found you. Nice to meet you. Can you tell me your name?
User 2: (User 2's name).
AMI : You're (User2's name). Let's shake hands.
User 2: OK.
User 2 and AMI shake hands.
AMI: How are you today?
User 2: Fine.
AMI: I'm fine too.
(TALKING)
AMI: What did you have for lunch?
User 2: I had a hamburger, potato chips, and a salad.
AMI: Sounds nice. What's your favorite food?
User 2: Hmm. Nothing particular. I like everything.
AMI: Good.
(PLAYING)
AMI: You look fine today. Let's play a game.
User 2: OK.
AMI: Good. How about a common sense quiz. I'll give you some questions and you have to answer. Is that OK?
User 2: OK.
AMI: The first world cup was held in Greece.
User 2: Correct.
AMI: Sorry. Wrong. Next. The reason for VDT syndrome is too much exercise.
User 2: No!
AMI: You're right. Good job.

User 2: Had enough?
AMI: OK. Let's do it again later.
... (The conversation continues until the user says, "AMI, let's stop talking.")

7.2 Facial Expression

The 3D facial expression shows the robot's emotional status synthesized in the motivation system, as described in section 5. These expressions make up for the limitations of the robot's mechanical face which has difficulty in expressing its emotions. These facial emotion expressions were implemented using 3D graphics. Our 3D graphical face is displayed on the LCD screen which located on the robot's chest. We developed two different facial expression programs. One is more face like version and the other is more artificial and abstract version. The facial expressions in our 3D graphical faces and the dimension of emotions are shown as Fig. 6.

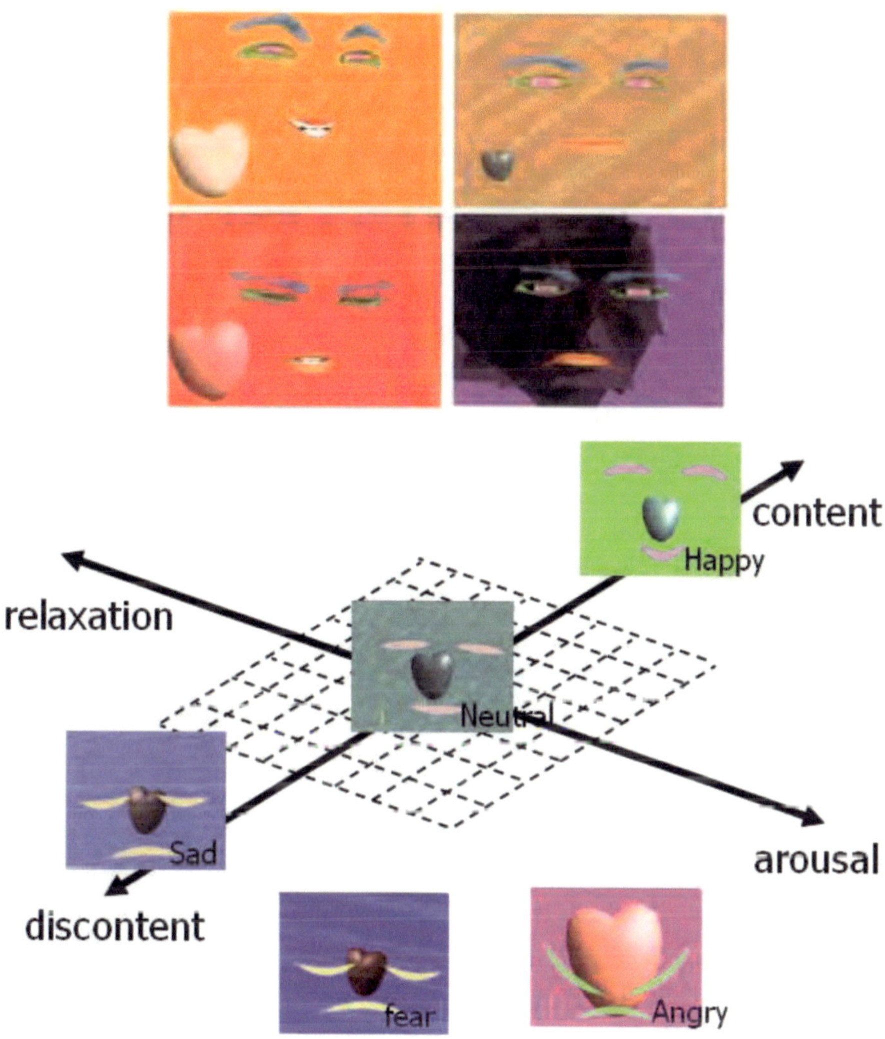

Figure 6. Graphical Facial Emotion Expressions and Dimension of Emotions

7.3 Emotional Gesture Expression

Gestures(or Motions) for our humanoids were generated to be human-like and friendly. Gestures are used to express its own emotions and to make interaction with humans more expressive. Therefore, expressions that would best attract the interest of humans were considered, and various interesting gestures were developed for our humanoids that would match the robot's dialogs and emotional statuses.

Humans tend to guess the emotional states of other people or some object from their body motions. Motions of a service robot are also important because they give strong impressions to a person. Most people think that robots act unnaturally and strangely. There are three types of functional disorders in communication methods between a human and a robot excluding speech. The details are in Table 3.

Limitations of conventional robots	**Functional disorders**
Motions to express internal state	Problem Conventional robots can not express their internal state ex) out of battery, emergency Solution Motions can be used for expressing internal state of a robot ex) no movement – out of battery slow movement – ready fast movement - emergency
Communication using sense of touch	Problem No reactions when a robot is touched Ex) An accident can be occurred even though someone tries to stop the robot. Solution A robot can express its internal state using motions when it touched by others Ex) When a person punishes a robot for its fault by hitting it, it trembles.
Eye Contact	Problem A robot which has no eyes looks dangerous ex) Humans usually feel that robots with no eyes are dangerous Solution A robot can look at a person of interest with sense of vision. ex) When a robot is listening to its master, it looks at his/her eyes.

Table 3. Limitations of conventional robots' interaction

We have to improve above functions of a robot to express its internal emotional state. As we can see Table 3, these functions can be implemented by using the channels of touch and vision. We focused the channel of vision perception-especially motion cues, so we studied about how to express emotions of a robot using motions such as postures, gestures and dances.

To generate emotional motions of a service robot, making an algorithm which can convert emotion to motions and describing motions quantitatively are necessary. We defined some parameters to generate emotional motions. These parameters are like in Table 4. We defined the parameters for the body part and the parameters for the two arms independently, so we can apply these parameters to a robot without considering whether it has two arms or not. Posture control and velocity control are very important to express emotional state using activities. These parameters are not absolute values, but relative values.

	Parameter	Joy	Sad	Anger	Disgust	Surprise
Body	Velocity	Fast	Slow	Fast	Slow	Slow
	Acceleration	Small	-	Large	Small	Large
	Direction	Possible turns	-	Forward / Backward	Backward	Backward / Stop
Arms	Position	Up	Down	Center	Center	Up
	Velocity	Fast	Slow	Fast	Normal	Fast
	Velocity change	Small	-	Large	Small	Small
	Shape	Arc	Line	Perpen-dicular	Perpen-dicular	Perpen-dicular
	Symmetry	Symme-trical	-	Unsymme-trical	-	-

Table 4. Parameters for emotional motions

To generate emotional gestures, we used the concept of Laban Movement Analysis, which is used for describing body movements (Toru Nakata, Taketoshi Mori, et al., 2002).

There are various parameters which are related to produce emotional motion generation. These parameters are related to generating natural emotional motions. To make natural motions of a robot, these parameters are used for expressing the intensity of the emotional state. According to the intensity of emotion, the number of parameters is changed to generate emotional motions. The higher intensity of an emotion is going to be expressed in a motion, the more parameters are going to be used for generating that motion.

We described the details of the parameters for emotional motions and we developed the emotional motions generating method. We defined 8 parameters and we can express 6 emotions by adjusting these parameters. The emotions we can express are joy, sad, neutral, surprise and disgust. We can make emotional motions with 5 levels of intensity by adjusting parameters in Table 2. We developed a simulator program to preview the generated motions before applying to the robot platform. The simulator is shown in Figure 7.

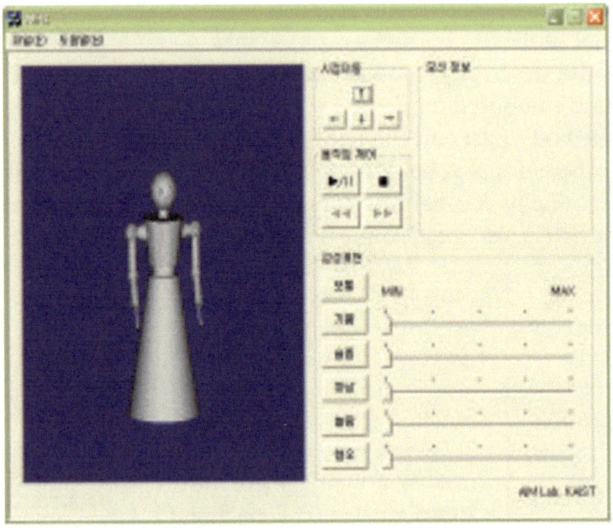

Figure 7. Simulator for emotional motion generation

We can preview a new generated motion using this simulator, so we can prevent some problems which can be occurred when we try to apply that motion to the robot. In this simulator, we can produce 6 emotional motions. Each emotional motion has 5 levels corresponding to the intensity of the emotion. Some examples of these emotional motions of our humanoid robots are shown in Fig. 8, Fig. 9, respectively.

Figure 8. Guesture expressions of AMI

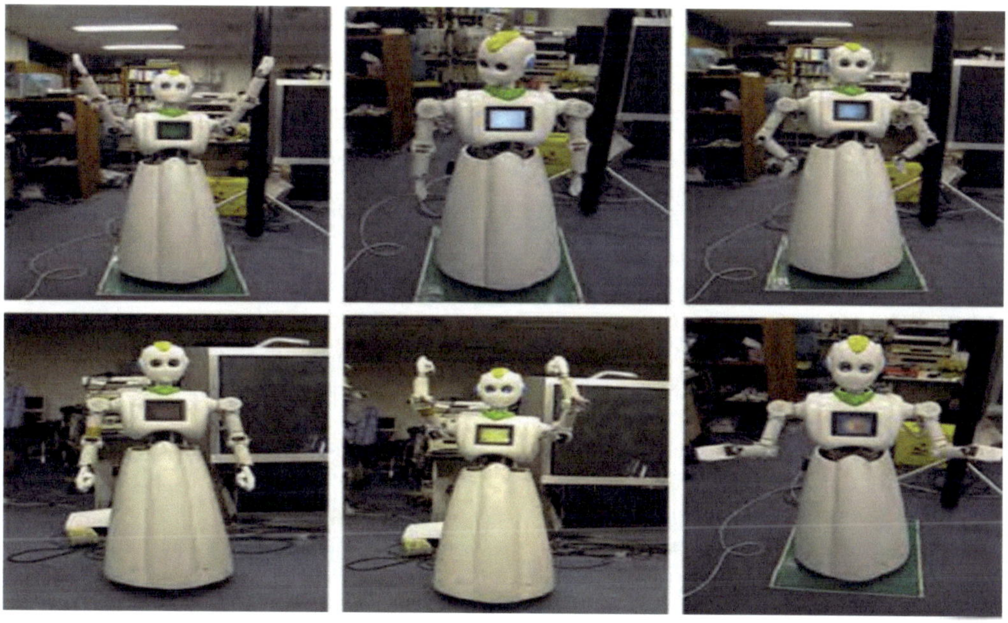

Figure 9. Guesture expressions of AMIET- joy, sad, anger, neutral, surprise and disgust in sequence

8. AIM Lab's Humanoid Robots

This section summarizes the study on design and development of the humanoid robots of AIM Lab to realize the enhanced interaction with humans. Especially, we have been focusing on building a new robot with the self-contained physical body, the intelligence which make the robot be autonomous, and the emotional communication capability toward a human symbiotic robot in AIM Lab, since 1999.

So far, the members of AIM Lab have developed autonomous robots called AMI and AMIET which have two wheeled mobile platform, anthropomorphic head, arms and hands. And also we have been developing software system which performs intelligent tasks using a unified control architecture based on behavior architecture and emotional communication interfaces. Humanoid robots, AMI, AMIET were released to the public in 2001, 2002 respectively.

AMIO is the biped humanoid robot which was developed recently. The developed robot consists of a self-contained body, head, two arms, with a two legged (biped) mechanism. Its control hardware includes vision and speech capabilities and various control boards such as motion controllers, with a signal processing board for several types of sensors. Using the developed robot, biped locomotion study and social interaction research were concurrently carried out.

An anthropomorphic shape is appropriate for a human robot interaction. Also it is very important that the robot has the stable mobility in dynamically changing and uncertain environment. Therefore, we decided the design of our robot as the mobile manipulation robot system which has upper torso, head, two arm, hand, and vehicle. In the first mechanical design stage of AMI, we consider the following factors for our robot to satisfy the requirements of these human symbiotic situations.

We considered the height of the robot firstly. The height of the robot should be appropriate to interact with human. Secondly, Manipulation capability and motion range of robot arm should be similar to human. Thirdly, Reliable mobility to move around household while ensuring human's safety is required. Fourthly, Information Screen to show helpful information to user through the network environments and to check the current status of robot and to show emotional expression of robot itself is required. In the consideration of these points, we tried to make our robot be more natural and intimate to human. And then we built the robots, AMI and AMIET as following Fig. 10.

The designed robot, AMI has a mouth to open and close in the case of speaking and making expressions, two eyes with CCD camera, and speaker unit to make sounds in his face. And his neck is equipped with two motors to track the moving target and implement active vision capabilities.

The developed torso part supports the robot's head and two arms, and includes the main computer system, arm and head controllers, and motor driving units. And also, LCD screen is attached to his breast to check the internal status of robot and to recover the limitations of mechanical face which has difficult in making emotional expressions.

We designed two symmetric arms which have five degrees of freedom each. Hand has six degrees of freedom and three fingers. Each finger has two motors. At the end of fingers, FSR(Force Sensing Register) sensors are located to sense the force in grasping object. The total length of AMI's arm with hand is 750[mm].

In case of AMIET, it is designed to make human-like motions with its two arms; each arm has 6 degrees of freedom (DOF), so it can imitate the motion of a human arm. Additionally, AMIET has a waist with 2 DOF to perform rotating and bending motions. Thus, AMIET can perform many human-like acts.

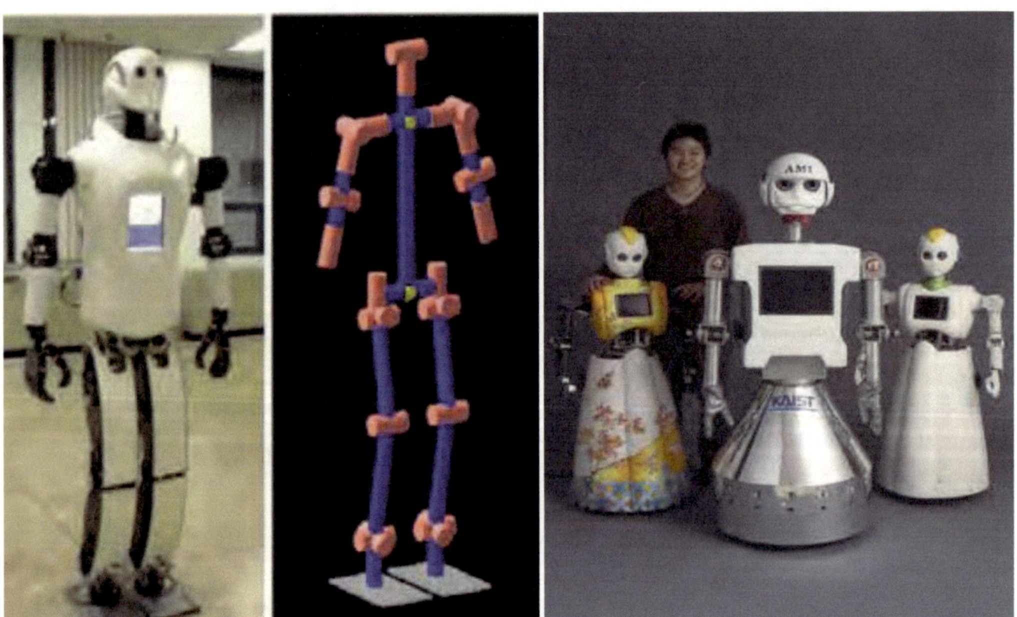

Figure 10. AIM Lab's Biped Humanoid Robots, AMIO, AMI and two AMIETs

AMI is 1550 mm tall. The breadth of the shoulder is 650 mm and the total weight is about 100 kg. Figure 13 shows the shape and DOFs of the assembled robot. AMIET has child-like height. Ami is 160 cm tall and AMIET is 130 cm tall. Differently with AMI, AMIET is designed by the concept of a child and her appearance is considered before when she is developed. So AMET could be felt friendly for humans.
A newly developed biped humanoid robot named AMI was designed and manufactured based on the dimensions of the human body. The lower part of the robot has two legs, which have 3, 1, and 2 degrees of freedom at the pelvis, knees, and ankles, respectively. This allows the robot to lift and spread its legs, and to bend forward at the waist. This structure, which was verified by previous research to be simple and stable for biped-walking robots, makes it possible for the robot to walk as humans walk. The shape and D.O.F arrangement of the developed robot, AMIO are shown in Fig. 10.

9. Experimental Results

To test the performance of the affective communication model, we conducted several experiments with humanoid robots, AMI, AMIET and AMIO. We confirmed that each subsystem satisfies its objectives. From our evaluation, we drew the graph in Fig. 12, which shows the subsystem's flow during a sample inter-action. The graph also shows the behavior system (finding, greeting, consoling and so on), the motivation system (robot's drives and emotions), and the perception system (the user's emotional status)
To evaluate the proposed communication model with the emotional memory, we compared three types of systems: one without an emotional or memory system, one with an emotional system but without a memory system, and one with both an emotional and memory system. Table 5 shows the results. The results suggest that the overall performance of the systems with an emotional memory is better than the system without it. The results clearly suggest that emotional memory helps the robot to synthesize more natural emotions and to reduce redundancy in conversational topics. Fig. 13 and Fig. 14 show the cases of the human and humanoid robot interaction experiments using the proposed affective communication model.

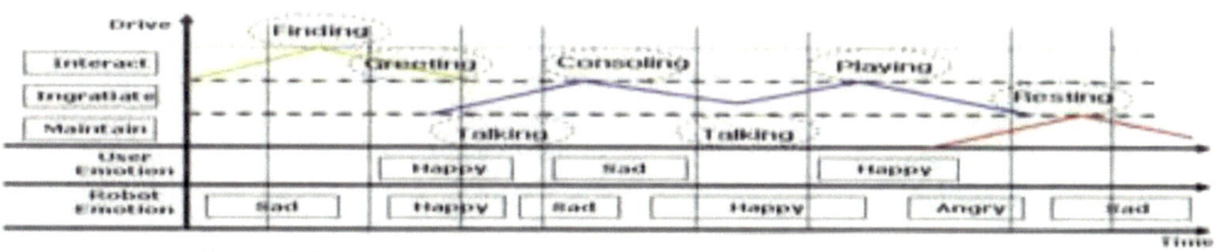

Figure 12 Work flow of the system

		without emotion. without memory	with emotion, without memory	with emotion, with memory
Natural		54%	70%	76%
Unnatural	Emotion mismatch	19%	10%	5%
	Redundancy	14%	6%	3%
	Other errors	13%	14%	16%

Table 5. Experimental results of Affective interactions with a robot

Figure 13 AMI's Affective Dialogue Interactions with Mr. Ryu in TV program

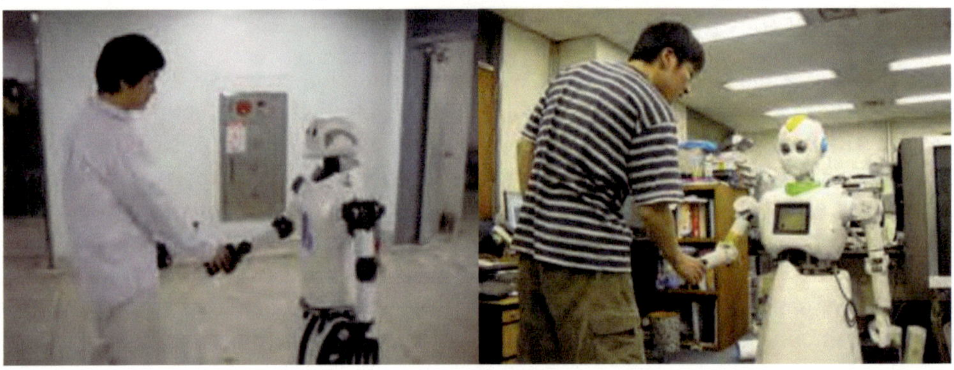

Figure 14 AMIO and AMIET shake hands with human in Affective Interaction Experiment

10. Conclusion

This chapter presented the affective communication model for humanoids that is designed to lead human robot interactions by recognizing human emotional status and expressing its emotion through multimodal emotion channels like a human, and behaves appropriately in response to human emotions.

We designed and implemented an affective human-robot communication model for a humanoid robot, which makes a robot communicate with a human through dialogue. Through this proposed model, a humanoid robot can communicate with humans by preserving emotional memories of users and topics, and it naturally engages in dialogue with humans.

With explicit emotional memories on users and topics, in the proposed system, we successfully improved the affective interaction between humans and robots. Previous sociable robots either ignored emotional memories or maintained them implicitly. Our research proves that explicit emotional memory can help high-level affective dialogue interactions.

In several experiments, the robots chose an appropriate conversation topic and behaved appropriately in response to human emotions. They could ask what the human is angry about and then make a joke to console him or give pleasure to him. Therefore, this robot is able to help human mentally and emotionally as a robot therapy function. The human partner perceives the robot to be more human-like and friendly, thus enhancing the interaction between the robot and human.

In the future, we plan to extend the robot's memory system to contain more various memories, such as visual objects or high level concepts. The robot's current memory cells are limited to conversational topics and users. Our future system will be capable of memorizing information on visual inputs and word segments, and connections between them.

To interact sociably with a human, we are going to concentrate on building a real humanoid robot in terms of thinking and feeling that can not only recognize and express emotions like a human, but also share emotional experience with humans while the robot is talking to users on many kinds of interesting and meaningful scenarios supported and updated dynamically from outside database systems such as worldwide web and network based contents server.

Acknowledgement

This research was partially supported by the Korea Ministry of Commerce, Industry and Energy(MOCIE) through the Brain Science Research Project and Health-Care Robot Project and by the Korea Ministry of Science and Technology(MOST) through AITRC program.

11. References

Armon-Jones, C. (1985): The social functions of emotions. R. Harre (ed.), *The Social Construction of Emotions*, Basil Blackwell, Oxford.

Arkin, R.C., Fujita, M., Takagi, T., Hasegawa, R. (2003): An Ethological and Emotional Basis for Human-Robot Interaction. *Robotics and Autonomous Systems*, 42.

Breazeal, C. and Scassellati, B. (1999), A context-dependent attention system for a social robot. In *Proceedints of the Sixteenth International Joint Conference on Artificial Intelligence* (IJCAI'99), pp.1146-1151.

C. Bartneck, M. Okada (2001), Robotic user interfaces, *Proceedings of the Human and Computer Conference*, 2001.

Cahn, J. (1990), Generating expression in synthesized speech, *Master's Thesis*, MIT Media Lab.

Ekman, P., Friesen, W.V. (1978): *Facial Action Coding System: Investigator's Guide.* Consulting Psychologists Press, Palo Alto, CA.

Huang, T.S., Chen, L.S., and Toa, H. (1998), Bimodal Emotion Recognition by Man and Machine. *ATR Workshop on Virtual Communication Environments.* 1998

Hyun S. Yang, Yong-Ho Seo, Yeong-Nam Chae, Il-Woong Jeong, Won Hyung Kang and Ju-Ho Lee (2006), Design and Development of Biped Humanoid Robot, AMI2, for Social Interaction with Humans, *proceedings of IEEE-RAS HUMANOIDS* 2006

Jung, H., Yongho Seo, Ryoo, M.S., Yang, H.S. (2004): Affective communication system with multimodality for the humanoid robot AMI. *proceedings of IEEE-RAS HUMANOIDS* 2004

Ledoux, J. (1996): *The Emotional brain: the mysterious under pinning of emotional life.* New York: Simon & Schuster

M. Lansdale, T. Ormerod (1994), *Understanding Interfaces*, Academic Press, New York

Naoko Tosa and Ryohei Nakatsu (1996), *Life-like Communication Agent - Emotion Sensing Character "MIC" & Feeling Session Character "MUSE"*, ICMCS, 1996

Nakatsu, R., Nicholson, J. and Tosa, N. (1999), Emotion Recognition and Its Application to Computer Agents with Spontaneous Interactive Capabilities, *Proc. of the IEEE Int. Workshop on Multimedia Signal Processing*, pp. 439-444 , 1999

Ledoux, J. (1996): *The Emotional brain: the mysterious under pinning of emotional life*. New York: Simon & Schuster.

Schlossberg, H. (1954): Three dimensions of emotion. *Psychology Review* 61

Shibata, T. at al. (2000): Emergence of emotional behavior through physical interaction between human and artificial emotional creature. *ICRA* (2000) 2868-2873

Sidner, C.L.; Lee, C.; Kidds, C.D.; Lesh, N.; Rich, C. (2005), Explorations in Engagement for Humans and Robots, *Artificial Intelligence*, May 2005

Toru Nakata, Taketoshi Mori & Tomomasa Sato (2002), Analysis of Impression of Robot Bodily Expression, *Journal of Robotics and Mechatronics*, Vol.14, No.1, pp.27--36, 2002

Yong-Ho Seo, Ho-Yeon Choi, Il-Woong Jeong, and Hyun S. Yang (2003), Design and Development of Humanoid Robot AMI for Emotional Communication and Intelligent Housework, *Proceedings of IEEE-RAS HUMANOIDS* 2003, pp.42.

Yoon, S.Y., Burke, R.C., Blumberg, B.M., Schneider, G.E. (2000): Interactive Training for Synthetic Characters. *AAAI* 2000

29

Communication Robots in Real Environments

Masahiro Shiomi, Takayuki Kanda, Hiroshi Ishiguro and Norihiro Hagita
ATR-IRC
Japan

1. Introduction

The development of robots is entering a new stage, where the focus is on interaction with people in their daily environments. We are interested in developing a communication robot that operates in everyday conditions and supports peoples' daily life through interactions with body movements and speech. The concept of the communication robot is now rapidly emerging and developing, with communication robots in the not-too-distant future likely to act as peers providing mental, communication, and physical support. Such interactive tasks are important for enabling robots to take part in human society.

Many robots have already been applied to various fields in real environments. We discuss the details of related works in the next section. Here, there are mainly two kinds of fields: closed and open. The difference between a closed and an open environment lies in the people who are interacting. In a closed environment, such as an elementary school or an office, robots interact with a limited group of people. On the contrary, we chose to work in an open environment because we expect that many people, in a wide-range of ages, will interact with robots. In line with this prospect, we have been developing a science museum guide robot that we believe to be a promising application.

There is a double benefit in choosing a science museum as the experiment field. On the one hand, visitors have the opportunity to interact with the robots and experience the advanced technologies by which they are made, which is the fundamental purpose of a science museum. Thus, we can smoothly deploy our research in a real environment.

On the other hand, in a science museum we are naturally targeting people who are interested in science and are unlikely to miss the chance to interact with our robots; thus this field is one of the best choices for collecting feedback and examining the interaction between people and the communication robot in various tasks. The need for extensive and accurate feedback goes back to our belief that interaction with humans through tasks is one of the communication robot's essential roles. This feedback is vital for developing the ability of the robots to act appropriately in a daily living environment.

In this chapter, we introduce recent research efforts in communication robots in real environments and describe an experiment in which a system using many ubiquitous sensors and humanoid robots -- Robovies -- guide the visitors at a science museum. In this setting, the Robovies interacted with the visitors and showed them around to exhibits according to information from ubiquitous sensors, such as the visitors' positions and movement histories.

Location	Purpose	People		Function		Interaction
		Wide age range	Many people	Navigation	Person identification	Human-like
Home	Entertainment [Fujita 1998]	✓	-	-	-	-
Hospital	Mental care [Shibata 2004]	✓	-	-	-	-
School	Language education [Kanda et al. 2004]	-	✓	-	✓	✓
Lobby	Interaction [Gockley et al. 2005]	✓	✓	-	✓	-
Office	Guiding [Asoh et al. 1997]	-	-	✓	✓	-
Museum	Guidance & navigation [Burgard et al. 1998] [Siegwart et al. 2003]	✓	✓	✓	✓	-
	Interaction & guidance [This paper]	✓	✓	✓	✓	✓

Table 1. Various field experiments with interactive robots

2. Related Works

Table 1 shows a comparison between our work and previous works based on the concept of the communication robot. Aibo [Fujita, 1998] and Paro [Shibata, 2004] had animal-like appearances - respectively, a dog and a seal. These robots provide entertainment or mental care to people through a human-pet style of interaction. Both studies indicated the effectiveness of interaction between people and pet-like robots.

In an elementary school, Robovie [Kanda et al. 2004] was used to assist with the language education of elementary school students. This research detailed the importance of using personal information in an interaction and the effectiveness of human-like interaction from a communication robot in a real environment. On the contrary, this research mainly focused on the interaction between people and a robot with a human-like body. In addition, Robovie only interacted with a limited group of people. As a result, it is not clear how a robot should operate in large-scale real environments where a wide variety of people visit.

Valerie [Gockley et al. 2005] is a robotic-receptionist. This research indicated the effectiveness of the robot's personality through long-term interaction with people. Valerie used functions of emotion expression and storytelling to represent her personality. In addition, Valerie interacted with many people using personal information such as name (from a magnetic card) in a large-scale environment. Valerie cannot, however, move around and has only simple interfaces such as a keyboard; therefore, this research did not address problems associated with navigation and human-like interaction.

Jijo-2 [Asoh et al. 1997], RHINO [Burgard et al. 1998], and RoboX [Siegwart et al. 2003] are traditional mobile robots, the developers of which designed robust navigation functions for real environments. In particular, RoboX and RHINO guided thousands of people in large-scale environments. Although these works represent the effectiveness of robust navigation functions in interactions between people and robots, their interaction functions are quite different from human-like interactions. In addition, these researchers mainly focused on the robustness of their systems, thus none of these reports evaluated the effectiveness of human-like interaction in large-scale real environments.

From surveying the related research, we consider it important to investigate the effectiveness of human-like interaction as well as to develop robust functions. Therefore, we designed our robots to interact with people using human-like bodies and personal information obtained via ubiquitous sensors.

3. System Configurations

3.1 The Osaka Science Museum

1) General settings:

Seventy-five exhibits were installed on the fourth floor of the Osaka Science Museum, and visitors could freely explore these exhibits. Figure 1 shows a map of the fourth floor of the museum. Generally, visitors walk in the counterclockwise direction from the entrance to the exit. The width and length of the Osaka Science Museum are 84[m] and 42[m], respectively.

2) Our experimental settings:

We installed the humanoid robots, RFID tag readers, infrared cameras and video cameras in the fourth floor of the Osaka Science Museum for an experiment. Visitors could freely interact with our robots similar to the other exhibits. Typically, in our experiment, visitors progress through the following steps:

• If a visitor decides to register as part of our project, personal data such as name, birthday, and age (under 20 or not) is gathered at the reception desk (Fig. 1, point A). The visitor receives a tag at the reception desk. The system binds those data to the ID of the tag and automatically produces a synthetic voice for the visitor's name.

• The visitors could freely experience the exhibits in the Osaka Science Museum as well as interact with our robots. Four robots were placed at positions B, C, and D on the fourth floor, as shown in Fig. 1. When leaving the exhibit, visitors returned their tags at the exit point (Fig. 1, point E).

3.2 Humanoid Robots

1) Robovie:

Figure 2 shows "Robovie," an interactive humanoid robot characterized by its human-like physical expressions and its various sensors. The reason we used humanoid robots is that a human-like body is useful to naturally control the attention of humans [Imai et al. 2001]. The human-like body consists of a head, a pair of eyes, and two arms. When combined, these parts can generate the complex body movements required for communication. We decided on a robot height of 120 cm to decrease the risk of scaring children. The diameter was 40 cm. The robot has two 4*2 DOFs (degrees of freedom) in its arms, 3 DOFs in its head, and a mobile platform. It can synthesize and produce a voice via a speaker. We also attached an RFID tag reader to Robovie [Kanda et al. 2004] that enables it to identify the individuals around it. Two of the four robots used in this experiment were Robovies.

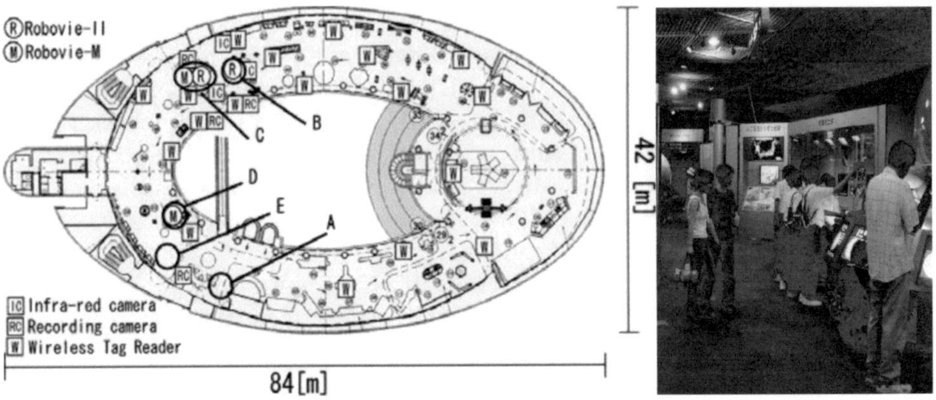

Figure 1. Map of the fourth floor of the Osaka Science Museum

2) Robovie-M:

Figure 3 shows a "Robovie-M" humanoid robot characterized by its human-like physical expressions. We decided on a height of 29 cm for this robot. Robovie-M has 22 DOFs and can perform two-legged locomotion, bow its head, and do a handstand. We used a personal computer and a pair of speakers to enable it to speak, since it was not originally equipped for this function. The two other robots in this experiment were Robovie-Ms.

Figure 2. Robovie

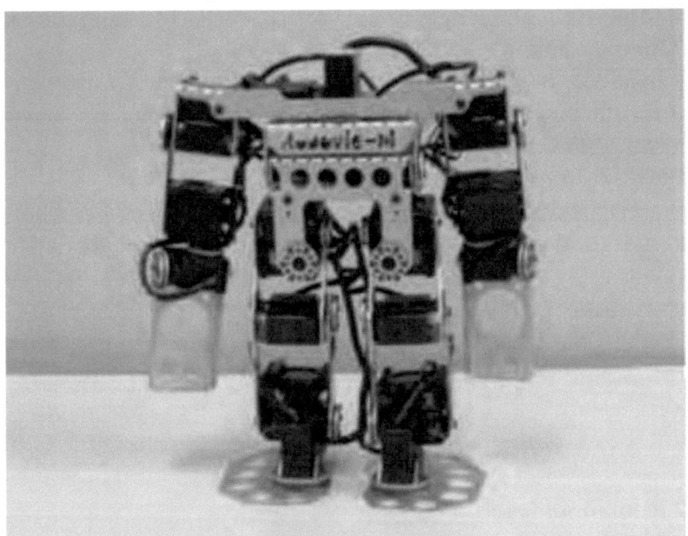

Figure 3. Robovie-M

3.3 Ubiquitous sensors in an environment:

On the fourth floor of the Osaka Science Museum, we installed 20 RFID tag readers (Spider-IIIA, RF-CODE), which included the two equipped on the Robovies, three infrared sensors, and four video cameras. All sensor data were sent to a central server's database via an Ethernet network.

In the following sections, we describe each type of sensor used.

1) RFID tag readers:

We used an active type of RFID tag. This technology enables easy identification of individuals, since detection is unaffected by the occurrence of occlusions, the detection area is wide, and the distance between the tag reader and the RFID tag can be roughly estimated. Such benefits are advantageous for large environments.

However, drawbacks of this approach include low accuracy over long distances and the inability to detect exact positions. We compensated for these shortcomings by installing many RFID tag readers in the environment.

To achieve approximate distance estimation, we set the RFID tag readers to have eight levels of sensitivity. Detection areas, however, are affected by the position of the RFID tag readers and reflections due to walls. Therefore, we measured each detection area prior to the experiment. We then attached the tag readers in positions two meters above the floor, and to successfully detect the tags we had to set the reader sensitivity level to at least five.

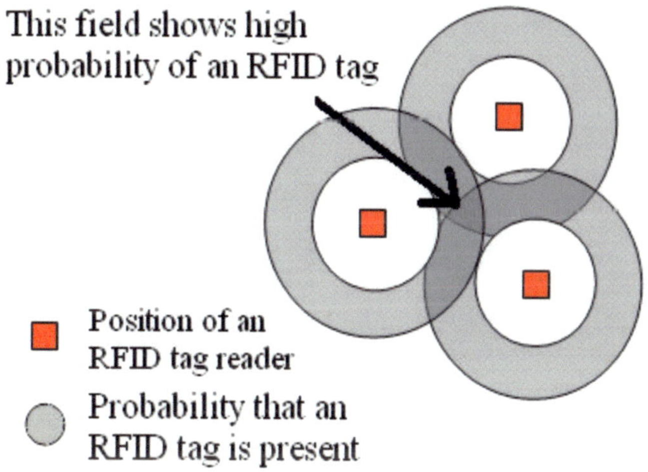

Figure 4. Detection fields of RFID tag

Figure 1 shows an example of how we positioned the tag readers. We placed them around particular exhibits, so that the system could detect whether visitors approached them. Moreover, since a tag reader's detection field has a torus shape, the system can estimate the tag position by superposing the circles calculated from the reader outputs (Fig. 4).

2) Infrared cameras:

We placed an infrared LED on top of a Robovie and attached infrared cameras to the ceiling to determine the robot's correct position. The system produces binary images from the infrared cameras and detects bright areas. It calculates absolute coordinates with a reference to the weighted center of the detection area and sends them to the database.

Infrared camera positions are shown in Fig. 1. The distance between the floor and ceiling is about 4 m. The width and height of images from an infrared camera are 320 and 240 pixels, respectively. One pixel represents about 1 cm^2 of area.

3) Video cameras:

The video camera positions are also shown in Fig. 1. The output images of each video camera are recorded onto a PC and used to analyze the data generated during the experiment.

4. Robot Behaviour

4.1 Locomotive robot

We used a Robovie as a locomotive robot that moved around in parts of the environment, interacted with visitors, and guided them to exhibits. Such behaviour can be divided into four types, the details of which are explained as follows.

1) Interaction with humans: Childlike interaction

The robot can engage in such childlike behaviour as handshaking, hugging, and playing the game of "rock, paper, scissors." Moreover, it has reactive behaviours such as avoidance and gazing at the touched part of its body, as well as patient behaviour such as solitary playing and moving back and forth. Figure 5 shows interaction scenes between the robot and visitors.

(a) A child touching the robot

(b) The robot hugging children

Figure 5. Scenes of interaction between visitors and the robot

2) Interaction with humans: Using information from RFID tags

The robots can detect RFID tag signals around themselves by using their RFID tag reader, which allows them to obtain personal data on visitors from their RFID tag IDs. Each robot can greet visitors by name, wish them a happy birthday, and so on. In addition, the system records the time that visitors spend on the fourth floor of the Osaka Science Museum. The robots can behave according to that duration of visit.

3) Guiding people to exhibits: Human guidance

The robot can guide people to four kinds of exhibits by randomly determining the target. Foe example, when bringing visitors to the telescope, the robot says, "I am taking you to an exhibit, please follow me!", and approaches the telescope. It suggests that the person look through it and then talks about its inventor.

4) Guiding people to exhibits: Using information from RFID tags

The RFID tags' data are also used for interaction. We used the amount of time that visitors spent near an exhibit to judge whether visitors tried it. For example, when an RFID-tagged visitor has stayed around the "magnetic power" exhibit longer than a predefined time, the system assumes that the visitor has already tried it. Thus, the robot says, "Yamada-san, thank you for trying 'magnetic power.' What did you think of it?" If the system assumes that the visitor has not tried it, the robot will ask, "Yamada-san, you didn't try 'magnetic power.' It's really fun, so why don't you give it a try?"

4.2 Robots that talk with each other

Two stationary robots (Robovie and Robovie-M) can casually talk about the exhibits as humans do with accurate timing because they are synchronized with each other through an Ethernet network. When taking, Robovie controlled the timing of Robovie-M's motion and speech. The topic itself is determined by data from RFID tags. By knowing a visitor's previous course of movement through the museum, the robots can try to interest the visitor in an exhibit he or she overlooked by starting a conversation about that exhibit. Figure 6 shows scenes of robots talking to each other.

(a) Two robots talking

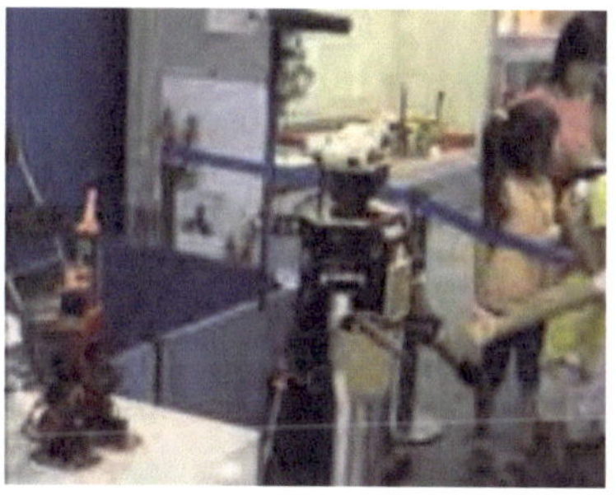

(b) Two robots talking to visitors

Figure 6. Scenes of robots talking to each other

4.3 A robot bidding farewell

This robot is positioned near the exit and, after requesting data from their RFID tags, says goodbye to the departing visitors. We used Robovie-M for this robot. It also reorients visitors on the tour who are lost by examining the visitor's movement history and time spent on the fourth floor of the Osaka Science Museum, which was recorded by the system. If visitors walk clockwise, they will immediately see this robot at the beginning and will be pointed in the right direction by the robot. Figure 7 shows a scene with this robot.

Figure 7. The robot bidding farewell

5. Results of Exhibition Experiment

We performed experiments to investigate the impressions made by robots on visitors to the fourth floor of the Osaka Science Museum during a two-month period. By the end of the two-month period, the number of visitors had reached 91,107, the number of subjects who wore RFID tags was 11,927, and the number of returned questionnaires was 2,891 (details are not discussed here). Details of questionnaires' results are described by Nomura et al [Nomura et al, 2005]. The results of the two-month experiment indicate that most visitors were satisfied with their interaction with our robots. In this section, we describe the observed interaction between visitors and our robots at the exhibition and how the ubiquitous sensors contributed to the entire system in the real environment.

5.1 Observations of interaction between visitors and the robots

1) Locomotive robot

- Often there were many adults and children crowded around the robot. In crowded situations, a few children simultaneously interacted with the robot.
- Similar to Robovie's free-play interaction in a laboratory [Ishiguro et al. 2001], children shook hands, played the paper-rock-scissors game, hugged, and so forth. Sometimes they imitated the robot's body movements, such as the robot's exercising.
- When the robot started to move to a different place (in front of an exhibit), some children followed the robot to the destination.
- After the robot explained a telescope exhibit, one child went to use the telescope. When she came back to the robot, another child used the telescope.
- Its name-calling behaviour attracted many visitors. They tried to show the RFID tags embedded in the nameplates to the robot. Often, when one visitor did this, several other visitors began showing the robots their nameplates, too, as if they were competing to have their names called.

We demonstrated that robots could provide visitors with the opportunity to play with and study science through exhibits they might have otherwise missed. In particular, this reminds us of the importance of making a robot move around, a capability that attracts people to interact with it. Moreover, as shown in the scene where children followed the locomotive robot, it drew their attention to the exhibit, although the exhibit (a telescope) was relatively unexciting. (The museum features many attractive exhibits for visitors to move and operate to gain an understanding of science, such as a pulley and a lever.)

2) Robots that talk to each other

- There were two types of typical visitor behaviours. One was simply listening to the robots' talk. For example, after listening to them, the visitors talked about the exhibit that was explained to them, and sometimes visited the exhibit.
- The other behaviour is to expect to have their name called. In this case, the visitors paid rather less attention to the robots' talk but instead showed their names to the robots, which is similar to the actions observed around the locomotive robot. Often, a visitor would leave the area in front of the robot just after his/her name was called.

One implication is that displaying a conversation between robots can attract people and convey information to them, even though the interactivity is very low. Such examples are also shown in other works [Hayashi et al. 2005] [Hayashi et al. 2007].

3) Robot bidding farewell

- There were two types of typical visitor behaviours. One was simply watching the robot's behaviour.
- The other was, again, to expect to have their name called. In this case, the visitors often showed their names to the robots.

Significantly, the effectiveness of the name-calling behaviour was again demonstrated, as seen in the children's behaviour when returning their RFID tags.

5.2 Contribution of the ubiquitous sensors

1) RFID tags and tag readers

As shown in the scenes of interaction with the robots, the application of RFID technology largely promoted the human-robot interaction, particularly with regard to the name-calling behaviour. However, the information obtained from the distributed RFID tag readers made a relatively small contribution to the system. Robots talked to the visitors about their exhibit-visiting experience, such as "You did not see the telescope exhibit, did you? It is very interesting. Please try it," based on the information from the RFID reader network, but it seemed to be less attractive and impressive to the visitors. This was also pointed out in our previous work [Koide et al. 2004]. Perhaps, robots are too novel for visitors, so they highly value the mere experience of interacting with the robots while paying less attention to the detailed services that they offer.

2) Other ubiquitous sensors

Regarding the ubiquitous sensors other than the RFID tags, their role was limited. The infrared camera supplied the exact position of the robot, which was very helpful in the crowded environment. We believe that there will be much useful information from ubiquitous sensors available for human-robot interaction, which should be incorporated in our future work. For example, Nabe et al. reported that the distance between a robot and a person was influenced by age [Nabe et al. 2006].

6. Conclusion

We have developed an interactive robot system that combines autonomous robots and ubiquitous sensors. The system guided visitors through a science museum with human-like interaction, such as calling their names in a free-play behaviour and explaining exhibits with voice and gestures. In a two-month exhibition, 91,107 people visited the Osaka Science Museum, 11,927 of whom wore RFID tags to participate in the field trial. The results from questionnaires revealed that almost all of the visitors evaluated these robots highly.

7. Acknowledgments

We wish to thank the staff at the Osaka Science Museum for their kind cooperation. We also wish to thank the following ATR members for their helpful suggestions and cooperation: Tatsuya Nomura, Hideaki Terauchi, Takugo Tasaki, Daniel Eaton, Toshihiko Shibata, Koutarou Hayashi, Masaaki Kakio, Taichi Tajika, and Fumitaka Yamaoka. This research was supported by the Ministry of Internal Affairs and Communications of Japan.

8. References

Asoh, H., Hayamizu, S., Hara, I., Motomura, Y., Akaho, S., and Matsui, T. Socially Embedded Learning of the Office-Conversant Mobile Robot Jijo-2, *Int. Joint Conf. on Artificial Intelligence* (IJCAI), 1997.

Burgard, W., Cremers, A. B, Fox, D., Hähnel, D., Lakemeyer, G., Schulz, D., Steiner, W., and Thrun, S. The Interactive Museum Tour-Guide Robot, *Proc. National Conference on Artificial Intelligence (AAAI)*, 1998.

Fujita, M. AIBO: Toward the era of digital creatures, *Int. J. Robot.* Res., vol. 20, no. 10, pp. 781–794, 2001.

Gockley, R., Bruce, A., Forlizzi, L., Michalowski, M., Mundell, A., Rosenthal S., Sellner, B., Simmons, R., Snipes, K., Schultz, Alan C., and Wang. J. Designing Robots for Long-Term Social Interaction, *IROS2005.* pp. 2199-2204, 2005

Hayashi, K., Kanda, T., Miyashita, T., Ishiguro, H., and Hagita, N. Robot Manzai – Robots' conversation as a passive social medium-, *IEEE International Conference on Humanoid Robots (Humanoids 2005)*, 2005.

Hayashi, K., Sakamoto D., Kanda T., Shiomi M., Ishiguro H., and Hagita N., Humanoid robots as a passive-social medium - a field experiment at a train station -, *ACM 2nd Annual Conference on Human-Robot Interaction (HRI 2007)*, 2007. (to appear)

Imai, M., Ono T., and Ishiguro, H. Physical Relation and Expression: Joint Attention for Human-Robot Interaction, *Proceedings of the 10th IEEE International Workshop on Robot and Human Communication (RO-MAN2001)*, pp. 512-517, 2001

Ishiguro, H., Imai, M., Maeda, T., Kanda, T., and Nakatsu, R. Robovie: an interactive humanoid robot, *Int. J. Industrial Robot*, Vol. 28, No. 6, pp. 498-503, 2001.

Kanda T., Ishiguro H., Imai M., Ono T., and Mase K. A constructive approach for developing interactive humanoid robots, *IEEE/RSJ International Conference on Intelligent Robots and Systems (IROS 2002)*, pp. 1265-1270, 2002.

Kanda, T., Hirano, T., Eaton, D., and Ishiguro, H. Interactive Robots as Social Partners and Peer Tutors for Children: A Field Trial, *Journal of Human Computer Interaction*, Vol. 19, No. 1-2, pp. 61-84, 2004.

Koide, Y., Kanda, T., Sumi, Y., Kogure, K., and Ishiguro, H. An Approach to Integrating an Interactive Guide Robot with Ubiquitous Sensors, *IROS2004*, pp. 2500-2505, 2004.

Nabe, S., Kanda, T., Hiraki, K., Ishiguro, H. Kogure, K., Hagita, N. Analysis of Human Behavior in an Open Field to Improve Communication Robots, *ACM 1st Annual Conference on Human-Robot Interaction (HRI2006)*, pp. 234-241, 2006.

Nomura, T., Tasaki, T., Kanda, T., Shiomi, M., Ishiguro, H., and Hagita, N. Questionnaire-Based Research on Opinions of Visitors for Communication Robots at an Exhibition in Japan, *International Conference on Human-Computer Interaction* (Interact 2005), 2005.

Siegwart, R. et al. Robox at Expo.02: A Large Scale Installation of Personal Robots. *Robotics and Autonomous Systems*, 42, 203-222, 2003

Shibata, T. An overview of human interactive robots for psychological enrichment, The *proceedings of IEEE* November 2004.

Neural Control of Actions Involving Different Coordinate Systems

Cornelius Weber[1], Mark Elshaw[2], Jochen Triesch[1] and Stefan Wermter[2]
[1]*Johann Wolfgang Goethe University*, [2]*University of Sunderland*
[1]*Germany*, [2]*UK*

1. Introduction

The human body has a complex shape requiring a control structure of matching complexity. This involves keeping track of several body parts that are best represented in different frames of reference (coordinate systems). In performing a complex action, representations in more than one system are active at a time, and switches from one set of coordinate systems to another are performed. During a simple act of grasping, for example, an object is represented in a purely visual, retina-centered coordinate system and is transformed into head- and body-centered representations. On the control side, 3-dimensional movement fields, found in the motor cortex, surround the body and determine the goal position of a reaching movement. A conceptual, object-centered coordinate space representing the difference between target object and hand position may be used for movement corrections near the end of grasping. As a guideline for the development of more sophisticated robotic actions, we take inspiration from the brain. A cortical area represents information about an object or an actuator in a specific coordinate system. This view is generalized in the light of population coding and distributed object representations. In the motor system, neurons represent motor "affordances" which code for certain configurations of object- and effector positions, while mirror neurons code actions in an abstract fashion.

One challenge to the technological development of a robotic / humanoid action control system is - besides vision - its complexity, another is learning. One must explain the cortical mechanisms which support the several processing stages that transform retinal stimulation into the mirror neuron and motor neuron responses (Oztop et al., 2006). Recently, we have trained a frame of reference transformation network by unsupervised learning (Weber & Wermter, 2006). It transforms between representations in two reference frames which may dynamically change their position to each other. For example the mapping between retinal and body-centered coordinates while the eyes may move. We will briefly but concisely present this self-organizing network in the context of grasping. We will also discuss mechanisms required for unsupervised learning such as requested slowness of neuronal response changes in those frames of reference that tend to remain constant during a task. This book chapter shall guide and inspire the development of sensory-motor control strategies for humanoids.

This book chapter is organized as follows. Section 2 reviews neurobiological findings; Section 3 reviews robotic research. Then, after motivating learning in Section 4, we will carefully introduce neural frame of reference transformations in Section 5, and in Section 6 present a model for their unsupervised learning. Section 7 discusses the biological context, the model's solution for visual routing, and open questions for motor control.

2. Neurobiology

2.1 Cortical Areas Involved in Sensory Motor Control

This section addresses some individual cortical areas, based primarily on macaque data (Luppino & Rizzolatti, 2000), highlighting the variety of frames of reference that exist in the cortex. In lower visual areas such as V1 and V2, neurons are responsive only to visual stimuli shown at a defined region in the visual field, the receptive field. They code in a retinal (eye-centered) coordinate frame. In higher visual areas the receptive fields become larger and can comprise half of the visual field. IT (infero temporal cortex) responses are for example dominated by the presence of an object. The retinal coordinate frame is unimportant, neither is any other spatial frame of reference. MT/MST (middle temporal / medial superior temporal) neurons respond to motion stimuli.

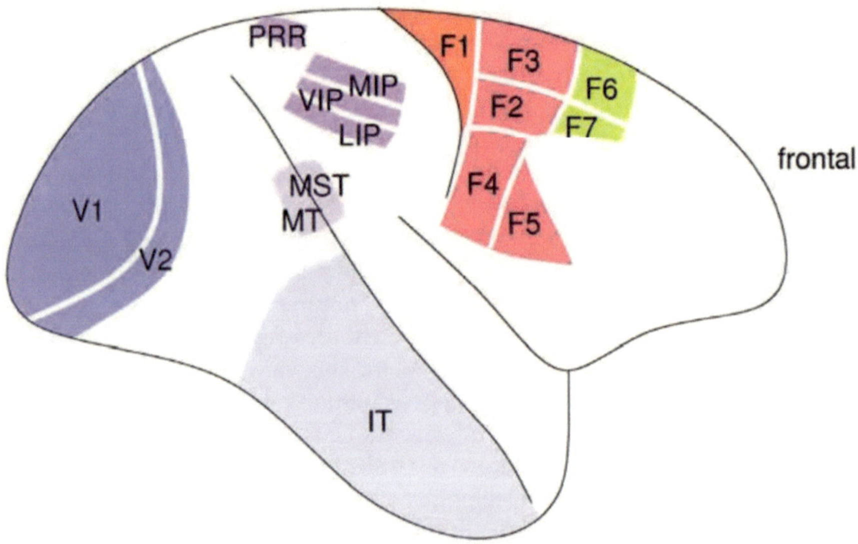

Figure 1. Cortical areas involved in visuomotor computations. The figure is schematic rather than faithful (drawn after Luppino and Rizzolatti (2000) and Van Essen et al. (1992))

Of specific interest to frame of reference transformations are the areas of the posterior parietal cortex (PPC):

- LIP (lateral intraparietal) neurons encode locations retinotopically i.e. in eye-centered coordinates (Duhamel et al., 1997).
- VIP (ventral IP) neurons encode locations in eye- and also in head-centered coordinates (Duhamel et al., 1997). Some cells show response fields that shift only partway with the eyes as gaze changes (Batista, 2002) ("intermediate reference frame"). Others have a head-centered response: the receptive field is fixed w.r.t. the head, but the response magnitude can be scaled depending on the position of the eyes. Because of this multiplicative interaction these are termed "gain fields".

Many parietal neurons, such as in LIP and VIP, respond when an eye movement brings the site of a previously flashed stimulus into the receptive field (Duhamel et al., 1992). Hence they predict reference frame changes caused by eye movement.

- MIP (medial IP) neurons represent reach plans in a retinal coordinate frame. For example, a neuron fires during reaching movements, but only when the eyes are centered at the reach target (Batista, 2002).
- PRR (parietal reach region) neurons code the next planned reaching movement. In the neighboring area 5 of the PPC, targets are coded w.r.t. both eye and hand (Buneo et al., 2002). Buneo et al. (2002) suggest that the transformation between these two reference frames may be achieved by subtracting hand location from target location, both coded in eye centered coordinates.

The motor cortex (for reviews see Rizzolatti et al. (2001); Luppino and Rizzolatti (2000); Graziano (2006)) is also called agranular cortex, because it misses the granular layer which in sensory areas receives the bottom-up sensory input. The motor areas can be subdivided into two groups. One group of motor areas connects to the spinal cord. These more caudally situated areas transform the sensory information into motor commands:

- F1 projections end directly on motor neurons. Neural activations are correlated with hand position, finger control, and velocity to perform actions such as grasping, with lesion studies showing that damage to this area prevents hand grasping (Fadiga & Craighero, 2003).

F2, F3 and parts of F4, F5 activate preformed medullar circuits and also send connections to Fl (Luppino & Rizzolatti, 2000).

- F2 encodes motor execution and motor preparation and is somatotopically organized. There are few visually responsive neurons (≈16%), mostly within parts of the forelimb representation (Fogassi et al., 1999).
- F3 encodes complete body movements. Stimulation evokes proximal and axial muscle movements; typically a combination of different joints. There are frequent responses to somato-sensory stimulation (Luppino & Rizzolatti, 2000).
- F4 is active at simple movements, e.g. toward mouth or body. It responds to sensory stimulation: 50% of neurons to somato-sensory 50% to somato-sensory and visual stimuli (Luppino & Rizzolatti, 2000). Visual receptive fields are 3-dimensional, located around the tactile receptive fields, such as on face, body or arms. Hence, there is are egocentric, body-part centered frames of reference.
- F5 controls hand and mouth. Neuronal firing is correlated with action execution, such as precision- or power grip, but not with its individual movements. Some neurons do not respond to vision, some respond to a sheer 3-dimensional presentation of objects (pragmatic representation of graspable objects), finally, "mirror neurons" respond to action observation. They will be described in more detail in Section 2.3.

The other group of motor areas do not have direct connections to the spinal cord, nor to Fl. These more frontally situated areas are involved in controlling the more caudal motor areas:

- F6 neurons display long leading activity during preparation of movement. They are modulated by the possibility of grasping an object. Stimulation, only with high currents, leads to slow complex movements which are restricted to the arm. Visual responses are common. It receives input from cingulate cortical areas which are associated with working memory, temporal planning of movements, motivation (Rizzolatti & Luppino, 2001).

- F7 displays visual responses. It may be involved in coding object locations in space for orienting and coordinated arm-body movements (Luppino & Rizzolatti, 2000).

Because of different investigation methods, there are different classification schemes of cortical areas. The primary motor cortex F1 is also referred to as MI. The dorsal premotor area (PMd) comprises F2 and F7. F3 is referred to as supplementary motor area (SMA), and F6 as pre-SMA. The ventral premotor area (PMv) comprises F4 and F5 (Matsumoto et al., 2006).

2.2 Key Aspects of Functionality

We summarize the following general observations:

- There are many frames of reference, accounting for the dynamic complexity of the body. Some frame of reference transformations are likely to depend on the correct functioning of certain others, while others may function in parallel as independent systems.
- Some neurons code in "intermediate" reference frames, such as between eye- and head-centered coordinates. Likewise, neurons with a constant receptive field position (in a certain reference frame) may have their responses modulated by, e.g. eye position ("gain fields").
- There is convergence: For example, an object position can be computed in a body-centered frame from visual input; the hand position can be computed from somato-sensory signals; from these two positions, a motor error (the difference between target object and hand) can be computed. On the other hand, when both hand and object are in sight, this difference can be read directly in retinal coordinates (Buneo et al., 2002). This redundancy may be used to align different frames of reference, or supervise the learning of one representation by the representation of another.

Frame of reference transformations enable the understanding of actions performed by others, as observed in the mirror neurons in F5. This is a prerequisite for social communication, and because of its importance we will discuss the mirror neuron system in the following.

2.3 Mirror Neurons

Rizzolatti and Arbib (1998) and Umilta et al. (2001) describe neurons located in the rostral region of a primate's inferior area, the F5 area (see Fig. 1), which are activated by the movement of the hand, mouth, or both. These neurons fire as a result of the action, not of the movements that are the components of this action. The recognition of motor actions depends on the presence of a goal, so the motor system does not solely control movement (Gallese & Goldman, 1998; Rizzolatti et al., 2002). A seen tool also activates regions in the premotor cortex, an effect which increases when subjects name the tool use (Grafton et al., 1997). The F5 neurons are organized into action categories such as 'grasping', 'holding' and 'tearing' (Gallese & Goldman, 1998; Rizzolatti & Arbib, 1998). More generally, the motor cortex is partly organized around action-defined categories (Graziano, 2006).

Certain grasping-related neurons fire when grasping an object, whether the grasping is performed by the hand, mouth, or both. This supports the view that these neurons do not represent the motor action but the actual goal of performing the grasping task. Within area F5 the pure motor neurons only respond to the performing of the action. Visuomotor neurons also respond to the presentation of an object ("canonical neurons") or when a monkey sees the action performed (mirror neurons) (Kohler et al., 2002; Rizzolatti & Arbib, 1998; Rizzolatti et al., 2001). The mirror neuron system indicates that the motor cortex is not only involved in the production of actions but in the action understanding from visual and auditory information (Rizzolatti et al., 2002; Rizzolatti & Luppino, 2001; Rizzolatti & Arbib, 1998) and so the observer has the same internal representation of action as the actor (Umilta et al., 2001).

These mirror neurons are typically found in area F5c and do not fire in response to the presence of the object or mimicking of the action. Mirror neurons require the action to interact with the actual object. They respond not only to the aim of the action but also how the action is carried out (Umilta et al., 2001). Already an understanding that the object is there without being visible causes the activation of the mirror neurons if the hand reaches for the object in the appropriate manner. This is achieved when primates are first shown the action being performed completely visible and then with the hand-object interaction hidden behind an opaque sliding screen (Umilta et al., 2001). Since these areas are active during both performance and recognition, when simply observing the action, there must be a set of mechanisms that suppress the movements to perform the action.

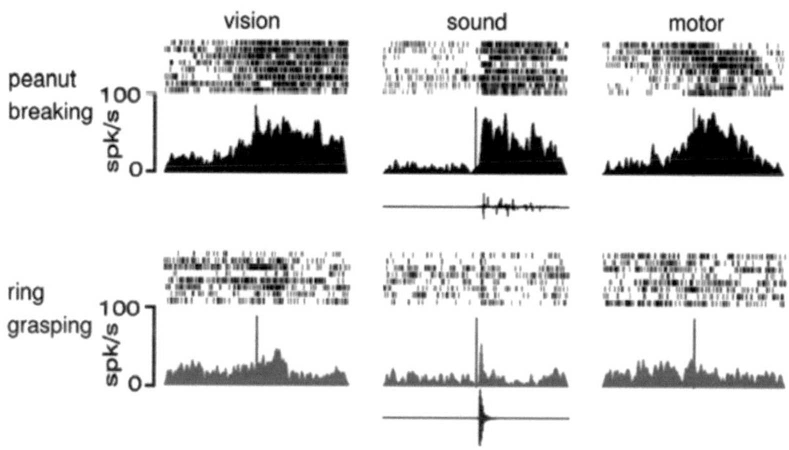

Figure 2. Responses of a macaque F5 mirror neuron during action recognition and performance (from Kohler et al. (2002), with permission)

Fig. 2 provides mirror neuron responses. The individual pictures show at the top a rastergram that represents the spikes during 10 trials, below which is a histogram (the central line represents onset of stimulus). Recognition is achieved through visual stimuli, left, or auditory stimuli, middle, for which oscillograms are shown below. The motor readings, right, are recordings of the primate when performing the action. It can be seen that this mirror neuron is active when the monkey breaks a peanut or observes someone performing this action, but not during a control action of ring grasping. These audiovisual mirror neurons have a role in the discrimination of different actions, and constitute together with Broca's area for language representation, a part of a "hearing-doing" system (Lahav et al., 2007).

2.4 Mirror Neurons and Imitation

One possible application for the mirror neuron system is imitation learning. According to Schaal et al. (2003) and Demiris and Hayes (2002) imitation learning is common to everyday life and is able to speed up the learning process. Imitation learning allows the observer to gain skills by creating an abstract representation of the teacher's behavior, understanding the aims of the teacher and creating the solution (Dillmann, 2003). Imitation requires the ability to understand the seen action and produce the appropriate motor primitives to recreate it (Buccino et al., 2004). The role of mirror neurons is to encode actions so they are understood or can be imitated, by gaining the reason for the action (Rizzolatti & Arbib, 1998; Sauser & Billard, 2005a).

A possible explanation for the ability to imitate is the internal vocabulary of actions that are recognized by the mirror neurons (Rizzolatti & Luppino, 2001). This ability to understand others' actions, beliefs, goals and expectations aids the inclusiveness of the group. This allows the observer to predict the actions and so determine if they are helpful, unhelpful, threatening, and to act accordingly (Gallese & Goldman, 1998; Gallese, 2005). It is argued by Demiris and Hayes (2002) that through the mirror neuron system when a primate or human watches an action they are to imitate they put themselves in the place of the demonstrator. Understanding the actions of the demonstrator comes from creating alternatives and choosing the most appropriate one. A requirement for imitation is to connect the sensory system with the motor system so that the multimodal inputs are linked to the appropriate actions.

2.5 Mirror Neurons and Frame of Reference Transformations

If mirror neurons are involved in imitation then they have to mediate between different coordinate systems in order to establish a mapping from the visual representation of another agent's action to an appropriate motor behavior. If the observed action is targeted at a specific object, then a visual representation of agent and object in retinal coordinates must be transferred into an object centered representation of the action. If the object-directed action is then to be imitated, it has to be transfered to the appropriate motor commands taking into account a potentially different relative position of the imitator to the object.

A very instructive example is that of gaze following. Gaze following is the skill of looking to an object because another agent turns to look at it. In this case the object-directed action is simply to look at it. Human infants seem to learn this behavior during their first two years of life presumably because they learn that whereever other people are looking there is typically something interesting to see – such as another person, an interesting object, or the other person's hands manipulating something (Triesch, Teuscher et al., 2006). Consider the situation of an infant and her mother facing each other. When the mother turns to her right to look at something, the infant will see the head of the mother turning. But how does the infant know that she needs to turn her own head to the left and how much she needs to turn to look at the same object? The question is not trivial and in fact infants seem to need more than 18 month to fully master this skill. The infant needs to learn to associate different locations and head poses of the mother with potential locations in space where the mother may be looking. These locations comprise the line of sight of the mother, which is represented, presumably, in an ego-centric reference frame of the infant. Recently, Triesch, Jasso and Deáak (2006) presented a simple model that demonstrates how such a mapping can be learnt with generic reinforcement learning mechanisms. Specific head poses of the mother become associated with gaze shifts of the infant that have a high probability of matching the location where the mother is looking. This learning process is driven by rewards the infant receives for looking at interesting objects, whose locations are predicted by the looking direction of the mother. Interestingly, the model develops a mirror-neuron-like premotor representation with a population of model neurons that become activated when the infant plans to look at a certain location or when the infant sees the mother looking in the direction of that location. The existence of a similar representation in the brain is the major prediction of the model.

The example demonstrates that the three-dimensional pose of the agent and the imitator and their interrelation need not be fully computed to achieve imitation behavior. On the other hand, Sauser and Billard (2005b) present a model (cf. Section 3.2) in which imitation is performed, only by constructing several consecutive frame of reference transformations, in order to map from an agent- to an imitator coordinate system. How much geometry has to be calculated for imitation?

In another view, mirror neurons represent actions in an abstract fashion, such as when activated by sound (cf. Fig. 2), but are in general unresponsive to the individual movements that make up an action. Hence, they are invariant with respect to the exact geometrical constellations. The demand for invariances makes geometrical frame of reference computations even more challenging, and encourages an alternative view in which the mere presence of features, such as a sound, are detected by mirror neurons with little involvement of geometry. Both views can be reconciled by the notion of convergence (cf. Section 2.2): the result of a geometrical frame of reference computation may be supported by, e.g., the target being successfully focused, or grasped. The error- or reinforcement signal at the end of a successfully completed action can give feedback to the learning of the geometrical calculations.

3. Action in Robotics

The complex geometry of humanoid robots complicates the mapping between sensors and actuators. Additionally, sensors (head, eyes/camera, ears/microphone) may be moved independently of the body on which the actuators are mounted. This is demonstrated in Fig. 3. The robot must know not only the visually seen direction of the object (blue dotted arrow) in retinal coordinates, but also the gaze direction (yellow arrow) which defines the retinal reference frame, before acting in the body-centered reference frame (defined by the long red arrow). Geometrical calculations are complicated by the fact that axes do not coalign, nor do they lie in the centers of sensors.

3.1 Coordinate Transformations in Robotics

There has been a lot of traditional and current research on coordinate transformations in robotics since traditional robotic models rely extensively on converting perceptual input to internal representations and external actions. For instance, at the most recent Intelligent Robotics and Systems international conference proceedings, 2006, there are 62 papers addressing "coordinate transformations" in some form. Coordinate transformations are used in traditional control modeling for instance for motion generation, ball catching, sound localization, visual tracking, simultaneous localization and mapping, path following, motion planning, cooperating manipulators, balancing tasks, dance step selection, or pedestrian tracking. If multiple audio visual and motor maps are all involved in controlling robot behavior these maps have to be coordinated. Most of the existing coordinate transformation approaches in robotics rely on traditional inverse dynamics / inverse kinematics models based on control approaches. However, for new intelligent robotics and in particular humanoid robots with many degrees of freedom researchers look for new and alternative solutions to coordinate transformations. For instance, one important cluster of work relates to the motion generation and motion interpretation for humanoid robots (Harada et al., 2006) and in order to grasp a cup with a handle from a table, the robotic head and the arm have to be coordinated (Tsay & Lai, 2006). For such tasks several different coordinate transformations are involved in a simple grasping action, e.g. including an object coordinate system, a palm coordinate system, gaze coordinate system, arm base coordinate system, and wrist coordinate system etc. Furthermore, if sound maps are involved, for instance different sizes of heads, ear shapes have an influence on the sound maps and coordinate transformations in reality are difficult to calibrate from sound to vision or motor maps (Hoernstein et al., 2006).

Figure 3. Grasping by a humanoid robot. The object is localized in a visual reference frame (yellow arrow points into gaze direction) by the direction of the dotted, blue arrow. The grasping is performed in a body-centered reference frame (long red arrow). The relation between visual and body-centered systems varies when the head turns

Particularly challenging are also coordinate transformations between different maps of different robots, the so-called multi robot map alignment problem. A new approach has been proposed to use the relative pose measurements of each of two robots to determine the coordinate transformation between two maps while assuming that the initial poses of the robots are not known (Zhou & Roumeliotis, 2006).

Besides these traditional approaches based on formal algebraic transformations there are more and more non-traditional approaches to motor control based on biomimetic or cognitive approaches. The general problem of movement control is to map from a cartesian task space to a solution in joint space. Especially for higher degrees of freedom robots with 30-40 degrees of freedom (similar to humans) the task of inverse dynamics of computing these coordinate transformations is very demanding. Therefore, different from traditional approaches some new bioinspired control models have been suggested to address computational efficiency with task independent motor control, only involving those joints which also participate in the task (Gu & Ballard, 2006). This approach by Gu and Ballard proposes that actions are planned in segments with equilibrium end points in joint space. Movements are done by moving between equilibrium points.

Another interesting novel approach has been proposed as a learning-based neurocontroller for a humanoid arm. This approach is based on a self-organizing neural network model and does not assume knowledge about the geometry of the manipulators. Essential is an action-perception cycle which helps to learn transformations from spatial movement to joint movements in the neural map (Asuni et al., 2006). Another new approach to biologically inspired cross modal mapping is further pursued for robotic eye-hand systems (Meng & Lee, 2006) where they incrementally construct mapping networks for eye hand coordination based on extended Kalman filters and radial basis function networks. The system converts the location of a target to an eye-centered coordinate system from which it is mapped via another network into a hand-based coordinate system. Then the difference between the actual and desired position can be computed to steer the motor commands.

3.2 Robot Mirror Neuron System Imitation Learning

The mirror neuron system's principles offer inspiration for developing robotic systems particularly with regards to imitation learning to allow robots to cope with complex environments by reducing the search space (Belpaeme et al., 2003; Triesch et al., 1999).

A mirror neuron-based approach for imitation that relies on learning the reference frame transformation is that of Sauser and Billard (2005b, 2005a). This approach for three-dimensional frames of reference transformations is based on a recurrent multi-layer neural network. This two layer neural network can create a non-linear composition from its inputs. The model includes an attractor network in the first layer with lateral weights, with the second layer containing neurons that receive inputs from a recurrent population but lack lateral connections. This network is able to represent two characteristics, direction and amplitude, in a population vector code. It is able to perform translation or rotation of vectorial activities. To perform non-linear transformation such as rotation there was the need for an intermediary population known as the gain field. A rotation around an axis the rotation is split into three transformations. When carrying out rotation for a reaching activity the target is observed using the visual system and represented using head-centered coordinate neurons. The angle between the head and body is represented in neurons that get proprioceptive information from the appropriate muscles receptors. To produce a movement to the target there is a need to take the head- centered coordinate representation and alter it so it is a body-centered frame of reference vector represented in a population of neurons.

However many of the approaches simplify the problem to get round the need to include a learnt approach for reference transformation. Takahashi et al. (2006) have put forward a mirror neuron-based model for learning behaviors and others' intentions that does not depend on a precise model of the world or coordinate transforms. This approach relies on reinforcement learning to perform activities including navigation and ball passing with in a modular approach. Once the observer robot has developed the appropriate behaviors through reinforcement learning, the observer watchers the behavior and maps the sensory information from the observer's position to that of the performers based on state variables created during reinforcement. This model is based on a modular learning system made up of behavior modules.

An additional robotics approach that uses imitation learning based on mirror neurons is that of Demiris and Hayes (2002) and Demiris and Johnson (2003) through behavior and forward models. A reference transformation is made in that the observer and the actor robots face each other. The behavior model gives information on the current state and the goal and produces the required motor commands. The forward model then creates the expected next state based on the output from the behavior model. The predicted state is compared with the actual state of the demonstrator to produce an error signal. A confidence value is created from the error signal and used to establish the confidence by which a particular action or series of actions is identified by the observer robot. The architecture not only allows the observer robot to produce the appropriate behavior but also to recognize the behavior being performed by the demonstrator.

A further mirror neuron-based approach is that of Billard and Matarić (2001) who use a hierarchy of neural networks and provides an abstract and high level depiction of the neurological structure that is the basis of the visuomotor pathways to examine the ability to reproduce human arm movements. The model consists of three parts for visual recognition, motor control and learning and uses seven modules. A module based on the temporal cortex processes visual information to identify the direction and orientation of the teacher's arms with reference to a point on the teacher's body. The temporal cortex model receives as input Cartesian coordinates for the limbs of the person demonstrating the action and this is transformed into the frame of reference of the observer. This transfer is supported by studies that have observed orientation-sensitive cells in the temporal cortex. The motor control is based on a hierarchical model with a spinal cord module at the lower level. Learning of movement occurs in the pre-motor cortex and cerebellum modules and learning creates links between the primary motor cortex, premotor cortex and the cerebellum and within the premotor cortex and the cerebellum. These modules use a dynamic recurrent associative memory architecture which is a fully connected recurrent network that enables time series and spatio-temporal data to be learnt using short-term memory. The model when tested on series of arm movements is found to reproduce all motions despite the noisy environment.

An additional mirror neuron system based approach for grounding is that of Tani et al. (2004). A recurrent neural network with parametric biases (RNNPB) learns to recognize and produce multiple behaviors with distributed coding using a self-organizing technique. The reference transformation takes the spatial coordinates of the actor's hands which are mapped to the robot's hands using centered cartesian coordinates without learning. In this approach, sections of spatio-temporal data of sensory-motor flow are depicted by using vectors of small dimensions. The nonlinear dynamical system is produced using a Jordan-type recurrent network that has parametric biases ***(PB)*** incorporated in the input layer function. Learning is achieved through the self-organizing mapping of the ***PB*** and the behavior representation. To reproduce the characteristics of the mirror neuron system, the RNNPB creates the appropriate dynamic pattern from fixed ***PB*** to learn and perform recognition by producing the PB from a target pattern. Movement patterns are learnt using the forward model by producing the ***PB*** vectors and a synaptic weight matrix. Following learning it is possible to produce sensory-motor series by using the forward dynamics of the RNNPB with the parameter biases fixed. When the network produces a behavior it operates in a closed loop where the prediction of the next action is fed back as an input.

As part of the MirrorBot project a mirror neuron-based docking action was generated using a 4-step model. First, feature detectors from the visual input of neurons in a "what" area are learnt unsupervised. Second, associator weights within and between the "what" area and a "where" area are learnt supervised. After training, these two levels visually localize an object in a camera-centered coordinate system. Third, weights to a robot motor output are trained by reinforcement learning to drive the robot to a position to grasp the object (Weber et al., 2004). Since the position of the camera was held fixed, the pixel coordinates could be directly used for motor control, and no dynamic reference frame transformation was needed. In an equivalent approach, Martinez-Marin and Duckett (2005) make the camera always focus at a grasp target, which allows to use the camera gaze angle directly to control the robot. In (Weber et al., 2006) finally, a fourth layer observes the self-performed actions and learns to perform and to predict them based on only visual input. This loosely mimics mirror neurons, but as visual recognition of other robots was not done, neurons are active only when observing self-performance. Another, higher level with additional language input learnt to associate words to actions (Wermter et al., 2005). In simulation, the network could perform and recognize three behaviors, docking ('pick'), wander ('go'), and move away ('lift').

4. Importance of Learning

Maybe one of the most fundamental questions we can ask about coordinate transformations is how the brain comes to know how to compute them. More precisely, to what extent is the ability to compute proper coordinate transformations already built into our brains at birth and can be seen as a product of our evolutionary heritage? And to what extent is it the product of experience dependent learning mechanisms, i.e. a product of our lifetime experience? This set of questions is a particular example of what is often called the nature/nurture debate (Elman et al., 1996).

Several pieces of evidence point to an important role of evolution in setting up proper sensorimotor coordinate mappings in our brains. Most important, maybe, is the fact that even newborn infants are robustly capable of computing certain sensorimotor transformations. For example they will readily turn their head in the direction of a salient visual stimulus, requiring them to map the location of a visually perceived object represented in retinotopic coordinates to the appropriate motor commands for turning the head. This finding is important, because it shows that some sensorimotor transformations are already in place minutes after birth, before there was much time for any experience dependent processes to learn such a mapping. The sensorimotor abilities of other species are even more striking. In some precocious species like gazelles, newborns will be able to run at high speeds a few hours after birth, successfully avoiding obstacles and navigating across uneven terrain.

The remarkable sensorimotor abilities of some species at birth suggest that much of the solution may be "hardwired" into our brains, but experience dependent learning processes must also play an important role in setting up and maintaining proper coordinate transformations. A chief reason for this is that the required coordinate transformations are not constant but change during the organism's development as the body grows and the geometry of sense organs and limbs matures. In fact, our ability to constantly adapt our sensorimotor coordination seems to be ubiquitous and it even persists into adulthood. A well-studied example is that of prism adaptation (Harris, 1965). When subjects wear glasses with wedge prisms that produce a sidewards shift of the visual scene, they will initially be quite inaccurate at a task like throwing darts at a target. The prism glasses introduce a bias to throw the dart to one side. Within a brief period, however, subjects are able to adapt their sensorimotor mappings to compensate for the effect of the prism glasses and their throws become accurate again. If the glasses are now taken away again, the subjects will again make errors – but this time in the opposite direction. Thus, our sensorimotor mappings can be viewed as being the product of adaptation processes that change the required coordinate transformations whenever needed. At this point it is clear that both nature and nurture play important roles in allowing us to become so adept at performing sophisticated coordinate transformations. The fact that our coordinate transformations are so adaptable suggests that humanoid robots may also benefit from the flexibility to learn and adapt their coordinate transformations during their "life-time".

5. Neural Frame of Reference Transformations

5.1 Neural Population Code

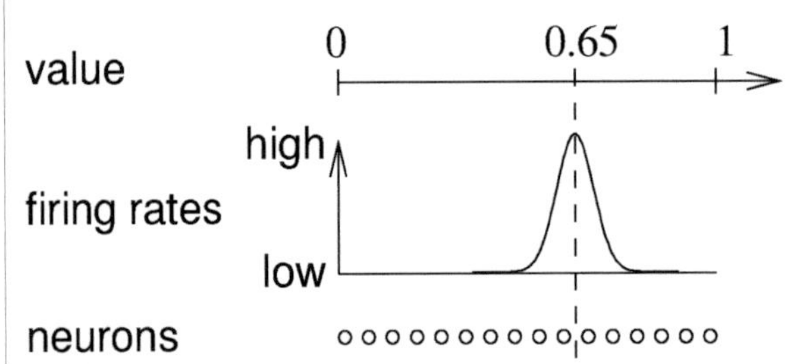

Figure 4. Neural population code. A scalar value, e.g. 0.65, is encoded by the activation rates of an array of neurons, the *population vector*

A continuous number is likely to be represented in the brain on an array of neurons, rather than in the firing rate of a single neuron. Fig. 4 visualizes how neurons may encode a continuous value by a neural population code. Each neuron codes for one value for which it will fire maximally, but it also responds to similar values. The set of neurons must cover all values that can be coded. The coded value can be read from the center of mass of the activation hill. One reason for such a code in the brain may be that a cortical neuron's firing is noisy and its firing rate is hard to adjust precisely. Many neurons for example decrease their firing rate within a few seconds of constant stimulus presentation, a phenomenon called firing rate adaptation (Blakemore & Campbell, 1969).

A second reason for population coding may be that sensory neurons are spatially distributed. A seen object activates a blob of neurons on the retina, so it is computationally straightforward to retain the positional information topographically in cortical visual areas. Furthermore, there is more information in such a code than just the value coded in the maximally active neuron. A wider, or narrower hill of activation may account for uncertainty in the network's estimate of the encoded value. If two estimates, e.g. in the dark an uncertain visual position estimation and a more precise sound-based estimation are combined, networks that perform Bayesian cue integration can combine the information to obtain an estimate that has higher saliency than an estimate based on one cue alone (Battaglia et al., 2003). A neural activation pattern not only encodes location information. A visual stimulus activates several hierarchically organized cortical areas, each analyzing a different aspect, such as color, shape, identity, etc. The pure location information of an object may be most beneficial in the motor cortex, but also there, information like object orientation, size and expected weight is relevant, e.g. if the object is to be grasped. A more complex shape of the neural activation pattern can represent such additional information. However, if the activation pattern becomes too complex, then information might interfere, and the coded variables may be read out incorrectly.

5.2 Principles of Frame of Reference Transformations

Coordinate transformations exist in several complexities. In the simple case, for example visual coordinates (x_1, x_2) representing a grasp target are transformed into arm angle coordinates (z_1, z_2) used for reaching the target. This can be described as a fixed (non-linear) mapping $f : \vec{x} \mapsto \vec{z}$, and assumes that only the arm moves. The body of the agent, its head, eyes, and the object are statically fixed (Ghahramani et al., 1996). The mapping gets more involved if the arm angle coordinates have redundant degrees of freedom, as is the case in human(oid)s (Asuni et al., 2006).

In this book chapter we are focusing on the more complex, *dynamic* coordinate transformations. In the example, the visual-motor mapping would be altered by other influences, such as the gaze direction $\vec{y}$, which is determined by the posture of eyes and head. Since the mapping is now also influenced by the values of (y_1, y_2), we may express it as $f : (\vec{x}, \vec{y}) \mapsto \vec{z}$. It is these dynamic transformations that are in the literature referred to as frame of reference transformations.

Our paramount example (for population variables $\vec{x}$ and $\vec{y}$ representing scalar variables) is that μ_x is the horizontal object position on the retina, and μ_y is the horizontal gaze angle (composed of eye- and head-angle). Then the body-centered horizontal position of the target is

$$\mu_z = \mu_x + \mu_y \tag{1}$$

This is challenging, because there is no other sensory input supplying μ_z and the computation is done with population codes.

For this scalar case, Fig. 5 shows how a neural frame of reference transformation can be performed. The scalar variables μ_x and μ_y define the centers of Gaussian hills of neural activations $\vec{x}$ and $\vec{y}$, each along one dimension. The outer product of these population codes is then represented on two dimensions, depicted as squares in Fig. 5. These two dimensions contain all information of $\vec{x}$ and $\vec{y}$, be it in a rather wasteful manner. The advantage is that for each constant μ_z, a diagonal line in that square represents all possible combinations of μ_x and μ_y for which Eq. 1 holds.

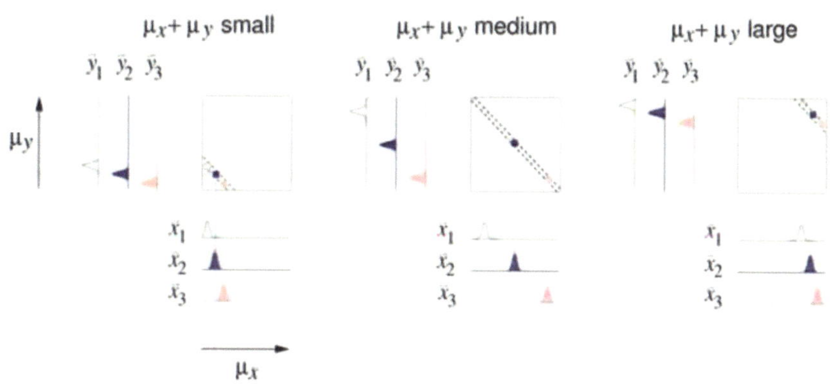

Figure 5. Schematic of a frame of reference transformation. Three different variations of the inputs are shown for each of three different results of the transformation, i.e. resulting in small, medium and large sums. The input is a pair of Gaussian population vectors, e.g. $(\vec{x}_2, \vec{y}_2)$, representing μ_x and μ_y, the to be transformed variables. Different combinations of μ_x and μ_y are possible that lead to the same sum; these combinations lie on a diagonal on the depicted squares. A neuron that responds to a given sum will retrieve its input from one such diagonal

A straightforward method to get the result is to plaster the square with neurons which give input to a one-dimensional array of output neurons. Each output neuron receives its input from a diagonal line with an offset corresponding to the result μ_z that it represents (Rossum & Renart, 2004).

Such networks have been termed "basis function networks" (Deneve et al., 2001) referring to the units on the middle (square) layer, and partially shifting receptive fields of these units in response to a change of their input have been reported, akin to responses in several parietal and premotor areas. The term "gain field architecture" (Sauser & Billard, 2005a) refers to the multiplication of the two inputs (a component of $\vec{x}$ times a component of $\vec{y}$) which effects the middle layer responses, a behavior also observed in some cortical areas (see Section 2).

A problem in such a two-layer neural network is to find an unsupervised learning scheme. This is important, because, in the above example, μ_z is not directly available, hence there is no supervisor. We will present in Section 6 a network that replaces the middle layer by enhanced synapses of the output layer neurons, and which allows unsupervised training.

It is also worthwhile mentioning that the body geometry may considerably constrain, and therefore simplify, the possible transformations. For example, if the head turns far right, then no object on the left side of the body is visible. Hence, one input such as μ_y already constrains the result μ_z without any knowledge of the other input μ_x (assuming that only visible objects are of interest). A simple transformation network without any middle layer that was trained in a supervised fashion takes advantage of this simplification (Weber et al., 2005).

6. A Mapping with Sigma-Pi Units

A standard connectionist unit i is activated by the sum of input activations x_j weighted by its weights w_{ij}:

$$a_i = \sum_j w_{ij} x_j \quad (2)$$

This net input a_i is then usually passed through a transfer function. A Sigma-Pi neuron i evaluates the weighted sum of multiplications

$$a_i = \sum_{j,k} w_{ijk} x_j y_k \quad (3)$$

As a specific case, the input vector $\vec{y}$ can be the same as $\vec{x}$, but in our example we have different input layers. In general, Sigma-Pi units may evaluate the product of more than just two terms.

The advantage of the Sigma-Pi neuron for our problem can be seen in Fig. 6. Consider a neuron that shall be active if, and only if, $\mu_x + \mu_y$ leads to a medium sum, as in Fig. 5 middle. We can construct it by assigning non-zero weights to the according combinations in the x- and y-input layers, as depicted by the small blobs on the diagonal of the square in Fig. 6, and zero weights away from the diagonal. This neuron will be activated according to

$$a_i = w_{ijj} x_j y_j + w_{ikk} x_k y_k + w_{ill} x_l y_l \quad (4)$$

so it will be activated by the selected combinations of x- and y-inputs. It will not be activated by different combinations, such as e.g. (x_j, y_k), because w_{ijk} is zero. Such a selective response is not feasible with one connectionist neuron.

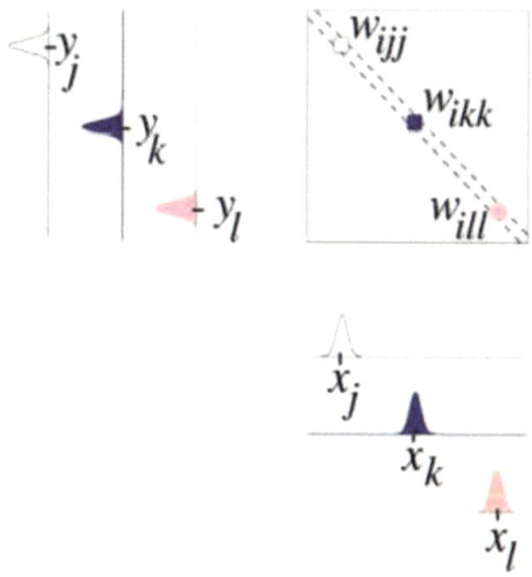

Figure 6. A Sigma-Pi neuron with non-zero weights along the diagonal will respond only to selected input combinations, such as (x_j, y_j) or (x_k, y_k) or (x_l, y_l). This corresponds to Fig. 5, middle, where $\mu_x + \mu_y$ has medium value

6.1 A Sigma-Pi SOM Learning Algorithm

The main idea for an algorithm to learn frame of reference transformations exploits that a representation of an object remains constant over time in some coordinate system while it changes in other systems. When we move our eyes, a retinal object position will change with the positions of the eyes, while the head-centered, or body centered, position of the object remains constant. In the algorithm presented in Fig. 7 we exploit this by sampling two input pairs (e.g. retinal object position and position of the eyes, at two time instances), but we "connect" both time instances by learning with the output taken from one instance with the input taken from the other. We assume that neurons on the output (map) layer respond invariantly while the inputs are varied. This forces them to adopt, e.g. a body centered representation. In unsupervised learning, one has to devise a scheme how to activate those neurons which do not see the data (the map neurons). Some form of competition is needed so that not all of these "hidden" neurons behave, and learn, the same. Winner-take-all is one of the simplest form of enforcing this competition without the use of a teacher. The algorithm uses this scheme (Fig. 7, step 2(c)) based on the assumption that exactly one object needs to be coded. The corresponding winning unit to each data pair will have its weights modified so that they resemble these data more closely, as given by the difference term in the learning rule (Fig. 7, step 4). Other neurons will not see these data, as they cannot win any more, hence the competition. They will specialize on a different region in data space. The winning unit will also activate its neighbors by a Gaussian activation function placed over it (Fig. 7, step 2(d)). This causes neighbors to learn similarly, and hence organizes the units to form a topographic map. Our Sigma-Pi SOM shares with the classical self-organizing map (SOM) (Kohonen, 2001) the concepts of winner-take-all, Gaussian activation, and a difference-based weight update. The algorithm is described in detail in

Weber and Wermter (2006). Source code is available at the ModelDB database: http://senselab.med.yale.edu/senselab/modeldb (Migliore et al., 2003).

1. Choose a random sum location μ_z.

   This sum location will be constant for the following two sample input pairs. If μ_z is a body-centered object position, then the following two sample pairs correspond, e.g., to two different views of the object, such as two different eye positions and corresponding retinal positions of the object.

2. (a) Sample randomly input pair (μ_x, μ_y) with $\mu_x + \mu_y = \mu_z$ and produce the corresponding population code $(\vec{x}, \vec{y})$

   E.g. eye position 1, and retinal object position 1.

   (b) Activate all map units $\{i\}$: $a_i = \sum_{j,k} w_{ijk} x_j y_k$

   Determine which map unit will be activated by co-occurrence of the given inputs, using the higher-order weights w_{ijk}.

   (c) Find the winning unit: $m = \text{argmax}_i a_i$

   This determines the center of the representation on the output layer which is constant for both views of the object.

   (d) Obtain activations: $z_i(\vec{x}, \vec{y}) = G_{xy}(|i-m|, \sigma)$

   A Gaussian G around the winning unit makes the network forgive small assignment errors, and organizes the units topographically.

   (e) Decrease the neighborhood interaction width σ

   The Gaussian is made more narrow, so after the units have learnt roughly, they can become more distinct.

3. Sample randomly input pair $(\bar{\mu}_x, \bar{\mu}_y)$ with $\bar{\mu}_x + \bar{\mu}_y = \mu_z$ and produce the corresponding population code $(\bar{x}, \bar{y})$

   E.g. eye position 2, and retinal object position 2. We may also compute the activation on the output layer from this input pair, in order to check whether this leads to the same (or nearby) winning unit as in Step 2(c). However, the asymmetric learning rule needs the postsynaptic activation of only one of the two data pairs.

4. Change weights incrementally: $\Delta w_{ijk} = \epsilon \underbrace{z_i(\vec{x}, \vec{y})}_{\text{post}} \underbrace{(\bar{x}_j \bar{y}_k - w_{ijk})}_{\text{pre}}$

   The post-synaptic learning term is the map activation resulting from the 1st data pair; only activated neurons will learn. The pre-synaptic term is the difference between the 2nd data pair and the current Sigma-Pi weight vector.

Repeat until the neighborhood interaction width σ is zero.

Figure 7. One iteration of the Sigma-Pi SOM learning algorithm

6.2 Results of the Transformation with Sigma-Pi Units

We have run the algorithm with two-dimensional location vectors μ_x and μ_y as relevant for example for a retinal object location and a gaze angle, since there are horizontal and vertical components. μ_z then encodes a two-dimensional body-centered direction. The corresponding inputs in population code $\vec{x}$ and $\vec{y}$ are each represented by 15 x 15 units. Hence each of the 15 x 15 units on the output layer has 15^4 = 50,625 Sigma-Pi connection parameters. For an unsupervised learnt mapping, it cannot be determined in advance exactly which neurons of the output layer will react to a specific input. A successful frame of reference transformation, in the case of our prime example Eq. 1, is achieved, if for different combinations (μ_x, μ_y) that belong to a given μ_z always the same output unit is activated, hence $\vec{z}$ will be constant. Fig. 8, left, displays this property for different $(\vec{x}, \vec{y})$ pairs. Further, different output units must be activated for a different sum μ'_z. Fig. 8, right, shows that different points on one layer, here together forming an "L"-shaped pattern, are mapped to different points on the output layer in a topographic fashion. Results are detailed in Weber and Wermter (2006).

The output $\vec{z}$ (or possibly, $\vec{a}$) is a suitable input to a reinforcement-learnt network. This is despite the fact that, before learning, $\vec{z}$ is unpredictable: the "L" shape of $\vec{a}$ in Fig. 8, right, might as well be oriented otherwise. However, after learning, the mapping is consistent. A reinforcement learner will learn to reach the goal region of the trained map (state space) based on a reward that is administered externally.

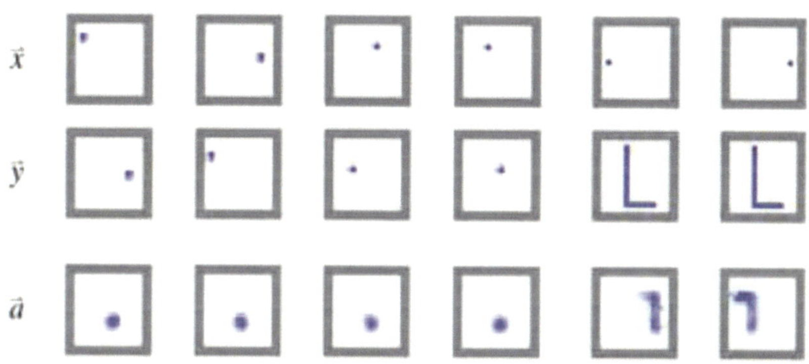

Fig. 8: Transformations of the two-dimensional Sigma-Pi network. Samples of inputs $\vec{x}$ and $\vec{y}$ given to the network are shown in the first two rows, and the corresponding network response a, from which $\vec{z}$ is computed, in the third row. Leftmost four columns: random input pairs are given under the constraint that they belong to the same sum value $\bar{\mu}_z$. The network response a is almost identical in all four cases. Rightmost two columns: when a more complex "L"-shaped test activation pattern is given to one of the inputs, a similar activation pattern emerges on the sum area. It can be seen that the map polarity is rotated by 180°.

6.3 An Approximate Cost Function

A cost function for the SOM algorithm does not strictly exist, but approximate ones can be stated, to gain an intuition of the algorithm. In analogy to Kaski (1997) we state (cf. Fig. 7):

$$E(\{W\},\{m(x,y)\}) = \frac{1}{2} \sum_{i,(x,y),j,k} G_{xy}(|i-m|,\sigma) \cdot (\bar{x}_j \bar{y}_k - w_{ijk})^2 \tag{5}$$

where the sum is over all units, data, and weight indices. The cost function is minimized by adjusting its two parameter sets in two alternating steps. The first step, winner-finding, is to minimize *E* w.r.t. the assignments $\{m(\vec{x}, \vec{y})\}$ (cf. Fig. 7, Step 2 (c)), assuming fixed weights:

$$m(x, y) = \text{argmin}_i \sum_{j,k} (w_{ijk} - x_j y_k)^2 \approx \text{argmax}_i \sum_{j,k} w_{ijk} x_j y_k \tag{6}$$

Minimizing the difference term and maximizing the product term can be seen as equivalent if the weights and data are normalized to unit length. Since the data are Gaussian activations of uniform height, this is approximately the case in later learning stages when the weights approach a mean of the data. The second step, weight-learning (Fig. 7, Step 4), is to minimize *E* w.r.t. the weights $\{\vec{w}_i\}$, assuming given assignments. When convergend, $\Delta w_{ijk} = 0$ and

$$w_{ijk} = \frac{\sum_{(x,y)} z_i(\vec{x}, \vec{y}) \cdot \bar{x}_j \bar{y}_k}{\sum_{(x,y)} z_i(\vec{x}, \vec{y})} \tag{7}$$

Hence, the weights of each unit reach the center of mass of the data assigned to it. Assignment uses $(\vec{x}, \vec{y})$ while learning uses a pair $(\bar{x}, \bar{y})$ of an "adjacent" time step, to create invariance. The many near-zero components of x and y keep the weights smaller than active data units.

7. Discussion

Sigma-Pi units lend themselves to the task of frame of reference transformations. Multiplicative attentional control can dynamically route information from a region of interest within the visual field to a higher area (Andersen et al, 2004). With an architecture involving Sigma-Pi weights activation patters can be dynamically routed, as we have shown in Fig. 8 b). In a model by Grimes and Rao (2005) the dynamic routing of information is combined with feature extraction. Since the number of hidden units to be activated depends on the inputs, they need an iterative procedure to obtain the hidden code. In our scenario only the position of a stereotyped activation hill is estimated. This allows us to use a simpler, SOM-like algorithm.

7.1 Are Sigma-Pi Units Biologically Realistic?

A real neuron is certainly more complex than a standard connectionist neuron which performs a weighted sum of its inputs. For example, there exists input, such as shunting inhibition (Borg-Graham et al., 1998; Mitchell & Silver, 2003), which has a multiplicative effect on the remaining input. However, such potentially multiplicative neural input often targets the cell soma or proximal dendrites (Kandel et al., 1991). Hence, multiplicative neural influence is rather about gain modulation than about individual synaptic modulation. A Sigma-Pi unit model proposes that for each synapse from an input neuron, there is a further input from a third neuron (or even a further "receptive field" from within a third neural layer). There is a debate about potential multiplicative mutual influences between synapses, happening particularly when synapses gather in clusters at the postsynaptic dendrites (Mel, 2006). It is a challenge to implement the transformation of our Sigma-Pi network with more established neuron models, or with biologically faithful models.

A basis function network (Deneve et al., 2001) relates to the Sigma-Pi network in that each each Sigma-Pi connection is replaced by a connectionist basis function unit - the intermediate layer built from these units then has connections to connectionist output units. A problem of this architecture is that by using a middle layer, unsupervised learning is hard to implement: the middle layer units would not respond invariantly when in our example, another view of an object is being taken. Hence, the connections to the middle layer units cannot be learnt by a slowness principle, because their responses change as much as the input activations do. An alternative neural architecture is proposed by Poirazi et al. (2003). They found that the complex input-output function of one hippocampal pyramidal neuron can be well modelled by a two-stage hierarchy of connectionist neurons. This could pave a way toward a basis function network in which the middle layer is interpreted as part of the output neurons' dendritic trees. Being parts of one neuron would allow the middle layer units to communicate, so that certain learning rules using slowness might be feasible.

7.2 Learning Invariant Representations with Slowness

Our unsupervised learnt model of Section 6 maps two fast varying inputs (retinal object position $\vec{x}$ and gaze direction $\vec{y}$) into one representation (body-centered object position $\vec{z}$) which varies slowly in comparison to the inputs. This parallels a well known problem in the visual system: the input changes frequently via saccades while the environment remains relatively constant. In order to understand the environment, the visual system needs to transform the "flickering" input into slowly changing neural representations - these encoding constant features of the environment.

Examples include complex cells in the lower visual system that respond invariantly to small shifts and which can be learnt with an "activity trace" that prevents fast activity changes (Földiák, 1991). With a 4-layer network reading visual input and exploiting slowness of response, Wyss et al. (2006) let a robot move around while turning a lot, and found place cells emerging on the highest level. These neurons responded when the robot was at a specific location in the room, no matter the robot's viewing direction.

How does our network relate to invariance in the visual system? The principle is very similar: in vision, certain complex combinations of pixel intensities denote an object, while each of the pixels themselves have no meaning. In our network, certain combinations of inputs $(\vec{x}, \vec{y})$ denote a $\vec{z}$, while $\vec{x}$ or $\vec{y}$ alone have no information. The set of inputs that lead to a given $\vec{z}$ is manageable, and a one-layer Sigma-Pi network can transform all possible input combinations to the appropriate output. In vision, the set of inputs that denotes one object is rather unmanageable; an object often needs to be recognized in novel view, such as a person with new clothes. Therefore, the visual system is multi level hierarchical and uses strategies such as de-composition of objects into different parts.

Computations like our network does may be realized in parts of the visual system. Constellations of input pixel activities that are always the same can be detected by simple feature detectors made with connectionist neurons; there is no use for Sigma-Pi networks. It is different if constellations need to be detected when transformed, such as when the image is rotated. This requires the detector to be invariant over the transformation, while distinguishing from other constellations. Rotation invariant object recognition, reviewed in Bishop (1995), but also the routing of visual information (Van Essen et al., 1994), as we show in Fig. 8 b), can easily be done with second order neural networks, such as Sigma-Pi networks.

7.3 Learning Representations for Action

We have seen above how slowness can help unsupervised learning of stable sensory representations. Unsupervised learning ignores the motor aspect, i.e. the fact that the transformed sensory representations only make sense if used for motor action. Cortical representations in the motor system are likely to be influenced by motor action, and not merely by passive observation. Learning to catch a moving object is unlikely to be guided by a slowness principle. Effects of action outcome that might guide learning are observed in the visual system. For example, neurons in V1 of rats can display reward contingent activity following presentation of a visual stimulus which predicts a reward (Shuler & Bear, 2006). In monkey V1, orientation tuning curves increased their slopes for those neurons that participated in a discrimination task, but not for other neurons that received comparable visual stimuli (Schoups et al., 2001). In the Attention-Gated Reinforcement Learning model, Roelfsema and Ooyen (2005) combine unsupervised learning with a global reinforcement signal and an "attentional" feedback signal that depends on the output units' activations. For 1-of-*n* choice tasks, these biologically plausible modifications render learning as powerful as supervised learning. For frame of reference transformations that extend into the motor system, unsupervised learning algorithms may analogously be augmented by additional information obtained from movement.

8. Conclusion

The control of humanoid robots is challenging not only because vision is hard, but also because the complex body structure demands sophisticated sensory-motor control. Human and monkey data suggest that movements are coded in several coordinate frames which are centered at different sensors and limbs. Because these are variable against each other, dynamic frame of reference transformations are required, rather than fixed sensory-motor mappings, in order to retain a coherent representation of a position, or an object, in space. We have presented a solution for the unsupervised learning of such transformations for a dynamic case. Frame of reference transformations are at the interface between vision and motor control. Their understanding will advance together with an integrated view of sensation and action.

9. Acknowledgements

We thank Philipp Wolfrum for valuable discussions. This work has been funded partially by the EU project MirrorBot, IST-2001-35282, and NEST-043374 coordinated by SW. CW and JT are supported by the Hertie Foundation, and the EU project PLICON, MEXT-CT-2006-042484.

10. References

Andersen, C.; Essen, D. van & Olshausen, B. (2004). Directed Visual Attention and the Dynamic Control of Information Flow. In *Encyclopedia of visual attention,* L. Itti, G. Rees & J. Tsotsos (Eds.), Academic Press/Elsevier.

Asuni, G.; Teti, G.; Laschi, C.; Guglielmelli, E. & Dario, P. (2006). Extension to end-effector position and orientation control of a learning-based neurocontroller for a humanoid arm. In *Proceedings of IROS,* pp. 4151-4156.

Batista, A. (2002). Inner space: Reference frames. *Curr Biol, 12,*11, R380-R383.
Battaglia, P.; Jacobs, R. & Aslin, R. (2003). Bayesian integration of visual and auditory signals for spatial localization. *Opt Soc Am A, 20, 7,*1391-1397.
Belpaeme, T.; Boer, B. de; Vylder, B. de & Jansen, B. (2003). The role of population dynamics in imitation. In *Proceedings of the 2nd international symposium on imitation in animals and artifacts,* pp. 171-175.
Billard, A. & Mataric, M. (2001). Learning human arm movements by imitation: Evaluation of a biologically inspired connectionist architecture. *Robotics and Autonomous Systems, 941,* 1-16.
Bishop, C. (1995). *Neural network for pattern recognition.* Oxford University Press.
Blakemore, C. & Campbell, F. (1969). On the existence of neurones in the human visual system selectively sensitive to the orientation and size of retinal images. *Physiol, 203,* 237-260.
Borg-Graham, L.; Monier, C. & Fregnac, Y. (1998). Visual input evokes transient and strong shunting inhibition in visual cortical neurons. *Nature, 393,* 369-373.
Buccino, G.; Vogt, S.; Ritzi, A.; Fink, G.; Zilles, K.; Freund, H.-J. & Rizzolatti, G. (2004). Neural circuits underlying imitation learning of hand actions: An event-related fMRI study. *Neuron, 42,* 323-334.
Buneo, C.; Jarvis, M.; Batista, A. & Andersen, R. (2002). Direct visuomotor transformations for reaching. *Nature, 416,* 632-636.
Demiris, Y. & Hayes, G. (2002). Imitation as a dual-route process featuring prediction and learning components: A biologically-plausible computational model. In *Imitation in animals and artifacts,* pp. 327-361. Cambridge, MA, USA, MIT Press.
Demiris, Y. & Johnson, M. (2003). Distributed, predictive perception of actions: A biologically inspired robotics architecture for imitation and learning. *Connection Science Journal, 15,* 4, 231-243.
Deneve, S.; Latham, P. & Pouget, A. (2001). Efficient computation and cue integration with noisy population codes. *Nature Neurosci, 4,* 8, 826-831.
Dillmann, R. (2003). Teaching and learning of robot tasks via observation of human performance. In *Proceedings of the IROS workshop on programming by demonstration,* pp. 4-9.
Duhamel, J.; Bremmer, R; Benhamed, S. & Graf, W. (1997). Spatial invariance of visual receptive fields in parietal cortex neurons. *Nature, 389,* 845-848.
Duhamel, J.; Colby, C. & Goldberg, M. (1992). The updating of the representation of visual space in parietal cortex by intended eye movements. *Science, 255,* 5040, 90-92.
Elman, J. L.; Bates, E.; Johnson, M.; Karmiloff-Smith, A.; Parisi, D. & Plunkett, K. (1996). *Rethinking innateness: A connectionist perspective on development.* Cambridge, MIT Press.
Fadiga, L. & Craighero, L. (2003). New insights on sensorimotor integration: From hand action to speech perception. *Brain and Cognition, 53,* 514-524.
Fogassi, L.; Raos, V.; Franchi, G.; Gallese, V.; Luppino, G. & Matelli, M. (1999). Visual responses in the dorsal premotor area F2 of the macaque monkey. *Exp Brain Res, 128,* 1-2,194-199.
Foldiak, P. (1991). Learning invariance from transformation sequences. *Neur Comp, 3,*194-200.

Gallese, V. (2005). The intentional attunement hypothesis. The mirror neuron system and its role in interpersonal relations. In *Biomimetic multimodal learning in a robotic systems,* pp. 19-30. Heidelberg, Germany, Springer-Verlag.

Gallese, V. & Goldman, A. (1998). Mirror neurons and the simulation theory of mind-reading. *Trends in Cognitive Science, 2,12,* 493-501.

Ghahramani, Z.; Wolpert, D. & Jordan, M. (1996). Generalization to local remappings of the visuomotor coordinate transformation. *Neurosci, 16,*21, 7085-7096.

Grafton, S.; Fadiga, L.; Arbib, M. & Rizzolatti, G. (1997). Premotor cortex activation during observation and naming of familiar tools. *Neuroimage, 6,*231-236.

Graziano, M. (2006). The organization of behavioral repertoire in motor cortex. *Annual Review Neuroscience, 29,*105-134.

Grimes, D. & Rao, R. (2005). Bilinear sparse coding for invariant vision. *Neur Comp, 17,*47-73.

Gu, X. & Ballard, D. (2006). Motor synergies for coordinated movements in humanoids. In *Proceedings of IROS,* pp. 3462-3467.

Harada, K.; Hauser, K.; Bretl, T. & Latombe, J. (2006). Natural motion generation for humanoid robots. In *Proceedings of IROS,* pp. 833-839.

Harris, C. (1965). Perceptual adaptation to inverted, reversed, and displaced vision. *Psychol Rev, 72, 6,* 419-444.

Hoernstein, J.; Lopes, M. & Santos-Victor, J. (2006). Sound localisation for humanoid robots - building audio-motor maps based on the HRTF. In *Proceedings of IROS,* pp. 1170-1176.

Kandel, E.; Schwartz, J. & Jessell, T. (1991). *Principles of neural science.* Prentice-Hall.

Kaski, S. (1997). *Data exploration using self-organizing maps.* Doctoral dissertation, Helsinki University of Technology. Published by the Finnish Academy of Technology.

Kohler, E.; Keysers, C.; Umilta, M.; Fogassi, L.; Gallese, V. & Rizzolatti, G. (2002). Hearing sounds, understanding actions: Action representation in mirror neurons. *Science, 297,* 846-848.

Kohonen, T. (2001). *Self-organizing maps* (3. ed., Vol. 30). Springer, Berlin, Heidelberg, New York.

Lahav, A.; Saltzman, E. & Schlaug, G. (2007). Action representation of sound: Audio motor recognition network while listening to newly acquired actions. *Neurosci, 27,* 2, 308-314.

Luppino, G. & Rizzolatti, G. (2000). The organization of the frontal motor cortex. *News Physiol Sci, 15,* 219-224.

Martinez-Marin, T. & Duckett, T. (2005). Fast reinforcement learning for vision-guided mobile robots. In *Proceedings of the IEEE International Conference on Robotics and Automation (ICRA 2005).*

Matsumoto, R.; Nair, D.; LaPresto, E.; Bingaman, W.; Shibasaki, H. & Ltiders, H. (2006). Functional connectivity in human cortical motor system: a cortico-cortical evoked potential study. *Brain, 130,1,*181-197.

Mel, B. (2006). Biomimetic neural learning for intelligent robots. In *Dendrites,* G. Stuart, N. Spruston, M. Hausser & G. Stuart (Eds.), (in press). Springer.

Meng, Q. & Lee, M. (2006). Biologically inspired automatic construction of cross-modal mapping in robotic eye/hand systems. In *Proceedings of IROS,* pp. 4742-4747.

Migliore, M.; Morse, T.; Davison, A.; Marenco, L.; Shepherd, G. & Hines, M. (2003). ModelDB Making models publicly accessible to support computational neuroscience. *Neuroin-formatics, 1*,135-139.

Mitchell, S. & Silver, R. (2003). Shunting inhibition modulates neuronal gain during synaptic excitation. *Neuron, 38*,433-445.

Oztop, E.; Kawato, M. & Arbib, M. (2006). Mirror neurons and imitation: A computationally guided review. *Neural Networks, 19*,254-271.

Poirazi, P.; Brannon, T. & Mel, B. (2003). Pyramidal neuron as two-layer neural network. *Neuron, 37*, 989-999.

Rizzolatti, G. & Arbib, M. (1998). Language within our grasp. *Trends in Neuroscience, 21, 5*, 188-194.

Rizzolatti, G.; Fogassi, L. & Gallese, V. (2001). Neurophysiological mechanisms underlying the understanding and imitation of action. *Nature Review, 2*, 661-670.

Rizzolatti, G.; Fogassi, L. & Gallese, V. (2002). Motor and cognitive functions of the ventral premotor cortex. *Current Opinion in Neurobiology, 12*,149-154.

Rizzolatti, G. & Luppino, G. (2001). The cortical motor system. *Neuron, 31*, 889-901.

Roelfsema, P. & Ooyen, A. van. (2005). Attention-gated reinforcement learning of internal representations for classification. *Neur Comp, 17*,2176-2214.

Rossum, A. van & Renart, A. (2004). Computation with populations codes in layered networks of integrate-and-fire neurons. *Neurocomputing, 58-60*, 265-270.

Sauser, E. & Billard, A. (2005a). Three dimensional frames of references transformations using recurrent populations of neurons. *Neurocomputing, 64*, 5-24.

Sauser, E. & Billard, A. (2005b). View sensitive cells as a neural basis for the representation of others in a self-centered frame of reference. In *Proceedings of the third international symposium on imitation in animals and artifacts, Hatfield, UK.*

Schaal, S.; Ijspeert, A. & Billard, A. (2003). Computational approaches to motor learning by imitation. *Transactions of the Royal Society of London: Series B, 358*, 537-547.

Schoups, A.; Vogels, R.; Qian, N. & Orban, G. (2001). Practising orientation identification improves orientation coding in VI neurons. *Nature, 412*, 549-553.

Shuler, M. & Bear, M. (2006). Reward timing in the primary visual cortex. *Science, 311*,1606-1609.

Takahashi, Y; Kawamata, T. & Asada, M. (2006). Learning utility for behavior acquisition and intention inference of other agents. In *Proceedings of the IEEE/RSJ IROS workshop on multi-objective robotics*, pp. 25-31.

Tani, J.; Ito, M. & Sugita, Y. (2004). Self-organization of distributedly represented multiple behavior schemata in a mirror system: Reviews of robot experiments using RNNPB. *Neural Networks, 17, 8-9*,1273-1289.

Triesch, J., Jasso, H. & Deak, G. (2006). Emergence of mirror neurons in a model of gaze following. In *Proceedings of the Int. Conf on Development and Learning (ICDL 2006).*

Triesch, J.; Teuscher, C; Deak, G. & Carlson, E. (2006). Gaze following: why (not) learn it? *Developmental Science, 9*, 2,125-147.

Triesch, J.; Wieghardt, J.; Mael, E. & Malsburg, C. von der. (1999). Towards imitation learning of grasping movements by an autonomous robot. In *Proceedings of the third gesture workshop (gw'97)*. Springer, Lecture Notes in Artificial Intelligence.

Tsay T. & Lai, C. (2006). Design and control of a humanoid robot. In *Proceedings of IROS*, pp. 2002-2007.

Umilta, M.; Kohler, E.; Gallese, V; Fogassi, L.; Fadiga, L.; Keysers, C. et al. (2001). I know what you are doing: A neurophysiological study. *Neuron, 31,*155-165.

Van Essen, D.; Anderson, C. & Felleman, D. (1992). Information processing in the primate visual system: an integrated systems perspective. *Science, 255,*5043, 419-423.

Van Essen, D.; Anderson, C. & Olshausen, B. (1994). Dynamic Routing Strategies in Sensory, Motor, and Cognitive Processing. In *Large scale neuronal theories of the brain,* pp.271-299. MIT Press.

Weber, C.; Karantzis, K. & Wermter, S. (2005). Grasping with flexible viewing-direction with a learned coordinate transformation network. In *Proceedings of Humanoids,* pp. 253-258.

Weber, C. & Wermter, S. (2007). A self-organizing map of Sigma-Pi units. *Neurocomputing,* 70, 2552-2560.

Weber, C.; Wermter, S. & Elshaw, M. (2006). A hybrid generative and predictive model of the motor cortex. *Neural Networks, 19,* 4, 339-353.

Weber, C.; Wermter, S. & Zochios, A. (2004). Robot docking with neural vision and reinforcement. *Knowledge-Based Systems, 17,*2-4,165-172.

Wermter, S.; Weber, C.; Elshaw, M.; Gallese, V & Pulvermuller, F. (2005). A Mirror Neuron Inspired Hierarchical Network for Action Selection. In *Biomimetic neural learning for intelligent robots,* S. Wermter, G. Palm & M. Elshaw (Eds.), pp. 162-181. Springer.

Wyss, R.; König, P. & Verschure, P. (2006). A model of the ventral visual system based on temporal stability and local memory. *PLoS Biology, 4,* 5, e120.

31

Towards Tutoring an Interactive Robot

Britta Wrede, Katharina J. Rohlfing, Thorsten P. Spexard and Jannik Fritsch
Bielefeld University, Applied Computer Science Group
Germany

1. Introduction

Many classical approaches developed so far for learning in a human-robot interaction setting have focussed on rather low level motor learning by imitation. Some doubts, however, have been casted on whether with this approach higher level functioning will be achieved (Gergeley, 2003). Higher level processes include, for example, the cognitive capability to assign meaning to actions in order to learn from the tutor. Such capabilities involve that an agent not only needs to be able to mimic the motoric movement of the action performed by the tutor. Rather, it understands the constraints, the means, and the goal(s) of an action in the course of its learning process. Further support for this hypothesis comes from parent-infant instructions where it has been observed that parents are very sensitive and adaptive tutors who modify their behaviour to the cognitive needs of their infant (Brand et al., 2002).

Figure 1. Imitation of deictic gestures for referencing on the same object

Based on these insights, we have started our research agenda on analysing and modelling learning in a communicative situation by analysing parent-infant instruction scenarios with automatic methods (Rohlfing et al., 2006). Results confirm the well known observation that parents modify their behaviour when interacting with their infant. We assume that these modifications do not only serve to keep the infant's attention but do indeed help the infant to understand the actual goal of an action including relevant information such as constraints and means by enabling it to structure the action into smaller, meaningful chunks. We were

able to determine first objective measurements from video as well as audio streams that can serve as cues for this information in order to facilitate learning of actions.
Our research goal is to implement such a mechanism on a robot. Our robot platform *Barthoc* (Bielefeld Anthropomorphic RoboT for Human-Oriented Communication) (Hackel et al., 2006) has a human-like appearance and can engage in human-like interactions. It encompasses a basic attention system that allows it to focus the attention on a human interaction partner, thus maintaining the system's attention on the tutor. Subsequently, it can engage in a grounding-based dialog to facilitate human robot interaction.
Based on our findings on learning in parent-infant interaction and Barthoc's functionality as described in this Chapter, our next step will be to integrate algorithms for detecting infant-directed actions that help the system to decide when to learn and when to stop learning (see Fig. 1). Furthermore, we will use prosodic measures and correlate them with the observed hand movements in order to help structuring the demonstrated action. By implementing our model of communication-based action acquisition on the robot-platform Barthoc we will be able to study the effects of tutoring in detail and to enhance our understanding of the interplay between representation and communication.

2. Related Work

The work plan of social robotics for the next future is to create a robot that can observe a task performed by a human (Kuniyoshi et al., 1994) and interpret the observed motor pattern as a meaningful behaviour in such a manner that the meanings or goals of actions can activate a motor program within the robot.
Within the teaching by showing paradigm (Kuniyoshi et al., 1994), the first step according to this work plan has been accomplished by focussing on mapping motor actions. Research has been done on perception and formation of internal representation of the actions that the robot perceives (Kuniyoshi et al., 1994), (Wermter et al., 2005). However, from the ongoing research we know that one of the greatest challenges for robotics is how to design the competence not only of imitating but of action understanding. From a developmental psychology perspective Gergely (2003) has pointed out that the so far pursued notion of learning lacks higher-level processes that include "understanding" of the semantics in terms of goal, means and constraints. What is meant by this critique is the point that robots learning from human partners not only should know *how* to imitate (Breazeal et al., 2002) (Demiris et al., 1996) and *when* to imitate (Fritsch et al., 2005) but should be able to come up with their own way of reproducing the achieved change of state in the environment. This challenge, however, is tightly linked to another challenge, occurring exactly because of the high degree of freedom of how to achieve a goal. This forms the complexity of human actions, and the robot has to cope with action variations, which means that when comparing across subjects, most actions typically appear variable at a level of task instruction. In other words, we believe that the invariants of action, which are the holy grail of action learning, will not be discovered by analyzing the "appearance" of a demonstrated action but only by looking at the higher level of semantics. One modality that is pre-destined for analyzing semantics is speech. We therefore advocate the use of multiple modalities, including speech, in order to derive the semantics of actions.
So far these points have barely been addressed in robotics: Learning of robots usually consists of uni-modal abstract learning scenarios involving generally the use of vision systems to track movements and transform observed movements to ones own morphology

("imitation"). In order for a robot to be able to learn from actions based on the imitation paradigm, it seems to be necessary to reduce the variability to a minimum, for example by providing another robot as a teacher (Weber et al., 2004).
We argue that information from communication, such as the coordination of speech and movements or actions, in learning situations with a human teacher can lighten the burden of semantics by providing an organization of presented actions.

3. Results from Parent-infant tutoring

In previous work (Rohlfing et al., 2006) we have shown that in parent-child interaction there is indeed a wealth of cues that can help to structure action and to assign meaning to different parts of the action. The studies were based on experimental settings where parents were instructed to teach the functions of ten different objects to their infants. We analysed multi-modal cues from the parents' hand movements on the one hand and the associated speech cues on the other hand when one particular object was presented.
We obtained objective measurements from the parents' hand movements - that can also be used by a robot in a human-robot interaction scenario - by applying automatic tracking algorithms based on 2D and 3D models that were able to track the trajectories of the hand movements based on movies from a monocular camera (Schmidt et al., 2006). A number of variables capturing objectively measurable aspects of the more subjectively defined variables as used by (Brand et al., 2002) were computed. Results confirmed that there are statistically significant differences between child-directed and adult-directed actions. First, there are more pauses in child-directed interaction, indicating a stronger structuring behaviour. Secondly, the roundness of the movements in child-directed action is less than in adult-directed interaction. We define roundness as the covered motion path (in meters) divided by the distance between motion on- and offset (in meters). This means that a round motion trajectory is longer and more common in an adult-adult interaction (Fritsch et al., 2005); similarly to the notion of "punctuation" in (Brand et al., 2002), an action performed towards a child, is less round because it consists of more pauses between single movements, where the movements are shorter and more straight resulting in simpler action chunks. Thirdly, the difference between the velocity in child-directed movements and adult-directed movements shows a strong trend towards significance when measured in 2D. However, measurements based on the 3D algorithms did not provide such a trend. This raises the interesting question whether parents are able to plan their movements by taking into account the perspective of their infant who will mainly perceive the movement in a 2D-plane.
In addition to these vision-based features, we analysed different speech variables derived from the videos. In general, we found a similar pattern as in the movement behaviour (see also (Rohlfing et al., 2006)): Parents made more pauses in relation to their speaking time when addressing their infants than when instructing an adult. However, we observed a significantly higher variance in this verbosity feature between subjects in the adult-adult condition, indicating that there is a stronger influence of personal style when addressing an adult. In more detail, we observed that the beginnings and endings of action and speech segments tend to coincide more often in infant directed interaction. In addition, when coinciding with an action end, the speech end is much stronger prosodically marked in infant directed speech than in adult directed speech. This could be an indication that the semantics of the actions in terms of goals and subgoals are much more stressed when

addressing an infant. Finally, we observed more instances of verbally stressed object referents and more temporal synchrony of verbal stress and "gestural stress", i.e. shaking or moving of the shown object. These findings match previous findings by (Zukow-Goldring, 2006).
From these results, we derived 8 different variables that can be used for (1) deciding whether a teaching behaviour is being shown (2) analysing the structure of the action and (3) assigning meaning to specific parts of the action (see Table1).

Variable	Detecting "when" to imitate	Detecting action end / (sub)goal	Detecting naming of object attribute (colour, place)
Motion roundness	+		
Motion velocity (2D)	+		
Motion pauses		+	
Speech pauses		+	
Coincidence of speech and movement end		+	
Prosodic emphasis of speech end coinciding with movement end		+	
Verbal stress			+
Synchrony of verbal stress and "shaking" movement			+

Table 1. Variables and their functions in analysing a human tutor's behaviour

In order for a robot to make use of these variables, it needs to be equipped with basic interaction capabilities so it is able to detect when a human tutor is interacting with it and when it is not addressed. While this may appear to be a trivial pre-condition for learning, the analysis of the social situation is generally not taken into account (or implicitely assumed) in imitation learning robots. Yet, to avoid that the robot will start to analyse any movements in its vicinity, it needs to be equipped with a basic attention system that enables it to focus its attention on an interaction partner or on a common scene, thus establishing so called joint attention. In the next section, we describe how such an attention system is realized on Barthoc.

4. The Robot Platform Barthoc

Our research is based on a robot that has the capabilities to establish a communication situation and can engage in a meaningful interaction.
We have a child-sized and an adult-sized humanoid robot Barthoc as depicted in Fig. 2 and Fig. 3. It is a torso robot that is able to move its upper body like a sitting human. The adult-sized robot corresponds to an adult person with the size of 75 cm from its waist upwards. The torso is mounted on a 65 cm high chair-like socket, which includes the power supply,

two serial connections to a desktop computer, and a motor for rotations around its main axis. One interface is used for controlling head and neck actuators, while the second one is connected to all components below the neck. The torso of the robot consists of a metal frame with a transparent cover to protect the inner elements. Within the torso all necessary electronics for movement are integrated. In total 41 actuators consisting of DC- and servo motors are used to control the robot. To achieve human-like facial expressions ten degrees of freedom are used in its face to control jaw, mouth angles, eyes, eyebrows and eyelids. The eyes are vertically aligned and horizontally controllable autonomously for object fixations. Each eye contains one FireWire colour video camera with a resolution of 640x480 pixels.

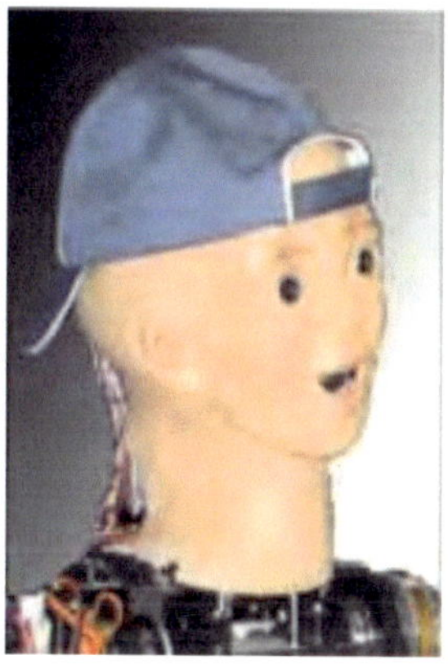

Figure 2. Child-sized Barthoc Junior

Besides facial expressions and eye movements the head can be turned, tilted to its side and slightly shifted forwards and backwards. The set of human-like motion capabilities is completed by two arms, mounted at the sides of the robot. With the help of two five finger hands both deictic gestures and simple grips are realizable. The fingers of each hand have only one bending actuator but are controllable autonomously and made of synthetic material to achieve minimal weight. Besides the neck two shoulder elements are added that can be lifted to simulate shrugging of the shoulders. For speech understanding and the detection of multiple speaker directions two microphones are used, one fixed on each shoulder element. This is a temporary solution. The microphones will be fixed at the ear positions as soon as an improved noise reduction for the head servos is available.

Figure 3. Adult-sized Barthoc

5. System Architecture

For the presented system a three layer architecture (Fritsch et al., 2005) is used consisting of a deliberative, an intermediate, and a reactive layer (see Fig. 4). The top deliberative layer contains the speech processing modules including a dialog system for complex user interaction. In the bottom layer reactive modules capable of adapting to sudden changes in the environment are placed. Since neither the deliberative layer dominates the reactive layer nor the reactive layer dominates the deliberative one, a module called Execution Supervisor (ESV) was developed (Kleinehagenbrock et al., 2004) located in the intermediate layer as well as a knowledge base. The ESV coordinates the different tasks of the individual modules by reconfiguring the parameters of each module. For example, the Actuator Interface for controlling the hardware is configured to receive movement commands from different

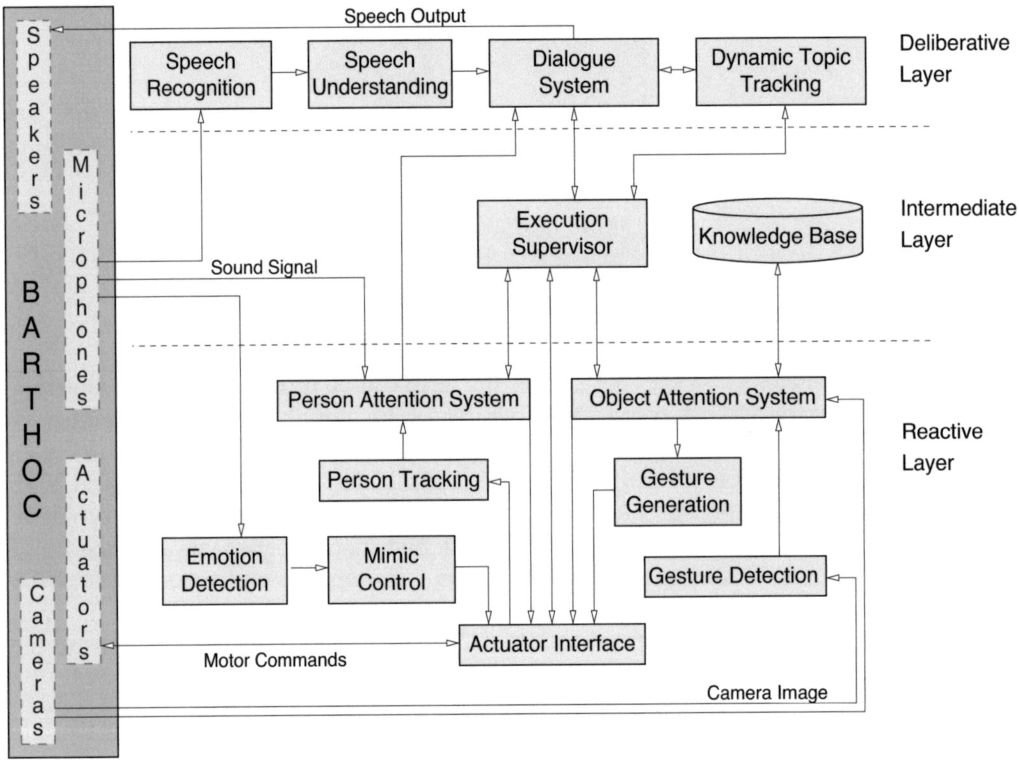

Figure 4. Three layer system architecture of Barthoc, representing the different modules connected by XCF

modules. The ESV can be described as a finite state machine. The different HRI abilities are represented as states and a message sent from a module to the ESV can result in a transition from state A to state B. For each transition the modules in the different layers are reconfigured. Additionally to the new configuration, data like an object label is exchanged between the modules. All data exchange via the ESV is based on the XML Communication Framework (Wrede et al., 2004) using four predefined XML structures, only. All XML data is designed in a human readable style for easy understanding of the internal system communication and efficient debugging.

Using a predefined set of XML structures (see Table 2) data exchange between the ESV and each module is automatically established after reading a configuration file. The file also contains the definition of the finite state machine and the transitions that can be performed. This makes the system easily extendable for new HRI capabilities, by simply changing the configuration file for adding new components without changing one line of source code. Due to the automatic creation of the XML interfaces with a very homogenous structure, fusing the data from the different modules is achieved easily. The system already contains modules for multiple person tracking with attention control (Lang et al., 2003; Fritsch et al., 2004) and an object attention system (Haasch et al., 2005) based on deictic gestures for learning new objects. Additionally an emotion classification based on the intonation of user utterances (Hegel et al., 2006) was added, as well as a Dynamic Topic Tracking (Maas et al., 2006) to follow the content of a dialog. In the next sections we detail how the human-robot interaction is performed by analysing not only system state and visual cues, but spoken language via the dialog system (Li et al., 2006) as well, delivered by the independent operating modules.

```
<MSG xmlns:xs="http://www.w3.org/2001/XMLSchema-instance" xs:type="event">
  <GENERATOR>PTA</GENERATOR>
  <TIMESTAMP>1145869461268</TIMESTAMP>
  <ID>
    <ORIGIN mod="PTA">3</ORIGIN>
  </ID>
  <NAME>CPFound</NAME>
  <STATE>PersonAlertness</STATE>
  <BESTBEFORE>1145869461568</BESTBEFORE>
  <DATA>
    <CPDATA>
      <ID>4</ID>
      <NAME>UNKNOWN</NAME>
  </CPDATA>
  </DATA>
</MSG>
```

```
<MSG xmlns:xs="http://www.w3.org/2001/XMLSchema-instance" xs:type="order">
  <GENERATOR>ESV</GENERATOR>
  <TIMESTAMP>1145870556599</TIMESTAMP>
  <ID>
    <ORIGIN mod="DLG">2</ORIGIN>
  </ID>
  <NAME>FocusCPFace</NAME>
  <STATE>PersonAttention</STATE>
  <DATA>
    <CPDATA>
      <ID>36</ID>
      <NAME>unknown</NAME>
    </CPDATA>
</DATA>
</MSG>
```

Table 2. Examples for data exchange from Person Tracking (PTA) to ESV to inform the system that a person was found (above) and data exchange from Dialog (DLG) via ESV to PTA with the order to focus the face of the current communication partner

6. Finding Someone to Interact with

In the first phase of an interaction, a potential communication partner has to be found and continuously tracked. Additionally, the HRI system has to cope not only with one but also with multiple persons in its surrounding, and thus, discriminating which person is currently attempting to interact with the robot and who is not. The Person Tracking and Attention System is solving both tasks, first finding and continuously tracking multiple persons in the robot's surrounding and secondly deciding to whom the robot will pay attention.

The multiple person tracking is based on the *Anchoring* approach originally introduced by Coradeschi & Saffiotti (2001) and can be described as the connection (Anchoring) between the sensor data (*Percept*) of a real world object and the software representation (*Symbol*) of this object during a fixed time period. To create a robust tracking we extended the tracking from a single to a multi modal approach not tracking a human as one Percept-Symbol relation but as two using a face detector and a voice detector. While the face detector is based on Viola & Jones (2001) the voice detector uses a Cross-Power Spectrum Phase to estimate multiple speaker directions from the signal runtime difference of the two microphones mounted on the robot's shoulders. Each modality (face and voice) is separately anchored and afterwards assigned to a so called *Person Anchor*. A Person Anchor can be initiated by a found face or voice or both if the distance in the real world is below an adjustable threshold. The Person Anchor will be kept and thus a person tracked as long as at least one of its *Component Anchors* (face and voice) is anchored. To avoid anchor losses due to singular misclassifications a Component Anchor will not be lost immediately after a missing Percept for a Symbol. Empirical evaluation showed that a temporal threshold of two seconds increases the robustness of the tracking while maintaining a high flexibility to a changing environment.

As we did not want to restrict the environment to a small interaction area in front of the robot, it is necessary to consider the limited field of view of the video cameras in the eyes of Barthoc. The robot reacts and turns towards people starting to speak outside the current field of view. This possibly results in another person getting out of view due to the robot's movement. To achieve this robot reaction towards real speakers but not towards TV or radio and to avoid loosing track of persons as they get out of view by the robot's movement, we extended the described Anchoring process by a simple but very efficient voice validation and a short term memory (Spexard et al., 2006). For the voice validation we decided to follow the example humans give us. If they encounter an unknown voice out of their field of view humans will possibly have a short look in the corresponding direction evaluating whether the reason for the voice raises their interest or not. If it does, they might change their attention to it, if not they will try to ignore it as long as it persists. Since we have no kind of voice classification any sound will be of the same interest for Barthoc and cause a short turn of its head to the corresponding direction looking for potential communication partners. If the face detection does not find a person there after an adjustable number of trials (lasting on average 2 seconds) although the sound source should be in sight the sound source is marked as not trustworthy. From here on, the robot does not look at it, as long as it persists. Alternatively, a re-evaluation of not trusted sound sources is possible after a given time, but experiments revealed that this is not necessary because the speaker verification is working reliable.

If a person is trusted by the Voice Validation and got out of view due to the robot's movement the short term memory will keep the person's position and return to it later according to the attention system. If someone gets out of sight because he is walking away the system will not return to the position. When a memorized communication partner re-enters the field of view, because the robot shifts its attention to him it is necessary to add another temporal threshold of three seconds since the camera needs approximately one second to adjust focus and gain for an acceptable image quality. The person remains tracked if a face is detected within this time span, otherwise the corresponding person is forgotten and the robot will not look at his direction again. In this case it is assumed that the person has left while the robot did not pay attention.

The decision to which person Barthoc currently pays attention is taken by current user behaviour as observed by the robot. The system is capable of classifying whether someone is standing still or passing by, it can recognize the current gaze of a person by the face detector and the Voice Anchor provides the information whether a person is speaking. Assuming that people look at the communication partner they are talking to the following hierarchy was implemented: Of the lowest interest are people passing by independently of the remaining information. Persons who are looking at the robot are more interesting than persons looking away. Taking into account that the robot might not see all people as they are out of its field of view a detected voice raises the interest. The most interest is paid to a person who is standing in front of, talking towards and facing the robot. It is assumed that this person wants to start an interaction and the robot will not pay attention to another person as long as these three conditions are fulfilled. Given more than one Person on the same interest level the robot's attention will skip from one person to the next one after an adjustable time span, which is currently four to six seconds. The order for the changes is determined by the order in which the people were first recognized by Barthoc.

Figure 5. Scenario: Interacting with Barthoc in a human-like manner

7. Communicating with Barthoc

When being recognized and tracked by the robot, a human interaction partner is able to use a natural spoken dialog system to communicate with the robot (Li et al., 2006). The dialog model is based on the concept of grounding (Clark, 1992) (Traum, 1994), where dialog contributions are interpreted in terms of adjacency pairs. According to this concept, each interaction starts with a *presentation* which is an account introduced by one participant. This presentation needs to be answered by the interlocutor, indicating that he has understood what has been said. This answer is termed *acceptance*, referring to the pragmatic function it plays in the interaction. Note that the concept of presentation and acceptance does not refer to the semantic content of the utterance. The term *acceptance* can also be applied to a negative answer. However, if the interlocutor did not understand the utterance, regardless of the reason (i.e. acoustically or semantically), his answer will be interpreted as a new presentation which needs to be answered by the speaker, before the original question can be answered. Once an acceptance is given, the semantic content of the two utterances are interpreted as *grounded*, that is, the propositional content of the utterances will be interpreted as true for this conversation and as known to both participants. This framework allows to interpret dialog interactions with respect to their pragmatic function.
Furthermore, the implementation of this dialog model allows to integrate verbal as well as non-verbal contributions. This means, given for example a vision-based head nod recognizer, a head nod would be interpreted as an acceptance. Also, the model can generate non-verbal feedback within this framework which means that instead of giving a verbal answer to a command, the execution of the command itself would serve as the acceptance of the presentation of the command.
With respect to the teaching scenario this dialog model allows us to frame the interaction based on the pragmatic function of verbal and non-verbal actions. Thus, it would be possible for the robot to react to the instructor's actions by non-verbal signals. Also, we can interpret the instructor's actions or sub-actions as separate contributions of the dialog to which the robot can react by giving signals of understanding or non-understanding. This way, we can establish an interaction at a fine grained level. This approach will allow us to test our hypotheses about signals that structure actions into meaningful parts such as sub-goals, means or constraints in an interactive situation by giving acceptance at different parts of the instructor's action demonstration.

8. Outlook

Modelling learning on a robot requires that the robot acts in a social situation. We have therefore integrated a complex interaction framework on our robot Barthoc that it is, thus, able to communicate with humans, more specifically, with tutors. This interaction framework enables the robot (1) to focus its attention on a human communication partner and (2) to interpret the communicative behaviour of its communication partner as communicative acts within a pragmatic framework of grounding.
Based on this interaction framework, we can now integrate our proposed learning mechanism that aims at deriving a semantic understanding of the presented actions. In detail, we can now make use of the above mentioned variables derived from the visual (hand tracking) and acoustic (intonation contour and stress detection) channel in order to chunk demonstrated actions into meaningful parts. This segmentation of the action can be

tested during interactions with human instructors. It will also allow us to analyse the effect of different segmentation strategies, which are reflected in the feedback behaviour of the robot, on the behaviour of the tutor.

9. References

Brand, R.J. ; Baldwin, D. A. & Ashburn, L.A. (2002). Evidence for 'motionese': modifications in mothers' infant-directed action. *Developmental Science* 5, 72-83 (2002).

Caradeschi, S. & Saffiotti, A. (2001). Perceptual anchoring of symbols for action. In : *Proc. Of the 17th IJCAI Conference*, pp. 407-412.

Clark, H.H., ed. (1992). *Arenas of Language Use*. University of Chicago Press.

Fritsch, J. ; Kleinehagenbrock, M. ; Lang, S. ; Fink, G.A. & Sagerer, G. (2004). Audiovisual person tracking with a mobile robot. In : *Proc. Int. Conf. on Intelligent Autonomous Systems*. pp. 898-906.

Fritsch, J. ; Hofemann, N. & Rohlfing K. (2005). Detecting 'when to imitate' in a social context with a human caregiver. In *Proc. IEEE ICRA, Workshop on The Social Mechanisms of Robot Programming by Demonstration*. April, Barcelona, Spain.

Fritsch, J. ; Kleinehagenbrock, M. ; Haasch, A. ; Wrede, S. & Sagerer, G. (2005b). A flexible infrastructure for the development of a robot companion with extensible HRI-capabilitites. In : *Proc. IEEE Int. Conf. On Robotics and Automation*, pp. 3419-3425.

Gergely, G. (2003): What should a robot learn from an infant? Mechanisms of action interpretation and observational learning in infancy. In *Proc. 3d International Workshop on Epigenetic Robotics*. August, Boston, USA.

Haasch, A.; Hofemann, N.; Fritsch, J. & Sagerer, G. (2005). A multi-modal object attention system for a mobile robot. In: *Proc. IEEE/RSJ Int. Conf. On Intelligent Robots and Systems*, pp. 1499-1504.

Hackel, M.; Schwope, M.; Fritsch, J.; Wrede, B. & Sagerer, G. (2006). Designing a sociable humanoid robot for interdisciplinary research. In: *Advanced Robotics*, 20(11): pp. 1219-1235.

Hegel, F.; Spexard, T.; Vogt, T.; Horstmann, G. & Wrede, B. (2006). Playing a different imitation game: Interaction with an empathic android robot. In: *Proc. IEEE-RAS Int. Conf. On Humanoid Robots*, pp. 56-61.

Kleinehagenbrock, M.; Fritsch, J. & Sagerer, G. (2004). Supporting Advanced Interaction Capabilities on a Mobile Robot with a Flexible Control System. In: *Proc. IEEE/RSJ In. Conf. On Intelligent Robots and Systems*, Vol. 3, pp. 3649-3655.

Kuniyoshi, Y., Inaba, M. & Inoue, H. (1994): Learning by watching: Extracting reusable task knowledge from visual observation of human performance. In: *IEEE Transactions on Robotics and Automation*, 10(6), 165-170.

Lang, S. ; Kleinehagenbrock, M. ; Hohenner, S. ; Fritsch, J. ; Fink, G.A. & Sagerer, G. (2003). Providing the basis for human-robot interaction : a multi-modal attention system for a mobile robot. In : Proc. Int. Conf. on Multimodal Interfaces, pp. 28-35.

Li, S. ; Wrede, B. & Sagerer, G. (2006). A computational model of multi-modal grounding. In : *Proc. ACL SIGdial workshop on discourse and dialog, in conjunction with COLING/ACL*, pp. 153-160.

Maas, J.F. ; Spexard, T. ; Fritsch, J. ; Wrede, B. & Sagerer, G. (2006). BIRON, what's the topic ? – A multi-modal topic tracker for improved human-robot interaction. In : *Proc. IEEE Int. Workshop on Robot and Human Interactive Communication*, pp : 26-32.

Rohlfing, K. ; Fritsch, J. ; Wrede, B. & Junmann, T. (2006). How can multimodal cues from child-directed interaction reduce learning complexity in robots, In : *Advanced Robotics*, 20(10) : 1183-1199.

J. Schmidt, J.; Kwolek, B. & Fritsch, J. (2006). Kernel Particle Filter for Real-Time 3D Body Tracking in Monocular Color Images. In: Proc. of Automatic Face and Gesture Recognition, pp. 567-572.

Spexard, T. ; Haasch, A. ; Fritsch, J. & Sagerer, G. (2006). Human-like person tracking with an anthropomorphic robot. In : *Proc. IEEE Int. Conf. on Robotics and Automation*, pp. 1286-1292.

Traum, D. ; Rickel, J. (1994). *A Computational Theory of Grounding in Natural Language Conversation*. Ph.D. Dissertation, University of Rochester.

Viola, P. & Jones, M. (2001). Rapid object detection using a boosted cascade of simple features. In *: Proc. Int. Conf. on Computer Vision and Pattern Recognition*, pp. 511-518.

Weber, C., Elshaw, M., Zochios, M. & Wermter, S. (2004). A multimodal hierarchial approach to robot learning by imitation. In L. Berthouze, H. Kozima, C. G. Prince, G. Sandini, G. Stojanov, G. Metta, and C. Balkenius (Eds.): *Proc. 4th International Workshop on Epigenetic Robotics: Modeling Cognitive Development in Robotic Systems*, pp. 131-134.

Wermter, S., Weber, C., Elshaw, M., Gallese, V. & Pulvermüller, F. (2005). Grounding neural robot language in action. In: Wermter, S., Palm, G. & Elshaw, M. (Eds.): *Biomimetic Neural Learning for Intelligent Robots*, pp. 162-181. Heidelberg: Springer.

Wrede, S. ; Fritsch, J. ; Bauckhage, C. & Sagerer, G. (2004). An XML based framework for cognitive vision architectures. In : *Proc. Int. Conf. on Pattern Recognition*, Vol. 1, pp. 757-760.

Zukow-Goldring, P. (2006). Assisted imitation: Affordances, effectivities, and the mirror system in early language development. In *From action to language*, M. A. Arbib (Ed.). Cambridge: CUP.

32

Intuitive Multimodal Interaction with Communication Robot Fritz

Maren Bennewitz, Felix Faber, Dominik Joho and Sven Behnke
University of Freiburg, Computer Science Institute
Germany

1. Introduction

One of the most important motivations for many humanoid robot projects is that robots with a human-like body and human-like senses could in principle be capable of intuitive multimodal communication with people. The general idea is that by mimicking the way humans interact with each other, it will be possible to transfer the efficient and robust communication strategies that humans use in their interactions to the man-machine interface. This includes the use of multiple modalities, such as speech, facial expressions, gestures, body language, etc. If successful, this approach yields a user interface that leverages the evolution of human communication and that is intuitive to naïve users, as they have practiced it since early childhood.

We work towards intuitive multimodal communication in the domain of a museum guide robot. This application requires interacting with multiple unknown persons. The testing of communication robots in science museums and on science fairs is popular, because the robots encounter there many new interaction partners, which have a general interest in science and technology. Here, we present the humanoid communication robot Fritz that we developed as successor to the communication robot Alpha (Bennewitz et al., 2005). Fritz uses speech, facial expressions, eye-gaze, and gestures to interact with people. Depending on the audio-visual input, our robot shifts its attention between different persons in order to involve them into an interaction. He performs human-like arm gestures during the conversation and also uses pointing gestures generated with eyes, its head, and arms to direct the attention of its communication partners towards the explained exhibits. To express its emotional state, the robot generates facial expressions and adapts the speech synthesis.

The remainder of the chapter is organized as follows. The next section reviews some of the related work. The mechanical and electrical design of Fritz is covered in Sec. 3. Sec. 4 details the perception of the human communication partners. Sec. 5 explains the robot's attentional system. The generation of arm gestures and of facial expressions is presented in Sec. 6 and 7, respectively. Finally, in the experimental section, we discuss experiences made during public demonstrations of our robot.

2. Related Work

Many research groups world-wide work on intuitive multimodal communication between humanoid robots and humans. Some example projects are the Leonardo robot at MIT

(Breazeal et al., 2004), Repliee Q2 at Osaka University (Matsui et al., 2005), and BARTHOC at Bielefeld University (Spexard et al., 2006).

Several systems exist that use different types of perception to sense and track people during an interaction and that use a strategy to decide which person gets the attention of the robot. Lang et al. apply an attention system in which only the person that is currently speaking is the person of interest (Lang et al., 2003). While the robot is focusing on this person, it does not look to another person to involve it into the conversation. Only if the speaking person stops talking for more than two seconds, the robot will show attention to another person.

Okuno et al. also follow the strategy to focus the attention on the person who is speaking (Okuno et al., 2002). They apply two different modes. In the first mode, the robot always turns to a new speaker, and in the second mode, the robot keeps its attention exclusively on one conversational partner. The system developed by Matsusaka et al. is able to determine the one who is being addressed to in the conversation (Matsusaka et al., 2001). Compared to our application scenario (museum guide), in which the robot is assumed to be the main speaker or actively involved in a conversation, in their scenario the robot acts as an observer. It looks at the person who is speaking and decides when to contribute to a conversation between two people.

The model developed by Thorisson focuses on turn-taking in one-to-one conversations (Thorisson, 2002). This model has been applied to a virtual character. Since we focus on how to decide which person in the surroundings of the robot gets its focus of attention, a combination of both techniques is possible.

In the following, we summarize the approaches to human-like interaction behavior of previous museum tour-guide projects. Bischoff and Graefe presented a robotic system with a humanoid torso that is able to interact with people using its arms (Bischoff & Graefe, 2004). This robot also acted as a museum tour-guide. However, the robot does not distinguish between different persons and does not have an animated face. Several (non-humanoid) museum tour-guide robots that make use of facial expressions to show emotions have already been developed. Schulte et al. used four basic moods for a museum tour-guide robot to show the robot's emotional state during traveling (Schulte, et al., 1999). They defined a simple finite state machine to switch between the different moods, depending on how long people were blocking the robot's way. Their aim was to enhance the robot's believability during navigation in order to achieve the intended goals. Similarly, Nourbakhsh et al. designed a fuzzy state machine with five moods for a robotic tour-guide (Nourbakhsh et al., 1999). Transitions in this state machine occur depending on external events, such as people standing in the robot's way. Their intention was to achieve a better interaction between the users and the robot. Mayor et al. used a face with two eyes, eyelids and eyebrows (but no mouth) to express the robot's mood using seven basic expressions (Mayor et al., 2002). The robot's internal state is affected by several events during a tour (e.g., a blocked path or no interest in the robot).

Most of the existing approaches do not allow continuous changes of facial expression. Our approach, in contrast, uses a bilinear interpolation technique in a two-dimensional state space (Ruttkay et al., 2003) to smoothly change the robot's facial expression.

3. Mechanical and Electrical Design of the Robot Fritz

Our humanoid robot Fritz has originally been designed for playing soccer in the RoboCup Humanoid League TeenSize class (Behnke et al., 2006). The left part of Fig. 1 shows him as goalie in the TeenSize Penalty Kick final at RoboCup 2006. Fritz is 120cm tall and has a total weight of about 6.5kg. His body has 16 degrees of freedom (DOF): Each leg is driven by five large digitally controlled Tonegawa PS-050 servos and each arm is driven by three digital Futaba S9152 servos.

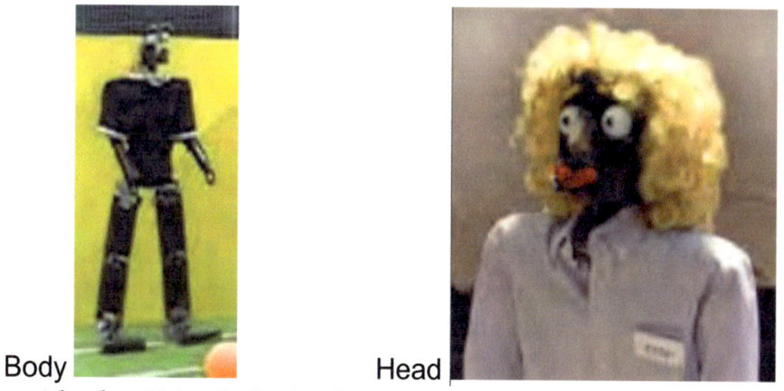

Figure 1. Humanoid robot Fritz. Left: Goalie at RoboCup 2006. Right: Communication head

For the use as communication robot, we equipped Fritz with a 16DOF head, shown in the right part of Fig. 1. The head is mounted on a 3DOF neck. The eyes are USB cameras that can be moved together in pitch and independently in yaw direction. Six servo motors animate the mouth and four servos animate the eyebrows.
The servo motors are controlled by a total of four ChipS12 microcontroller boards, which are connected via RS-232 to a main computer. We use a standard PC as main computer. It runs computer vision, speech recognition/synthesis, and behavior control.

4. Perception of Communication Partners

To detect and track people in the environment of our robot, we use the two cameras and a stereo microphone. In order to keep track of persons even when they are temporarily outside the robot's field of view, the robot maintains a probabilistic belief about the people in its surroundings.

4.1 Visual Detection and Tracking of People

Our face detection system is based on the AdaBoost algorithm and uses a boosted cascade of Haar-like features (Viola & Jones, 2004). Whenever a new observation is made it must be determined to which person, that has already been detected by the robot, the newly detected face belongs. To solve this data association problem, we apply the Hungarian Method using a distance-based cost function. We use a Kalman filter to track the position of a face over time. Fig. 2 shows three snapshots during face tracking. As indicated by the differently colored boxes, all faces are tracked correctly.

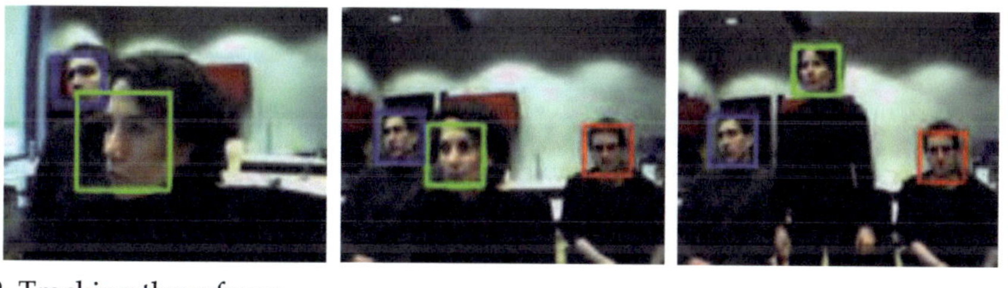

Figure 2. Tracking three faces

To account for false classifications of face/non-face regions and association failures, we apply a probabilistic technique. We use a recursive Bayesian update scheme (Moravec & Elfes, 1985) to compute the existence probability of a face. In this way, the robot can also estimate whether a person outside the current field of view is still there. When a person's probability falls below a threshold, the robot makes a reconfirming gaze to this person.

4.2 Speaker Localization

In addition to the visual perception of the persons around the robot, we implemented a speaker localization system that uses a stereo microphone. We apply the Cross-Power Spectrum Phase Analysis (Giuliani et al., 1994) to calculate the spectral correlation measure between the left and the right microphone channel. Using the corresponding delay, the relative angle between the speaker and the microphones can be calculated.
The person in the robot's belief that has the minimum distance to the sound source angle gets assigned the information that it has spoken. If the angular distance between the speaker and all persons is greater than a certain threshold, we assume the speaker to be a new person, who just entered the scene.

5. Attentional System

It is not human-like to fixate a single conversational partner all the time when there are other people around. Fritz shows interest in different persons in his vicinity and shifts his attention between them so that they feel involved into the conversation. We currently use three different concepts in order to change the robot's gaze direction.

5.1 Focus of Attention

In order to determine the focus of attention of the robot, we compute an importance value for each person in the belief. It currently depends on the time when the person has last spoken, on the distance of the person to the robot (estimated using the size of the bounding box of its face), and on its position relative to the front of the robot. The resulting importance value is a weighted sum of these three factors. In the future, we plan to consider further aspects to determine the importance of persons like, for example, waving with hands.
The robot focuses its attention always on the person who has the highest importance, which means that it keeps eye-contact with this person. Of course, the focus of attention can change during a conversation with several persons.

5.2 Attentiveness to a Speaker

If a person that is outside the current field of view, which has not been detected so far, starts to speak, the robot reacts to this by turning towards the corresponding direction. In this way, the robot shows attentiveness and also updates its belief about the people in its surrounding.

5.3 Gazes outside the Focus of Attention

Since the field of view of the robot is constrained, it is important that the robot changes its gaze direction to explore the environment and to update its belief about it. Our robot regularly changes its gaze direction and looks in the direction of other faces, not only to the most important one. This reconfirms that the people outside the field of view are still there and involves them into the conversation.

5.4 Example

Fig. 3 illustrates an experiment that was designed to show how the robot shifts its attention from one person to another if it considers the second one to be more important. In the situation depicted here, person 2 was talking to the robot. Since person 2 had the highest importance, the robot initially focused its attention on person 2 but also looked to person 1 at time steps 10 and 21, to signal awareness and to involve him/her into the conversation. When looking to person 1 at time step 21, the robot then noticed that this person had come very close. This yielded a higher importance value for this person and the robot shifted its attention accordingly.

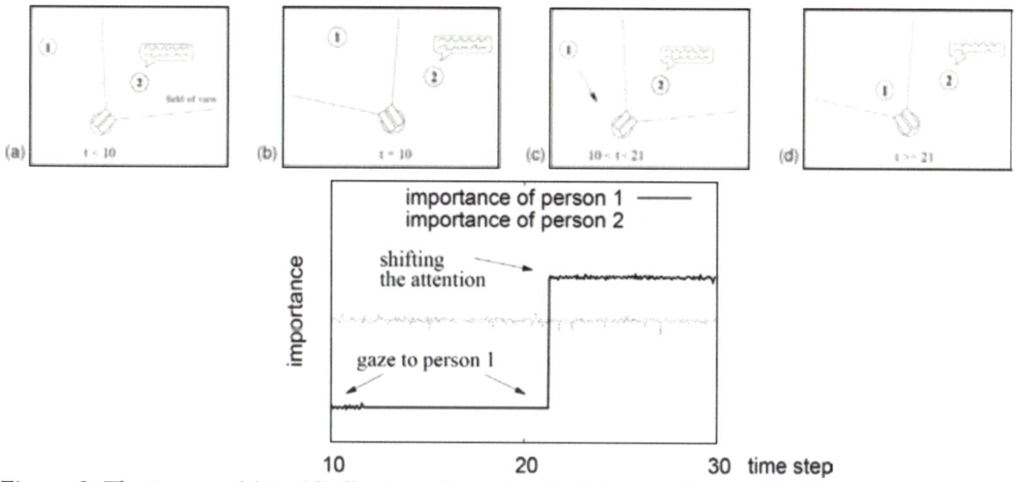

Figure 3. The images (a) to (d) illustrate the setup in this experiment. The lower image shows the evolution of the importance values of two people. See text for a detailed explanation

6. Arm and Head Gestures

Our robot uses arm and head movements to generate gestures, and to appear livelier. The gestures are generated online. Arm gestures consist of a preparation phase, where the arm moves slowly to a starting position, the stroke phase that carries the linguistic meaning, and a retraction phase, where the hand moves back to a resting position (MacNeill, 1992). The stroke is synchronized to the speech synthesis module.

6.1 Symbolic Gestures

Greeting  Inquiring 

Figure 4. Fritz performing two symbolic gestures with its arms

Symbolic gestures are gestures in which the relation between form and content is based on social convention. They are culture-specific.

- **Greeting Gesture:** The robot performs a single-handed gesture while saying hello to newly detected people. As shown in the left part of Fig. 4 it raises its hand, stops, and lowers it again.
- **Come Closer Gesture:** When the robot has detected persons farther away than the normal conversation distance (1.5-2.5m), he animates the people to come closer. Fig. 5 shows that the robot moves both hands towards the people in the preparation phase and towards its chest during the stroke.
- **Inquiring Gesture:** While asking certain questions, the robot performs an accompanying gesture, shown in the right part of Fig. 4. It moves both elbows outwards to the back.
- **Disappointment Gesture:** When the robot is disappointed (i.e., because it did not get an answer to a question), it carries out a gesture to emphasize its emotional state. During the stroke it moves both hands quickly down.
- **Head Gestures:** To confirm or disagree, the robot nods or shakes its head, respectively.

Figure 5. Fritz asks a person to come closer

6.2 Batonic Gestures

Humans continuously gesticulate to emphasize their utterances while talking to each other. Fritz also makes small emphasizing gestures with both arms when he is speaking longer sentences.

6.3 Pointing Gestures

To draw the attention of communication partners towards objects of interest, our robot performs pointing gestures. While designing the pointing gesture for our robot, we followed the observation made by Nickel et al. that people usually move the arm in such a way that, in the poststroke hold, the hand is in one line with the head and the object of interest (Nickel et al., 2004). This is illustrated in Fig. 6.

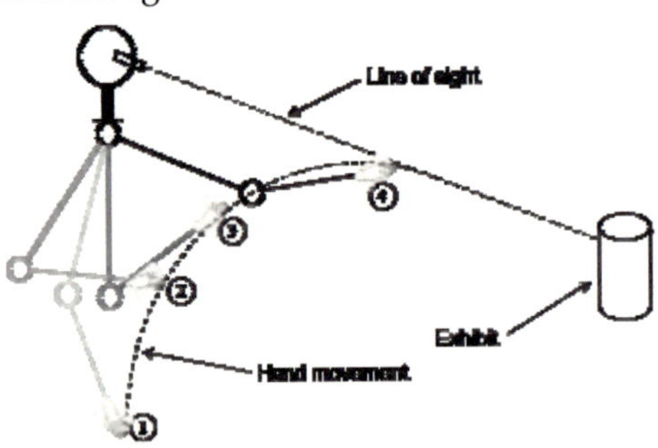

Figure 6. Side view of the arm movement during a pointing gesture

When the robot wants to draw the attention to an object, it simultaneously moves the head and the eyes in the corresponding direction and points in the direction with the respective arm while uttering the object name.

6.4 Non-Gestural Arm Movements

While standing, people typically move unconsciously with their arms and do not keep completely still. Our robot also performs such minuscule movements with its arms. The arms move slowly, with low amplitude in randomized oscillations.

7. Expression of Emotions

Showing emotions plays an important role in inter-human communication. During an interaction, the perception of the mood of the conversational partner helps to interpret his/her behavior and to infer intention. To communicate the robot's mood we use a face with animated mouth and eyebrows that displays facial expressions and also synthesize speech according to the current mood. The robot's mood is computed in a two-dimensional space, using six basic emotional expressions (joy, surprise, fear, sadness, anger, and disgust). Here, we follow the notion of the Emotion Disc (Ruttkay et al., 2003).

7.1 Facial Expressions

Fig. 7 shows the six basic facial expressions of our robot. As parameters for an expression we use the height of the mouth corners, the mouth width, the mouth opening angle, and the angle and height of the eye-brows.

To influence the emotional state of our robot, we use behaviors that react to certain events. For example, if no one is interested in the robot, it is getting more and more sad, if someone then talks to it, the robot's mood changes to a mixture of surprise and happiness. Each behavior submits its request in which direction and with which intensity it wants to change the robot's emotional state. After all behaviors submitted their requests, the resulting vector is computed by the sum of the individual requests. We allow any movement within the circle described by the Emotion Disc.

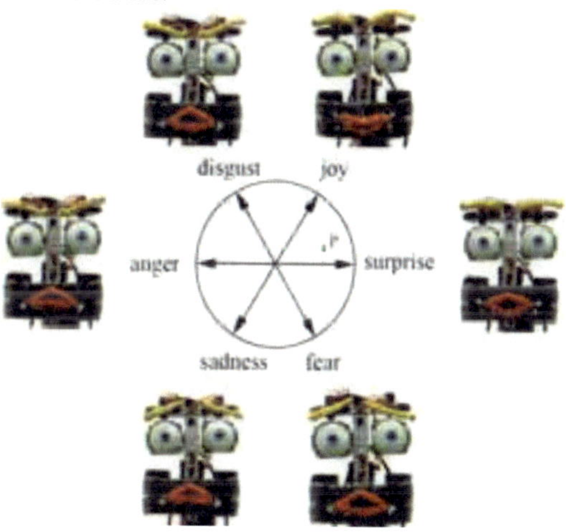

Figure 7. The two-dimensional space in which we compute the robot's emotional state. The images show the six basic facial expressions

The parameters P' for the facial expression corresponding to a certain point P in the two-dimensional space are calculated by linear interpolation between the parameters E'_i and E'_{i+1}, the adjacent basic expressions:

$$P' = l(p) \cdot \left(\alpha(p) \cdot E'_i + (1 - \alpha(p)) \cdot E'_{i+1} \right). \tag{1}$$

Here, $l(p)$ is the length of the vector p that leads from the origin (corresponding to the neutral expression) to P, and $\alpha(p)$ denotes the normalized angular distance between p and the vectors corresponding to the two neighboring basic expressions. This technique allows continuous changes of the facial expression.

7.2 Emotional Speech Synthesis

In combination with facial expressions, we use emotional speech to express the robot's mood. Most speech synthesis systems do not support emotional speech directly; neither does the Loquendo TTS system that we use. However, in this system, we can influence the parameters average pitch, speed, and volume and thereby express emotional speech.
Cahn proposed a mapping of emotional states to the relative change of several parameters of a speech synthesis system (Cahn, 1989). She carried out experiments to show that test persons were able to recognize the emotion category of several synthesized sample sentences. In the mapping, she used the same six basic emotions that constitute the axes of the Emotion Disc. We use her mapping for the parameters average pitch, speech rate and loudness to set the parameters average pitch, speed and volume of our speech synthesizer.
The mapping of emotional states to the relative change of the speech parameters can be seen in Tab. 1. Let **M** be such a mapping matrix, and **e** be an emotion intensity vector of the six basic emotions. Then we can compute the three speech parameters as a vector **s**, as follows:

$$\mathbf{s} = \mathbf{d} + \mathbf{S}\,\mathbf{M}\,\mathbf{e}. \tag{2}$$

The three-element vector **d** contains the default values for the parameters and **S** is a diagonal matrix used to scale the result of the mapping, thereby allowing for an adaptation of the mapping to the characteristics of the synthesizer system. The emotion intensity vector contains only two non-zero entries, $l(p)\cdot\alpha(p)$ and $l(p)\cdot(1 - \alpha(p))$, that correspond to the influence factors of the two adjacent basic expressions of the current mood (see Fig. 7 and Eq. 1).

	joy	surprise	fear	sadness	anger	disgust
pitch	-3	0	10	0	-5	0
speed	2	4	10	-10	8	-3
volume	0	5	10	-5	10	0

Table 1. Mapping of emotions to the relative change of the speech parameters (**M** in Eq. 2)

Emotions influence many more characteristics of speech, e.g. breathiness, precision of articulation, and hesitation pauses. Hence, the three parameters used in our system can only roughly approximate emotional speech. In spite of these limitations, we experienced that even such simple adjustments can, in conjunction with facial expressions, contribute to the emotional expressiveness.

8. Public Demonstrations

To evaluate our system, we tested our communication robots Alpha and Fritz in two public demonstrations. In this section, we report the experiences we made during these exhibitions. We chose a scenario in which the robot presents four of its robotic friends. We placed the exhibits on a table in front of the robot. Our communication robot interacted multimodally with the people and had simple conversations with them. For speech recognition and speech synthesis, we used the Loquendo software. Our dialog system is realized as a finite state machine. Fig. 8 illustrates a simple version. With each state, a different grammar of phrases is associated that the recognition system should be able to recognize. The dialog system generates some small talk and allows the user to select which exhibits should be explained and to what level of detail.

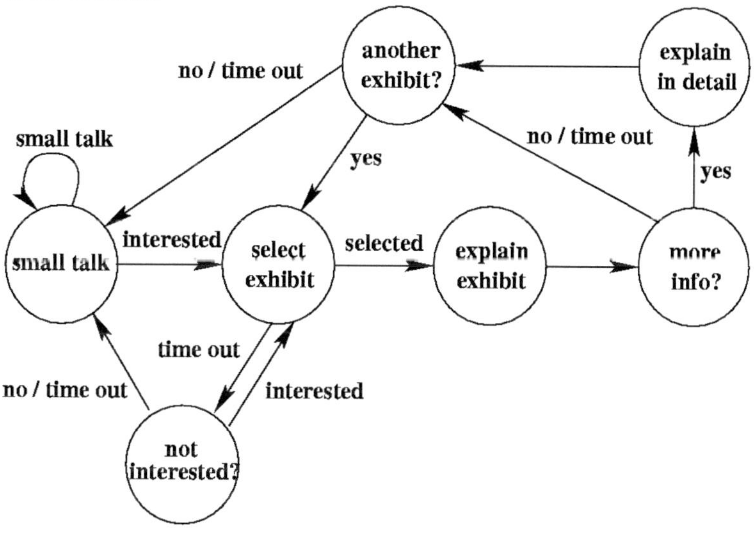

Figure. 8. Finite state machine controlling the flow of the dialog

8.1 Two-Day Demonstration at the Science Fair 2005 in Freiburg

The first demonstration was made using the robot Alpha, the predecessor of Fritz. We exhibited Alpha during a two-day science fair of Freiburg University in June 2005. In contrast to Fritz, Alpha did not use emotional speech and performed pointing gestures with his arms but not any other human-like gestures.

At the science fair, we asked the people who interacted with the robot to fill out questionnaires about their interaction-experiences with Alpha (see (Bennewitz et al., 2005) for more details). Almost all people found the eye-gazes, gestures, and the facial expression human-like and felt that Alpha was aware of them. The people were mostly attracted and impressed by the vivid human-like eye movements. To evaluate the expressiveness of the pointing gestures, we carried out an experiment in which the people had to guess the target of the pointing gestures. The result was that 91% of the gestures were correctly interpreted. However, one limitation that was obvious is that speech recognition does not work sufficiently well in noisy environments, even when using close-talking microphones. To account for this problem, in our current system, the robot asks for an affirmation when the speech recognition system is not sure about the recognized phrase.

8.2 Three-Day Demonstration at the Science Days 2006 in the Europapark Rust

Figure 9. Fritz presenting its robot friends to visitors at the Science Days

In October 2006, we exhibited Fritz for three days at the Science Days in the Europapark Rust (see Fig. 9). Since the people at the previous exhibition were most attracted by the human-like behavior, we augmented the number of arm gestures as explained in Sec. 6. In general, the gestures served their purpose. However, the *come closer* gesture did not always have the desired result. In the beginning of the interaction, some people were still too shy and barely wanted to come closer to the robot. This effect is not uncommon even for human museum guides starting a tour. As soon as the visitors became more familiar with the robot, their shyness vanished and they choose a suitable interaction distance by themselves.

In contrast to the exhibition of Alpha, where toddlers often were afraid of the robot and hid behind their parents, we did not observe such a behavior with Fritz. This is probably due to the different sizes and appearances of the robots. The kids found Fritz apparently very exciting. Most of them interacted several times with the robot. At the end, some of them knew exactly what the robot was able to do and had fun in communicating with Fritz.

When there were people around Fritz but nobody started to talk to the robot, Fritz proactively explained to the people what he is able to do. While speaking, he performed gestures with his head and arms so that after the explanation the people had a good idea about the capabilities of the robot. This idea resulted from lessons learned of the first exhibition where people often did not know about what the robot is actually able to do and what not.

Due to the severe acoustical conditions, speech recognition did not always work well. The affirmation request helped only if the correct phrase was the most likely one. Hence, for the next exhibition, we plan to employ an auditory front-end that focuses on the fundamental frequency of the speaker, in order to separate it from background noise.

A video of the demonstration can be downloaded from http://www.NimbRo.net.

9. Conclusions

In this chapter, we presented our humanoid communication robot Fritz. Fritz communicates in an intuitive, multimodal way. He employs speech, an animated face, eye-gaze, and gestures to interact with people. Depending on the audio-visual input, our robot shifts its attention between different communication partners in order to involve them into an interaction. Fritz performs human-like arm and head gestures, which are synchronized to the speech synthesis.

He generates pointing gestures with its head, eyes, and arms to direct the attention of its communication partners towards objects of interest. Fritz changes its emotional state according to the number of people around him and the dialog state. Its emotional state is communicated by facial expressions and emotional speech synthesis. We tested the described multimodal dialog system during two public demonstrations outside our lab. The experiences made indicate that the users enjoyed interacting with the robot. They treated him as an able communication partner, which was sometimes difficult, as its capabilities are limited.

The experienced problems were mainly due to perception deficits of the robot. While speech synthesis works fairly well, robust speech recognition in noisy environments is difficult. This is problematic, because the users expect the robot to understand speech at least as well as it talks. Similarly, while the robot is able to generate gestures and emotional facial expressions, its visual perception of the persons around it is limited to head position and size. To reduce this asymmetry between action generation and perception, we currently work on head posture estimation from the camera images and on the visual recognition of gestures.

10. Acknowledgment

Funding for this project is provided by the DFG (Deutsche Forschungsgemeinschaft), grant BE 2556/2-1,2.

11. References

Behnke, S.; Schreiber, M.; Stückler, J.; Strasdat, H. & Bennewitz M. (2006). NimbRo TeenSize 2006 team description, In: *RoboCup 2006 Humanoid League Team Descriptions*, Bremen, Germany

Bennewitz, M.; Faber, F.; Joho, D.; Schreiber, M. & Behnke, S. (2005). Towards a humanoid museum guide robot that interacts with multiple persons, *Proceedings of the IEEE/RSJ International Conference on Humanoid Robots (Humanoids)*, pp. 418-423, Tsukuba, Japan

Bischoff, R. & Graefe, V. (2004). Hermes - a versatile personal robot assistant. *Proc. IEEE - Special Issue on Human Interactive Robots for Psychological Enrichment*, 92(11):1759-1779

Breazeal, C; Brooks, A.; Gray, J.; Hoffman, G.; Kidd, C.; Lee, H.; Lieberman, J.; Lockerd, A. & Mulanda, D. (2004). Humanoid robots as cooperative partners for people. *International Journal of Humanoid Robots*,1(2):1-34

Cahn, J. (1989). *Generating expression in synthesized speech*, Master thesis, Massachusetts Institute of Technology

Giuliani, D.; Omologo, M. & Svaizer P. (1994). Talker localization and speech recognition using a microphone array and a cross-powerspectrum phase analysis, *Proceedings of International Conference on Spoken Language Processing (ICSLP)*, pp. 1243-1246, Yokohama, Japan

Lang, S.; Kleinehagenbrock, M.; Hohenner, S.; Fritsch, J.; Fink, G. & Sagerer, G. (2003). Providing the basis for human-robot-interaction: A multi-modal attention system for a mobile robot, *Proceedings of IEEE International Conference on Multimodal Interfaces*, pp. 28-35, Vancouver, Canada

MacNeill, D. (1992). *Hand and Mind: What Gestures Reveal about Thought*, University of Chicago Press

Matsui, D.; Minato, T.; MacDorman, K. F. & Ishiguro, H. (2005). Generating natural motion in an android by mapping human motion, *Proceedings of IEEE/RSJ International Conference on Intelligent Robots and Systems (IROS)*, pp. 1089-1096, Edmonton, Canada

Matsusaka, Y.; Fujie, S. & Kobayashi, T. (2001). Modeling of conversational strategy for the robot participating in the group conversation, *Proc. of the European Conference on Speech Communication and Technology (Eurospeech)*, pp. 2173-2176, Aalborg, Denmark

Mayor, L.; Jensen, B.; Lorotte, A. & Siegwart, R. (2002). Improving the expressiveness of mobile robots, *Proceedings of IEEE International Workshop on Robot and Human Interactive Communication (ROMAN)*, pp. 325-330, Berlin, Germany

Moravec, H. & Elfes, A. (1985). High resolution maps from wide angle sonar, *Proceedings of IEEE International Conference on Robotics & Automation (ICRA)*, pp. 116-121, St. Louis

Nickel, K.; Seemann, E. & Stiefelhagen, R. (2004). 3D-Tracking of heads and hands for pointing gesture recognition in a human-robot interaction scenario, *Proceedings of International Conference on Face and Gesture Recognition (FGR)*, pp. 565-570, Seoul, Korea

Nourbakhsh, I.; Bobenage, J.; Grange, S.; Lutz, R.; Meyer, R. & Soto, A. (1999). An affective mobile robot educator with a full-time job. *Artificial Intelligence,* 114(1-2):95-124

Okuno, H.; Nakadai, K. & Kitano, H. (2002). Social interaction of humanoid robot based on audio-visual tracking, *Proceedings of IEEE International Conference on Industrial and Engineering Applications of Artificial Intelligence and Expert Systems (IEA/AIE)*, pp. 725-735, Cairns, Australia

Ruttkay, Z.; Noot, H. & Hagen, P. (2003). Emotion Disc and Emotion Squares: Tools to explore the facial expression space. *Computer Graphics Forum,* 22(1):49-53

Schulte, J.; Rosenberg, C. & Thrun, S. (1999). Spontaneous short-term interaction with mobile robots in public places, *Proceedings of IEEE International Conference on Robotics and Automation (ICRA)*, pp. 658-663, Detroit, Michigan

Spexard, T.; Haasch, A.; Fritsch, J. & Sagerer G. (2006). Human-like person tracking with an anthropomorphic robot, *Proceedings of IEEE International Conference on Robotics and Automation (ICRA)*, pp. 1286-1292, Orlando, Florida

Thorisson, K. R. (2002). Natural turn-taking needs no manual: Computational theory and model, from perception to action, In: *Multimodality in Language and Speech Systems*, pp. 173-207, Kluwer Academic Publishers

Viola, P. & Jones, M.J. (2004). Robust Real-Time Face Detection. *International Journal of Computer Vision,* 57(2):137-154

33

Hierarchical Reactive Control for Soccer Playing Humanoid Robots

Sven Behnke, Jörg Stückler, Hauke Strasdat and Michael Schreiber
University of Freiburg, Computer Science Institute
Germany

1. Introduction

What drives thousands of researchers worldwide to devote their creativity and energy to make robots kick a ball into a goal? The answer lies not only in the fascination of the soccer game, but rather in the quest to advance the fields of artificial intelligence research and robotics. AI researchers started to investigate games early-on. Already in the 1950th, Simon predicted that computers would be able to win against the human chess world champion within ten years (Simon & Newell , 1958). Playing chess was viewed as the epitome of intelligence. The dominant view at that time was that human intelligence could be simulated by manipulating symbols. While the chess world champion was defeated by a machine in 1997 (Newborn, 1997), human intelligence is still far from being understood.
The basis for intelligent action is the perception of the world. Already this seemingly easy task frequently exceeds the capabilities of current computer systems. Perceptual processes, which interpret the flood of stimuli streaming into our senses and make it accessible for behavior control, are mostly unconscious. Hence, we are not aware of the difficulties involved. The performance of our perceptual system becomes clear only when trying to solve the same task with machines. This applies to behavior control as well. Human locomotion, for example, does not seem to be problematic. That walking and running on two legs is not an easy task becomes clear only when one tries to implement it on a real robot.
Based on these observations, a view on intelligence has established itself over the last two decades that does not rely on manipulating symbols, but emphasizes the interaction of an agent with its environment (Brooks 1990; Pfeifer & Scheier 1999). The term embodiment stresses the importance of having a body as the physical basis for intelligence. Situatedness of an agent in a rich environment enables feedback from its actions to its sensory signals. The complexity of the interaction is increased significantly when the environment does not only contain passive objects, but other agents as well.

1.1 RoboCup Competitions

Motivated by the success in the chess domain, the RoboCup Federation organizes since 1997 international robotic soccer competitions. Similar competitions are organized by FIRA. The long-term goal of RoboCup is to develop by the year 2050 a team of humanoid soccer robots that wins against the FIFA world champion (Kitano & Asada, 2000). The soccer game was selected for the competitions, because, as opposed to chess, multiple players of one team must cooperate in a dynamic environment. The players must interpret sensory signals in real-time, select appropriate actions, and execute them. The soccer competitions do not test isolated components, but during a match two systems compete with each other. The number of goals scored is an objective performance measure for comparing systems that implement a large variety of approaches to robot construction, perception, and behavior control. The presence of opponent teams, which continuously improve their system, raises the bar every year. Such a challenge problem focuses the effort of many research groups worldwide and facilitates the exchange of ideas.

The RoboCup championships grew to the most important robotic competition worldwide. In the last RoboCup, which took place in June 2006 in Bremen, Germany, 440 teams from 36 countries competed, not only in RoboCupSoccer, but also in RoboCupRescue, RoboCupJunior, and RoboCup@home. The total number of participants was more than 2.600.

1.2 Humanoid League

Figure 1. Some of the RoboCup 2006 Humanoid League participants

The RoboCupSoccer competitions are held in five leagues for simulated, wheeled, four-legged, and biped robots. The Humanoid League was established in 2002. Here, robots with a human-like body plan compete with each other. Fig. 1 shows some participants of the 2006 competition. The players must have two legs, two arms, a head, and a trunk. Size restrictions make sure that the center of mass of the robots is not too low, that their feet are not too large, and so on. The participants are grouped in two size classes: KidSize (<60cm) and TeenSize (>80cm). The humanoid robots must be able to walk on two legs, and they must be fully autonomous. They may communicate with each other via a wireless network. Help from outside the field is not permitted, neither by humans nor by computers.

Because the construction and the control of humanoid robots is significantly more complex than that of wheeled robots, initially, there were only preliminary competitions held, but no soccer games played, in the Humanoid League. The robots had to footrace around a pole and faced each other in penalty kicks. Since 2005, 2 vs. 2 soccer games take place in the KidSize class with rules derived from the FIFA laws. Some simplifications apply, however. For example, the offside rule is not observed and key objects are color-coded.

The complexity of playing soccer games is much higher than the complexity of kicking penalties. The ball might be at any position on the field and the robots need to search for it if they lost track of its position. The robots must also perceive at least the two goals and the other players. Higher-level behaviors require self-localization on the field. The distances to walk are much longer. Hence, the walking speed must be higher. As two robots play together, there is need for coordination. While some teams use one dedicated goalie and one field player, other teams use two field players. This makes dynamic role assignment necessary. Last, but not least, in soccer games robots of the two teams interact physically when going for the ball. This disturbs walking and leads to falls. The robots need to get up from the ground by themselves in order to continue play. As a result of these difficulties, in the RoboCup 2006 competition, only a fraction of the teams able to play penalty kick was able to play decent soccer games.

The other RoboCupSoccer leagues have been facing the complexity of soccer games for some years now. There, tools for structured behavior engineering have been developed. For example, Jaeger and Christaller proposed the Dual Dynamics architecture (Jaeger & Christaller, 1998), which has been used in the MiddleSize League. The architecture distinguishes elementary behaviors, which implement a target dynamics, and complex behaviors, which control the activation of elementary behaviors. Another tool used in the MiddleSize League is the BAP-framework (Utz et al., 2005), which allows for specifying hierarchical, event-driven, behavior-based control systems. In the Four-Legged League, the German Team developed XABSL (Lötzsch et al., 2004). It allows for XML-based specification of hierarchies of behavior modules that contain state machines for decision making. State transitions are modeled as decision trees. Parts of the German Team system are used now in the Humanoid League by Darmtadt Dribblers, Humanoid Team Humboldt, and BreDoBrothers. Another example for a behavior architecture used in more than one league is the architecture proposed by Laue and Röfer (Laue & Röfer, 2005), which combines action selection and potential field motion planning. It was used to control SmallSize and Aibo soccer robots.

To implement the behavior control software for the humanoid soccer robots of our team NimbRo, we used a framework that supports a hierarchy of reactive behaviors (Behnke & Rojas, 2001). This framework has been originally developed for the FU-Fighters SmallSize robots. It was later adapted to the FU-Fighters MiddleSize robots and also used by CMU in the Four-Legged League. We adapted it for the control of soccer playing humanoid robots by extending the agent-hierarchy to: joint - body part - player - team. The lowest levels of this hierarchy contain position control of individual joints and kinematic interfaces for body parts. At the next level, basic skills like omnidirectional walking, kicking, and getting-up behaviors are implemented. These are used at the player level by soccer behaviors like searching for the ball, approaching the ball, avoiding obstacles, and defending the goal. Finally, on the team level, the players communicate via a wireless network to share information about the world state and to negotiate roles like attacker and defender.

The remainder of this chapter is organized as follows. In the next section, we describe our robots. We cover mechanical design, electronics, and perception. Sec. 3 describes our behavior control framework. The implementation of basic skills is covered in Sec. 4. Sec. 5 explains the design of our soccer behaviors. Finally, we present the results of using the soccer behaviors at RoboCup 2006.

2. NimbRo 2006 Robots

2.1 Mechanical Design

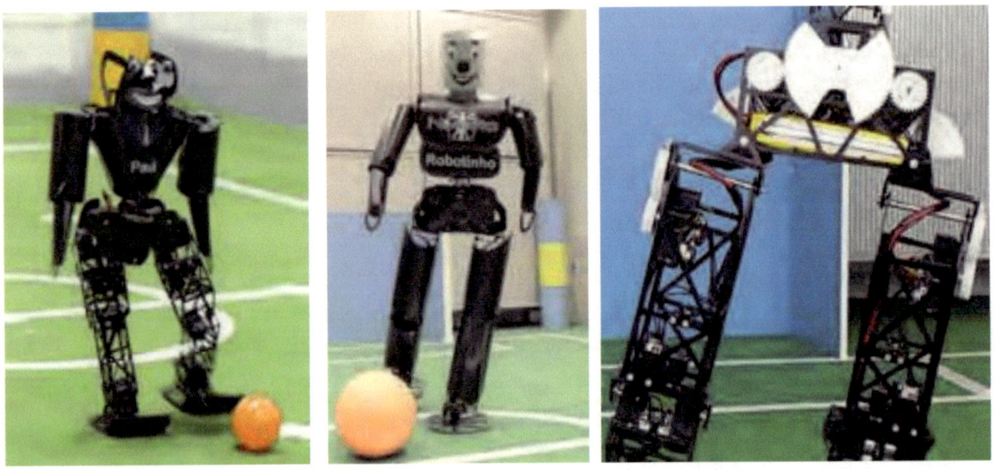

(a) (b) (c)

Figure 2. NimbRo 2006 robots: (a) KidSize robot Paul; (b) TeenSize robot Robotinho; (c) close-up of Robotinho's mechanics

Fig. 2 shows Paul, one of our 2006 KidSize robots, and Robotinho, our 2006 TeenSize robot. As can be seen, the robots have human-like proportions. Their mechanical design focused simplicity, robustness, and weight reduction. The KidSize robots have a height of 60cm and a weight of only 2.9kg, including batteries. They are driven by 24 Dynamixel actuators: 8 per leg, 3 in each arm, and two in the trunk. For the leg and the trunk joints, we use the DX-117 actuators (66g, 37kg·cm). Three orthogonal axes constitute the 3DOF hip joint. For the hip pitch and roll axes, we use two of these actuators in parallel. The actuators are coupled in a master-slave configuration. This doubles the torque and lowers operating temperatures. The master-slave pair of actuators has the same interface as the single actuators used for all other joints. Two orthogonal servos form the 2DOF ankle joint. One servo drives the knee joint. The trunk joints are in the pitch and yaw axes. The arms do not need to be as strong as the legs. They are powered by DX-113 actuators (58g, 10.2kg·cm). Two orthogonal servos constitute the shoulder joint and one servo drives the elbow joint.

The TeenSize robot Robotinho is 100cm tall and has a total weight of about 5kg. Its 21 DOF are driven by a total of 33 DX-117 actuators. The additional joint is the roll axis in the trunk. All joints in the legs and the trunk, except for the yaw axes, are driven by two parallel actuators. The hip and trunk yaw axes are reinforced by external 2:1 spur gears. The hip and trunk roll axes are reduced by 3:1, resulting in a holding torque of 222kg·cm at 16V.

The skeleton of the robots is constructed from aluminum extrusions with rectangular tube cross section. In order to reduce weight, we removed all material not necessary for stability. The feet and the forearms are made from sheets of carbon composite material. The elasticity of the feet and the carpet, the robots walk on, helps to maintain non-degenerate foot-ground contact, even when the supporting foot is not parallel to the ground. Robotinho's head is made of lightweight foam. The upper part of the KidSize robots and the entire body of the TeenSize robot is protected by a layer of foam and an outer shell of thin carbon composite material.

2.2 Electronics

Our soccer robots are fully autonomous. They are powered by high-current Lithium-polymer rechargeable batteries, which are located in their hip. Four Kokam 910mAh cells are used for the KidSize robots. Robotinho has four Kokam 3200mAh cells. The batteries last for about 25 minutes of operation. The Dynamixel actuators have a RS-485 differential half-duplex interface. Each robot is equipped with a CardS12 microcontroller board, which manages the detailed communication with all Dynamixels. These boards feature the Motorola MC9S12D64 chip, a 16-bit controller belonging to the popular HCS12 family. We clock it with 32MHz. It has 4kB RAM, 64kB flash, two serial interfaces, CAN bus, 8 timers, 8 PWM channels, and 16 A/D converters.

The Dynamixel actuators can be configured in a flexible way. Not only target positions are sent to the actuators, but also parameters of the control loop, such as the compliance. In the opposite direction, the current positions, speeds, loads, temperatures, and voltages are read back. In addition to these joint sensors, each robot is equipped with an attitude sensor, located in the trunk. It consists of a dual-axis accelerometer (ADXL203, ±1.5g) and two gyroscopes (ADXRS 300, ±300 °/s). The four analog sensor signals are digitized with A/D converters of the HCS12 and are preprocessed by the microcontroller. The microcontroller communicates with the Dynamixels at 1MBaud and with a main computer via a RS-232 serial line at 115KBaud. Every 12ms, target positions and compliances for the actuators are sent from the main computer to the HCS12 board, which distributes them to the actuators. The microcontroller sends the preprocessed sensor readings back. This allows keeping track of the robot's state in the main computer.

We use a Pocket PC as main computer, which is located in the upper part of the robots. The FSC Pocket Loox 720 has a weight of only 170g, including the battery. It features a 520MHz XScale processor PXA-272, 128MB RAM, 64MB flash memory, a touch-sensitive display with VGA resolution, Bluetooth, wireless LAN, a RS-232 serial interface, and an integrated 1.3 MPixel camera. This computer runs behavior control, computer vision, and wireless communication. It is equipped with a Lifeview FlyCam CF 1.3M that has been fitted to an ultra-wide-angle lens. Robotinho's FlyCam lens also serves as nose. It looks in forward direction. For the KidSize robots, we took the integrated camera out of the Pocket PC and connected it via an extension cable. This camera uses the QuickCapture feature of the XScale chipset. Images of size 640×480 can be captured at 15fps using DMA. The camera is fitted to a wide-angle converter. Located above the Pocket PC, it looks in forward direction. The FlyCam is looking in backward direction in the KidSize robots.

2.3 Perception

Our robots need information about themselves and the situation on the soccer field to act successfully.

- **Proprioception:** The readings of accelerometers and gyros are fused to estimate the robot's tilt in roll and pitch direction. The gyro bias is automatically calibrated and the low-frequency components of the tilt estimated from the accelerometers are combined with the integrated turning rates to yield an estimate of the robot's attitude that is insensitive to short linear accelerations. As described above, joint angles, speeds, and loads are also available. Temperatures and voltages are monitored to notify the user in case of overheating or low batteries.

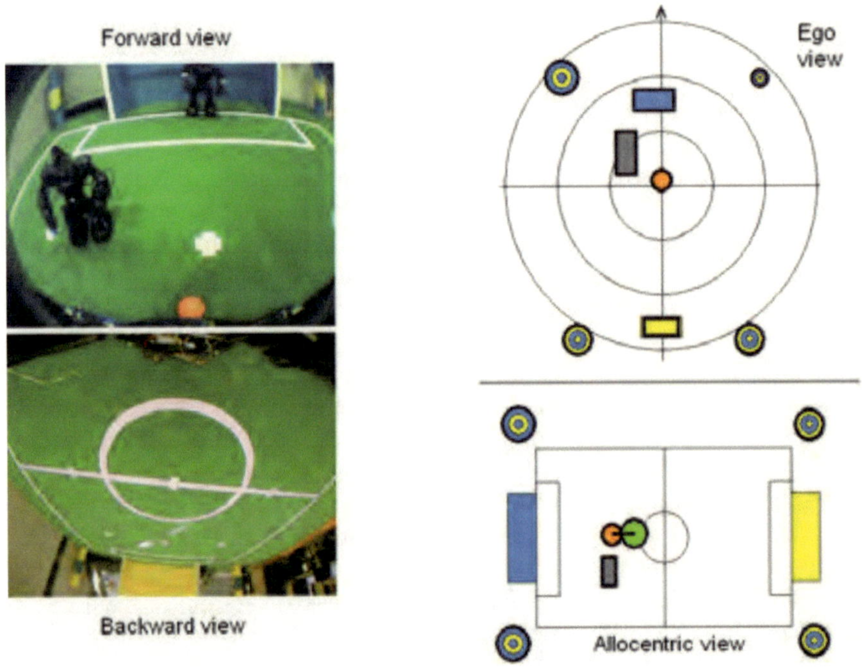

Figure 3. Left: Images of the two cameras mounted on the robot. Upper right: Egocentric coordinates of key objects (ball, goals, corner poles, and obstacle) detected in the image. Lower right: Localization of the robot, the ball, and the obstacle on the soccer field

- **Visual Object Detection:** The only source of information about the environment for our robots is their camera. The wide field of view of the cameras allows the robots to see their own feet and objects above the horizon at the same time (see left part of Fig. 3). Our computer vision software detects the ball, the goals, the corner poles, and other players based on their color in YUV space. Using a look-up table, the colors of individual pixels are classified into color-classes that are described by ellipses in the UV-plane. In a multistage process, we discard insignificant colored pixels and detect colored objects. We estimate their coordinates in an egocentric frame (distance to the robot and angle to its orientation), based on the inverted projective function of the camera. We correct first for the lens distortion and invert next the affine projection from the ground plane to the camera plane. The estimated egocentric coordinates of the key objects are illustrated in the upper right part of Fig. 3. Here, the objects detected by both cameras are fused, based on their confidence. The objects are also combined with previous observations, which are adjusted by a motion model, if the robot is moving. This yields a robust egocentric world representation.
- **Ball Tracking:** The limited computing power of the Pocket PC does not allow for processing all images at the frame rate of the camera (15fps). Because the ball is the

most important object on the field, we implemented a tracking procedure for it. If the ball could be detected in the previous frame, a small window is placed at the predicted ball position. Only this window is analyzed for every frame. Every third frame is processed entirely to detect the other objects. The full frame is also processed if the vision system looses track of the ball.

Fig. 4 illustrates a typical problem when processing images captured from a walking robot. The walking induces camera motion that causes motion blur in the image. Because the orange of the ball blends with the green carpet to a brownish color, we use such a candidate color to detect blurred balls. In this case, however, it is important to make sure that a brownish color blob is surrounded by green carpet, in order prevent false positive detections caused by brownish objects outside the field.

(a) (b)

Figure 4. The orange ball behind a white line on a green field. (a) Clean image captured from a standing robot. (b) Same situation with motion blur due to humanoid walking movements. The green blends with the orange to a brownish color

- **Self-Localization:** The relative coordinates suffice for many relative behaviors like positioning behind the ball while facing the goal. To keep track of non-visible goals or to communicate about moving objects with other team members, we need the robot coordinates in an allocentric frame ((x, y) -position on the field and orientation θ). We solve self-localization by triangulation over pairs of landmark observations, i.e. detected goals and corner poles. When observing more than two landmarks, the triangulation results are fused based on their confidence. Again, the results of self-localization are integrated over time and a motion model is applied. The lower-right of Fig. 3 illustrates the resulting allocentric representation.

3. Behavior Architecture

We control the robots using a framework that supports a hierarchy of reactive behaviors (Behnke & Rojas, 2001). This framework allows for structured behavior engineering. Multiple layers that run on different time scales contain behaviors of different complexity. When moving up the hierarchy, the speed of sensors, behaviors, and actuators decreases. At the same time, they become more abstract. This is illustrated in Fig. 5.

The framework forces the behavior engineers to define abstract sensors that are aggregated from faster, more basic sensors. One example for such an abstract sensor is the robot's attitude that is computed from the readings of accelerometers and gyros. Abstract actuators give higher-level behaviors the possibility to configure lower layers in order to eventually influence the state of the world. One such abstract actuator is the desired walking direction, which configures the gait engine, described below, implemented in the lower control levels.

The behaviors within one layer of the behavior framework are activated according to the current state of its sensors. Activation is indicated by an activation factor in the interval [0, 1]. Each active behavior can manipulate the actuators in its layer. If multiple behaviors try to manipulate the same actuator, the actuator is set to the weighted sum of desired values, where the activation factors are used as weights. To prevent conflicting behaviors from being active at the same time, behaviors can inhibit other behaviors. If an inhibiting behavior is not completely active, the inhibited behaviors share the remaining activation, such that the activation factors sum to one.

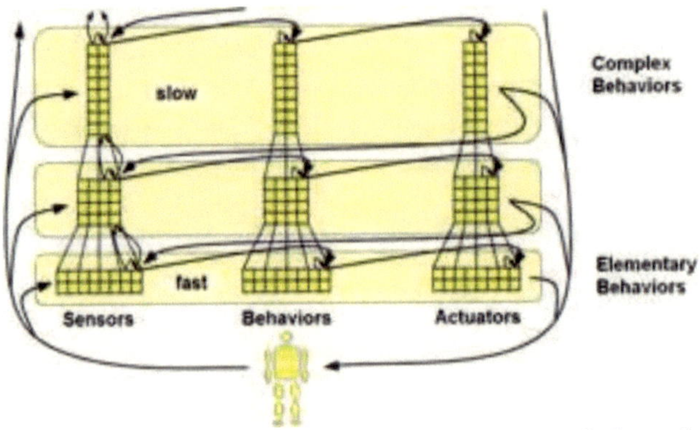

Figure 5. Sketch of the hierarchical framework for reactive control. Complex behaviors are evaluated less often than elementary behaviors. They make decisions based on aggregated fast sensors or rely on slow physical sensors. Complex behaviors use slow actuators to configure lower levels or to directly influence the environment

The control hierarchy of our soccer robots is arranged in an agent hierarchy:

- multiple joints (e.g. left knee) constitute a body part (e.g. left leg),
- multiple body parts constitute a player (e.g. field player), and
- multiple players constitute a team.

In our system, two teams can be controlled simultaneously. The behavior framework manages all but the motor control loop within the Dynamixel actuators, which has been implemented by Robotis. The behaviors on the lower level in the framework implement basic skills which generate target positions for individual joints at a rate of 83.3Hz.
To abstract from the individual joints, we implemented here a kinematic interface for the body parts. The leg interface, for example, allows to independently change leg extension η, leg angle θ_{Leg}, and foot angle θ_{Foot}, as illustrated in Fig. 6. A detailed description of the kinematic leg interface is given in (Behnke, 2006).

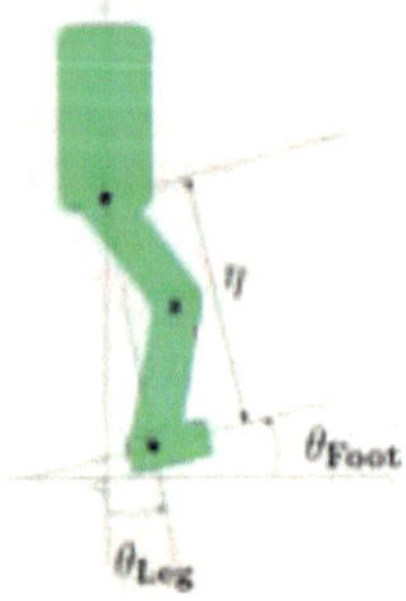

Figure 6. Kinematic interface to a leg

4. Basic Skills

Several basic skills use this kinematic interface. Fundamental for playing soccer are the abilities to walk and to kick. As body contact between the physical agents is unavoidable, the capability of getting up after a fall is also essential. To act as a goalkeeper, the robot must be able to perform special motions. The basic skills are implemented on the body part layer. Fig. 12 illustrates the inhibitory structure of the basic skills and the interface that they provide for the next higher level of our behavior control system.

4.1 Omnidirectional Walking

Omnidirectional locomotion is a concept that has proven to be advantageous in dynamic environments and in restricted spaces. The ability to move into any direction, irrespective of the orientation, and to control the rotational speed at the same time has advantages in many domains, including RoboCupSoccer. Omnidirectional drives are used by most teams in wheeled leagues, and omnidirectional walking is heavily used in the Four-legged League. It is much easier to position robots for kicking and to outmaneuver opponents when using omnidirectional locomotion.

We use the leg interface to implement omnidirectional walking for our humanoid soccer robots. Shifting the weight from one leg to the other, shortening of the leg not needed for support, and leg motion in walking direction are the key ingredients of this gait. In contrast to the low-frequency gait of our 2005 robots (Behnke, 2006), we were able to increase the step frequency significantly to 3.45Hz for the KidSize robots and to 2.44Hz for Robotinho.

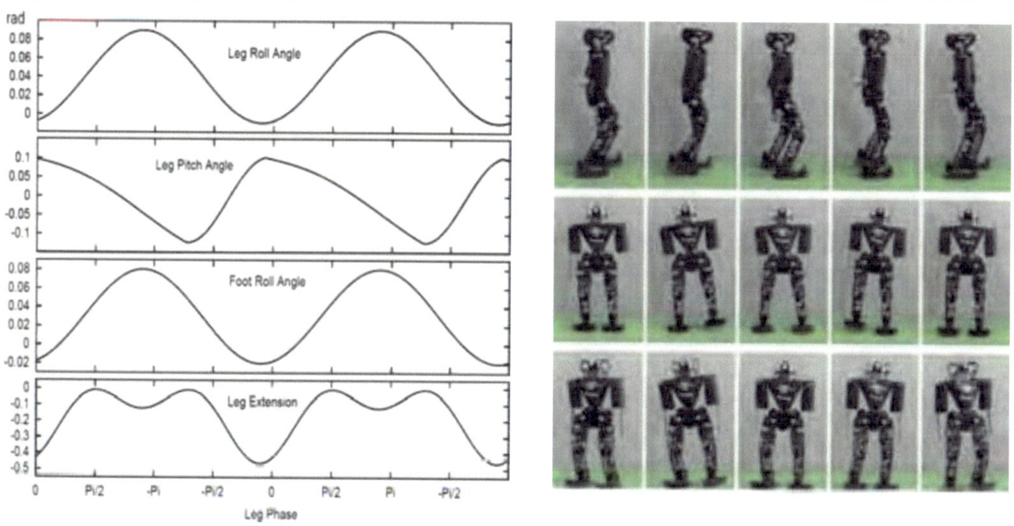

Figure 7. Trajectories for forward walking of KidSize robots (left) and resulting robot motion during forward, lateral, and rotational walking (right).

Fig. 7 shows in its left part the trajectories generated for forward walking. Note that the leg is not only shortening during swing, but also in the middle of the stance phase. Walking forward, to the side, and rotating on the spot are generated in a similar way. The three basic walking directions can be smoothly combined. The robots are able to walk in every direction and to change their heading direction at the same time. The gait target vector (v_x, v_y, v_θ) can be changed continuously while the robot is walking. This makes it possible to correct for deviations in the actual walking direction and to account for changes in the environment by using visual feedback. When using this flexible gait, the maximal forward walking speed of the robots is approx. 25cm/s. The right part of Fig. 7 shows image sequences of the robot Franz walking forward, laterally, and turning. Behaviors of the upper level can control the gait target vector with an actuator that enforces maximal speeds and accelerations.

4.2 Kicking

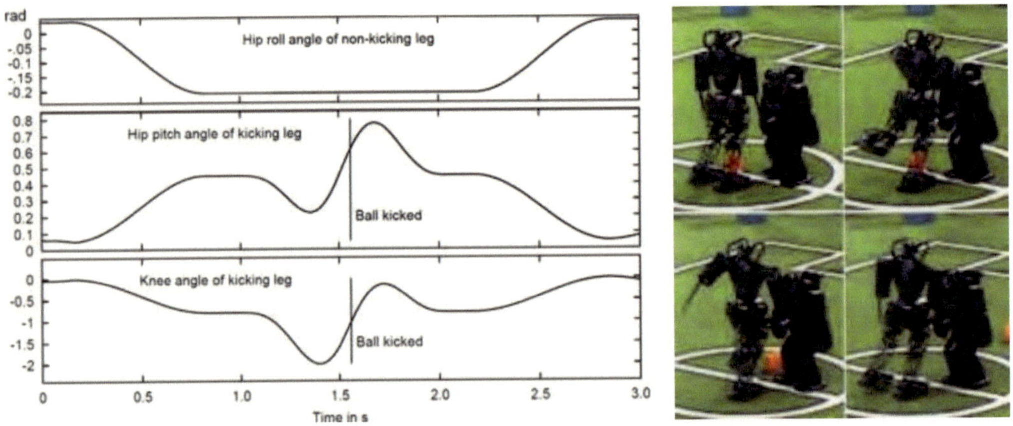

Figure 8. Trajectories for kicking (left) and resulting robot motion (right)

In addition to walking, we implemented kicking. An actuator allows behaviors in the upper level to trigger kicks with both, the left and the right leg. Fig. 8 shows some of the trajectories generating the kicking motion. After inhibiting the walking behavior and coming to a stop, the robot moves its weight to the non-kicking leg (see hip roll angle). Then, it shortens the kicking leg, swings it back, and accelerates forward. The kicking leg reaches its maximal speed when it comes to the front of the robot. At this point, the hip pitch joint and the knee both move the foot forward and the ball is kicked. The kicking movement continues with deceleration of the foot and slow motion back to the bipedal stand. The resulting kick can be seen in the right part of Fig. 8.

4.3 Getting up from the Floor

Since in soccer games physical contact between the robots is unavoidable, the walking patterns are disturbed and the robots might fall. Hence, they must be able to detect the fall, to assess their posture on the ground, and to get back into an upright posture. After falling, the robot's center of mass (COM) projection to the ground is outside the convex hull spanned by the foot-contact points. Additional support points like knees, elbows, and hands must be used in order to move the COM back inside the foot polygon. Using their attitude sensors, the robots detect a fall, classify the prone or supine posture and trigger the corresponding getting-up sequence. We designed the getting-up sequences in the simulator using sinusoidal trajectories (Stückler et al., 2006). Fig. 9 illustrates the four phases of getting up from the prone and the supine posture. The getting-up sequences work very reliably. Under normal circumstances, i.e. appropriate battery voltage, the routines worked with 100 successes in 100 tests.

I. Lift the trunk and bring the forearms under the shoulders.
II. Move the COM projection as close as possible to the leading edges of the feet by bending in the spine, the hip pitch and the knee joints.
III. Straighten the arms to let the robot tip over the leading edges of the feet.
IV. Bring the body into an upright posture.

I. Move the upper body into a sit-up posture and move the arms into a supporting position behind the back.
II. Move into a bridge-like position using the arms as support.
III. Move the COM over the feet by swinging the upper body to the front.
IV. Bring the body into an upright posture.

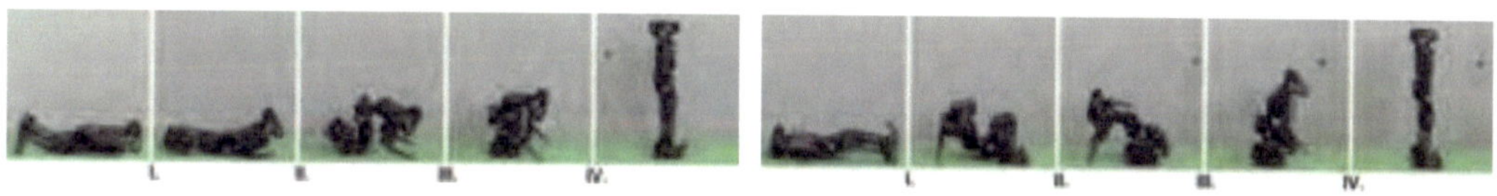

Figure 9. Standing up from the supine posture (left) and the prone posture (right)

4.4 Goalkeeper Motions

The goalkeeper is capable of diving into both directions or to bend forward with spread arms. Fig. 10 shows Franz diving to the left. First, it moves its COM and turns its upper body towards the left while shortening the legs. As soon as it tips over its left foot, it starts straightening its body again. While doing so it is sliding on its hands and elbows. The fully extended robot covers the entire goal half. After the dive Franz gets up again, as described above.

Figure 10. Diving motion of the goalkeeper

5. Playing Soccer

The next higher level of our behavior control framework contains soccer behaviors which are executed at a rate of 41.7Hz. They build on the basic skills and have been designed for 2 vs. 2 soccer games.

5.1 Representation of the Game Situation

The soccer behaviors require knowledge of the current game situation. The visual perception supplies relative distance, angle, and perceptual confidence for the ball, the own goal, the opponent goal, and the nearest obstacle. In the attacking role, the relative position and confidence of the opponent goal is used as the target to kick at (ball-target). The decision for the kicking leg is made at every time step, depending on the relative position of the ball and the line from ball to ball-target, which we denote as ball to target-line. If the robot has to approach the ball-to-target-line from the right, it kicks with the left leg, and vice versa. To avoid oscillations it is important that the decision may only be changed if the distance of the robot to the ball-to-target-line exceeds a threshold.

To kick the ball with the chosen leg, the robot has to position itself behind the ball with lateral and sagittal offsets, δ_l and δ_s that depend on the distance between the legs and the length of the feet. To generate smoothly approaching trajectories, the sagittal offset is increased by an amount δ_a that is proportional to the angle between the robot's heading direction and the ball-target. The ball approach is illustrated in Fig. 11.
When playing as defensive field player, the own goal is used as ball-target, such that the position behind the ball is set to a position between ball and own goal. The distance kept to the ball depends on the distance to the own goal. A threshold for the minimal distance to the goal lets the robot stay out of its goal, as long as the ball is still far away. If the ball and the robot are near the goal, the robot keeps behind the ball at a minimum distance.

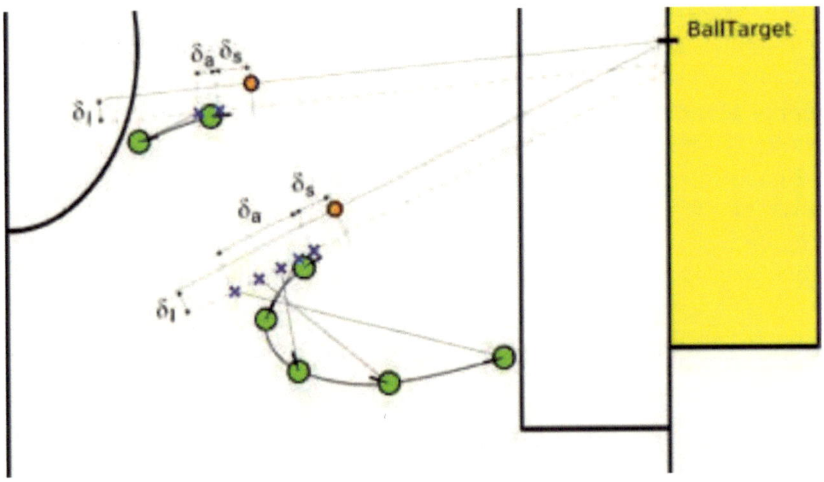

Figure 11. Two examples showing sequences of robot poses, target positions behind the ball (blue crosses), and ball positions while approaching the ball with the left leg as kicking leg

The robot maintains additional hypotheses about the relative ball location that are used for searching the ball. If a kick is triggered, one hypothesis is set in front of the robot at a distance depending on kick strength. The confidence of the hypothesis is discounted by the time since the kick started. Its relative position is altered according to the motion model.
Additionally, hypotheses are maintained for the perceptions of the ball by other players on the field. The confidences of these hypotheses depend on the self-localization and ball perception confidences of the other players and the self-localization confidence of the robot itself.

5.2 Soccer Behaviors

According to the current game situation, behaviors like searching the ball, positioning behind the ball, or avoiding obstacles are activated. These behaviors are implemented on the player level and use the actuator interface that the basic skills of the lower layer provide. For example, they set the gait target vector or trigger a kick. Fig. 12 illustrates the inhibitory structure of the soccer behaviors and the actuator interface used for configuring the basic skills.

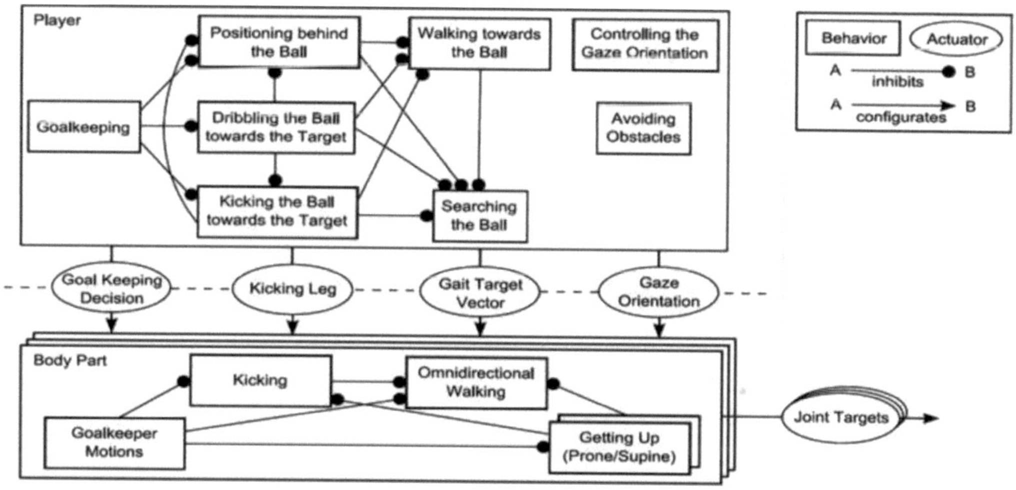

Figure 12. Behaviors on the body part and player levels of the behavior hierarchy

- **Searching the Ball:** Exploring the environment for the ball is always active, but inhibited by behaviors that activate when the ball has been perceived with a certain confidence. If a ball hypothesis with confidence over a certain threshold exists, the robot walks towards the most confident hypothesis. Otherwise, it turns towards the most confident hypothesis for a short time. If the ball still is not visible, it starts to walk around the center circle in a constant distance in order to inspect all parts of the field.
- **Walking towards the Ball:** The robot walks straight towards the ball, if it perceives the ball. The own goal must be either not visible or far away to avoid scoring an own goal. This behavior controls the gait target velocity to keep the robot near the ball, e.g. if visual perception fails to detect the opponent goal. The behavior inhibits searching the ball.
- **Positioning behind the Ball:** If the ball and the ball-target are perceived, the robot positions itself behind the ball, facing towards the ball-target. The robot is positioning on the behind-ball-position by controlling the gait target velocity. If the distance to the target position is large, the robot rotates towards the target position, such that it can approach it by mainly combining forward walking with turning. If it is near the target position, the robot aligns itself towards the ball-target. For intermediate distances, the gait rotation is interpolated linearly between both alignment targets. The behavior also handles the case when the ball is located between the robot and the behind ball-position. Here, the robot walks around the ball by walking towards the target position but avoiding the ball-to target-line. When playing as defensive field player, the robot rotates towards the ball at any distance. It does not avoid the ball-to-target-line, because the ball-target is the own goal. This behavior inhibits walking towards the ball, such that the inhibited behavior may only be active, if the ball-target has not been perceived. It also inhibits searching the ball.
- **Kicking the Ball towards the Target:** This behavior is activated as soon as the behind-ball position has been reached with a certain precision in angle to the ball-target and in distance to the target position. If the precision conditions hold, a kick is triggered. Obviously, ball and ball-target must be perceived and the own goal must not be in front of the robot. If the ball comes into a kicking position by chance, the behavior initiates a

kick with the corresponding leg. As the robot has to come to a complete stop before the kicking motion can be executed, the robot can cancel the kick, if the ball moves away in the meantime. This behavior inhibits searching the ball, walking towards the ball, and positioning behind the ball.

- **Dribbling the Ball towards the Target:** If positioning behind the ball was not successful for a longer time, or the game started with a kick-off for the player, the robot activates dribbling the ball towards the ball-target for some time. Additional preconditions for activation are that the ball and ball-target are perceived and the angle towards the ball-target is small. Dribbling is performed by steering towards the ball. The forward walking speed is inversely related to the angle to the ball. In combination with positioning behind the ball, the robot is kept behind the ball, facing the ball-target when dribbling. Dribbling inhibits searching the ball, walking towards the ball, and positioning behind the ball. As we want the decision for dribbling to be strict, it also inhibits kicking the ball towards the target.
- **Avoiding Obstacles:** After a fall, the robot needs valuable time to get back on its feet. The main reason for our robots to fall is physical contact with other robots. Hence, obstacle avoidance is an important feature. The visual perception supplies the behavior with the nearest obstacle. If it is detected closely in front of the robot, obstacle avoidance is activated by a factor that interpolates linearly between a minimum and a maximum distance for the obstacle. The avoidance sets the gait target actuator to a constant and a variable part of the direction from obstacle to robot. The strength of the variable part depends on the distance to the obstacle, similar to the activation factor. If the ball is between obstacle and robot, the variable avoidance is weakened, such that the robot moves more aggressively behind the ball. A stuck situation is indicated by a resulting gait target vector that is small in length for a longer time. In this case, the robot may sidestep the obstacle, if the ball is not between the obstacle in the front and the robot and is perceived on one side of the obstacle. The action is cancelled, if either the preconditions for sidestepping do not hold anymore or a certain amount of time has elapsed since sidestepping has been activated. The deactivation of sidestepping after some time is important, because the decision for the sidestep direction is made only once on activation.
- **Controlling the Gaze Direction:** Although the robot has wide-angled views to the front and the rear, it cannot perceive objects on the sides. Thus, a gaze control behavior is always active and primarily keeps the ball within an angular range of $\pm\pi/4$ by twisting the upper trunk with the trunk yaw joint. If the ball is not visible or within range and the robot is localized, it aligns the upper body with the line between the goals to keep the localization landmarks visible. This is achieved by keeping the angle to the line within the angular range of $\pm\pi/4$. The twisting of the trunk is limited to $\pm\pi/4$.
- **Goalkeeping:** The goalkeeper's objective apparently is to keep the ball out of the own goal. While the ball is visible and is not in kicking distance to the robot, the goalkeeping behavior is active. Otherwise, the robot behaves like a field player and tries to shoot the ball towards the opponent goal. Hence, goalkeeping inhibits positioning behind the ball, kicking the ball, and dribbling the ball towards the target. Walking towards the ball and searching for the ball is not activated when playing as a goalkeeper. The goalkeeper stands still until it reacts on the ball. Balls close to the robot let it react immediately. It uses the ball angle to determine the appropriate type of motion (diving

left/right or bending forward). To achieve fast reaction on an approaching ball, the visual perception supplies the difference of the ball position between the last two images. The magnitude of this vector is interpreted as approaching speed. The goalkeeper does not react on small speeds. The type of the goalkeeper motion is determined by the intersection point of the moving ball direction and the goal line. At kick-off, the goalkeeper is placed in the goal. After a diving motion, it gets up and repositions itself in the goal while facing the opponent goal.

5.3 Team Behaviors

The importance of team behaviors is still low in the Humanoid League, as only two players per team have competed so far. In Bremen 2006, most teams assigned one player to keep the goal clear and used the other player as field player. In our team, the players share perceptions via wireless communication. The ball perceptions communicated by other players are used for search. For the soccer play with two field players, we implemented simple but effective role negotiation between the players. As soon as one of our players has control of the ball, the other player goes to a defensive position between the ball and the own goal.

A player takes control of the ball, if it is close to the ball and perceived it with high confidence. It loses control, if the ball gets too far away or has low confidence. The thresholds for taking and losing control implement hysteresis to prevent oscillations of the control state.

6. RoboCup 2006 Results

Our robots performed well at RoboCup 2006, where 21 teams from eleven countries competed in the Humanoid League. In the 2 vs. 2 soccer round robin, the KidSize robots played 2 games and scored 12:0 goals. In the quarter final, they won 6:1 against team RO-PE from Singapore. They met the German-Japanese team Darmstadt Dribblers and Hajime in the semi-final. Our robots won 6:2. The final game was between our robots and Team Osaka, as in 2005. Our robots played well in the first half and scored a lead of 4:0. Fig. 13(a) shows one of the shots. After a goal directly from kick-off, the score at halftime was 4:1. Due to hardware problems of our robots, Team Osaka was able to reach a draw of 4:4 after regular playing time. As we already had taken the available two substitutions, we needed to continue playing with impaired robots in the extra time. The final score was 9:5 for Team Osaka.

Our KidSize robots also kicked penalties very reliably. In the Penalty Kick competition they scored in 31 of 34 attempts. In the KidSize Penalty Kick final (Fig. 13(c)) our robots won 8:7 against Team Osaka.

In the technical challenge, our KidSize robot Gerd was one of the two robots able to walk across the rough terrain (Fig. 13(b)). Our KidSize robots also scored in the passing challenge. Our TeenSize robot Robotinho used a simplified version of the KidSize behaviors. It also reached the final of its Penalty Kick competition (Fig. 13(d)).

In the overall Best Humanoid ranking, our KidSize robots came in second, next only to the titleholder, Team Osaka. Videos showing the performance of our robots at RoboCup 2006 can be found at http://www.NimbRo.net.

(a) (b) (c) (d)

Figure 13. RoboCup 2006: (a) 2 vs. 2 Soccer final NimbRo vs. Team Osaka. (b) NimbRo robot Gerd walking over rough terrain. (c) KidSize Penalty Kick final NimbRo vs. Team Osaka. (d) TeenSize Penalty Kick final NimbRo vs. Team Osaka

7. Conclusions

This chapter described the design of the behavior control software for our humanoid soccer robots, which successfully took part as team NimbRo at the RoboCup 2006 competitions. We implemented the control software in a framework that supports a hierarchy of reactive behaviors. This structure restricts interactions between the system variables and thus reduces the complexity of behavior engineering.

A kinematic interface for body parts made it possible to abstract from individual joints when implementing basic skills like omnidirectional walking. These basic skills made it possible to abstract from body parts when implementing more complex soccer behaviors. At this player level, our humanoid robots are very similar to wheeled or four-legged soccer robots. Finally, at the team level, the players are coordinated through role negotiation.

Playing soccer with humanoid robots is a complex task, and the development has only started. So far, there has been significant progress in the Humanoid League, which moved in its few years from remotely controlled robots to soccer games with fully autonomous humanoids. Indeed, the Humanoid League is currently the most dynamic RoboCupSoccer league. We expect to see the rapid progress continue as more teams join the league. Many research issues, however, must be resolved before the humanoid robots reach the level of play shown in other RoboCupSoccer leagues. For example, the humanoid robots must maintain their balance, even when disturbed. Postural reflexes that are trigged by deviations from the normal walking patterns are one way to minimize the number of falls (Renner & Behnke, 2006).

In the next years, the speed of walking must be increased significantly. We work on automatic gait optimization to increase both speed and stability. At higher speeds, running will become necessary. We recently started to explore this direction. The visual perception of the soccer world must become more robust against changes in lighting and other interferences. We continuously improve our computer vision software to make it more reliable.

The 2006 competition has shown that most teams were able to kick penalties, but that soccer games are much richer and more interesting. In the team leader meeting after the competition, the majority voted for abandoning penalty kick as a separate competition. Instead, the KidSize teams will focus on soccer games. Unfortunately, most teams do not feel ready to increase the number of players to more than two per team. This limits the possibilities for team play.

As the basic skills of the humanoid soccer robots improve every year, teams will be able to focus on the more complex soccer behaviors and on team play. This will make structured behavior engineering a key factor for success.

8. Acknowledgment

Funding for this project is provided by the DFG (Deutsche Forschungsgemeinschaft), grant BE 2556/2-1,2,4.

9. References

Behnke, S. & Rojas, R. (2001). A Hierarchy of Reactive Behaviors Handles Complexity, In: *Balancing Reactivity and Social Deliberation in Multi-Agent Systems,* pp. 125-136, Springer, LNCS 2103

Behnke, S. (2006). Online Trajectory Generation for Omnidirectional Biped Walking, *Proceedings of IEEE International Conference on Robotics and Automation (ICRA'06),* pp. 1597-1603, Orlando, Florida

Brooks, R. A. (1990). Elephants Don't Play Chess. *Robotics and Autonomous Systems,* 6:3-15

Jaeger, H. & Christaller, T. (1998). Dual dynamics: Designing behavior systems for autonomous robots. *Artificial Life and Robotics,* 2(3):108-112

Kitano, H. & Asada, M. (2000). The RoboCup Humanoid Challenge as the millennium challenge for advanced robotics. *Advanced Robotics,* 13(8):723-737

Laue, T. & Röfer, T. (2005). A Behavior Architecture for Autonomous Mobile Robots Based on Potential Fields, In: *RoboCup 2004: Robot Soccer World Cup VIII,* pp. 122-133, Springer, LNCS 3276

Lötzsch, M.; Bach, J; Burkhard, H.-D. & Jüngel, M. (2004). Designing Agent Behavior with the Extensible Agent Behavior Specification Language XABSL, In: *RoboCup 2003: Robot Soccer World Cup VII,* pp. 114-124, Springer, LNCS 3020

Newborn, M. (1997). *Kasparov versus Deep Blue: Computer chess comes of age,* Springer

Pfeifer, R. & Scheier, C. (1999). *Understanding Intelligence,* The MIT Press, Cambridge MA

Renner, R. & Behnke, S. (2006). Instability Detection and Fall Avoidance for a Humanoid using Attitude Sensors and Reflexes, *Proceedings of IEEE/RSJ International Conference on Intelligent Robots and Systems (IROS),* pp. 2967-2973, Beijing, China

Simon, H. A. & Newell, A. (1958). Heuristic Problem Solving: The Next Advance in Operations Research. *Operations Research,* 6(1):1-10

Stückler, J.; Schwenk, J. & Behnke, S. (2006). Getting Back on Two Feet: Reliable Standing-up Routines for a Humanoid Robot, *Proceedings of the 9th International Conference on Intelligent Autonomous Systems (IAS-9),* pp. 676-685, Tokyo, Japan

Utz, H.; Kraetzschmar, G.; Mayer, G. & Palm, G. (2005). Hierarchical Behavior Organization, *Proceedings of the 2005 IEEE/RSJ International Conference on Intelligent Robots and Systems (IROS)*, pp. 2598-2605, Edmonton, Canada